Nuclear Receptors

PROTEINS AND CELL REGULATION

Volume 8

Series Editors: Professor Anne Ridley
Ludwig Institute for Cancer Research
and Department of Biochemistry and Molecular Biology
University College London
London
United Kingdom

Professor Jon Frampton
Professor of Stem Cell Biology
Institute for Biomedical Research,
Birmingham University Medical School,
Division of Immunity and Infection
Birmingham
United Kingdom

Aims and Scope

Our knowledge of the ways in which a cell communicates with its environment and how it responds to information received has reached a level of almost bewildering complexity. The large diagrams of cells to be found on the walls of many a biologist's office are usually adorned with parallel and interconnecting pathways linking the multitude of components and suggest a clear logic and understanding of the role played by each protein. Of course this two-dimensional, albeit often colourful representation takes no account of the three-dimensional structure of a cell, the nature of the external and internal milieu, the dynamics of changes in protein levels and interactions, or the variations between cells in different tissues.

Each book in this series, entitled "*Proteins and Cell Regulation*", will seek to explore specific protein families or categories of proteins from the viewpoint of the general and specific functions they provide and their involvement in the dynamic behaviour of a cell. Content will range from basic protein structure and function to consideration of cell type-specific features and the consequences of diseaseassociated changes and potential therapeutic intervention. So that the books represent the most up-to-date understanding, contributors will be prominent researchers in each particular area. Although aimed at graduate, postgraduate and principal investigators, the books will also be of use to science and medical undergraduates and to those wishing to understand basic cellular processes and develop novel therapeutic interventions for specific diseases.

For further volumes:
http://www.springer.com/series/6442

Nuclear Receptors

Current Concepts and Future Challenges

Edited by

Chris M. Bunce
University of Birmingham, Birmingham, United Kingdom

Moray J. Campbell
Roswell Park Cancer Institute, Buffalo, USA

Springer

Editors
Chris M. Bunce
University of Birmingham
School of Biosciences
Birmingham
Edgbaston
United Kingdom B15 2TT

Moray J. Campbell
Roswell Park Cancer Institute
Dept. Pharmacology &
Therapeutics
Elm & Carlton Streets
Buffalo NY 14263
USA

ISBN 978-90-481-3302-4 e-ISBN 978-90-481-3303-1
DOI 10.1007/978-90-481-3303-1
Springer Dordrecht Heidelberg London New York

Library of Congress Control Number: 2010921599

Printed on acid-free paper

Springer is part of Springer Science+Business Media (www.springer.com)

FOREWORD

Organisms have come up with many clever mechanisms for sensing and reacting to the environment. For example, our powerful nervous system with its complex neuronal networks can sense, process and respond in a fraction of a second and cells can sense the outside world with an impressive array of surface receptors that trigger intricate phosphorylation cascades. However metazoan animals have evolved an additional form signal perception mediated by *nuclear receptors*. These transcription factors are able to bypass the complex 'second messenger' signalling pathways of surface receptors and directly link gene transcription to the availability of ligands coming from the surrounding environment.

Starting from 1985, when the first cDNA encoding the human glucocorticoid receptor was cloned, there has been an exponential growth in the knowledge covering the nuclear receptor field with more than a hundred thousand publications on this topic. As PhD students starting projects in the field we found that the available textbooks were not up to date and the task of mining the literature to compile a more contemporary picture proved to be very difficult. This new book offers a solid background in nuclear receptor biology, written by many of the leaders in the field, along with the most recent advances and will surely be an indispensable tool for researchers entering into the field.

As described in the ensuing pages nuclear receptors can sense steroid hormones, retinoids, dietary lipids, xenobiotics and many other ligands and quickly elicit a response. These actions are subject to an elegant regulation by a cohort of co-regulators, mainly co-repressor and co-activator complexes, that epigenetically modulate the chromatin environment through histone deacetylases and histone acetyltransferases that, respectively, repress or promote transcription. Not surprisingly, nuclear receptors play a key role in normal physiology, adaptation to environmental conditions and in the development of diseases such as cancer where they are commonly exploited as therapeutical targets.

The widespread action of nuclear receptors easily explains why they are key hubs in cellular signaling within animals and why so many research projects focus on nuclear receptor biology, making it a very dynamic and constantly changing field of research. As PhD students, we have been motivated by curiosity and enthusiasm in this kind of research. We hope that the readers of this book will be driven by the same passion.

Sebastiano Battaglia and Pedro Veliça Phd students with the editors, funded by the NucSys Marie Curie Research Training Network.

Sebastiano Battaglia
Pedro Veliça

CONTENTS

LIST OF CONTRIBUTORS

Trevor K. Archer Laboratory of Molecular Carcinogenesis, National Institute of Environmental Health Sciences, Research Triangle Park, NC, USA, archer1@niehs.nih.gov

Ferdinando Auricchio Department of General Pathology, II University of Naples, Via L. De Crecchio 7, 80138 Naples, Italy, ferdinando.auricchio@unina2.it

François Bonneton Institut de Génomique Fonctionnelle de Lyon, Université de Lyon, Université Lyon 1, CNRS, INRA, Ecole Normale Supérieure de Lyon, Lyon cedex, France

F.J. Bruggeman Regulatory Networks Group, Netherlands Institute for Systems Biology, University of Amsterdam, Amsterdam, The Netherlands; Life Sciences, Institute for Mathematics and Computer Science (CWI), Amsterdam 1098 SJ, The Netherlands; Molecular Cell Physiology, Vrije Universiteit, Amsterdam 1081 HV, The Netherlands

Chris M. Bunce School of Biosciences, University of Birmingham, Edgbaston, Birmingham, B15 2TT, UK

Craig J. Burd Laboratory of Molecular Carcinogenesis, National Institute of Environmental Health Sciences, Research Triangle Park, NC, USA

Moray J. Campbell Department of Pharmacology and Experimental Therapeutics, Roswell Park Cancer Institute, Buffalo, NY 14263, USA, Moray.Campbell@RoswellPark.org

Carsten Carlberg Life Sciences Research Unit, University of Luxembourg, L-1511 Luxembourg City, Luxembourg; Department of Biosciences, University of Kuopio, FIN-70211 Kuopio, Finland, carsten.carlberg@uni.lu

Gabriella Castoria Department of General Pathology, II University of Naples, Via L. De Crecchio 7, 80138 Naples, Italy

Chawnshang Chang George H. Whipple Lab for Cancer Research, Departments of Urology and Pathology, University of Rochester Medical Center, Rochester, NY 14642, USA, chang@urmc.rochester.edu

Weidong Chen Department of Gene Regulation and Drug Discovery, Beckman Research Institute, City of Hope National Medical Center, 1500 E. Duarte Road, Duarte, CA 91010, USA

Xiaosong Chen Department of Gene Regulation and Drug Discovery, Beckman Research Institute, City of Hope National Medical Center, 1500 E. Duarte Road, Duarte, CA 91010, USA

Christine Chomienne CRAS Inserm UMR-S-718, Hôpital Saint Louis, Institut Universitaire d'Hématologie, Avenue Claude Vellefaux, 75010 Paris, France; Université Paris- Diderot UMR-S-718, 75010, Paris, France, christine.chomienne@sls.aphp.fr

John A. Cidlowski National Institute of Environmental Health Sciences, NC 27709, USA, cidlows1@niehs.nih.gov

Audrey Cras Inserm UMR-S-718, Hôpital Saint Louis, Institut Universitaire d'Hématologie, Avenue Claude Vellefaux, 75010 Paris, France; Université Paris-Diderot UMR-S-718, 75010, Paris, France

James T. Dalton GTx Inc., Memphis, TN 38163, USA, jdalton@gtxinc.com

Douglas Forrest NIDDK, Clinical Endocrinology Branch, National Institutes of Health, Bethesda, MD 20892-1772, USA

Wenqing Gao School of Pharmacy and Pharmaceutical Sciences, Department of Pharmaceutical Sciences, University at Buffalo SUNY, Buffalo, NY 14260, USA

Fabien Guidez Department of Medical and Molecular Genetics, King's College London, Guy's Hospital, London SE1 9RT, UK

Jan-Åke Gustafsson Department of Biosciences and Nutrition, Novum, Huddinge SE-141 57, Sweden

Martin Hewison Department of Orthopaedic Surgery, David Geffen School of Medicine UCLA, Los Angeles, CA 90095, USA, mhewison@mednet.ucla.edu

Wendong Huang Department of Gene Regulation and Drug Discovery, Beckman Research Institute, City of Hope National Medical Center, 1500 E. Duarte Road, Duarte, CA 91010, USA

Sander Kersten Nutrigenomics Consortium, TI Food and Nutrition, Nieuwe Kanaal 9A, 6709 PA Wageningen, The Netherlands; Nutrition, Metabolism and

Genomics Group, Division of Human Nutrition, Wageningen University, Bomenweg 2, 6703 HD Wageningen, The Netherlands, sander.kersten@wur.nl

Curtis Klaassen Department of Pharmacology, Toxicology and Therapeutics, University of Kansas Medical Center, Kansas City, KS, USA

A. Kolodkin Molecular Cell Physiology, Vrije Universiteit, Amsterdam 1081 HV, The Netherlands

Vincent Laudet Institut de Génomique Fonctionnelle de Lyon, Université de Lyon, Université Lyon 1, CNRS, INRA, Ecole Normale Supérieure de Lyon, Lyon Cedex, France

Yifen Lee George H. Whipple Lab for Cancer Research, Departments of Urology and Pathology, University of Rochester Medical Center, Rochester, NY 14642, USA

Hong Liu NIDDK, Clinical Endocrinology Branch, National Institutes of Health, Bethesda, MD 20892-1772, USA

Hong Lu Department of Pharmacology, Toxicology and Therapeutics, University of Kansas Medical Center, Kansas City, KS, USA

Orla Maguire Department of Pharmacology and Therapeutics, Roswell Park Cancer Institute, Elm & Carlton Streets, Buffalo, NY 14263, USA

Gabriel Markov Institut de Génomique Fonctionnelle de Lyon, Université de Lyon, Université Lyon 1, CNRS, INRA, Ecole Normale Supérieure de Lyon, Lyon Cedex, France; USM 501 – Evolution des Régulations Endocriniennes. Muséum National d'Histoire Naturelle, Paris, France

David Martín Institute of Evolutionary Biology (CSIC-UPF), 08003 Barcelona, Spain, david.martin@ibe.upf-csic.es

Antimo Migliaccio Department of General Pathology, II University of Naples, Via L. De Crecchio 7, 80138 Naples, Italy

M. Moné Molecular Cell Physiology, Vrije Universiteit, Amsterdam 1081 HV, The Netherlands

Stefan Nilsson Karo Bio AB, Novum, Huddinge SE-141 57, Sweden; Department of Biosciences and Nutrition, Novum, Huddinge SE-141 57, Sweden

Robert H. Oakley Laboratory of Signal Transduction, Department of Health and Human Services, National Institute of Environmental Health Sciences, National Institutes of Heath, NC 27709, USA

K. Rybakova Molecular Cell Physiology, Vrije Universiteit, Amsterdam 1081 HV, The Netherlands

Linda M. Sanderson Nutrigenomics Consortium, TI Food and Nutrition, Nieuwe Kanaal 9A, 6709 PA Wageningen, The Netherlands; Nutrition, Metabolism and Genomics Group, Division of Human Nutrition, Wageningen University, Bomenweg 2, 6703 HD Wageningen, The Netherlands

Liu Su George H. Whipple Lab for Cancer Research, Departments of Urology and Pathology, University of Rochester Medical Center, Rochester, NY 14642, USA

Björn Vennström Department of Cell and Molecular Biology, Karolinska Institute, S-171 77 Stockholm, Sweden, Bjorn.Vennstrom@ki.se

Yandong Wang Department of Gene Regulation and Drug Discovery, Beckman Research Institute, City of Hope National Medical Center, 1500 E. Duarte Road, Duarte, CA 91010, USA

H.V. Westerhoff Molecular Cell Physiology, Vrije Universiteit, Amsterdam 1081 HV, The Netherlands; Manchester Centre for Integrative Systems Biology, Manchester Interdisciplinary Biocentre, University of Manchester, Manchester, UK

Shaozhen Xie George H. Whipple Lab for Cancer Research, Departments of Urology and Pathology, University of Rochester Medical Center, Rochester, NY 14642, USA

CHAPTER 1

NUCLEAR RECEPTORS AN INTRODUCTORY OVERVIEW

CHRIS M. BUNCE[1] AND MORAY J. CAMPBELL[2]

[1] *School of Biosciences, University of Birmingham, Edgbaston, Birmingham, B15 2TT, UK*
[2] *Department of Pharmacology and Experimental Therapeutics, Roswell Park Cancer Institute, Buffalo, NY, 14263, USA*

Abstract: This semi historical account of the origins of the nuclear receptor field and our current understanding of the biology these receptors is aimed at both the novice and experienced nuclear receptor biologist. For the novice it will provide a background scaffold for the forthcoming chapters. For the most experienced it will serve as a timely reminder of what fun we have had and how far this vitally important field has come since the first report of a nuclear rector in the mid 1980s.

1.1. RECEPTORS INSIDE CELLS!

In 1665 Robert Hooke reported his observations of tree bark using an early primitive microscope and gave birth to cell biology. Nineteen years later Antony van Leeuwenhoek observed protozoa as singular living cells. These seminary observations laid the foundations of the 'cell theory' proposed one and a half centuries later in 1839 by Schwann and Schleiden. At the core of this theory was the thesis that all life forms are made from one or more cells, that cells arise from pre-existing cells and that the cell is the smallest form of life. This principle of the cell as the 'unit of life' was immediately and almost universally accepted and remains a cornerstone of biology to this day.

At school students still learn about the cell as a unit of life capable of performing all the sustaining functions of the individual (in the case of protozoans) or becoming specialised 'differentiated' cells that perform discrete functions that collectively sustain and propagate metazoans. The cell, as a living unit, has a plasma membrane that serves as a physical interface between its inner functional machinery and its external environment. It's an easy step from there to conceptualise cell surface receptors inserted into these membranes to sense the environment, and thence to communicate the sensory experience to the inner cell, thereby eliciting appropriate physiological and transcriptional responses. Examples include the use of cell surface receptors by protozoans to sense chemorepellants, and the use of sophisticated and highly specific

C.M. Bunce, M.J. Campbell (eds.), *Nuclear Receptors*, Proteins and Cell Regulation 8, 1–13. DOI 10.1007/978-90-481-3303-1_1,

surface antigen receptors by lymphocytes that evolved only relatively recently in the jawed vertebrates.

It is not surprising therefore that amongst the student bodies in our universities and amongst our biological colleagues there is a relatively high level of appreciation of cell surface receptors and how they translate extracellular signals into intracellular messengers such as Ca^{2+}, cAMP and IP_3. However, the appreciation that not all the environmental sensing machinery of cells is on the 'outside' has much less of a stronghold within the general biological consciousness. It is our hope that this book will at least in part help re-address this imbalance in appreciation and understanding.

1.1.1. Nuclear Receptors

Collectively the family of receptors that this book addresses are most frequently termed 'Nuclear Receptors' (NRs). This is because many of the functions mediated by these receptors depend on their translocation to the nucleus either bound to ligand or in a ligand free state that is subsequently able to bind ligand when within the nucleus. As will be revealed these receptors commonly function within complex transcriptional complexes that selectively regulate target gene expression.

Herein lays part of the reason why NRs are less appreciated as elements of the sensory apparatus of cells. Nuclear receptors are both receptors and transcription factors. Indeed a great deal of our understanding of gene transcription has arisen from the study of NR- mediated transactivation and transrepression. Thus there is a tendency for biologists interested in cell signalling to 'pigeonhole' NRs as transcription factors rather than receptors. Anecdotally, this phenomenon was acutely illustrated in a recent teaching meeting at one of the contributing authors' institutions when a senior member of staff with a renowned research group in cell signalling, expressed surprise at the inclusion of lectures on nuclear receptor signalling in a module entitled 'Cellular signalling by hormones and neurotransmitters'! On another level nuclear receptors have evolved in metazoan animals they do not occur in bacteria, protozoa, yeast or in plants. Thus large cohorts of biologists are not exposed to nuclear receptor signalling in their day by day academic experiences.

The aim of the current book was in part to address these discrepancies. However quite apart from generating a volume for students and researchers wishing to better familiarise themselves with nuclear receptors our purpose here was also to provide a platform that will provoke cross-fertilisation of ideas and emerging concepts amongst researchers already within the field. As throughout the history of nuclear receptor biology new and exciting twists in the story keep appearing. Whilst giving a sound background in nuclear receptor biology we hope that this book highlights some of the new challenges arising from more recently acquired insights. Ultimately we hope that young researchers establishing themselves in this arena will find this volume of use in formulating the questions behind future investigations that will continue to sustain the evolution of this vital area of modern biology.

As a note of caution, be advised that this volume is not an exhaustive overview of modern nuclear receptor biology. It has not been possible for us to cover all aspects of this rapidly evolving field. A number of authors that we approached could not contribute because of recent or ongoing commitments to other projects. As with all rapidly developing research areas the field of nuclear receptors is subject to ongoing debate and conceptual development. In this regard it would also be erroneous to assume that all the contributors fully endorse all that is written outside of their own contribution. However in our role as editors we have satisfied ourselves that this volume gives an accurate snapshot of the field today.

1.2. HISTORICAL ASPECTS

1.2.1. The Early Years

Diverse small lipophilic hormone-like molecules including steroids, retiniods, thyroid hormones and vitamin D_3 were purified and characterised long before they were recognised as nuclear receptor ligands. They were isolated and identified because of their abilities to modulate development (including metamorphosis), cell differentiation and physiological processes and because of associations with human disease processes. Although for some time it was possible to measure selective binding of these 'hormones' within cells it was not until the mid 1980s that a bewilderingly explosive advance was made in the understanding of the receptors that perceive these signals.

In 1985 the primary structure of a cDNA encoding a human glucocorticoid receptor was described that encoded a protein that bound glucocorticoids similarly to native glucocorticoid receptors [1]. This was quickly followed by the cloning of a rat cDNA that when expressed in receptor negative, i.e. non steroid binding cells, restored steroid responses via hormone-dependent transcriptional enhancers [2]. At the same time cDNA encoding the first human estrogen receptor was also cloned [3]. In the latter study it was noted that the cloned estrogen receptor had extensive homology with the v-*erb-A* oncogene of the avian erythroblastosis virus. Similarly, a follow-up study by the group that had identified the glucocorticoid receptor also reported homologies with v-*erb-A* [4]. These observations quickly lead to the identification of the c-*erbA* locus as the gene encoding the human and chicken thyroid hormone receptors [5, 6]. Thus in just 2 years the field went from being unaware of how cells perceived lipophilic signalling hormones to having identified three receptors that displayed cross species conservation. Most remarkable was the realisation that chemically distinct hormones were demonstrably signalling via highly related receptors providing the first evidence of a new receptor family.

Based on the fact that the first receptors in this family to be cloned were steroid receptors the term *Steroid Hormone Receptor*-family or -superfamily was coined. However, as it became clear that not all ligands are true steroids, coupled with an increased understanding of the biology of these receptors the alternative description of *Nuclear Receptors* has become more widely adopted.

In the modern laboratory it is easy to overlook that these seminal observations were being made at a time when technologies that we now take for granted were only just becoming established. There is no doubt that the accelerated understanding of nuclear receptor biology from this point in time represents a tangible illustration of the impact these technologies have had on modern biology.

In 1987 the cloning of the avian vitamin D_3 receptor was described using newly available receptor-specific monoclonal antibodies to recover the appropriate cDNA from a chicken intestinal lambda gt11 expression library [7]. The identification that glucocorticoid, estrogen and thyroid hormone receptors and now vitamin D_3 were structurally related, prompted researchers to seek additional family members. The growing applicability of molecular biology allowed the searching of DNA for homologous sequences by using relatively conserved nuclear receptor sequences as low stringency hybridisation probes to identify candidate novel nuclear receptors and the subsequent cloning of those sequences. These approaches rapidly led to the identification of human mineralocorticoid receptor, and human retinoic acid receptors [8–10].

In less than a decade from the publishing of the cloning of the glucocorticoid receptor, 30 or more nuclear receptors had been identified (reviewed in [11]). These included a drosophila receptor for ecdysone, an insect hormone that orchestrates metamorphosis [12], indicating the early origin of this receptor superfamily during the development of metazoan animals. Toward the end of the 1990s the completion of the *C. elegans* genome project identified greater than two hundred candidate nuclear receptors, further emphasising the early origin of this receptor family [13]. By this time the number of receptors identified in mammals, birds and insects had risen to around seventy in total and so the potential that *C. elegans* alone had more than two hundred receptors appeared as a biological anomaly that remains poorly understood to this day. Despite this, interest in nuclear receptors within the *C. elegans* field has been surprisingly slow to develop; perhaps because few of the available *C. elegans* nuclear receptor mutants have phenotypes that can be measured. This possibly implies redundancy between receptors in this species that renders the study of their biology more difficult. One exception is the nuclear receptor DAF-12 that has been implicated in regulating the lifespan of *C. elegans*. Recently a metabolomics approach has identified steroids present in *C. elegans* that are identical to steroids present in humans including pregnenalone [14]. Treating with *C.* elegans with pregnenalone extended the lifespan of wt worms but not that of DAF-12 mutants strongly indicating that pregnenalone or a metabolite thereof is the activating ligand of DAF-12 [14].

For the seventy or so nuclear receptors identified in insects, birds and animals by the start of the twenty-first century less than half had assigned activating ligands. Those that did not became known as ‘orphan receptors’. Since that time several orphan receptors have become ‘adopted’ following the identification of intrinsic or xenobiotic ligands.

1.2.2. Key Concepts Begin to Emerge

By the mid 1990s some general concepts about nuclear receptor biology had emerged that remain more or less embedded in the current literature. Many of these concepts can be observed in their nascence in the excellent volume 'Nuclear Hormone Receptors' edited by Malcolm Parker and Published in 1991. They also were succinctly described in the 1995 review of David Mangelsdorf; 'The Nuclear Receptor Superfamily: The Second Decade' [15].

1.2.2.1. *Primary structure of nuclear receptors*

The primary structure of nuclear receptors became commonly presented as having five or six contiguous regions of variable degrees of homology running from the N-terminus to the C-terminus and nominated as A/B, C, D, E and F (Figure 1.1). This representation of nuclear receptors arose from early comparisons made of the human and chicken estrogen receptor amino acid sequences that identified three regions of high homology (A, C and E) interspersed with regions of lesser homology (B, D, and F) [16]. Retrospectively it is now recognised that these early structural classifications were foretellers of the functional compartmentalisation of nuclear receptor proteins. Krust et al. [16] identified the C region as entirely conserved between human and chicken estrogen receptor amino acid sequences and this region was later recognised as the most highly conserved across the nuclear receptor family. The region encodes a DNA binding domain (DBD) that is composed of two zinc

```
hER1    1   MTMTLHTKASGMALLHQIQGNELEPLNRPQLKIPLERPLGEVYLDSSKPAVYNYPEGAAY 60    A
cER1    1   ***********VT*******T***T*S**********S*SDM*VE*N*TG*F******T* 60

hER1   61   EFNAAAAANAQVYGQTGLPYGPGSEAAAFGSNGLGGFPPLNSVSPSPLMLLHPPPQLSPF 120   B
cER1   61   D*G----TT*P***S*T*S*A*T**S--***SS*A**HS**N*P***VVF*QTA****** 114

hER1  121   LQPHGQQVPYYLENEPSGYTVREAGPPAFYRPNSDNRRQGGRERLASTNDKGSMAMESAK 180
cER1  115   IHH*S**********QGSFGM***A*******S*****HSI***MS***E***LS***T* 174

hER1  181   ETRYCAVCNDYASGYHYGVWSCEGCKAFFKRSIQGHNDYMCPATNQCTIDKNRRKSCQAC 240   C
cER1  175   ************************************************************ 234

hER1  241   RLRKCYEVGMMKGGIRKDRRGGRMLKHKRQRDDGEGRGEVGSAGDMRAANLWPSPLMIKR 300   D
cER1  235   **********************E*M*Q****EEQDS*NGEA*STEL**PT**T***VV*H 294

hER1  301   SKKNSLALSLTADQMVSALLDAEPPILYSEYDPTRPFSEASMMGLLTNLADRELVHMINW 360
cER1  295   N****P******E*******E*****V******N***N*****T**************** 354

hER1  361   AKRVPGFVDLTLHDQVHLLECAWLEILMIGLVWRSMEHPGKLLFAPNLLLDRNQGKCVEG 420
cER1  355   ************************************************************ 414   E

hER1  421   MVEIFDMLLATSSRFRMMNLQGEEFVCLKSIILLNSGVYTFLSSTLKSLEEKDHIHRVLD 480
cER1  415   ***********AA*************************************R*Y****** 474

hER1  481   KITDTLIHLMAKAGLTLQQQHQRLAQLLLILSHIRHMSNKGMEHLYSMKCKNVVPLYDLL 540
cER1  475   ************S**S*****R**************************N************* 534

hER1  541   LEMLDAHRLHAPTSRGGASVEETDQSHLATAGSTSSHSLQKYYITG-EAEGFPATV 595      F
cER1  535   ************AA*SA*PM**ENRNQ*T**-PA******SF**NSK*E*SMQN*I 589
```

Figure 1.1. Re-representation of the comparison of the human (hER) and chicken (cER) estrogen receptor amino acid sequences as originally described by Krust and colleagues in 1986 [16]. The *shaded* and *non shaded* areas denote regions of identified higher and lesser homology that were the foretellers of the structure/function relationships of the nuclear receptor family

fingers. This DNA binding motif is characteristic of nuclear receptors and sets them apart from other DNA binding proteins [17, 18]. Indeed it was the exclusivity of the sequences encoding these DBDs that allowed for the cloning of novel nuclear receptors using moderate stringency hybridisation approaches.

The second most conserved region of Nuclear Receptors is the region denoted E which transpired to include the ligand binding domain. The DBD (region C) and the ligand binding domain (region E) are separated by a variable region D which acts as a hinge region conferring flexibility on the receptor that allows it to shape-change ligand upon binding. The N-terminal A/B and C-terminal F variable domains encode a number of effector functions that will be described in later chapters.

1.2.2.2. Different receptor classes bind DNA in differing ways

Nuclear receptors became broadly divided into classes based on their binding to DNA as either homodimers, heterodimers or as monomers. The receptors of the classical steroids, the glucocorticoid, mineralocorticoid, progesterone androgen and estrogen receptors (GR, MR, PR, AR and ER) were found to bind DNA as homodimers and formed the 'Group I' receptors. The Group II and III receptors bind DNA as predominantly as heterodimers. The group II receptors were distinct from group III receptors in as much as they had defined ligands whereas group III receptors were orphan receptors. This distinction is becoming less clear as a number of receptors once considered as orphan receptors, have now been assigned ligands or putative ligands. The group II receptors originally included the receptors of triidothyronine (thyroid hormone) all-trans retinoic acid, $1\alpha,25(OH)_2$-vitamin D_3 and the ecdysone receptor. This has grown to include receptors of eicosinoids and other lipid derived mediators. In 1990 Mangelsdorf et al described a novel receptor that bound 9-cis retinoic acid that they named the Retinoid X Receptor [19]. This discovery marked a major breakthrough in understanding as RXRs became understood as common heterodimeric partners of both group II and III receptors.

The group IV receptors bind DNA as monomers and until recently were almost exclusively orphan receptors. Indeed some group III and IV orphan receptors have been proposed to be ligand independent mediators of gene transcription. However time will tell. For example the monomeric binding Nor1 receptor had been considered to be ligand independent until the cyclopentenone prostaglandin PGA_2 was recently described as an activating ligand in mice [20]. The importance of group IV receptors in many systems has been difficult to ascertain possibly because of overlapping 'promiscuous' functions. For example Nor1 deficient mice display only mild phenotypes. Similarly, mice lacking the related receptor Nur77 lack overt defects. However, a recent report demonstrated that abrogation of both receptors leads to the development of acute myeloid leukaemia [21]. This somewhat startling observation serves as a timely reminder of the central importance of nuclear receptors in tissue homeostasis and a powerful illustration of how much is still to be elucidated in this exciting and rapidly moving field.

1.2.2.3. A unifying nomenclature system

By the late 1990s more than 300 bonefide or potential, nuclear receptor sequences had been identified. Increasingly a crisis of nomenclature was arising which was most acute in the case of orphan receptors that could not be named by virtue of their activating ligand. As in other superfamilies of proteins different researchers began to define the same nuclear receptors using differing titles. To combat this problem Vincent Laudet, Johan Auwerx, Jan-Ake Gustafsson and Walter Wahli formed the Nuclear Receptors Nomenclature Committee and together with the INTERNATIONAL Union of Pharmacology Committee on Receptor Nomenclature and Drug Classification (NC-IUPHAR) formed a system for unifying the nomenclature of Nuclear Receptors. The system is similar to that used for the cytochrome P450 family [22] and in principle is based on an evolutionary tree as constructed using sequence identities. In this way subfamilies become identified that are given Arabic numerals. Nuclear receptors showing grouping of evolutionary homology within a subfamily are further identified by the use of capitol letters and individual receptors within these clusters identified with a further Arabic numeral [23]. The value of this system is illustrated by the nuclear receptor designated as NR1H2 that has been variously described as UR, OR-1, NER-1 RIP15 and LXRβ. The Nomenclature committee proposed that all new sequences be submitted prior to publication to receive an official tag. The committee also proposed that all publications concerning nuclear receptor studies should include the use of the new official name at least once in the summary and again in the introduction. The system was not meant to remove the use of trivial or functional names but to provide clarity about which receptors each study concerned. As the field progressed the nomenclature system had to develop to encompass a number of nuclear receptor isoforms that arise from alternative RNA splicing. Thus under this system, the thyroid hormone receptor isoforms TRα1, TRα2 and TRα3 are defined as NR1A1a, NR1A1b and NR1A1c respectively. The role of splice variant isoforms and the extra level of intricacy that they bring to the nuclear receptor signalling network will be touched upon in later chapters.

1.2.2.4. Different receptor classes bind at selective response elements

By the late 1980s to the early 1990s it became apparent that the ability of nuclear receptors to regulate gene expression in a selective fashion is largely mediated by the recognition of and binding to specific DNA sequences initially known as hormone response elements (HREs) [24, 25]. The response elements for many nuclear receptors utilise a common consensus sequence often described as AGGTCA although as will be illustrated in later chapters variations do occur. For dimeric nuclear receptors HREs are composed of two repeats of the concensus sequence separated by small but variable numbers of non-consensus nucleotides. Response element consensus sequences can variably exist as direct repeats (AGGTCA...AGGTCA), inverted repeats (AGGTCA...ACTGGA) or everted repeats (ACTGGA...AGGTCA). By 1995 the concept had arisen that these configurations and the spacing between them compliment the physical nature of the nuclear receptor dimers that bind them and thereby contribute strongly to selectivity of gene regulation (reviewed by Forman and Evans in 1995 [26]). Attempts to verify these concepts by structural biologists have until recently been hampered by the need to use nuclear receptor DNA binding domain subfragments in order to derive good enough crystals. However, a recent study by Chandra et al. has reported structures of an intact nuclear receptor heterodimer binding to its selective response element [27]. This ground breaking study and others that are bound to follow, will provide a more intimate insight into the complex actions of nuclear receptors and no doubt spawn a new wave of investigations.

1.2.2.5. Nuclear receptors can both trans-activate and trans-repress gene expression via binding to response elements

By the end of the 1990s studies of the retinoic acid and thyroid hormone receptors had demonstrated that heterodimeric receptors are able bind to their REs both in the presence and absence of ligand. Furthermore, when bound in the absence of ligand NRs were observed to repress gene expression actively whereas in the presence of ligand gene expression was activated. A concerted research effort involving many laboratories across the globe has now revealed that RE-bound NRs serve to recruit immense protein holo-complexes that bridge the DNA-bound NR complex with the basal transcriptional machinery at the start site of the targeted gene. Importantly these complexes differ in their composition and function depending on whether the associated NRs have bound ligand.

A seminal discovery in this arena was the identification that repressive complexes associated with unliganded NRs contain key proteins termed co-repressors. The first of these to be described were NCOR1 (nuclear receptor co-repressor [28]) and NCOR2/SMRT (silencing mediator for retinoid and thyroid hormone receptor [29]) and perhaps they remain the most studied and best understood. It soon became apparent that co-repressor complexes containing NCOR2/SMRT or NCOR1 also contain histone deacetylases (HDACs). These enzymes modify local histones permitting local changes to chromatin structure and configuration that contribute to gene repression. In contrast complexes associated with DNA-bound and liganded NR dimers contain co-activator proteins rather than co-repressors. Some of these co activators and other proteins in the complex have associated histone acetyl transferase (HAT) activities that result in the reverse effect of HDAC activity. A frenetic period of research ensued aimed at understanding the structure and function relationships of both co-repressive and co-activating NR complexes. A number of reviews written at the turn of the century endeavoured to pull this information together into comprehensible models. To single out any one of these is perhaps a disservice to the others. However, at the risk of being disingenuous the reader is referred to the excellent review of Christopher Glass and Michael Rosenfeld which was timely published in January 2000 [30].

Research in this area continues apace and a picture is evolving of a highly dynamic regulation of gene expression by NRs. From studies of estrogen receptor and vitamin D receptor regulated genes it appears that NR-mediated transcription upon ligand exposure is cyclical with levels of transcribed mRNA fluctuating rather than rising steadily [31–33]. A recent study from one of the contributory authors of this book has provided mechanistic insight to this process. During the 1α, 25-dihydroxyvitamin D_3 induced expression of the *p21* gene the fluctuating rise of mRNA was mirrored by cyclical binding of VDR to vitamin D HREs (VDREs) distal to the transcription start site (TSS). The model that arises from the study is that 1α,25-dihydroxyvitamin D_3 regulates VDR-mediated *p21* gene transcription via inducing cyclical chromatin looping by the bridging of the distal VDRE with the *p21* TSS. Thus in addition to local histone modifications, NR associated complexes also regulate gene expression by engendering larger scale mechanical changes in chromatin structure [34].

1.3. TOWARDS A MORE INTEGRATED VIEW OF NR SIGNALLING

To date NR research has been largely introspective with groups being focused on a single or a few NRs of choice. Despite this it has emerged that NRs do not work in isolation and considerable cross talk occurs between the various NRs expressed within a given cell. Intuitively this is appropriate since normal cell function must depend on the integration of simultaneously perceived signals. However, this must also include the integration of signals received by cell surface receptors and NRs expressed by the same cell.

Considerable insight into the possible mechanisms of such integration has come from studies of inflammation. It has been recognised for some time that certain nuclear receptor ligands are able to repress the expression of inflammatory response genes. This gave rise to the concept of ligand dependent gene repression that flew against the established concept of ligand dependant gene transcription. In the autumn of 2005 the groups of Michael Rosenfeld and Christopher Glass published two collaborative studies that provided new mechanistic insights to these processes.

One study investigated the complex interplay between cell surface toll like receptor (TLR) signalling and multiple NRs in the regulation of inflammatory response genes. The study revealed that liganded glucocorticoid receptors repressed a large set of inflammatory response genes by disrupting p65 containing transcriptional complexes required for signalling via the toll like receptors TLR4 and TLR9 but not for TLR3. In contrast the ligand dependent transrepression of TLR signalling by PPARγ and LXR was observed to be p65-independent [35]. The other study considered the repression of NF-κB activity by ligands of the nuclear receptor PPARγ and identified that the process involved the ligand dependent SUMOylation of PPARγ. This protein modification of PPARγ resulted in the targeting of the receptor not to conventional PPAR response elements (PPREs), but to NCor-HDAC complexes at NF- κB reponse elements within inflammatory gene promoters. As a result these corepressor complexes were protected from ubiquitinylation thereby preventing their proteasomal degradation and resulting in the maintenance of the promoter in a repressed state [36].

These studies and others reveal that NRs not only interact with each other to regulate gene transcription but also with other transcriptional complexes including those that mediate gene transcription in response to messages first perceived at the cell surface. What emerges is a highly complex role for NRs in regulating cellular responses to hormonal, environmental and xenobiotic cues. The picture lightly sketched here is that NRs have the ability to dominate over cell surface signalling and thus act to police responses by cell surface receptors. This is almost certainly not the case. It is just as likely that cell surface receptor signalling regulates NR activity. One mode of regulation that is receiving growing attention is via NR phosphorylation. A growing number of reports indicate receptor and site specific phosphorylations of NRs that regulate the sensitivity of ligand responses, subcellular localisation, DNA binding, protein stability and protein-protein interactions (reviewed in [37]). The possibility for signalling crosstalk by posttranslational modifications of NRs becomes still more complex when one also considers that NRs have been demonstrated direct targets for acetylation and arguably glycosylation.

1.4. NON GENOMIC ACTIONS OF NRS

No introductory overview of NR biology would be complete without reference to non genomic actions of NRs. This broad and ill defined term relates to a large and still growing body of literature that could be perhaps better described under the term 'extranuclear activities of nuclear receptors'. When reading this diverse literature it is important to distinguish those studies that show NR proteins having extra-nuclear signalling properties from others that describe non-genomic actions of recognised NR ligands mediated by cellular targets other than their cognitive NR. For example thyroid hormones have been demonstrated to bind integrin alphaVbeta3 and activate mitogen activated protein kinase (MAPK) pathways [38]. However this is not a non-genomic action of either TRα or TRβ nuclear receptor proteins.

In contrast a recent study identified evidence that an oncogenic dominant negative TRβ (PV) directly interacts with the regulatory p85alpha subunit of phosphatidylinositol 3-kinase (PI3K) and thereby activates PI3K-downstream signalling pathways [39]. Although an activity of a mutant TRβ, it has yet to be determined if this activity is integral to the normal activity of these receptors. Nonetheless, it is interesting to note that NCOR1 was found to regulate PV-activated PI3K signalling by competitive binding to the C-terminal SH2 domain of p85alpha.

It is also noteworthy that other NRs have been identified as putative ligand dependent regulators of PI3K activity. For example high dose corticosteroids resulted in enhanced PI3K signalling via a mechanism that was antagonised by GR-antagonists. The use of inhibitors, antagonist and analyses of the dose and time dependency of this action indicated that stimulation of PI3K was a rapid, non-transcriptional effect of GR [40]. Separate studies also demonstrated ligand induced association of estrogen receptor (ERα) with p85alpha and the consequent activation PI3K [41, 42].

Furthermore the in vivo vascular protective effect of oestrogen following ischaemia and reperfusion injury in mice was abolished in the presence PI3K inhibitors [41].

More recently it has been demonstrated that retinoic acid nuclear receptors (RARs) also bind to the p85 subunit of PI3K in SH-SY5Y neuroblastoma cells and again regulate PI3K activity in a ligand dependent manner [43]. The use of RAR-null cells appeared to confirm that the ligand dependent activation of PI3K was indeed mediated by NRs. However the mechanism was described as distinct from that of the GR and ER mediated affects. Whereas in the aforementioned studies, GR and ER association with p85 was driven by ligand, the association of RARα with p85 was detected in the absence of exogenously added ligand. In this scenario addition of ligand appeared to increase the association of the catalytic sub unit of PI3K (p110) with p85 [43].

It could be argued that the few studies of non-genomic actions of NRs quoted here are flawed because they used mutated receptors, tumour cell lines rather than normal cells and non-physiological doses of either natural or pharmaceutical ligands. Equally these criticisms could be rolled out to discredit the many other studies that we do not have the space to mention. However, similar criticisms could have also been made of the many studies that have helped establish much of the central dogma of NR signalling. It should be noted that in the recent study by Masiá S et al. the association of RARα with p85 was independent of applied natural or pharmacological ligands [43]. Furthermore, there is no evidence that this association requires either RARα or p85 mutations. Thus with the caveat that this was an in vitro observation in a tumour cell line it adds to accumulating evidence that PI3K represents a potential interface between cell surface receptor mediated signalling and extranuclear NR mediated signalling.

1.5. HERE WE GO

The pages and chapters that follow represent the best understanding and beliefs of researchers currently studying NR biology. The authors collectively have a varied background, experience and standing in the field and no doubt would debate strenuously what is important and what is not. What we hope is that you the reader will share with us an enthusiasm and excitement about the field and the importance it has for understanding the organisation, homeostasis and evolution of metazoan animals such as ourselves.

REFERENCES

1. Hollenberg, S. M. et al. (1985). Primary structure and expression of a functional human glucocorticoid receptor cDNA. *Nature* 318, 635–641.
2. Miesfeld, R. et al. (1986). Genetic complementation of a glucocorticoid receptor deficiency by expression of cloned receptor cDNA. *Cell* 46, 389–399.
3. Greene, G. L. et al. (1986). Sequence and expression of human estrogen receptor complementary DNA. *Science* 231, 1150–1154.

4. Weinberger, C., Hollenberg, S. M., Rosenfeld, M. G., and Evans, R. M. (1985). Domain structure of human glucocorticoid receptor and its relationship to the v-erb-A oncogene product. *Nature* 318, 670–672.
5. Weinberger, C. et al. (1986). The c-erb-A gene encodes a thyroid hormone receptor. *Nature* 324, 641–646.
6. Sap, J. et al. (1986). The c-erb-A protein is a high-affinity receptor for thyroid hormone. *Nature* 324, 635–640.
7. McDonnell, D. P., Mangelsdorf, D. J., Pike, J. W., Haussler, M. R., and O'Malley, B. W. (1987). Molecular cloning of complementary DNA encoding the avian receptor for vitamin D. *Science* 235, 1214–1217.
8. Arriza, J. L. et al. (1987). Cloning of human mineralocorticoid receptor complementary DNA: Structural and functional kinship with the glucocorticoid receptor. *Science* 237, 268–275.
9. Giguere, V., Ong, E. S., Segui, P., and Evans, R. M. (1987). Identification of a receptor for the morphogen retinoic acid. *Nature* 330, 624–629.
10. Petkovich, M., Brand, N. J., Krust, A., and Chambon, P. (1987). A human retinoic acid receptor which belongs to the family of nuclear receptors. *Nature* 330, 444–450.
11. Reichel, R. R. and Jacob, S. T. (1993). Control of gene expression by lipophilic hormones. *FASEB J* 7, 427–436.
12. Koelle, M. R. et al. (1991). The Drosophila EcR gene encodes an ecdysone receptor, a new member of the steroid receptor superfamily. *Cell* 67, 59–77.
13. Sluder, A. E., Mathews, S. W., Hough, D., Yin, V. P., and Maina, C. V. (1999). The nuclear receptor superfamily has undergone extensive proliferation and diversification in nematodes. *Genome Res* 9, 103–120.
14. Broue, F., Liere, P., Kenyon, C., and Baulieu, E. E. (2007). A steroid hormone that extends the lifespan of Caenorhabditis elegans. *Aging Cell* 6, 87–94.
15. Mangelsdorf, D. J. et al. (1995). The nuclear receptor superfamily: The second decade. *Cell* 83, 835–839.
16. Krust, A. et al. (1986). The chicken oestrogen receptor sequence: Homology with v-erbA and the human oestrogen and glucocorticoid receptors. *EMBO J* 5, 891–897.
17. Berg, J. M. (1989). DNA binding specificity of steroid receptors. *Cell* 57, 1065–1068.
18. Klug, A. and Schwabe, J. W. (1995). Protein motifs 5. Zinc fingers. *FASEB J* 9, 597–604.
19. Mangelsdorf, D. J., Ong, E. S., Dyck, J. A., and Evans, R. M. (1990). Nuclear receptor that identifies a novel retinoic acid response pathway. *Nature* 345, 224–229.
20. Kagaya, S. et al. (2005). Prostaglandin A2 acts as a transactivator for NOR1 (NR4A3) within the nuclear receptor superfamily. *Biol Pharm Bull* 28, 1603–1607.
21. Mullican, S. E. et al. (2007). Abrogation of nuclear receptors Nr4a3 and Nr4a1 leads to development of acute myeloid leukemia. *Nat Med* 13, 730–735.
22. Nebert, D. W. et al. (1987). The P450 gene superfamily: Recommended nomenclature. *DNA* 6, 1–11.
23. Nuclear Receptors Nomenclature Committee. A unified nomenclature system for the nuclear receptor superfamily. *Cell* 97, 161–163 (1999).
24. Umesono, K., Giguere, V., Glass, C. K., Rosenfeld, M. G., and Evans, R. M. (1988). Retinoic acid and thyroid hormone induce gene expression through a common responsive element. *Nature* 336, 262–265.
25. Truss, M., Chalepakis, G., Slater, E. P., Mader, S., and Beato, M. (1991). Functional interaction of hybrid response elements with wild-type and mutant steroid hormone receptors. *Mol Cell Biol* 11, 3247–3258.
26. Forman, B. M. and Evans, R. M. (1995). Nuclear hormone receptors activate direct, inverted, and everted repeats. *Ann NY Acad Sci* 761, 29–37.
27. Chandra, V. et al. (2008). Structure of the intact PPAR-gamma-RXR- nuclear receptor complex on DNA. *Nature* 456, 350–356.
28. Horlein, A. J. et al. (1995). Ligand-independent repression by the thyroid hormone receptor mediated by a nuclear receptor co-repressor. *Nature* 377, 397–404.

29. Chen, J. D. and Evans, R. M. (1995). A transcriptional co-repressor that interacts with nuclear hormone receptors. *Nature* 377, 454–457.
30. Glass, C. K. and Rosenfeld, M. G. (2000). The coregulator exchange in transcriptional functions of nuclear receptors. *Genes Dev* 14, 121–141.
31. Metivier, R. et al. (2003). Estrogen receptor-alpha directs ordered, cyclical, and combinatorial recruitment of cofactors on a natural target promoter. *Cell* 115, 751–763.
32. Kim, S., Shevde, N. K., and Pike, J. W. (2005). 1,25-Dihydroxyvitamin D3 stimulates cyclic vitamin D receptor/retinoid X receptor DNA-binding, co-activator recruitment, and histone acetylation in intact osteoblasts. *J Bone Miner Res* 20, 305–317.
33. Banwell, C. M. et al. (2006). Altered nuclear receptor corepressor expression attenuates vitamin D receptor signaling in breast cancer cells. *Clin Cancer Res* 12, 2004–2013.
34. Saramaki, A. et al. (2009). Cyclical chromatin looping and transcription factor association on the regulatory regions of the p21 (CDKN1A) gene in response to 1alpha,25-dihydroxyvitamin D3. *J Biol Chem* 284, 8073–8082.
35. Ogawa, S. et al. (2005). Molecular determinants of crosstalk between nuclear receptors and toll-like receptors. *Cell* 122, 707–721.
36. Pascual, G. et al. (2005). A SUMOylation-dependent pathway mediates transrepression of inflammatory response genes by PPAR-gamma. *Nature* 437, 759–763.
37. Weigel, N. L. and Moore, N. L. (2007). Steroid receptor phosphorylation: A key modulator of multiple receptor functions. *Mol Endocrinol* 21, 2311–2319.
38. Bergh, J. J. et al. (2005). Integrin alphaVbeta3 contains a cell surface receptor site for thyroid hormone that is linked to activation of mitogen-activated protein kinase and induction of angiogenesis. *Endocrinology* 146, 2864–2871.
39. Furuya, F., Lu, C., Guigon, C. J., and Cheng, S. Y. (2009). Nongenomic activation of phosphatidylinositol 3-kinase signaling by thyroid hormone receptors. *Steroids* 74, 628–634.
40. Hafezi-Moghadam, A. et al. (2002). Acute cardiovascular protective effects of corticosteroids are mediated by non-transcriptional activation of endothelial nitric oxide synthase. *Nat Med* 8, 473–479.
41. Simoncini, T. et al. (2000). Interaction of oestrogen receptor with the regulatory subunit of phosphatidylinositol-3-OH kinase. *Nature* 407, 538–541.
42. Castoria, G. et al. (2001). PI3-kinase in concert with Src promotes the S-phase entry of oestradiol-stimulated MCF-7 cells. *EMBO J* 20, 6050–6059.
43. Masia, S., Alvarez, S., de Lera, A. R., and Barettino, D. (2007). Rapid, nongenomic actions of retinoic acid on phosphatidylinositol-3-kinase signaling pathway mediated by the retinoic acid receptor. *Mol Endocrinol* 21, 2391–2402.

CHAPTER 2

WHAT DOES EVOLUTION TEACH US ABOUT NUCLEAR RECEPTORS?

GABRIEL MARKOV[1,2], FRANÇOIS BONNETON[1], AND VINCENT LAUDET[1]

[1] *Institut de Génomique Fonctionnelle de Lyon; Université de Lyon; Université Lyon 1; CNRS; INRA; Ecole Normale Supérieure de Lyon; Lyon cedex, France*

[2] *USM 501 – Evolution des Régulations Endocriniennes. Muséum National d'Histoire Naturelle, Paris, France*

Abstract: In this chapter we first summarise the current knowledge about the phylogenetic spectrum of nuclear receptors (NRs). Then, we discuss how studying their diversity can be helpful to make insights about their evolution. Significant attention is paid to the evolution of ligand-binding ability. Recent evolutionary and functional data have challenged the traditional concept of ligand, providing a more complex view of the mechanisms by which the transcriptional activity of NRs can be modulated. Finally, we argue that the evolutionary analysis of NRs has contributed to a conceptual shift of our understanding of nuclear receptors, from highly specific endocrine regulators to a promiscuous metabolic rheostat.

2.1. INTRODUCTION

Nuclear receptors (NRs) are classically defined as ligand-activated transcription factors that allow the regulation of target genes by small lipophilic molecules such as hormones (e.g. thyroid hormones or steroids), morphogen (e.g. retinoic acid) or dietary components (e.g. fatty acids). All built with a similar organization, NRs are nevertheless regulated by a wide diversity of compounds and are implicated in a tremendous diversity of physiological and metabolic processes. The question of the origin of such a system has received much attention because of its intrinsic interest, but also because it provides a nice experimental and conceptual framework to understand the origin of complex regulatory systems. Indeed, the NR family is a model of choice to address evolutionary issues because these proteins are present, with various physiological roles, in a broad range of well-studied organisms. Due to their importance, sequence, structural and physiological data have all been used to answer evolutionary questions. Notably, since NRs are involved in the integration

C.M. Bunce, M.J. Campbell (eds.), *Nuclear Receptors*, Proteins and Cell Regulation 8, 15–29. DOI 10.1007/978-90-481-3303-1_2,

of genomic and environmental processes, they are a critical link to understand the molecular basis of phenotypic plasticity.

2.2. NRs PHYLOGENY AND CLASSIFICATION

Before addressing questions about NR evolution, it is necessary to place the current knowledge on a solid phylogenetic framework. Indeed, classification is the first step required prior to any evolutionary analysis, since it is only by knowing the relationships between taxa (either proteins or organisms) that one can safely propose evolutionary hypothesis to be tested. The presence of two functionally conserved domains, the DNA-binding domain (DBD) and the ligand-binding domain (LBD) have proved highly informative for tree reconstruction, since it allows the generation of robust phylogenies [1–3].

The phylogeny of NRs provided a framework to establish a nomenclature for the family, a particularly useful tool given the increasing number of new NRs sequences coming from many different organisms [4]. According to this nomenclature, the NR family is divided into six subfamilies (NR1 to NR6). Each of these subfamilies is a robust monophyletic group, in which all receptors clustered in a subfamily originate from a single ancestor. The precise relationships between the six subfamilies are still unclear, blurring our views about the origin of the family itself (see below). It is interesting to note that two subfamilies (I and IV) cluster receptors that are able to interact with RXR in vertebrates, suggesting that this feature is not common to the whole family (Figure 2.3) [5]. Similarly, in vertebrates, subfamily III clusters members that are able to dimerize on palindromic elements. In contrast to this link between evolutionary history and DNA binding activity, no link between the ligand binding ability and the phylogeny has been detected. Steroid receptors are present in different subfamilies (I and III), and strongly related receptors (e.g. within the NR1 family), bind molecules as different as thyroid hormones (NR1A), retinoids (NR1B) or prostaglandins (NR1C).

Also noteworthy is the fact that the family contains proteins that lack one of the two conserved domains. In the official nomenclature, these proteins are artificially gathered into a specific subfamily (NR0) that has no biological meaning in itself (that is, all its members do not share a specific ancestral receptor [4]). Without a LBD, the protein is not a receptor, although it can act as a classical ligand-independent transcription factor. The well-known gap segmentation gene *knirps*, a transcriptional repressor in insects, is a good example of the NR0A group [6]. Without a DBD (group NR0B), the protein cannot act as a transcription factor. For example, the Small Heterodimer Partner (SHP; NR0B2) is an orphan corepressor of various transcription factors, including nuclear receptors [7] whereas DAX-1 (NR0B1) plays a role in sex determination mechanisms in mammals. These two paralogues are distantly related to the TLL group within subfamily II. Members of the NR0 subfamily provide interesting examples of how protein domains are reshuffled during evolution, a major source of molecular innovation [8]. In this view, proteins are composed of modules that follow their own evolutionary path and the phylogeny of a given protein is not necessarily identical to the phylogenies of its constituent modules. In the NR family, this remains an exception.

2.3. NR COMPLEXITY IS NOT LIMITED TO VERTEBRATES

In addition to an established phylogeny, the understanding of NR evolution requires a better knowledge of the phylogenetic distribution of receptors in a various sets of metazoans [9, 10]. This also allows discovery of a much more diverse family than originally anticipated by studying only NRs from common model organisms such as either human, mouse or *Drosophila*.

Figure 2.1 summarises the current knowledge on the phylogenetic distribution of NRs, with the addition of some notable events concerning specific model organisms. It should be underlined that the two best-studied non-vertebrate models, i.e. *Drosophila* and *Caenorhabditis*, are members of the group ecdysozoa, which cannot be taken as a picture of the 'ancestral condition' of NR functioning. This figure shows that the last common ancestor of all bilaterian animals, *Urbilateria*, most likely possessed about 22–25 receptors, and subsequently complex events of gene loss, gene duplication and domain shuffling occurred. Some receptors such as NR3A/ER or NR1A/TR, were lost in tunicates and ecdyzosoans, indicating that such animals should not be taken as representatives of a 'primitive state'. On the contrary, the physiological regulatory networks of these organisms are certainly as much derived from earlier organisms as vertebrate networks are. In vertebrates, the number of receptors and their phylogenetic relationships fit very well with whole genome duplication events [11]. There are also many examples of duplication of one peculiar receptor. The amphioxus *Branchiostoma floridae* has ten copies of the NR1H, due to lineage-specific duplication [12], and, more spectacular, the NR2A of *Caenorhabditis elegans* has about 250 copies [13, 14]. Concerning the nematode receptors, it should be mentioned that many of them have diverged considerably, with some receptors that are linked to no clear group, for example the 'NR1K' of *Onchocerca volvulus* [15]. Such sequences may be very divergent forms of NRs that are otherwise well conserved among metazoans. It should be possible to test this exciting hypothesis using recent data about the genomes of *Brugia malayi* [16], *Meloidogyne incognita* [17], and *Pristionchus pacificus* [18]. The very different life styles of these species may also allow correlation between biology (nutrition, ecology, reproduction) and NR duplications. The finding of NRs with 2 DBDs in various lophotrochozoans and in *Daphnia* opens other promising research fields, suggesting once again parallel losses in nematodes, insects and maybe vertebrates [19].

In short, the emerging complete picture of metazoan NRs indicates that, in spite of the conservation of some basic mechanisms, receptors present in various metazoan phyla are extremely diverse. The characterization of NRs throughout the whole metazoan biodiversity thus offers a view on the real flexibility of the NR structural

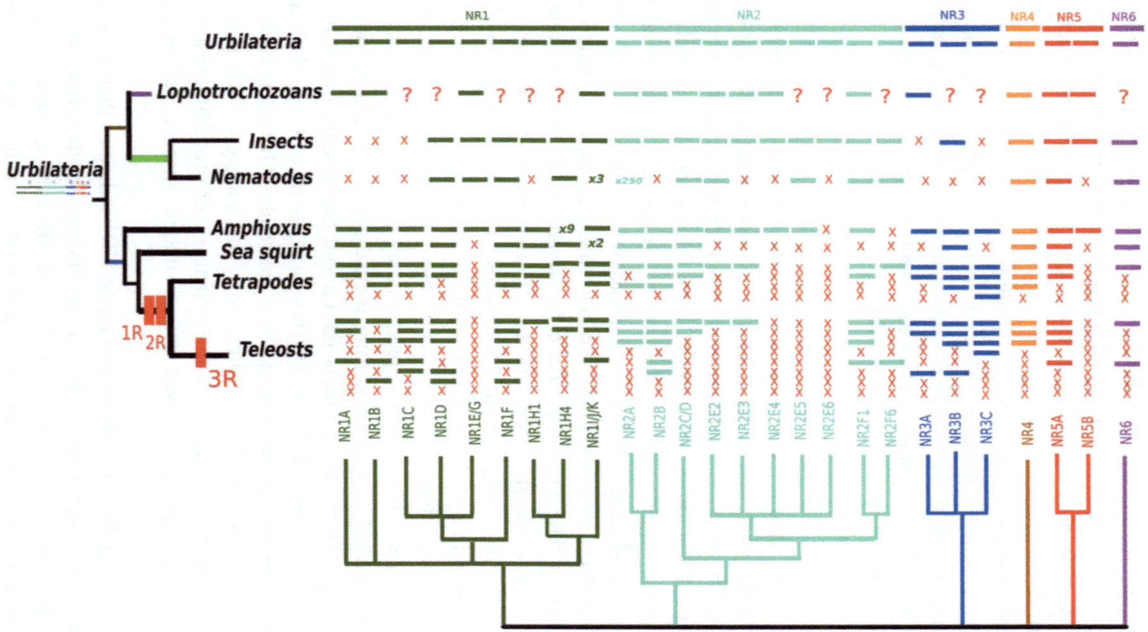

Figure 2.1. Phylogeny of the NR family and NR contain of some bilaterian genomic models. The following species were taken as representative of their taxon: *Tribolium castaneum* for insects [64], *Caenorhabditis elegans* for nematodes [10], *Homo sapiens* for tetrapodes [10] and *Danio rerio* for teleosts [11]. For lophotrochozoans, given the lack of extensive genomic search for NRs in the litterature, we merged partial data about various species [10, 65]. The first line represents the inferred NR contain of *Urbilateria*, the last common ancestor of all bilaterian animals [10]. Orthologous receptors are on the same column. Lost receptors are indicated by *red crosses*, whereas missing data are indicated by *red question mark*; lineage-specific duplications are indicated by « x*N* », where *N* is the number of paralogs. The three rounds or vertebrate whole-genome duplications that led to four paralog groups in tetrapodes and eight paralog groups in teleosts are indicated by 1R, 2R and 3R

modules. NRs are conserved proteins that were prone to some changes during evolution and a future challenge is to better understand to what extent these changes have contributed to phenotypic plasticity.

2.4. NR-LIKE ARE FOUND THROUGHOUT THE TREE OF LIFE

The origin of NRs is still unknown. A decade ago, a PCR screen indicated that NRs are found only in metazoans [9]. Recently it was confirmed that NRs are absent from the genome of the unicellular *Monosiga brevicollis*, which belongs to choanoflagellates, the sister group of animals [20]. The metazoan-specific DBD of nuclear receptors contains two C4-zinc fingers that are structurally related to the GATA C4-zinc fingers, which are found in all eukaryotes. It is possible that Nuclear Receptor DBD arose by duplication of a single ancestral C4-Zinc finger. By contrast, the LBD of nuclear receptors share no similarity with other domains outside animals. However, NR-like proteins were identified recently in the budding yeast. The heterodimeric transcription factors Oaf1/Pip2 are bound and regulated by fatty acids (oleate) through a mechanism that is very similar to PPAR/RXR [21]. Even more surprising is the suggestion, based on structure predictions, deletion and mutation analysis, that these proteins contain a LBD with a NR folding. Since the sequence identity is not significant, it is impossible to determine whether these resemblances are due to either homology or homoplasic evolution. Similarly, the yeast transcription factors Pdr1p/Pdr3p are regulated by xenobiotics, like the PXR nuclear receptor [22]. All these four proteins contain a zinc-finger DBD (Zn6Cys2) and orthologs of the putative LBD were found in other ascomycetes [21] (Figure 2.2). Therefore, fungi can use ligand-regulated transcription factors that share many functional and structural characteristics with NRs of animals (Figure 2.2).

It is interesting to recall that other ligand-activated transcription factors exist in animals. Indeed, the aryl hydrocarbon receptor (AHR) is a member of the bHLH-PAS family that contains a DBD of the basic-Helix-Loop-Helix type (bHLH) and a Per-Arnt-Sim (PAS) domain involved in the ligand binding activity [23] (Figure 2.2). Like NRs, the AHR can be bound and activated by a wide diversity of small lipophilic ligands that can act as either signalling molecules or xenobiotics (including dioxin).

Unfortunately, structural data of the AHR LBD are currently not yet available. Cross-talk interactions exist between NRs and AHR in the steroids and retinoids pathways of mammals [24, 25]. Similar relationships have been found in insects between the bHLH-PAS protein Methoprene-tolerant (Met) and the ecdysone receptor, which is a heterodimer between two nuclear receptors: ECR (NR1H) and USP (NR2B) [26]. Finally, NR coactivators such as the p160 proteins (SRC1, TIF2 and ACTR) are also bHLH-PAS proteins. Therefore, the possibility of interactions between NRs and bHLH-PAS pathways might well be a common theme in animals. These examples show that any information obtained for one of the two groups of receptor can lead to interesting suggestions for the other group, despite totally different DBD and LBD.

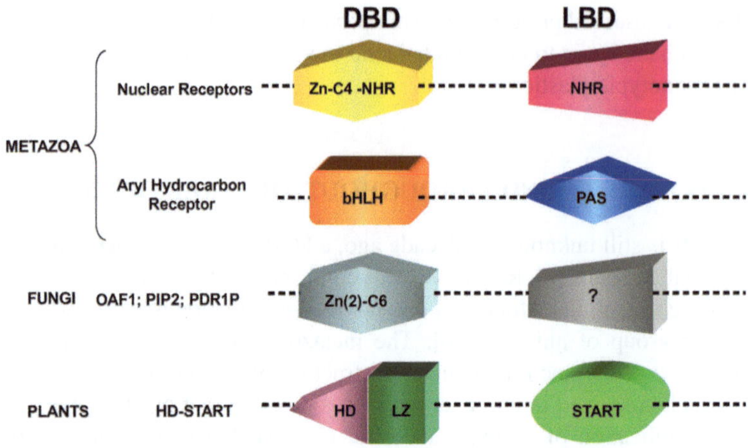

Figure 2.2. NR and NR-like transcription factors in eukaryotes. The domain structures of some known metazoan ligand-activated transcription factors are compared. They all possess a DBD and a LBD, but from different domain families, so they are not homologous

Plants also possess proteins that combine a DBD and a LBD: the HD-START family (Figure 2.2). These transcription factors contain a homeodomain associated with a leucine zipper important for dimerisation. The Steroidogenic Acute Regulatory-related lipid Transfer (START) domain was first identified in mammals as a lipid-sterol binding motif [27]. The crystal structure of several mammalian START domains revealed a hydrophobic tunnel that allows the transport of sterol or phospholipids [28]. The HD and START domains are combined with various other protein motifs in eukaryotes and prokaryotes. However, surprisingly, only within the plant kingdom are they associated together [29]. Several of these HD-START proteins are implicated in cell differentiation during plant development, possibly by linking the lipid metabolic state of the cell to the regulation of transcription [29]. Unfortunately, START sequences of plants are not closely related to those of animals, and their structure is unknown, as well as their ligands. Hydrophobic steroid hormones are very similar in plants and animals but the identified receptors are totally different [30]. Nevertheless, the plant-specific HD-START family may have an analogous role to the one of NRs in animals.

Finally, it is important to recall that what seems to be an exception in eukaryotes is actually the rule in eubacteria, where the regulation of transcription is based on proteins that bind to specific DNA sequences in a ligand-dependent manner [31]. One of the best example of this mechanism is the *lac* operon, the first discovered case of gene regulation. The repressor (lacI) undergoes a conformational transition in response to bound ligands that increases or decreases its affinity for the operator DNA of the *lac* operon [32]. NRs act fundamentally in a similar allosteric fashion. The structure of the repressor revealed a modular structure with four domains: a DBD of the HLH type; a hinge region; the LBD forming a sugar binding pocket;

a C-terminal helix important for tetramerization [33]. This general organisation is familiar to those studying NRs, which are usually organised in 4–5 domains: A/B, C (DBD), D (hinge), E (LBD) and sometimes a F domain.

Beyond these simplistic comparisons, the key point is that ligand-dependent gene regulation is probably a very ancient mechanism. In that perspective, the mode of action of NRs could be qualified as 'primitive', in the sense that it would represent an example of an old and widespread mechanism already present in the late common ancestor to prokaryotes and eukaryotes.

2.5. WHAT IS A NR-LIGAND?

In biochemistry, a ligand is a substance that is able to bind to and form a complex with a biomolecule in a biological context. In a narrower sense, it is a signal-triggering molecule binding to a site on a target protein, by intermolecular forces such as ionic bonds, hydrogen bonds and Van der Waals forces. The docking (association) is usually reversible (dissociation). Actually, irreversible covalent binding between a ligand and its target molecule is rare in biological systems.

The nature of the NR ligand is a question that has been strongly biased by the history of the discovery of NRs. The first NR ligands were classical hormones of the endocrinology field, like steroid hormones or thyroid hormones, hence the name often given to the family: the steroid/thyroid hormone receptor family. When the receptor for ecdysone was cloned, it was clear for everyone in the field that NRs were high affinity receptors (at the nanomolar range) for very specific compounds with a hormonal function [34]. The identification of the first orphan receptors in no way changed this paradigm and many pharmaceutical companies performed high-throughput screens on orphan receptors in order to discover new hormonal ligands. Nevertheless, the discovery of the metabolic receptors and, prominently, of the PPARs that were shown to bind a wide diversity of compounds (including fatty acids with an affinity in the micromolar range), provided the first clue that the situation was more complex than previously expected. This view is reinforced by data showing that 9-cis RA may only be a pharmacological ligand of RXR, at least in mammals, and that its ligands are rather fatty acids [35]. The characterization of xenobiotic regulators such as PXR and CAR have also much broadened our view since these receptors bind to an extremely wide variety of unrelated compounds such as rifampicine or RU486.

In addition, several intriguing results have highlighted the tremendous diversity of interactions that can exist between NRs and small molecules. We are now far from the key/lock model of a stable and simple interaction between a hormone (the key) and a receptor (the lock)! Among recent observations that modified these views on ligand binding it is interesting to mention the ability of ligands to bind unique ligand binding pockets; e.g. FMOC-Leu on PPARγ [36], and the existence of a still much discussed second ligand binding site in an estrogen receptor [36]; the regulation of a receptor activity by gas (NO and CO), which controls the redox status of a heme molecule that is permanently bound to the E75 ligand binding pocket [38]; the several cases of structural ligands that are small molecules, often fatty acids, that are bound in the ligand binding pockets of USP and HNF4 [39, 40]. To finish on this rapid panorama of unusual binding modes one has to recall that classical receptors such as estrogen receptors are in fact promiscuous, since they recognize a large number of exogenous compounds (the endocrine disruptors) that can regulate their activity in very subtle ways. If some of these compounds are indeed artificial (BPA, DDT), others occur naturally (phytoestrogens) suggesting that may have been part of the ancestral regulatory system of NR activity by food. The historic idea of highly selective ligands controlling NR activities is now inconsistent with the fact that several endogenous ligands of estrogen receptors exist: 5α-androstane-3β,17β-diol is a natural agonist of ERβ [41] and 17OH-cholesterol, a naturally occurring steroid compound, is an endogenous antagonist of ERβ, [42].

All these data suggest that, at their origin, NRs were probably not hormonal receptors with high affinity for very specific compounds. Rather this is a feature that was acquired later during evolution. We propose that NRs instead act as a sensor, by interacting with a wide variety of compounds, to transfer, as transcriptional activity, subtle metabolic balances in the respective amounts of various compounds. Viewed in this manner, there is a continuum between classical hormones, endogenous regulators, exogenous regulators including pharmacological ligands, food derivatives, endocrine disruptors and even structural ligands that are permanently bound to the receptors.

2.6. EVOLUTION OF LIGAND BINDING

There are some well-known examples about refinement of the ligand-binding activity within closely related receptors. For the RAR/NR1B, which are retinoic-acid receptors in vertebrates, it was shown that the ability to bind slightly different molecules was acquired through definite mutations in the LBP [43].

The RXR/USP receptor, that binds 9cis-RA in metazoans, underwent at least two major shifts during its evolution in insects. First, a loss of the ligand-binding ability, after the split between grasshopers (where USP can bind 9cis-RA) and more derived insects, such as coleoptera, hemiptera and mecopterida. This first shift seems to be due only to punctual mutations that prevent the formation of a ligand binding pocket. Secondly, during the emergence of the mecopterida clade, a new and large ligand-binding pocket appeared, allowing the binding of a structural ligand. Insertions in the LBD of USP were probably responsible of this second evolutionary shift [44]. Therefore, all the possible situations regarding ligand binding (classical ligand, structural ligand, real orphan) occur for USP-RXR. Currently, it is the only NR exhibiting such plasticity.

The careful analysis of the ligand-binding specificity of PXR in mammals also provides an example of shifts in NR ligand-binding ability. In vitro and cell culture binding assays between various vertebrate species showed that the biliary salt receptor PXR underwent a broadening of its specificity, being able to bind a various set of androstanes, pregnanes, C27 bile alcohol sulfates and some xenobiotics in the common vertebrate ancestor, whereas its specificity is restricted to C24 bile acids in amniotes [45, 46].

Despite numerous examples of transitions from one ligand to another, the original acquisition of ligand-binding ability remains unknown. The observation, that there are no obvious correlations between known ligand specificity and phylogenetic relationships between the receptors (see Figure 2.3a), led to the proposition that the evolution of the ligand-binding specificity of NRs involved several independent gains and losses of ligand-binding ability ([9], reviewed in [47, 48]). According to this notion, the LBD, which is conserved in all NRs, irrespective of the fact that they have or not a ligand, should rather be seen as an allosteric domain controlling the activity of the transcription factor. It is precisely because this conformational change

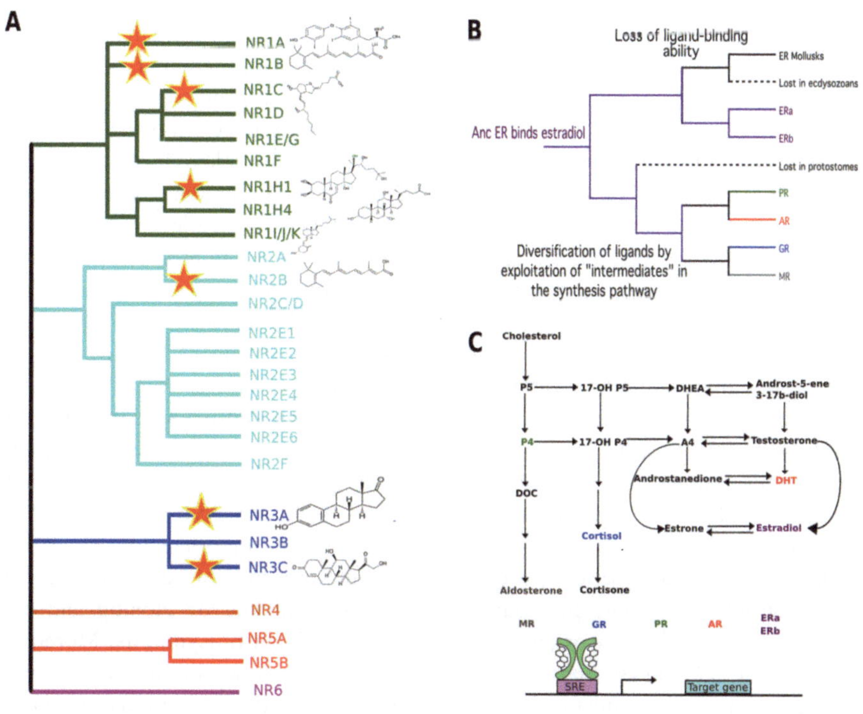

Figure 2.3. Different models about the evolution of ligand-binding ability. **a**: a general model for the whole family, independent acquisition of ligand-binding ability. Ligand for some human receptors are indicated on the tree. The lack for obvious correlations between known ligand specificity and phylogenetic relationships between the receptors led to the proposition that the evolution of the ligand-binding specificity of NRs involved several independent gains, that are here indicated by red stars **b**: a model for NR3A and NR3C. Gene losses, duplications and shifts in ligand-binding ability are indicated on a phylogenetic tree, with colors refering to ligands indicated in **c**. **c**: steroid biosynthesis pathway in human. Ligand and their receptors are indicated with the same colour

is conserved that the LBD sequence is conserved. It is interesting to note that several reports indicate that the conformational change can be triggered by several processes in addition to ligand binding [49]. Indeed, phosphorylation or protein-protein interactions can induce conformational changes. Thus, ligand binding appears as just one possible trigger.

An hypothesis, based on reconstruction of ancestral sequences at internal nodes of the evolutionary tree and functional characterization of the 'resurrected' receptor suggests that the ancestral steroid receptor was liganded and that orphan receptors secondarily lost the ability to bind a ligand [50, 51]. According to the ligand exploitation model, the terminal product in a biosynthetic pathway is the first compound for which a receptor evolves; selection for this hormone also selects for the synthesis of intermediates (see Figure 2.3b, c), and duplicated receptors then evolved affinity for these intermediates. This model accounts for the divergence observed in ligand specificity of the steroid receptors, namely AR/NR3C4, GR/NR3C1, MR/NR3C2, PR/NR3C3 and ERs/NR3A [51, 52]. It also suggests that ligands for some 'orphan' receptors may be found among intermediates in the synthesis of ligands for evolutionary related receptors. The ligand exploitation model considers that 17β-estradiol (E2) was the ligand of the ancestral receptor and that E2 was present and active in a wide range of metazoans before the diversification of steroid receptors [51]. However this axiom is not supported by the current data (see for example a discussion in [53] and [54]). There are clear evidences that different steroids are synthesized in vertebrates, insects and nematodes. For example, moulting is controlled by dafachronic acids in *C. elegans* [55] and by ecdysteroids (ecdysone and related compounds) in insects [56]. In humans, the main active steroids are dihydroxytestosterone (DHT), progesterone (P4), cortisol and aldosterone that bind the four members of the NR3C group (MR, GR, AR, PR), and E2 that binds to ERs (Figure 2.3b). Even within vertebrates, there are some variations in the identity of active steroid hormones. For example, in teleosts, there are two different active androgens: DHT and 11-ketotestosterone (11KT) [57], while aldosterone is not present [58]. Strikingly, despite these known variations of steroids identity among vertebrates, many authors have searched for the presence and putative roles of 'human'-type steroids such as estradiol or progesterone, in all metazoan groups. It is important to realize that, to date, none of the biochemical evidences for the presence of vertebrate steroids in lophotrochozoans and cnidarians has been substantiated by cloning and biochemical characterization of enzymes responsible of their biosynthesis. Moreover, many of the techniques that were used to detect endocrine activity are prone to artefacts and misidentification [53, 54]. Thus, the identity of steroids present in non-model 'invertebrates' is still an open and important question. This evolutionary variability in ligands may be found for other hormonal systems. Indeed, the recent characterization of the thyroid hormone receptor signalling in amphioxus has shown that the ligand of amphioxus TR is not T3 itself, but TRIAC, a derivative of T3 [59, 60].

A way to reconcile the different views could be to suppose that the ancestral receptor was not an orphan but rather a sensor that was able to bind with low affinity a wide range of molecules, probably provided by food [61], and that ligand binding specificity with high affinity evolved sometimes through selection of synthesis pathways for molecules that had a strong positive effect on animal physiology, and thus became endogenous synthesized hormones. This view is consistent with the fact that low affinity ligands are now known for a large set of receptors, which were previously thought to be orphans but are in fact sensors (e.g. NR2A/HNF4 in mammals [62]). Their abundance makes the hypothesis of an ancestral sensor more parsimonious than previous speculations (Figure 2.3a) that only distinguished between high affinity ligand binding receptors and orphans with no ligand binding ability at all. Some evolutionary studies at a subfamily scale [63] also reinforce the view that it may be a continuum between true orphan, sensors able to bind many molecules with low specificity and true endocrine receptors, which bind specifically one signalling molecule only.

2.7. CONCLUSION: EVOLUTION AS A REFLECTION FRAME TO UNDERSTAND NRs

The evolutionary story of NR is far from being fully elucidated but important recent progress has occurred leading to a radical shift in our view of NR signalling in recent years. NRs appear as very dynamic at the evolutionary level, being able to become adapted to a wide variety of physiological, metabolic and developmental roles. These molecules are thus now very promising evolutionary models. Striking conclusions from recent studies are that (i) non-usual genetic models such as lophotrochozoans and cnidarians continue to reveal provocative genomic and functional insight (ii) there is no obligatory co-linearity, at an evolutionary scale, between a given receptor and ligand, and (iii) physiological roles of NRs cannot be fully understood without an integrative view, taking into account the genetic environment in which receptors are evolving.

REFERENCES

1. Laudet, V., Hänni, C., Coll, J., Catzeflis, F., and Stéhelin, D. (1992). Evolution of the nuclear receptor gene superfamily. *EMBO J* 11(3), 1003–1013.
2. Amero, S. A., Kretsinger, R. H., Moncrief, N. D., Yamamoto, K. R., and Pearson, W. R. (1992). The origin of nuclear receptor proteins: A single precursor distinct from other transcription factors. *Mol Endocrinol* 6(1), 3–7.
3. Laudet, V. (1997). Evolution of the nuclear receptor superfamily: Early diversification from an ancestral orphan receptor. *J Mol Endocrinol* 19(3), 207–226.
4. Nuclear Receptors Nomenclature Committee (1999). A unified nomenclature system for the nuclear receptor superfamily. *Cell* 97(2), 161–163.
5. Brelivet, Y., Kammerer, S., Rochel, N., Poch, O., and Moras, D. (2004). Signature of the oligomeric behaviour of nuclear receptors at the sequence and structural level. *EMBO Rep* 5(4), 423–429, DOI10.1038/sj.embor.7400119.
6. Arnosti, D. N., Gray, S., Barolo, S., Zhou, J., and Levine, M. (1996). The gap protein knirps mediates both quenching and direct repression in the *Drosophila* embryo. *EMBO J* 15(14), 3659–3666.

7. Båvner, A., Sanyal, S., Gustafsson, J. A., and Treuter, E. (2005). Transcriptional corepression by SHP: Molecular mechanisms and physiological consequences. *Trends Endocrinol Metab* 16(10), 478–488, DOI10.1016/j.tem.2005.10.005.
8. Miyata, T. and Suga, H. (2001). Divergence pattern of animal gene families and relationship with the cambrian explosion. *Bioessays* 23(11), 1018–1027, DOI10.1002/bies.1147.
9. Escriva, H., Safi, R., Hänni, C., Langlois, M. C., Saumitou-Laprade, P., Stehelin, D., Capron, A., Pierce, R., and Laudet, V. (1997). Ligand binding was acquired during evolution of nuclear receptors. *Proc Natl Acad Sci U S A* 94(13), 6803–6808.
10. Bertrand, S., Brunet, F. G., Escriva, H., Parmentier, G., Laudet, V., and Robinson-Rechavi, M. (2004). Evolutionary genomics of nuclear receptors: From twenty-five ancestral genes to derived endocrine systems. *Mol Biol Evol* 21(10), 1923–1937, DOI10.1093/molbev/msh200.
11. Bertrand, S., Thisse, B., Tavares, R., Sachs, L., Chaumot, A., Bardet, P. L., Escrivà, H., Duffraisse, M., Marchand, O., Safi, R., Thisse, C., and Laudet, V. (2007). Unexpected novel relational links uncovered by extensive developmental profiling of nuclear receptor expression. *PLoS Genet* 3(11), e188, DOI10.1371/journal.pgen.0030188.
12. Schubert, M., Brunet, F., Paris, M., Bertrand, S., Benoit, G., and Laudet, V. (2008). Nuclear hormone receptor signaling in amphioxus. *Dev Genes Evol*, DOI10.1007/s00427-008-0251-y.
13. Sluder, A. E. and Maina, C. V. (2001). Nuclear receptors in nematodes: Themes and variations. *Trends Genet* 17(4), 206–213.
14. Robinson-Rechavi, M., Maina, C. V., Gissendanner, C. R., Laudet, V., and Sluder, A. (2005). Explosive lineage-specific expansion of the orphan nuclear receptor HNF4 in nematodes. *J Mol Evol* 60(5), 577–586, DOI10.1007/s00239-004-0175-8.
15. Yates, R. A., Tuan, R. S., Shepley, K. J., and Unnasch, T. R. (1995). Characterization of genes encoding members of the nuclear hormone receptor superfamily from *Onchocerca volvulus*. *Mol Biochem Parasitol* 70(1–2), 19–31.
16. Ghedin, E., Wang, S., Spiro, D., Caler, E., Zhao, Q., Crabtree, J., Allen, J. E., Delcher, A. L., Guiliano, D. B., Miranda-Saavedra, D., Angiuoli, S. V., Creasy, T., Amedeo, P., Haas, B., El-Sayed, N. M., Wortman, J. R., Feldblyum, T., Tallon, L., Schatz, M., Shumway, M., Koo, H., Salzberg, S. L., Schobel, S., Pertea, M., Pop, M., White, O., Barton, G. J., Carlow, C. K. S., Crawford, M. J., Daub, J., Dimmic, M. W., Estes, C. F., Foster, J. M., Ganatra, M., Gregory, W. F., Johnson, N. M., Jin, J., Komuniecki, R., Korf, I., Kumar, S., Laney, S., Li, B. W., Li, W., Lindblom, T. H., Lustigman, S., Ma, D., Maina, C. V., Martin, D. M. A., McCarter, J. P., McReynolds, L., Mitreva, M., Nutman, T. B., Parkinson, J., Peregrín-Alvarez, J. M., Poole, C., Ren, Q., Saunders, L., Sluder, A. E., Smith, K., Stanke, M., Unnasch, T. R., Ware, J., Wei, A. D., Weil, G., Williams, D. J., Zhang, Y., Williams, S. A., Fraser-Liggett, C., Slatko, B., Blaxter, M. L., and Scott, A. L. (2007). Draft genome of the filarial nematode parasite *Brugia malayi*. *Science* 317(5845), 1756–1760, DOI10.1126/science.1145406.
17. Abad, P., Gouzy, J., Aury, J. M., Castagnone-Sereno, P., Danchin, E. G. J., Deleury, E., Perfus-Barbeoch, L., Anthouard, V., Artiguenave, F., Blok, V. C., Caillaud, M. C., Coutinho, P. M., Dasilva, C., Luca, F. D., Deau, F., Esquibet, M., Flutre, T., Goldstone, J. V., Hamamouch, N., Hewezi, T., Jaillon, O., Jubin, C., Leonetti, P., Magliano, M., Maier, T. R., Markov, G. V., McVeigh, P., Pesole, G., Poulain, J., Robinson-Rechavi, M., Sallet, E., Ségurens, B., Steinbach, D., Tytgat, T., Ugarte, E., van Ghelder, C., Veronico, P., Baum, T. J., Blaxter, M., Bleve-Zacheo, T., Davis, E. L., Ewbank, J. J., Favery, B., Grenier, E., Henrissat, B., Jones, J. T., Laudet, V., Maule, A. G., Quesneville, H., Rosso, M. N., Schiex, T., Smant, G., Weissenbach, J., and Wincker, P. (2008). Genome sequence of the metazoan plant-parasitic nematode *Meloidogyne incognita*. *Nat Biotechnol* 26(8), 909–915, DOI10.1038/nbt.1482.
18. Dieterich, C., Clifton, S. W., Schuster, L. N., Chinwalla, A., Delehaunty, K., Dinkelacker, I., Fulton, L., Fulton, R., Godfrey, J., Minx, P., Mitreva, M., Roeseler, W., Tian, H., Witte, H., Yang, S. P., Wilson, R. K., and Sommer, R. J. (2008). The *Pristionchus pacificus* genome provides a unique perspective on nematode lifestyle and parasitism. *Nat Genet* 40(10), 1193–1198, DOI10.1038/ng.227.

19. Wu, W., Niles, E. G., Hirai, H., and LoVerde, P. T. (2007). Evolution of a novel subfamily of nuclear receptors with members that each contain two DNA binding domains. *BMC Evol Biol* 7, 27, DOI10.1186/1471-2148-7-27.
20. King, N., Westbrook, M. J., Young, S. L., Kuo, A., Abedin, M., Chapman, J., Fairclough, S., Hellsten, U., Isogai, Y., Letunic, I., Marr, M., Pincus, D., Putnam, N., Rokas, A., Wright, K. J., Zuzow, R., Dirks, W., Good, M., Goodstein, D., Lemons, D., Li, W., Lyons, J. B., Morris, A., Nichols, S., Richter, D. J., Salamov, A., Sequencing, J. G. I., Bork, P., Lim, W. A., Manning, G., Miller, W. T., McGinnis, W., Shapiro, H., Tjian, R., Grigoriev, I. V., and Rokhsar, D. (2008). The genome of the choanoflagellate *Monosiga brevicollis* and the origin of metazoans. *Nature* 451(7180), 783–788, DOI10.1038/nature06617.
21. Phelps, C., Gburcik, V., Suslova, E., Dudek, P., Forafonov, F., Bot, N., MacLean, M., Fagan, R. J., and Picard, D. (2006). Fungi and animals may share a common ancestor to nuclear receptors. *Proc Natl Acad Sci U S A* 103(18), 7077–7081, DOI10.1073/pnas.0510080103.
22. Thakur, J. K., Arthanari, H., Yang, F., Pan, S. J., Fan, X., Breger, J., Frueh, D. P., Gulshan, K., Li, D. K., Mylonakis, E., Struhl, K., Moye-Rowley, W. S., Cormack, B. P., Wagner, G., and Näär, A. M. (2008). A nuclear receptor-like pathway regulating multidrug resistance in fungi. *Nature* 452(7187), 604–609, DOI10.1038/nature06836.
23. Furness, S. G. B., Lees, M. J., and Whitelaw, M. L. (2007). The dioxin (aryl hydrocarbon) receptor as a model for adaptive responses of bHLH/PAS transcription factors. FEBS Lett 581(19):3616–3625, DOI10.1016/j. febslet. 200704.011.
24. Murphy, K. A., Quadro, L., and White, L. A. (2007). The intersection between the aryl hydrocarbon receptor (AHR)- and retinoic acid-signaling pathways. *Vitam Horm* 75, 33–67, DOI10.1016/S0083-6729(06)75002-6.
25. Ohtake, F., Baba, A., Takada, I., Okada, M., Iwasaki, K., Miki, H., Takahashi, S., Kouzmenko, A., Nohara, K., Chiba, T., Fujii-Kuriyama, Y., and Kato, S. (2007). Dioxin receptor is a ligand-dependent E3 ubiquitin ligase. *Nature* 446(7135), 562–566, DOI10.1038/nature05683.
26. Konopova, B. and Jindra, M. (2007). Juvenile hormone resistance gene methoprene-tolerant controls entry into metamorphosis in the beetle *Tribolium castaneum. Proc Natl Acad Sci U S A* 104(25), 10,488–10,493, DOI10.1073/pnas.0703719104.
27. Soccio, R. E. and Breslow, J. L. (2003). Star-related lipid transfer (start) proteins: Mediators of intracellular lipid metabolism. *J Biol Chem* 278(25), 22,183–22,186, DOI10.1074/jbc.R300003200.
28. Tsujishita, Y. and Hurley, J. H. (2000). Structure and lipid transport mechanism of a star-related domain. *Nat Struct Biol* 7(5), 408–414, DOI10.1038/75192.
29. Schrick, K., Nguyen, D., Karlowski, W. M., and Mayer, K. F. X. (2004). Start lipid/sterol-binding domains are amplified in plants and are predominantly associated with homeodomain transcription factors. *Genome Biol* 5(6), R41, DOI10.1186/gb-2004-5-6-r41.
30. Kushiro, T., Nambara, E., and McCourt, P. (2003). Hormone evolution: The key to signalling. *Nature* 422(6928), 122, DOI10.1038/422122a.
31. Beckett, D. (2001). Regulated assembly of transcription factors and control of transcription initiation. *J Mol Biol* 314(3), 335–352, DOI10.1006/jmbi.2001.5134.
32. Jacob, F. and Monod, J. (1961). Genetic regulatory mechanisms in the synthesis of proteins. *J Mol Biol* 3, 318–356.
33. Lewis, M. (2005). The *lac* repressor. *C R Biol* 328(6), 521–548, DOI10.1016/j.crvi.2005.04.004.
34. Koelle, M. R., Talbot, W. S., Segraves, W. A., Bender, M. T., Cherbas, P., and Hogness, D. S. (1991). The *Drosophila EcR* gene encodes an ecdysone receptor, a new member of the steroid receptor superfamily. *Cell* 67(1), 59–77.
35. Calléja, C., Messaddeq, N., Chapellier, B., Yang, H., Krezel, W., Li, M., Metzger, D., Mascrez, B., Ohta, K., Kagechika, H., Endo, Y., Mark, M., Ghyselinck, N. B., and Chambon, P. (2006). Genetic and pharmacological evidence that a retinoic acid cannot be the RXR-activating ligand in mouse epidermis keratinocytes. *Genes Dev* 20(11), 1525–1538, DOI10.1101/gad.368706.
36. Rocchi, S., Picard, F., Vamecq, J., Gelman, L., Potier, N., Zeyer, D., Dubuquoy, L., Bac, P., Champy, M. F., Plunket, K. D., Leesnitzer, L. M., Blanchard, S. G., Desreumaux, P., Moras, D., Renaud, J. P.,

and Auwerx, J. (2001). A unique PPARgamma ligand with potent insulin-sensitizing yet weak adipogenic activity. *Mol Cell* 8(4), 737–747.

37. Wang, Y., Chirgadze, N. Y., Briggs, S. L., Khan, S., Jensen, E. V., and Burris, T. P. (2006). A second binding site for hydroxytamoxifen within the coactivator-binding groove of estrogen receptor beta. *Proc Natl Acad Sci U S A* 103(26), 9908–9911.
38. Thummel, C. S. (2005). Powered by gas–a ligand for a fruit fly nuclear receptor. Cell 122(2): 151–153, DOI10.1016/j.cell.2005.07.007.
39. Clayton, G. M., Peak-Chew, S. Y., Evans, R. M., and Schwabe, J. W. (2001). The structure of the ultraspiracle ligand-binding domain reveals a nuclear receptor locked in an inactive conformation. *Proc Natl Acad Sci U S A* 98(4), 1549–1554, DOI10.1073/pnas.041611298.
40. Dhe-Paganon, S., Duda, K., Iwamoto, M., Chi, Y. I., and Shoelson, S. E. (2002). Crystal structure of the HNF4 alpha ligand binding domain in complex with endogenous fatty acid ligand. *J Biol Chem* 277(41), 37,973–37,976, DOI10.1074/jbc.C200420200.
41. Weihua, Z., Lathe, R., Warner, M., and Gustafsson, J. A. (2002). An endocrine pathway in the prostate, ERbeta, AR, 5alpha-androstane-3beta,17beta-diol, and CYP7B1, regulates prostate growth. *Proc Natl Acad Sci U S A* 99(21), 13,589–13,594, DOI10.1073/pnas.162477299.
42. Umetani, M., Domoto, H., Gormley, A. K., Yuhanna, I. S., Cummins, C. L., Javitt, N. B., Korach, K. S., Shaul, P. W., and Mangelsdorf, D. J. (2007). 27-hydroxycholesterol is an endogenous serm that inhibits the cardiovascular effects of estrogen. *Nat Med* 13(10), 1185–1192, DOI10.1038/nm1641.
43. Escriva, H., Bertrand, S., Germain, P., Robinson-Rechavi, M., Umbhauer, M., Cartry, J., Duffraisse, M., Holland, L., Gronemeyer, H., and Laudet, V. (2006). Neofunctionalization in vertebrates: The example of retinoic acid receptors. *PLoS Genet* 2(7), e102, DOI10.1371/journal.pgen.0020102.
44. Iwema, T., Billas, I. M. L., Beck, Y., Bonneton, F., Nierengarten, H., Chaumot, A., Richards, G., Laudet, V., and Moras, D. (2007). Structural and functional characterization of a novel type of ligand-independent RXR-USP receptor. *EMBO J* 26(16), 3770–3782, DOI10.1038/sj.emboj.7601810.
45. Krasowski, M. D., Yasuda, K., Hagey, L. R., and Schuetz, E. G. (2005). Evolution of the pregnane X receptor: Adaptation to cross-species differences in biliary bile salts. *Mol Endocrinol* 19(7), 1720–1739, DOI10.1210/me.2004-0427.
46. Reschly, E. J., Bainy, A. C. D., Mattos, J. J., Hagey, L. R., Bahary, N., Mada, S. R., Ou, J., Venkataramanan, R., and Krasowski, M. D. (2007). Functional evolution of the vitamin D and pregnane X receptors. *BMC Evol Biol* 7, 222, DOI10.1186/1471-2148-7-222.
47. Escriva, H., Delaunay, F., and Laudet, V. (2000). Ligand binding and nuclear receptor evolution. *Bioessays* 22(8), 717–727, DOI3.0.CO;2-I.
48. Baker, M. E. (2003). Evolution of adrenal and sex steroid action in vertebrates: A ligand-based mechanism for complexity. *Bioessays* 25(4), 396–400, DOI10.1002/bies.10252.
49. Popov, V. M., Wang, C., Shirley, L. A., Rosenberg, A., Li, S., Nevalainen, M., Fu, M., and Pestell, R. G. (2007). The functional significance of nuclear receptor acetylation. Steroids 72(2):221–230, DOI10.1016/j.steroids.2006.12.001.
50. Thornton, J. W. (2001). Evolution of vertebrate steroid receptors from an ancestral estrogen receptor by ligand exploitation and serial genome expansions. *Proc Natl Acad Sci U S A* 98(10), 5671–5676, DOI10.1073/pnas.091553298.
51. Thornton, J. W., Need, E., and Crews, D. (2003). Resurrecting the ancestral steroid receptor: Ancient origin of estrogen signaling. *Science* 301(5640), 1714–1717, DOI10.1126/science.1086185.
52. Bridgham, J. T., Carroll, S. M., and Thornton, J. W. (2006). Evolution of hormone-receptor complexity by molecular exploitation. *Science* 312(5770), 97–101, DOI10.1126/science.1123348.
53. Lafont, R. and Mathieu, M. (2007). Steroids in aquatic invertebrates. *Ecotoxicology* 16(1), 109–130, DOI10.1007/s10646-006-0113-1.
54. Markov, G. V., Paris, M., Bertrand, S., and Laudet, V. (2008). The evolution of the ligand/receptor couple: A long road from comparative endocrinology to comparative genomics. *Mol Cell Endocrinol* 293(1–2), 5–16, DOI10.1016/j.mce.2008.06.011.
55. Rottiers, V., Motola, D. L., Gerisch, B., Cummins, C. L., Nishiwaki, K., Mangelsdorf, D. J., and Antebi, A. (2006). Hormonal control of *C. elegans* dauer formation and life span by a rieske-like oxygenase. Dev Cell 10(4):473–482, DOI10.1016/j.devcel.2006.02.008.

56. Rewitz, K. F., Rybczynski, R., Warren, J. T., and Gilbert, L. I. (2006). The halloween genes code for cytochrome P450 enzymes mediating synthesis of the insect moulting hormone. *Biochem Soc Trans* 34(Pt 6), 1256–1260, DOI10.1042/BST0341256.
57. Lokman, P. M., Harris, B., Kusakabe, M., Kime, D. E., Schulz, R. W., Adachi, S., and Young, G. (2002). 11-oxygenated androgens in female teleosts: Prevalence, abundance, and life history implications. *Gen Comp Endocrinol* 129(1), 1–12.
58. Bury, N. R., Sturm, A., Rouzic, P. L., Lethimonier, C., Ducouret, B., Guiguen, Y., Robinson-Rechavi, M., Laudet, V., Rafestin-Oblin, M. E., and Prunet, P. (2003). Evidence for two distinct functional glucocorticoid receptors in teleost fish. *J Mol Endocrinol* 31(1), 141–156.
59. Paris, M., Escriva, H., Schubert, M., Brunet, F., Brtko, J., Ciesielski, F., Roecklin, D., Vivat-Hannah, V., Jamin, E. L., Cravedi, J. P., Scanlan, T. S., Renaud, J. P., Holland, N. D., and Laudet, V. (2008). Amphioxus postembryonic development reveals the homology of chordate metamorphosis. *Curr Biol* 18(11), 825–830, DOI10.1016/j.cub.2008.04.078.
60. Paris, M. and Laudet, V. (2008). The history of a developmental stage: Metamorphosis in chordates. *Genesis* 46(11), 657-672, DOI10.1002/dvg.20443.
61. Ben-Shlomo, I. and Hsueh, A. J. W. (2005). Three's company: Two or more unrelated receptors pair with the same ligand. *Mol Endocrinol* 19(5), 1097–1109, DOI10.1210/me.2004-0451.
62. Wisely, G. B., Miller, A. B., Davis, R. G., Thornquest, A. D., Johnson, R., Spitzer, T., Sefler, A., Shearer, B., Moore, J. T., Miller, A. B., Willson, T. M., and Williams, S. P. (2002). Hepatocyte nuclear factor 4 is a transcription factor that constitutively binds fatty acids. *Structure* 10(9), 1225–1234.
63. Reschly, E. J. and Krasowski, M. D. (2006). Evolution and function of the NR1I nuclear hormone receptor subfamily (VDR, PXR, and CAR) with respect to metabolism of xenobiotics and endogenous compounds. *Curr Drug Metab* 7(4), 349–365.
64. Bonneton, F., Brunet, F. G., Kathirithamby, J., and Laudet, V. (2006). The rapid divergence of the ecdysone receptor is a synapomorphy for mecopterida that clarifies the strepsiptera problem. *Insect Mol Biol* 15(3), 351–362, DOI 10.1111/j.1365-2583.2006.00654.x.
65. Wu, W., Niles, E. G., El-Sayed, N., Berriman, M., and LoVerde, P. T. (2006). *Schistosoma mansoni* (Platyhelminthes, Trematoda) nuclear receptors: Sixteen new members and a novel subfamily. *Gene* 366, 303–315.

CHAPTER 3

FUNCTIONS OF NUCLEAR RECEPTORS IN INSECT DEVELOPMENT

DAVID MARTÍN
Institute of Evolutionary Biology (CSIC-UPF), 08003 Barcelona, Spain

Abstract: Nuclear hormone receptors form a large family of ligand activated transcription factors that regulate the transcription of target genes involved in a variety of important biological processes in animal development. Intense research over the past years has led to the characterization of the complete set of nuclear hormone receptors in several invertebrate genomes, including those of six insect species. In these organisms, extensive studies have defined critical roles for nuclear hormone receptors in controlling embryonic development, molting, metamorphosis, reproduction and metabolism. In this chapter, I review our understanding of the biological functions of all insect nuclear hormone receptors.

3.1. INTRODUCTION

Nuclear hormone receptors (NRs) are transcription factors that control global changes in gene transcription upon binding lipophilic ligands and, hence, exert key roles in a wide range of physiological processes during development. Members of this family are exclusively found in metazoans, since no NRs have been isolated in plants, fungi or unicellular eukaryotes [1]. NRs function through ligand binding due to their remarkable modular architecture of 4–5 structural domains (Figure 3.1). The two more important domains are the highly conserved DNA binding domain (DBD or C domain) and the ligand binding domain (LBD or E domain). The DBD is composed by two C_4-zinc fingers and contains the P-box, which is responsible for the binding to DNA, and the D-box that facilitates the heterodimerization of the receptor. The LBD, usually less conserved and located C-terminal of the DBD, is made up by 11–13 α-helices that form a hydrophobic pocket where the ligand can bind. Importantly, within the helix 12 it is located the activation function 2 domain (AF-2), that interacts with different coactivator proteins ligand-dependently [2]. Furthermore, there is a second AF-1 domain in the N-terminus A/B region that functions in a ligand-independent way. Between the DBD and the LBD, there is a hinge region (D domain) with different functional characteristics. Finally, at the C-terminus end of the receptor, it is located the F domain, variable in size and even completely absent in several NRs (Figure 3.1).

C.M. Bunce, M.J. Campbell (eds.), *Nuclear Receptors*, Proteins and Cell Regulation 8, 31–61. DOI 10.1007/978-90-481-3303-1_3,

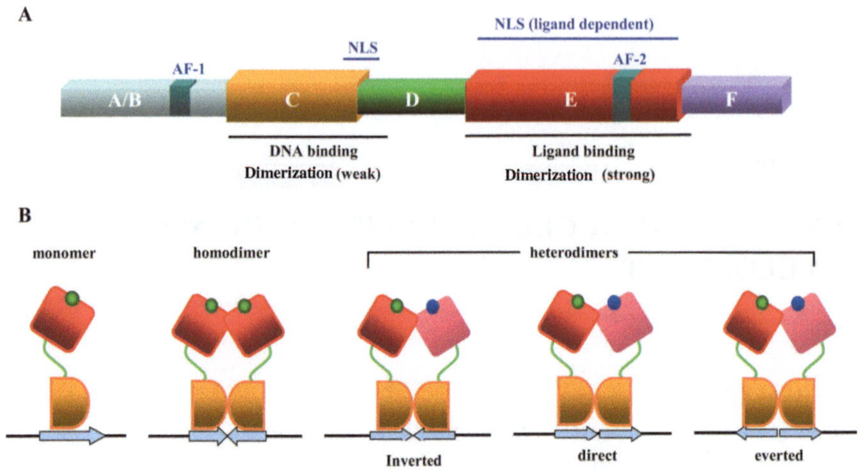

Figure 3.1. (**a**) Schematic representation of structural and functional modular organization of nuclear receptors. *White letters* indicate each domain. The typical nuclear receptor contains a variable N-terminal domain (**A/B**), a conserved DNA binding domain (**C**), a variable hinge region (**D**), a conserved ligand-binding domain (**E**) and a highly variable C-terminal region (**F**). The most relevant domain functions are depicted above and below the scheme. NLS, nuclear localization signal; AF-1 and AF-2, activation function domain-1 and -2 respectively. (**b**) Nuclear receptors bind the DNA as monomers, homodimers or heterodimers to single as well as directed, inverted or everted repeats of the core sequence 5′-RGGTCA-3′, in which R is a purine; represented by a *blue arrow*. The specificity for the different hormone response elements is given by the number of nucleotides spaced between the two core elements

Some NRs function as monomers through binding to a single hormone-response element (5′-RGGTCA-3′, in which R is a purine), whereas others can bind DNA as homodimers or heterodimers to direct, inverted or everted single repeats separated with a variable number of nucleotides, therefore increasing their regulatory complexity (Figure 3.1). Moreover, the activity of the NRs is also regulated by different post-transcriptional modifications such as phosphorylation, ubiquitination, sumoylation or acetylation.

Although NRs are defined as ligand-dependent transcription factors, most of them are still orphans, since they lack an identified ligand. In fact, in insects only three out of 24 NRs (EcR, E75 and HR51) have a known ligand. Conversely, in humans, about 50% of the 48 NRs are liganded receptors. Currently, it is not known whether the rest of the NRs have natural ligands or are true orphans.

3.2. NUCLEAR RECEPTORS IN INVERTEBRATES

NRs have been identified in several invertebrates, including cnidarians, annelids and molluscs. During the last years, however, large-scale sequencing of entire genomes has provided an incredibly detailed knowledge of the entire NR repertoire in insects and nematodes, both belonging to the Ecdysozoa, a proposed clade of animals that share the developmental trait of molting [3]. Currently, six complete insect genomes, those of the dipterans *Drosophila melanogaster*, *Anopheles gambiae* and *Aedes aegypti*, the lepidopteran *Bombyx mori*, the coleopteran *Tribolium castaneum* and the hymenopteran *Apis mellifera*, along with that of the nematode *Caenorhabditis elegans*, have been sequenced. Detailed analysis of these genomes has revealed the entire set of NRs in these species [4, 5, 6, 7, 8] (Table 3.1). The availability of these sequenced genomes and the fact that insects and *C. elegans* are amenable to detailed genetic analysis due to the power of well-established genetic and genomic methodologies, have allowed to use these organisms as outstanding models to study NR functions.

Using a detailed phylogenetic approach, Bertrand et al. [9] have shown that the ancestral NR set in the Urbilateria, the ancestor of bilaterians, was composed by 25 proteins, and that gene duplication as well as gene loss played relevant roles in the evolution of the NRs repertoires in the different bilaterian lineages. Whereas humans have 48 NRs, insects have 19–22 and *C. elegans* has 284, mainly due to a lineage-specific duplication burst of the *HNF4* gene [10]. Based on extensive phylogenetic analysis using the DBD and LBD of the different NRs, this family has been divided into six subfamilies. Insects have representatives in all six subfamilies (Table 3.1). Furthermore, there are also 2–3 atypical receptors that lack the whole LBD. These receptors (knirps, knirps-related and eagle) have been included in a new and not related subfamily (Table 3.1).

In this chapter I provide an overview of the biological functions of all insect NRs, as models of invertebrate organisms. In insects, about 40% of their NRs are devoted to detect and transduce one specific hormonal signal, the steroidal 20-hydroxyecdysone (20E). The remaining receptors exert multiple roles related with a wide variety of physiological processes, from embryonic patterning and organogenesis to metabolism and detoxification. Throughout the chapter, differences between NR functions between more primitive hemimetabolous insects and more derived holometabolous insects are discussed. In addition, I also provide information about the biological functions of the NR homologs of *C. elegans*, presenting examples of interesting functional conservation with their insect counterparts.

3.3. INSECT NUCLEAR RECEPTORS: THE 20-HYDROXYECDYSONE-MEDIATED PATHWAY

Most NRs act as dimers or interact temporarily with each other to control transcription of target genes. These protein–protein interactions generate complex cross-regulatory networks. The main network in insect development is related with the transcriptional response to ecdysteroids, specific steroid hormones of insects. In fact, 9 out of 24 insect NRs are involved in the signalling pathway activated by the most active ecdysteroid, 20E. But, why is 20E so important in insect development?

Table 3.1. Nuclear hormone receptors in insects and the nematode *C. elegans*. Complete set of insect nuclear receptors found in the genomes of the dipterans *Drosophila melanogaster*, *Aedes aegypti* and *Anopheles gambiae*, the lepidopteran *Bombyx mori*, the hymenopteran *Apis mellifera* and the coleopteran *Tribolium castaneum*. Conserved nuclear receptors of humans as well as the nematode *C. elegans* are included. Subfamily classification and nuclear receptor nomenclature according to Ruau et al. [170]. Nuclear receptor accession numbers for *D. melanogaster*, *A. aegypti* and *A. mellifera* are from GenBank (http://www.ncbi.nlm.nih.gov/), *A. gambiae* are from VectorBase (http://www.vectorbase.org/), *B. mori* are from Silkworm Genome Database (http://silkworm.genomics.org.cn/) and *T. castaneum* are from Beetlebase (htpp://bioinformaticas.ksu.edu/BeetleBase/)

Nomenclature	General name	Human orthologue	*D. melanogaster*	*A. aegypti*	*A. gambiae*	*B. mori*	*A. mellifera*	*T. castaneum*	*C. elegans*
NR1D3	E75	REV-ERBA	CG8127	AM397060	AGAP012223	BGIBMGA006839	GB11364	TC_12440	NHR-85
NR1E1	E78	REV-ERVBA	CG18023	AM773442	AGAP006571	BGIBMGA003895	GB30226	TC_03935	SEX-1
NR1F4	HR3	RORB	CG33183	AF230281	AGAP009002	BGIBMGA009688	GB10650	TC_08909	NHR-23
NR1H1	ECR	FXR/LXR	CG1765-PA CG1765-PB	AY345989 AF305214	AGAP012211 AGAP012201	BGIBMGA006767	GB30298	TC_12112	–
NR1J1	HR96	VDR	CG11783	AM773443	AGAP012210	BGIBMGA010783	GB10331	TC_10645	DAF-12 NHR-8 NHR-48
NR2A4	HNF4	HNF4A	CG9310	AF059026	AGAP004224	BGIBMGA001947	GB11424	TC_08726	NHR-64 NHR-69 NHR-49
NR2B4	USP	RXR	CG4380	AF305213	AGAP002095	BGIBMGA006183	GB16648	TC_14027	–
NR2D1	HR78	TR2	CG7199	BN001173	AGAP008384	BGIBMGA013363	GB18358	TC_04598	NHR-41
NR2E2	TLL	TLX	CG1378	BN001174	AGAP000819	BGIBMGA002000	GB20053	TC_00441	NHR-67
NR2E3	HR51	PNR	CG16801	BN001175	AGAP009890	BGIBMGA013855	GB10077	TC_09378	–
NR2E4	DSF	TLX	CG9019	BN001178	AGAP009575	BGIBMGA010370	GB14217	TC_01069	–
NR2E5	HR83	–	CG10296	AM773444	AGAP002234	–	GB17656	TC_10460	FAX-1
NR2E6	Nameless	–	–	AM773445	AGAP001348	BGIBMGA000670	GB17775	TC_13148	–

(Continued)

Table 3.1. (Continued)

NR2F3	SVP	COUP-TF	CG11502	AF303224	AGAP002544	BGIBMGA001391	GB17100	TC_01722	UNC-55
NR3B4	ERR	ERRb	CG7404	BN001176	AGAP001743	BGIBMGA004890	GB11125	TC_09140	–
NR4A4	HR38	NURR1	CG1864	AF165528	AGAP008334	BGIBMGA002964	GB17814	TC_13146	NHR-6
NR5A3	FTZ-F1	LRH-1/SF-1	CG4059	AF274870	AGAP005661	BGIBMGA000716	GB16873	TC_02550	NHR-25
NR5B1	HR39	LRH-1/SF-1	CG8676	BM001177	AGAP009399	BGIBMGA007914	GB11634	TC_14986	–
NR6A1	HR4	GCNF	CG16902	AM773447	AGAP004693	BGIBMGA007888	GB16863	TC_00543	NHR-91
NR0A1	KN1	–	CG4717	–	–	–	GB15945	–	–
NR0A2	KNRL	–	CG4761	BN001172	AGAP010438	–	GB13710	TC_03412	–
NR0A3	EG	–	CG7383	–	–	BGIBMGA014624	GB18215	TC_03409	–

Insects present different types of developmental strategies. In primitive ametabolous and hemimetabolous insects, their juvenile forms, the nymphs, are morphologically similar to the adults, only differing by colour, wing and genitalia details. In these insects, growth and maturation occur simultaneously throughout successive nymphal stages. Conversely, holometabolous insects have more complicated life cycles that include larval periods, a pupal stage and finally adulthood. Unlike in more primitive insects, these stages do not represent developmental forms that gradually evolve towards the adult stage. Importantly, during the pupal stage there is a complete metamorphosis in which almost all larval structures are destroyed and replaced by the structures of the adult organism. Remarkably, 20E controls development transitions in all types of development. Periodic pulses of 20E at the end of each nymphal/larval instar act as developmental timer triggering the molt to the next stage, and hence allowing the animal to accommodate internal growth. Furthermore, during the metamorphic pupal stage of holometabolous insects, 20E also induces the destruction of larval tissues and the formation of the adult body plan [11]. Thus, 20E controls a remarkable variety of key developmental processes, from molting, morphogenesis and apoptosis to metabolism and reproduction.

From a molecular point of view, 20E acts upon binding to a heterodimer of two NRs, the ecdysone receptor (EcR) and the Retinoid X Receptor (RXR) homolog, ultraspiracle (USP) [12]. The activated receptor elicits cascades of gene expression that mediate and amplify the hormonal signal. Molecular characterization of these genes, revealed that they also encode several NRs, namely E75, E78, HR3, HR4, FTZ-F1, HR39 and HR78 (reviewed in [7, 13, 11]). All the factors that form the functional 20E-dependent pathway are expressed throughout the major 20E-regulated developmental transitions in hemimetabolous and holometabolous insects, revealing in most cases repeated cascades of NRs expression, which suggests temporal cross-regulatory interactions between them (Figure 3.2). Below, I present the description of the functional knowledge of the heterodimeric 20E receptor and the NRs that form the hierarchical 20E-dependent signalling pathway.

3.3.1. The Functional 20-Hydroxyecdysone Receptor. EcR (NR1H1) and USP/RXR (NR2B4)

As previously mentioned, the functional 20E receptor is composed by the heterodimer of EcR and USP/RXR [14, 12, 15]. Whereas EcR is most similar to the vertebrate liver X receptor (LXR) and the farnesoid X receptor (FXR), USP/RXR is closest to the retinoid X Receptor (RXR). Given the functional relevance of these two NRs, both genes have been cloned in a remarkably large number of insect and arthropod species. Although the ECR-USP/RXR is the functional receptor of 20E in all insect analyzed, it has been recently shown in *D. melanogaster* that EcR would homodimerize or interact with another unknown NR, in a USP/RXR independent way, for the specific 20E-dependent activation of *glue* genes at mid-third larval instar [16]. EcR and USP/RXR have different expression patterns throughout development,

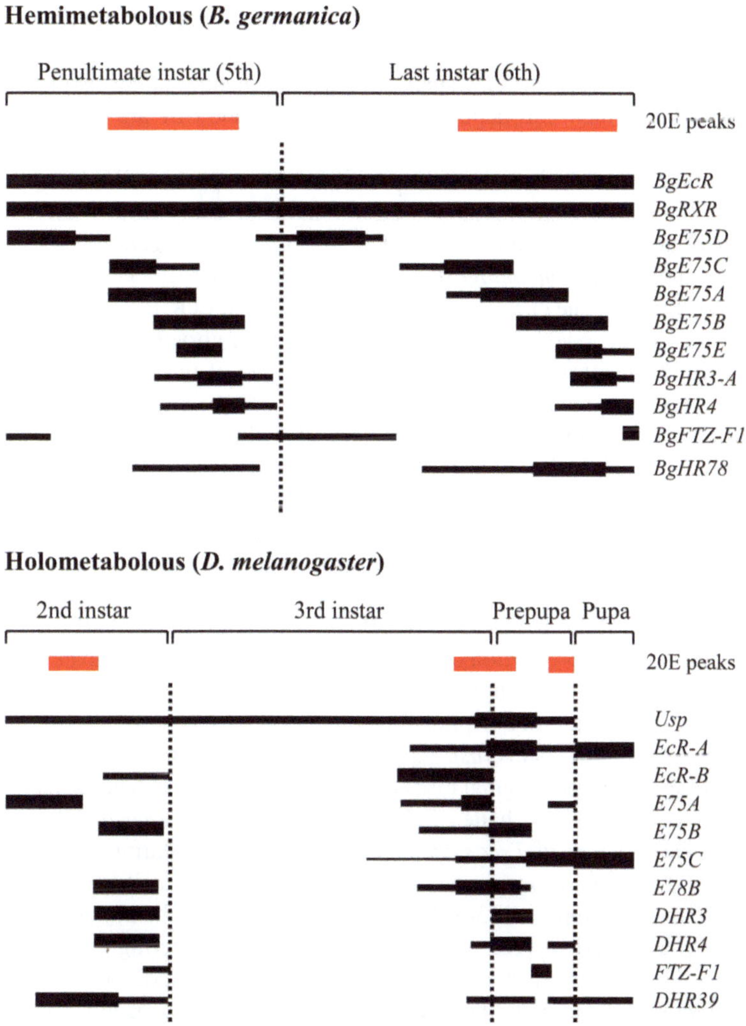

Figure 3.2. Schematic representation comparing the circulating ecdysteroids and mRNA expression of the nuclear receptors that form the 20-hydroxyecdysone-dependent pathway between the last two nymphal instars of the hemimetabolous insect *Blattella germanica* (*upper panel*) and the second and third larval instars as well as the prepupal stage of the holometabolous insect *Drosophila melanogaster*. The approximate times of 20-hydroxyecdysone pulses are shown in *red boxes* at the *top of each panel*. The transitions between different stages are marked by *vertical dotted lines*. The diagrams of mRNA expression, where the width of the box represents approximate levels of expression, are based on Sullivan and Thummel [18], Maestro et al. [17], Cruz et al. [37,60,68], Mané-Padrós et al. [46] and Mané-Padrós et al. (unpublished results)

from a housekeeping like pattern in the hemimetabolous model insect, the cockroach *Blattella germanica* [17], to more dynamic patterns in holometabolous species, such as *D. melanogaster* [18] (Figure 3.2). Usually, the marked changes in the expression of both receptors mirror their 20E-dependency. In fact, EcR and USP are induced by 20E in cultured organs of many species [19, 20, 21, 22, 23, 24, 25].

By using mutational analysis as well as double-stranded RNA (dsRNA) mediated RNA interference (RNAi) methodologies, functional analysis of EcR and USP/RXR have been carried out in the hemimetabolous *B. germanica* and in the holometabolous *D. melanogaster*, *T. castaneum* and *A. mellifera*. In *D. melanogaster*, the *ecr* gene encodes three isoforms, EcR-A, EcR-B1 and EcR-B2 [14, 26] whereas *usp* encodes only one [27, 28]. During embryogenesis, EcR is required for germ band retraction and head involution, two of the most important morphogenetic processes in the fly embryogenesis [29]. Conversely, *usp* mutants present a different embryonic phenotype to that of EcR, suggesting that EcR heterodimerize with a different partner during mid-embryogenesis [30]. During post-embryonic development, EcR isoforms display different regulatory functions although all related with the control of 20E-dependent transitions. *EcR-A* mutants, for example, arrest at early to mid-pupal stage, indicating that this protein is necessary after the formation of the pupal body plan and before the differentiation of adult structures [31]. *EcR-B1* and *EcR-B2* double mutants arrest during early larval development, and the small proportion of specimens that progress fail to pupate [32]. Finally *EcR-B1* single mutants do not pupate [33]. Furthermore, USP is required in late third instar larvae for the transcriptional response to the ecdysteroid pulse that induces the pupal formation. Thus, *usp* mutants are not able to initiate metamorphosis, and processes such as imaginal discs differentiation, adult midgut formation and larval tissue destruction are halted [34]. Similar to *D. melanogaster*, TcEcR or TcUSP/RXR knockdowns of *T. castaneum* arrest development during larval and pupal stages [35]. In the honeybee *A. mellifera*, the reduction of AmUSP/RXR levels also affects pupal development [36]. On the other hand, in the hemimetabolous insect *B. germanica*, RNAi in vivo of BgEcR or BgUSP/RXR during embryonic development results in the impairment of the germ band formation, due to the inhibition of proliferation of the cells that are fated to form the germ band (Maestro et al., unpublished results). Furthermore, during nymphal development, the reduction of BgEcR or BgUSP/RXR impairs molting to the next stage. Remarkably, all arrested knockdown nymphs show duplication of all structures of ectodermic origin, such as mouthparts, hypopharinge, tracheal system and cuticle layers, as well as impaired proliferation of the follicular epithelium that surrounds the immature oocytes. Furthermore, during adult development, both receptors are required for autophagic death of the prothoracic gland (the tissue responsible for the synthesis of ecdysteroids during nymphal development) and for choriogenesis [37, 38]. Although development is very different between hemimetabolous and holometabolous insects, a common feature is that the reduction of EcR or USP/RXR clearly inhibits the expression of 20E-dependent genes, affecting the functionality of the 20E-dependent pathway. Finally, it is interesting to note that, although the nematode *C. elegans* do molt several times during its life cycle, it does not have EcR and USP homologs in the genome [39].

3.3.2. E75 (NR1D3)

The E75 gene is present in all insect genomes sequenced. In addition, homologs have been identified in many other insects. Interestingly, whereas some E75 isoforms present the typical domain structure, others lack one (E75B) or both (E75D) zinc-fingers in the DBD, which renders them incapable of DNA binding. As its vertebrate homologs RevErbα and RevErbβ, E75 acts as a transcriptional repressor [40]. The expression of most E75 isoforms correlates well with the increase of circulating ecdysteroids (Figure 3.2). Furthermore, tissue and cell cultures show that most E75 isoforms are directly induced by 20E [19, 41, 42, 43, 44].

In *D. melanogaster*, homozygote flies with a loss-of-function mutation of all E75 isoforms die as late embryos or during a prolonged first larval instar. Conversely, isoform-specific mutants display different phenotypes. E75B mutants are viable, whereas E75C mutants die as pharate adults or within few days after adult eclosion. Finally, E75A mutants die throughout development, from embryo to pharate adult. Remarkably, a fraction of second instar E75A mutant larvae shows very low levels of circulating ecdysteroids during the stage, suggesting that this receptor acts in a feed-forward loop to stimulate ecdysteroid synthesis [45]. Similarly, in *B. germanica*, BgE75 is also necessary to increase the ecdysteroid titer at the end of the last nymphal instar. The reason of the ecdysteroid shortage is the premature degeneration of the prothoracic glands [46]. Surprisingly, knockdown last instar nymphs that do not molt initiate the adult-specific developmental program properly, which indicates that BgE75 links molting and developmental progression in hemimetabolous insects. In *T. castaneum*, TcE75 knockdown larvae are unable to pupariate [47].

From a detailed genetic analysis of the E75 mutants/knockdowns, it has been shown that E75 plays a central role in the 20E-signaling pathway (Figure 3.3). It interacts with another 20E-dependent NR, HR3, impairing the activity of the later [48]. Remarkably, this protein–protein interaction is modulated by diatomic gases. *D. melanogaster* E75 contains a heme group within its LBD that functions as a prosthetic group binding NO and CO. CO binding interferes with the ability of E75 to interact with DHR3, allowing the latter to induce transcription of target genes [49]. The functional interaction between E75 and HR3 is evolutionary conserved since RevErbβ and RAR-related orphan receptor (RORα), the vertebrate homolog of HR3, work antagonistically to activate *Bmal1* expression, an important part of the circadian clock machinery [50].

On the other hand, the nematode *C. elegans* has one E75 homolog, NHR-85 [39]. This gene is expressed in the developing vulva, the hypodermis and in specialized epithelia of the rectum and excretory duct. RNAi of *nhr-85* has revealed that is required for the development and function of the egg-laying machinery [51].

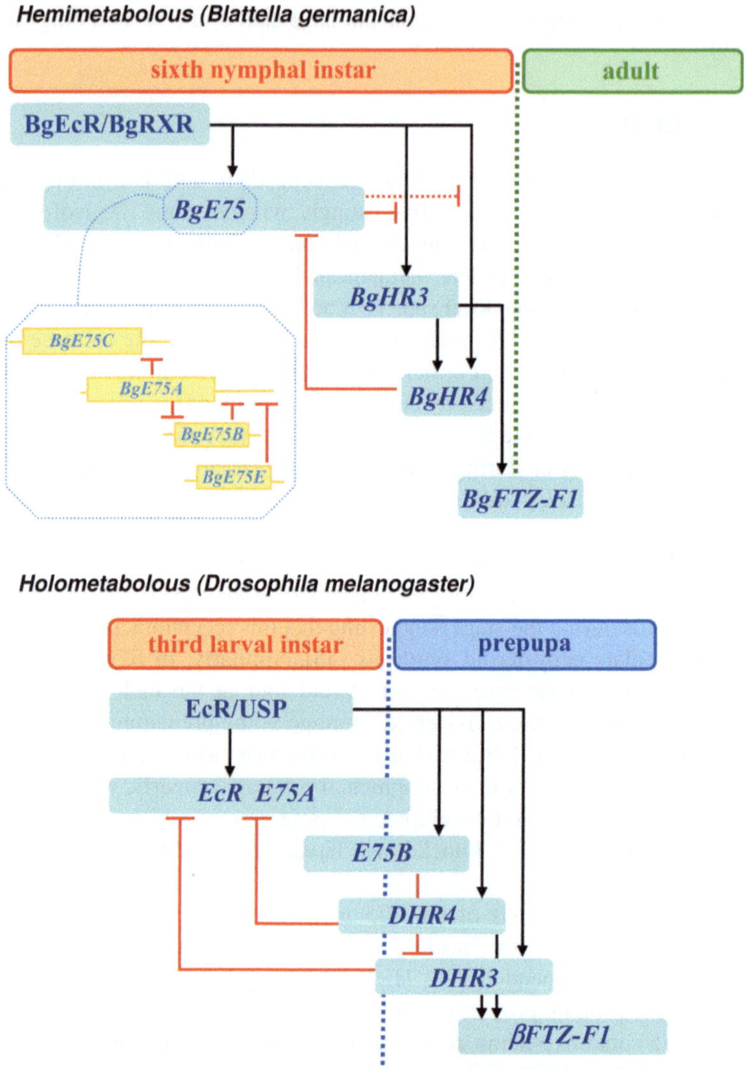

Figure 3.3. Schematic representation summarizing the similarities and differences of the regulatory cross-interactions between the nuclear receptors that form the 20-hydroxyecdysone-dependent signalling pathway in the hemimetabolous insect *Blattella germanica* (*upper panel*) and the holometabolous insect, *Drosophila melanogaster* (*lower panel*). In the case of *B. germanica* it is represented the transition between the last nymphal instar and the adult stage, whereas in *D. melanogaster*, the panel represents the transition between the last larval instar and the prepupal stage. The transitions between stages are represented by *vertical dotted lines*. The *blue* and *yellow boxes* represent the duration of the expression of each nuclear receptor. *Black arrows* represent inductive effects and *red lines* represent repressive effects. The complex regulatory interactions between different isoforms of the nuclear receptor BgE75 in *B. germanica* during the sixth nymphal instar is represented within a *dotted blue box*. See main text for references

3.3.3. E78 (NR1E1)

E78 is an *E75*-related gene most similar in sequence to human Rev-Erbα and to *C. elegans* SEX-1. In *D. melanogaster*, the mRNA of this receptor is detected in concert with larval ecdysteroid pulses [18]. Furthermore 20E induces its expression in organ cultures [52]. Conversely to E75, however, *D. melanogaster* E78 mutants are viable and fertile, although some defects are detected on larval polytene chromosomal puffing as well as in the morphology of the dorsal chorionic appendages of the eggs [53, 54]. In *T. castaneum*, TcE78 is not involved in any 20E-dependent process [47]. In *C. elegans*, SEX-1 regulates sex determination and dosage compensation by inhibiting *xol-1*, an important sex determining gene [55].

3.3.4. HR3 (NR1F4)

HR3 is the insect homolog of the retinoid-related orphan receptor (ROR), and *C. elegans* NHR-23. In all insects, HR3 shows a restricted expression pattern paralleling the ecdysteroid titer (Figure 3.2). Consequently, HR3 is induced by 20E in cultured organs [56, 57, 20, 58, 59, 60]. Although HR3 is defined as an orphan receptor, the GAL4-LBD ligand-trap system for in vivo detection of NR ligands developed in *D. melanogaster* shows a clear activation of the GAL4-HR3 sensor in the fat body of metabolically active third instar larvae, suggesting that HR3 can function as a nutrient or lipid sensor at this stage [61]. In fact, RORα and RORβ bind cholesterol derivatives and all-trans retinoic acid respectively [62, 63].

In the fruitfly, *DHR3* mutants die during embryonic development exhibiting defects in the nervous system [64]. During post-embryonic development, *DHR3* mutants that are rescued to the last larval instar die at the prepupal stage displaying defects in the tracheal system, gas bubble translocation and head eversion. Furthermore, it is also necessary for bristle, wing and cuticular development during pupal development [65]. In *B. germanica*, RNAi in vivo shows that BgHR3 is crucial to complete the ecdysis process. Along with BgEcR and BgRXR, cockroach nymphs with reduced BgHR3 levels arrest development presenting duplication in all structures of ectodermic origin [60]. The role of HR3 in molting is also conserved in the coleopteran *T. castaneum* [47]. On the other hand, the expression of HR3 in the fat body and ovary during the adult stage of *A. aegypti* and *B. mori*, suggests an active role of this factor in other 20E-dependent processes, such as oogenesis and vitellogenesis [58, 59]. Interestingly, the function of HR3 is evolutionary conserved, since during development of *C. elegans*, NHR-23 also controls molting. The expression of *nhr-23* oscillates with each molting cycle and RNAi provokes molting impairment with associated defects of cuticle shedding [66].

From a mechanistic point of view, it is important to note that most of the HR3 mutant/knockdown phenotypes are due to the absence of another important NR, FTZ-F1, a direct target of HR3 [67], [65, 68]. However, as it has been previously described, the HR3-dependent induction of FTZ-F1 is negatively regulated by E75, hence defining a key 20E-dependent cross-regulatory network

[48]. Thus, in hemimetabolous and holometabolous insects, the precise timing of FTZ-F1 activation is controlled by the accumulation of HR3 and the decline of E75, which occurs when the ecdysteroid pulse declines at the end of each instar (Figure 3.3).

3.3.5. HR4 (NR6A6)

HR4 is the homolog of the vertebrate germ cell nuclear factor (GCNF), a potent transcriptional repressor [69]. *HR4* gene is directly induced by 20E in all insect species analyzed [70, 71, 72]. Remarkably, its transcript always appears at the same time or just a little after HR3 and before FTZ-F1, suggesting that HR4 is another important member of the 20E-dependent genetic network (Figure 3.2) [73, 74, 70, 75, 18].

In the fruitfly, a P-insertion located in this gene, causes embryonic lethality, with cuticular defects [76]. Furthermore, disruption of DHR4 function during post-embryonic development revealed two different phenotypes: first, DHR4 third instar mutants display premature pupariation, resulting in small animals; and second, DHR4 mutants arrested development at early metamorphosis [72]. In *B. germanica*, BgHR4 is required during all developmental transitions to complete the ecdysis process. As happens with BgHR3, BgHR4 knockdowns arrest development with duplicated ectoderm-derived structures (Figure 3.4) (Mané-Padrós et al., unpublished results). The same molting impairment is observed in *T. castaneum* last instar larvae treated with dsRNA against TcHR4 [47]. Interestingly, whereas HR4 is required for all nymphal-nymphal and nymphal-adult transitions in *B. germanica*, it is not required for essential functions before the pupal stage in *D. melanogaster* [72]. From these results, it is tempting to speculate that the *HR4* gene has evolved from an ancestral function related with the control of molting in hemimetabolous insects to a new holometabolous-specific role in coordinating growth and maturation by mediating the endocrine response to the attainment of the critical weight during the larval stage. Despite these functional differences, however, a common regulatory feature of HR4 between holometabolous and hemimetabolous insects is that it acts as a potent transcriptional repressor of several 20E-dependent genes at the end of the last larval/nymphal instars, such as E75A, E75B, HR3 and IMP-L1 among others ([72]; Mané-Padrós et al., unpublished results). A second common characteristic is that HR4 is required at the end of the instar for proper FTZ-F1 activation. In summary, HR3 and HR4 function together as a transcriptional switch at the larval-prepupal transition in holometabolous insects, and nymphal-adult transition in hemimetabolous insects by downregulating the early regulatory response to 20E and also by inducing the FTZ-F1 receptor (Figure 3.3). One homolog of HR4 has been found in *C. elegans* genome, NHR-91, that is expressed in the developing vulva, seam cells, espermatheca, excretory cells and anterior neurons. Although RNAi experiments have showed no obvious phenotypes, recently it has been suggested that NHR-49 together with the ATP-binding cassette protein E would function in vulvae development and molting [77].

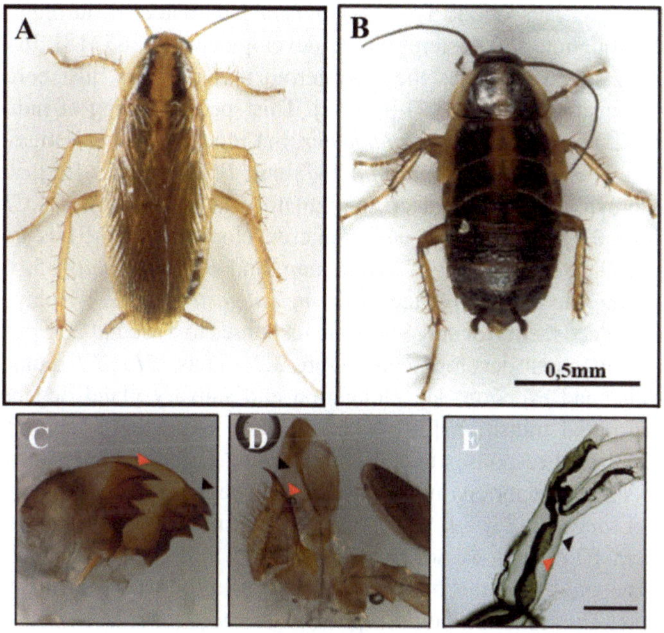

Figure 3.4. Effect of RNAi-mediated knockdown of BgHR4 on the imaginal molt of the hemimetabolous insect, *Blattella germanica*. The same phenotype is observed in nymphs with reduced amounts of BgEcR, BgRXR and BgFTZ-F1 [37, 38, 68], indicating that these receptors are part of the cross-regulated 20-hydroxyecdysone-dependent pathway. Sixth instar female nymphs are treated with dsRNA targeted to BgHR4 at the onset of the instar and left until the imaginal molt, 8 days later. (**a**) Wild-type animal 1 day after the imaginal molt showing a normal winged adult shape. (**b**) BgHR4 knockdown nymph at the same time point arrests development without molting. (**c–e**) The arrested nymph shows duplication of cuticular structures (nymphal structures are indicated with a *black arrowhead* and adult ones with a *red arrowhead*), like double mandibles (**c**), double laciniae within the maxilla (**d**), and two superimposed thrachea (**e**). Scale bars: 5 mm in (**a**) and (**b**); 500 μm in (**c–e**)

3.3.6. FTZ-F1 (NR5A3)

FTZ-F1 is one of the most studied receptors of the 20E-signaling pathway. This receptor, together with the steroidogenic factor 1 (SF-1), the liver receptor homolog 1 (LRH-1) and the *C. elegans* NHR-25 forms the NR5 subfamily. FTZ-F1 binds to DNA as a monomer with high specificity due to the presence of a specific C-terminal extension of its DBD, named the FTZ-F1 box [78, 79].

In *D. melanogaster*, the *ftz-f1* gene generates two isoforms differing only in their N-terminal sequences. The αFTZ-F1 isoform, from maternal origin, is only present during early embryonic development, where it regulates early segmentation through the interaction with the homeotic gene *ftz* [80, 81]. This interaction, mediated through the AF-2 domain within the LBD of αFTZ-F1, has allowed the evolution of FTZ from its ancestral homeotic function in short-germ band insects to a new role in segmentation in more derived long germ band insects, such as

D. melanogaster [82, 83]. The second isoform, βFTZ-F1, is detected during late embryogenesis and throughout the postembryonic development [84, 85, 18]. In all these stages, *βFTZ-F1* is expressed as the ecdysteroid titer declines, just before each developmental transition (Figure 3.2) [85, 18]. This specific timing of induction has been observed, not only in *D. melanogaster*, but in all insects investigated so far, *M. sexta* [75], *B. mori* [86], *A. aegypti* [87] and *B. germanica* [68], and is the result of the previously described cross-regulatory interplay between E75, HR3 and HR4 (Figure 3.3). During post-embryonic development FTZ-F1 is necessary for molting in *D. melanogaster*, *T. castaneum*, and *B. germanica* [85, 68, 47]. Furthermore, during the prepupal stage of *D. melanogaster*, βFTZ-F1 plays a crucial role as a competence factor for stage-specific responses to a second 20E pulse that triggers the onset of pupal development (Figure 3.2). Thus, *βFtz-F1* mutants show defects in adult head eversion, leg elongation and salivary gland degeneration at the prepupal-pupal transition [88]. The role as a competence factor for stage-specific transcriptional responses to 20E is conserved in the holometabolous insects since it has been also reported during the previtellogenic-vitellogenic transition in the dipteran *A. aegypti* [87, 89, 90]. Conversely, in more primitive insects, such as *B. germanica*, FTZ-F1 does not play any role in transcriptional competence, but controls the timing of the onset of ecdysteroid production during the intermolt periods [68]. This function could represent an ancestral role of FTZ-F1 during post-embryonic development that has been probably lost in more derived insects. In fact, holometabolous animals utilize a new timer mechanism based on the nutritional and circadian-dependent release of the prothoracicotropic hormone produced in the brain [91]. In *C. elegans*, *nhr-25* is functionally conserved. Along with nhr-23, this receptor is periodically expressed during larval development and it is necessary to complete the molting process [92, 93]. These results show that the functional axis HR3/nhr-23-FTZ-F1/nhr-25 is well conserved in the Ecdysozoa in relation to molting. However, from a regulatory point of view, nhr-25 expression is not dependent of nhr-23 activity, suggesting that the HR3-mediated induction of FTZ-F1 is a regulatory process that has been acquired in insects.

3.3.7. HR39 (NR5B1)

HR39 is the result of the duplication of the *ftz-f1* gene that occurred early during protostomian evolution [94], and hence is most related to SF-1 and LRH-1. In *D. melanogaster*, *DHR39* is induced by 20E and its expression is always inversely related to that of βFTZ-F1 (Figure 3.2), suggesting crossregulatory interactions between both factors [57, 95, 18]. In fact, through binding to the same response element, DHR39 repress and βFTZ-F1 activates transcription of target genes, such as the *alcohol dehydrogenase* and *ftz* [96, 97]. Although *DHR39* null mutants are viable and fertile [98], it has been recently shown that it is essential for sexual development [99]. *DHR39* mutant females are sterile due to the absence of sperm-storing spermathecae and glandular paraovaria. It is also necessary for activation of spermathecal secretion in females as well as to repress male-specific courtship genes [99]. On the other hand, in *T. castaneum*, TcHR39 is necessary for normal development during larval and pupal stages [47].

3.4. INHIBITING THE 20E-SIGNALLING PATHWAY

Throughout development the precise timing of the 20E responses requires not only the positive actitivity of the 20E-dependent NRs but also the participation of negative transcriptional regulators. Three such NRs, HR78, HR38 and Seven-up (SVP) attenuate the activity of the 20E-signalling pathway by inhibiting the normal functioning of the EcR-USP/RXR heterodimer. Below I present a detailed description of these NRs.

3.4.1. HR78 (NR2D1)

This receptor is the insect homolog of the vertebrate orphan receptors Testis related 2 (TR2) and TR4. Although *HR78* expression is induced by 20E in cultured organs and its mRNA levels oscillate with each molting cycle [100,74,18], several results show that it is a negative modulator of EcR-USP/RXR action. First, in *D. melanogaster*, DHR78 binds to a subset of EcR/USP binding sites [100]. Second, DHR78 inhibits the 20E-dependent activation of a reported gene through binding site competition with the 20E heterodimeric receptor [101]. Finally, in *B. mori*, BmHR78 heterodimerizes with BmUSP in vitro [102].

Functional analysis of DHR78, carried out in *D. melanogaster,* confirm its repressive role of 20E signalling [103, 104]. Whereas *DHR78* null mutants eclose without problems, they show post-embryonic growth defects, dying as small third instar larvae with clear tracheal defects. These mutants fail to activate properly the 20E-dependent regulatory hierarchy that occurs at mid instar and that prepares the larvae for metamorphosis. Interestingly, the absence of any phenotypic effect upon ectopic expression of DHR78 indicates that its activity is regulated post-transcriptionally [103]. This has been recently confirmed by characterizing a new DHR78-interacting corepressor, Moses (Middleman of 78 signalling). Biochemical and genetic data show that levels of Moses regulate the activity and the stability of DHR78, therefore maintaining the correct rate of larval growth [105]. The role of HR78 in controlling growth is reminiscent with that of its vertebrate homolog. TR4 mutant mice display early lethality and clear growth retardation [106]. The function of HR78 has been also analyzed in *T. castaneum*. In this insect, 40% of the TcHR78 last larva knockdowns are unable to proceed to the pupal stage and die at the end of the larval instar. Furthermore, knockdown animals that are able to progress to the adult stage present significant reduction in the number of laid eggs, defining possible roles of TcHR78 in reproduction and/or egg development [47]. On the other hand, the genome of *C. elegans* has one HR78 homolog, HNR-41 [39]. RNAi in developing worms does not reveal clear functions for this receptor, although it seems to be related to molting and dauer formation [51].

3.4.2. HR38 (NR4A4)

This receptor is a member of the small NR4 subfamily group together with the vertebrate's NGFI-B and Nurr1 as well as with *C. elegans* NHR-6. *HR38* is widely expressed throughout development and binds DNA as monomer or heterodimer with USP/RXR [100, 107, 108, 109, 110]. Similar to HR78, HR38 is a negative modulator of the 20E-signaling pathway. It exerts its repressive role by interacting with USP/RXR, hence disrupting the functional 20E heterodimer. One example of such inhibition is found in adult females of *A. aegypti* where AaHR38 interacts with AaUSP at the previtellogenic stage preventing the formation of AaEcR-AaUSP heterodimer, and maintaining inhibited the 20E-signaling pathway. After feeding, the repressive action of AaHR38 is halted and the increase of circulating ecdysteroids promotes the heterodimerization of AaEcR-AaUSP and the activation of 20E-dependent genes [110]. The HR38-USP/RXR heterodimer can also silence 20E action by competing with EcR-USP/RXR for binding to selected response elements in target genes, as happens in the *ng-1* and *ng-2* genes of *D. melanogaster* [109].

Aside from the inhibitory function, DHR38 may be part of a new ecdysteroid signaling pathway in *D. melanogaster*. The heterodimer of DHR38 with ligand-activated USP and/or RXR receptors can respond not only to 20E but also to a wide variety of different ecdysteroids, such as α-ecdysone, 3-Epi-20E and 3-dehydromakisterone A. This response, however, does not involve direct binding of these ligands to DHR38 since the crystal structure of the receptor revealed the lack of a conventional ligand binding pocket and a co-activator binding site as well [111]. Similarly, the vertebrate Nurr1 protein is a ligand independent transcriptional activator that presents a similar atypical ligand binding pocket [112]. Taken together, these results indicate that the heterodimer HR38-USP/RXR could play an important physiological role at very specific stages of development, by transducing a new ecdysteroid signalling pathway through an atypical mechanism.

DHR38 is also important for correct synthesis of the insect cuticle. Thus, pupae lacking DHR38 present fragility and rupturing of the adult cuticle [108]. In fact, DHR38 is necessary for the correct activation of several important epidermis genes, namely *AcP65A* cuticle gene and the *DOPA-decarboxylase* [113, 114]. Confirming these results, *DHR38* mutants die as pharate adults showing significant reduction in several cuticle gene expression [115]. Finally, RNAi of HR38 in *T. castaneum* showed that this factor is vital to complete larval-pupal and pupal-adult transitions [47]. Conversely to the described roles of HR38 in insects, *C. elegans* NHR-6, expressed in the posterior and anterior spermatheca during the third and fourth larval stages and in two chemosensory neurons, is necessary for ovulation [51].

3.4.3. Seven-up (NR2F3)

Along with Hepatocyte nuclear factor 4 (HNF4) and Tailless (Tll), Svp is the most conserved receptor. It shares a 90% identity in the DBD and LBD with its vertebrate homolog, chicken ovalbumin upstream promoter transcription factor 1 (COUP-TF1).

Svp is a potent repressor that is widely expressed throughout development [116, 117, 74, 118]. In *D. melanogaster* and *A. aegypti*, SVP represses the 20E-dependent transactivation mediated by EcR-USP [119, 118]. In the mosquito, AaSVP can even interact directly with AaUSP, as has been demonstrated by two-hybrid and GST-pull down analysis [89, 90]. This interaction has an important role during the end of the vitellogenic cycle, when the 20E titer is low, inhibiting the 20E-dependent transactivation of target genes such as *vitellogenin* [89, 90]. The repressive role of SVP is also observed in *D. melanogaster*, where ectopic overexpression of this receptor at different larval and pupal stages leads to lethality at the onset of metamorphosis [119].

In addition to the 20E-related functions, SVP exerts multiple roles in a wide variety of developmental processes. In *D. melanogaster*, for example, is required during embryogenesis for the correct development of four photoreceptor cells, R1, R3, R4 and R6. The absence of SVP forces these cells to adopt a R7 fate whereas ectopic expression in R7 cells transforms their identity to an R1 or R6 fate [116, 120]. SVP is also necessary for the proper formation of the embryonic central and peripheral nervous systems, as well as the fat body, Malpighian tubules and the different cardioblasts in the dorsal vessel [121, 122, 123, 124]. Furthermore, during late development it controls the proliferation and differentiation in many different neuroblast lineages [125]. In the herbivorous coleopteran *Callosobruchus maculates*, CmSVP is involved in the negative regulation of counter defense genes that help to cope with plant defense compounds [126]. In *T. castaneum*, TcSVP is required to complete the metamorphic process [47]. In adult females of *B. germanica*, BgSVP induces juvenile hormone production during the first days of the vitellogenic cycle (Borras, F, and Martín, D., unpublished results).

Similar to insects, the *C. elegans* homolog UNC-55 is related with neural differentiation. This receptor, expressed in the ventral motor neurons of the worm, modifies the genetic program that produces the correct synaptic pattern that identifies these type of neurons [127].

3.5. 20E-INDEPENDENT NUCLEAR RECEPTORS

The remaining 11 insect NRs are not directly involved in the 20E-dependent signalling pathway. Instead, they play key roles in different physiological processes during embryonic and post-embryonic development, mainly related with embryogenesis, neural development, metabolism and detoxification. Below, I describe the current knowledge about the biological function of these receptors. The order in which are described is based on the subfamily classification.

3.5.1. HR96 (NR1J1)

This receptor is most similar to the vertebrate vitamin D receptor (VDR) and two xenobiotic receptors related with detoxification responses, CAR and SXR. In

D. melanogaster, DHR96 mRNA is present throughout development and, surprisingly, its basal levels of expression are upregulated by 20E in cultured larval organs [100, 18]. Recently, the analysis of *DHR96* mutants has confirmed the important role in the coordination of the transcriptional response to two xenobiotic compounds, Phenobarbital and the pesticide DTT. Furthermore, DHR96 also controls different metabolic and stress-related genes, defining this receptor as a crucial regulator of multiple xenobiotic-stress induced responses [128]. In *C. elegans*, there are three homologs, DAF-12, NHR-8 and NHR-48. NHR-8 is expressed in the gut and is proposed to function as a xenobiotic sensor, similar to DHR96 [129]. DAF-12, the best characterized NR in the worm, links environmental information with multiple physiological processes, such as, dauer formation, life span, metabolism and heterochronicity [130].

3.5.2. HNF4 (NR2A4)

HNF4, together with SVP and TLL, is the most conserved NR, and two homolog genes are found in mammals, HNF4α and HNF4γ. Remarkably, a surprising burst of gene duplications in *C. elegans* produced 269 HNF4-derived receptors. Given the strong conservation of HNF4 between insects and vertebrates, it is not surprising that the expression pattern is also well conserved. During the embryogenesis of *D. melanogaster*, for example, HNF4 is present in the developing fat body, Malpighian tubules and the midgut during organogenesis [131]. Likewise, in rat and mouse, HNF4 mRNA can be detected in the liver (fat body homologue), kidney (Malpighian tubules analogue) and intestine (midgut analogue) [132]. HNF4 is also expressed in the fat body, gut and ovaries of *B. mori* [133] and *A. aegypti* [134]. In *D. melanogaster*, the in vivo ligand sensor system has shown that the activity of GAL4-HNF4, along with GAL4-DHR3 and GAL4-DHR38, is very high in the embryonic yolk (main source of nutrients and lipids in the embryo) and in the fat body of metabolically active third instar larvae. Conversely, the activity is clearly halted in non-feeding prepupae, suggesting that HNF4 operates as a metabolic sensor [61]. Recently, this hypothesis has been confirmed analyzing *dHNF4* null mutant larvae, which are sensitive to starvation [135]. Although starved mutant larvae consume glycogen normally, they are unable to use stored fat in the midgut and fat body for energy. Furthermore, a detailed microarray analysis has shown that dHNF4 is necessary for normal activation of genes involved in lipid catabolism and β-oxidation required for energy production [135]. Similarly, mammalian HNF4 binds different fatty acids and is required for the transcriptional regulation of genes involved in the hepatic metabolism of lipids [136, 137]. On the other hand, in *T. castaneum*, the reduction of HNF4 expression by RNAi in vivo, dramatically affects reproduction and egg maturation of adult females [47].

In *C. elegans*, the two more similar genes are *nhr-64* and *nhr-69*, although they do not display any obvious phenotypes during development [51]. During the last years another HNF4-derived gene, *nhr-49*, has been functionally analyzed. Similar to the metabolic role of its mammalian and insect counterparts, *nhr-49* knockdown nematodes present elevated fat content and short life span. The high-fat content is due to reduce expression of enzymes in fatty-acid β-oxidation and the shortened life span resulted from the impaired expression of Δ_9-desaturases [138].

3.5.3. Tailless (NR2E2)

The *tailless* gene is one of the most conserved NRs. *tll* has one vertebrate homolog, *tailless related* (*Tlx*). The function of this gene has been extensively studied in relation with its function during the embryonic development of *D. melanogaster*. In the fruitfly, patterning of the early embryo depends on a complex network of transcriptional regulation. Maternal and zygotic transcription factors interact to progressively subdivide the embryo into different spatial domains. At mid-blastoderm stage, several of these factors are expressed in highly defined gradients, bands and stripes that prefigure the body plan of the organism [139]. In this context, TLL controls the specification of the most anterior and posterior embryonic structures, defining it as a gap gene [140, 141, 142]. *tll* mutants present abnormal development of parts of the head and the anterior brain and absence of the posterior terminal domain comprising the eighth abdominal segment, telson and the posterior gut [141, 142, 143]. Furthermore, ectopic expression of *tll* throughout the entire embryo at the blastoderm stage, force the trunk and abdomen regions to adopt a terminal fate [144]. The effect of *tll* during embryonic development is achieved by acting as activator and repressor of other patterning genes. It is required for the activation of *hunchback*, *brachyenteron* and *forkhead* at the posterior region and also for repressing the expression of the gap genes *knirps*, *krüppel* and *giant* [145, 146, 147, 148, 149]. Recently, however, it has been shown that TLL functions exclusively as a transcriptional repressor and that its activation effect is mediated by repressing some repressor factors [150]. On the other hand, in *T. castaneum*, which represents a more ancestral form of embryogenesis, posterior *Tc-tll* expression occurs during a short time ceasing before tll-target genes known from *D. melanogaster* are activated, suggesting that TcTLL is not a direct regulator of segmentation genes at the posterior end of the embryo [151]. It is clear, then, that the function of *tll* has experienced a shift during the evolution of insect embryogenesis, from short germ band to long germ band type.

TLL also exerts important functions during later stages of development. It is expressed in the developing forebrain of *D. melanogaster*, where is required for the normal formation of the protocerebral neuroblasts and in the eye formation [152, 143, 153]. Remarkably, the functions associated with brain and eye development are evolutionary conserved, since its mammalian ortholog Tlx also controls these processes [154]. In *C. elegans* the *tll* homolog, *NHR-67*, presents a highly restricted expression pattern in only four head neurons and is necessary for normal development. Knockdown larvae are uncoordinated and show slow growing and defects in vulval development during the adult stage. Interestingly, *NHR-67* knockdown worms show incomplete shedding of the larval cuticle, defining possible nematode-specific functions of NHR-67 in relation with molting [51].

3.5.4. Dissatisfaction (NR2E4)

The *dissatisfaction* gene seems to be the result of a duplication event of the *tll* gene occurred after the separation between arthropods and vertebrates. In insects, this gene is related with the control of complex behaviours. DSF acts as repressor upon binding as a monomer to different direct and inverted repeats of the hormonal response element. The repressive activity of DSF is located in the LBD and hinge regions [155].

Dsf is expressed both in males and females during all developmental stages in a very limited number of different neurons [156]. Correlating with its neuronal expression, *dsf* mutation affects sex-specific courtship behaviours and neuronal differentiation in both sexes. Virgin *dsf* female mutants resist males during courtship and copulation and also fail to lay mature eggs. *dsf* mutant males court and attempt copulation with both mature males and females but they have problems copulating because of reduced abdominal curling. All these phenotypes are consequence of impaired differentiation of the sex-specific abdominal neurons that innervate the uterine muscles in females and in the ventral abdominal muscles of males [157, 156].

3.5.5. HR51 (NR2E3), HR83 (NR2E5) and NR2E6

HR51 and HR83 were recently identified thanks to the sequencing of the *D. melanogaster* genome, although both are present in all the other insect genomes. Both are homologs to the vertebrate photoreceptor-specific nuclear receptor (PNR), and *C. elegans* FAX-1. Unfortunately, the information about these genes is scarce. Recently, however, two independent *DHR51* alleles have been characterized in *D. melanogaster* [158]. Consistent with the highly restricted expression of DHR51 in mushroom body neurons and in a small number of other cells of the central nervous system, *DHR51* mutants displayed behavioural and sensory defects, including compromised fertility and coordination and wing expansion failure [158]. Furthermore, it has been shown that DHR51 is a thiolate heme-binding protein and, as E75, binds NO and CO which suggests that this factor can also be a gas or a heme sensor [159]. On the other hand, it has been impossible to detect HR83 mRNA in any stage of development [18], suggesting that its expression is tissue restricted. Although there is no information of the function of this receptor, it is tempting to speculate that it could exert conserved functions in neuronal development similar to those of HR51, PNR and FAX-1. *PNR* expression is restricted to the retina and controls the development of photoreceptors [160, 161]. Similarly, *FAX-1* is also expressed in a subset of neurons and is required for axon pathfinding and for the expression of neuropeptides in a number of interneurons [162]. Finally, NR2E6 is new insect's specific NR of the NR2E subgroup. It is found in all insect genomes but in that of *D. melanogaster*

(Table 3.1). In *T. castaneun*, the only species in which NR2E6 function has been studied, RNAi reveal no phenotypes in any stage of development [47].

3.5.6. Estrogen-Related Receptor (NR3B4)

The estrogen-related receptor (ERR) is the only member of the NR3 subfamily that is present in insects. Along with ERR, this subfamily comprises the receptors for estrogens (ER), androgens (AR), glucocorticoids (GR), progesterone (PG) and mineralocorticoids (MR). The *ERR* gene is expressed during mid-embryogenesis and the third larval instar of *D. melanogaster* [18]. Unfortunately, the information on ERR function is almost non-existent. Almost all data come from the analysis of the activation pattern of the GAL4-ERR LBD sensor [61]. GAL4-ERR is activated during mid-embryogenesis and in the mid-third larval instar, paralleling the expression of the *ERR* gene. The activation at mid-third instar is relevant in that it coincides with a global switch in gene expression that prepares the larva for metamorphosis. Given that neither EcR nor USP are activated at this precise moment, it is possible that a yet undefined ERR-specific hormonal signal activates the mentioned genetic reprogramming through this receptor. Obtaining functional data requires, however, the generation of specific ERR mutants and/or knockdowns.

3.5.7. The NR0 Subfamily

This subfamily includes three atypical NRs, knirps, knirps-related and eagle. All three, which do not have vertebrate homologs, lack their LBDs. In *D. melanogaster*, *knirps* and *knirps-related* genes derive from a recent gene duplication event. During embryonic development, at the blastodermal stage, *knirps* and *knirps-related* are expressed in the anterior and posterior domains. The posterior expression of *knirps* is involved in the regulation of pair-rule genes. In fact, the absence of this receptor in the posterior domain results in the deletion of abdominal segments one to seven, hence the classification of this gene as an abdominal gap gene [163]. Furthermore, elimination of the expression of *knirps* and *knirps-related* genes results in defects in head morphogenesis and in tracheal development during later development [164, 165]. Interestingly, the head defects are not observed in single mutants revealing the functional redundancy of both genes. Conversely, *T. castaneum* has a single homolog that is crucial for head segmentation but exerts a minor role during abdominal patterning, which indicates that in more primitive insects, this gene does not function as a canonical gap gene [166].

On the other hand, in *D. melanogaster*, the *eagle* gene is expressed in only four neuronal lineages and also transiently in the embryonic gonad [167, 168]. Whereas *eagle* mutants show a correct number of neuronal cells, there is a dramatic reduction in the number of serotonin cells and abnormal axon projections of few neurons as well, indicating that eagle is vital for the terminal differentiation of selected neuronal lineages [167, 169].

ACKNOWLEDGEMENTS

I thank to all members of my laboratory for their work and critical comments on the manuscript. I also want to aopologize to authors whose work has not been cited owing to length restrictions. Research in the lab is supported by the Spanish Ministry of Education and Science (projects BMC2002-03222 and BFU2006-13212) and the Generalitat de Catalunya (2001 SGR 003245).

REFERENCES

1. Laudet, V. and Bonneton, F. (2005). In Gilbert, L. I., Iatrou, K., Gill, S. S., eds., Comprehensive molecular insect science Vol. 3. Elsevier, Oxford/Amsterdam, 287–318.
2. Glass, C. K. and Rosenfeld, M. G. (2000). The coregulator exchange in transcriptional functions of nuclear receptors. *Genes Dev* 14, 121–141.
3. Aguinaldo, A. M., Turbeville, J. M., Linford, L. S., Rivera, M. C., Garey, J. R., Raff, R. A., and Lake, J. A. (1997). Evidence for a clade of nematodes, arthropods and other moulting animals. *Nature* 387, 489–493.
4. Bonneton, F., Chaumot, A., and Laudet, V. (2008). Annotation of Tribolium nuclear receptors reveals an increase in evolutionary rate of a network controlling the ecdysone cascade. *Insect Biochem Mol Biol* 38, 416–429.
5. Cheng, D., Xia, Q., Duan, J., Wei, L., Huang, C., Li, Z., Wang, G., and Xiang, Z. (2008). Nuclear receptors in Bombyx mori: Insights into genomic structure and developmental expression. *Insect Biochem Mol Biol* 38, 1130–1137.
6. Cruz, J., Sieglaff, D. H., Arensburger, P., Atkinson, P. W., and Raikhel, A. S. (2009). Nuclear receptors in the mosquito Aedes aegypti: Annotation, hormonal regulation and expression profiling. *FEBS J* 276, 1233–1254.
7. King-Jones, K. and Thummel, C. S. (2005). Nuclear receptors – a perspective from *Drosophila*. *Nat Rev Genet* 6, 311–323.
8. Velarde, R. A., Robinson, G. E., and Fahrbach, S. E. (2006). Nuclear receptors of the honey bee: Annotation and expression in the adult brain. *Insect Mol Biol* 15, 583–595.
9. Bertrand, S., Brunet, F. G., Escriva, H., Parmentier, G., Laudet, V., and Robinson-Rechavi, M. (2004). Evolutionary genomics of nuclear receptors: From twenty-five ancestral genes to derived endocrine systems. *Mol Biol Evol* 21, 1923–1937.
10. Maglich, J. M., Sluder, A., Guan, X., Shi, Y., McKee, D. D., Carrick, K., Kamdar, K., Willson, T. M., and Moore, J. T. (2001). Comparison of complete nuclear receptor sets from the human, *Caenorhabditis elegans* and *Drosophila* genomes. *Genome Biol* 2, research0029.1–0029.7.
11. Thummel, C. S. (1995). From embryogenesis to metamorphosis: The regulation and function of *Drosophila* nuclear receptor superfamily members. *Cell* 83, 871–877.
12. Yao, T. P., Segraves, W. A., Oro, A. E., McKeown, M., and Evans, R. M. (1992). *Drosophila* ultraspiracle modulates ecdysone receptor function via heterodimer formation. *Cell* 71, 63–72.
13. Riddiford, L. M., Cherbas, P., and Truman, J. W. (2000). Ecdysone receptors and their biological actions. *Vitam Horm* 60, 1–73.
14. Koelle, M. R., Talbot, W. S., Segraves, W. A., Bender, M. T., Cherbas, P., and Hogness, D. S. (1991). The *Drosophila* EcR gene encodes an ecdysone receptor, a new member of the steroid receptor superfamily. *Cell* 67, 59–77.
15. Yao, T. P., Forman, B. M., Jiang, Z., Cherbas, L., Chen, J. D., McKeown, M., Cherbas, P., and Evans, R. M. (1993). Functional ecdysone receptor is the product of EcR and Ultraspiracle genes. *Nature* 366, 476–479.
16. Costantino, B. F., Bricker, D. K., Alexandre, K., Shen, K., Merriam, J. R., Antoniewski, C., Callender, J. L., Henrich, V. C., Presente, A., and Andres, A. J. (2008). A novel ecdysone receptor

mediates steroid-regulated developmental events during the mid-third instar of *Drosophila*. *PLoS Genet* 4, e1000102.

17. Maestro, O., Cruz, J., Pascual, N., Martín, D., and Bellés, X. (2005). Differential expression of two RXR/ultraspiracle isoforms during the life cycle of the hemimetabolous insect Blattella germanica (Dictyoptera, Blattellidae). *Mol Cell Endocrinol* 238, 27–37.
18. Sullivan, A. A. and Thummel, C. S. (2003). Temporal profiles of nuclear receptor gene expression reveal coordinate transcriptional responses during *Drosophila* development. *Mol Endocrinol* 17, 2125–2137.
19. Karim, F. D. and Thummel, C. S. (1992). Temporal coordination of regulatory gene expression by the steroid hormone ecdysone. *EMBO J* 11, 4083–4093.
20. Jindra, M. and Riddiford, L. M. (1996). Expression of ecdysteroid-regulated transcripts in the silk gland of the wax moth, *Galleria mellonella*. *Dev Genes Evol* 206, 305–314.
21. Jindra, M., Malone, F., Hiruma, K., and Riddiford, L. M. (1996). Developmental profiles and ecdysteroid regulation of the mRNAs for two ecdysone receptor isoforms in the epidermis and wings of the tobacco hornworm, *Manduca sexta*. *Dev Biol* 180, 258–272.
22. Jindra, M., Huang, J. Y., Malone, F., Asahina, M., and Riddiford, L. M. (1997). Identification and mRNA developmental profiles of two ultraspiracle isoforms in the epidermis and wings of *Manduca sexta*. *Insect Mol Biol* 6, 41–53.
23. Wang, S. F., Li, C., Zhu, J., Miura, K., Miksicek, R. J., and Raikhel, A. S. (2000). Differential expression and regulation by 20-hydroxyecdysone of mosquito ultraspiracle isoforms. *Dev Biol* 218, 99–113.
24. Wang, S. F., Li, C., Sun, G., Zhu, J., and Raikhel, A. S. (2002). Differential expression and regulation by 20-hydroxyecdysone of mosquito ecdysteroid receptor isoforms A and B. *Mol CellEndocrinol* 196, 29–42.
25. Verras, M., Gourzi, P., Zacharopoulou, A., and Mintzas, A. C. (2002). Developmental profiles and ecdysone regulation of the mRNAs for two ecdysone receptor isoforms in the Mediterranean fruit fly *Ceratitis capitata*. *Insect Mol Biol* 11, 553–565.
26. Talbot, W. S., Swyryd, E. A., and Hogness, D. S. (1993). *Drosophila* tissues with different metamorphic responses to ecdysone express different ecdysone receptor isoforms. *Cell* 73, 1323–1337.
27. Henrich, V. C., Sliter, T. J., Lubahn, D. B., MacIntyre, A., and Gilbert, L. I. (1990). A steroid/thyroid hormone receptor superfamily member in *Drosophila melanogaster* that shares extensive sequence similarity with a mammalian homologue. *Nucleic Acids Res* 18, 4143–4148.
28. Oro, A. E., McKeown, M., and Evans, R. M. (1990). Relationship between the product of the *Drosophila* ultraspiracle locus and the vertebrate retinoid X receptor. *Nature* 347, 298–301.
29. Kozlova, T. and Thummel, C. S. (2003). Essential roles for ecdysone signaling during *Drosophila* mid-embryonic development. *Science* 301, 1911–1914.
30. Perrimon, N., Engstrom, L., and Mahowald, A. P. (1985). Developmental genetics of the 2C-D region of the *Drosophila* X chromosome. *Genetics* 111, 23–41.
31. Davis, M. B., Carney, G. E., Robertson, A. E., and Bender, M. (2005). Phenotypic analysis of EcR-A mutants suggests that EcR isoforms have unique functions during *Drosophila* development. *Dev Biol* 282, 385–396.
32. Schubiger, M., Wade, A. A., Carney, G. E., Truman, J. W., and Bender, M. (1998). *Drosophila* EcR-B ecdysone receptor isoforms are required for larval molting and for neuron remodeling during metamorphosis. *Development* 125, 2053–2062.
33. Bender, M., Imam, F. B., Talbot, W. S., Ganetzky, B., and Hogness, D. S. (1997). *Drosophila* ecdysone receptor mutations reveal functional differences among receptor isoforms. *Cell* 91, 777–788.
34. Hall, B. L. and Thummel, C. S. (1998). The RXR homolog ultraspiracle is an essential component of the *Drosophila* ecdysone receptor. *Development* 125, 4709–4717.

35. Tan, A. and Palli, S. R. (2008b). Ecdysone [corrected] receptor isoforms play distinct roles in controlling molting and metamorphosis in the red flour beetle, *Tribolium castaneum*. *Mol Cell Endocrinol* 291, 42–49.
36. Barchuk, A. R., Figueiredo, V. L., and Simões, Z. L. (2008). Downregulation of ultraspiracle gene expression delays pupal development in honeybees. *J Insect Physiol* 54, 1035–1040.
37. Cruz, J., Mané-Padrós, D., Bellés, X., and Martín, D. (2006). Functions of the ecdysone receptor isoform-A in the hemimetabolous insect *Blattella germanica* revealed by systemic RNAi in vivo. *Dev Biol* 297, 158–171.
38. Martín, D., Maestro, O., Cruz, J., Mané-Padrós, D., and Bellés, X. (2006). RNAi studies reveal a conserved role for RXR in molting in the cockroach *Blattella germanica*. *J Insect Physiol* 52, 410–416.
39. Sluder, A. E. and Maina, C. V. (2001). Nuclear receptors in nematodes: Themes and variations. *Trends Genet* 17, 206–213.
40. Jiang, C., Lamblin, A. F., Steller, H., and Thummel, C. S. (2000). A steroid-triggered transcriptional hierarchy controls salivary gland cell death during *Drosophila* metamorphosis. *Mol Cell* 5, 445–455.
41. Zhou, B., Hiruma, K., Jindra, M., Shinoda, T., Segraves, W. A., Malone, F., and Riddiford, L. M. (1998). Regulation of the transcription factor E75 by 20-hydroxyecdysone and juvenile hormone in the epidermis of the tobacco hornworm, Manduca sexta, during larval molting and metamorphosis. *Dev Biol* 193, 127–138.
42. Pierceall, W. E., Li, C., Biran, A., Miura, K., Raikhel, A. S., and Segraves, W. A. (1999). E75 expression in mosquito ovary and fat body suggests reiterative use of ecdysone-regulated hierarchies in development and reproduction. *Mol Cell Endocrinol* 150, 73–89.
43. Swevers, L., Eystathioy, T., and Iatrou, K. (2002). The orphan nuclear receptors BmE75A and BmE75C of the silkmoth Bombyx mori: Hornmonal control and ovarian expression. *Insect Biochem Mol Biol* 32, 1643–, 1652.
44. Dubrovsky, E. B., Dubrovskaya, V. A., and Berger, E. M. (2004). Hormonal regulation and functional role of *Drosophila* E75A orphan nuclear receptor in the juvenile hormone signaling pathway. *Dev Biol* 268, 258–270.
45. Bialecki, M., Shilton, A., Fichtenberg, C., Segraves, W. A., and Thummel, C. S. (2002). Loss of the ecdysteroid-inducible E75A orphan nuclear receptor uncouples molting from metamorphosis in *Drosophila*. *Dev Cell* 3, 209–220.
46. Mané-Padrós, D., Cruz, J., Vilaplana, L., Pascual, N., Bellés, X., and Martín, D. (2008). The nuclear hormone receptor BgE75 links molting and developmental progression in the direct-developing insect *Blattella germanica*. *Dev Biol* 315, 147–160.
47. Tan, A. and Palli, S. R. (2008a). Identification and characterization of nuclear receptors from the red flour beetle, *Tribolium castaneum*. *Insect Biochem Mol Biol* 38, 430–439.
48. White, K. P., Hurban, P., Watanabe, T., and Hogness, D. S. (1997). Coordination of *Drosophila* metamorphosis by two ecdysone-induced nuclear receptors. *Science* 276, 114–117.
49. Reinking, J., Lam, M. M., Pardee, K., Sampson, H. M., Liu, S., Yang, P., Williams, S., White, W., Lajoie, G., Edwards, A., and Krause, H. M. (2005). The *Drosophila* nuclear receptor E75 contains heme and is gas responsive. *Cell* 122, 195–207.
50. Guillaumond, F., Dardente, H., Giguère, V., and Cermakian, N. (2005). Differential control of Bmal1 circadian transcription by REV-ERB and ROR nuclear receptors. *J Biol Rhythms* 20, 391–403.
51. Gissendanner, C. R., Crossgrove, K., Kraus, K. A., Maina, C. V., and Sluder, A. E. (2004). Expression and function of conserved nuclear receptor genes in *Caenorhabditis elegans*. *Dev Biol* 266, 399–416.
52. Stone, B. L. and Thummel, C. S. (1993). The *Drosophila* 78C early late puff contains E78, an ecdysone-inducible gene that encodes a novel member of the nuclear hormone receptor superfamily. *Cell* 22, 307–320.
53. Russell, S. R., Heimbeck, G., Goddard, C. M., Carpenter, A. T., and Ashburner, M. (1996). The *Drosophila* Eip78C gene is not vital but has a role in regulating chromosome puffs. *Genetics* 144, 159–170.

54. Bryant, Z., Subrahmanyan, L., Tworoger, M., LaTray, L., Liu, C. R., Li, M. J., van den Neg, G., and Ruohola-Baker, H. (1999). Characterization of differentially expressed genes in purified *Drosophila* follicle cells: Toward a general strategy for cell type-specific developmental analysis. *Proc Natl Acad Sci USA* 96, 5559–5564.
55. Carmi, I., Kopczynski, J. B., and Meyer, B. J. (1998). The nuclear hormone receptor SEX-1 is an X-chromosome signal that determines nematode sex. *Nature* 396, 168–173.
56. Palli, S. R., Hiruma, K., and Riddiford, L. M. (1992). An ecdysteroid-inducible *Manduca* gene similar to the *Drosophila* DHR3 gene, a member of the steroid hormone receptor superfamily. *Dev Biol* 150, 306–318.
57. Horner, M. A., Chen, T., and Thummel, C. (1995). Ecdysteroid regulation and DNA binding properties of *Drosophila* nuclear family members. *Dev Biol* 168, 490–502.
58. Kapitskaya, M. Z., Li, C., Miura, K., Segraves, W., and Raikhel, A. S. (2000). Expression of the early-late gene encoding the nuclear receptor HR3 suggests its involvement in regulating the vitellogenic response to ecdysone in the adult mosquito. *Mol Cell Endocrinol* 160, 25–37.
59. Eystathioy, T., Swevers, L., and Iatrou, K. (2001). The orphan nuclear receptor BmHR3 of *Bombyx mori*: Hormonal control, ovarian expression and functional properties. *Mech Dev* 103, 107–115.
60. Cruz, J., Martin, D., and Belles, X. (2007). Redundant ecdysis regulatory functions of three nuclear receptor HR3 isoforms in the direct-developing insect *Blattella germanica. Mech Dev* 124, 180–189.
61. Palanker, L., Necakov, A. S., Sampson, H. M., Ni, R., Hu, C., Thummel, C. S., and Krause, H. M. (2006). Dynamic regulation of *Drosophila* nuclear receptor activity in vivo. *Development* 133, 3549–3562.
62. Kallen, J. A., Schlaeppi, J. M., Bitsch, F., Geisse, S., Geiser, M., Delhon, I., and Fournier, B. (2002). X-ray structure of the hRORalpha LBD at 1.63 A: Structural and functional data that cholesterol or a cholesterol derivative is the natural ligand of RORalpha. *Structure* 10, 1697– 1707.
63. Stehlin-Gaon, C., Willmann, D., Zeyer, D., Sanglier, S., Van Dorsselaer, A., Renaud, J. P., Moras, D., and Schüle, R. (2003). All-trans retinoic acid is a ligand for the orphan nuclear receptor ROR beta. *Nat Struct Biol* 10, 820–825.
64. Carney, G. E., Wade, A. A., Sapra, R., Goldstein, E. S., and Bender, M. (1997). DHR3, an ecdysone-inducible early-late gene encoding a *Drosophila* nuclear receptor, is required for embryogenesis. *Proc Natl Acad Sci USA* 94, 12024–12029.
65. Lam, G., Hall, B. L., Bender, M., and Thummel, C. S. (1999). DHR3 is required for the prepupal-pupal transition and differentiation of adult structures during *Drosophila* metamorphosis. *Dev Biol* 212, 204–216.
66. Kostrouchova, M., Krause, M., Kostrouch, Z., and Rall, J. E. (2001). Nuclear hormone receptor CHR3 is a critical regulator of all four larval molts of the nematode *Caenorhabditis elegans. Proc Natl Acad Sci USA* 98, 7360–7365.
67. Lam, G. T., Jiang, C., and Thummel, C. S. (1997). Coordination of larval and prepupal gene expression by the DHR3 orphan receptor during *Drosophila* metamorphosis. *Development* 124, 1757–1769.
68. Cruz, J., Nieva, C., Mané-Padrós, D., Martín, D., and Bellés, X. (2008). Nuclear receptor BgFTZ-F1 regulates molting and the timing of ecdysteroid production during nymphal development in the hemimetabolous insect *Blattella germanica. Dev Dyn* 237, 3179–3191.
69. Zhang, Y. and Dufau, M. L. (2004). Gene silencing by nuclear orphan receptors. *Vitam Horm* 68, 1–48.
70. Hiruma, K. and Riddiford, L. M. (2001). Regulation of transcription factors MHR4 and betaFTZ-F1 by 20-hydroxyecdysone during a larval molt in the tobacco hornworm, *Manduca sexta. Dev Biol* 232, 265–274.
71. Chen, J. H., Turner, P. C., and Rees, H. H. (2002). Molecular cloning and induction of nuclear receptors from insect cell lines. *Insect Biochem Mol Biol* 32, 657–667.
72. King-Jones, K., Charles, J. P., Lam, G., and Thummel, C. S. (2005). The ecdysone-induced DHR4 orphan nuclear receptor coordinates growth and maturation in *Drosophila. Cell* 121, 773–784.

73. Charles, J. P., Shinoda, T., and Chinzei, Y. (1999). Characterization and DNA-binding properties of GRF, a novel monomeric binding orphan receptor related to GCNF and betaFTZ-F1. *Eur J Biochem* 266, 181–190.
74. Mouillet, J. F., Bousquet, F., Sedano, N., Alabouvette, J., Nicolaï, M., Zelus, D., Laudet, V., and Delachambre, J. (1999). Cloning and characterization of new orphan nuclear receptors and their developmental profiles during *Tenebrio* metamorphosis. *Eur J Biochem* 265, 972–981.
75. Weller, J., Sun, G. C., Zhou, B., Lan, Q., Hiruma, K., and Riddiford, L. M. (2001). Isolation and developmental expression of two nuclear receptors, MHR4 and betaFTZ-F1, in the tobacco hornworm, *Manduca sexta*. *Insect Biochem Mol Biol* 31, 827–837.
76. Bourbon, H. M., Gonzy-Treboul, G., Peronnet, F., Alin, M. F., Ardourel, C., Benassayag, C., Cribbs, D., Deutsch, J., Ferrer, P., Haenlin, M., Lepesant, J. A., Noselli, S., and Vincent, A. (2002). A P-insertion screen identifying novel X-linked essential genes in *Drosophila*. *Mech Dev* 110, 71–83.
77. Zhao, Z., Fang, L. L., Johnsen, R., and Baillie, D. L. (2004). ATP-binding cassette protein E is involved in gene transcription and translation in *Caenorhabditis elegans*. *Biochem Biophys Res Commun* 323, 104–111.
78. Li, L. -A., Chiang, E. F.-L., Chen, J.-C., Hsu, N.-C., Chen, Y.-J., and Chung, B.-C. (1999). Function of steroidogenic factor 1 domains in nuclear localization, transactivation, and interaction with transcription factor TFIIB and c-Jun. *Mol Endocrinol* 13, 1588–1598.
79. Ueda, H., Sun, G.-C., Murata, T., and Hirose, S. (1992). A novel DNA-binding motif abuts the zinc finger domain of insect nuclear hormone receptor FTZ-F1 and mouse embryonal long terminal repeat-binding protein. *Mol Cell Biol* 12, 5667–5672.
80. Guichet, A., Copeland, J. W., Erdélyi, M., Hlousek, D., Závorszky, P., Ho, J., Brown, S., Percival-Smith, A., Krause, H. M., and Ephrussi, A. (1997). The nuclear receptor homologue Ftz-F1 and the homeodomain protein Ftz are mutually dependent cofactors. *Nature* 385, 548–552.
81. Yu, Y., Li, W., Su, K., Yussa, M., Han, W., Perrimon, N., and Pick, L. (1997). The nuclear hormone receptor Ftz-F1 is a cofactor for the *Drosophila* homeodomain protein Ftz. *Nature* 385, 552– 555.
82. Alonso, C. R., Maxton-Kuechenmeister, J., and Akam, M. (2001). Evolution of Ftz protein function in insects. *Curr Biol* 11, 1473–1478.
83. Löhr, U., Yussa, M., and Pick, L. (2001). *Drosophila* fushi tarazu. A gene on the border of homeotic function. *Curr Biol* 11, 1403–1412.
84. Ueda, H., Sonoda, S., Brown, J. L., Scott, M. P., and Wu, C. (1990). A sequence-specific DNA-binding protein that activates fushi tarazu segmentation gene expression. *Genes Dev* 4, 624– 635.
85. Yamada, M., Murata, T., Hirose, S., Lavorgna, G., Suzuki, E., and Ueda, H. (2000). Temporally restricted expression of transcription factor betaFTZ-F1: Significance for embryogenesis, molting and metamorphosis in *Drosophila melanogaster*. *Development* 127, 5083–5092.
86. Sun, G. C., Hirose, S., and Ueda, H. (1994). Intermittent expression of BmFTZ-F1, a member of the nuclear hormone receptor superfamily during development of the silkworm *Bombyx mori*. *Dev Biol* 162, 426–437.
87. Li, C., Kapitskaya, M. Z., Zhu, J., Miura, K., Segraves, W., and Raikhel, A. S. (2000). Conserved molecular mechanism for the stage specificity of the mosquito vitellogenic response to ecdysone. *Dev Biol* 224, 96–110.
88. Broadus, J., McCabe, J. R., Endrizzi, B., Thummel, C. S., and Woodard, C. T. (1999). The *Drosophila* beta FTZ-F1 orphan nuclear receptor provides competence for stage-specific responses to the steroid hormone ecdysone. *Mol Cell* 3, 143–149.
89. Zhu, J., Chen, L., and Raikhel, A. S. (2003a). Posttranscriptional control of the competence factor βFTZ-F1 by juvenile hormone in the mosquito *Aedes aegypti*. *Proc Natl Acad Sci USA* 100, 13338–13343.
90. Zhu, J., Miura, K., Chen, L., and Raikhel, A. S. (2003b). Cyclicity of mosquito vitellogenic ecdysteroid-mediated signaling is modulated by alternative dimerization of the RXR homologue ultraspiracle. *Proc Natl Acad Sci USA* 100, 544–549.
91. Gilbert, L. I. (2004). Halloween genes encode P450 enzymes that mediate steroid hormone biosynthesis in *Drosophila melanogaster*. *Mol Cell Endocrinol* 215, 1–10.

92. Asahina, M., Ishihara, T., Jindra, M., Kohara, Y., Katsura, I., and Hirose, S. (2000). The conserved nuclear receptor Ftz-F1 is required for embryogenesis, moulting and reproduction in *Caenorhabditis elegans*. *Genes Cells* 5, 711–723.
93. Gissendanner, C. R. and Sluder, A. E. (2000). nhr-25, the *Caenorhabditis elegans* ortholog of ftz-f1, is required for epidermal and somatic gonad development. *Dev Biol* 221, 259–272.
94. De Mendonça, R. L., Bouton, D., Bertin, B., Escriva, H., Noël, C., Vanacker, J. M., Cornette, J., Laudet, V., and Pierce, R. J. (2002). A functionally conserved member of the FTZ-F1 nuclear receptor family from *Schistosoma mansoni*. *Eur J Biochem* 269, 5700–5711.
95. Huet, F., Ruiz, C., and Richards, G. (1995). Sequential gene activation by ecdysone in *Drosophila melanogaster*: The hierarchical equivalence of early and early late genes. *Development* 121, 1195–1204.
96. Ayer, S., Walker, N., Mosammaparast, M., Nelson, J. P., Shilo, B. Z., and Benyajati, C. (1993). Activation and repression of *Drosophila* alcohol dehydrogenase distal transcription by two steroid hormone receptor superfamily members binding to a common response element. *Nucleic Acids Res* 21, 1619–1627.
97. Ohno, C. K., Ueda, H., and Petkovich, M. (1994). The *Drosophila* nuclear receptors FTZ-F1 alpha and FTZ-F1 beta compete as monomers for binding to a site in the fushi tarazu gene. *Mol Cell Biol* 14, 3166–3175.
98. Horner, M. and Thummel, C. S. (1997). Mutations in the DHR39 orphan receptor gene havno ee ffect on viability. *Dros Info Serv* 80, 35–37.
99. Allen, A. K. and Spradling, A. C. (2008). The Sf1-related nuclear hormone receptor Hr39 regulates *Drosophila* female reproductive tract development and function. *Development* 135, 311–321.
100. Fisk, G. J. and Thummel, C. S. (1995). Isolation, regulation, and DNA-binding properties of three *Drosophila* nuclear hormone receptor superfamily members. *Proc Natl Acad Sci USA* 92, 10604–10608.
101. Zelhof, A. C., Yao, T. P., Evans, R. M., and McKeown, M. (1995b). Identification and characterization of a *Drosophila* nuclear receptor with the ability to inhibit the ecdysone response. *Proc Natl Acad Sci USA* 92, 10477–10481.
102. Hirai, M., Shinoda, T., Kamimura, M., Tomita, S., and Shiotsuki, T. (2002). *Bombyx mori* orphan receptor, BmHR78: cDNA cloning, testis abundant expression and putative dimerization partner for *Bombyx* ultraspiracle. *Mol Cell Endocrinol* 189, 201–211.
103. Fisk, G. J. and Thummel, C. S. (1998). The DHR78 nuclear receptor is required for ecdysteroid signaling during the onset of *Drosophila* metamorphosis. *Cell* 93, 543–555, Genetics 144, 159–170.
104. Astle, J., Kozlova, T., and Thummel, C. S. (2003). Essential roles for the Dhr78 orphan nuclear receptor during molting of the *Drosophila* tracheal system. *Insect Biochem Mol Biol* 33, 1201–1209.
105. Baker, K. D., Beckstead, R. B., Mangelsdorf, D. J., and Thummel, C. S. (2007). Functional interactions between the Moses corepressor and DHR78 nuclear receptor regulate growth in *Drosophila*. *Genes Dev* 21, 450–464.
106. Collins, L. L., Lee, Y. F., Heinlein, C. A., Liu, N. C., Chen, Y. T., Shyr, C. R., Meshul, C. K., Uno, H., Platt, K. A., and Chang, C. (2004). Growth retardation and abnormal maternal behavior in mice lacking testicular orphan nuclear receptor 4. *Proc Natl Acad Sci USA* 101, 15058–15063.
107. Sutherland, J. D., Kozlova, T., Tzertzinis, G., and Kafatos, F. C. (1995). *Drosophila* hormone receptor 38: A second partner for *Drosophila* USP suggests an unexpected role for nuclear receptors of the nerve growth factor-induced protein B type. *Proc Natl Acad Sci USA* 92, 7966–7970.
108. Kozlova, T., Pokholkova, G. V., Tzertzinis, G., Sutherland, J. D., Zhimulev, I. F., and Kafatos, F. C. (1998). *Drosophila* hormone receptor 38 functions in metamorphosis: A role in adult cuticle formation. *Genetics* 149, 1465–1475.
109. Crispi, S., Giordano, E., D'Avino, P. P., and Fúria, M. (1998). Cross-talking among *Drosophila* nuclear receptors at the promiscuous response element of the ng-1 and ng-2 intermolt genes. *J Mol Biol* 275, 561–574.

110. Zhu, J., Miura, K., Chen, L., and Raikhel, A. S. (2000). AHR38, a homolog of NGFI-B, inhibits formation of the functional ecdysteroid receptor in the mosquito *Aedes aegypti*. *EMBO J* 19, 253–262.
111. Baker, K. D., Shewchuk, L. M., Kozlova, T., Makishima, M., Hassell, A., Wisely, B., Caravella, J. A., Lambert, M. H., Reinking, J. L., Krause, H., Thummel, C. S., Willson, T. M., and Mangelsdorf, D. J. (2003). The *Drosophila* orphan nuclear receptor DHR38 mediates an atypical ecdysteroid signaling pathway. *Cell* 113, 731–742.
112. Wang, Z., Benoit, G., Liu, J., Prasad, S., Aarnisalo, P., Liu, X., Xu, H., Walter, N. P., and Perlmann, T. (2003). Structure and function of Nurr1 identifies a class of ligand-independent nuclear receptors. *Nature* 423, 555–560.
113. Bruey-Sedano, N., Alabouvette, J., Lestradet, M., Hong, L., Girard, A., Gervasio, E., Quennedey, B., and Charles, J. P. (2005). The *Drosophila* ACP65A cuticle gene: Deletion scanning analysis of cis-regulatory sequences and regulation by DHR38. *Genesis* 43, 17–27.
114. Davis, M. M., Yang, P., Chen, L., O'Keefe, S. L., and Hodgetts, R. B. (2007). The orphan nuclear receptor DHR38 influences transcription of the DOPA decarboxylase gene in epidermal and neural tissues of *Drosophila melanogaster*. *Genome* 50, 1049–1060.
115. Kozlova, T., Lam, G., and Thummel, C. S. (2009). *Drosophila* DHR38 nuclear receptor is required for adult cuticle integrity at eclosion. *Dev Dyn* 238, 701–707.
116. Mlodzik, M., Hiromi, Y., Weber, U., Goodman, C. S., and Rubin, G. M. (1990). The *Drosophila* seven-up gene, a member of the steroid receptor gene superfamily, controls photoreceptor cell fates. *Cell* 60, 211–224.
117. Broadus, J. and Doe, C. Q. (1995). Evolution of neuroblast identity: Seven-up and prospero expression reveal homologous and divergent neuroblast fates in *Drosophila* and *Schistocerca*. *Development* 121, 3989–3996.
118. Miura, K., Zhu, J., Dittmer, N. T., Chen, L., and Raikhel, A. S. (2002). A COUP-TF/Svp homolog is highly expressed during vitellogenesis in the mosquito *Aedes aegypti*. *J Mol Endocrinol* 29, 223–238.
119. Zelhof, A. C., Yao, T. P., Chen, J. D., Evans, R. M., and McKeown, M. (1995a). Seven-up inhibits ultraspiracle-based signaling pathway in vitro and in vivo. *Mol Cell Biol* 15, 6736–6745.
120. Hiromi, Y., Mlodzik, M., West, S. R., Rubin, G. M., and Goodman, C. S. (1993). Ectopic expression of seven-up causes cell fate changes during ommatidial assembly. *Development* 118, 1123–1135.
121. Hoshizaki, D. K., Blackburn, T., Price, C., Ghosh, M., Miles, K., Ragucci, M., and Sweis, R. (1994). Embryonic fat-cell lineage in *Drosophila melanogaster*. *Development* 120, 2489–2499.
122. Kerber, B., Fellert, S., and Hoch, M. (1998). Seven-up, the *Drosophila* homolog of the COUP-TF orphan receptors, controls cell proliferation in the insect kidney. *Genes Dev* 12, 1781–1786.
123. Lo, P. C. and Frasch, M. (2001). A role for the COUP-TF-related gene seven-up in the diversification of cardioblast identities in the dorsal vessel of *Drosophila*. *Mech Dev* 104, 49–60.
124. Sudarsan, V., Pasalodos-Sanchez, S., Wan, S., Gampel, A., and Skaer, H. (2002). A genetic hierarchy establishes mitogenic signalling and mitotic competence in the renal tubules of *Drosophila*. *Development* 129, 935–944.
125. Maurange, C., Cheng, L., and Gould, A. P. (2008). Temporal transcription factors and their targets schedule the end of neural proliferation in *Drosophila*. *Cell* 133, 891–902.
126. Ahn, J. E., Guarino, L. A., and Zhu-Salzman, K. (2007). Seven-up facilitates insect counter-defense by suppressing cathepsin B expression. *FEBS J* 274, 2800–2814.
127. Zhou, H. M. and Walthall, W. W. (1998). UNC-55, an orphan nuclear hormone receptor, orchestrates synaptic specificity among two classes of motor neurons in *Caenorhabditis elegans*. *J Neurosci* 18, 10438–10444.
128. King-Jones, K., Horner, M. A., Lam, G., and Thummel, C. S. (2006). The DHR96 nuclear receptor regulates xenobiotic responses in *Drosophila*. *Cell Metab* 4, 37–48.
129. Lindblom, T. H., Pierce, G. J., and Sluder, A. E. (2001). A *C. elegans* orphan nuclear receptor contributes to xenobiotic resistance. *Curr Biol* 11, 864–868.

130. Magner, D. B. and Antebi, A. (2008). *Caenorhabditis elegans* nuclear receptors: Insights into life traits. *Trends Endocrinol Metab* 19, 153–160.
131. Zhong, W., Sladek, F. M., and Darnell, J. E., Jr. (1993). The expression pattern of a *Drosophila* homolog to the mouse transcription factor HNF-4 suggests a determinative role in gut formation. *EMBO J* 12, 537–544.
132. Sladeck, F. M. (1994). In Tronche, F., Yaniv, M., eds., Transcriptional regulation of liver-specific genes. R.G. Landes Co, Austin, TX, 207–230.
133. Swevers, L. and Iatrou, K. (1998). The orphan receptor BmHNF-4 of the silkmoth Bombyx mori: Ovarian and zygotic expression of two mRNA isoforms encoding polypeptides with different activating domains. *Mech Dev* 72, 3–13.
134. Kapitskaya, M. Z., Dittmer, N. T., Deitsch, K. W., Cho, W. L., Taylor, D. G., Leff, T., and Raikhel, A. S. (1998). Three isoforms of a hepatocyte nuclear factor-4 transcription factor with tissue- and stage-specific expression in the adult mosquito. *J Biol Chem* 273, 29801–29810.
135. Palanker, L., Tennessen, J. M., Lam, G., and Thummel, C. S. (2009). *Drosophila* HNF4 regulates lipid mobilization and beta-oxidation. *Cell Metab* 9, 228–239.
136. Dhe-Paganon, S., Duda, K., Iwamoto, M., Chi, Y. I., and Shoelson, S. E. (2002). Crystal structure of the HNF4 alpha ligand binding domain in complex with endogenous fatty acid ligand. *J Biol Chem* 277, 37973–37976.
137. Hayhurst, G. P., Lee, Y. H., Lambert, G., Ward, J. M., and Gonzalez, F. J. (2001). Hepatocyte nuclear factor 4alpha (nuclear receptor 2A1) is essential for maintenance of hepatic gene expression and lipid homeostasis. *Mol Cell Biol* 21, 1393–1403.
138. Van Gilst, M. R., Hadjivassiliou, H., Jolly, A., and Yamamoto, K. R. (2005). Nuclear hormone receptor NHR-49 controls fat consumption and fatty acid composition in *C. elegans*. *PLoS Biol* 3, e53.
139. Pankratz, M. J. and Jackle, H. (1993). Blastoderm segmentation. In Bate, M., Martinez-Arias, A., (ed.), The development of *Drosophila melanogaster*. Cold Spring Harbour Laboratory Press, Cold Spring Harbour, NY.
140. Casanova, J. (1990). Pattern formation under the control of the terminal system in the *Drosophila* embryo. *Development* 110, 621–628.
141. Pignoni, F., Baldarelli, R. M., Steingrímsson, E., Diaz, R. J., Patapoutian, A., Merriam, J. R., and Lengyel, J. A. (1990). The *Drosophila* gene tailless is expressed at the embryonic termini and is a member of the steroid receptor superfamily. *Cell* 62, 151–163.
142. Strecker, T. R., Merriam, J. R., and Lengyel, J. A. (1988). Graded requirement for the zygotic terminal gene, tailless, in the brain and tail region of the *Drosophila* embryo. *Development* 102, 721–734.
143. Younossi-Hartenstein, A., Green, P., Liaw, G. J., Rudolph, K., Lengyel, J., and Hartenstein, V. (1997). Control of early neurogenesis of the *Drosophila* brain by the head gap genes tll, otd, ems, and btd. *Dev Biol* 182, 270–283.
144. Steingrímsson, E., Pignoni, F., Liaw, G. J., and Lengyel, J. A. (1991). Dual role of the *Drosophila* pattern gene tailless in embryonic termini. *Science* 254, 418–421.
145. Pankratz, M. J., Hoch, M., Seifert, E., and Jäckle, H. (1989). Krüppel requirement for knirps enhancement reflects overlapping gap gene activities in the *Drosophila* embryo. *Nature* 341, 337–340.
146. Weigel, D., Jürgens, G., Küttner, F., Seifert, E., and Jäckle, H. (1989). The homeotic gene fork head encodes a nuclear protein and is expressed in the terminal regions of the *Drosophila* embryo. *Cell* 57, 645–658.
147. Kraut, R. and Levine, M. (1991). Spatial regulation of the gap gene giant during *Drosophila* development. *Development* 111, 601–609.
148. Kispert, A., Herrmann, B. G., Leptin, M., and Reuter, R. (1994). Homologs of the mouse *Brachyury* gene are involved in the specification of posterior terminal structures in *Drosophila*, *Tribolium*, and *Locusta*. *Genes Dev* 8, 2137–2150.

149. Margolis, J. S., Borowsky, M. L., Steingrímsson, E., Shim, C. W., Lengyel, J. A., and Posakony, J. W. (1995). Posterior stripe expression of hunchback is driven from two promoters by a common enhancer element. *Development* 121, 3067–3077.
150. Morán, E. and Jiménez, G. (2006). The tailless nuclear receptor acts as a dedicated repressor in the early *Drosophila* embryo. *Mol Cell Biol* 26, 3446–3454.
151. Schröder, R., Eckert, C., Wolff, C., and Tautz, D. (2000). Conserved and divergent aspects of terminal patterning in the beetle *Tribolium castaneum*. *Proc Natl Acad Sci USA* 97, 6591–6596.
152. Rudolph, K. M., Liaw, G. J., Daniel, A., Green, P., Courey, A. J., Hartenstein, V., and Lengyel, J. A. (1997). Complex regulatory region mediating tailless expression in early embryonic patterning and brain development. *Development* 124, 4297–4308.
153. Daniel, A., Dumstrei, K., Lengyel, J. A., and Hartenstein, V. (1999). The control of cell fate in the embryonic visual system by atonal, tailless and EGFR signaling. *Development* 126, 2945–2954.
154. Land, P. W. and Monaghan, A. P. (2003). Expression of the transcription factor, tailless, is required for formation of superficial cortical layers. *Cereb Cortex* 13, 921–931.
155. Pitman, J. L., Tsai, C. C., Edeen, P. T., Finley, K. D., Evans, R. M., and McKeown, M. (2002). DSF nuclear receptor acts as a repressor in culture and in vivo. *Dev Biol* 245, 315–328.
156. Finley, K. D., Edeen, P. T., Foss, M., Gross, E., Ghbeish, N., Palmer, R. H., Taylor, B. J., and McKeown, M. (1998). Dissatisfaction encodes a tailless-like nuclear receptor expressed in a subset of CNS neurons controlling *Drosophila* sexual behavior. *Neuron* 21, 1363–1374.
157. Finley, K. D., Taylor, B. J., Milstein, M., and McKeown, M. (1997). Dissatisfaction, a gene involved in sex-specific behavior and neural development of *Drosophila melanogaster*. *Proc Natl Acad Sci USA* 94, 913–918.
158. Sung, C., Wong, L. E., Chang, S. L. Q., Nguyen, E., Lazaga, N., Ganzer, G., McNabb, S. L., and Robinow, S. (2009). The unfulfilled/DHR51 gene of *Drosophila* melanogaster modulates wing expansion and fertility. *Dev Dyn* 238, 171–182.
159. de Rosny, E., de Groot, A., Jullian-Binard, C., Borel, F., Suarez, C., Le Pape, L., Fontecilla-Camps, J. C., and Jouve, H. M. (2008). DHR51, the *Drosophila melanogaster* homologue of the human photoreceptor cell-specific nuclear receptor, is a thiolate heme-binding protein. *Biochemistry* 47, 13252–13260.
160. Kobayashi, M., Takezawa, S., Hara, K., Yu, R. T., Umesono, Y., Agata, K., Taniwaki, M., Yasuda, K., and Umesono, K. (1999). Identification of a photoreceptor cell-specific nuclear receptor. *Proc Natl Acad Sci USA* 96, 4814–4819.
161. Haider, N. B., Jacobson, S. G., Cideciyan, A. V., Swiderski, R., Streb, L. M., Searby, C., Beck, G., Hockey, R., Hanna, D. B., Gorman, S., Duhl, D., Carmi, R., Bennett, J., Weleber, R. G., Fishman, G. A., Wright, A. F., Stone, E. M., and Sheffield, V. C. (2000). Mutation of a nuclear receptor gene, NR2E3, causes enhanced S cone syndrome, a disorder of retinal cell fate. *Nat Genet* 24, 127–131.
162. Much, J. W., Slade, D. J., Klampert, K., Garriga, G., and Wightman, B. (2000). The fax-1 nuclear hormone receptor regulates axon pathfinding and neurotransmitter expression. *Development* 127, 703–712.
163. Nauber, U., Pankratz, M. J., Kienlin, A., Seifert, E., Klemm, U., and Jäckle, H. (1988). Abdominal segmentation of the *Drosophila* embryo requires a hormone receptor-like protein encoded by the gap gene knirps. *Nature* 336, 489–492.
164. Chen, C. K., Kühnlein, R. P., Eulenberg, K. G., Vincent, S., Affolter, M., and Schuh, R. (1998). The transcription factors KNIRPS and KNIRPS RELATED control cell migration and branch morphogenesis during *Drosophila* tracheal development. *Development* 125, 4959–4968.
165. González-Gaitán, M., Rothe, M., Wimmer, E. A., Taubert, H., and Jäckle, H. (1994). Redundant functions of the genes knirps and knirps-related for the establishment of anterior *Drosophila* head structures. *Proc Natl Acad Sci USA* 91, 8567–8571.
166. Cerny, A. C., Grossmann, D., Bucher, G., and Klingler, M. (2008). The *Tribolium* ortholog of knirps and knirps-related is crucial for head segmentation but plays a minor role during abdominal patterning. *Dev Biol* 321, 284–294.

167. Higashijima, S., Shishido, E., Matsuzaki, M., and Saigo, K. (1996). Eagle, a member of the steroid receptor gene superfamily, is expressed in a subset of neuroblasts and regulates the fate of their putative progeny in the *Drosophila* CNS. *Development* 122, 527–536.
168. Rothe, M., Nauber, U., and Jäckle, H. (1989). Three hormone receptor-like *Drosophila* genes encode an identical DNA binding finger. *EMBO J* 8, 3087–3094.
169. Lundell, M. J. and Hirsh, J. (1998). Eagle is required for the specification of serotonin neurons and other neuroblast 7-3 progeny in the *Drosophila* CNS. *Development* 125, 463–472.
170. Ruau, D., Duarte, J., Ourjdal, T., Perrière, G., Laudet, V., and Robinson-Rechavi, M. (2004). Update of NUREBASE: Nuclear hormone receptor functional genomics. *Nucleic Acids Res* 32(Database issue), D165–D167.

CHAPTER 4

THE GLUCOCORTICOID RECEPTOR

ROBERT H. OAKLEY AND JOHN A. CIDLOWSKI

Laboratory of Signal Transduction, Department of Health and Human Services, National Institute of Environmental Health Sciences, National Institutes of Heath, Research Triangle Park, NC 27709, USA

Abstract: Glucocorticoids are essential for life and play critical roles in many diverse physiological processes including metabolism, development, growth, inflammation, and apoptosis. Drugs that mimic the actions of glucocorticoids are widely used to treat diseases such as cancer, inflammation, and autoimmune disorders. The effects of glucocorticoids and their synthetic derivatives are mediated by the glucocorticoid receptor (GR, NR3C1), a member of the nuclear receptor superfamily of ligand-dependent transcription factors. Upon binding hormone, the GR translocates into the nucleus where it regulates gene expression by direct binding to DNA and/or through interactions with other transcription factors. The GR is derived from a single gene, yet recent work has demonstrated that alternative processing of this gene generates an astonishing array of GR isoforms with unique expression, functional, and gene regulatory profiles. Here, we discuss the molecular and cellular mechanisms of GR signaling, and the potential role for GR isoforms in regulating the specificity and sensitivity of glucocorticoid responsiveness in healthy and diseased tissues.

4.1. INTRODUCTION

Glucocorticoids are endogenous hormones essential for life that are released by the adrenal cortex in a circadian manner and in response to stress. The secretion of these hormones is under control of the hypothalamic-pituitary-adrenal (HPA) axis. In response to physical or emotional stress, the hypothalamus releases corticotrophin releasing hormone (CRH) which acts on the anterior pituitary to stimulate the synthesis and secretion of adrenocorticotropin hormone (ACTH). ACTH then acts on the adrenal cortex to stimulate the production and secretion of glucocorticoids. Only about 10% of circulating glucocorticoid is free, as the majority of the hormone is transported in the blood bound to corticosteroid binding globulin. Glucocorticoids act on nearly all tissues and cells to regulate many diverse physiological processes that function to maintain homeostasis in the face of stressful perturbations. Named for their metabolic effects converting proteins and lipids into glucose during times of low blood sugar, glucocorticoids also play critical roles in immune

C.M. Bunce, M.J. Campbell (eds.), *Nuclear Receptors*, Proteins and Cell Regulation 8, 63–89. DOI 10.1007/978-90-481-3303-1_4,

function, the inflammatory response, central nervous system function, skeletal growth, reproduction, development, and apoptosis among others [1, 2]. Additionally, in a classic negative feedback loop, glucocorticoids act on the hypothalamus and anterior pituitary to inhibit the synthesis and secretion of CRH and ACTH.

Because glucocorticoids influence so many facets of human physiology, dysregulation of glucocorticoid production can lead to severe health consequences. Addison's disease is characterized by a deficiency in glucocorticoids and results most commonly from destruction of the adrenal cortex caused by autoimmune reaction or infection (tuberculosis) and more rarely from genetic abnormalities in steroid biosynthesis [3]. Symptoms include impaired stress resistance, lymphoid tissue hypertrophy, weight loss, hypoglycemia, and fatigue. Patients with Addison's disease are successfully treated with glucocorticoid replacement therapy. In contrast, Cushing's syndrome is characterized by excessive glucocorticoid production due commonly to pituitary adenomas or ectopic ACTH-producing tumors [4]. The clinical manifestations that accompany this disorder include central obesity, muscle wasting, hyperglycemia, thinning of the skin, hypertension, osteoporosis, depression, and diabetes. Interestingly, metabolic syndrome shares many of the features of Cushing's syndrome suggesting that glucocorticoids may be involved in this pathology [5]. Surgical removal of the tumor or administration of glucocorticoid antagonists are the primary treatment options for patients with Cushing's syndrome.

Glucocorticoids are one of the most widely prescribed therapeutic agents in the world today, particularly for their powerful anti-inflammatory and immunosuppressive actions [6]. Synthetic glucocorticoids are indispensable for the treatment of acute and chronic inflammatory diseases such as asthma, rheumatoid arthritis, and ulcerative colitis. They are also used also to prevent organ transplant rejection and to treat cancers of the lymphoid system such as leukemias, lymphomas, and myelomas. For their effects on development, glucocorticoids are administered to stimulate maturation of the lung in premature babies. The therapeutic benefit of glucocorticoids, however, is limited by serious side effects that accompany chronic administration of these drugs [6, 7]. These adverse effects mimic the symptoms of Cushing's syndrome and include abdominal obesity, muscle wasting, skin thinning, osteoporosis, hyperglycemia, growth retardation in children, and increased susceptibility to infection.

Both the physiological and pharmacological actions of glucocorticoids are mediated by the glucocorticoid receptor (GR, NR3C1). The GR belongs to the nuclear receptor superfamily of intracellular proteins that function as ligand-dependent transcription factors [8]. Upon binding hormone, the receptor enhances or represses expression of target genes leading to specific changes in cellular function. Consistent with the widespread and essential actions of glucocorticoids, the GR is expressed in almost every cell of the body and its ablation is incompatible with life as GR deficient mice die shortly after birth due to respiratory failure [9]. Cellular responsiveness to glucocorticoids, however, is not uniform [10–12]. Sensitivity to glucocorticoids varies among individuals and among tissues of the same individual. Tissue-specific glucocorticoid resistance frequently develops with long-term exposure to glucocorticoids in patients with rheumatoid arthritis, asthma, Crohn's disease, ulcerative colitis, and osteoarthritis. Conversely, hypersensitivity to glucocorticoids is observed in some tissues in patients with visceral obesity-related insulin resistance, diabetes mellitus type II, and essential hypertension. Alterations in glucocorticoid sensitivity are also observed in the same normal tissue during development and in the same cell during the cell cycle [13, 14]. Moreover, glucocorticoids selectively kill some cells, such as osteoblasts and thymocytes, but protect other cells such as hepatocytes from death. How glucocorticoids exert these cell-type specific effects and how cells alter their sensitivity to hormone under physiological and pathological conditions is not well understood.

The GR is derived from a single gene, and the prevailing assumption over the last several decades has been that a single receptor isoform is responsible for the myriad and diverse actions of glucocorticoids. Changes in the expression of this receptor or in its capacity to function as a transcription factor may account for the unique glucocorticoid signaling properties within and between cells [15, 16]. An additional explanation for the cellular differences in glucocorticoid specificity and sensitivity, however, comes from recent studies that have challenged the simple one gene-one receptor paradigm by revealing the presence of an astonishing array of GR isoforms that arise from alternative processing of the GR gene [17–19]. The combination of alternative splicing and alternative translation initiation generates multiple receptor subtypes with distinct expression, functional, and gene regulatory profiles. Such processing events are strategically utilized in humans to turn an estimated 30,000 genes into well over 100,000 functionally diverse proteins [20, 21]. Consequently, glucocorticoid responsiveness and specificity of action will be determined in large measure by the complement of the various GR isoforms in a particular cell. In the following chapter, we review the classic GR signaling pathway with emphasis on the origin and properties of these GR subtypes and their potential contribution to the regulation and dysregulation of glucocorticoid signaling.

4.2. CLASSIC GR AND ITS SIGNALING PATHWAY

4.2.1. Domain Structure of GR

Like other members of the nuclear receptor superfamily, the GR is a modular protein comprised of three major domains: an amino-terminal transactivation domain (NTD), a central DNA binding domain (DBD), and a carboxyl-terminal ligand binding domain (LBD) (Figure 4.1) [16, 22]. Separating the DBD and LBD is a flexible region of the molecule referred to as the hinge region. The NTD, which is the most variable among family members, contains the ligand-independent transcriptional activation function (AF1) that plays a major role in gene activation by associating with various coregulators and with the basal transcriptional machinery. The DBD is the most conserved domain across the nuclear receptor family and contains two zinc finger motifs that recognize and bind target DNA sequences. Three amino acids comprising the 'P-box' are found in the first zinc finger and are important for target site

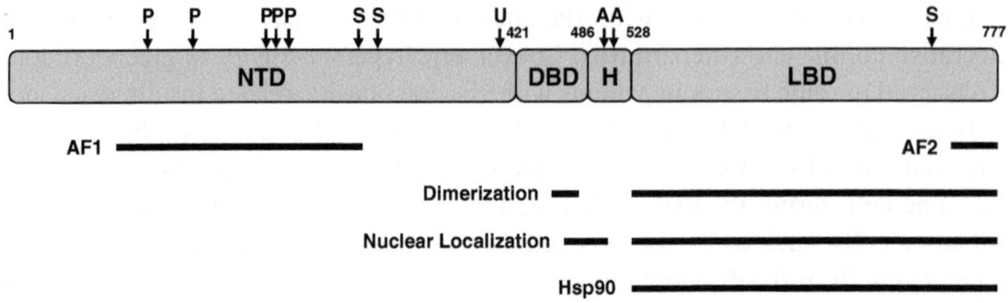

Figure 4.1. GR domain structure and sites of post-translational modification. The GR is comprised of an amino-terminal transactivation domain (NTD), DNA binding domain (DBD), hinge region (H), and ligand binding domain (LBD). Regions involved in transactivation (AF1 and AF2), dimerization, nuclear localization, and hsp90 binding are indicated. Also shown is the location of residues post-translationally modified by phosphorylation (P) (Ser-113, Ser-141, Ser-203, Ser-211, and Ser-226); ubiquitination (U) (Lys-419); sumoylation (S) (Lys-277, Lys-293, and Lys-703); and acetylation (A) (Lys-494 and Lys-495). Numbers are for the human glucocorticoid receptor

discrimination, and 5 amino acids comprising the 'D-box' are found in the second zinc finger and are important for receptor dimerization. The LBD not only contains the residues that mediate glucocorticoid binding but also houses a second transcriptional activation function (AF2). AF2 is exposed by ligand binding and modulates transcription via interactions with coactivators and corepressors. Sequences important for receptor dimerization and for interacting with chaperone proteins such as heat shock protein 90 (hsp90) are also found in the LBD. Finally, two nuclear localization signals have been identified in the receptor, one of which spans the junction of the DBD and hinge region and the other resides within the LBD.

4.2.2. Hormone Binding and Nuclear Translocation

In the absence of hormone, the GR is found predominantly in the cytoplasm of cells as part of a multiprotein complex that includes chaperone proteins (hsp90, hsp70, and p23) and immunophilins of the FK506 binding family (FKBP51 and FKBP52) (Figure 4.2) [23, 24]. These proteins assist in proper folding of the GR and maintain the receptor in a transcriptionally inactive conformation that favors high affinity ligand binding [25]. As lipophilic molecules, glucocorticoids readily diffuse across the plasma membrane into the cell. Bioavailability of cortisol, the most abundant endogenous glucocorticoid in humans, is exquisitely controlled inside the cell by two enzymes working in an opposing manner [26]. 11β-Hydroxysteroid dehydrogenase type II (11β-HSD2) converts the active glucocorticoid cortisol to the metabolite cortisone, whereas 11β-hydroxysteroid dehydrogenase type I (11β-HSD1) converts cortisone to the active cortisol. The relative levels of these two enzymes vary in a cell-type specific manner, with 11β-HSD1 found at high levels in liver, adipose tissue, and brain and 11β-HSD2 expressed at high levels in kidney, colon, and reproductive organs. Because glucocorticoid action is amplified by increases in 11β-HSD1

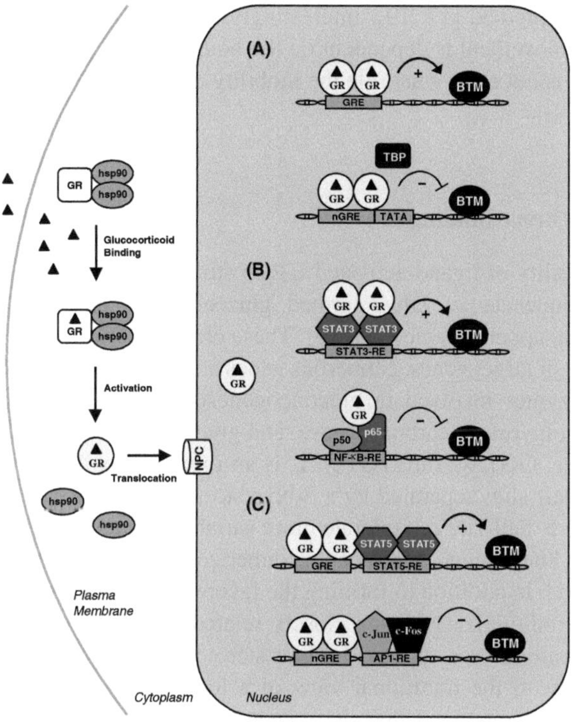

Figure 4.2. GR signaling pathway. The unliganded GR resides in the cytoplasm of cells in a complex with chaperone proteins such as hsp90. Upon binding glucocorticoids (▲), the receptor undergoes a change in conformation, dissociates from accessory proteins, and translocates into nucleus where it regulates gene expression in 3 primary ways. (**a**), GR binds directly to DNA and enhances (*cis*-activation) or inhibits (*cis*-repression) transcription of target genes. (**b**), GR interacts with DNA-bound transcription factors without itself binding to DNA and enhances (*trans*-activation) or inhibits (*trans*-repression) transcription of target genes. (**c**), GR binds directly to DNA and interacts with transcription factors bound to neighboring sites and enhances (*cis*-activation) or inhibits (*cis*-repression) transcription of target genes. NPC, nuclear pore complex. BTM, basal transcription machinery. TBP, TATA-box binding protein

activity and reduced by increases in 11β-HSD2 activity, expression of these enzymes will contribute to tissue-specific differences in glucocorticoid responsiveness. Upon binding hormone, the GR undergoes a conformational change that results in the dissociation of the chaperone proteins and the exposure of the nuclear localization signals. The exposed nuclear localization signals are recognized by importins and the receptor is translocated rapidly into the nucleus via nuclear pores [27]. Once in the nucleus, the activated receptor regulates positively or negatively the expression of target genes by binding directly to specific sequences of DNA (*cis*-activation or *cis*-repression) and/or by binding other transcription factors and modulating their activity (*trans*-activation or *trans*-repression). Global gene expression analyses indicate that up to 20% of the genome is regulated by glucocorticoids, with similar numbers being induced and repressed [17, 19]. Interestingly, GR is highly mobile in the nucleus and its rate of movement is dependent on the bound ligand as well as the cell-type [28], however it is not clear what role the mobility of nuclear GR plays in the regulation of gene expression.

4.2.3. GR Signaling by Direct Binding to DNA

Cis-activation refers to the ability of ligand-activated GR to stimulate gene expression by binding specific sequences of DNA termed glucocorticoid responsive elements (GREs) (Figure 4.2a, upper scheme) [29, 30]. These elements are typically found in the promoter regions of target genes. Numerous genes are regulated by these positive GREs including enzymes involved in gluconeogenesis, such as tyrosine aminotransferase, phosphoenolpyruvate carboxykinase, and glucose-6-phosphatase. The consensus GRE sequence, GGTACAnnnTGTTCT, is an imperfect palindrome that is comprised of 2 6-bp half-sites separated by a 3-bp spacer. The $3'$ half-site is highly conserved whereas the $5'$ half-site can tolerate more variability. Each half-site binds one subunit of the GR homodimer, and receptor dimerization is required for high affinity DNA binding [31]. In addition to forming the favored homodimers, GR has also been shown to heterodimerize with the closely related mineralocorticoid and androgen receptors, permitting cross talk with other steroid receptor signaling pathways [32, 33]. In contrast to the traditional view of a long-lasting GR-GRE association, recent work has demonstrated that the receptor interacts only briefly with target sites, rapidly cycling on and off the promoter every few seconds [34]. This rapid exchange may permit more precise control of gene expression by allowing the receptor to sample a large number of potential binding sites and interacting proteins and by introducing additional points of regulation by chaperones and the proteasome [35].

GR binding to the GRE induces a conformational change in the receptor that promotes the coordinated recruitment of various coactivators that are necessary for stimulating transcription of the target gene [22, 36–38]. Many of the coactivators interact initially with the AF2 helix of GR via a conserved LXXLL motif or nuclear receptor box. Three general classes of coactivators have been shown to assemble on the DNA-bound GR. The ATP-dependent chromatin remodeling complex BRG1 (SWI/SNF) is thought to be recruited at an early point and mediates large noncovalent disruptions in chromatin structure. Next to associate are coactivators such as CBP, p300, and the SRC/p160 family of proteins that modify chromatin structure locally through their intrinsic histone acetyltransferase (HAT) activity. Transfer of acetyl groups to certain histone lysine residues results in the unwinding of chromatin, which facilitates access of the basal transcription machinery to the regulatory region of the target gene. The third class of coactivators recruited to the receptor includes components of the DRIP/TRAP complex that directly contact the basal transcription machinery and assist in its recruitment to the newly accessible promoter. Adding further complexity to the regulation is that many of the various coactivators themselves recruit other coactivators with HAT or methyltransferase activity for further remodeling of chromatin structure. The specific coactivators assembled by GR at target genes will determine the consequent gene induction profile and glucocorticoid response, and this assembly will depend on multiple factors including the promoter context, the nature of the bound glucocorticoid, cell-type, and the expression level and activity of the coactivators themselves.

In some promoter contexts, the ligand-bound GR binds a more heterogeneous response element and suppresses gene activation by a process termed *cis*-repression (Figure 4.2a, lower scheme) [39]. These response elements are referred to as negative GREs (nGREs) and bear little resemblance to the classic, positive GRE. Additionally, the nGREs typically contain only one conserved half-site and bind GR with lower affinity than a positive GRE, suggesting a GR monomer may be sufficient to mediate the negative regulation at least in some promoter contexts. Multiple mechanisms have been proposed to underlie the nGRE-dependent repression [39]. For some genes, such as osteocalcin, GR binding to the nGRE interferes sterically with the binding of positive-acting transcription factors to their cognate response elements. The nGRE in the osteocalcin promoter overlaps the TATA box, and binding of GR precludes binding of the TATA-box binding protein resulting in decreased transcription [40]. The situation is more complex for other genes negatively regulated by glucocorticoids, as both GR binding to DNA and interactions with transcription factors on adjacent sites appear to be required (Figure 4.2c, lower scheme). An nGRE of this nature, often called a composite GRE, is found in the proliferin promoter where the site for GR binding lies adjacent to an activator protein-1 (AP-1) site [41]. DNA-bound GR interacts with the neighboring AP-1 complex and, depending on the subunit composition of AP-1, can either repress or enhance gene expression. A similar composite GRE, comprised of discrete adjacent binding sites for GR and AP-1, mediates the glucocorticoid-dependent repression of CRH [42]. Finally, since nGREs correspond poorly to the GRE consensus sequence, it has been proposed that GR binding to DNA may not be required at all for the negative regulation to occur. In fact, several recent reports support a model in which GR targets DNA-bound positive transcription factors without itself binding to the promoter [43–45]. This 'tethering' mechanism of repression (described in more detail below) appears to mediate the glucocorticoid-dependent inhibition of many target genes including gonadotropin-releasing hormone receptor, via the interaction of GR with the DNA-bound Oct1 transcription factor [43], and proopiomelanocortin, via the interaction of GR with the DNA-bound orphan transcription factor Nur77 [45].

4.2.4. GR Signaling by Interactions with Other Transcription Factors

The ability of activated GR to physically associate with other transcription factors and repress their activity on glucocorticoid responsive promoters without itself binding to DNA is referred to as *trans*-repression (Figure 4.2b, lower scheme). The two most extensively studied examples of GR-mediated *trans*-repression involve the transcription factors AP-1 and nuclear factor-κB (NF-κB). These two proteins are central mediators of the inflammatory and immune response, and their inhibition by GR is thought to underlie the major anti-inflammatory and immunosuppressive actions of glucocorticoids [46–48]. Both AP-1 and NF-κB are activated by stress signals such as proinflammatory cytokines, bacterial and viral infection agents, and proapoptotic stimuli. In the nucleus, activated AP-1 and NF-κB bind their cognate response elements and induce the expression of many proinflammatory genes including cytokines, cell adhesion molecules, and enzymes involved in tissue destruction and the synthesis of inflammatory mediators such as COX-2 and iNOS. Glucocorticoids are known to antagonize the actions of AP-1 and NF-κB in many different ways. For example, GR stimulates expression of the inhibitor IκB via a positive GRE to sequester NF-κB in the cytoplasm [49]. The receptor also induces expression of MAPK phosphatase 1 (MKP-1) which dephosphorylates c-Jun N-terminal kinase (JNK) and thereby prevents activation of the c-Jun component of AP-1 [50, 51]. Furthermore, GR upregulates the expression of the RNA binding protein tristetraprolin which destabilizes the mRNA of many AP-1 and NF-κB induced genes [52].

The primary way, however, in which hormone-bound GR is thought to repress the activity of AP-1 and NF-κB is through a direct interaction with the c-Jun subunit of AP-1 and the p65 subunit of NF-κB [46, 47]. GR monomers appear to be sufficient for the antagonism as receptor mutants deficient in dimerization retain the inhibitory capacity [53]. Interestingly, the antagonism is reciprocal as both AP-1 and NF-κB inhibit the activity of GR on glucocorticoid-responsive genes. How the association of GR with these two proteins disrupts their ability to signal is controversial as evidence exists for the involvement of multiple mechanisms [47, 48, 54], suggesting the mode of action may vary in a cell-type and/or signal-dependent manner. For example, the receptor has been reported to prevent AP-1 and NF-κB from binding target sites on DNA by sequestering them in the cytoplasm and/or nucleus. More recent studies, however, have shown that GR does not interfere with the capacity of these proteins to bind DNA. Instead, the activated GR is tethered to the DNA-bound AP-1 and NF-κB and appears to impair transcription through alterations in the recruitment of coregulators and/or interference with the activity of the basal transcription machinery. For example, GR has been shown to repress NF-κB activity on several toll-like receptor genes by disrupting the interaction of p65 with the promoter-specific coactivator IRF3 [55]. In addition, the association of GR with NF-κB impairs the phosphorylation of the C-terminal domain of RNA polymerase II at the IL-8 and ICAM-1 promoters, perhaps through recruitment of a corepressor with phosphatase or kinase-inhibitory activity [56]. Furthermore, the GR-dependent recruitment of the coactivator GRIP1 potentiated receptor-mediated repression of AP-1 and NF-κB on target genes, suggesting coactivators can function as corepressors depending on the promoter context [57, 58]. Histone deacetylases (HDACs) form a prominent class of co-repressors by removing acetyl groups from chromatin. GR has been shown to interact with HDAC2 and to repress the HAT activity of NF-κB, leading to the proposal that GR-mediated antagonism of NF-κB is accomplished by recruitment of HDAC to target genes [59].

In contrast to the inhibitory effects of GR on AP-1 and NF-κB, the physical association of the receptor with other transcription factors can enhance their activity on target genes (Figure 4.2b, upper scheme). This process of *trans*-activation has been described for members of the signal transducer and activator of transcription (STAT) family [60]. STAT transcription factors are activated by a range of cytokines through induction of the Janus kinase pathway (JAK) and tyrosine phosphorylation. These proteins bind response elements and regulate genes that effect cell differentiation, survival, and apoptosis. The GR has been shown to interact with several members of the STAT family, including STAT3 and STAT5, and to synergistically enhance their activity on target genes in a promoter-dependent fashion [60]. For example, the GR has been shown to associate with STAT3 at target gene promoters that lack identifiable GREs, such as γ-fibrinogen and α2-macroglobulin, and to superinduce their expression [61–63]. In addition, the association of GR with STAT5 bound to DNA results in the synergistic activation of β-casein and toll-like receptor 2 transcription [64–67]. However, the observed synergy at these two STAT5-responsive promoters may also require GR binding to DNA [66, 67], and therefore more accurately reflect a case of *cis*-activation via a composite GRE (Figure 4.2c, upper scheme). The precise role by which GR synergistically activates STAT-regulated genes is not clear, but data suggest a role for GR enhancing STAT nuclear localization [68], prolonging the promoter occupancy of STAT by inhibiting its tyrosine dephosphorylation [61, 69], and promoting the co-utilization of certain coactivators [60]. Interestingly, the synergism of these two transcription factors is not always reciprocal. STATs were found to modulate the transcriptional activity of GR on GRE-driven reporters but the effects were inhibitory or stimulatory depending on the particular STAT isoform [63, 64, 70, 71].

4.3. GR POST-TRANSLATIONAL MODIFICATIONS AND GLUCOCORTICOID SIGNALING

4.3.1. Phosphorylation of GR

The GR protein is a substrate for several types of post-translational modifications that regulate glucocorticoid signaling by modulating the levels and/or transcriptional activity of the receptor. Phosphorylation was one of the earliest identified post translational modification of GR [72]. The human GR is phosphorylated at 5 serine residues (Ser-113, Ser-141, Ser-203, Ser-211, and Ser-226) located in AF1 of the NTD (Figure 4.1) [73]. Some of the sites are phosphorylated in the absence of hormone, whereas others become phosphorylated with glucocorticoid binding. The major kinases mediating receptor phosphorylation include mitogen protein kinases (MAPK) [74], cyclin-dependent kinase (CDK) [74], glycogen synthase kinase-3 (GSK-3) [75], and JNK [76, 77]. Phosphorylation has been shown to affect the expression level of GR by promoting degradation of the receptor protein, as wild-type GR degrades approximately three times faster than a phosphorylation deficient mutant following glucocorticoid exposure [78]. Interestingly, the unliganded, hypophosphorylated GR has been shown to interact more favorably with the protein TSG101 which appears to protect the ligand-free receptor from proteasome-dependent degradation [79]. The trafficking of GR both into and out of the nucleus is also regulated by receptor phosphorylation, and impaired nuclear import by p38 MAPK-directed phosphorylation of GR has been implicated in glucocorticoid resistance [80].

The phosphorylation status of the receptor has also been shown to modulate its transcriptional activity. For example, GR mutants deficient in phosphorylation were compromised in their ability to stimulate transcription from some glucocorticoid-responsive promoters [78]. In addition, studies have shown that phosphorylation of Ser-203 and Ser-211 is necessary for full transcriptional enhancement by GR [74, 81, 82], whereas JNK phosphorylation of Ser-226 leads to reduced GR transcriptional activity in part by enhancing receptor export from the nucleus [76, 77, 83]. Furthermore, the pattern of GR recruitment to promoters of endogenous target genes, as well as the magnitude of the ensuing activation or repression, was shown to depend on the specific receptor residues that were phosphorylated and to vary in a gene specific fashion [83, 84]. The molecular mechanisms underlying these phosphorylation-specific effects on GR activity is still under investigation but may involve differences in cofactor recruitment. Indeed, phosphorylation of Ser-211 has been shown to enhance the interaction of GR with the coactivator MED14 (DRIP150) [83]. Thus, the pattern of GR phosphorylation will confer distinct gene regulatory profiles on the receptor, allowing the magnitude and specificity of glucocorticoid action to be controlled with fine-tuned precision.

4.3.2. Ubiquitination of GR

Another important post-translational modification of the GR is the covalent attachment of a chain of ubiquitin moieties. Ubiquitin is a 76 amino acid protein that when attached to specific lysine residues of target proteins marks them for degradation by the proteasome. Evidence for a functional role of the ubiquitin-proteasome degradation pathway in the regulation of GR was revealed by treatment of cells with the proteasome inhibitor MG-132. Proteasome inhibition not only enhanced GR expression by blocking glucocorticoid-dependent down-regulation of the receptor but also increased transcriptional activity of GR on reporter genes [85, 86]. The mechanism underlying the increase in glucocorticoid responsiveness is not clear but may involve an elevation in the concentration of activated GR in the nucleus and/or proteasome-dependent modulations in receptor-cofactor interactions [85]. Direct ubiquitination of GR has been demonstrated [86], and analysis of the human, rat, and mouse GR primary sequences revealed a conserved PEST (Pro, Glu, Ser, and Thr) motif located just upstream of the DBD (Figure 4.1) [86]. This motif, which is important for substrate recognition by E2/E3 enzymes involved in the ubiquitylation process, appears to direct the ubiquitination of Lys-426 in mouse GR (Lys-419 in humans). Confirming the importance of ubiquitination in GR signaling, mutation of this lysine generated a receptor that was resistant to glucocorticoid-induced degradation and displayed enhanced transcriptional activity [86]. Additionally, a recent study identified the carboxyl-terminus of heat shock protein 70-interacting protein (CHIP) as an E3 ligase for GR and demonstrated that alterations in CHIP expression modulate hormone-dependent down-regulation of GR and cellular responsiveness to glucocorticoids [87].

4.3.3. Sumoylation of GR

Sumoylation describes a post-translational modification in which a small, ubiquitin-related modifier (SUMO) peptide is covalently attached to lysine residues of target proteins and modulates various functions including protein stability, subcellular localization, and transcriptional activity. The SUMO-conjugating E2 enzyme Ubc9 is known to interact with GR [88], and the receptor can serve as a substrate for modification by three SUMO isoforms (SUMO-1, SUMO-2, and SUMO-3) in a manner that does not require ligand [89–92]. Three consensus SUMO attachment sites have been identified at Lys-277, Lys-293, and Lys-703 of the human GR (Figure 4.1). Sumoylation of the receptor dramatically promotes its degradation [90]. In addition, attachment of SUMO proteins to GR inhibits its transcriptional activity on many, but not all, promoters sensitive to glucocorticoids, suggesting the observed effects are dependent on the promoter context [89–93]. Interestingly, the two proximal sites of SUMO attachment coincide exactly with so called ‘synergy control motifs’ (SC motifs). The SC motifs, which are found in many different transcription factors, limit the synergistic transcriptional output of ligand-activated GR from promoters containing multiple GREs but not single GREs, and this process is dependent on SUMO modification [89, 94]. Adding further complexity to this regulation, the extent to which the transcriptional synergy was inhibited was also dependent on the particular SUMO isoform attached, as SUMO-2 and SUMO-3 were most efficacious. How attachment of SUMO proteins inhibits GR activity is not known but some studies have implicated the recruitment of inhibitory factors [93], such as the transcriptional corepressor DAXX that interacts with the SUMO protein on GR [95]. Finally, a recent report has shown functional interdependence between GR phosphorylation and sumoylation [92]. JNK-mediate phosphorylation of GR facilitated subsequent sumoylation at the two acceptor lysines in the NTD of the receptor and repressed GR activity on responsive genes in a promoter-specific fashion.

4.3.4. Acetylation of GR

Acetylation is a common modification for regulating the function and activity of many different proteins, including histones and transcription factors. However, an appreciation for direct acetylation of nuclear receptors as a means for regulating their activity has only recently emerged. Both the estrogen receptor alpha and the androgen receptor are directly acetylated with consequent alterations in hormone-induced signaling [96, 97]. The identified nuclear receptor acetylation motif (KXKK) is conserved in human GR at amino acids 492–495 which is located within the nuclear localization signal NL1 spanning the DBD and hinge region (Figure 4.1). In response to glucocorticoids, the GR was shown recently to be acetylated and the level of acetylation was reduced upon mutation of Lys-494 and Lys-495 [98]. Functionally, acetylation of GR appears to impair its antagonism of NF-κB, since deacetylation of GR by histone deacetylase 2 (HDAC2) promotes receptor binding to NF-κB and repression of downstream target genes [98]. Interestingly, HDAC2 expression and activity is reduced in smokers and patients with chronic obstructive pulmonary disease that are known to be resistant to the anti-inflammatory actions of glucocorticoids. A defect in GR deacetylation caused by deficient HDAC2 might compromise the ability of GR to repress NF-κB and account for the observed resistance. In support of this hypothesis, overexpression of HDAC2 in macrophages isolated from these patients restored GR-dependent repression of NF-κB activity [98].

4.4. GR SPLICE VARIANTS AND GLUCOCORTICOID SIGNALING

4.4.1. GR Gene and Alternative Processing at 5′ End of Primary Transcript

The human GR gene, which was the first nuclear receptor cloned and founding member of the superfamily, is located on chromosome 5q31-32 [99–101]. The gene covers a region of approximately 80 kb and is comprised of 9 exons (Figure 4.3) [102, 103]. The receptor NTD is encoded primarily by exon 2, the DBD is encoded by exons 3 and 4, and the hinge region and LBD are encoded by exons 5–9. The 5′ untranslated region (5′UTR) of the mature GR mRNA is derived from multiple alternative exon 1s and the proximal part of exon 2, whereas the large 3′ untranslated region (3′UTR) comes from the distal portion of exon 9. Alternative splicing of the GR primary transcript generates heterogeneity in the mature GR mRNA and several unique protein isoforms.

A total of 9 alternative exon 1s have been identified in the human GR gene (exons 1A, 1B, 1C, 1D, 1E, 1F, 1H, 1I, and 1 J) [104–106]. Each exon 1 appears to be driven by its own unique promoter, and collectively they give rise to 13 splice variants all joined to a common acceptor site in exon 2. Differing only in their 5′ UTRs, the variant mRNAs encode the same GR protein. Since even small reductions in GR expression can have major health consequences [107, 108], the presence of multiple promoters controlled by different transcription factors likely serves to tightly regulate GR expression. A number of recent studies have shown that the various promoters are utilized in a tissue- and cell-type specific manner and are regulated differentially by glucocorticoids [109]. For example, exons 1B and 1C are ubiquitously expressed whereas a splice variant of exon 1A (1A3) is found primarily in cells of hematopoietic origin [104]. In response to glucocorticoids, most cells negatively autoregulate expression of the GR by reduced transcription of the GR

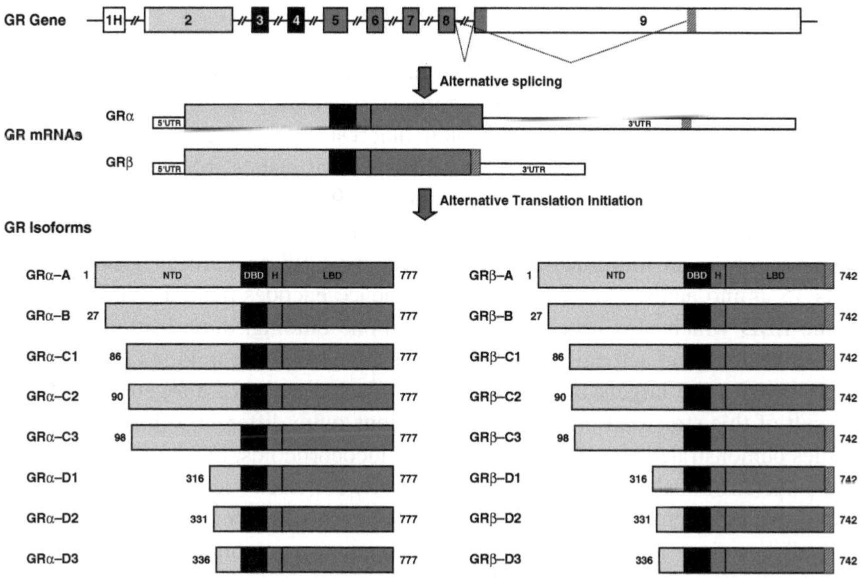

Figure 4.3. Alternative processing of the GR gene generates multiple GR isoforms. The human GR gene is comprised of 9 exons. Alternative splicing at the 3′ end of the primary transcript generates GRα and GRβ mRNAs, which encode GRα and GRβ proteins differing only at their carboxyl-terminus. Alternative translation initiation from 8 different AUG start codons derived from exon 2 generates additional protein isoforms with progressively shorter NTDs. Numbers shown denote the first and last residues for the human GR isoforms. *Shading* indicates that the NTD is encoded primarily by exon 2, the DBD is encoded by exons 3 and 4, and the hinge (H) and LBD are encoded by exons 5–9. For simplicity, only the most proximal of 9 alternate exon 1s (1H) is shown

gene and/or by proteasome-mediated degradation of the receptor protein [15, 86]. However, in human leukemia cells sensitive to glucocorticoid-induced apoptosis, glucocorticoids positively autoregulate GR levels in part by increased usage of the 1A and 1I promoters, the only two promoters with predicted GREs [104, 105].

The GR promoter employed by the cell can also regulate alternative splicing that occurs elsewhere in the receptor primary transcript. Usage of the 1B promoter enhances the occurrence of an alternative splicing event at the 3′ end that generates a truncated form of the receptor, termed GR-P (described below), that does not bind hormone and has been implicated in glucocorticoid resistance [110]. Finally, heterogeneity in the 5′UTR may affect the stability, export, and translation of the GR mRNA. Indeed, GR transcripts possessing the 1A3 exonic sequences show increased translation from a second start codon [111]. The resulting amino-terminal truncated receptor, termed GR-B, has greater transcriptional activity on reporter genes than the full length receptor [112]. By regulating the expression and composition of GR variants within cells, the diverse array of promoters, alternative exons, and resulting mRNAs with distinct 5′UTRs are likely to play a major role controlling the sensitivity and specificity of glucocorticoid action.

4.4.2. Splice Variant GRβ

Alternative splicing in exon 9 at the 3′ end of the human GR primary transcript generates two different GR protein isoforms termed GRα and GRβ [99, 102, 103] (Figure 4.3). GRα is the classic GR protein that results from the end of exon 8 being joined to the beginning of exon 9. The GRβ variant results from the end of exon 8 being joined to downstream sequences in exon 9. The two resulting isoforms are identical through amino acid 727 but then diverge at the carboxyl-terminus, with GRα containing an additional 50 amino acids and GRβ an additional, nonhomologous 15 amino acids. The GRα-specific sequence encodes helix 11 and 12 (AF2) of the LBD which play crucial roles in hormone binding, dimerization, coactivator recruitment, and activation of gene expression [113]. The unique GRβ sequence is predicted to be largely disordered and is missing helix 12 altogether [114]. As a result of this structural change, GRβ functions quite differently than GRα. Early studies demonstrated that GRβ did not bind glucocorticoids, resided predominantly in the nucleus of cells, and was unable to directly activate or repress glucocorticoid-responsive reporter genes [103, 115, 116]. However, when co-expressed with GRα, the splice variant functioned as a dominant negative inhibitor of GRα on genes both positively and negatively regulated by glucocorticoids [103, 115–118]. Several mechanisms have been reported to underlie the antagonism including competition for GRE binding, the formation of transcriptionally deficient GRα/GRβ heterodimers, and competition for needed coactivators such as the p160 coactivator glucocorticoid receptor interacting protein-1 (GRIP-1) [116, 119].

The ability of GRβ to antagonize GRα suggests that expression of this splice variant may play an important role regulating cellular sensitivity to glucocorticoids. GRβ is detected in most tissues and cell lines, but it is generally expressed at much lower levels than GRα [103, 115, 120–122], leading some to question its proposed role as a physiological inhibitor of GRα [123]. High levels of GRβ, however, have been observed in certain cell types such as neutrophils and specific epithelial cells, including those lining the terminal bronchiole of the lung, forming the outer layer of Hassall's corpuscles in the thymus, and lining the bile duct in the liver [121, 124]. Additionally, expression of GRβ can be selectively increased in a variety of cells in response to proinflammatory cytokines and in peripheral blood monocytes exposed to microbial superantigens [124–128]. In these cases, the enhanced expression of GRβ is accompanied by glucocorticoid resistance presumably through more effective antagonism of GRα. Levels of GRβ are also elevated in glucocorticoid resistant forms of asthma, rheumatoid arthritis, ulcerative colitis, nasal polyposis, systemic lupus erythematosus, acute lymphoblastic leukemia, and chronic lymphocytic leukemia [129–137]. Interestingly, methotrexate, an effective drug for treating autoimmune and inflammatory diseases, has been reported to selectively increase GRα at the expense of GRβ thereby improving glucocorticoid sensitivity of lymphocytes [138]. The molecular basis for altered expression of GRβ is not known but recent work has implicated the possible involvement of the alternative splicing factor serine-arginine rich protein p30c (SRp30c) [139]. This splicing protein was shown to be necessary for the specific generation of GRβ transcripts over GRα transcripts in neutrophils, and its expression was induced by agents that stimulated a selective increase in GRβ.

Elevated GRβ levels also result from a naturally occurring ATTTA to GTTTA polymorphism (A3669G) in exon 9 that corresponds to position 3,669 in the 3′UTR of the GRβ mRNA [140]. The nucleotide substitution disrupts an mRNA destabilization motif (AUUUA) leading to increased stability of the GRβ mRNA and enhanced protein expression [140, 141]. The frequency of having one or more copies of the A3669G allele has been determined in several populations and ranges from approximately 15 to 30% [140, 142, 143]. In a study of approximately 300 subjects of European origin, the A3669G allele was found to be associated with reduced central obesity in women and a more favorable lipid profile in men [142], suggesting that the increase in GRβ might antagonize some of the undesirable effects of GRα on fat distribution and lipid metabolism. The A3669G polymorphism also appears to attenuate the immunosuppressive actions of GRα. For example, individuals with the A3669G allele are reported to have an elevated risk of the autoimmune disease rheumatoid arthritis and a reduced risk of *Staphylococcus aureus* nasal infection [140, 144]. Moreover, in a recent study of almost 8,000 subjects in the Netherlands, persons homozygous for the A3669G polymorphism were associated with a more active proinflammatory phenotype, and had a 2.2-fold increased risk of myocardial infarction and a 2.8-fold increased risk of coronary heart disease [143]. Since inflammation is involved in the pathogenesis of rheumatoid arthritis, myocardial infarction, and coronary heart disease, GRβ-mediated inhibition of the anti-inflammatory actions of GRα might contribute to the increased incidence of these diseases. Alternatively, GRβ might contribute to the etiology of these diseases by as yet unidentified roles in cell signaling.

Over the last several years, new functions of GRβ have emerged from studies that have challenged the canonical view that GRβ does not bind glucocorticoids and is transcriptionally inactive apart from its effects on GRα. GRβ was shown to bind the synthetic glucocorticoid antagonist/partial agonist RU486 with a Kd of approximately 100 nM [145]. Additionally, it was demonstrated that GRβ functioned in a manner similar to GRα to repress the basal activity of the interleukin 5 and interleukin 13 promoters through the recruitment of histone deacetylase 1 [146]. Moreover, genome-wide microarray analysis performed in cells selectively expressing GRβ found the splice variant to alter the expression of over 5,000 genes, with approximately equal numbers being induced or repressed [145]. Less than 20% of these genes were commonly regulated by ligand-activated GRα, indicating that GRβ possesses its own distinct gene regulatory capacity. Interestingly, RU486 behaved as an antagonist of the unliganded activity of GRβ and abolished most of the GRβ-mediated changes in gene expression. Thus, when evaluated on endogenous promoters in their native chromatin context, GRβ was found to have direct effects on gene expression both in the absence and presence of RU486. These findings indicate that GRβ can function as a bonafide transcription factor and raise the possibility that GRβ not only contributes to alterations in glucocorticoid sensitivity by genomic effects distinct from its inhibition of GRα but also plays an important gene regulatory role in other biological and pathological processes.

4.4.3. Other GR Splice Variants

Alternative splicing of the GR primary transcript generates several additional receptor isoforms with unique properties. GRγ, derived from the use of an alternative splice donor site in the intron between exons 3 and 4, contains an insertion of a single arginine residue between the two zinc fingers of the DBD that compromises its transcriptional activity on reporter genes [147]. This splice variant was initially discovered in glucocorticoid resistant small cell lung carcinoma cells and corticotroph adenomas and shows a widespread tissue distribution [147, 148]. The level of GRγ expression has recently been correlated with glucocorticoid resistance in childhood acute lymphoblastic leukemia [149]. Two GR splice variants with altered LBDs were discovered in glucocorticoid resistant human multiple myeloma cells [150]. GR-A is missing middle exons 5–7, which encode the amino-terminal half of the LBD, due to alternative splicing linking the end of exon 4 to the beginning of exon 8. GR-P is missing the terminal exons 8 and 9, which encode the carboxyl-terminal half of the LBD, and results from a failure to splice at the exon 7/8 boundary. As expected, neither of these receptor isoforms binds glucocorticoids. The GR-P variant is widely expressed in normal tissues and found in many hematological malignancies [110, 151], and enhanced levels of GR-P have been associated with glucocorticoid resistance in multiple myeloma cells [152]. GR-P also appears to be the predominant receptor variant expressed in glucocorticoid resistant cells derived from ACTH-secreting small cell lung cancer [153]. Interestingly, GR-P has been shown to repress the transcriptional activity of GRα in certain cell types but stimulate its activity in others [151].

4.5. GR TRANSLATIONAL ISOFORMS AND GLUCOCORTICOID SIGNALING

While the generation of distinct GR isoforms by alternative splicing has been recognized for many years, only recently was it demonstrated that an additional cohort of receptor subtypes was produced by alternative translation initiation [17, 18, 112]. Specifically, eight GR isoforms with progressively shorter NTDs are generated from one GRα mRNA transcript by alternative translation initiation (Figure 4.3). All eight AUG start codons are located in exon 2 and are 100% conserved in human, monkey, rat, and mouse. The names of the translational isoforms and position of the initiator methionine are as follows: GRα-A (Met-1), GRα-B (Met-27), GRα-C1 (Met-86), GRα-C2 (Met-90), GRα-C3 (Met-98), GRα-D1 (Met-316), GRα-D2 (Met-331), and GRα-D3 (Met-336). Notably, the GRβ mRNA also contains the identical start codons and would be expected to generate a similar complement of subtypes (Figure 4.3). GRα-A is the classic full-length 777 amino acid protein that is generated from the first initiator AUG codon. Production of the shorter isoforms from internal start codons involves both ribosomal leaky scanning and ribosomal shunting mechanisms [17]. All eight GRα isoforms are widely expressed, however the relative levels of the different isoforms vary significantly between and within tissues [17, 19]. For example, the GRα-C isoforms were expressed at low levels in the liver but at high levels in the pancreas, lung, and colon. The GRα-D subtypes were most abundant in spleen and bladder but expressed at low levels in heart and pancreas. Within the same tissue, GRα-B was found to be more abundant than GRα-A in thymus and colon but not in other tissues examined.

The GRα translational isoforms possess both common and unique properties. Consistent with all 8 isoforms having an intact DBD and LBD, no significant differences were observed in the ability of the subtypes to bind glucocorticoids or in the capacity of the activated receptors to bind GREs [19]. Additionally, all 8 isoforms were localized in the nucleus of cells following glucocorticoid treatment. Surprisingly, however, the subcellular distribution of the isoforms differed in the absence of hormone [17]. In contrast to the predominant cytoplasmic location of GRα-A, GRα-B, and GRα-C isoforms, the GRα-D isoforms were found in the nucleus of cells. This result suggests that sequences in the NTD may play a previously unappreciated role in nuclear localization, nuclear export, and/or cytoplasmic retention. Owing perhaps to its unique nuclear distribution in the absence of hormone, the unactivated GRα-D3 isoform also exhibited constitutive binding to certain GRE-containing promoters, an event not observed for the other receptor subtypes [19].

The transcriptional activity of the GRα translational isoforms was initially evaluated on several glucocorticoid-responsive reporter genes [17]. In response to glucocorticoids, all 8 isoforms induced gene expression with similar potencies but to different extents. The GRα-C3 isoform was the most active (approximately 200% of GRα-A), whereas the three GRα-D subtypes were the least active (approximately 50% of GRα-A). In agreement with these findings, similar isoform-selective effects were observed on endogenous genes commonly regulated by the GR subtypes in response to hormone [19]. These distinct transcriptional profiles appear to be due to isoform-selective recruitment patterns of transcriptional factors and cofactors [19]. GRα-C3 was found to recruit significantly higher amounts of the coactivators CBP and p300 and the active RNA polymerase II compared to the other isoforms, whereas a significantly lower amount of each protein was recruited by GRα-D. Structural differences in the NTD are presumed to account for the unique recruitment patterns, as this region of the receptor contains the powerful AF1 domain (amino acids 77-262) which plays a critical role in the recruitment of coregulators and the basal transcription machinery [36]. The GRα-D isoforms are missing the entire AF1 region and the GRα-C3 isoform is missing the proximal portion. Interestingly, and in contrast to the isoform-selective effects on gene induction, no significant differences have been observed so far in the ability of these GR translational isoforms to repress NF-κB mediated activation of both reporter and endogenous genes [19].

The transcriptional properties of the GRα isoforms were further investigated by performing a whole genome microarray analysis on U2OS osteosarcoma cells stably expressing wild-type GRα, GRα-A, GRα-B, GRα-C3, or GRα-D3 [19]. Each isoform acted as a potent regulator of gene expression, with GRα-A regulating the largest number of genes (3,451) and GRα-D3 the smallest number (1,761). The total number of genes regulated by at least one GRα isoform was approximately 6,500. Remarkably, less than 500 of these genes were commonly regulated by all receptor isoforms, indicating that the majority of the target genes were selectively regulated by different receptor isoforms. The unique gene regulatory profiles of the various GR isoforms were also shown to translate into functional differences in cellular responsiveness to glucocorticoids [19]. In particular, the receptor subtypes exhibited distinct capabilities to induce apoptosis in U2OS cells treated with glucocorticoids. Cells expressing GRα-C3 were killed earlier and to a greater extent than cells expressing the other isoforms. In contrast, cells expressing GRα-D3 were much more resistant to the apoptosis-inducing actions of glucocorticoids than cells expressing the other subtypes. The molecular mechanism underlying this functional divergence involves differences among the isoforms in their ability to recruit transcriptional proteins and induce the proapoptotic enzymes granzyme A and caspase-6 that are necessary for cell death [19].

Research into the biological significance of the GR translational isoforms is still in its infancy. A critical goal of future studies will be to determine the contribution each isoform makes in an endogenous setting to glucocorticoid signaling in both healthy and diseased tissues. Because each isoform possesses a unique gene regulatory profile, the cellular complement of isoforms will likely play a major role in determining both the specificity and sensitivity of glucocorticoid responsiveness. Are the GRα-D isoforms more prevalent in glucocorticoid resistant tissues? Is the GRα-C3 isoform more abundant in tissues displaying hypersensitivity to glucocorticoids? Do the relative levels of the GR subtypes change under physiological or pathophysiological conditions and, if so, what operative factors control isoform selective expression? These are but a few of the many important questions waiting to be answered in this exciting new area in GR signaling.

4.6. CONCLUSION

Glucocorticoids act through the GR to regulate many diverse physiological processes and their synthetic counterparts are mainstays in the treatment of inflammation, autoimmune disorders, and cancer. The traditional view that glucocorticoids exert their pleiotropic effects through one receptor isoform has dramatically changed in recent years with the discovery of additional GR subtypes arising from alternative processing of the single GR gene. Alternative splicing and alternative translation initiation generate a heterogeneous population of GR isoforms with unique expression, functional, and gene regulatory profiles. The potential for these isoforms to undergo post-translational modifications and to function as monomers, homodimers, and heterodimers on both common and unique sets of genes provides cells with a wealth of possibilities for controlling a wide range of functions with fine-tuned precision. Moreover, alterations in the relative levels of the GR subtypes may underlie pathologies characterized by hyposensitivity or hypersensitivity to glucocorticoids. Of great future interest will be the possibility of developing synthetic glucocorticoids that selectively target one or more of the GR isoforms. Such molecules hold promise for better patient care by maintaining the powerful immunosuppressive actions of glucocorticoids while minimizing their harmful side effects.

REFERENCES

1. Barnes, P. J. (1998). Anti-inflammatory actions of glucocorticoids: Molecular mechanisms. *Clin Sci (Lond)* 94, 557–572.
2. Sapolsky, R. M., Romero, L. M., and Munck, A. U. (2000). How do glucocorticoids influence stress responses? Integrating permissive, suppressive, stimulatory, and preparative actions. *Endocr Rev* 21, 55–89.
3. Lovas, K. and Husebye, E. S. (2003). Replacement therapy in Addison's disease. *Expert Opin Pharmacother* 4, 2145–2149.
4. Shibli-Rahhal, A., Van Beek, M., and Schlechte, J. A. (2006). Cushing's syndrome. *Clin Dermatol* 24, 260–265.
5. Friedman, T. C., Mastorakos, G., Newman, T. D., Mullen, N. M., Horton, E. G., Costello, R., Papadopoulos, N. M., and Chrousos, G. P. (1996). Carbohydrate and lipid metabolism in endogenous hypercortisolism: Shared features with metabolic syndrome X and NIDDM. *Endocr J* 43, 645–655.
6. Rhen, T. and Cidlowski, J. A. (2005). Antiinflammatory action of glucocorticoids – new mechanisms for old drugs. *N Engl J Med* 353, 1711–1723.
7. Miner, J. N., Hong, M. H., and Negro-Vilar, A. (2005). New and improved glucocorticoid receptor ligands. *Expert Opin Investig Drugs* 14, 1527–1545.
8. Evans, R. M. (1988). The steroid and thyroid hormone receptor superfamily. *Science* 240, 889–895.
9. Cole, T. J., Blendy, J. A., Monaghan, A. P., Krieglstein, K., Schmid, W., Aguzzi, A., Fantuzzi, G., Hummler, E., Unsicker, K., and Schutz, G. (1995). Targeted disruption of the glucocorticoid receptor gene blocks adrenergic chromaffin cell development and severely retards lung maturation. *Genes Dev* 9, 1608–1621.
10. Lamberts, S. W., Huizenga, A. T., de Lange, P., de Jong, F. H., and Koper, J. W. (1996). Clinical aspects of glucocorticoid sensitivity. *Steroids* 61, 157–160.
11. Chrousos, G. P. and Kino, T. (2005). Intracellular glucocorticoid signaling: A formerly simple system turns stochastic. *Sci STKE* 2005, pe48.
12. Kino, T., De Martino, M. U., Charmandari, E., Mirani, M., and Chrousos, G. P. (2003). Tissue glucocorticoid resistance/hypersensitivity syndromes. *J Steroid Biochem Mol Biol* 85, 457–467.
13. Gorovits, R., Ben-Dror, I., Fox, L. E., Westphal, H. M., and Vardimon, L. (1994). Developmental changes in the expression and compartmentalization of the glucocorticoid receptor in embryonic retina. *Proc Natl Acad Sci U S A* 91, 4786–4790.
14. Hsu, S. C. and DeFranco, D. B. (1995). Selectivity of cell cycle regulation of glucocorticoid receptor function. *J Biol Chem* 270, 3359–3364.
15. Oakley, R. H. and Cidlowski, J. A. (1993). Homologous down regulation of the glucocorticoid receptor: The molecular machinery. *Crit Rev Eukaryot Gene Expr* 3, 63–88.
16. Oakley, R. H. and Cidlowski, J. A. 2001 The glucocorticoid receptor: Expression, function, and regulation of glucocorticoid responsiveness. In: Goulding N. J., Flower R. J. eds. Glucocorticoids. Birkhauser Verlag, Basel, 55–80

17. Lu, N. Z. and Cidlowski, J. A. (2005). Translational regulatory mechanisms generate N-terminal glucocorticoid receptor isoforms with unique transcriptional target genes. *Mol Cell* 18, 331–342.
18. Lu, N. Z. and Cidlowski, J. A. (2006). Glucocorticoid receptor isoforms generate transcription specificity. *Trends Cell Biol* 16, 301–307.
19. Lu, N. Z., Collins, J. B., Grissom, S. F., and Cidlowski, J. A. (2007). Selective regulation of bone cell apoptosis by translational isoforms of the glucocorticoid receptor. *Mol Cell Biol* 27, 7143–7160.
20. Maniatis, T. and Tasic, B. (2002). Alternative pre-mRNA splicing and proteome expansion in metazoans. *Nature* 418, 236–243.
21. Southan, C. (2004). Has the yo-yo stopped? An assessment of human protein-coding gene number. *Proteomics* 4, 1712–1726.
22. Kumar, R. and Thompson, E. B. (2005). Gene regulation by the glucocorticoid receptor: Structure: Function relationship. *J Steroid Biochem Mol Biol* 94, 383–394.
23. Grad, I. and Picard, D. (2007). The glucocorticoid responses are shaped by molecular chaperones. *Mol Cell Endocrinol* 275, 2–12.
24. Pratt, W. B. and Toft, D. O. (1997). Steroid receptor interactions with heat shock protein and immunophilin chaperones. *Endocr Rev* 18, 306–360.
25. Nemoto, T., Ohara-Nemoto, Y., Denis, M., and Gustafsson, J. A. (1990). The transformed glucocorticoid receptor has a lower steroid-binding affinity than the nontransformed receptor. *Biochemistry* 29, 1880–1886.
26. Seckl, J. R. (2004). 11beta-hydroxysteroid dehydrogenases: Changing glucocorticoid action. *Curr Opin Pharmacol* 4, 597–602.
27. Freedman, N. D. and Yamamoto, K. R. (2004). Importin 7 and importin alpha/importin beta are nuclear import receptors for the glucocorticoid receptor. *Mol Biol Cell* 15, 2276–2286.
28. Schaaf, M. J. and Cidlowski, J. A. (2003). Molecular determinants of glucocorticoid receptor mobility in living cells: The importance of ligand affinity. *Mol Cell Biol* 23, 1922–1934.
29. Beato, M. (1989). Gene regulation by steroid hormones. *Cell* 56, 335–344.
30. Freedman, L. P. (1992). Anatomy of the steroid receptor zinc finger region. *Endocr Rev* 13, 129–145.
31. Drouin, J., Sun, Y. L., Tremblay, S., Lavender, P., Schmidt, T. J., de Lean, A., and Nemer, M. (1992). Homodimer formation is rate-limiting for high affinity DNA binding by glucocorticoid receptor. *Mol Endocrinol* 6, 1299–1309.
32. Chen, S., Wang, J., Yu, G., Liu, W., and Pearce, D. (1997). Androgen and glucocorticoid receptor heterodimer formation. A possible mechanism for mutual inhibition of transcriptional activity. *J Biol Chem* 272, 14087–14092.
33. Trapp, T. and Holsboer, F. (1996). Heterodimerization between mineralocorticoid and glucocorticoid receptors increases the functional diversity of corticosteroid action. *Trends Pharmacol Sci* 17, 145–149.
34. McNally, J. G., Muller, W. G., Walker, D., Wolford, R., and Hager, G. L. (2000). The glucocorticoid receptor: Rapid exchange with regulatory sites in living cells. *Science* 287, 1262–1265.
35. Stavreva, D. A., Muller, W. G., Hager, G. L., Smith, C. L., and McNally, J. G. (2004). Rapid glucocorticoid receptor exchange at a promoter is coupled to transcription and regulated by chaperones and proteasomes. *Mol Cell Biol* 24, 2682–2697.
36. Lonard, D. M. and O'Malley, B. W. (2005). Expanding functional diversity of the coactivators. *Trends Biochem Sci* 30, 126–132.
37. Jenkins, B. D., Pullen, C. B., and Darimont, B. D. (2001). Novel glucocorticoid receptor coactivator effector mechanisms. *Trends Endocrinol Metab* 12, 122–126.
38. Rosenfeld, M. G. and Glass, C. K. (2001). Coregulator codes of transcriptional regulation by nuclear receptors. *J Biol Chem* 276, 36865–36868.
39. Dostert, A. and Heinzel, T. (2004). Negative glucocorticoid receptor response elements and their role in glucocorticoid action. *Curr Pharm Des* 10, 2807–2816.

40. Meyer, T., Gustafsson, J. A., and Carlstedt-Duke, J. (1997). Glucocorticoid-dependent transcriptional repression of the osteocalcin gene by competitive binding at the TATA box. *DNA Cell Biol* 16, 919–927.
41. Diamond, M. I., Miner, J. N., Yoshinaga, S. K., and Yamamoto, K. R. (1990). Transcription factor interactions: Selectors of positive or negative regulation from a single DNA element. *Science* 249, 1266–1272.
42. Malkoski, S. P. and Dorin, R. I. (1999). Composite glucocorticoid regulation at a functionally defined negative glucocorticoid response element of the human corticotropin-releasing hormone gene. *Mol Endocrinol* 13, 1629–1644.
43. Chandran, U. R., Warren, B. S., Baumann, C. T., Hager, G. L., and DeFranco, D. B. (1999). The glucocorticoid receptor is tethered to DNA-bound Oct-1 at the mouse gonadotropin-releasing hormone distal negative glucocorticoid response element. *J Biol Chem* 274, 2372–2378.
44. Subramaniam, N., Cairns, W., and Okret, S. (1998). Glucocorticoids repress transcription from a negative glucocorticoid response element recognized by two homeodomain-containing proteins, Pbx and Oct-1. *J Biol Chem* 273, 23567–23574.
45. Martens, C., Bilodeau, S., Maira, M., Gauthier, Y., and Drouin, J. (2005). Protein-protein interactions and transcriptional antagonism between the subfamily of NGFI-B/Nur77 orphan nuclear receptors and glucocorticoid receptor. *Mol Endocrinol* 19, 885–897.
46. Necela, B. M. and Cidlowski, J. A. (2004). Mechanisms of glucocorticoid receptor action in noninflammatory and inflammatory cells. *Proc Am Thorac Soc* 1, 239–246.
47. Newton, R. and Holden, N. S. (2007). Separating transrepression and transactivation: A distressing divorce for the glucocorticoid receptor? *Mol Pharmacol* 72, 799–809.
48. Smoak, K. A. and Cidlowski, J. A. 2008 Glucocorticoid signaling in health and disease. In: Rey A. D., Chrousos G., Besedovsky H. eds. The hypothalamus-pituitary-adrenal axis. Elsevier, Amsterdam, 33–53
49. De Bosscher, K., Vanden Berghe, W., and Haegeman, G. (2003). The interplay between the glucocorticoid receptor and nuclear factor-kappaB or activator protein-1: molecular mechanisms for gene repression. *Endocr Rev* 24, 488–522.
50. Caelles, C., Gonzalez-Sancho, J. M., and Munoz, A. (1997). Nuclear hormone receptor antagonism with AP-1 by inhibition of the JNK pathway. *Genes Dev* 11, 3351–3364.
51. Hirasawa, N., Sato, Y., Fujita, Y., Mue, S., and Ohuchi, K. (1998). Inhibition by dexamethasone of antigen-induced c-Jun N-terminal kinase activation in rat basophilic leukemia cells. *J Immunol* 161, 4939–4943.
52. Smoak, K. and Cidlowski, J. A. (2006). Glucocorticoids regulate tristetraprolin synthesis and posttranscriptionally regulate tumor necrosis factor alpha inflammatory signaling. *Mol Cell Biol* 26, 9126–9135.
53. Reichardt, H. M., Kaestner, K. H., Tuckermann, J., Kretz, O., Wessely, O., Bock, R., Gass, P., Schmid, W., Herrlich, P., Angel, P., and Schutz, G. (1998). DNA binding of the glucocorticoid receptor is not essential for survival. *Cell* 93, 531–541.
54. Kassel, O. and Herrlich, P. (2007). Crosstalk between the glucocorticoid receptor and other transcription factors: Molecular aspects. *Mol Cell Endocrinol* 275, 13–29.
55. Ogawa, S., Lozach, J., Benner, C., Pascual, G., Tangirala, R. K., Westin, S., Hoffmann, A., Subramaniam, S., David, M., Rosenfeld, M. G., and Glass, C. K. (2005). Molecular determinants of crosstalk between nuclear receptors and toll-like receptors. *Cell* 122, 707–721.
56. Nissen, R. M. and Yamamoto, K. R. (2000). The glucocorticoid receptor inhibits NFkappaB by interfering with serine-2 phosphorylation of the RNA polymerase II carboxy-terminal domain. *Genes Dev* 14, 2314–2329.
57. Rogatsky, I., Luecke, H. F., Leitman, D. C., and Yamamoto, K. R. (2002). Alternate surfaces of transcriptional coregulator GRIP1 function in different glucocorticoid receptor activation and repression contexts. *Proc Natl Acad Sci U S A* 99, 16701–16706.

58. Rogatsky, I., Zarember, K. A., and Yamamoto, K. R. (2001). Factor recruitment and TIF2/GRIP1 corepressor activity at a collagenase-3 response element that mediates regulation by phorbol esters and hormones. *EMBO J* 20, 6071–6083.
59. Ito, K., Barnes, P. J., and Adcock, I. M. (2000). Glucocorticoid receptor recruitment of histone deacetylase 2 inhibits interleukin-1beta-induced histone H4 acetylation on lysines 8 and 12. *Mol Cell Biol* 20, 6891–6903.
60. Rogatsky, I. and Ivashkiv, L. B. (2006). Glucocorticoid modulation of cytokine signaling. *Tissue Antigens* 68, 1–12.
61. Lerner, L., Henriksen, M. A., Zhang, X., and Darnell, J. E., Jr. (2003). STAT3-dependent enhanceosome assembly and disassembly: Synergy with GR for full transcriptional increase of the alpha 2-macroglobulin gene. *Genes Dev* 17, 2564–2577.
62. Takeda, T., Kurachi, H., Yamamoto, T., Nishio, Y., Nakatsuji, Y., Morishige, K., Miyake, A., and Murata, Y. (1998). Crosstalk between the interleukin-6 (IL-6)-JAK-STAT and the glucocorticoid-nuclear receptor pathway: Synergistic activation of IL-6 response element by IL-6 and glucocorticoid. *J Endocrinol* 159, 323–330.
63. Zhang, Z., Jones, S., Hagood, J. S., Fuentes, N. L., and Fuller, G. M. (1997). STAT3 acts as a co-activator of glucocorticoid receptor signaling. *J Biol Chem* 272, 30607–30610.
64. Stocklin, E., Wissler, M., Gouilleux, F., and Groner, B. (1996). Functional interactions between Stat5 and the glucocorticoid receptor. *Nature* 383, 726–728.
65. Stoecklin, E., Wissler, M., Moriggl, R., and Groner, B. (1997). Specific DNA binding of Stat5, but not of glucocorticoid receptor, is required for their functional cooperation in the regulation of gene transcription. *Mol Cell Biol* 17, 6708–6716.
66. Lechner, J., Welte, T., Tomasi, J. K., Bruno, P., Cairns, C., Gustafsson, J., and Doppler, W. (1997). Promoter-dependent synergy between glucocorticoid receptor and Stat5 in the activation of beta-casein gene transcription. *J Biol Chem* 272, 20954–20960.
67. Hermoso, M. A., Matsuguchi, T., Smoak, K., and Cidlowski, J. A. (2004). Glucocorticoids and tumor necrosis factor alpha cooperatively regulate toll-like receptor 2 gene expression. *Mol Cell Biol* 24, 4743–4756.
68. Cella, N., Groner, B., and Hynes, N. E. (1998). Characterization of Stat5a and Stat5b homodimers and heterodimers and their association with the glucocortiocoid receptor in mammary cells. *Mol Cell Biol* 18, 1783–1792.
69. Wyszomierski, S. L., Yeh, J., and Rosen, J. M. (1999). Glucocorticoid receptor/signal transducer and activator of transcription 5 (STAT5) interactions enhance STAT5 activation by prolonging STAT5 DNA binding and tyrosine phosphorylation. *Mol Endocrinol* 13, 330–343.
70. Biola, A., Lefebvre, P., Perrin-Wolff, M., Sturm, M., Bertoglio, J., and Pallardy, M. (2001). Interleukin-2 inhibits glucocorticoid receptor transcriptional activity through a mechanism involving STAT5 (signal transducer and activator of transcription 5) but not AP-1. *Mol Endocrinol* 15, 1062–1076.
71. De Miguel, F., Lee, S. O., Onate, S. A., and Gao, A. C. (2003). Stat3 enhances transactivation of steroid hormone receptors. *Nucl Recept* 1, 3.
72. Zhou, J. and Cidlowski, J. A. (2005). The human glucocorticoid receptor: One gene, multiple proteins and diverse responses. *Steroids* 70, 407–417.
73. Ismaili, N. and Garabedian, M. J. (2004). Modulation of glucocorticoid receptor function via phosphorylation. *Ann NY Acad Sci* 1024, 86–101.
74. Krstic, M. D., Rogatsky, I., Yamamoto, K. R., and Garabedian, M. J. (1997). Mitogen-activated and cyclin-dependent protein kinases selectively and differentially modulate transcriptional enhancement by the glucocorticoid receptor. *Mol Cell Biol* 17, 3947–3954.
75. Rogatsky, I., Waase, C. L., and Garabedian, M. J. (1998). Phosphorylation and inhibition of rat glucocorticoid receptor transcriptional activation by glycogen synthase kinase-3 (GSK-3). Species-specific differences between human and rat glucocorticoid receptor signaling as revealed through GSK-3 phosphorylation. *J Biol Chem* 273, 14315–14321.

76. Itoh, M., Adachi, M., Yasui, H., Takekawa, M., Tanaka, H., and Imai, K. (2002). Nuclear export of glucocorticoid receptor is enhanced by c-Jun N-terminal kinase-mediated phosphorylation. *Mol Endocrinol* 16, 2382–2392.
77. Rogatsky, I., Logan, S. K., and Garabedian, M. J. (1998). Antagonism of glucocorticoid receptor transcriptional activation by the c-Jun N-terminal kinase. *Proc Natl Acad Sci U S A* 95, 2050–2055.
78. Webster, J. C., Jewell, C. M., Bodwell, J. E., Munck, A., Sar, M., and Cidlowski, J. A. (1997). Mouse glucocorticoid receptor phosphorylation status influences multiple functions of the receptor protein. *J Biol Chem* 272, 9287–9293.
79. Ismaili, N., Blind, R., and Garabedian, M. J. (2005). Stabilization of the unliganded glucocorticoid receptor by TSG101. *J Biol Chem* 280, 11120–11126.
80. Irusen, E., Matthews, J. G., Takahashi, A., Barnes, P. J., Chung, K. F., and Adcock, I. M. (2002). p38 Mitogen-activated protein kinase-induced glucocorticoid receptor phosphorylation reduces its activity: Role in steroid-insensitive asthma. *J Allergy Clin Immunol* 109, 649–657.
81. Miller, A. L., Webb, M. S., Copik, A. J., Wang, Y., Johnson, B. H., Kumar, R., and Thompson, E. B. (2005). p38 Mitogen-activated protein kinase (MAPK) is a key mediator in glucocorticoid-induced apoptosis of lymphoid cells: Correlation between p38 MAPK activation and site-specific phosphorylation of the human glucocorticoid receptor at serine 211. *Mol Endocrinol* 19, 1569–1583.
82. Wang, Z. and Garabedian, M. J. (2003). Modulation of glucocorticoid receptor transcriptional activation, phosphorylation, and growth inhibition by p27Kip1. *J Biol Chem* 278, 50897–50901.
83. Chen, W., Dang, T., Blind, R. D., Wang, Z., Cavasotto, C. N., Hittelman, A. B., Rogatsky, I., Logan, S. K., and Garabedian, M. J. (2008). Glucocorticoid receptor phosphorylation differentially affects target gene expression. *Mol Endocrinol* 22(8), 1754–1766
84. Blind, R. D. and Garabedian, M. J. (2008). Differential recruitment of glucocorticoid receptor phospho-isoforms to glucocorticoid-induced genes. *J Steroid Biochem Mol Biol* 109, 150–157.
85. Deroo, B. J., Rentsch, C., Sampath, S., Young, J., DeFranco, D. B., and Archer, T. K. (2002). Proteasomal inhibition enhances glucocorticoid receptor transactivation and alters its subnuclear trafficking. *Mol Cell Biol* 22, 4113–4123.
86. Wallace, A. D. and Cidlowski, J. A. (2001). Proteasome-mediated glucocorticoid receptor degradation restricts transcriptional signaling by glucocorticoids. *J Biol Chem* 276, 42714–42721.
87. Wang, X. and DeFranco, D. B. (2005). Alternative effects of the ubiquitin-proteasome pathway on glucocorticoid receptor down-regulation and transactivation are mediated by CHIP, an E3 ligase. *Mol Endocrinol* 19, 1474–1482.
88. Gottlicher, M., Heck, S., Doucas, V., Wade, E., Kullmann, M., Cato, A. C., Evans, R. M., and Herrlich, P. (1996). Interaction of the Ubc9 human homologue with c-Jun and with the glucocorticoid receptor. *Steroids* 61, 257–262.
89. Holmstrom, S., Van Antwerp, M. E., and Iniguez-Lluhi, J. A. (2003). Direct and distinguishable inhibitory roles for SUMO isoforms in the control of transcriptional synergy. *Proc Natl Acad Sci U S A* 100, 15758–15763.
90. Le Drean, Y., Mincheneau, N., Le Goff, P., and Michel, D. (2002). Potentiation of glucocorticoid receptor transcriptional activity by sumoylation. *Endocrinology* 143, 3482–3489.
91. Tian, S., Poukka, H., Palvimo, J. J., and Janne, O. A. (2002). Small ubiquitin-related modifier-1 (SUMO-1) modification of the glucocorticoid receptor. *Biochem J* 367, 907–911.
92. Davies, L., Karthikeyan, N., Lynch, J. T., Sial, E. A., Gkourtsa, A., Demonacos, C., and Krstic-Demonacos, M. (2008). Cross talk of signaling pathways in the regulation of the glucocorticoid receptor function. *Mol Endocrinol* 22, 1331–1344.
93. Holmstrom, S. R., Chupreta, S., So, A. Y., and Iniguez-Lluhi, J. A. (2008). Sumo-mediated inhibition of glucocorticoid receptor synergistic activity depends on stable assembly at the promoter but not on daxx. *Mol Endocrinol* 22, 2061–2075
94. Iniguez-Lluhi, J. A. and Pearce, D. (2000). A common motif within the negative regulatory regions of multiple factors inhibits their transcriptional synergy. *Mol Cell Biol* 20, 6040–6050.

95. Lin, D. Y., Huang, Y. S., Jeng, J. C., Kuo, H. Y., Chang, C. C., Chao, T. T., Ho, C. C., Chen, Y. C., Lin, T. P., Fang, H. I., Hung, C. C., Suen, C. S., Hwang, M. J., Chang, K. S., Maul, G. G., and Shih, H. M. (2006). Role of SUMO-interacting motif in Daxx SUMO modification, subnuclear localization, and repression of sumoylated transcription factors. *Mol Cell* 24, 341–354.
96. Fu, M., Wang, C., Reutens, A. T., Wang, J., Angeletti, R. H., Siconolfi-Baez, L., Ogryzko, V., Avantaggiati, M. L., and Pestell, R. G. (2000). p300 and p300/cAMP-response element-binding protein-associated factor acetylate the androgen receptor at sites governing hormone-dependent transactivation. *J Biol Chem* 275, 20853–20860.
97. Wang, C., Fu, M., Angeletti, R. H., Siconolfi-Baez, L., Reutens, A. T., Albanese, C., Lisanti, M. P., Katzenellenbogen, B. S., Kato, S., Hopp, T., Fuqua, S. A., Lopez, G. N., Kushner, P. J., and Pestell, R. G. (2001). Direct acetylation of the estrogen receptor alpha hinge region by p300 regulates transactivation and hormone sensitivity. *J Biol Chem* 276, 18375–18383.
98. Ito, K., Yamamura, S., Essilfie-Quaye, S., Cosio, B., Ito, M., Barnes, P. J., and Adcock, I. M. (2006). Histone deacetylase 2-mediated deacetylation of the glucocorticoid receptor enables NF-kappaB suppression. *J Exp Med* 203, 7–13.
99. Hollenberg, S. M., Weinberger, C., Ong, E. S., Cerelli, G., Oro, A., Lebo, R., Thompson, E. B., Rosenfeld, M. G., and Evans, R. M. (1985). Primary structure and expression of a functional human glucocorticoid receptor cDNA. *Nature* 318, 635–641.
100. Francke, U. and Foellmer, B. E. (1989). The glucocorticoid receptor gene is in 5q31-q32 [corrected]. *Genomics* 4, 610–612.
101. Theriault, A., Boyd, E., Harrap, S. B., Hollenberg, S. M., and Connor, J. M. (1989). Regional chromosomal assignment of the human glucocorticoid receptor gene to 5q31. *Hum Genet* 83, 289–291.
102. Encio, I. J. and Detera-Wadleigh, S. D. (1991). The genomic structure of the human glucocorticoid receptor. *J Biol Chem* 266, 7182–7188.
103. Oakley, R. H., Sar, M., and Cidlowski, J. A. (1996). The human glucocorticoid receptor beta isoform. Expression, biochemical properties, and putative function. *J Biol Chem* 271, 9550–9559.
104. Breslin, M. B., Geng, C. D., and Vedeckis, W. V. (2001). Multiple promoters exist in the human GR gene, one of which is activated by glucocorticoids. *Mol Endocrinol* 15, 1381–1395.
105. Presul, E., Schmidt, S., Kofler, R., and Helmberg, A. (2007). Identification, tissue expression, and glucocorticoid responsiveness of alternative first exons of the human glucocorticoid receptor. *J Mol Endocrinol* 38, 79–90.
106. Turner, J. D. and Muller, C. P. (2005). Structure of the glucocorticoid receptor (NR3C1) gene 5′ untranslated region: Identification, and tissue distribution of multiple new human exon 1. *J Mol Endocrinol* 35, 283–292.
107. King, L. B., Vacchio, M. S., Dixon, K., Hunziker, R., Margulies, D. H., and Ashwell, J. D. (1995). A targeted glucocorticoid receptor antisense transgene increases thymocyte apoptosis and alters thymocyte development. *Immunity* 3, 647–656.
108. Pepin, M. C., Pothier, F., and Barden, N. (1992). Impaired type II glucocorticoid-receptor function in mice bearing antisense RNA transgene. *Nature* 355, 725–728.
109. Turner, J. D., Schote, A. B., Macedo, J. A., Pelascini, L. P., and Muller, C. P. (2006). Tissue specific glucocorticoid receptor expression, a role for alternative first exon usage? *Biochem Pharmacol* 72, 1529–1537.
110. Russcher, H., Dalm, V. A., de Jong, F. H., Brinkmann, A. O., Hofland, L. J., Lamberts, S. W., and Koper, J. W. (2007). Associations between promoter usage and alternative splicing of the glucocorticoid receptor gene. *J Mol Endocrinol* 38, 91–98.
111. Pedersen, K. B., Geng, C. D., and Vedeckis, W. V. (2004). Three mechanisms are involved in glucocorticoid receptor autoregulation in a human T-lymphoblast cell line. *Biochemistry* 43, 10851–10858.
112. Yudt, M. R. and Cidlowski, J. A. (2001). Molecular identification and characterization of a and b forms of the glucocorticoid receptor. *Mol Endocrinol* 15, 1093–1103.
113. Bledsoe, R. K., Montana, V. G., Stanley, T. B., Delves, C. J., Apolito, C. J., McKee, D. D., Consler, T. G., Parks, D. J., Stewart, E. L., Willson, T. M., Lambert, M. H., Moore, J. T., Pearce, K. H., and

Xu, H. E. (2002). Crystal structure of the glucocorticoid receptor ligand binding domain reveals a novel mode of receptor dimerization and coactivator recognition. *Cell* 110, 93–105.

114. Yudt, M. R., Jewell, C. M., Bienstock, R. J., and Cidlowski, J. A. (2003). Molecular origins for the dominant negative function of human glucocorticoid receptor beta. *Mol Cell Biol* 23, 4319–4330.
115. Bamberger, C. M., Bamberger, A. M., de Castro, M., and Chrousos, G. P. (1995). Glucocorticoid receptor beta, a potential endogenous inhibitor of glucocorticoid action in humans. *J Clin Invest* 95, 2435–2441.
116. Oakley, R. H., Jewell, C. M., Yudt, M. R., Bofetiado, D. M., and Cidlowski, J. A. (1999). The dominant negative activity of the human glucocorticoid receptor beta isoform. Specificity and mechanisms of action. *J Biol Chem* 274, 27857–27866.
117. Fruchter, O., Kino, T., Zoumakis, E., Alesci, S., De Martino, M., Chrousos, G., and Hochberg, Z. (2005). The human glucocorticoid receptor (GR) isoform {beta} differentially suppresses GR{alpha}-induced transactivation stimulated by synthetic glucocorticoids. *J Clin Endocrinol Metab* 90, 3505–3509.
118. Hauk, P. J., Goleva, E., Strickland, I., Vottero, A., Chrousos, G. P., Kisich, K. O., and Leung, D. Y. (2002). Increased glucocorticoid receptor Beta expression converts mouse hybridoma cells to a corticosteroid-insensitive phenotype. *Am J Respir Cell Mol Biol* 27, 361–367.
119. Charmandari, E., Chrousos, G. P., Ichijo, T., Bhattacharyya, N., Vottero, A., Souvatzoglou, E., and Kino, T. (2005). The human glucocorticoid receptor (hGR) beta isoform suppresses the transcriptional activity of hGRalpha by interfering with formation of active coactivator complexes. *Mol Endocrinol* 19, 52–64.
120. de Castro, M., Elliot, S., Kino, T., Bamberger, C., Karl, M., Webster, E., and Chrousos, G. P. (1996). The non-ligand binding beta-isoform of the human glucocorticoid receptor (hGR beta): tissue levels, mechanism of action, and potential physiologic role. *Mol Med* 2, 597–607.
121. Oakley, R. H., Webster, J. C., Sar, M., Parker, C. R., Jr., and Cidlowski, J. A. (1997). Expression and subcellular distribution of the beta-isoform of the human glucocorticoid receptor. *Endocrinology* 138, 5028–5038.
122. Pujols, L., Mullol, J., Roca-Ferrer, J., Torrego, A., Xaubet, A., Cidlowski, J. A., and Picado, C. (2002). Expression of glucocorticoid receptor alpha- and beta-isoforms in human cells and tissues. *Am J Physiol Cell Physiol* 283, C1324–C1331.
123. Carlstedt-Duke, J. (1999). Glucocorticoid Receptor beta: View II. *Trends Endocrinol Metab* 10, 339–342.
124. Strickland, I., Kisich, K., Hauk, P. J., Vottero, A., Chrousos, G. P., Klemm, D. J., and Leung, D. Y. (2001). High constitutive glucocorticoid receptor beta in human neutrophils enables them to reduce their spontaneous rate of cell death in response to corticosteroids. *J Exp Med* 193, 585–593.
125. Orii, F., Ashida, T., Nomura, M., Maemoto, A., Fujiki, T., Ayabe, T., Imai, S., Saitoh, Y., and Kohgo, Y. (2002). Quantitative analysis for human glucocorticoid receptor alpha/beta mRNA in IBD. *Biochem Biophys Res Commun* 296, 1286–1294.
126. Tliba, O., Cidlowski, J. A., and Amrani, Y. (2006). CD38 expression is insensitive to steroid action in cells treated with tumor necrosis factor-alpha and interferon-gamma by a mechanism involving the up-regulation of the glucocorticoid receptor beta isoform. *Mol Pharmacol* 69, 588–596.
127. Webster, J. C., Oakley, R. H., Jewell, C. M., and Cidlowski, J. A. (2001). Proinflammatory cytokines regulate human glucocorticoid receptor gene expression and lead to the accumulation of the dominant negative beta isoform: A mechanism for the generation of glucocorticoid resistance. *Proc Natl Acad Sci U S A* 98, 6865–6870.
128. Hauk, P. J., Hamid, Q. A., Chrousos, G. P., and Leung, D. Y. (2000). Induction of corticosteroid insensitivity in human PBMCs by microbial superantigens. *J Allergy Clin Immunol* 105, 782–787.
129. Chikanza, I. C. (2002). Mechanisms of corticosteroid resistance in rheumatoid arthritis: A putative role for the corticosteroid receptor beta isoform. *Ann NY Acad Sci* 966, 39–48.
130. Hamid, Q. A., Wenzel, S. E., Hauk, P. J., Tsicopoulos, A., Wallaert, B., Lafitte, J. J., Chrousos, G. P., Szefler, S. J., and Leung, D. Y. (1999). Increased glucocorticoid receptor beta in airway cells of glucocorticoid-insensitive asthma. *Am J Respir Crit Care Med* 159, 1600–1604.

131. Hamilos, D. L., Leung, D. Y., Muro, S., Kahn, A. M., Hamilos, S. S., Thawley, S. E., and Hamid, Q. A. (2001). GRbeta expression in nasal polyp inflammatory cells and its relationship to the anti-inflammatory effects of intranasal fluticasone. *J Allergy Clin Immunol* 108, 59–68.
132. Honda, M., Orii, F., Ayabe, T., Imai, S., Ashida, T., Obara, T., and Kohgo, Y. (2000). Expression of glucocorticoid receptor beta in lymphocytes of patients with glucocorticoid-resistant ulcerative colitis. *Gastroenterology* 118, 859–866.
133. Koga, Y., Matsuzaki, A., Suminoe, A., Hattori, H., Kanemitsu, S., and Hara, T. (2005). Differential mRNA expression of glucocorticoid receptor alpha and beta is associated with glucocorticoid sensitivity of acute lymphoblastic leukemia in children. *Pediatr Blood Cancer* 45, 121–127.
134. Leung, D. Y., Hamid, Q., Vottero, A., Szefler, S. J., Surs, W., Minshall, E., Chrousos, G. P., and Klemm, D. J. (1997). Association of glucocorticoid insensitivity with increased expression of glucocorticoid receptor beta. *J Exp Med* 186, 1567–1574.
135. Piotrowski, P., Burzynski, M., Lianeri, M., Mostowska, M., Wudarski, M., Chwalinska-Sadowska, H., and Jagodzinski, P. P. (2007). Glucocorticoid receptor beta splice variant expression in patients with high and low activity of systemic lupus erythematosus. *Folia Histochem Cytobiol* 45, 339–342.
136. Shahidi, H., Vottero, A., Stratakis, C. A., Taymans, S. E., Karl, M., Longui, C. A., Chrousos, G. P., Daughaday, W. H., Gregory, S. A., and Plate, J. M. (1999). Imbalanced expression of the glucocorticoid receptor isoforms in cultured lymphocytes from a patient with systemic glucocorticoid resistance and chronic lymphocytic leukemia. *Biochem Biophys Res Commun* 254, 559–565.
137. Sousa, A. R., Lane, S. J., Cidlowski, J. A., Staynov, D. Z., and Lee, T. H. (2000). Glucocorticoid resistance in asthma is associated with elevated in vivo expression of the glucocorticoid receptor beta-isoform. *J Allergy Clin Immunol* 105, 943–950.
138. Goecke, I. A., Alvarez, C., Henriquez, J., Salas, K., Molina, M. L., Ferreira, A., and Gatica, H. (2007). Methotrexate regulates the expression of glucocorticoid receptor alpha and beta isoforms in normal human peripheral mononuclear cells and human lymphocyte cell lines in vitro. *Mol Immunol* 44, 2115–2123.
139. Xu, Q., Leung, D. Y., and Kisich, K. O. (2003). Serine-arginine-rich protein p30 directs alternative splicing of glucocorticoid receptor pre-mRNA to glucocorticoid receptor beta in neutrophils. *J Biol Chem* 278, 27112–27118.
140. Derijk, R. H., Schaaf, M. J., Turner, G., Datson, N. A., Vreugdenhil, E., Cidlowski, J., de Kloet, E. R., Emery, P., Sternberg, E. M., and Detera-Wadleigh, S. D. (2001). A human glucocorticoid receptor gene variant that increases the stability of the glucocorticoid receptor beta-isoform mRNA is associated with rheumatoid arthritis. *J Rheumatol* 28, 2383–2388.
141. Schaaf, M. J. and Cidlowski, J. A. (2002). AUUUA motifs in the 3′UTR of human glucocorticoid receptor alpha and beta mRNA destabilize mRNA and decrease receptor protein expression. *Steroids* 67, 627–636.
142. Syed, A. A., Irving, J. A., Redfern, C. P., Hall, A. G., Unwin, N. C., White, M., Bhopal, R. S., and Weaver, J. U. (2006). Association of glucocorticoid receptor polymorphism A3669G in exon 9beta with reduced central adiposity in women. *Obesity (Silver Spring)* 14, 759–764.
143. van den Akker, E. L., Koper, J. W., van Rossum, E. F., Dekker, M. J., Russcher, H., de Jong, F. H., Uitterlinden, A. G., Hofman, A., Pols, H. A., Witteman, J. C., and Lamberts, S. W. (2008). Glucocorticoid receptor gene and risk of cardiovascular disease. *Arch Intern Med* 168, 33–39.
144. van den Akker, E. L., Nouwen, J. L., Melles, D. C., van Rossum, E. F., Koper, J. W., Uitterlinden, A. G., Hofman, A., Verbrugh, H. A., Pols, H. A., Lamberts, S. W., and van Belkum, A. (2006). Staphylococcus aureus nasal carriage is associated with glucocorticoid receptor gene polymorphisms. *J Infect Dis* 194, 814–818.
145. Lewis-Tuffin, L. J., Jewell, C. M., Bienstock, R. J., Collins, J. B., and Cidlowski, J. A. (2007). The human glucocorticoid receptor {beta} (hGR{beta}) Binds RU-486 and is transcriptionally active. *Mol Cell Biol* 27, 2266–2282.
146. Kelly, A., Bowen, H., Jee, Y. K., Mahfiche, N., Soh, C., Lee, T., Hawrylowicz, C., and Lavender, P. (2008). The glucocorticoid receptor beta isoform can mediate transcriptional repression by recruiting histone deacetylases. *J Allergy Clin Immunol* 121(203–208), e201.

147. Ray, D. W., Davis, J. R., White, A., and Clark, A. J. (1996). Glucocorticoid receptor structure and function in glucocorticoid-resistant small cell lung carcinoma cells. *Cancer Res* 56, 3276–3280.
148. Rivers, C., Levy, A., Hancock, J., Lightman, S., and Norman, M. (1999). Insertion of an amino acid in the DNA-binding domain of the glucocorticoid receptor as a result of alternative splicing. *J Clin Endocrinol Metab* 84, 4283–4286.
149. Beger, C., Gerdes, K., Lauten, M., Tissing, W. J., Fernandez-Munoz, I., Schrappe, M., and Welte, K. (2003). Expression and structural analysis of glucocorticoid receptor isoform gamma in human leukaemia cells using an isoform-specific real-time polymerase chain reaction approach. *Br J Haematol* 122, 245–252.
150. Moalli, P. A., Pillay, S., Krett, N. L., and Rosen, S. T. (1993). Alternatively spliced glucocorticoid receptor messenger RNAs in glucocorticoid-resistant human multiple myeloma cells. *Cancer Res* 53, 3877–3879.
151. de Lange, P., Segeren, C. M., Koper, J. W., Wiemer, E., Sonneveld, P., Brinkmann, A. O., White, A., Brogan, I. J., de Jong, F. H., and Lamberts, S. W. (2001). Expression in hematological malignancies of a glucocorticoid receptor splice variant that augments glucocorticoid receptor-mediated effects in transfected cells. *Cancer Res* 61, 3937–3941.
152. Krett, N. L., Pillay, S., Moalli, P. A., Greipp, P. R., and Rosen, S. T. (1995). A variant glucocorticoid receptor messenger RNA is expressed in multiple myeloma patients. *Cancer Res* 55, 2727–2729.
153. Gaitan, D., DeBold, C. R., Turney, M. K., Zhou, P., Orth, D. N., and Kovacs, W. J. (1995). Glucocorticoid receptor structure and function in an adrenocorticotropin-secreting small cell lung cancer. *Mol Endocrinol* 9, 1193–1201.

CHAPTER 5

ESTROGEN RECEPTORS: THEIR ACTIONS AND FUNCTIONAL ROLES IN HEALTH AND DISEASE

STEFAN NILSSON[1,2] AND JAN-ÅKE GUSTAFSSON[2]

[1] *Karo Bio AB, Novum, Huddinge SE-141 57, Sweden*

[2] *Department of Biosciences and Nutrition, Novum, Huddinge SE-141 57, Sweden*

Abstract: Our view of estrogen signalling has undergone a paradigm shift over the recent 10–15 years with the discovery of a second estrogen receptor, ERβ, in 1995 and the finding that estrogens play an important role also in male physiology. Aromatase deficient patients and aromatase knock-out mice have highlighted the importance of estrogens in development and metabolic homeostasis while ER knock-out mice, ERα–/– and ERβ–/–, have shown that both ERs are of physiological importance and that ERα and ERβ have distinct and non-overlapping functions in the body. The uses of ER subtype-selective ligands in various animal models have further substantiated the distinctive physiological roles of ERα and ERβ and shown that they, in many contexts, are antagonistic against one another. Structural studies of the ligand-binding domains of ERα and ERβ have provided in-depth information on ligand recognition, receptor activation, and recruitment of coregulators. The cloning of coregulators and chromatin modulators together with sophisticated methodology to study gene regulation has significantly increased our understanding of cellular and target gene responses to estrogens, SERMs, and ER subtype-selective ligands.

This book chapter will review our current understanding of the mechanisms of ERα- and ERβ-dependent estrogen signalling, the role of ERα and ERβ in health and disease, and the potential clinical uses of ERα- and ERβ-selective pharmaceuticals.

5.1. INTRODUCTION AND HISTORICAL PERSPECTIVE

In the late nineteenth century two scientists made the observation that female sex steroid hormones played an essential role in promoting mammary tumour growth [1, 2]. A little more than half a century later the pioneering endocrinologists, Jensen and Jacobsen [3, 4], arrived at the conclusion that the biological effects of 17β-estradiol (E2) could be mediated through a specific receptor rather than through enzymatic metabolism of estradiol itself. A few years later the groups of Gorski and Jensen isolated and characterized an estrogen-binding protein from the uterus and proposed a model for its mechanism of action [5–7]. For many years this protein and

C.M. Bunce, M.J. Campbell (eds.), *Nuclear Receptors*, Proteins and Cell Regulation 8, 91–141. DOI 10.1007/978-90-481-3303-1_5,

its role in the effects of estrogens were extensively studied in several laboratories [8–12], and in 1986, two groups reported the cloning of the estrogen receptor (ER) [13–15]. Like the glucocorticoid receptor [16] the estrogen receptor was also found to interact with specific DNA sequences, estrogen response elements (ERE), which function as receptor-selective transcriptional enhancer elements, located upstream of or in the vicinity of estrogen-sensitive target genes [17–21]. Following biochemical and mutational analysis of the cloned ER the primary structure and modular organization of the receptor protein as well as its genomic organization was established [22, 23]. A comparison of the primary sequence and domain structure, and function, of ER with other intracellular receptors characterized at the time, suggested the existence of a large family of proteins [the nuclear steroid and thyroid hormone receptor (NR) family] with many similarities in their overall structures and hormone-responsive modes of action [24–30]. Until 1995, it was assumed that there was only one ER responsible for mediating all of the physiological and pharmacological effects of natural and synthetic estrogens and antiestrogens. However, early in 1995 a second ER, ERβ (NR3A2; [31]), was cloned from a rat prostate cDNA library [32], rapidly followed by the cloning of ERβ from also other species [33–36], including several isoforms (reviewed in [37–40]). The ER cloned in 1986 was consequently renamed ERα (NR3A1; [31]).

Similar to the other members of the ligand-activated nuclear receptor family, the transcriptional regulation of target genes by the ERs is dependent on conformational changes of the transcription activation domains contained within the N- and C-terminal regions of the receptors [41, 37], and their ability to interact with coregulator proteins [42, 43].

ERα and ERβ are expressed in most cell types in the body: epithelium, endothelium, stroma, smooth and skeletal muscle, bone, cartilage, hematopoietic cells, neurons and glia [44–46]. In some organs, both receptor subtypes are expressed at similar levels, sometimes in different cell types within the organ, whereas in others, one or the other subtype predominates. ERα is primarily expressed in uterus, prostate (stroma), ovary (theca cells), testes (Leydig cells), epididymis, bone, breast, liver, kidney, white adipose tissue, and various regions of the brain. ERβ is predominantly expressed in colon, prostate (epithelium), testis, ovary (granulosa cells), bone marrow, salivary gland, vascular endothelium, lung, bladder, and certain regions of the brain [47, 48].

The discovery of ERβ has changed the concepts of estrogen signalling [49] and has revealed a wider role for estrogens in physiology including not only the reproductive functions in females but also the growth, development, and homeostasis of organs such as, but not limited to, the mammary gland, bone, the cardiovascular system, the immune system, and the CNS, in females as well as in males [50, 39, 51]. Estrogen and its cognate receptors are also involved in a number of different pathological conditions, for example, various cancers, osteoporosis, disorders of the CNS, cardiovascular disease, endometriosis, insulin resistance, autoimmune diseases, and obesity [39, 52].

Common regimens in use to treat symptoms of the menopause and post-menopausal osteoporosis are 17β-estradiol, esterified estrogens or conjugated equine estrogens, each in combination with a progestin, for example, medroxyprogesterone acetate (MPA), to avoid increased risk of endometrial or uterine cancer in women with an intact uterus. The awareness of undesired effects (e.g. resumption of monthly bleedings, breast tenderness, and headaches) or health risks (breast cancer, endometrial cancer, venous thromboembolism, ovarian cancer, asthma, and gall bladder disease) with existing hormone replacement therapy (HRT) calls for alternatives with improved safety profile [53–56].

Tamoxifen, the first selective estrogen receptor modulator (SERM) developed for the endocrine treatment of ER-positive breast cancer, has been successfully used in the clinic for more than 30 years [57]. However, the side-effect profile of tamoxifen and rate of breast cancer recurrence have spurred the search for and the development of additional SERMs and SERDs (selective estrogen receptor down-regulators). Today these agents are in clinical use or in clinical development for prevention or treatment of breast cancer or other pathological conditions associated with ER activity such as, for example prevention of the development of osteroporosis [58–62]. Aromatase inhibitors represent still another alternative strategy for blocking estrogen signalling in breast cancer, and have been reported to be superior to tamoxifen with respect to disease-free survival although not in terms of side-effect profile and overall survival [63–69].

5.2. PRIMARY STRUCTURE, ISOFORMS AND POLYMORPHISMS OF ERs

The ERα and ERβ proteins in humans are encoded by genes spanning approximately 450 kb of chromosome 6 and 260 kb of chromosome 14, respectively, each comprised of eight coding exons interrupted by long introns [70]. The primary structure and modular organization of the two ERs are similar to the other members of the nuclear receptor family, comprising five distinguishable domains, A/B, C, D, E, and F [28] (Figure 5.1). The size of the wild-type ERα protein is 595 amino acids with a molecular weight of 66.2 kDa, and the wild-type ERβ protein is 530 amino acids long with a molecular weight of 59.2 kDa [71]. Several ERα and ERβ splicing variants with either truncations or insertions in their C-terminal domains have been described [72, 39, 73, 40]. Single nucleotide polymorphisms within coding sequences or in regions surrounding ESR1 (ERα gene) or ESR2 (ERβ gene), respectively, have been reported, some of which have been associated with breast and ovarian cancer, prostate cancer, bulimia, cardiovascular disease, lipid and apoprotein levels, type 2 diabetes, venous ulcers, pelvic organ prolapse, or osteoporosis [72, 74, 75, 39, 76, 77–82].

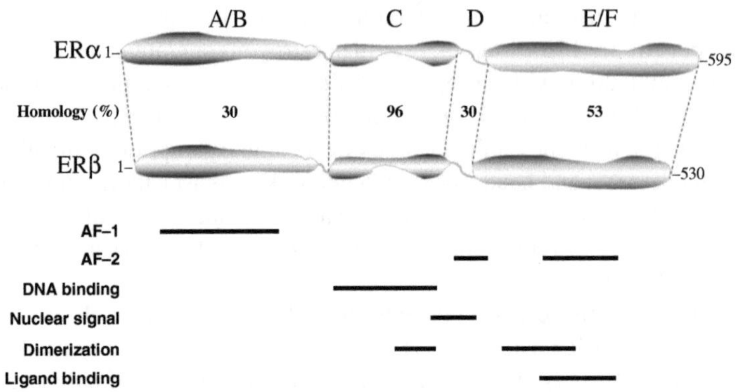

Figure 5.1. The domain structure of the estrogen receptors, ERα and ERβ. The N-terminal (A/B) domain with its ligand-independent, autonomous transcription activation function-1 (AF-1), including sites for phosphorylation by the MAPK pathway. The conserved C-domain mediates sequence-specific DNA binding. The hinge (D) domain contains the nuclear translocation signal and is target for multiple postranslational modifications such as acetylation, sumoylation, and ubiqutination. The C-terminal, multifunctional E/F domain, is involved in ligand binding, receptor homo- and heterodimerization, and ligand-dependent transcription activation (AF-2). Both AF-1 and AF-2 provide surfaces for protein-protein interactions with coregulators. Shown is also the degree of homology of the various receptor domains between ERα and ERβ

5.3. FUNCTIONAL DOMAINS OF ERα AND ERβ

5.3.1. The N-Terminal A/B Domain

The N-terminal domains of ERα and ERβ encode amino acid sequences involved in interactions between ER-domains and between ER and coactivators [83–85]. In addition to the segments in the N-terminal domain of ERα that promote coactivator interactions and hormone-independent transcription activation (AF-1) (Figure 5.1), there is a segment at the beginning of the N-terminus that has an inhibitory function on hormone-independent gene activation. Removal of this inhibitory sequence, which dynamically interacts with a hydrophobic surface in the C-terminal ligand-binding domain, enhances hormone-independent activation of reporter gene expression in mammalian cells [86, 87, 85]. The N-termini of both ERs are targets for phosphorylation by kinases in growth factor pathways [88–90, 84, 91], leading to stimulation of AF-1 activity [92, 93].

5.3.2. The DNA-Binding C Domain

The 70–80 amino acid-long DNA binding domains (DBD) of ERα and ERβ are very similar with 95% homology at the amino acid sequence level [94]. The DBD plays a pivotal role in the binding of the ERs to specific sequences on DNA and is part of the receptor dimerization surface. A general characteristic of all DBDs in the nuclear receptor family is the two zinc-binding motifs, in which a zinc atom is coordinated by four cysteine residues. At the stem of the first zinc finger is the P-box, which is involved in sequence-specific DNA recognition and binding [95, 20]. The crystal structure of ERα DBD dimers bound to a consensus ERE showed that the side chains of four amino acids in the P box, make direct and indirect contacts with nucleotides in the ERE [95, 39]. The beginning of the second motif, which contains the distal box (the D-box), provides the surface for ERs to dimerize head-to-head on DNA and contributes to the cooperative binding of two ER monomers to the ERE [95, 39].

Although the amino acid sequences of the P- and D-boxes in ERα and ERβ, respectively, are identical and, thus, one might expect that ERα and ERβ would have similar affinities to EREs, chip-on-chip genome landscape analysis of ERα and ERβ DNA interactions in intact chromatin has shown that, in addition to regions where there is a high degree of overlap between the two ER subtypes, there are also regions preferentially bound by ERα or ERβ, respectively [96]. In addition, there are regions in chromatin that bind ERα only in the presence of ERβ, suggesting that ERβ may function as a recruiter of ERα. ERα subtype-specific DNA-binding regions showed distinct properties; they had an overrepresentation of TA-rich motifs and Forkhead (FoxA1) transcription factor binding sites while ERβ-bound regions had a predominance of classical estrogen response elements and GC-rich motifs. Furthermore, ERβ binding regions were generally located closer to gene transcription start sites than the ERα-bound regions [96]. Direct binding of ERα to chromatin sites requires the presence of FoxA1, which facilitates recruitment of ERα and chromatin remodelling; silencing of FoxA1 expression blocks ERα from associating with chromatin and subsequent modulation of target gene transcription [97–100]. Approximately 50–60% of ERα binding sites overlap with sites occupied by the FoxA1 transcription factor [101].

5.3.3. The Hinge (D) Domain

Until recently not much was known about the function of the hinge domain except that part of the nuclear translocation signal is endoded by this domain. In recent publications, however, it was shown that this domain is target for acetylation of lysine (K) residues (K266, 268, 302 and 303) by the p300 acetylase and that this posttranslational modification enhances the DNA binding activity and hormone sensitivity of ERα as well as its transcriptional activity [102, 103]. The same lysine residues susceptible for acetylation are also targets for sumoylation, which is important for hormone-dependent activation of transcription [104], as well as for ubiquitination, targeting the receptor for proteasome degradation [105].

5.3.4. The Ligand Binding E Domain

The ligand binding domain (LBD) is a multifunctional region participating in nuclear translocation, ligand recognition and binding, receptor homo/hetero dimerization,

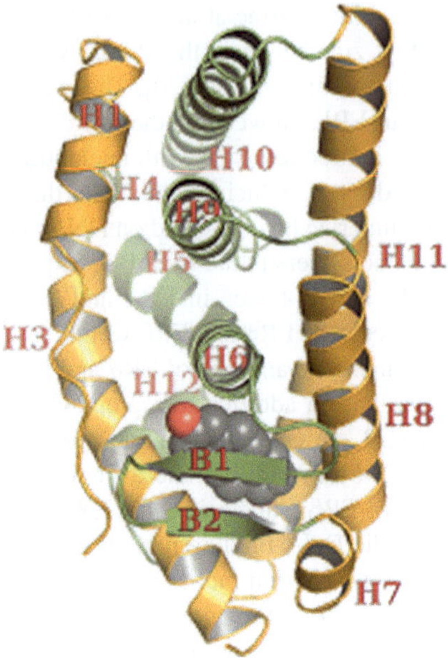

Figure 5.2. The overall configuration of ER-LBD with helices H1-H12 and the two-stranded anti-parallel β-sheet, B1 and B2, displayed. Shown in grey is 17β-estradiol bound in the ligand binding cavity within the interior of the LBD

protein–protein interaction with coregulators, and transcription activation. The E-domains of ERα and ERβ are 59% homologous at the amino acid level [94] and both are folded into a three-layered, anti-parallel α-helical sandwich comprised of 12 α-helices (H1-12) and a small two-stranded anti-parallel β-sheet (Figure 5.2; kindly provided by Dr Björn Kauppi, Karo Bio AB) [106–109].

The ligand binding cavity is completely hidden from the external environment and occupies a relatively large portion of the hydrophobic core of the ER LBD [106]. Ligand recognition is achieved by specific hydrogen bonds and a combination of the hydrophobic complementarity of the binding cavity with the shape and non-polar character of the ligand (Figure 5.3; kindly provided by Dr Björn Kauppi, Karo Bio AB) [106, 109]. Except for two amino acids, Leu384 and Met421 in ERα corresponding to Met336 and Ile373 in ERβ, respectively [110], the amino acids lining the hormone binding cavities of ERα and ERβ are identical. The wide variety of structurally distinct natural and synthetic ligands that bind to ERα and ERβ [111–116, 109, 62] can be attributed to the size and plasticity of their respective ligand-binding cavities, with an internal accessible volume considerably larger than the bound ligand [110, 109].

In contrast to AF-1 in the N-terminal AB domain, the transcription activation function (AF-2) in the LBD is ligand-dependent and constitutes a hydrophobic

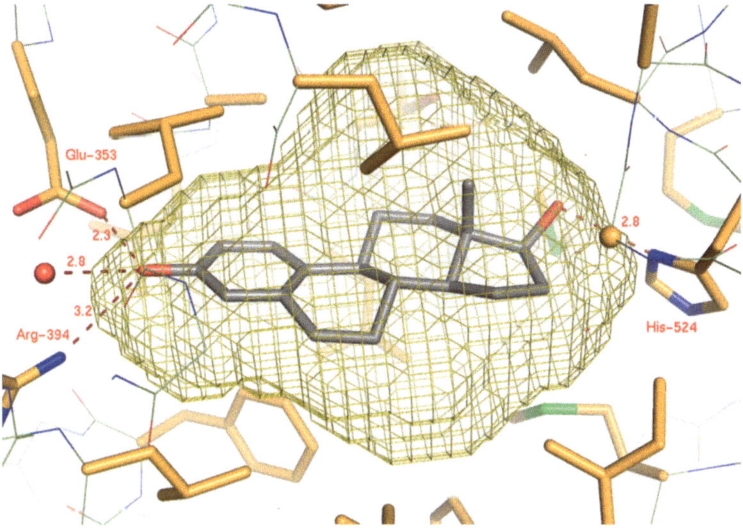

Figure 5.3. 17β-estradiol (E2) bound in the ERα ligand binding cavity. Ligand recognition is achieved by specific hydrogen bonds with amino acid residues lining the receptor cavity in combination with the overall shape complementarity of the pocket to the nonpolar character of E2. The phenolic hydroxyl in the A-ring of E2 is coordinated by glutamic acid (Glu-353) in H3, a water molecule, and arginine (Arg-394) in H6. The OH-group in the D-ring of E2 makes a hydrogen bond with His-524 in H11 via a water molecule. The remainder of E2, in-between the two hydroxyl groups, participates in a number of hydrophobic contacts. Displayed is also the internal accessible volume, which is considerably larger than the bound 17β-estradiol

coactivator: receptor docking surface formed by H12 and amino acids in H3, H4 and H5. Key player in the function of AF-2 in transcriptional regulation of target genes is H12, which, in a ligand-dependent fashion acts as a dynamic molecular switch by adopting different positions, discriminating between coactivator and corepressor binding ('agonist'- or 'antagonist' conformation, respectively) (Figure 5.4a, b; kindly provided by Dr Björn Kauppi, Karo Bio AB) [106, 107, 117–120, 85, 121–123, 109], thereby facilitating or silencing transcription activation, respectively. The amino acid sequence in H12, important for AF-2 activity, is conserved between ERα and ERβ [110], with similar efficiency in promoting interaction with the coactivator SRC-1 and in stimulating transcription activation [124]. However, also amino acid sequences outside the AF-2 core sequence may affect the specificity and strength of coactivator interaction in a ligand and ER subtype-specific manner.

Nuclear receptors form strong dimers that are essential for their function as transcription factors. The major ER dimerization interface involves hydrophobic amino acids at the N-terminal end of H11 of each monomer, that intertwine to form a rigid backbone, and H8 of one monomer interacting with parts of H9 and H10 from the neighbouring monomer [106, 125].

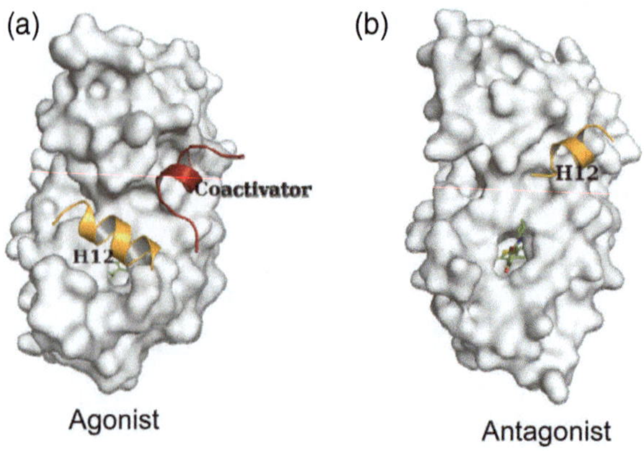

Figure 5.4. Agonist and antagonist position, respectively, of helix 12 (H12). H12 has a key role in the function of the ligand-dependent AF-2 transcription activation module of the ERs, acting as a dynamic switch by assuming different positions depending on bound ligand: (**a**) Agonist conformation of H12 that allows interaction with coactivators via their LXXLL motifs and the ER coactivator docking surface, (**b**) In the presence of ligands with a bulky side-chain, H12 orients along the hydrophobic coactivator cleft, foiling the AF-2 surface, thereby inhibiting a productive interaction with coactivators

Although ER can form stable dimers in the absence of bound ligand, agonists stabilize the dimers and coactivator peptides increase this stabilization [126, 127]. Previously it was shown that coactivator NR-box binding to agonist activated ER also locks the ligand within the ligand binding cavity, substantially decreasing the rate of ligand dissociation from the receptor as well as increasing the agonist potency in transcription activation [128].

5.3.5. The F-Domain

The structure and function of the F-domain in other members of the NR family suggested a modulatory role for receptor transcription activity and recruitment of coactivators and corepressors [129, 130]. This is clearly the case with hepatocyte nuclear receptor-4α, where a naturally occurring mutation in the F-domain co-segregates with the development of diabetes and impaired insulin secretion [131].

The structure and function of the F-domain of ERα has been analysed by deletion and point mutation studies [132–135]. Mutations that deleted the F-domain or targeted a predicted α-helical structure in the F-domain, reduced or abrogated the agonist activity of 4-hydroxytamoxifen but increased its effectiveness in suppressing E2-stimulated gene transcription. The ERα F-domain has also been reported to inhibit receptor dimerization and interaction with the coregulator RIP140 (receptor interacting protein of 140 kDa) in the absence of E2 [136]. Furthermore, the F-domain of ERα was shown to be essential for E2-dependent activation of GC-rich (SP1 containing) promoters while dispensable in the presence of antiestrogens [137]. Thus, the F-domain of ERα seems to exhibit multiple modulatory functions, which affect the agonist/antagonist effectiveness of antiestrogens and the transcriptional activity of the liganded ER. The length, amino acid sequence and predicted secondary structure of the ERβ F-domain are quite different from the F-domain in ERα and its role in ERβ activity and function remains elusive.

5.4. LIGAND-INDUCED CONFORMATION AND SURFACE MORPHOLOGY: EFFECTS AND CONSEQUENCES

AF-2 is located in a conformational dynamic region in ER strongly affected by the presence of ligands in the binding pocket. Different classes of ER ligands induce unique surface conformations [138–143] influencing the orientation of H12 and, thus, the formation and shape of the coregulator protein interaction site [106, 107, 110, 119, 123, 109]. Agonist ligands cause H12 to orient itself across the LBD, sealing the ligand within the cavity, and complete the AF-2 core surface, thereby stabilizing receptor conformation optimal for efficient interaction with α-helical nuclear receptor (NR)-box motifs (LXXLL) in coactivator proteins ('agonist' conformation), which in turn facilitate transcription activation (Figure 5.4a; kindly provided by Dr Björn Kauppi, Karo Bio AB) [144–146, 109]. In contrast, antagonists, for example tamoxifen, raloxifene and ICI 164,384, do not promote positioning of H12 to complete the AF-2 coactivator recruitment site but instead orient H12 along the hydrophobic coactivator groove, foiling the AF-2 surface ('antagonist' conformation) (Figure 5.4b; kindly provided by Dr Björn Kauppi, Karo Bio AB) [106, 107, 110, 119, 147, 148, 123, 109]. In the H12 'antagonist' conformation other surfaces are created that are accessible for corepressors to interact with the receptor, thereby antagonizing agonist-dependent transcriptional activation [149–151, 118, 120, 85, 121, 152, 122, 153, 154].

Based on crystallographic information two mechanisms by which ligands can induce the 'antagonist' conformation have been reported: (i) active antagonism induced by antagonists with bulky side-chain extensions, which sterically prevent H12 to adopt the 'agonist' conformation [106, 107, 119, 123] and (ii) passive antagonism induced by ligands (without bulky side-chain substituents), which fit nicely within the cavity boundaries but fail to make the appropriate contacts with amino acids lining the ligand binding cavity, thereby destabilizing the 'agonist' orientation of H12 [155]. Genistein is an example of a ligand that induces destabilization of H12 'agonist' orientation in ERβ [110]. This destabilization is, however, more imaginary as H12 in the ERβ-genistein complex, co-crystallized with a coactivator fragment containing an LXXLL motif, is in a conformation similar to that observed for other ER-agonist complexes [156], and consistent with its almost full agonism in promoting gene transcription activation in cells [157].

Although the core amino acid sequence of the AF-2 coactivator docking surface of ERα and ERβ is conserved [124] there are differences in their recognition and recruitment of NR-box-containing proteins or peptides, in response to various ligands. By the use of fragments of coactivator proteins encoding their NR-boxes, single NR-box peptides, or artificial peptides, ERα and ERβ subtype-selective interactions were demonstrated [158, 128, 159, 160–163, 143]. It can be concluded from these studies that the interaction of coactivators with the ERs is multifactorial, determined by the structure of the agonist and by the NR-box sequence including its flanking sequences, and that additional ligand-induced receptor surfaces outside the AF-2 core required to interact with coactivators are not identical between ERα and ERβ.

Coactivators such as SRC-1 and GRIP-1 not only interact separately with AF-1 or AF-2, respectively, but they also promote association of AF-1 with AF-2 in a ligand-dependent fashion, integrating the N-terminal and C-terminal receptor functions. Both in the presence of agonist (E2) and antagonist (tamoxifen), respectively, coactivators enhance the AF-1 – AF-2 interaction, but in a cooperative fashion in the presence of agonist only. Lack of cooperativity in the presence of antagonist is likely due to disruption of the AF-2 core coactivator binding surface. [164, 165, 83, 166, 124, 167, 87]. ER ligands and their ability to modulate receptor conformation also affect ER binding kinetics to various ERE sequences on DNA, with consequences for target gene selection and transcription regulation [168–170].

5.5. GENE REGULATION BY ERα AND ERβ IN RESPONSE TO LIGANDS

Modulation of target gene expression by ERα and ERβ is multifactorial, including ligand binding, binding to promoter regions on DNA, and interaction with coregulators (chromatin remodelers, coactivators, and corepressors) [26, 41, 171–175, 42, 176]. Various mechanisms for transcriptional regulation of estrogen-sensitive genes have been described for both ERs (reviewed in [37, 177]). The classical mechanism is by binding to EREs on DNA, either as homodimers (ERα:ERα or ERβ:ERβ) or heterodimers (ERα:ERβ) [172, 173, 166]. Other reported mechanisms of ER-dependent modulation of gene transcription is binding to non-ERE response elements on DNA or through tethering onto other transcription factors bound to their specific sites on DNA.

An example of a non-classical DNA binding site for the ERs is the electrophile response element (EpRE) found in the promoter of the quinone reductase gene [178]. Both ERα and ERβ bind directly to the EpRE sequence and activate gene expression in the presence of antiestrogens but not E2, which actually antagonizes the effect of antiestrogens on this type of response element. Furthermore, ERβ is a more potent activator than ERα on an electrophile response element [179].

The SP (specificity protein) transcription factors are a family of structurally related proteins (SP1–SP8) that bind GC-rich response elements on DNA and that can activate gene expression in complex with the ERs in a ligand-dependent manner [180, 181]. It has previously been shown that both ER subtypes physically interact with the C-terminal domain of the SP1 transcription factor on GC-rich sites but that only ERα activated ER/SP1 driven reporter gene expression in an E2- and antiestrogen-dependent fashion [182, 181]. On the human RARα-1 promoter, however, ERβ activated gene expression in complex with SP1 in the presence of antiestrogens while E2 blocked this activity [180]. Furthermore, various structural classes of ER ligands, on the one hand, differentially activate ERα/SP-dependent gene activation from GC-rich promoters and, on the other hand, selectively use different SP proteins in complex with ERα (ERα/SP1, ERα/SP3, and ERα/SP4, respectively) to regulate gene expression [183].

The activator protein-1 (AP-1) is another example where the ERs may affect gene expression through tethering onto a transcription factor on DNA [184–186]. A broad spectrum of ER ligands, from agonists to antagonists, activate genes through an ERα/AP-1 complex on AP-1 response elements [187]. In contrast, gene expression mediated by ERβ/AP-1 on AP-1 sites is stimulated only by antagonists while agonists silence that response.

In the presence of E2, ERα and ERβ both interfere with cytokine gene expression by blocking the activity of the transcription factor NFκB (nuclear factor kappa B; heterodimer of p50 and p65) [188]. In addition, ERα but not ERβ is inhibitory also in the presence of antiestrogens [189, 190, 125, 191–194]. ERα inhibits NF-kB activity in the cytoplasm by affecting upstream NF-kB signalling steps as well as in the nucleus by directly affecting NF-κB DNA binding or transcriptional activation [191, 193]. ERα suppresses NFκB-dependent expression of MCP-1 (monocyte chemoattractant protein-1) and the cytokine IL-8 at the transcriptional level by displacement of CBP (cAMP response element-binding protein) from NFκB while the expression of IL-6 was inhibited by ERα by displacement of the p65 monomer of NFκB [195]. Deletion analysis of ERα has shown that inhibition of NF-kB activity requires the LBD, the DBD and the D region of ERα, but not the AF-1 domain. Point mutations in the LBD identified amino acids within the ERα major dimerization interface to be important for transrepression of NF-kB activity [125]. Furthermore, amino acids within helices that constitute the AF-2 of the ERs were also shown to be important for E2-dependent inhibition of NF-kB-mediated gene activation [125].

The relative importance of the two activation functions AF-1 and AF-2 for control of target gene transcription activity by the ERs is both cell-type and gene promoter-context specific. In HeLa cells the integrity of the ligand-dependent AF-2 seems to play a major role, in HepG2 and chicken embryo fibroblast cells the ligand-independent AF-1 was reported to have a dominant role, and in CHO K1 and HEK 293 cells both AF-1 and AF-2 seem equally important and necessary for synergistic and maximum transcriptional activity [196, 157, 124, 87]. Consequently, in cells where AF-2 is the major contributor to transcription activation, SERMs act more like pure antagonists whereas in cells where AF-1 plays the major role, SERMs may display more estrogen-like effects.

Estrogen-dependent target gene regulation by ERα: ERβ heterodimers was shown to require binding of E2 to only one of the partners in the heterodimeric complex; the AF2 core of both receptors must, however, be intact to allow stable interaction with a coactivator protein, for transcriptional activity to occur [166, 160, 197]. The relative level of ERα and ERβ in a cell appears to be an important determinant of the response to agonists and antagonists [173, 198]. Gradual increase in the ERβ/ERα ratio was shown to silence the ERα-mediated partial agonist response to tamoxifen and, moreover, the presence of ERβ potentiated the antagonism of tamoxifen on gene expression, apoptosis, and cell cycling [173, 198, 199]. Thus, there is a 'yin yang' relationship between ERα and ERβ; ERβ acting as a dominant modulator of ERα transcriptional activity in cells where ERα and ERβ are co-expressed. Furthermore, heterodimerization of ERα and ERβ may result in receptor activities distinct from that of homodimers, including differences in the selection of target genes, [200–203].

Cellular responses to ER ligands are restricted in their duration, allowing cells to adapt to changes in the microenvironment and to sense fluctuations in presence and levels of agonist/antagonist ligands. ER-mediated transcriptional activation of target genes is a sequential, coordinated cyclical process including binding of ligand-bound ER to target gene promoters, recruitment of coregulators, chromatin remodelling, recruitment and assembly of basal transcription machinery, initiation of transcription, RNA elongation and splicing, and termination [118, 204–206, 42, 207, 176]. As shown in studies with live cells, the presence of either agonists (E2) or antagonists (tamoxifen) causes redistribution of ERα from a diffuse nucleoplasmic appearance to nuclear matrix-bound foci within minutes, and that SRC-1 colocalizes with E2-bound ERα, but not with tamoxifen-bound ERα, in these foci [208, 209]. These studies also showed that preformed ERα coactivator complexes undergo very rapid molecular changes even in the continuous presence of agonist, suggesting that agonist-activated ERα and its interaction with coregulators and DNA is a highly dynamic process. The models described for ERα regulated transcription of target genes suggest that ERα, together with a number of coregulators, cycles on and off the promoters in response to E2 [118, 204] and that ubiquitination and proteasome-mediated degradation of the receptor is essential for the on/off cycling process, each round of productive transcription completed by proteasome-mediated clearance of ubiquinated ERα [210, 211, 206, 207]. Also methylation/demethylation of CpG islets within target gene promoters are integral components of ER-dependent gene regulation, as demonstrated for the pS2 gene in a breast cancer cell line [212, 213]

Recruitment and binding of ER to estrogen sensitive target genes often involves the FoxA1 factor and ERE elements at distances > 100 kb from putative transcription start sites [97, 101]. Initiation of estrogen-dependent target gene transcription is then achieved by physical contact between protein factors at the distal enhancer with protein factors at the TATA-box proximal promoter region, facilitated by chromosomal looping [214].

5.6. RAPID NON-GENOMIC AND MEMBRANE-MEDIATED MECHANISMS

Other proposed and potentially very important mechanisms of estrogen-mediated signalling involve indirect non-genomic functions of ER and/or the action of membrane-bound ERs or receptors of non-nuclear receptor origin [215–225]. Examples of rapid, non-transcription-dependent estrogen signalling are changes in intracellular calcium [226, 227], increases in cAMP [228, 229], modulation of calcium and potassium channels [230, 231], phosphorylation of CREB (cAMP response element binding protein) [232], and activation of MAPK (mitogen-activated protein kinase) [233, 234, 227, 235]. As proof for the existence of a membrane-bound ER, both ER-specific antibodies and a non-permeable E2-BSA complex have been used to exclude the involvement of the nuclear ER [236–239].

E2 exerts vaso- and atheroprotective properties through rapid stimulation of endothelial NO synthase and the production of nitric oxide (NO) [240–245]. This effect of E2 has been suggested to be mediated by membrane-associated ERα localized to caveolae, which leads to activation of $G_{\alpha i}$, and downstream kinase signalling pathways [242, 243, 245]. ER46, a naturally occurring splice variant of ERα is efficiently targeted to caveolae in the membrane of endothelial cells and was found to be more efficient than the full-length wild-type ERα in mediating E2-stimulated eNOS activation [152, 245]. Furthermore, palmitoylation of Cys^{447} in the ligand binding domain of ERα seems to target the receptor to the plasma membrane and may be involved in E2-induced rapid activation of the MAPK/ERK signalling pathway [246, 152, 247, 245, 248]. Also ERβ has been associated with nongenomic pathways enhancing nitric oxide release in human endothelial cells in response to the β_1 adrenergic receptor blocker nebivolol [249].

SERMs like raloxifene, tamoxifen and acolbifene (EM-652), but not the pure antagonist ICI 182,780, stimulate rapid production of nitric oxide in endothelial cells, suggesting that SERMs, acting directly on endothelial cells in the vessel wall by non-genomic mechanisms, may have cardio-protective effects, [250–252].

5.7. PHYSIOLOGICAL IMPORTANCES OF ESTROGENS

The final step in the biosynthesis of estrogens is catalyzed by the cytochrome P-450 aromatase enzyme ($P450_{AROM}$). The enzyme is located in the endoplasmic reticulum of estrogen-producing cells and is expressed in a variety of tissues such as ovaries (granulosa cells), placenta, testis (Sertoli and Leydig cells), adipose tissue, liver, bone, breast, muscle (including vascular smooth muscle), endothelial cells, and brain [253–257]. Until the early 1990s it was a general perception that E2 is essential only for women's development and health and of no significant importance for men. However, clinical cases of estrogen resistance in a man diagnosed with a non-functional ERα or male patients with aromatase deficiency have dramatically and convincingly demonstrated the paramount importance of estrogens in also human male physiology [258, 259].

5.7.1. Breast Tissue

The importance of estrogens in the development of female breast tissue is well documented [260–264]. Aromatase deficient females, unable to convert C_{19} steroids to estrogens, show no signs of breast development at the onset of puberty. However, estrogen replacement therapy results in normal pre- and postpubertal breast development [265–267, 51].

Female ERα knock-out (ERKO) mice have lost their capacity to develop mammary gland tissue beyond the embryonic and fetal stages despite elevated levels of circulating E2 [268, 46]. In contrast, female ERβ knock-out (BERKO) mice have normal breast development and lactate normally [269, 270, 46]. Studies in a combined aromatase transgenic and ERα-deficient mouse model (aromatase/ERKO) confirmed the important role of ERα in the development of the mammary gland but also its role in inducing mammary hyperplasia in response to estradiol [271]. In contrast, by the use of an ERβ-selective agonist it could be concluded that ERβ is non-mammotrophic [272]. Furthermore, treatment of HER-2/neu x aromatase double transgenic mice with THC (tetrahydrochrysene), a combined ERα agonist/ERβ antagonist, lead to increased expression of ERα, complete downregulation of ERβ, increased mammary proliferation, and increased incidence of tumorigenesis, strengthening the different and opposing roles of ERα and ERβ in mammary tissue [273].

5.7.2. Urogenital Tract

ERα and ERβ are both expressed in uterus, ovary, testes, and prostate, but with different cellular localization. In ovary, ERβ is primarily expressed in the granulosa cells while ERα is mainly found in thecal cells [46]. In prostate ERα resides primarily in the stromal compartment and ERβ mainly in the glandular epithelium [46, 274, 275]. In the urinary bladder and urethra ERβ is the dominant estrogen receptor [276, 44, 45, 277].

Absence of ERα expression results in infertility in both male and female mice; in females due to hypoplastic, estrogen resistant uteri and cystic ovarian phenotype with no ovulatory capacity, and in males primarily caused by disturbed fluid reabsorption in the efferent ductules [268, 278–280]. Fertility in female mice deficient in ERβ expression is compromised but not absent, explained by reduced ovulatory capacity due to an attenuated response to FSH-mediated granulosa cell differentiation and the subsequent LH surge, required for follicle rupture and ovulation [269, 279, 46, 281]. Another plausible explanation for this subfertile phenotype is failure in the communication between the granulosa and theca cell layers of the ovaries and by insufficient blood supply to the follicles, impairing their maturation, an explanation supported by restoration of the fertility to 100% following surgical ovarian wedge resection [282]. Male BERKO mice show no overt abnormalities and reproduce normally [269, 279].

Treatment of hypophysectomised or gonadotropin-releasing hormone antagonist-treated female rats with an ERβ-selective agonist resulted in stimulation of early folliculogenesis, a decrease in follicular atresia, and stimulation of late follicular growth, followed by an increase in the number of ovulated oocytes [283]. Treatment with an ERα-selective agonist had, however, no effects on follicular development and ovulation [283]. In contrast, ERα-selective but not ERβ-selective agonist treatment, stimulate uterine growth and weight, confirming findings in ER knock-out mice [284, 272, 285, 283, 115, 286].

Estrogens play an important physiological role in prostate development, and immunohistochemical studies revealed that ERβ is the predominant ER subtype in prostate, located in the epithelial cells along the ductal network of the prostate while ERα has been localized primarily to the stromal compartment of the prostate [287–290, 275]. In ERβ–/– mice the level of androgen receptor (AR) in the prostate is elevated. Exposure of wild-type mice to 5α-androstane-3β, 17β-diol, an ERβ-selective estrogenic metabolite of dihydrotestosterone, caused a decrease in the level of AR in the prostate [289], suggesting that the AR-gene is an ERβ target in the prostate. ERβ deficient mice display signs of prostatic hyperplasia with aging [289], suggesting that ERβ may protect against abnormal prostate growth.

Aromatase deficient female patients were reported with ambiguous genitalia at birth, a phenotype that was further pronounced at puberty, and with polycystic ovaries, characterized by a disproportionate number of atretic follicles. Replacement with estrogen in affected female patients led to normalized gonadotropin and androgen levels, resolution of the ovarian cysts, and menarche [266, 267, 51]. Similar to female patients, aromatase knock-out (ArKO) female mice have low serum estrogen and high testosterone and gonadotropin levels [291]. Female ArKO mice also display genital anomalies with underdeveloped external genitalia and uteri and with ovaries containing numerous follicles arrested before ovulation. No corpora lutea were present [291].

Aromatase activity for production of estrogen in the testis has been extensively characterized [292]. Defective estrogen production or estrogen insensitivity was reported to cause macroorchidism or oligozoospermia and/or decreased sperm viability or motility in male patients diagnosed with aromatase or ERα deficiency [258, 267, 293, 259, 51]. Male ArKO mice were initially fertile but developed progressive infertility due to arrested spermatogenesis [294]. Thus estrogen has a crucial role in male germ cell development and for male fertility [295].

5.7.3. Skeletal Homeostasis

The importance of estrogen for women's bone physiology is evident clinically from the occurrence of osteoporosis in postmenopausal women [296], and from the evidence that estrogens protect postmenopausal women from bone loss and the development of osteoporosis [297, 298, 296, 299, 67]. Estrogens may also play a more important role than testosterone in maintenance of bone mass in ageing men. This conclusion comes from the positive correlation between bone mineral density (BMD) and plasma estradiol concentrations rather than to testosterone levels in aging men [300–306].

In female but not in male ERKO mice there was a significant decrease in the length and size of the femur while decreased BMD and bone mineral content (BMC) was more pronounced in ERKO males than in females [268]. In male BERKO mice the bone phenotype was unaffected while there was a masculinization of the long bones (femur) in the females [307]. Lack of ERβ expression in the female mice led to increased length of the femur, thicker cortical bone, and increased size of the vertebrae, approaching the corresponding characteristics of wild-type male mice. Trabecular architecture and BMD were, however, unaffected in the male and female BERKO mice. Following ovariectomy a loss in trabecular BMD in wild-type and BERKO female mice was observed suggesting an important role for ERα in the maintenance of trabecular BMD and architecture [307]. The decrease in length and size of the femur in female ERKO mice occurs despite plasma levels of E2 10-fold higher than normal. The increased femur length in female BERKO mice and the decreased length of the femur in female ERKO mice may be indicative of an important role for ERβ in growth plate fusion and bone length in females [308, 309]. Furthermore, female BERKO mice are partially protected against age-related trabecular bone loss compared to wild-type female mice, and this is thought to be in part, an effect of increased ERα expression and increased osteoblast differentiation and activity [310]. Additional support for an important role of ERα in bone homeostasis was obtained by the protection from ovariectomy-induced loss of bone mineral density in the rat by treatment with ERα-selective agonists [284, 115, 286].

Estrogen deficiency is known to result in a marked increase in bone resorption by the osteoclast, which is not compensated for by osteoblast-mediated bone formation. Acute changes in estrogen levels have a direct effect on immune functions, cytokine levels, the balance of the RANK/RANKL/OPG system, and osteoblast and osteoclast functions [311, 312]. Selective ablation of ERα expression in osteoclasts in female mice suggested that the osteoprotective effect of estrogen is mediated by ERα in the osteoclast; osteoclast ERα controls the life span of the osteoclast in response to estrogens and SERMs by upregulation of FAS ligand in the osteoclast that in turn leads to programmed cell death of the osteoclast [313]. Another study on the bone protective role of estrogen suggested a paracrine mechanism; estrogen and SERM activated ERα affects osteoclast survival through upregulation of FAS ligand in the osteoblast (not the osteoclast) that subsequently leads to apoptosis of pre-osteoclasts [314]. Osteoclast formation and the production of osteoclastogenic cytokines were studied on peripheral blood mononuclear cells and T-cells isolated from pre- and postmenopausal women, diagnosed with or without osteoporosis [315]. The result of this study confirms that estrogen deficiency increases the number of osteoclast precursor cells, stimulates osteoclast formation and activity, and increases the level of the pro-osteoclastogenic cytokines, TNFα and RANKL. Furthermore, it was reported that T-cells and monocytes from women with osteoporosis produced more TNFα than the healthy women involved in the study and that the activity of T-cells in the osteoporotic women was higher than in the healthy postmenopausal women [315].

Support for the importance of ERα in also male bone physiology was obtained from examination of the ERα deficient male patient [258]. Despite elevated levels of estrogens, he suffered from low BMD and continuous linear growth because of unfused epiphyses. Whether ERβ has a similar role in human bone development and homeostasis as it does in mice needs further investigation [309].

Male and female patients with aromatase deficiency have delayed bone maturation, increased bone turnover, low BMD, and tall stature due to unfused epiphyses [266, 316, 267, 317–319, 259, 51, 306]. Estrogen replacement therapy of both female and male aromatase deficient patients resulted in growth spurt and bone maturation and cessation of linear growth with concomitant closure of the epiphyses, reduced bone turn-over, and increased bone mineral density.

5.7.4. Metabolic Effects

The role of estrogen and estrogen receptors in the control of energy and glucose homeostasis has been investigated in detail [320–323]. Decline in estrogen levels as a natural consequence of aging may be associated with dysregulation of lipid and glucose metabolism [324]. In postmenopausal women, estrogen replacement improves the serum lipid/cholesterol profile, increases the level of apolipoprotein A1, and lowers plasma levels of Lp(a) [325–328]. Estrogen also has beneficial effects on carbohydrate metabolism including insulin sensitivity [329–335].

Studies in female ob/ob mice showed that long-term E2 treatment improved glucose tolerance and insulin sensitivity [336].

An additional aspect of and support for the importance of E2 in metabolic control in also males was reported from characterization of PPARα knock-out mice [337]. Male PPARα-/- mice develop massive hepatic and cardiac lipid accumulation and die from hypoglycemia when metabolically stressed by etomoxir inhibition of carnitine palmitoyltransferase I. E2 treatment not only rescued male PPARα–/– mice from death due to stress-induced hypoglycemia but also significantly decreased hepatic and cardiac lipid accumulation [337].

Both female and male ERα knock-out mice are diabetic [338–340], with profound hepatic insulin resistance [340] and radically reduced cell membrane levels of GLUT4 in skeletal muscle [341]. Ovariectomy of ERα–/– mice lowered the blood glucose and insulin levels compared to sham-ovx ERKO mice and E2 administration reversed this effect [342], suggesting that ERβ may be pro-diabetogenic. In support of this notion, ERβ deficient mice exhibit improved insulin sensitivity and glucose tolerance [342, 341, 343, 344]. Barros et al. [341, 343] have proposed that the pro-diabetogenic effect of ERβ is due to suppressive effect of ERβ on muscle GLUT4 expression while. Foryst-Ludwig et al. [344] have suggested a blocking effect by ERβ of PPARγ transcriptional activity in adipocytes. Both male and female ERKO mice are obese, with adipocyte hyperplasia and hypertrophy and elevated levels of leptin [338, 345, 339]. Ovariectomy of female ERKO mice resulted in normalization of body weight that was counteracted by the administration of E2 [342], suggesting that ERβ has an adipogenic effect. BERKO mice did not show an increase in total body fat but perhaps a tendency for a leaner body composition than wild-type mice [345]. However, under high fat diet conditions also BERKO mice showed increased body weight and gain in fat mass, explained by unopposed PPARγ activity in adipocytes [344].

Besides direct effects of ERα in peripheral target organs the CNS plays an important role in the regulation of metabolic homeostasis. Ablation of ERα in the ventromedial nucleus in the brain of female mice and rats resulted in the development of obesity, hyperphagia, impaired glucose tolerance, and reduced energy expenditure, a phenotype characteristic for the metabolic syndrome [346].

In further support for the role of ERα in metabolic control, chronic treatment of ovariectomized rats with the ERα-selective agonists, PPT (propyl pyrazole triol) or 16α-LE_2, resulted in 50% reduction of total plasma cholesterol levels and completely prevented the weight gain observed in ovariectomized vehicle control animals [284, 286].

As in ERα-/- mice, the male ERα-negative patient was characterized by hyperinsulinemia, impaired glucose clearance, elevated glycosylated haemoglobin, and axillary acanthosis nigricans [258, 347]. Whether ERβ contributes to his diabetogenic phenotype is not known.

Male and female ArKO mice are obese with perturbed expression of fatty acid-metabolizing enzymes, sexually dimorphic hepatic steatosis, elevated fasting glucose levels, impaired glucose clearance, hyperinsulinemia, and insulin resistance [348–350, 51]. Obesity in these mice develops over time and becomes apparent at the age of 10–12 weeks while glucose intolerance does not become evident until the age of 18 weeks [351, 352]. Estrogen replacement therapy increased the rate of glucose clearance to that of wild type mice while placebo treated ArKO mice remained diabetic.

Long term treatment with E2 initiated at birth of the ArKO mice and continued till the age of 36 weeks prevented the development of obesity and type 2 diabetes in these mice [349, 353]. Interestingly, the effect of E2 replacement on glucose tolerance and insulin sensitivity was similar to or better than the effect of the PPARα agonist bezafibrate and the PPARγ agonist pioglitazone, respectively, in ArKO mice [352]. In agreement with findings in ERKO and BERKO mice [342, 341, 26, 344], treatment of ArKO mice with the ERβ-selective agonist DPN (diarylpropionitrile) markedly suppressed the expression of GLUT4 and its presence in the plasma membrane of skeletal muscle cells [341, 26].

Gender differences in the effects of estrogen on target organs such as liver, adipocytes, and skeletal muscle, and on perturbed metabolic control and sexually dimorphic hepatic steatosis in male ArKO mice, are also influenced by differences in the signalling in the CNS. Aging male but not female ArKO mice were reported with significant loss of dopaminergic neurons in the arcuate nucleus, an area of great importance for ERα-mediated estrogen regulation of energy uptake, storage, and mobilisation [350, 354–356].

Male and female aromatase deficient patients suffer from impaired carbohydrate metabolism, hyperinsulinemia, and dyslipidemia, abnormalities that were reversed towards normal levels by low-dose oral or transdermal estrogen replacement therapy [266, 347, 317, 318, 357, 51, 356, 358].

In conclusion, the reported phenotypic observations of ERKO, BERKO, ArKO, and aromatase or ERα deficient patients, as well as the use of ER subtype-selective ligands, suggests a profound role of ERα in the maintenance of metabolic control while ERβ may have an opposite effect, especially when unopposed, i.e. in the absence of agonist-bound ERα, resulting in a perturbed energy balance and increased risk for the development of diabetes, particularly in a situation of high-energy/fat diet and low physical activity.

5.7.5. The Cardiovascular System

Both estrogen receptors, ERα and ERβ, are expressed in the cardiovascular system and mediate the effects of estrogen by both rapid non-genomic and longer-term genomic pathways [359–363]. Examples of non-genomic effects are vasodilatation as a consequence of estrogen regulation of ion-channel function [364] and nitric oxide (NO) synthesis [365]. Examples of long-term genomic effects are modulation of e.g. prostaglandin synthase, NO synthase and endothelin gene expression [360, 366], regulation of AT1 receptor density on vascular smooth muscle cells [367], and inhibition of injury induced vascular intimal thickening [368, 369].

Women's risk to develop cardiovascular disease has been correlated to their levels of endogenous or supplemented estrogens [362, 370]. That the level of estrogen is important for the function and maintenance of men's cardiovascular system has recently become evident [371]. A male aromatase-deficient patient, diagnosed with precocious alteration of coronary morphology and function, was treated with E2, which resulted in disappearance of carotid atherosclerosis [372]. Inhibition of the conversion of testosterone to estrogen by the aromatase inhibitor anastrozole impaired normal vascular relaxation in healthy young men [373] and male ArKO mice showed impaired vasorelaxation in response to endothelium-dependent acetylcholine-induced release of NO [374]. Castrated male mice develop aortic fatty streak lesions that could be attenuated by administration of testosterone or E2. Simultaneous treatment with testosterone and the aromatase inhibitor anastrozole blunted the effect of testosterone [375]. Thus, data from the various studies support an important role for testosterone aromatization and E2 in cardiovascular function and atherogenic protection in males.

Sex steroid hormones and their cognate receptors are critical determinants of cardiovascular gender differences [370, 376]. Women's risk to develop cardiovascular disease is lower than for men before mid life but rises sharply after menopausal transition, a consequence attributed to the loss of female sex steroid hormones at the time of menopause. Gender differences also exist in cardiac electrophysiological function [377] and in blood pressure, which is higher in men than in women at similar ages [378]. It has also been reported that men have a higher prevalence of left ventricular hypertrophy than women [379, 380, 359], hypothesized to be an effect of lower levels of circulating estrogens in men than in women [380] but possibly also to gender-specific differences in estrogen receptor levels and in the regulation of endogenous gene expression in cardiac myocytes in response to estrogens [359, 360]. Other reported difference is the rapid response to estrogen following acetylcholine-induced coronary arterial constriction in men and women with coronary artery disease [381]. In the female patients, administration of estrogen attenuated the acetylcholine induced constriction while there was no response to estrogen in the male patients. Furthermore, coronary blood flow was significantly enhanced in the presence of estrogen in the female but not in the male patients. A plausible explanation for these gender differences is that the vascular endothelium produces more nitric oxide in response to estrogen in women than in men [381].

The role of ERα and ERβ, respectively, in mediating the effects of estrogen on the function and integrity of the cardiovascular system has been studied in various animal models including mice in which the ERs have been disrupted [382]. ERα but not ERβ was reported to accelerate re-endothelialization following carotid artery injury in response to E2 and to inhibit smooth muscle cell proliferation and matrix deposition [383, 384]. Treatment of ovariectomized, spontaneously hypertensive rats with E2 or the ERα-selective 16α-LE2 agonist attenuated cardiac hypertrophy, increased cardiac output, and left ventricular stroke volume, implicating a favourable effect of ERα on cardiac hypertrophy and myocardial contractility [385]. However, also ERβ has been reported to mediate E2-dependent inhibition of cardiac hypertrophy and fibrosis, induced by angiotensin II treatment [386]. ERα is the primary ER subtype responsible for mediating the well documented atheroprotective effects of estrogens [387, 388], as demonstrated in various animal models of atherosclerosis [389]. In female mice deficient in low density lipoprotein receptor, the ERα-selective agonist PPT but not the ERβ-selective agonist WAY-200070 protected against the development of atherosclerosis by activation of cyclooxygenase 2 (COX-2) and increased production of the prostacyclin PGI_2 [390]. ERβ, on the other hand, was crucial for maintenance of normal vasodilatation and blood pressure in both male and female mice [391], and for cardiac protection following ischemia/reperfusion injury in female but not male mice [392, 393].

Support for an important role of ERα in the function of the cardiovascular system in humans comes from characterization of the male ERα-deficient patient [258], who was reported with intimal thickening of the common carotid arteries despite elevated circulating levels of estrogen and with an absence of endothelium-dependent vasodilatation of the carotid arteries following ischaemic cuff occlusion [394, 395].

5.7.6. The Central Nervous System and the Hypothalamo-Pituitary Axis

Estrogens are involved in the development of the central nervous system (CNS) and in modulation of a variety of its functions such as learning, memory, cognition, fine motor skills, temperature regulation, mood, and reproductive functions [396–399]. Various brain structures and neurotransmitter systems are involved in the effects of estrogens: the serotonergic, dopaminergic, catecholaminergic, cholinergic, and γ-aminobutyric acidergic systems [400–402, 396, 397].

Both ERα and ERβ are found in various regions of the human and rodent brain, including the hypothalamus, hippocampus, cerebral cortex, midbrain, brainstem, and forebrain, with differential presence of the two ER subtypes in certain areas of the brain but also areas where they seem to overlap [403, 404, 405, 46, 52, 399, 406].

Besides CNS characterization of ERα and ERβ knock-out mice [268, 46, 52], the use of ER subtype-selective ligands help to unravel the specific functions of ERα and ERβ in the brain [407]. The ERα-selective agonist, PPT, increased progesterone receptor mRNA in arcuate and ventromedial nuclei of the hypothalamus, enhanced sexual receptivity in ovariectomized rats, and prevented the increase in tail skin temperature in the morphine addicted rat hot flush model [284, 408]. In the ovariectomized rat hot flush model both PPT and the ERβ-selective agonist, DPN, reversed the increase in tail skin temperature [409]. PPT but not DPN indirectly promoted sprouting of dorsal root ganglion nociceptors, suggesting that ERα may have implications for female pain syndromes [410]. E2 was shown to have an antidepressant effect in various strains of ovariectomized mice but lack of effect in ERβ deficient (BERKO) mice, suggesting a specific role of ERβ in mood and behaviour [411]. The ERβ-selective agonists, WAY-200070 and DPN, showed efficacy in animal models of anxiety and depression, having effects on the dopamine and the serotonergic systems while the ERα-selective agonist, PPT, was without effect [412, 413, 408, 414–416]. Cognitive processes are influenced by E2, and in ovariectomized wild-type but not BERKO mice, E2 and DPN showed improved recognition task performance, implicating a more important role of ERβ than ERα in cognitive processes [417]. In mice hippocampal synaptic plasticity and hippocampus-dependent memory tasks were reported to be specifically modulated by activation of ERβ in the presence of DPN while activation of ERα by PPT was ineffective [418]. Cerebellar plasticity and motor learning was promoted by ERβ in response to E2 [419]. In wild-type and ArKO mice DPN protected from acoustic trauma and increased the expression of the neuroprotective peptide BDNF (brain-derived neurotrophic factor), implicating important neuroprotective role for ERβ in the auditory system [420]. Treatment of the male estrogen deficient ArKO mice with an ERα- and an ERβ-selective agonist, respectively, protected against dopaminergic neuronal cell death; the ERα-selective agonist specifically having anti-apoptotic efficacy in the arcuate nucleus (involved in hormone and metabolic regulation) and the ERβ-selective agonist specifically in the medial preoptic area (involved in modulation of behaviour) [355].

The hypothalamo-pituitary axis (HPA) regulates overall endocrine homeostasis in the body. Estrogens, through effects on the HPA, modulate the expression and secretion of several hormones from the anterior pituitary gland, such as LH, FSH, GH and PRL [268]. Although serum levels of LH and FSH are directly controlled by hypothalamic gonadotropin-releasing hormone (GnRH) neurons, it is the circulating plasma levels of estrogen, other sex steroids, and the inhibin glycoproteins that are the most important physiological determinants of serum gonadotropin levels [268, 421, 422]. From studies in ER subtype and ERα neuron-specific knock-out mice and by the use of ER subtype-selective agonists and antagonists the conclusion was that ERα is responsible for mediating the positive feed-back of estrogen to generate the pre-ovulatory gonadotropin surge but that ERβ may act as a negative modifier [423–426].

As a result of estrogen deficiency, male but not wild-type or female ArKO mice displayed a progressive decrease in dopaminergic neurons in the medial preoptic area due to apoptosis, explaining the disrupted sexual behaviour of male ArKO mice that became severe upon aging [354, 355]. The number of dopaminergic neurons in the arcuate nucleus of male ArKO mice was also significantly decreased, probably explaining the elevated levels of prolactin and gonadotropins observed in these mice [354, 355].

Female and male patients with aromatase deficiency are reported to have elevated levels of LH and FSH, elevated circulating plasma levels of androgens but very low levels of estradiol and estrone [427, 267, 259, 51]. Therapy with conjugated estrogens in both male and female aromatase deficient patients resulted in normalization of gonadotropin and testosterone levels. Clinical data on the male patient with the ERα nonsense mutation [258, 293] also showed increased serum LH and FSH levels despite normal levels of testosterone and high estrogen levels. As expected, transdermal ethinyl-estradiol therapy of this man did not have any effect on lowering of serum LH or FSH levels, suggesting an important function of ERα for proper regulation of gonadotropin levels also in men.

5.8. WOMEN'S HEALTH INITIATIVE AND BEYOND

The WHI (Women's Health Initiative) study on women at the age of 50–79 years (average 63 years), arrived at the conclusion that the overall health risks exceeded the benefits of HRT and that the use of conjugated equine estrogens (CEE) with or without a gestagen combination should not be recommended for chronic disease prevention in postmenopausal women, including cardiovascular disease [428, 429]. However, this has created debate and controversies that following an extensive review of non-clinical, human observational, and human clinical trial data have led to a recommendation to start HRT earlier than in women > 60 years of age to avoid serious side effects and health risks [430]. Ideally estrogen therapy with or without progestagens should start already at the menopausal transition to avoid damages to the cardiovascular and other organ systems that otherwise may develop and that cannot be repaired by hormone therapy initiated several years after menopause [431, 432, 430, 188].

Our view of estrogen signalling has undergone a paradigm shift over the recent 10–15 years and in the near future optional treatment strategies to generalized HRT will emerge. Today selective therapies, targeting specific symptoms and health issues with ER subtype-selective drugs or optimized balanced SERMs with improved safety profile, are becoming a reality [113, 284, 272, 433–437, 413, 52, 438–440, 441, 412, 62].

5.9. ESTROGEN RECEPTORS AND CANCER

There are numerous publications suggesting an important role of the ERs in various forms of cancer: adrenocortical carcinoma [442], bladder cancer [443], breast cancer [59, 444, 445, 60, 446], cholangiocarcinoma [447–449], chronic myeloid leukemia [450], colon cancer [451–454], lung cancer [455, 456], medulloblastoma [457], melanoma [458], multiple myeloma [459, 460], ovarian cancer [461], prostate cancer [289, 462–464], and rhabdomyosarcoma [465].

5.9.1. Prostate Cancer

Both androgens and estrogens are critical for normal prostate growth and development [466, 467, 290, 468, 274, 275]. Studies on ERβ knock-out mice showed increased epithelial proliferation, decreased apoptosis, and accumulation of incompletely differentiated cells that eventually lead to prostatic epithelial hyperplasia [462]. The metabolite of 5α-dihydrotestosterone, 5α-androstane-3β, 17β-diol, is a ligand of ERb in the prostate and inhibits proliferation and migration of prostate cancer cells [469, 470, 220], and an ERβ- but not an ERα-selective agonist reverses the hyperplasia seen in ArKO mouse prostate [464]. In another hyperplastic prostate model, luteinizing hormone receptor knock-out (LuRKO) mice, testosterone but not 5α-dihydrotestosterone induced hyperplasia and inflammation that was further pronounced by aromatase inhibition and treatment with the pure ER antagonist ICI 182,780. The ERβ-selective agonist, DPN, however, prevented inflammation and the development of hyperplasia in the testosterone–treated LuRKO mice [471]. SERBA-1, a novel ERβ-selective agonist, was reported to have desired effects in a model of benign prostatic hyperplasia without feed-back effects on testosterone and dihydrotestosterone levels [472].

In many types of cancers, including prostate cancer, the expression of ERβ decreases over time due to epigenetic silencing [473–476]. Co-treatment of prostate cancer cells with inhibitors of DNA methyltransferase and histone deacetylase, respectively, resulted in reduced cell proliferation, increased cell death by apoptosis and a marked re-expression of ERβ [476]. Treatment of LNCap cells with the histone deacetylase inhibitor valproic acid (VPA) and the phytoestrogen tectorigenin, respectively, caused a significant increase in ERβ expression, accompanied by an antiproliferative effect. In contrast, silencing of ERβ expression resulted in resumed LNCap proliferation including upregulation of genes involved in cell cycle regulation [475].

The response to various balanced and ER subtype-selective ligands on cell growth/viability and expression of the chromosomal translocation gene product TMPRSS2-ERG (TMPRSS2-ERG containing prostate cancers are more aggressive than other forms of prostate cancer), was assessed in the androgen receptor negative but ERα and ERβ expressing NCI-H660 prostate cancer cell line [477]. Both E2 and the ERβ-selective agonist, DPN, decreased the growth and the expression of the TMPRSS2-ERG fusion gene product in these cells while the ERα-selective agonist, PPT, resulted in stimulation of cell growth and TMPRSS2-ERG expression, as compared to vehicle control cells [477].

In summary, a plethora of scientific data points to ERβ-selective agonists as a new therapeutic strategy against prostate cancer.

5.9.2. Breast Cancer

ERβ has emerged as an important and potential prognostic marker for predicting response to endocrine breast cancer therapy [478] but similar to prostate cancer, loss of ERβ expression is a common step in tumor progression [479–481].

Lack of proliferative response of the normal, mouse mammary epithelial cell line HC11 to E2 was explained as an integrated result of E2 acting on ERα and ERβ, respectively, where the ERα-mediated cell growth was balanced out by the apoptotic effect mediated by ERβ [482]. However, upon selective knock-down of ERβ, E2 stimulated cell proliferation and rendered the cells capable to grow in soft agar. In contrast, selective knock-down of ERα resulted in E2 dependent cell death by ERβ stimulation of apoptotic pathways. Furthermore, following treatment of the HC11 cells with the ERα-selective agonist, PPT, only a proliferative response was observed while treatment with the ERβ-selective agonist, DPN, resulted in a decrease in cell number due to apoptosis [482], thus, confirming previous reports of the respective roles of ERα and ERβ in mammary tissue [272, 271, 273].

Expression of ERβ in the T47D and MCF-7 breast cancer cell lines, respectively, was reported to inhibit E2-dependent cell growth and to potentiate the anti-proliferative and apoptotic effect of tamoxifen [445, 199]. A xenograft model of T47D cells showed that ERβ expression in T47D cells resulted in reduced tumor growth and a reduction in the number of intratumoral blood vessels as compared to non-ERβ expressing T47D cells [483]. Treatment of xenograft models of ERα and ERβ expressing MCF-7 cells with the ERβ-selective herbal extract MF101 and the phytoestrogen liquiritigenin, respectively, did not stimulate tumor formation while large tumors were developed in animals treated with DES (diethylstilbestrol) and E2, respectively [439, 446].

In conclusion, like in prostate cancer, ERb-selective agonists may be a promising treatment strategy for breast cancer but this can only be established through appropriate clinical trials.

5.10. DEVELOPMENT OF ESTROGEN RECEPTOR SUBTYPE-SELECTIVE LIGANDS

Ligands with ER subtype-selective affinity and biological activity have been reported [113, 284, 272, 285, 390, 283, 115, 413, 482, 385, 341, 472, 355, 464, 471, 414, 410, 286, 412, 418, 420, 407, 426, 477]. A herbal extract with roughly similar affinity to both ERα and ERβ but with ERβ-selective transcriptional and biological activity has been described [439, 446]. The ERβ-selective transcriptional activity of this herbal extract was explained by the specific formation of ERβ-coactivator complexes, which could not be formed with herbal extract-bound ERα. Thus, subtype-selective receptor affinity is not the only determinant to achieve ERα- or ERβ-selective transcriptional and biological effects.

In summary, there are at least two major options for the design and development of ERα- and ERβ-selective ligands: (i) explore the subtle differences between the ERα and ERβ ligand binding cavities for development of ligands with high ER subtype-selective affinity, and (ii) develop subtype-selective ligands based on their ability to selectively modulate receptor conformation and recruitment of coregulator proteins [484, 439, 446].

5.11. CONCLUDING REMARKS

Today we have to consider the existence of two estrogen receptors, ERα and ERβ, in our understanding of the action of natural and synthetic estrogens and SERMs. Studies on knock-out mice, ERα–/–, ERβ–/–, and aromatase–/–, and patients diagnosed with aromatase deficiency or absence of ERα expression have led to awareness of the physiological importance of estrogens also in males and of novel targets for therapeutic interventions in both women and men [258, 436, 437, 259, 51].

Our knowledge of mechanisms involved in the regulation of estrogen-sensitive genes has increased tremendously. Target gene responses to agonists and antagonists are restricted in their duration by a cyclic on/off process of ER on promoters, thereby allowing the cell to sense and adapt to fluctuations in hormone levels [118, 204, 206, 213]. Furthermore, the relative expression levels of coactivators and corepressors as well as the ratio of ERα to ERβ expressed in cells and tissues determine the response to agonists and antagonists [173, 485, 482, 199].

Selective therapies, targeting specific symptoms and health issues, with ERα- and ERβ-selective drugs, respectively, or with improved, balanced or ER subtype-selective SERMs are becoming a reality [413, 417, 433–435, 438, 439, 440, 412, 62]. However, therapeutic intervention with ERα-selective agonists may lead to increased serious risks such as, but not limited to, venous thromboembolism and the development of breast and endometrial cancer [284, 285, 283, 115, 271, 286, 441] while treatment with an ERβ-selective agonist may increase the risk for the development of type 2 diabetes [342, 341, 26, 344]. However, to get a more complete understanding of the potential benefits and health risks with ER subtype-selective agonist therapy, additional studies in animal models are needed. For example, how much increased is the risk for the development of type 2 diabetes with an ERβ-selective agonist in the presence of also ERα and at levels of 17β-estradiol normally found in postmenopausal women? Eventually, only appropriate clinical trials can show the benefits and risks of ER subtype-selective therapies.

REFERENCES

1. Beatson, G. T. (1896). On the treatment of inoperable cases of carcinoma of the mamma: Suggestions for a new method of treatment with illustrative cases. *Lancet* 2, 104–107.
2. Boyd, S. (1900). On oophorectomy in cancer of the breast. *BMJ* 2, 1161–1167.
3. Jensen, E. V. and Jacobsen, H. I. (1962). Basic guides to the mechanism of estrogen action. *Rec Prog Horm Res* 18, 387–414.
4. Jensen, E. V. (2005). The contribution of "alternative approaches" to understanding steroid hormone action. *Mol Endocrinol* 19, 1439–1442.
5. Toft, D. O. and Gorski, J. (1966). A receptor molecule for estrogens: Isolation from the rat uterus and preliminary characterization. *Proc Natl Acad Sci USA* 5, 1574–1581.
6. Jensen, E. V., Sujuki, T., Kawashima, T., Stumpf, W. E., Jungblut, P. W., and DeSombre, E. R. (1968). A two-step mechanism for the interaction of estradiol with rat uterus. *Proc Natl Acad Sci USA* 59, 632–638.
7. Jensen, E. V. and DeSombre, E. R. (1973). Estrogen-receptor interaction. Estrogenic hormones effect transformation of specific receptor proteins to a biochemically functional form. *Science* 182, 126–134.
8. O'Malley, B. W., McGuire, W. L., and Middleton, P. A. (1968). Altered gene expression during differentiation: Population changes in hybridisable RNA after stimulation of the chick oviduct with estrogen. *Nature* 218, 1249–1251.
9. Chamness, G. C. and McGuire, W. L. (1972). Estrogen receptor in the rat uterus. Physiological forms and artefacts. *Biochemistry* 11, 2466–2472.
10. Eckert, R. L., Mullick, A., Rorke, E. A., and Katzenellenbogen, B. S. (1984). Estrogen receptor synthesis and turnover in MCF-7 breast cancer cells measured by a density shift technique. *Endocrinology* 114, 629–637.
11. Gorski, J., Welshons, W., and Sakai, D. (1984). Remodeling the estrogen receptor model. *Mol Cell Endocrinol* 36, 11–15.
12. Miller, M. A., Mullick, A., Greene, G. L., and Katzenellenbogen, B. S. (1985). Characterization of the subunit nature of nuclear estrogen receptors by chemical cross-linking and dense amino acid labeling. *Endocrinology* 117, 515–522.
13. Walter, P., Green, S., Greene, G., Krust, A., Bornert, J.-M., Jeltsch, J.-M., Staub, A., Jensen, E., Scrace, G., Waterfield, M., and Chambon, P. (1985). Cloning of the human estrogen receptor cDNA. *Proc Natl Acad Sci* 82, 7889–7893.
14. Green, S., Walter, P., Kumar, V., Krust, A., Bornert, J.-M., Argos, P., and Chambon, P. (1986). Human oestrogen receptor cDNA: Sequence, expression and homology to v-erb-A. *Nature* 320, 134–139.

15. Greene, G. L., Gilna, P., Waterfield, M., Baker, A., Hort, Y., and Shine, J. (1986). Sequence and expression of human estrogen receptor complementary DNA. *Science* 231, 1150–1154.
16. Payvar, F., DeFranco, D., Firestone, G. L., Edgar, B., Wrange, Ö., Okret, S., Gustafsson, J.-Å., and Yamamoto, K. R. (1983). Sequence-specific binding of glucocorticoid receptor to MTV DNA at sites within and upstream of the transcribed region. *Cell* 35, 381–392.
17. Walker, P., Germond, J.-E., Brown-Luedi, M., Givel, F., and Wahli, W. (1984). Sequence homologies in the region preceding the transcription initiation site of the liver estrogen-responsive vitellogenin and apo-VLDLII genes. *Nucleic Acids Res* 12, 8611–8626.
18. Klein-Hitpass, L., Schorpp, M., Wagner, U., and Ryffel, G. U. (1986). An estrogen-responsive element derived from the 5' flanking region of the xenopus vitellogenin A2 gene functions in transfected human cells. *Cell* 46, 1053–1061.
19. Seiler-Tuyns, Walker P, Martinez E, Mérillat A-M, Givel F, Wahli W (1986). Identification of estrogen-responsive DNA sequences by transient expression experiments in a human breast cancer cell line. *Nucleic Acid Res* 14, 8755–8770.
20. Umesono, K. and Evans, R. M. (1989). Determinants of target gene specificity for steroid/thyroid hormone receptors. *Cell* 57, 1139–1146.
21. Klinge, C. M. (2001). Estrogen receptor interaction with estrogen response elements. *Nucl Acids Res* 29, 2905–2919.
22. Kumar, V., Green, S., Stack, G., Berry, M., Jin, J.-R., and Chambon, P. (1987). Functionl domains of the human estrogen receptor. *Cell* 51, 941–951.
23. Ponglikitmongkol, M., Green, S., and Chambon, P. (1988). Genomic organization of the human oestrogen receptor gene. *EMBO J* 7, 3385–3388.
24. Giguere, V., Yang, N., Segui, P., and Evans, R. M. (1988). Identification of a new class of steroid hormone receptors. *Nature* 331, 91–94.
25. Evans, R. M. (1988). The steroid and thyroid hormone receptor superfamily. *Science* 240, 889–895.
26. Beato, M. (1989). Gene regulation by steroid hormones. *Cell* 56, 335–344.
27. Tsai, M.-J. and O'Malley, B. W. (1994). Molecular mechanisms of action of steroid/thyroid receptor superfamily members. *Annu Rev Biochem* 63, 451–486.
28. Gronemeyer, H. and Laudet, V. (1995). Transcription factors 3: Nulear receptors. *Protein Profile* 2, 1167–1322.
29. Mangelsdorf, D. J., Thummel, C., Beato, M., Herrlich, P., Schütz, G., Umesono, K., Blumberg, B., Kastner, P., Mark, M., Chambon, P., and Evans, R. M. (1995). The nuclear receptor superfamily: The second decade. *Cell* 83, 835–839.
30. Katzenellenbogen, J. A. and Katzenellenbogen, B. S. (1996). Nuclear hormone receptors: Ligand activated regulators of transcription and diverse cell responses. *Chem Biol* 3, 529–536.
31. Nuclear Receptors Nomenclature Committee (1999). A unified nomenclature system for the nuclear receptor superfamily. *Cell* 97, 161–163.
32. Kuiper, G. G. J. M., Enmark, E., Pelto-Huikko, M., Nilsson, S., and Gustafsson, J.-Å. (1996). Cloning of a novel estrogen receptor expressed in rat prostate and ovary. *Proc Natl Acad Sci USA* 93, 5925–5930.
33. Mosselman, S., Polman, J., and Dijkema, R. (1996). ERβ: Identification and characterization of a novel human estrogen receptor. *FEBS Lett* 392, 49–53.
34. Todo, T., Adachi, S., and Yamauchi, K. (1996). Molecular cloning and characterization of Japanese eel estrogen receptor cDNA. *Mol Cell Endocrinol* 119, 37–45.
35. Tremblay, G. B., Tremblay, A., Copeland, N. G., Gilbert, D. J., Jenkins, N. A., Labrie, F., and Giguère, V. (1997). Cloning, chromosomal localization, and functional analysis of the murine estrogen receptor β. *Mol Endocrinol* 11, 352–365.
36. Tchoudakova, A., Pathak, S., and Callard, G. V. (1999). Molecular cloning of an estrogen receptor beta subtype from the goldfish, *Carassius auratus*. *Gen Comp Endocrinol* 113, 388–400.
37. Nilsson, S. and Gustafsson, J.-Å. (2002a). Estrogen receptor action. *Crit Rev Eukaryot Gene Expr* 12, 237–257.

38. Matthews, J. and Gustafsson, J.-Å. (2003). Estrogen signalling: A subtle balance between ERα and ERβ. *Mol Interv* 3, 281–292.
39. Ascenzi, P., Bocedi, A., and Marino, M. (2006). Structure.function relationship of estrogen receptor α and β: Impact on human health. *Mol Aspects Med* 27, 299–402.
40. Zhao, C., Dahlman-Wright, K., and Gustafsson, J.-Å. (2008). Estrogen receptor β: An overview and update. *Nucl Recept Signal* 6, 1–10.
41. Beekman, J. M., Allan, G. F., Tsai, S. Y., Tsai, M.-J., and O'Malley, B. W. (1993). Transcriptional activation by the estrogen receptor requires a conformational change in the ligand binding domain. *Mol Endocrinol* 7, 1266–1274.
42. Hall, J. M. and McDonnell, D. P. (2005). Coregulators in nuclear estrogen receptor action: From concept to therapeutic targeting. *Mol Interv* 5, 343–357.
43. Lonard, D. M. and O'Malley, B. W. (2006). The expanding cosmos of nuclear receptor coactivators. *Cell* 125, 411–414.
44. Gustafsson, J.-Å. (1999). Estrogen receptor-β- a new dimension in estrogen mechanism of action. *J Endocrinol* 163, 379–383.
45. Taylor, A. H. and Al-Azzawi, F. (2000). Immunolocalization of estrogen receptor beta in human tissues. *J Mol Endocrinol* 24, 145–155.
46. Weihua, Z., Andersson, S., Cheng, G., Simpson, E. R., Warner, M., and Gustafsson, J.-Å. (2003). Update on estrogen signaling. *FEBS Lett* 546, 17–24.
47. Dahlman-Wright, K., Cavailles, V., Fuqua, S. A., Jordan, V. C., Katzenellenbogen, J. A., Korach, K. S., Maggi, A., Muramatsu, M., Parker, M. G., and Gustafsson, J.-Å. (2006). International union of pharmacology. LXIV. Estrogen receptors. *Pharmacol Rev* 58, 773–781.
48. Morani, A., Warner, M., and Gustafsson, J.-Å. (2008). Biological functions and clinical implications of oestrogen receptors alpha and beta in epithelial cells. *J Intern Med* 264, 128–142.
49. Gustafsson, J.-Å. (2000). Novel aspects of estrogen action. *J Soc Gynecol Investig* 7, S.
50. Oettel, M. (2002). Is there a role for estrogens in the maintenance of men's health? *Aging Male* 5, 248–257.
51. Jones, M. E., Boon, W. C., McInnes, K., Maffei, L., Carani, C., and Simpson, E. R. (2007a). Recognizing rare disorders: Aromatase deficiency. *Nat Clin Pract Endocrinol Metab* 3, 414–421.
52. Deroo, B. J. and Korach, K. S. (2006). Estrogen receptors and human disease. *J Clin Invest* 116, 561–570.
53. Nilsson, S. and Gustafsson, J.-Å. (2002b). Biological role of estrogen and estrogen receptors. *Crit Rev Biochem Mol Biol* 37, 1–28.
54. Albertazzi, P. (2007). Non-estrogenic approaches for the treatment of climacteric symptoms. *Climacteric* 10(suppl 2), 115–120.
55. Vitiello, D., Naftoilin, F., and Taylor, H. S. (2007). Menopause: Developing a rational treatment plan. *Gynecol Endocrinol* 23, 682–691.
56. Nelson, H. D. (2008). Menopause. *Lancet* 371, 760–770.
57. Jordan, V. C. (2003). Tamoxifen: A most unlikely pioneering medicine. *Nat Rev Drug Discov* 2, 205–213.
58. Burger, H. G. (2000). Selective oestrogen receptor modulators. *Horm Res* 53, 25–29.
59. Howell, S. J., Johnston, S. R. D., and Howell, A. (2004). The use of selective estrogen receptor modulators and selective estrogen receptor down-regulators in breast cancer. *Best Pract Res Clin Endocrinol Metab* 18, 47–66.
60. Johnston, S. R. D. (2005). Selective oestrogen receptor modulators and downregulators for breast cancer – have they lost their way? *Breast Cancer Res* 7, 119–130.
61. Draper, M. W. (2006). An update on raloxifene. *Int J Gynecol Cancer* 16, 502–503.
62. Shelly, W., Draper, M. W., Krishnan, V., Wong, M., and Jaffe, R. B. (2008). Selective estrogen receptor modulators: An update on recent clinical findings. *Obstet Gynecol Surv* 63, 163–181.
63. Dellapasqua, S. and Castiglione-Gertsch, M. (2005). Adjuvant endocrine therapies for post-menopausal women with early breast cancer: Standards and not. *Breast* 14, 555–563.
64. Felson, D. T. and Cummings, S. R. (2005). Aromatase inhibitors and the syndrome of arthralgias with estrogen deprivation. *Arthritis Rheum* 52, 2594–2598.

65. Mouridsen, H. T. and Robert, N. J. (2005). The role of aromatase inhibitors as adjuvant therapy for early breast cancer in postmenopausal women. *Eur J Cancer* 41, 1678–1689.
66. Birge, S. J. (2007). Aromatase inhibitors. *Menopause* 14, 971–972.
67. Hadji, P. (2008). Menopausal symptoms and adjuvant therapy-associated adverse events. *Endocr Relat Cancer* 15, 73–90.
68. Lin, N. U. and Winer, E. P. (2008). Advances in adjuvant endocrine therapy for postmenopausal women. *J Clin Oncol* 26, 798–805.
69. Ponzone, R., Mininanni, P., Cassina, E., Pastorino, F., and Sismondi, P. (2008). Aromatase inhibitors for breast cancer: Different structures, same effects. *Endocr Relat Cancer* 15, 27–36.
70. Sand, P., Luckhaus, C., Schlurmann, K., Götz, M., and Deckert, J. (2002). Untangling the human estrogen receptor gene structure. *J Neural Transm* 109, 567–583.
71. Ogawa, S., Inoue, S., Watanabe, T., Hiroi, H., Orimo, A., Hosoi, T., Ouchi, Y., and Muramatsu, M. (1998). The complete primary structure of human estrogen receptor ß (hERß) and its heterodimerization with ERα in vivo and in vitro. *Biochem Biophys Res Comm* 243, 122–126.
72. Herynk, M. H. and Fuqua, S. A. W. (2004). Estrogen receptor mutations in human disease. *Endocrine Rev* 25, 869–898.
73. Leung, Y.-K., Mak, P., Hassan, S., and Ho, S.-H. (2006). Estrogen receptor (ER)-β isoforms: A key to understanding ER-β signalling. *Proc Natl Acad Sci USA* 103, 13162–13167.
74. Nilsson, M., Naessén, S., Dahlman, I., Lindén Hirschberg, A., Gustafsson, J.-Å., and Dahlman-Wright, K. (2004). Association of estrogen receptor β polymorphisms with bulimic disease in women. *Mol Psychiatry* 9, 28–34.
75. Peter, I., Shearman, A. M., Vasan, R. S., Zucker, D. R., Schmid, C. H., Demissie, S., Cupples, L. A., Kuvin, J. T., Karas, R. H., Mendelsohn, M. E., Housman, D. E., and Benjamin, E. J. (2005). Association of estrogen receptor β gene polymorphisms with left ventricular mass and wall thickness in women. *Am J Hypertens* 18, 1388–1395.
76. Thellenberg-Karlsson, C., Lindström, S., Malmer, B., Wiklund, F., Augustsson-Bälter, K., Adami, H.-O., Stattin, P., Nilsson, M., Dahlman-Wright, K., Gustafsson, J.-Å., and Grönberg, H. (2006). Estrogen receptor β polymorphism is associated with prostate cancer risk. *Clin Cancer Res* 12, 1936–1941.
77. Gallagher, C. J., Langefeld, C. D., Gordon, C. J., Campbell, J. K., Mychalecky, J. C., Bryer-Ash, M., Rich, S. S., Bowden, D. W., and Sale, M. M. (2007). Association of the estrogen receptor-alpha gene with the metabolic syndrome and its component in African-American families: The insulin resistance atherosclerosis family study. *Diabetes* 56, 2135–2141.
78. Gennari, L., De Paola, V., Merlotti, D., Martini, G., and Nuti, R. (2007). Steroid hormone receptor gene polymorphisms and osteoporosis: A pharmacogenomic review. *Expert Opin Pharmacother* 8, 537–553.
79. Ashworth, J. J., Smyth, J. V., Pendleton, N., Horan, M., Payton, A., Worthington, J., Ollier, W. E., and Ashcroft, G. S. (2008). Polymorphisms spanning the 0 N exon and promoter of the estrogen receptor-beta (ERβ) gene ESR2 are associated with venous ulceration. *Clin Genet* 73, 55–61.
80. Chen, H.-Y., Wan, L., Chung, Y.-W., Chen, W.-H., Tsai, F.-J., and Tsai, C.-H. (2008). Estrogen receptor beta gene haplotype is associated with pelvic organ prolapse. *Eur J Obstet Gynecol Reprod Biol* 138, 105–109.
81. Klos, K. L., Boerwinkle, E., Ferrell, R. E., Turner, S. T., and Morrison, A. C. (2008). ESR1 polymorphism is associated with plasma lipid and apolipoprotein levels in Caucasians of the Rochester Family Heart Study. *J Lipid Res* 49, 1701–1706.
82. Leigh Pearce, C., Near, A. M., Butler, J. L., Van Den Berg, D., Bretsky, P., Conti, D. V., Stram, D. O., Pike, M. C., Hirschhorn, J. N., and Wu, A. H. (2008). Comprehensive evaluation of ESR2 variation and ovarian cancer risk. *Cancer Epidemiol Biomarkes Prev* 17, 393–396.
83. Webb, P., Nguyen, P., Shinsako, J., Anderson, C., Feng, W., Nguyen, M. P., Chen, D., Huang, S. M., Subramanian, S., McKinerney, E., Katzenellenbogen, B. S., Stallcup, M. R., and Kushner, P. J. (1998). Estrogen receptor activation function 1 works by binding p160 coactivator proteins. *Mol Endocrinol* 12, 1605–1618.

84. Tremblay, A., Tremblay, G. B., Labrie, F., and Giguère, V. (1999a). Ligand-independent recruitment of SRC-1 to estrogen receptor β through phosphorylation of activation function AF-1. *Mol Cell* 3, 513–519.
85. Métivier, R., Stark, A., Fluoriot, G., Hübner, M. R., Brand, H., Penot, G., Manu, D., Denger, S., Reid, G., Kos, M., Russell, R. B., Kah, O., Pakdel, F., and Gannon, F. (2002). A dynamic structural model for estrogen receptor-α activation by ligands, emphasizing the role of interactions between distant A and E domains. *Mol Cell* 10, 1019–1032.
86. Métivier, R., Petit, F. G., Valotaire, Y., and Pakdel, F. (2000). Function of N-terminal transactivation domain of the estrogen receptor requires a potential α-helical structure and is negatively regulated by the A domain. *Mol Endocrinol* 14, 1849–1871.
87. Métivier, R., Penot, G., Flouriot, G., and Pakdel, F. (2001). Synergism between ERα transactivation function 1 (AF-1) and AF-2 mediated by steroid receptor coactivator protein-1: Requirement for the AF-1 α-helical core and for a direct interaction between the N- and C-terminal domains. *Mol Endocrinol* 15, 1953–1970.
88. Kato, S., Endoh, H., Masuhiro, Y., Kitamoto, T., Uchiyama, S., Sasaki, H., Masushige, S., Gotoh, Y., Nishida, E., Kawashima, H., Metzger, D., and Chambon, P. (1995). Activation of the estrogen receptor through phosphorylation by mitogen- activated protein kinase. *Science* 270(1491-), 1494.
89. Bunone, G., Briand, P.-A., Miksicek, R. J., and Picard, D. (1996). Activation of the unliganded estrogen receptor by EGF involves the MAP kinase pathway and direct phosphorylation. *EMBO J* 15, 2174–2183.
90. Ignar-Trowbridge, D. M., Pimentel, M., Parker, M. G., McLachlan, J. A., and Korach, K. S. (1996). Peptide growth factor cross-talk with the estrogen receptor requires the A/B domain and occurs independently of protein kinase C or estradiol. *Endocrinology* 137, 1735–1744.
91. Lannigan, D. A. (2003). Estrogen receptor phosphorylation. *Steroids* 68, 1–9.
92. Katzenellenbogen, B. S. (1996). Estrogen receptors: Bioactivities and interactions with cell signaling pathways. *Biol Reprod* 54, 287–293.
93. Cenni, B. and Picard, D. (1999). Ligand-independent activation of steroid receptors: New roles for old players. *TEM* 10, 41–46.
94. Enmark, E., Pelto-Huikko, M., Grandien, K., Lagercrantz, S., Lagercrantz, J., Fried, G., Nordenskjold, M., and Gustafsson, J.-Å. (1997). Human estrogen receptor β- gene structure, chromosomal localisation and expression pattern. *J Clin Endocrinol Met* 82, 4258–4265.
95. Schwabe, J. W. R., Chapman, L., Finch, J. T., Rhodes, D., and Neuhaus, D. (1993). DNA recognition by the oestrogen receptor: From solution to the crystal. *Structure* 1, 187–204.
96. Liu, Y., Gao, H., Marstrand, T. T., Ström, A., Valen, E., Sandelin, A., Gustafsson, J.-Å., and Dahlman-Wright, K. (2008a). The genome landscape of ERα- and ERβ-binding DNA regions. *Proc Natl Acad Sci USA* 105, 2604–2609.
97. Carroll, J. S., Liu, X. S., Brodsky, A. S., Li, W., Meyer, C. A., Szary, A. J., Eeckhoute, J., Shao, W., Hestermann, E. V., Geistlinger, T. R., Fox, E. A., Silver, P. A., and Brown, M. (2005). Chromosome-wide mapping of estrogen receptor binding reveals long-range regulation requiring the forkhead protein FoxA1. *Cell* 122, 33–43.
98. Laganière, J., Deblois, G., Lefebvre, C., Bataille, A. R., Robert, F., and Giguère, V. (2005). Location analysis of estrogen receptor α target promoters reveals that FOXA1 defines a domain of the estrogen response. *Proc Natl Acad Sci USA* 102, 11651–11656.
99. Calnan, D. R. and Brunet, A. (2008). The FoxO code. *Oncogene* 27, 2276–2288.
100. Ho, K. K., Myatt, S. S., and Lam, E. W. (2008). Many forks in the path: Cycling with FoxO. *Oncogene* 27, 2300–2311.
101. Lupien, M., Eeckhoute, J., Meyer, C. A., Wang, Q., Zhang, Y., Li, W., Carroll, J. S., Liu, X. S., and Brown, M. (2008). FoxA1 translates epigenetic signatures into enhancer-driven lineage-specific transcription. *Cell* 132, 958–970.
102. Wang, C., Fu, M., Angeletti, R. H., Siconolfi-Baez, L., Reutens, A. T., Albanese, C., Lisanti, M. P., Katzenellenbogen, B. S., Kato, S., Hopp, T., Fuqua, S. A. W., Lopez, G. N., Kushner, P. J., and Pestell, R. G. (2001). Direct acetylation of the estrogen receptor α hinge region by p300 regulates transactivation and hormone sensitivity. *J Biol Chem* 276, 18375–18383.

103. Kim, M. Y., Woo, E. M., Chong, Y. T. E., Homenko, D. R., and Kraus, W. L. (2006). Acetylation of estrogen receptor α by p300 at lysines 266 and 268 enhances the deoxyribonucleic acid binding and transcriptional activities of the receptor. *Mol Endocrinol* 20, 1479–1493.
104. Sentis, S., Romancer, M. L., Bianchin, C., Rostan, M.-C., and Corbo, L. (2005). Sumoylation of the estrogen receptor α hinge region regulates its transcriptional activity. *Mol Endocrinol* 19, 2671–2684.
105. Berry, N. B., Fan, M., and Nephew, K. P. (2008). Estrogen receptor-α hinge-region lysine 302 and 303 regulate receptor degradation by the proteasome. *Mol Endocrinol* 22, 1535–1551.
106. Brzozowski, A. M., Pike, A. C. W., Dauter, Z., Hubbard, R. E., Bonn, T., Engström, O., Öhman, L., Greene, G. L., Gustafsson, J.-Å., and Carlquist, M. (1997). Molecular basis of agonism and antagonism in the oestrogen receptor. *Nature* 389, 753–758.
107. Shiau, A. K., Barstad, D., Loria, P. M., Cheng, L., Kushner, P. J., Agard, D. A., and Greene, G. L. (1998). The structural basis of estrogen receptor/coactivator recognition and the antagonism of this interaction by tamoxifen. *Cell* 95, 927–937.
108. Moras, D. and Gronemeyer, H. (1998). The nuclear receptor ligand binding domain: Structure and function. *Curr Opin Cell Biol* 10, 384–391.
109. Pike, A. C. W. (2006). Lessons learnt from structural studies of the estrogen receptor. *Best Pract Res Clin Endocrinol Metab* 20, 1–14.
110. Pike, A. C. W., Brzozowski, A. M., Hubbard, R. E., Bonn, T., Thorsell, A.-G., Engström, O., Ljunggren, J., Gustafsson, J.-Å., and Carlquist, M. (1999). Structure of the ligand-binding domain of oestrogen receptor beta in the presence of a partial agonist and a full antagonist. *EMBO J* 18, 4608–4618.
111. Kuiper, G. G. J. M., Carlsson, B., Grandien, K., Enmark, E., Häggblad, J., Nilsson, S., and Gustafsson, J.-Å. (1997). Comparison of the ligand binding specificity and transcript tissue distribution of estrogen receptors α and β. *Endocrinol* 138, 863–870.
112. Stauffer, S. R., Coletta, C. J., Tedesco, R., Nishiguchi, G., Carlson, K., Sun, J., Katzenellenbogen, B. S., and Katzenellenbogen, J. A. (2000). Pyrazole ligands: Structure-affinity/activity relationships and estrogen receptor-α-selective agonists. *J Med Chem* 43, 4934–4947.
113. Meyers, M. J., Sun, J., Carlson, K. E., Marriner, G. A., Katzenellenbogen, B. S., and Katzenellenbogen, J. A. (2001). Estrogen receptor-β potency-selective ligands: Structure-activity relationship studies of diarylpropionitriles and their acetylene and polar analogues. *J Med Chem* 44, 4230–4251.
114. Sun, J., Huang, Y. R., Harrington, W. R., Sheng, S., Katzenellenbogen, J. A., and Katzenellenbogen, B. S. (2002). Antagonists selective for estrogen receptor α. *Endocrinology* 143, 941–947.
115. Hillish, A., Peters, O., Kosemund, D., Müller, G., Walter, A., Schneider, B., Reddersen, G., and Elger, W.F. (2004). Dissecting physiological roles of estrogen receptor α and β with potent selective ligands from structure-based design. *Mol Endocrinol* 18, 1599–1609.
116. Kim, S., Wu, J. Y., Birzin, E. T., Frisch, K., Chan, W., Pai, L. Y., Yang, Y. T., Mosley, R. T., Fitzgerald, P. M., Sharma, N., Dahllund, J., Thorsell, A. G., DiNinno, F., Rohrer, S. P., Schaeffer, J. M., and Hammond, M. L. (2004). Estrogen receptor ligands. II. Discovery of benzoxathiins as potent, selective estrogen receptor alpha modulators. *J Med Chem* 47, 2171–2217.
117. Pike, A. C. W., Brzozowski, A. M., and Hubbard, R. E. (2000). A structural biologist's view of the oestrogen receptor. *J Steroid Biochem Mol Biol* 74, 261–268.
118. Shang, Y., Hu, X., DiRenzo, J., Lazar, M. A., and Brown, M. (2000). Cofactor dynamics and sufficiency in estrogen receptor-regulated transcription. *Cell* 103, 843–852.
119. Pike, A. C. W., Brzozowski, A. M., Walton, J., Hubbard, R. E., Thorsell, A.-G., Li, Y.-L., Gustafsson, J.-Å., and Carlquist, M. (2001). Structural insights into the mode of action of a pure antiestrogen. *Structure* 9, 145–153.
120. Marimuthu, A., Feng, W., Tagami, T., Nguyen, H., Jameson, J. L., Fletterick, R. J., Baxter, J. D., and West, B. L. (2002). TR surfaces and conformations required to bind nuclear receptor corepressors. *Mol Endocrinol* 16, 271–286.
121. Xu, H. E., Stanley, T. B., Montana, V. G., Lambert, M. H., Shearer, B. G., Cobb, J. E., McKee, D. D., Galardi, C. M., Plunket, K. D., Nolte, R. T., Parks, D. J., Moore, J. T., Kliewer, S. A.,

Willson, T. M., and Stimmel, J. B. (2002). Structural basis for antagonist-mediated recruitment of nuclear co-repressors by PPARα. *Nature* 415, 813–817.

122. Webb, P., Nguyen, P., and Kushner, P. J. (2003). Differential SERM effects on corepressor binding dictate ERα activity in vivo. *J Biol Chem* 278, 6912–6920.
123. Wu, Y.-L., Yang, X., Ren, Z., McDonnell, D. P., Norris, J. D., Willson, T. M., and Greene, G. L. (2005). Structural basis for an unexpected mode of SERM-mediated ER antagonism. *Mol Cell* 18, 413–424.
124. Cowley, S. M. and Parker, M. G. (1999). A comparison of transcriptional activation by ER alpha and ER beta. *J Steroid Biochem Mol Biol* 69, 165–175.
125. Valentine, J. E., Kalkhoven, E., White, R., Hoare, S., and Parker, M. G. (2000). Mutations in the estrogen receptor ligand binding domain discriminate between hormone-dependent transactivation and transrepression. *J Biol Chem* 275, 25322–25329.
126. Tamrazi, A., Carlson, K. E., Daniles, J. R., Hurth, K. M., and Katzenellenbogen, J. A. (2002). Estrogen receptor dimerization: Ligand binding regulates dimer affinity and dimer dissociation rate. *Mol Endocrionol* 16, 2706–2719.
127. Tamrazi, A., Carlson, K. E., Rodriguez, A. L., and Katzenellenbogen, J. A. (2005). Coactivator proteins as determinants of estrogen receptor structure and function: Spectroscopic evidence for a novel coactivator-stabilized receptor conformation. *Mol Endocrinol* 19, 1516–1528.
128. Gee, A. C., Carlson, K. E., Martini, P. G. V., Katzenellenbogen, B. S., and Katzenellenbogen, J. A. (1999). Coactivator peptides have a differential stabilizing effect on the binding of estrogens and antiestrogens with the estrogens receptor. *Mol Endocrinol* 13, 1912–1923.
129. Sladek, F. M., Ruse, M. D., Nepomuceno, L., Huang, S.-M., and Stallcup, M. R. (1999). Modulation of transcriptional activation and coactivator interaction by a splicing variation in the F domain of nuclear receptor hepatocyte nuclear receptor 4α1. *Mol Cell Biol* 19, 6509–6522.
130. Ruse, M. D., Privalsky, M. L., and Sladek, F. M. (2002). Competitive cofactor recruitment by orphan receptor hepatocyte nuclear factor 4α1: Modulation by the F domain. *Mol Cell Biol* 22, 1626–1638.
131. Hani, E. H., Suaud, L., Boutin, P., Chèvre, J.-C., Durand, E., Philippi, A., Demenais, F., Vionnet, N., Furuta, H., Velho, G., Bell, G. I., Laine, B., and Froguel, P. (1998). A missense mutation in hepatocyte nuclear receptor-4α, resulting in a reduced transactivation activity, in human late-onset non-insulin-dependent diabetes mellitus. *J Clin Invest* 101, 521–526.
132. Montano, M. M., Muller, V., Trobaugh, A., and Katzenellenbogen, B. S. (1995). The carboxy-terminal F domain of the human estrogen receptor: Role of the transcriptional activity of the receptor and the effectiveness of antiestrogens as estrogen antagonists. *Mol Endocrinol* 9, 814–825.
133. Montano, M. M., Ekena, K., Krueger, K. D., Keller, A. L., and Katzenellenbogen, B. S. (1996). Human estrogen receptor ligand activity inversion mutants: Receptors that interpret antiestrogens as estrogens and estrogens as antiestrogens and discriminate among different antiestrogens. *Mol Endocrinol* 10, 230–242.
134. Schwartz, J. A., Zhong, L., Deighton-Collins, S., Zhao, C., and Skafar, D. F. (2002). Mutations targeted to a predicted helix in the extreme carboxy-terminal region of the human estrogen receptor-α alter its response to estradiol and 4-hydroxytamoxifen. *J Biol Chem* 277, 13202–13209.
135. Skafar, D. F. and Koide, S. (2006). Understanding the human estrogen receptor-alpha using targeted mutagenesis. *Mol Cell Endocrinol* 246, 83–90.
136. Peters, G. A. and Khan, S. A. (1999). Estrogen receptor domains E and F: Role in dimerization and interaction with coactivator RIP-140. *Mol Endocrinol* 13, 286–296.
137. Kim, K., Thu, N., Saville, B., and Safe, S. (2003). Domains of estrogen receptor α (ERα) required for ERα/Sp1-mediated activation of GC-rich promoters by estrogens and antiestrogens in breast cancer cells. *Mol Endocrinol* 17, 804–817.
138. McDonnell, D. P., Clemm, D. L., Hermann, T., Goldman, M. E., and Pike, J. W. (1995). Analysis of estrogen receptor function in vitro reveals three distinct classes of antiestrogens. *Mol Endocrinol* 9, 659–669.
139. Paige, L. A., Christensen, D. J., Grøn, H., Norris, J. D., Gottlin, E. B., Padilla, K. M., Chang, C. Y., Ballas, L. M., Hamilton, P. T., McDonnell, D. P., and Fowlkes, D. M. (1999). Estrogen receptor

(ER) modulators each induce distinct conformational changes in ER alpha and ER beta. *Proc Nat Acad Sci USA* 96, 3999–4004.

140. Wijayaratne, A. L., Nagel, S. C., Paige, L. A., Christensen, D. J., Norris, J. D., Fowlkes, D. M., and McDonnell, D. P. (1999). Comparative analyses of mechanistic differences among antiestrogens. *Endocrinology* 140, 5828–5840.
141. Schaufele, F., Chang, C. Y., Liu, W., Baxter, J. D., Nordeen, S. K., Wan, Y., Day, R. N., and McDonnell, D. P. (2000). Temporally distinct and ligand-specific recruitment of nuclear receptor-interacting peptides and cofactors to subnuclear domains containing the estrogen receptor. *Mol Endocrinol* 14, 2024–2039.
142. Huang, H. J., Norris, J. D., and McDonnell, D. P. (2002). Identification of a negative regulatory surface within estrogen receptor alpha provides evidence in support of a role for corepressors in regulating cellular responses to agonists and antagonists. *Mol Endocrinol* 16, 1778–1792.
143. Iannone, M. A., Simmons, C. A., Kadwell, S. H., Svoboda, D. L., Vanderwall, D. E., Deng, S.-J., Consler, T. G., Shearin, J., Gray, J. G., and Pearce, K. H. (2004). Correlation between in vitro peptide binding profiles and cellular activities for estrogen receptor-modulating compounds. *Mol Endocrinol* 18, 1064–1081.
144. Heery, D. M., Kalkhoven, E., Hoare, S., and Parker, M. G. (1997). A signature motif in transcriptional co-activators mediates binding to nuclear receptors. *Nature* 387, 733–736.
145. Mak, H. Y., Hoare, S., Henttu, P. M., and Parker, M. G. (1999). Molecular determinants of the estrogen receptor-coactivator interface. *Mol Cell Biol* 19, 3895–3903.
146. Wärnmark, A., Treuter, E., Gustafsson, J. A., Hubbard, R. E., Brzozowski, A. M., and Pike, A. C. (2002). Interaction of TIF2 NR-box peptides with the coactivator binding site of ERα. *J Biol Chem* 277, 21862–21868.
147. Nagy, L. and Schwabe, J. W. R. (2004). Mechanism of the nuclear receptor molecular switch. *Trends Biochem Sci* 29, 317–324.
148. Nettles, K. W. and Greene, G. L. (2005). Ligand control of coregulator recruitment to nuclear receptors. *Ann Rev Physiol* 67, 309–333.
149. Hu, X. and Lazar, M. A. (1999). The CoRNR motif controls the recruitment of corepressors by nuclear hormone receptors. *Nature* 402, 93–96.
150. Nagy, L., Kao, H.-Y., Love, J. D., Li, C., Banayo, E., Gooch, J. T., Krishna, V., Chatterjee, K., Evans, R. M., and Schwabe, J. W. R. (1999). Mechanism of corepressor binding and release from nuclear hormone receptors. *Genes Dev* 13, 3209–3216.
151. Perissi, V., Staszewski, L. M., McInerney, E. M., Kurokawa, R., Krones, A., Rose, D. W., Lambert, M. H., Milburn, M. V., Glass, C. K., and Rosenfeld, M. G. (1999). Molecular determinants of nuclear receptor-corepressor interaction. *Genes Dev* 13, 3198–3208.
152. Li, Y., Lambert, M. H., and Xu, H. E. (2003a). Activation of nuclear receptors: A perspective from structural genomics. *Structure* 11, 741–746.
153. Kong, E. H., Heldring, N., Gustafsson, J.-Å., Treuter, E., Hubbard, R. E., and Pike, A. C. W. (2005). Delineation of a unique protein-protein interaction site on the surface of the estrogen receptor. *Proc Natl Acad Sci USA* 102, 3593–3598.
154. Heldring, N., Pike, A., Andersson, S., Matthews, J., Cheng, G., Hartman, J., Tujague, M., Ström, A., Treuter, E., Warner, M., and Gustafsson, J.-Å. (2007). Estrogen receptors: How do they signal and what are their targets. *Physiol Rev* 87, 905–931.
155. Shiau, A. K., Barstad, D., Radek, J. T., Meyers, M. J., Nettles, K. W., Katzenellenbogen, B. S., Katzenellenbogen, J. A., Agard, D. A., and Greene, G. L. (2002). Structural characterization of a subtype-selective ligand reveals a novel mode of estrogen receptor antagonism. *Nat Struct Biol* 9, 359–364.
156. Manas, E. S., Xu, Z. B., Unwalla, R. J., and Somers, W. S. (2004). Understanding the selectivity of genistein for human estrogen receptor-β using x-ray crystallography and computational methods. *Structure* 12, 2197–2207.
157. Barkhem, T., Carlsson, B., Nilsson, Y., Enmark, E., Gustafsson, J.-Å., and Nilsson, S. (1998). Differential response of estrogen receptor α and estrogen receptor β to partial estrogen agonists/antagonists. *Mol Pharmacol* 54, 105–112.

158. Chang, C.-Y., Norris, J. D., Grøn, H., Paige, L. A., Hamilton, P. T., Kenan, D. J., Fowlkes, D., and McDonnell, D. P. (1999). Dissection of the LXXLL nuclear receptor-coactivator interaction motif using combinatorial peptide libraries: Discovery of peptide antagonists of estrogen receptors α and β. *Mol Cell Biol* 19, 8226–8239.
159. Norris, J. D., Paige, L. A., Christensen, D. J., Chang, C.-Y., Huacani, M. R., Fan, D., Hamilton, P. T., Fowlkes, D. M., and McDonnell, D. P. (1999). Peptide antagonists of the human estrogen receptor. *Science* 285, 744–746.
160. Hall, J., Chang, C.-Y., and McDonnell, D. P. (2000). Development of peptide antagonists that target estrogen receptor β-coactivator interactions. *Mol Endocrinol* 14, 2010–2023.
161. Bramlett, K. S., Wu, Y., and Burris, T. P. (2001). Ligands specify coactivator nuclear receptor (NR) box affinity for estrogen receptor subtypes. *Mol Endocrinol* 15, 909–922.
162. Wong, C.-W., Komm, B., and Cheskis, B. J. (2001). Structure-function evaluation of ERα and β interplay with SRC family coactivators. ER selective ligands. *Biochemistry* 40, 6756–6765.
163. Brady, H., Doubleday, M., Gayo-Fung, L. M., Hickman, M., Khammungkhune, S., Kois, A., Lipps, S., Pierce, S., Richard, N., Shevlin, G., Sutherland, M. K., Anderson, D. W., Bhagwhat, S. S., and Stein, B. (2002). Differential response of estrogen receptors α and β to SP500263, a novel potent selective estrogen receptor modulator. *Mol Pharm* 61, 562–568.
164. McInerney, E. M., Tsai, M.-J., O'Malley, B. W., and Katzenellenbogen, B. S. (1996). Analysis of estrogen receptor transcriptional enhancement by a nuclear receptor coactivator. *Proc Natl Acad Sci* 93, 10069–10073.
165. Onate, S. A., Boonyaratanakornkit, V., Spencer, T. E., Tsai, S. Y., Tsai, M. J., Edwards, D. P., and O'Malley, B. W. (1998). The steroid receptor coactivator-1 contains multiple receptor interacting and activation domains that cooperatively enhance the activation function 1 (AF1) and AF2 domains of steroid receptors. *J Biol Chem* 273, 12101–12108.
166. Tremblay, G. B., Tremblay, A., Labrie, F., and Giguere, V. (1999b). Dominant activity of activation function 1 (AF-1) and differential stoichiometric requirements for AF-1 and-2 in the estrogen receptor alpha-beta heterodimeric complex. *Mol Cell Biol* 19, 1919–1927.
167. Benecke, A., Chambon, P., and Gronemeyer, H. (2000). Synergy between estrogen receptor alpha activation functions AF1 and AF2 mediated by transcription intermediary factor TIF2. *EMBO Rep* 1, 151–157.
168. Cheskis, B. J., Karathanasis, S., and Lyttle, C. R. (1997). Estrogen receptor ligands modulate its interaction with DNA. *J Biol Chem* 272, 11384–11391.
169. Hall, J. M. and Korach, K. S. (2002). Analysis of the molecular mechanisms of human estrogen receptors α and β reveals differential specificity in target promoter regulation by xenoestrogens. *J Biol Chem* 277, 44455–44461.
170. Krieg, A. J., Krieg, S. A., Ahn, B. S., and Shapiro, D. J. (2004). Interplay between estrogen response element sequence and ligands controls in vivo binding of estrogen receptor to regulated genes. *J Biol Chem* 279, 5025–5034.
171. Katzenellenbogen, J., O'Malley, B. W., and Katzenellenbogen, B. S. (1996). Tripartite steroid hormone receptor pharmacology: Interaction with multiple effector sites as a basis for the cell- and promoter-specific action of these hormones. *Mol Endocrinol* 10, 119–131.
172. Pettersson, K., Grandien, K., Kuiper, G. G., and Gustafsson, J.-Å. (1997). Mouse estrogen receptor β forms estrogen response element-binding heterodimers with estrogen receptor α. *Mol Endocrinol* 11, 1486–1496.
173. Hall, J. M. and McDonnell, D. P. (1999). The estrogen receptor beta-isoform (ERbeta) of the human estrogen receptor modulates ERalpha transcriptional activity and is a key regulator of the cellular response to estrogens and antiestrogens. *Endocrinology* 140, 5566–5578.
174. Collingwood, T. N., Urnov, F. D., and Wolffe, A. P. (1999). Nuclear receptors: Coactivators, corepressors and chromatin remodeling in the control of transcription. *J Mol Endocrinol* 1999(23), 255–275.
175. Kraus, W. L. and Wong, J. (2002). Nuclear receptor-dependent transcription with chromatin. Is it all about enzymes? *Eur J Biochem* 269, 2275–2283.

176. Lonard, D. M. and O'Malley, B. W. (2007). Nuclear receptor coregulators:–judges, juries, and executioners of cellular regulation. *Cell* 27, 691–700.
177. Katzenellenbogen, B. S. and Frasor, J. (2004). Therapeutic targeting in the estrogen receptor hormonal pathway. *Semin Oncol* 31, 28–38.
178. Montano, M. M., Bianco, N. R., Deng, H., Wittmann, B. M., Chaplin, L. C., and Katzenellenbogen, B. S. (2005). Estrogen receptor regulation of quinine reductase in breast cancer: Implications for estrogen-induced breast tumor growth and the therapeutic uses of tamoxifen. *Front Biosci* 10, 1440–1461.
179. Montano, M. M., Jaiswal, A. K., and Katzenellenbogen, B. S. (1998). Transcriptional regulation of the human quinine reductase gene by antiestrogen-liganded estrogen receptor-α and estrogen receptor-β. *J Biol Chem* 273, 25443–25449.
180. Zou, A., Marschke, K. B., Arnold, K. E., Berger, E. M., Fitzgerald, P., Mais, D. E., and Allegretto, E. (1999). Estrogen receptor β activates the human retinoic acid receptor α-1 promoter in response to tamoxifen and other estrogen receptor antagonists, but not in response to estrogen. *Mol Endocrinol* 13, 418–430.
181. Safe, S. and Kim, K. (2004). Nuclear receptor-mediated transactivation through interaction with Sp proteins. *Prog Nucl Acid Res Mol Biol* 77, 1–36.
182. Saville, B., Wormke, M., Wang, F., Nguyen, T., Enmark, E., Kuiper, G., Gustafsson, J.-A., and Safe, S. (2000). Ligand-, cell-, and estrogen receptor subtype (alpha/beta)-dependent activation at GC-rich (Sp1) promoter elements. *J Biol Chem* 275, 5379–5387.
183. Wu, F., Khan, S., Wu, Q., Barhoumi, R., Burghardt, R., and Safe, S. (2008). Ligand structure-dependent activation of estrogen receptor alpha/Sp by estrogens and xenoestrogens. *J Steroid Biochem Mol Biol* 110, 104–115.
184. Kushner, P. J., Agard, D. A., Greene, G. L., Scanlan, T. S., Shiau, A. K., Uht, R. M., and Webb, P. (2000). Estrogen receptor pathways to AP-1. *J Steroid Biochem Mol Biol* 74, 311–317.
185. Maruyama, S., Fujimoto, N., Asano, K., and Ito, A. (2001). Suppression by estrogen receptor β of AP-1 mediated transactivation through estrogen receptor α. *J Steroid Biochem Mol Biol* 78, 177–184.
186. Teyssier, C., Belguise, K., Galtier, F., and Chalbos, D. (2001). Characterization of the physical interaction between estrogen receptor α and jun proteins. *J Biol Chem* 276, 36361–36369.
187. Paech, K., Webb, P., Kuiper, G. G. J. M., Nilsson, S., Gustafsson, J.-Å., Kushner, P. J., and Scanlan, T. S. (1997). Differential ligand activation of estrogen receptors ERα and ERβ at AP1 sites. *Science* 277, 1508–1510.
188. Xing, D., Nozell, S., Chen, Y.-F., Hage, F., and Oparil, S. (2009). Estrogen and mechanisms of vascular protection. *Arterioscler Thromb Vasc Biol* 29, 289–295.
189. Stein, B. and Yang, M. X. (1995). Repression of the interleukin-6 promoter by estrogen receptor is mediated by NFκB and C/EBPβ. *Mol Cell Biol* 15, 4971–4979.
190. Galien, R. and Garcia, T. (1997). Estrogen receptor impairs interleukin-6 expression by preventing protein binding on the NF-kappaB site. *Nucleic Acids Res* 25, 2424–2429.
191. Biswas, D. K., Singh, S., Shi, Q., Pardee, A. B., and Iglehart, J. D. (2005). Crossroads of estrogen receptor and NF-κB signaling. *Sci STKE* 288, pe27.
192. Chadwick, C. C., Chippari, S., Matelan, E., Borges-Marcucci, L., Eckert, A. M., Keith, J. C., Albert, L. M., Leathbury, Y., Harris, H. A., Bhat, R. A., Ashwell, M., Trybulski, E., Winneker, R. C., Adelman, S. J., Steffan, R. J., and Harnish, D. C. (2005). Identification of pathway-selective estrogen receptor ligands that inhibit NF-κB transcriptional activity. *Proc Natl Acad Sci USA* 102, 2543–2548.
193. Kalaitzidis, D. and Gilmore, T. D. (2005). Transcription factor cross-talk: Estrogen receptor and NF-κB. *Trend Endocrinol Metab* 16, 46–52.
194. De Bosscher, K., Vanden Berghe, W., and Haegeman, G. (2006). Cross-talk between nuclear receptors and nuclear factor κB. *Oncogene* 25(6868-), 6886.
195. Nettles, K. W., Gil, G., Nowak, J., Métivier, R., Sharma, V. B., and Greene, G. L. (2008). CBP is a dosage-dependent regulator of nuclear factor-κB suppression by the estrogen receptor. *Mol Endocrinol* 22, 263–272.

196. Berry, M., Metzger, D., and Chambon, P. (1990). Role of the two activating domains of the oestrogen receptor in the cell-type and promoter-context dependent agonistic activity of the anti-oestrogen 4-hydroxytamoxifen. *EMBO J* 9, 2811–2818.
197. Margeat, E., Poujol, N., Boulahtouf, A., Chen, Y., Müller, J. D., Gratton, E., Cavailles, V., and Royer, C. A. (2001). The human estrogen receptor α dimer binds a single SRC-1 coactivator molecule with an affinity dictated by agonist structure. *J Mol Biol* 306, 433–442.
198. Pettersson, K., Delaunay, F., and Gustafsson, J.-Å. (2000). Estrogen receptor β acts as a dominant regulator of estrogen signalling. *Oncogene* 19, 4970–4978.
199. Hodges-Gallagher, L., Valentine, C. D., Bader, S. E., and Kushner, P. J. (2008). Estrogen receptor beta increases the efficacy of antiestrogens by effects on apoptosis and cell cycling in breast cancer. *Breast Cancer Res Treat* 109, 241–250.
200. Lindberg, M. K., Movérare, S., Skrtic, S., Gao, H., Dahlman-Wright, K., Gustafsson, J.-Å., and Ohlsson, C. (2003). Estrogen receptor (ER)-β reduces ERα-regulated gene transcription, supporting a "ying yang" relationship between ERα and ERβ in mice. *Mol Endocrinol* 17, 203–208.
201. Monroe, D. G., Getz, B. J., Johnsen, S. A., Riggs, B. L., Khosla, S., and Spelsberg, T. C. (2003). Estrogen receptor isoform-specific regulation of endogenous gene expression in human osteoblastic cell lines expressing either ERα or ERβ. *J Cell Biochem* 90, 315–326.
202. Stossi, F., Barnett, D. H., Frasor, J., Komm, B., Lyttle, C. R., and Katzenellenbogen, B. S. (2004). Transcriptional profiling of estrogen-regulated gene expression via estrogen receptor α or estrogen receptor β in human osteosarcoma cells: Distinct and common target genes for these receptors. *Endocrinology* 145, 3473–3486.
203. Monroe, D. G., Secreto, F. J., Subramaniam, M., Getz, B. J., Khosla, S., and Spelsberg, T. C. (2005). Estrogen receptor α and β heterodimers exert unique effects on estrogen- and tamoxifen-dependent gene expression in human U2OS osteosarcoma cells. *Mol Endocrinol* 19, 1555–1568.
204. Métivier, R., Penot, G., Hübner, M. R., Reid, G., Brand, H., Koš, M., and Gannon, F. (2003). Estrogen receptor-α directs ordered, cyclical, and combinatorial recruitment of factors on a natural target promoter. *Cell* 115, 751–763.
205. Reid, G., Denger, S., Kos, M., and Gannon, F. (2002). Human estrogen receptor-α: Regulation by synthesis, modification and degradation. *Cell Mol Life Sci* 59, 821–831.
206. Reid, G., Hübner, M. R., Métivier, R., Brand, H., Denger, S., Manu, D., Beaudouin, J., Ellenberg, J., and Gannon, F. (2003). Cyclic, proteasome-mediated turnover of unliganded and liganded ERα on responsive promoters is an integral feature of estrogen signalling. *Mol Cell* 11, 695–707.
207. Perissi, V. and Rosenfeld, M. G. (2005). Controlling nuclear receptors: The circular logic of cofactor cycles. *Nat Rev Mol Cell Biol* 6, 542–554.
208. Stenoien, D. L., Mancini, M. G., Patel, K., Allegretto, E. A., Smith, C. L., and Mancini, M. A. (2000). Subnuclear trafficking of estrogen receptor-α and steroid receptor coactivator-1. *Mol Endocrinol* 14, 518–534.
209. Stenoien, D. L., Nye, A. C., Mancini, M. G., Patel, K., Dutertre, M., O'Malley, B. W., Smith, C. L., Belmont, A. S., and Mancini, M. A. (2001a). Ligand-mediated assembley and real-time cellular dynamics of estrogen receptor α-coactivator complexes in living cells. *Mol Cell Biol* 21, 4402–4412.
210. Stenoien, D. L., Patel, K., Mancini, M. G., Dutertre, M., Smith, C. L., O'Malley, B. W., and Mancini, M. A. (2001b). FRAP reveals that mobility of oestrogen receptor-α is ligand- and proteasome-dependent. *Nat Cell Biol* 3, 15–23.
211. Muratani, M. and Tansey, W. P. (2003). How the ubiquitin-proteasome system controls transcription. *Nat Rev Mol Cell Biol* 4, 192–201.
212. Metivier, R., Gallais, R., Tiffoche, C., Le Peron, C., Jurkowska, R. Z., Carmouche, R. P., Ibberson, D., Barath, P., Demay, F., Reid, G., Benes, V., Jeltsch, A., Gannon, F., and Salbert, G. (2008). Cyclical DNA methylation of a transcriptionally active promoter. *Nature* 452, 45–50.
213. Reid, G., Gallais, R., and Métivier, R. (2009). Marking time: The dynamic role of chromatin and covalent modification in transcription. *Int J Biochem Cell Biol* 41, 155–163.
214. Carroll, J. S. and Brown, M. (2006). Estrogen receptor target gene: An evolving concept. *Mol Endocrinol* 20, 1707–1714.

215. Ho, K. J. and Liao, J. K. (2002a). Non-nuclear actions of estrogen: New targets for prevention and treatment of cardiovascular disease. *Mol Interv* 2, 219–228.
216. Ho, K. J. and Liao, J. K. (2002b). Nonnuclear actions of estrogen. *Arterioscler Thromb Vasc Biol* 22, 1952–1961.
217. Razandi, M., Pedram, A., Merchenthaler, I., Greene, G. L., and Levin, E. R. (2004). Plasma membrane estrogen receptors exist and functions as dimmers. *Mol Endocrinol* 18, 2854–2865.
218. Björnström, L. and Sjöberg, M. (2005). Mchanisms of estrogen receptor signaling: Convergence of genomic and nongenomic actions on target genes. *Mol Endocrinol* 19, 833–842.
219. Pedram, A., Razandi, M., and Levin, E. R. (2006). Nature of functional estrogen receptors at the plasma membrane. *Mol Endocrinol* 20, 1996–2009.
220. Oliveira, A. G., Coelho, P. H., Guedes, F. D., Mahecha, G. A. B., Hess, R. A., and Oliveira, C. A. (2007). 5α-androstane-3β,17β-diol (3β-diol), an estrogenic metabolite of 5α-dihydrotestosterone, is a potent modulator of estrogen receptor ERβ expression in the ventral prostate of adult rats. *Steroid* 72, 914–922.
221. Filardo, E. J., Quinn, J. A., and Sabo, E. (2008). Association of the membrane estrogen receptor, GPR30, with breast tumor metastasis and transactivation of the epidermal growth factor receptor. *Steroids* 73, 870–873.
222. McDevitt, M. A., Glidewell-Kenney, C., Jimenez, M. A., Ahearn, P. C., Weiss, J., Jameson, J. L., and Levine, J. E. (2008). New insights into the classical and non-classical actions of estrogen: Evidence from estrogen receptor knock-out and knock-in mice. *Mol Cell Endocrinol* 290, 24–30.
223. Raz, L., Khan, M. M., Mahesh, V. B., Vadlamudi, R. K., and Brann, D. W. (2008). Rapid estrogen signaling in the brain. *Neurosignals* 16, 140–153.
224. Vasudevan, N. and Pfaff, D. W. (2008). Non-genomic actions of estrogen and their interaction with genomic actions in the brain. *Front Neroendocrinol* 29, 238–257.
225. Pandey, D. P., Lappano, R., Albanito, L., Madeo, A., Maggiolini, M., and Picard, D. (2009). Estrogenic GPR30 signalling induces proliferation and migration of breast cancer cells through CTGF. *EMBO J* 28, 523–532.
226. Morley, P., Whitfield, J. F., Vanderhyden, B. C., Tsang, B. K., and Schwartz, J. L. (1992). A new, nongenomic estrogen action: The rapid release of intracellular calcium. *Endocrinology* 131, 1305–1312.
227. Improta-Brears, T., Whorton, A. R., Codazzi, F., York, J. D., Meyer, T., and McDonnell, D. P. (1999). Estrogen-induced activation of mitogen-activated protein kinase requires mobilization of intracellular calcium. *Proc Natl Acad Sci USA* 96, 4686–4691.
228. Aronica, S. M., Kraus, W. L., and Katzenellenbogen, B. S. (1994). Estrogen action via cAMP signaling pathway: Stimulation of adenylate cyclase and cAMP-regulated gene transcription. *Proc Natl Acad Sci USA* 91, 8517–8521.
229. Watters, J. J. and Dorsa, D. M. (1998). Transcriptional effects of estrogen on neuronal neurotensin gene expression involve cAMP/protein kinase A-dependent signaling mechanisms. *J Neurosci* 18, 6672–6680.
230. Mermelstein, P. G., Becker, J. B., and Surmeier, D. J. (1996). Estradiol reduces calcium currents in rat neostriatal neurons via a membrane receptor. *J Neurosci* 16, 595–604.
231. Nadal,A, Rovira JM, Laribi O, Leon-quinto T, Andreu E, Ripoll C, Soria B (1998). Rapid insulinotropic effect of 17beta-estradiol via a plasma membrane receptor. *FASEB J* 12, 1341–1348.
232. Zhou, Y., Watters, J. J., and Dorsa, D. M. (1996). Estrogen rapidly induces the phosphorylation of the cAMP response element binding protein in rat brain. *Endocrinology* 137, 2163–2166.
233. Migliaccio, A., Di Domenico, M., Castoria, G., de Falco, A., Bontempo, P., Nola, E., and Auricchio, F. (1996). Tyrosine kinase/ p21ras/MAP-kinase pathway activation by estradiol-receptor complex in MCF-7 cells. *EMBO J* 15, 1292–1300.
234. Watters, J. J., Campbell, J. S., Cunningham, M. J., Krebs, E. G., and Dorsa, D. M. (1997). Rapid membrane effects of steroids in neuroblastoma cells: Effects of estrogen on mitogen activated protein kinase signalling cascade and c-fos immediate early gene transcription. *Endocrinology* 138, 4030–4033.

235. Singer, C. A., Figueroa-Masot, X. A., Batchelor, R. H., and Dorsa, D. M. (1999). The mitogen-activated protein kinase pathway mediates estrogen neuroprotection after glutamate toxicity in primary cortical neurons. *J Neurosci* 19, 2455–2463.
236. Pappas, T. C., Gametchu, B., and Watson, C. S. (1995). Membrane estrogen receptors identified by multiple antibody labeling and impeded-ligand binding. *FASEB J* 9, 404–410.
237. Watson, C. S., Pappas, T. C., and Gametchu, B. (1995). The other estrogen receptor in the plasma membrane: Implications for the actions of environmental estrogens. *Environ Health Perspect* 103(Suppl 7), 41–50.
238. Razandi, M., Pedram, A., Greene, G. L., and Levin, E. R. (1999). Cell membrane and nuclear estrogen receptors (ERs) originate from a single transcript: Studies of ERalpha and ERbeta expressed in Chinese hamster ovary cells. *Mol Endocrinol* 13, 307–319.
239. Stevis, P. E., Deecher, D. C., Suhadolnik, L., Mallis, L. M., and Frail, D. E. (1999). Differential effects of estradiol and estradiol-BSA conjugates. *Endocrinology* 140, 5455–5458.
240. Kim, H. P., Lee, J. Y., Jeong, J. K., Bae, S. W., Lee, H. K., and Jo, I. (1999). Nongenomic stimulation of nitric oxide release by estrogen is mediated by estrogen receptor alpha localized in caveolae. *Biochem Biophys Res Commun* 263, 257–262.
241. Chambliss, K. L., Yuhanna, I. S., Mineo, C., Liu, P., German, Z., Sherman, T. S., Mendelsohn, M. E., Anderson, R. G., and Shaul, P. W. (2000). Estrogen receptor alpha and endothelial nitric oxide synthase are organized into a functional signaling module in caveolae. *Circ Res* 87, E44–E52.
242. Wyckoff, M. H., Chambliss, K. L., Mineo, C., Yuhanna, I. S., Mendelsohn, M. E., Mumby, S. M., and Shaul, P. W. (2001). Plasma membrane estrogen receptors are coupled to endothelial nitric-oxide synthase through $G\alpha_i$. *J Biol Chem* 276, 27071–27076.
243. Chambliss, K. L. and Shaul, P. W. (2002). Estrogen modulation of endothelial nitric oxide synthase. *Endocrine Rev* 23, 665–686.
244. Razandi, M., Oh, P., Pedram, A., Schnitzer, J., and Levin, E. R. (2002). ERs associate with and regulate the production of caveolin: Implications for signaling and cellular actions. *Mol Endocrinol* 16, 100–115.
245. Kim, K. H. and Bender, J. R. (2005). Rapid, estrogen-receptor-mediated signaling: Why is the endothelium so special? *Sci STKE* 288, pe28.
246. Li, L., Haynes, P., and Bender, J. R. (2003b). Plasma membrane localization and function of the estrogen receptor α variant (ER46) in human endothelial cells. *Proc Natl Acad Sci USA* 100, 4807–4812.
247. Acconcia, F., Ascenzi, P., Fabozzi, G., Visca, P., and Marino, M. (2004). S-palmitoylation modulates human estrogen receptor-α function. *Biochem Biophys Res Commun* 316, 878–883.
248. Pedram, A., Razandi, M., Sainson, R. C. A., Kim, J. K., Hughes, C. C., and Levin, E. R. (2007). A conserved mechanism for steroid receptor translocation to the plasma membrane. *J Biol Chem* 282, 22278–22288.
249. Hillebrand, U., Lang, D., Telgmann, R. G., Hagedorn, C., Reuter, S., Kliche, K., Stock, C. M., Oberleithner, H., Pavenstädt, H., Büssemaker, E., and Hausberg, M. (2009). Nebivolol decreases endothelial cell stiffness via the estrogen receptor beta: A nano-imaging study. *J Hypertens* 27, 517–526.
250. Simoncini, T. and Genazzani, A. R. (2000). Raloxifene acutely stimulates nitric oxide release from human endothelial cells via an activation of endothelial nitric oxide synthase. *J Clin Endocrinol Metab* 85, 2966–2969.
251. Simoncini, T., Genazzani, A. R., and Liao, J. K. (2002a). Nongenomic mechanisms of endothelial nitric oxide synthase activation by the selective estrogen receptor modulator raloxifene. *Circulation* 105, 1368–1373.
252. Simoncini, T., Varone, G., Fornari, L., Mannella, P., Luisi, M., Labrie, F., and Genazzani, A. R. (2002b). Genomic and nongenomic mechanisms of nitric oxide synthesis induction in human endothelial cells by a fourth-generation selective estrogen receptor modulator. *Endocrinology* 143, 2052–2061.

253. Harada, N., Utsumi, T., and Takagi, Y. (1993). Tissue-specific expression of the human aromatase cytochrome P-450 gene by alternative use of multiple exons 1 and promoters, and switching of tissue-specific exons 1 in carcinogenesis. *Proc Natl Acad Sci USA* 90, 11312–11316.
254. Brodie, A., Inkster, S., and Yue, W. (2001). Aromatase expression in the human male. *Mol Cell Endocrinol* 178, 23–28.
255. Conley, A. and Hinshelwood, M. (2001). Mammalian aromatases. *Reproduction* 121, 685–695.
256. Simpson, E. R., Clyne, C., Rubin, G., Boon, W. C., Robertson, K., Britt, K., Speed, C., and Jones, M. (2002). Aromatase-a brief overview. *Annu Rev Physiol* 64, 93–127.
257. Zirilli, L., Rochira, V., Diazzi, C., Caffagni, G., and Carani, C. (2008). Human models of aromatase deficiency. *J Steroid Biochem Mol Biol* 109, 212–218.
258. Smith, E. P., Boyd, J., Frank, G. R., Takahashi, H., Cohen, R. M., Specker, B., Williams, T. C., Lubahn, D. B., and Korach, K. S. (1994). Estrogen resistance caused by a mutation in the estrogen-receptor gene in a man. *New Engl J Med* 331, 1056–1061.
259. Jones, M. E., Boon, W. C., Proietto, J., and Simpson, E. R. (2006). Of mice and men: The evolving phenotype of aromatase deficiency. *Trends Endocrinol Metab* 17, 53–62.
260. Hennighausen, L. and Robinson, G. W. (2001). Signaling pathways in mammary gland development. *Dev Cell* 1, 467–475.
261. Hennighausen, L. and Robinson, G. W. (2005). Information networks in the mammary gland. *Nat Rev Mol Cell Biol* 6, 715–725.
262. De Muinck Keizer-Schrama, S. M. (2007). Introduction and management of puberty in girls. *Horm Res* 68, 80–83.
263. Molyneux, G., Regan, J., and Smalley, M. J. (2007). Mammary stem cells and breast cancer. *Cell Mol Life Sci* 64, 3248–3260.
264. Speirs, V. and Walker, R. A. (2007). New perspectives into the biological and clinical relevance of oestrogen receptors in the human breast. *J Pathol* 211, 499–506.
265. Conte, F. A., Grumbach, M. M., Ito, Y., Fisher, C. R., and Simpson, E. R. (1994). A syndrome of female pseudohermaphrodism, hypergonadotropic hypogonadismm, and multicystic ovaries associated with missense mutations in the gene encoding aromatase (P450arom). *J Clin Endocrinol Metab* 78, 1287–1292.
266. Morishima, A., Grumbach, M. M., Simpson, E. R., Fisher, C., and Qin, K. (1995). Aromatase deficiency in male and female siblings caused by a novel mutation and the physiological role of estrogens. *J Clin Endocrinol Metab* 80, 3689–3698.
267. MacGillivray, M. H., Morishima, A., Conte, F., Grumbach, M., and Smith, E. R. (1998). Pediatric endocrinology update: An overview. The essential roles of estrogens in pubertal growth, epiphyseal fusion and bone turnover: Lessons from Mutations in the genes for aromatase and the estrogen receptor. *Horm Res* 49(Suppl 1), 2–8.
268. Couse, J. F. and Korach, K. S. (1999). Estrogen receptor null mice: What have we learned and where will they lead us? *Endocr Reviews* 20, 358–417.
269. Krege, J. H., Hodgin, J. B., Couse, J. F., Enmark, E., Warner, M., Mahler, J. F., Sar, M., Korach, K., Gustafsson, J.-Å., and Smithies, O. (1998). Generation and reproductive phenotypes of mice lacking estrogen receptor β. *Proc Natl Acad Sci USA* 95, 15677–15682.
270. Förster, C., Mäkelä, S., Waärri, A., Kietz, S., Becker, D., Hultenby, K., Warner, M., and Gustafsson, J.-Å. (2002). Involvement of oestrogen receptor β in terminal differentiation of mammary gland epithelium. *Proc Natl Acad Sci USA* 99, 15578–15583.
271. Tekmal, R. R., Liu, Y.-G., Nair, H. B., Jones, J., Perla, R. P., Lubahn, D. B., Korach, K. S., and Kirma, N. (2005). Estrogen receptor alpha is required for mammary development and the induction of mammary hyperplasia and epigenetic alternations in the aromatase transgenic mice. *J Steroid Biochem Mol Biol* 95, 9–15.
272. Harris, H. A., Albert, L. M., Leathurby, Y., Malamas, S., Mewshaw, R. E., Miller, C. P., Kharhode, Y. P., Marzolf, J., Komm, B. S., Winneker, R. C., Frail, D. E., Henderson, R. A., Zhu, Y., and Keith, J. C., Jr (2003). Evaluation of an estrogen receptor-β agonist in animal models of human disease. *Endocrinology* 144, 4241–4249.

273. Tekmal, R. R., Nair, H. B., Perla, R. P., and Kirma, N. (2007). HER-2/neu x aromatase double transgenic mice model: The effects of aromatase overexpression on mammary tumorigenesis. *J Steroid Biochem Mol Biol* 106, 111–118.
274. Prins, G. S. and Korach, K. S. (2008). The role of estrogens and estrogen receptors in normal prostate growth and disease. *Steroids* 73, 233–244.
275. Ellem, S. J. and Risbridger, G. P. (2009). The dual, opposing roles of estrogen in the prostate. *Ann N Y Acad Sci* 1155, 174–186.
276. Saunders, P. T. K., Maguire, S. M., Gaughan, J., and Miller, M. R. (1997). Expression of oestrogen receptor beta (β) in multiple rat tissues visualised by immunohistochemistry. *J Endocrinol* 154, R13–R16.
277. Tincello, D. G., Taylor, A. H., Spurling, S. M., and Bell, S. C. (2009). Receptor isoforms that mediate estrogen and progestagen action in the female lower urinary tract. *J Urol* 181, 1474–1482.
278. Couse, J. F., Bunch, D. O., Lindzey, J., Schomberg, D. W., and Korach, K. S. (1999). Prevention of the polycystic ovarian phenotype and characterization of ovulatory capacity in the estrogen receptor-α knockout mouse. *Endocrinology* 140, 5855–5865.
279. Dupont, S., Krust, A., Gansmuller, A., Dierich, A., Chambon, P., and Mark, M. (2000). Effect of single and compound knockout of estrogen receptors α (ERα) and β (ERβ) on mouse reproductive phenotypes. *Development* 127, 4277–4291.
280. Hess, R. A., Bunick, D., and Bahr, J. (2001). Oestrogen, its receptors and function in the male reproductive tract – a review. *Mol Cell Endocrinol* 178, 29–38.
281. Couse, J. F., Yates, M. M., Deroo, B. J., and Korach, K. S. (2005). Estrogen receptor-β is critical to granulosa cell differentiation and the ovulatory response to gonadotropins. *Endocrinology* 146, 3247–3262.
282. Inzunza, J., Morani, A., Cheng, G., Warner, M., Hreinsson, J., Gustafsson, J.-Å., and Hovatta, O. (2007). Ovarian wedge resection restore fertility in the estrogen receptor β knockout (ERβ-/-) mice. *Proc Natl Acad Sci USA* 104, 600–605.
283. Hegele-Hartung, C., Siebel, P., Peters, O., Kosemund, D., Müller, G., Hillisch, A., Walter, A., Kraetschmer, J., and Fritzemeier, K.-H. (2004). Impact of isotype-selective estrogen receptor agonists on ovarian function. *Proc Natl Acad Sci USA* 101, 5129–5134.
284. Harris, H. A., Katzenellenbogen, J. A., and Katzenellenbogen, B. S. (2002). Characterization of the biological roles of the estrogen receptors, ERα and ERβ, in estrogen target tissues in vivo through the use of an ERα-selective ligand. *Endocrinology* 143, 4172–4177.
285. Frasor, J., Barnett, D. H., Danes, J. M., Hess, R., Parlow, A. F., and Katzenellenbogen, B. S. (2003). Response-specific and ligand dose-dependent modulation of estrogen receptor (ER) α activity by ERβ in the uterus. *Endocrinology* 144, 3159–3166.
286. Hertrampf, T., Seibel, J., Laudenbach, U., Fritzemeier, K. H., and Diel, P. (2008). Analysis of the effects of oestrogen receptor α (ERα)- and ERβ-selective ligands given in combination to ovariectomized rats. *Br J Pharmacol* 153, 1432–1437.
287. Chang, W. Y. and Prins, G. S. (1999). Estrogen receptor-β: Implications for the prostate gland. *Prostate* 40, 115–124.
288. Mäkelä, S., Strauss, L., Kuiper, G., Valve, E., Salmi, S., Santti, R., and Gustafsson, J.-Å. (2000). Differential expression of estrogen receptors α and β in adult rat accessory sex glands and lower urinary tract. *Mol Cell Endocrinol* 164, 109–116 (published erratum appears in Mol Cell Endocrinol 2000; 170: 217–229).
289. Weihua, Z., Mäkelä, S., Andersson, L. C., Salmi, S., Saji, S., Webster, J. I., Jensen, E. V., Nilsson, S., Warner, M., and Gustafsson, J.-Å. (2001). A role for estrogen receptor β in the regulation of growth of the ventral prostate. *Proc Natl Acad Sci USA* 98, 6330–6335.
290. Prins, G. S., Huang, L., Birch, L., and Pu, Y. (2006). The role of estrogens in normal and abnormal development of the prostate. *Ann N Y Acad Sci* 1089, 1–13.
291. Fisher, C. R., Graves, K. H., Parlow, A. F., and Evans, E. R. (1998). Characterization of mice deficient in aromatase (ArKO) because of targeted disruption of the cyp19 gene. *Proc Natl Acad Sci USA* 95, 6965–6970.

292. Lambard, S., Silandre, D., Delalande, C., Denis-Galeraud, I., Boutguiba, S., and Carreau, S. (2005). Aromatase in testis: Expression and role in male reproduction. *J Steroid Biochem Mol Biol* 95, 63–69.
293. Rochira, V., Balestrieri, A., Madeo, B., Baraldi, E., Faustini-Fustini, M., Granata, A. R. M., and Carani, C. (2001). Congenital estrogen deficiency: In search of the estrogen role in human male reproduction. *Mol Cell Endocrinol* 178, 107–115.
294. Robertson, K. M., O'Donnell, L., Jones, M. E., Meachem, S. J., Boon, W. C., Fisher, C. R., Graves, K. H., McLachlan, R. I., and Simpson, E. R. (1999). Impairment of spermatogenesis in mice lacking a functional aromatase (cyp 19) gene. *Proc Natl Acad Sci USA* 96, 7986–7991.
295. O'Donnell, L., Robertson, K. M., Jones, M. E., and Simpson, E. R. (2001). Estrogen and spermatogenesis. *Endocrine Rev* 22, 289–318.
296. Barrett-Connor, E., Wehren, L. E., Siris, E. S., Miller, P., Chen, Y.-T., Abbott, T. A., 3rd, Berger, M. L., Santora, A. C., and Sherwood, L. M. (2003). Recency and duration of postmenopausal hormone therapy: Effects on bone mineral density and fracture risk in the National Osteoporosis Risk Assessment (NORA) study. *Menopause* 10, 412–419.
297. Lindsay, R., Hart, D. M., Aitken, J. M., MacDonald, E. B., Anderson, J. B., and Clarke, A. C. (1976). Long-term prevention of postmenopausal osteoporosis by oestrogen. Evidence for an increased bone mass after delayed onset of oestrogen treatment. *Lancet* 1, 1038–1041.
298. Cauley, J. A., Seeley, D. G., Ensrud, K., Ettinger, B., Black, D., and Cummings, S. R. (1995). Estrogen replacement therapy and fractures in older women. Study of Osteoporotic Fractures Research Group. *Ann Intern Med* 122, 9–16.
299. Fitzpatrick, L. A. (2006). Estrogen therapy for postmenopausal osteoporosis. *Arq Bras Endcrinol Metabol* 50, 705–719.
300. Slemenda, C. W., Longcope, C., Zhou, L., Hui, S. L., Peacock, M., and Johnston, C. C. (1997). Sex steroids and bone mass in older men: Positive associations with serum estrogens and negative associations with androgens. *J Clin Invest* 100, 1755–1759.
301. Greendale, G. A., Edelstein, S., and Barrett-Connor, E. (1997). Endogenous sex steroids and bone mineral density in older women and men: The Rancho Bernardo Study. *J Bone Miner Res* 12, 1833–1843.
302. Barrett-Connor, E., Mueller, J. E., von Muhlen, D. G., Laughlin, G. A., Schneider, D. L., and Sartoris, D. J. (2000). Low levels of estradiol are associated with vertebral fractures in older men, but not women: The Rancho Bernardo study. *J Clin Endocrinol Metab* 85, 219–223.
303. Khosla, S., Melton, L. J., 3rd, and Riggs, B. L. (2002). Clinical review 144: Estrogen and the male skeleton. *J Clin Endocrinol Metab* 87, 1443–1450.
304. Gennari, L., Nuti, R., and Bilezikian, J. P. (2004). Aromatase activity and bone homeostasis in men. *J Clin Endocrinol Metab* 89, 5898–5907.
305. Gennari, L. and Bilezikian, J. P. (2007). Osteoporosis in men: Pathophysiology and treatment. *Curr Rheumatol Rep* 9, 71–77.
306. Khosla, S. (2008). Estrogen and bone: Insights from estrogen-resistant, aromatase deficient, and normal men. *Bone* 43, 414–417.
307. Windahl, S., Vidal, O., Andersson, G., Gustafsson, J.-Å., and Ohlsson, C. (1999). Increased cortical bone mineral content but unchanged trabecular bone mineral density in adult female estrogen receptor-β-/- mice. *J Clin Invest* 104, 895–901.
308. Chagin, A. S., Lindberg, M. K., Andersson, N., Moverare, S., Gustafsson, J.-Å., Sävendahl, L., and Ohlsson, C. (2004). Estogen receptor-β inhibits skeletal growth and has the capacity to mediate growth plate fusion in female mice. *J Bone Miner Res* 19, 72–77.
309. Chagin, A. S. and Sävendahl, L. (2007). Oestrogen receptors and linear bone growth. *Acta Paediatr* 96, 1275–1279.
310. Windahl, S. H., Hollberg, K., Vidal, O., Gustafsson, J.-Å., Ohlsson, C., and Andersson, G. (2001). Female estrogen receptor β-/- mice are partially protected against age-related trabecular bone loss. *J Bone Miner Res* 16, 1388–1398.
311. Clowes, J. A., Riggs, B. L., and Khosla, S. (2005). The role of the immune system in the pathophysiology of osteoporosis. *Immunol Rev* 208, 207–227.

312. Novack, D. V. (2007). Estrogen and bone: Osteoclasts take center stage. *Cell Metab* 6, 254–256.
313. Nakamura, T., Imai, Y., Matsumoto, T., Sato, S., Takeuchi, K., Igarashi, K., Harada, Y., Azuma, Y., Krust, A., Yamamoto, Y., Nishina, H., Takeda, S., Takayanagi, H., Metzger, D., Kanno, J., Takaoka, K., Martin, J., Chambon, P., and Kato, S. (2007). Estrogen prevents bone loss via estrogen receptor α and induction of Fas ligand in osteoclasts. *Cell* 130, 811–823.
314. Krum, S. A., Miranda-Carboni, G. A., Hauschka, P. V., Carroll, J. S., Lane, T. F., Freedman, L. P., and Brown, M. (2008). Estrogen protects bone by inducing Fas ligand in osteoblasts to regulate osteoclast survival. *EMBO J* 27, 535–545.
315. D`Amelio, P., Grimaldi, A., Di Bella, S., Brianza, S. Z. M., Cristofaro, M. A., Tamone, C., Giribaldi, G., Ulliers, D., Pescarmona, G. P., and Isaia, G. (2008). Estrogen deficiency increases osteoclastogenesis up-regulating T cells activity: A key mechanism in osteoporosis. *Bone* 43, 92–100.
316. Bilezikian, J. P., Morishima, A., Bell, J., and Grumbach, M. M. (1998). Increased bone mass as a result of estrogen therapy in a man with aromatase deficiency. *N Engl J Med* 339, 599–603.
317. Herrmann, B. L., Saller, B., Janssen, E. O., Gocke, P., Bockisch, A., Sperling, H., Mann, K., and Broecker, M. (2002). Impact of estrogen replacement therapy in a male with congenital aromatase deficiency caused by a novel mutation in the CYP19 gene. *J Clin Endocrinol Metab* 87, 5476–5484.
318. Herrmann, B. L., Janssen, O. E., Hahn, S., Broecker-Preuss, M., and Mann, K. (2005). Effects of estrogen replacement therapy on bone and glucose metabolism in a male with congenital aromatase deficiency. *Horm Metab Res* 37, 178–183.
319. Bouillon, R., Bex, M., Vanderschueren, D., and Boonen, S. (2004). Estrogens are essential for male pubertal periosteal bone expansion. *J Clin Endocrinol Metab* 89, 6025–6029.
320. Louet, J.-F., LeMay, C., and Mauvais-Jarvis, F. (2004). Antidiabetic actions of estrogen:–insights from human and genetic mouse models. *Curr Atheroscler Rep* 6, 180–185.
321. Godsland, I. F. (2005). Oestrogens and insulin secretion. *Diabetologia* 48, 2213–2220.
322. Kaaja, R. J. and Greer, I. A. (2005). Manifestations of chronic diseases during pregnancy. *JAMA* 294, 2751–2757.
323. Ropero, A. B., Alonso-Magdalena, P., Quesada, I., and Nadal, A. (2008). The role of estrogen receptors in the control of energy and glucose homeostasis. *sTERoids* 73, 874–879.
324. Carr, M. C. (2003). The emergence of the metabolic syndrome with menopause. *J Clin Endocrinol Metab* 88, 2404–2411.
325. Brinton, E. A. (1996). Oral estrogen replacement therapy in postmenopausal women selectively raises levels and production rates of lipoprotein A-I and lowers hepatic lipase activity without lowering the fractional catabolic rate. *Arterioscler Thromb Vasc Biol* 16, 431–440.
326. Nathan, L. and Chaudhuri, G. (1997). Estrogens and atherosclerosis. *Annu Rev Pharmacol Toxicol* 37, 477–515.
327. Zysow, B. R., Kauser, K., Lawn, R. M., and Rubanyi, G. M. (1997). Effects of estrus cycle, ovariectomy, and treatment with estrogen, tamoxifen, and progesterone on apolipoprotein(a) gene expression in transgenic mice. *Arterioscler Thromb Vasc Biol* 17, 1741–1745.
328. Godsland, I. F. (2001). Effects of postmenopausal hormone replacement therapy on lipid, lipoprotein, and apolipoprotein (a) concentrations: Analysis of studies published from 1974–2000. *Fertil Steril* 75, 898–915.
329. Andersson, B., Mattsson, L. A., Hahn, L., Marin, P., Lapidus, L., Holm, G., Bengtsson, B. A., and Bjorntorp, P. (1997). Estrogen replacement therapy decreases hyperandrogenicity and improves glucose homeostasis and plasma lipids in postmenopausal women with noninsulin-dependent diabetes mellitus. *J Clin Endocrinol Metab* 82, 638–643.
330. Brussaard, H. E., Gevers Leuven, J. A., Frolich, M., Kluft, C., and Krans, H. M. (1997). Short-term oestrogen replacement therapy improves insulin resistance, lipids and fibrinolysis in postmenopausal women with NIDDM. *Biabetologia* 40, 843–849.
331. Fineberg, S. E. (2000). Glycaemic control and hormone replacement therapy: Implications of the postmenopausal estrogen/progestogen intervention (PEPI) study. *Drugs Aging* 17, 453–461.

332. Ferrara, A., Karter, A. J., Ackerson, L. M., Liu, J. Y., and Selby, J. V. (2001). Hormone replacement therapy is associated with better glycemic control in women with type 2 diabetes: The Northern California Kaiser Permanente Diabetes Registry. *Diabetes Care* 24, 1144–1150.
333. Friday, K. E., Dong, C., and Fontenot, R. U. (2001). Conjugated equine estrogen improves glycemic control and blood lipoproteins in postmenopausal women with type 2 diabetes. *J Clin Endocrinol Metab* 86, 48–52.
334. Crespo, C. J., Smit, E., Snelling, A., Sempos, C. T., and Andersen, R. E. (2002). Hormone replacement therapy and its relationship to lipid and glucose metabolism in diabetic and nondiabetic postmenopausal women: Results from the Third National Health and Nutrition Examination Survey (NHANES III). *Diabetes Care* 25, 1675–1680.
335. Kanaya, A. M., Herrington, D., Vittinghoff, E., Lin, F., Grady, D., Bittner, V., Cauley, J. A., and Barrett-Connor, E. (2003). Glycemic effects of postmenopausal hormone therapy: The Heart and Estrogen/progestin replacement study. *Ann Intern Med* 138, 1–9.
336. Gao, H., Bryzgalova, G., Hedman, E., Khan, A., Efendic, S., Gustafsson, J.-Å., and Dahlman-Wright, K. (2006). Long-term administration of estradiol decreases expression of hepatic lipogenic genes and improves insulin sensitivity in ob/ob mice: A possible mechanism is through direct regulation of signal transducer and activator of transcripton 3. *Mol Endocrinol* 20, 1287–1299.
337. Djouadi, F., Weinheimer, C. J., Saffitz, J. E., Pitchford, C., and Bastin, J. (1998). A gender-related defect in lipid metabolism and glucose homeostasis in peroxisome proliferators-activated receptor α-deficient mice. *J Clin Invest* 102, 1083–1091.
338. Heine, P. A., Taylor, J. A., Iwamoto, G. A., Lubahn, D. B., and Cooke, P. S. (2000). Increased adipose tissue in male and female estrogen receptor-α knockout mice. *Proc Natl Acad Sci USA* 97, 12729–12734.
339. Cooke, P. S., Heine, P. A., Taylor, J. A., and Lubahn, D. B. (2001). The role of estrogen and estrogen receptor-α in male adipose tissue. *Mol Cell Endocrinol* 178, 147–154.
340. Bryzgalova, G., Gao, H., Ahren, B., Zierath, J. R., Galuska, D., Steiler, T. L., Dahlman-Wright, K., Nilsson, S., Gustafsson, J.-Å., Efendic, S., and Khan, A. (2006). Evidence that oestrogen receptor-α plays an important role in the regulation of glucse homeostasis in mice: Insulin sensitivity in the liver. *Diabetologia* 49, 588–597.
341. Barros, R. P., Machado, U. F., Warner, M., and Gustafsson, J. Å. (2006a). Muscle GLUT4 regulation by estrogen receptors ERβ and ERα. *Proc Natl Acad Sci USA* 103, 1605–1608.
342. Naaz, A., Zakroczymski, M., Heine, P., Taylor, J., Saunders, P., Lubahn, D., and Cooke, P. S. (2002). Effect of ovariectomy on adipose tissue of mice in the absence of estrogen receptor alpha (ERα): A potential role for estrogen receptor beta (ERβ). *Horm Metab Res* 34, 758–763.
343. Barros, R. P., Machado, U. F., and Gustafsson, J. Å. (2006b). Estrogen receptors: New players in diabetes mellitus. *Trends Mol Med* 12, 425–431.
344. Foryst-Ludwig, A., Clementz, M., Hohmann, S., Hartge, M., Sprang, C., Frost, N., Krikov, M., Bhanot, S., Barros, R., Morani, A., Gustafsson, J. A., Unger, T., and Kintscher, U. (2008). Metabolic actions of estrogen receptor beta (ERβ) are mediated by a negative cross-talk with PPARγ. *PLoS Genet* 4, e1000108.
345. Ohlsson, C., Hellberg, N., Parini, P., Vidal, O., Bohlooly, M., Rudling, M., Lindberg, M. K., Warner, M., Angelin, B., and Gustafsson, J.-Å. (2000). Obesity and disturbed lipoprotein profile in estrogen receptor-alpha-deficient male mice. *Biochem Biophys Res Commun* 278, 640–645.
346. Musatov, S., Chen, W., Pfaff, D. W., Mobbs, C. V., Yang, X.-J., Clegg, D. J., Kaplitt, M. G., and Ogawa, S. (2007). Silencing of estrogen receptor α in the ventromedial nucleus of hypothalamus leads to metabolic syndrome. *Proc Natl acad Sci USA* 104, 2501–2506.
347. Grumbach, M. M. and Auchus, R. J. (1999). Estrogen: Consequences and implications of human mutations in synthesis and action. *J Clin Endocrinol Metab* 84, 4677–4694.
348. Jones, M. E. E., Thorburn, A. W., Britt, K. L., Hewitt, K. N., Wreford, N. G., Proietto, J., Oz, O. K., Leury, B. J., Robertson, K. M., Yao, S., and Simpson, E. R. (2000). Aromatase-deficient (ArKO) mice have a phenotype of increased adiposity. *Proc Natl Acad Sci USA* 97, 12735–12740.

349. Nemoto, Y., Toda, K., Ono, M., Fujikawa-Adachi, K., Saibara, T., Onishi, S., Enzan, H., Okada, T., and Shizuta, Y. (2000). Altered expression of fatty acid-metabolizing enzymes in aromatase-deficient mice. *J Clin Invest* 105, 1819–1825.
350. Hewitt, K. N., Boon, W. C., Murata, Y., Jones, M. E. E., and Simpson, E. R. (2003). The aromatase knockout mouse presents with a sexually dimorphic disruption to cholesterol homeostasis. *Endocrinology* 144, 3895–3903.
351. Jones, M. E. E., Thorburn, A. W., Britt, K. L., Hewitt, K. N., Misso, M. L., Wreford, N. G., Proietto, J., Oz, O. K., Leury, B. J., Robertson, K. M., Yao, S., and Simpson, E. R. (2001). Aromatase-deficient (ArKO) mice accumulate excess adipose tissue. *J Steroid Biochem Mol Biol* 79, 3–9.
352. Takeda, K., Toda, K., Saibara, T., Nakagawa, M., Saika, K., Onishi, T., Sugiura, T., and Shizuta, Y. (2003). Progressive development of insulin resistance phenotype in male mice with complete aromatase (CYP19) deficiency. *J Endocrinol* 176, 237–246.
353. Toda, K., Takeda, K., Akira, S., Saibara, T., Okada, T., Onishi, S., and Shizuta, Y. (2001). Alternations in hepatic expression of fatty-acid metabolizing enzymes in ArKO mice and their reversal by the treatment with 17β-estradiol or a peroxisome proliferator. *J Steroid Biochem Mol Biol* 79, 11–17.
354. Hill, R. A., Pompolo, S., Jones, M. E. E., Simpson, E. R., and Boon, W. C. (2004). Estrogen deficiency leads to apoptosis in dopaminergic neurons in the medial preoptic area and arcuate nucleus of male mice. *Mol Cell Neurosci* 27, 466–476.
355. Hill, R. A., Chow, J., Fritzemeier, K., Simpson, E. R., and Boon, W. C. (2007). Fas/FasL-mediated apoptosis in the arcuate nucleus and medial preoptic area of male ArKO mice is ameliorated by selective estrogen receptor alpha and estrogen receptor beta agonist treatment, respectively. *Mol Cell Neurosci* 36, 146–157.
356. Jones, M. E. E., McInnes, K. J., Boon, W. C., and Simpson, E. R. (2007b). Estrogen and adiposity-utilizing models of aromatase deficiencies to explore the relationship. *J Steroid Biochem Mol Biol* 106, 3–7.
357. Simpson, E., Jones, M., Misso, M., Hewitt, K., Hill, R., Maffei, L., Carani, C., and Boon, W. C. (2005). Estrogen, a fundamental player in energy homeostasis. *J Steroid Biochem Mol Biol* 95, 3–5.
358. Maffei, L., Rochira, V., Zirilli, L., Antunez, P., Aranda, C., Fabre, B., Simone, M. L., Pignatti, E., Simpson, E. R., Houssami, S., Clyne, C. D., and Carani, C. (2007). A novel compound heterozygous mutation of the aromatase gene in an adult man: Reinforced evidence on the relationship between congenital oestrogen deficiency, adiposity and the metabolic syndrome. *Clin Endocrinol* 67, 218–224.
359. Grohé, C., Kahlert, S., Löbbert, K., Stimpel, M., Karas, R. H., Vetter, H., and Neyses, L. (1997). Cardiac myocytes and fibroblasts contain functional estrogen receptors. *FEBS Lett* 416, 107–112.
360. Grohé, C., Kahlert, S., Löbbert, K., and Vetter, H. (1998). Expression of oestrogen receptor alpha and beta in rat heart: Role of local oestrogen synthesis. *J Endocrinol* 156, R1–R7.
361. Foegh, M. L. and Ramwell, P. W. (1998). Cardiovascular effects of estrogen: Implications of the discovery of the estrogen receptor subtypeβ. *Curr Opinion Nephr Hypert* 7, 83–89.
362. Mendelsohn, M. E. and Karas, R. H. (1999). The protective effects of estrogen on the cardiovascular system. *N Engl J Med* 340, 1801–1811.
363. Mendelsohn, M. E. (2002). Genomic and nongenomic effects of estrogen in the vasculature. *Am J Cardiol* 90(suppl), 3F–6F.
364. Valverde, M. A., Rojas, P., Amigo, J., Cosmelli, D., Orio, P., Baharnonde, M. L., Mann, G. E., Vergara, C., and Latorre, R. (1999). Acute activation of Maxi-K channels (hSlo) by estradiol binding to the β subunit. *Science* 285, 1929–1931.
365. Chen, Z., Yuhanna, I. S., Galcheva-Gargova, Z., Karas, R. H., Mendelsohn, M. E., and Shaul, P. W. (1999). Estrogen receptor alpha mediates the nongenomic activation of endothelial nitric oxide synthase by estrogen. *J Clin Invest* 103, 401–406.
366. Mikkola, T., Viinikka, L., and Ylikorkala, O. (1998). Estrogen and postmenopausal estrogen/progestin therapy: Effect on endothelium-dependent prostacyclin, nitric oxide and endothelin-1 production. *Eur J Obstet Gynecol & Reprod Biol* 79, 75–82.

367. Nickening, G., Bäumer, A. T., Grohè, C., Kahlert, S., Strehlow, K., Rosenkranz, S., Stäblein, A., Beckers, F., Smits, J. F. M., Daemen, M. J. A. P., Vetter, H., and Böhm, M. (1998). Estrogen modulates AT1 receptor gene expression in vitro and in vivo. *Circulation* 97, 2197–2201.
368. Akishita, M., Ouchi, Y., Miyoshi, H., Kozaki, K., Inoue, S., Ishikawa, M., Eto, M., Toba, K., and Orimo, H. (1997). Estrogen inhibits cuff-induced intimal thickening of rat femoral artery: Effects on migration and proliferation of vascular smooth muscle cells. *Atherosclerosis* 130, 1–10.
369. Krom, Y. D., Pires, N. M. M., Jukema, J. W., de Vries, M. R., Frants, R. R., Havekes, L. M., van Dijk, K. W., and Quax, P. H. A. (2007). Inhibition of neointima formation by local delivery of estrogen receptor alpha and beta specific agonists. *Cardiovasc Res* 73, 217–226.
370. Mendelsohn, M. E. and Karas, R. H. (2005). Molecular and cellular basis of cardiovascular gender differences. *Science* 308, 1583–1587.
371. Mendelsohn, M. E. and Rosano, G. M. C. (2003). Hormonal regulation of normal vascular tone in males. *Circ Res* 93, 1142–1145.
372. Maffei, L., Murata, Y., Rochira, V., Tubert, G., Aranda, C., Vazquez, M., Clyne, C. D., Davis, S., Simpson, E. R., and Carani, C. (2004). Dysmetabolic syndrome in a man with a novel mutation of the aromatase gene: Effects of testosterone, alendronate, and estradiol treatment. *J Clin Endocrinol Metab* 89, 61–70.
373. Lew, R., Komesaroff, P., Williams, M., Dawood, T., and Sudhir, K. (2003). Endogenous estrogens influence endothelial function in young men. *Circ Res* 93, 1127–1133.
374. Kimura, M., Sudhir, K., Jones, M., Simpson, E., Jefferis, A.-M., and Chin-Dusting, J. P. F. (2003). Impaired acetylcholine-induced release of nitric oxide in the aorta of male aromatase-knockout mice: Regulation of nitric oxide production by endogenous sex hormones in males. *Circ Res* 93, 1267–1271.
375. Nathan, L., Shi, W., Dinh, H., Mukherjee, T. K., Wang, X., Lusis, A. J., and Chaudhuri, G. (2001). Testosterone inhibits early atherogenesis by conversion to estradiol: Critical role of aromatase. *Proc Natl Acad Sci USA* 98, 3589–3593.
376. Vitale, C., Miceli, M., and Rosano, G. M. C. (2007). Gender-specific chracteristics of atherosclerosis in menopausal women: Risk factors, clinical course and strategies for prevention. *Climacteric* 10(suppl 2), 16–20.
377. Larsen, J. A. and Kadish, A. H. (1998). Effects of gender on cardiac arrhythmias. *J Cardiovasc Electrophysiol* 9, 655–664.
378. Reckelhoff, J. F. (2001). Gender differences in the regulation of blood pressure. *Hypertension* 37, 1199–1208.
379. Marcus, R., Krause, L., Weder, A. B., Dominguez-Meja, A., Schork, N. J., and Julius, S. (1994). Sex-specific determinants of increased left ventricular mass in the Tecumseh Blood Pressure Study. *Circulation* 90, 928–936.
380. Gardin, J. M., Wagenknecht, L. E., Anton-Culver, H., Anton-Culver, H., Flack, J., Gidding, S., Kurosaki, T., Wong, N. D., and Manolio, T. A. (1995). Relationship of cardiovascular risk factors to echocardiographic left ventricular mass in healthy young black and white adult men and women. The CARDIA study. Coronary Artery Risk Development in Young Adults. *Circulation* 92, 380–387.
381. Collins, P., Rosano, G. M. C., Sarrel, P. M., Ulrich, L., Adamopoulos, S., Beale, C. M., McNeill, J. G., and Poole-Wilson, A. (1995). 17β-estradiol attenuates acetylcholine-induced coronary arterial constriction in women but not men with coronary heart disease. *Circulation* 92, 24–30.
382. Arias-Loza, P. A., Jazbutyte, V., and Pelzer, T. (2008). Genetic and pharmacological strategies to determine the function of estrogen receptor α and estrogen receptor β in the cardiovascular system. *Gend Med* 5(Suppl A), S34-S45.
383. Brouchet, L., Krust, A., Dupont, S., Chambon, P., Bayard, F., and Arnal, J. F. (2001). Estradiol accelerates reendothelialization in mouse carotid artery through estrogen receptor-α but not estrogen receptor-β. *Circulation* 103, 423–428.
384. Pare, G., Krust, A., Karas, R., Dupont, S., Aronovitz, M., Chambon, P., and Mendelsohn, M. E. (2002). Estrogen receptor-α mediates the protective effects of estrogen against vascular injury. *Circ Res* 90, 1087–1092.

385. Pelzer, T., Jazbutyte, V., Hu, K., Segerer, S., Nahrendorf, M., Nordbeck, P., Bonz, A. W., Muck, J., Fritzemeier, K.-H., Hegele-Hartung, C., Ertl, G., and Neyses, L. (2005). The estrogen receptor-α agonist 16α-LE2 inhibits cardiac hypertrophy and improves hemodynamic function in estrogen-deficient spontaneously hypertensive rats. *Cardiovasc Res* 67, 604–612.
386. Pedram, A., Razandi, M., Lubahn, D., Liu, J., Vannan, M., and Levin, E. R. (2008). Estrogen inhibits cardiac hypertrophy: Role of estrogen receptor-β to inhibit calcineurin. *Endocrinology* 149, 3361–3369.
387. Hanke, H., Kamenz, J., Hanke, S., Spieβ, J., Lenz, C., Brehme, U., Bruck, B., Finking, G., and Hombach, V. (1999). Effect of 17-β estradiol on pre-existing atherosclerotic lesions: Role of the endothelium. *Atherosclerosis* 147, 123–132.
388. Rosenfeld, M. E., Kauser, K., Martin-McNulty, B., Polinksy, P., Schwartz, S. M., and Rubanyi, G. M. (2002). Estrogen inhibits the initiation of fatty streaks throughout the vasculature but does not inhibit intra-plaque hemorrhage and the progression of established lesions in apolipoprotein E deficient mice. *Atherosclerosis* 164(251), 259.
389. Hodgin, J. B. and Maeda, N. (2002). Minireview: Estrogen and mouse models of atherosclerosis. *Endocrinology* 143, 4495–4501.
390. Egan, K. M., Lawson, J. A., Fries, S., Koller, B., Rader, D. J., Smyth, E. M., and FitzGerald, G. A. (2004). COX-2-derived prostacyclin confers atheroprotection on female mice. *Science* 306, 1954–1957.
391. Zhu, Y., Bian, Z., Lu, P., Karas, R. H., Bao, L., Cox, D., Hodgin, J., Shaul, P. W., Thorén, P., Smithies, O., Gustafsson, J.-Å., and Mendelsohn, M. E. (2002). Abnormal vascular function and hypertension in mice deficient in estrogen receptor β. *Science* 295, 505–507.
392. Gabel, S. A., Walker, V. R., London, R. E., Steenbergen, C., Korach, K. S., and Murphy, E. (2005). Estrogen receptor beta mediates gender differences in ischemia/reperfusion injury. *J Mol Cell Cardiol* 38, 289–297.
393. Nikolic, I., Liu, D., Bell, J. A., Collins, J., Steenbergen, C., and Murphy, E. (2007). Treatment with an estrogen receptor-beta-selective agonist is cardioprotective. *J Mol Cell Cardiol* 42, 769–780.
394. Sudhir, K., Chou, T. M., Messina, L. M., Hutchison, S. J., Korach, K. S., Chatterjee, K., and Rubanyi, G. M. (1997a). Endothelial dysfunction in a man with disruptive mutation in oestrogen-receptor gene. *The Lancet* 349, 1146–1147.
395. Sudhir, K., Chou, T. M., Chatterjee, K., Smith, E. P., Williams, T. C., Kane, J. P., Malloy, M. J., Korach, K. S., and Rubanyi, G. M. (1997b). Premature coronary artery disease associated with a disruptive mutation in the estrogen receptor gene in a man. *Circulation* 96, 3774–3777.
396. McEwen, B. S. and Alves, S. E. (1999). Estrogen actions in the central nervous system. *Endocrine Rev* 20, 279–307.
397. Sherwin, B. B. (2003). Estrogen and cognitive functioning in women. *Endocrine Rev* 24, 133–151.
398. McCarthy, M. M. (2008). Estradiol and the developing brain. *Physiol Rev* 88, 91–134.
399. Spencer, J. L., Waters, E. M., Romeo, R. D., Wood, G. E., Milner, T. A., and McEwen, B. S. (2008). Uncovering the mechanisms of estrogen effects on hippocampal function. *Front Neuroendocrinol* 29, 219–237.
400. Di Paolo, T. (1994). Modulation of brain dopamine transmission by sex steroids. *Rev Neurosci* 5, 27–42.
401. Grattan, D. R., Rocca, M. S., Strauss, K. I., Sagrillo, C. A., Selmanoff, M., and McCarthy, M. M. (1996). GABAergic neuronal activity and mRNA levels for both forms of glutamic acid decarboxylase (GAD_{65} and GAD_{67}) are reduced in the diagional band of Broca during the afternoon proestrus. *Brain Res* 733, 46–55.
402. Saunders-Pullman, R., Gordon-Elliott, J., Parides, M., Fahn, S., Saunders, H. R., and Bressman, S. (1999). The effect of estrogen replacement on early Parkinson's disease. *Neurology* 52, 1417–1421.
403. Shughrue, P. J., Lane, M. V., and Merchenthaler, I. (1997). Comparative distribution of estrogen receptor-α and-β mRNA in the rat central nervous system. *J Comp Neurol* 388, 507–525.
404. Laflamme, N., Nappi, R. E., Drolet, G., Labrie, C., and Rivest, S. (1998). Expression and neuropeptidergic characterization of estrogen receptors (ERα and ERβ) throughout the rat brain: Anatomical evidence of distinct roles of each subtype. *J Neurobiol* 36, 357–378.

405. Wang, L., Andersson, S., Warner, M., and Gustafsson, J.-Å. (2003). Estrogen receptor (ER)β knock-out mice reveal a role for ERβ in migration of cortical neurons in the developing brain. *Proc Natl Acad Sci USA* 100, 703–708.
406. Weiser, M. J., Foradori, C. D., and Handa, R. J. (2008). Estrogen receptor beta in the brain: From form to function. *Brain Res Rev* 57, 309–320.
407. Morissette, M., Le Saux, M., D'Astous, M., Jourdain, S., Al Sweidi, S., Morin, N., Estrada-Camarena, E., Mendez, P., Garcia-Segura, L. M., and Di Paolo, T. (2008). Contribution of estrogen receptors alpha and beta to the effects of estradiol in the brain. *Steroid Biochem Mol Biol* 108, 327–338.
408. Walf, A. A. and Frye, C. A. (2005). ERβ-selective estrogen receptor modulators produce antianxiety behavior when administrated systemic to ovariectomized rats. *Neuropsychopharmacology* 30, 1598–1609.
409. Bowe, J., Li, X. F., Kinsey-Jones, J., Heyerick, A., Brain, S., Milligan, S., and O'Byrne, K. (2006). The hop phytoestrogen, 8-prenylnaringenin, reverses the ovariectomy-induced rise in skin temperature in an animal model of menopausal hot flushes. *J Endocrinol* 191, 399–405.
410. Chakrabarty, A., Blacklock, A., Svojanovsky, S., and Smith, P. G. (2008). Estrogen elicits dorsal root ganglion axon sprouting via a rennin-angiotensin system. *Endocrinology* 149, 3452–3460.
411. Rocha, B. A., Fleischer, R., Schaeffer, J. M., Rohrer, S. P., and Hickey, G. J. (2005). 17β-estradiol-induced antidepressant-like effect in the forced swim test is absent in estrogen receptor-β knockout (BERKO) mice. *Psychopharmacology* 179, 637–643.
412. Hughes, Z. A., Liu, F., Platt, B. J., Dwyer, J. M., Pulicicchio, C. M., Zhang, G., Schechter, L. E., Rosenzweig-Lipson, S., and Day, M. (2008). WAY-200070, a selective agonist of estrogen receptor beta as a potential anxiolytic/antidepressant agent. *Neuropharm* 54, 1136–1142.
413. Walf, A. A., Rhodes, M. E., and Frye, C. A. (2004). Antidepressant effects of ERβ-selective estrogen receptor modulators in the forced swim test. *Pharmacol Biochem Behav* 78, 523–529.
414. Walf, A. A. and Frye, C. A. (2007). Administration of estrogen receptor beta-specific selective estrogen receptor modulators to the hippocampus decrease anxiety and depressive behavior of ovariectomized rats. *Pharmacol Biochem Behav* 86, 407–414.
415. Walf, A. A. and Frye, C. A. (2008). Rapid and estrogen receptor beta mediated actions in the hippocampus mediate some functional effects of estrogen. *Steroids* 73, 997–1007.
416. Weiser, M. J., Wu, T. J., and Handa, R. J. (2009). Estrogen receptor-β agonist diarylpropionitrile: Biological activities of R- and S-enantiomers on behaviour and hormonal response to stress. *Endocrinology* 150, 1817–1825.
417. Walf, A. A., Koonce, C. J., and Frye, C. A. (2008). Estradiol and diarylpropionitrile administration to wild type, but not estrogen receptor beta knockout, mice enhances performance in the object recognition and object placement tasks. *Neurobiol Learn Mem* 89, 513–521.
418. Liu, F., Day, M., Muñiz, L. C., Bitran, D., Arias, R., Revilla-Sanchez, R., Grauer, S., Zhang, G., Kelley, C., Pulito, V., Sung, A., Mervis, R. F., Navarra, R., Hirst, W. D., Reinhart, P. H., Marquis, K. L., Moss, S. J., Pangalos, M. N., and Brandon, N. J. (2008b). Activation of estrogen receptor-β regulates hippocampal synaptic plasticity and improves memory. *Nat Neurosci* 11, 334–343.
419. Andreescu, C. E., Milojkovic, B. A., Haasdijk, E. D., Kramer, P., De Jong, F. H., Krust, A., De Zeeuw, C. I., and De Jeu, M. T. G. (2007). Estradiol improves cerebellar memory formation by activating estrogen receptor β. *J Neurosci* 27, 10832–10839.
420. Meltser, I., Tahera, Y., Simpson, E., Hultcrantz, M., Charitidi, K., Gustafsson, J.-Å., and Canlon, B. (2008). Estrogen receptor β protects against acoustic trauma in mice. *J Clin Invest* 118, 1563–1570.
421. Renier, M. A., Vereecken, A., and Buytaert, P. (1998). Inhibins, activins and follistatins: A review of complex regulators of the reproductive system. *Eur J Contracep Reprod Health Care* 3, 129–135.
422. Lockwood, G. M., Muttukrishna, S., and Ledger, W. L. (1998). Inhibins and activins in human ovulation, conception and pregnancy. *Hum Reprod Update* 4, 284–295.
423. Wintermantel, T. M., Campbell, R. E., Porteous, R., Bock, D., Gröne, H.-J., Todman, M. G., Korach, K. S., Greiner, E., Pérez, C. A., Schütz, G., and Herbison, A. E. (2006). Definiton of estrogen receptor pathway critical for estrogen positive feed-back to gonadotropin-releasing hormone neurons and fertility. *Neuron* 52, 271–280.

424. Adachi, S., Yamada, S., Takatsu, Y., Matsui, H., Kinoshita, M., Takase, K., Suguira, H., Ohtaki, T., Matsumoto, H., Uenoyama, Y., Tsukamura, H., Inoue, K., and Maeda, K.-I. (2007). Involvement of anteroventral periventricular metastin/kisspeptin neurons in estrogen positive feedback action on luteinizing hormone release in female rats. *J Reprod Dev* 53, 367–378.
425. Roa, J., Aguilar, E., Dieguez, C., Pinilla, L., and Tena-Sempere, M. (2008a). New frontiers in kisspeptin/GPR54 physiology as fundamental gatekeepers of reproduction function. *Front Neuroendocrinoln* 29, 48–69.
426. Roa, J., Vigo, E., Castellano, J. M., Gaytan, F., Navarro, V. M., Aguilar, E., Dijcks, F. A., Ederveen, A. G. H., Pinilla, L., van Noort, P. I., and Tena-Sempere, M. (2008b). Opposite roles of estrogen receptor (ER)α and ERβ in the modulation of luteinizing hormone responses to kisspeptin in the female rat: Implications for the generation of the preovulatory surge. *Endocrinology* 149, 1627–1637.
427. Carani, C., Qin, K., Simoni, M., Faustini-Fustini, M., Serpente, S., Boyd, J., Korach, K. S., and Simpson, E. R. (1997). Effect of testosterone and estradiol in a man with aromatase deficiency. *New Engl J Med* 337, 91–95.
428. Writing Group for the Women's Health Initiative Investigators (2002). Risk and benefits of estrogen plus progestin in healthy postmenopausal women: Principle results from the women's health initiative randomized controlled trial. *JAMA* 288, 321–333.
429. The Women's Health Initiative Steering Committee (2004). Effects of conjugated equine estrogen in postmenopausal women with hysterectomy: The women's health initiative randomized controlled trial. *JAMA* 291, 1701–1712.
430. Phillips, L. S. and Langer, R. D. (2005). Postmenopausal hormone therapy: Critical reappraisal and a unified hypothesis. *Fertil Steril* 83, 558–566.
431. Clarkson, T. B. and Appt, S. E. (2005). Controversies about HRT – lessons from monkey models. *Maturitas* 51, 64–74.
432. Lam, P. M., Chung, T. K., and Haines, C. (2005). Where are we with postmenopausal hormone therapy in 2005? *Gynecol Endocrinol* 21, 248–256.
433. Harris, H. A. (2006a). The unexpected science of estrogen receptor-β selective agonists: A new class of anti-inflammatory agents? *Nucl Recept Signal* 4, e012.
434. Harris, H. A. (2006b). Preclinical characterization of selective estrogen receptor β agonists: new insights into their therapeutic potential. Ernst Schering Found Symp Proc. pp 149–161.
435. Harris, H. A. (2007). Estrogen receptor-β: Recent lessons from in vivo studies. *Mol Endocrinol* 21, 1–13.
436. Gustafsson, J.-Å. (2003). What pharmacologists can learn from recent advances in estrogen signalling. *Trends Pharmacol Sci* 24, 479–485.
437. Gustafsson, J.-Å. (2006). ERβ scientific visions translate to clinical uses. *Climacteric* 9, 156–160.
438. Craft, R. M. (2007). Modulation of pain by estrogens. *Pain* 132, S3–S12.
439. Cvoro, A., Paruthiyil, S., Jones, J. O., Tzagarakis-Foster, C., Clegg, N. J., Tatomer, D., Medina, R. T., Tagliaferri, M., Schaufele, F., Scanlan, T. S., Diamond, M. I., Cohen, I., and Leitman, D. C. (2007). Selective activation of estrogen receptor-β transcriptional pathways by an herbal extract. *Endocrinology* 148, 538–547.
440. Straub, R. H. (2007). The complex role of estrogens in inflammation. *Endocrine Rev* 28, 521–574.
441. L'Hermite, M., Simoncini, T., Fuller, S., and Genazzani, A. R. (2008). Could transdermal estradiol + progesterone be a safer postmenopausal HRT? A review. *Maturitas* 60, 185–201.
442. Montanaro, D., Maggiolini, M., Recchia, A. G., Sirianni, R., Aquila, S., Barzon, L., Fallo, F., Andò, S., and Pezzi, V. (2005). Antiestrogens upregulate estrogen receptor β expression and inhibit adrenocortical H295R cell proliferation. *J Mol Endocrinol* 35, 245–356.
443. Shen, S. S., Smith, C. L., Hsieh, J.-T., Yu, J., Kim, I. Y., Jian, W., Sonpavde, G., Ayala, G. E., Younes, M., and Lerner, S. P. (2006). Expression of estrogen receptors-α and-β in bladder cancer cell lines and human bladder tumor tissue. *Cancer* 106, 2610–2616.
444. Paruthiyil, S., Parmar, H., Kerekatte, V., Cunha, G. R., Firestone, G. L., and Leitman, D. C. (2004). Estrogen receptor β inhibits human breast cancer cell proliferation and tumor formation by causing a G_2 cell cycle arrest. *Cancer Res* 64, 423–428.

445. Ström, A., Hartman, J., Foster, J. S., Kietz, S., Wimalasena, J., and Gustafsson, J.-Å. (2004). Estrogen receptor β inhibits 17β-estradiol-stimulated proliferation of the breast cancer cell line T47D. *Proc Natl Acad Sci USA* 101, 1566–1571.

446. Mersereau, J. E., Levy, N., Staub, R. E., Baggett, S., Zogric, T., Chow, S., Ricke, W. A., Tagliaferri, M., Cohen, I., Bjeldanes, L. F., and Leitman, D. C. (2008). Liquiritigenin is a plant-derived highly selective estrogen receptor β agonist. *Mol Cell Endocrinol* 283, 49–57.

447. Sampson, L. K., Vicker, S. M., Ying, W., and Phillips, J. O. (1997). Tamoxifen-mediated growth inhibition of human cholangiocarcinoma. *Cancer Res* 57, 1743–1749.

448. Vickers, S. M., Jhala, N. C., Ahn, E.-Y., McDonald, J. M., Pan, G., and Bland, K. I. (2002). Tamoxifen (TMX)/Fas induced growth inhibition of human cholangiocarcinoma (HCC) by gamma interferon (IFN-γ). *Annals Surg* 235, 872–878.

449. Alvaro, D., Barbaro, B., Franchitto, A., Onori, P., Glaser, S. S., Alpini, G., Francis, H., Marucci, L., Sterpetti, P., Ginanni-Corradini, S., Muda, A. O., Dostal, D. E., De Santis, A., Attili, A. F., Benedetti, A., and Gaudio, E. (2006). Estrogens and insulin-like growth factor 1 modulate neoplastic cell growth in human cholangiocarcinoma. *Am J Pathol* 169, 877–888.

450. Shim, G.-J., Wang, L., Andersson, S., Nagy, N., Kis, L. L., Zhang, Q., Mäkelä, S., Warner, M., and Gustafsson, J.-Å. (2003). Disruption of the estrogen receptor β gene in mice causes myeloproliferative disease resembling chronic myeloid leukemia with lymphoid blast crisis. *Proc Natl Acad Sci USA* 100, 6694–6699.

451. Nelson, H. D., Humphrey, L. L., Nygren, P., Teutsch, S. M., and Allan, J. D. (2002). Postmenopausal hormone replacement therapy: Scientific review. *JAMA* 288, 872–881.

452. Wada-Hiraike, O., Imamov, O., Hiraike, H., Hultenby, K., Schwend, T., Omoto, Y., Warner, M., and Gustafsson, J.-Å. (2006). Role of estrogen receptor β in colonic epithelium. *Proc Natl Acad Sci USA* 103, 2959–2964.

453. Kennelly, R., Kavanagh, D. O., Hogan, A. M., and Winter, D. C. (2008). Oestrogen and the colon: Potential mechanisms for cancer prevention. *Lancet Oncol* 9, 385–391.

454. Raju, J., Bielecki, A., Caldwell, D., Lok, E., Taylor, M., Kapal, K., Curran, I., Cooke, G. M., Bird, R. P., and Mehta, R. (2009). Soy isoflavones modulate azoxymethane-induced rat colon carcinogenesis exposed pre- and postnatally and inhibit growth of DLD-1 human colon adenocarcinoma cells by increasing the expression of estrogen receptor-β. *J Nutr*, 474–481.

455. Alì, G., Donati, V., Loggini, B., Servadio, A., Dell'Omodame, M., Prati, M. C., Camacci, T., Lucchi, M., Melfi, F., Mussi, A., and Fontanini, G. (2008). Different estrogen receptor beta expression in distinct histologic subtypes of lung adenocarcinoma. *Hum Pathol* 39, 1465–1473.

456. Zhang, G.-F., Liu, X., Farkas, A. M., Parwani, A. V., Lathrop, L., Lenzner, D., Land, S. R., and Srinivas, H. (2009). Estrogen receptor β functions through nongenomic mechanisms in lung cancer cells. *Mol Endocrinol* 23, 145–156.

457. Belcher, S. M., Ma, X., and Le, H. H. (2009). Blockade of estrogen receptor signaling inhibits growth and migration of medulloblastoma. *Endocrinology* 150, 1112–1121.

458. Tanemura, A., van Hoesel, A. Q., Mori, T., Yu, T., and Hoon, D. S. B. (2007). The role of estrogen receptor in melanoma. *Expert Opin Ther Targets* 11, 1639–1648.

459. Sola, B. and Renoir, J.-M. (2007). Estrogenic or antiestrogenic therapies for multiple myeloma. *Mol Cancer* 6, 59.

460. Renoir, J.-M., Bouclier, C., Seguin, A., Marsaud, V., and Sola, B. (2008). Antioestrogen-mediated cell cycle arrest and apoptosis induction in breast cancer and multiple myeloma cells. *J Mol Endocrinol* 40, 101–112.

461. Chan, K. K. L., Wei, N., Liu, S. S., Xiao-Yun, L., Cheung, A. N., and Ngan, H. Y. S. (2008). Estrogen receptor subtypes in ovarian cancer. *Obstet Gynecol* 111, 144–151.

462. Imamov, O., Morani, A., Shim, G.-J., Omoto, Y., Thulin-Andersson, C., Warner, M., and Gustafsson, J.-Å. (2004). Estrogen receptor β regulates epithelial cellular differentiation in the mouse ventral prostate. *Proc Natl Acad Sci* 101, 9375–9380.

463. Mobley, J. A., L'Esperance, J. O., Wu, M., Friel, C. J., Hanson, R. H., and Ho, S.-M. (2004). The novel estrogen 17α-20Z-21-[(4-amino)phenyl]-19-norpregna-1,3,5(10),20-tetraene-3,17β-diol

induces apoptosis in prostate cancer cell lines at nanomolar concentrations in vitro. *Mol Cancer Ther* 3, 587–595.

464. McPherson, S. J., Ellem, S. J., Simpson, E. R., Patchev, V., Fritzemeier, K.-H., and Risbridger, G. P. (2007). Essential role for estrogen receptor β in stromal-epithelial regulation of prostatic hyperplasia. *Endocrinology* 148, 566–574.
465. Greenberg, J. A., Somme, S., Russnes, H. E., Durbin, A. D., and Malkin, D. (2008). The estrogen receptor pathway in rhabdomyosarcoma: A role for estrogen receptor-β in proliferation and response to the antiestrogen 4'OH-tamoxifen. *Cancer Res* 68, 3476–3485.
466. Ho, S.-M., Leung, Y.-K., and Chung, I. (2006). Estrogens and antiestrogens as etiological factos and therapeutics for prostate cancer. *Ann N Y Acad Sci* 1089, 177–193.
467. Bosland, M. C. (2006). Sex steroids and prostate carcinogenesis integrated, multifactorial working hypothesis. *Ann N Y Acad Sci* 1089, 168–176.
468. Carruba, G. (2007). Estrogen and prostate cancer: An eclipsed truth in an androgen-dominated scenario. *J Cell Biochem* 102, 899–911.
469. Weihua, Z., Lathe, R., Warner, M., and Gustafsson, J.-Å. (2002). An endocrine pathway in the prostate, ERβ, AR, 5α-androstane-3β,17β-diol, and CYP7B1, regulates prostate growth. *Proc Natl Acad Sci USA* 99, 13589–13594.
470. Guerini, V., Sau, D., Scaccianoce, E., Rusmini, P., Ciana, P., Maggi, A., Martini, P. G. V., Katzenellenbogen, B. S., Martini, L., Motta, M., and Poletti, A. (2005). The androgen derivative 5α-androstane-3β,17β-diol inhibits prostate cancer cell migration through activation of the estrogen receptor β subtype. *Cancer Res* 65, 5445–5453.
471. Savolainen, S., Pakarainen, T., Huhtaniemi, I., Poutanen, M., and Mäkelä, S. (2007). Delay of postnatal maturation sensitizes the mouse prostate to testosterone induced pronounced hyperplasia: Protective role of estrogen receptor β. *Am J Pathol* 171, 1013–1022.
472. Norman, B. H., Dodge, J. A., Richardson, T. I., Borromeo, P. S., Lugar, C. W., Jones, S. A., Chen, K., Wang, Y., Durst, G. L., Barr, R. J., Montrose-Rafizadeh, C., Osborne, H. E., Amos, R. M., Guo, S., Boodhoo, A., and Krishnan, V. (2006). Benzopyrans are selective estrogen receptor β agonists with novel activity in models of benign prostatic hyperplasia. *J Med Chem* 49, 6155–6157.
473. Sasaki, M., Tanaka, Y., Perinchery, G., Dharia, A., Kotcherguina, I., Fujimoto, S. I., and Dahiya, R. (2002). Methylation and inactivation of estrogen, progesterone, and androgen receptors in prostate cancer. *J Natl Cancer Inst* 94, 384–390.
474. Zhu, X., Leav, I., Leung, Y.-K., Wu, M., Liu, Q., Gao, Y., McNeal, J. E., and Ho, S.-M. (2004). Dynamic regulation of estrogen receptor-β expression by DNA methylation during prostate cancer development and metastasis. *Am J Pathol* 164, 2003–2012.
475. Stettner, M., Kaulfuβ, S., Burfeind, P., Schweyer, S., Strauss, A., Ringert, R. H., and Thelen, P. (2007). The relevance of estrogen receptor-β expression to the antiproliferative effects observed with histone deacetylase inhibitors and phytoestrogens in prostate cancer treatment. *Mol Cancer Ther* 6, 2626–2633.
476. Walton, T. J., Seth, G. L., McArdle, S. E., Bishop, M. C., and Rees, R. C. (2008). DNA methylation and histone deacetylation inhibition co-operate to re-express estrogen receptor beta and induce apoptosis in prostate cancer cell-lines. *Prostate* 68, 210–222.
477. Setlur, S. R., Mertz, K. D., Hoshida, Y., Demichelis, F., Lupien, M., Perner, S., Sboner, A., Pawitan, Y., Andrén, O., Johnson, L. A. et al. (2008). Estrogen-dependent signaling in a molecular distinct subclass of aggressive prostate cancer. *J Natl Cancer Inst* 100, 815–825.
478. Honma, N., Horii, R., Iwase, T., Saji, S., Younes, M., Takubo, K., Matsuura, M., Ito, Y., Akiyama, F., and Sakamoto, G. (2008). Clinical importance of estrogen receptor-β evaluation in breast cancer patients treated with adjuvant tamoxifen therapy. *J Clin Oncol* 26, 3727–3734.
479. Bardin, A., Boulle, N., Lazennec, G., Vignon, F., and Pujol, P. (2004). Loss of ERβ expression as a common step in estrogen-dependent tumor progression. *Endocr Relat Cancer* 11, 537–551.
480. Lin, C. Y., Ström, A., Li Kong, S., Kietz, S., Thomsen, J. S., Tee, J. B., Vega, V. B., Miller, L. D., Smeds, J., Bergh, J., and Gustafsson, J.-Å. (2007). Inhibitory effects of estrogen receptor beta on specific hormone-responsive gene expression and association with disease outcome in primary breast cancer. *Breast Cancer Res* 9, R25.

481. Fox, E. M., Davis, R. J., and Shupnik, M. A. (2008). ERβ in breast cancer – onlooker, passive player, or active protector. *Steroids* 73, 1039–1051.

482. Helguero, L. A., Hedengren Faulds, M., Gustafsson, J.-Å., and Haldosén, L.-A. (2005). Estrogen receptors alfa (ERα) and beta (ERβ) differentially regulate proliferation and apoptosis of the normal murine mammary epithelial cell line HC11. *Oncogene* 24, 6605–6616.

483. Hartman, J., Lindberg, K., Morani, A., Inzunza, J., Ström, A., and Gustafsson, J.-Å. (2006). Estrogen receptor β inhibits angiogenesis and growth of T47D breast cancer xenografts. *Cancer Res* 66, 11207–11213.

484. Nettles, K. W., Sun, J., Radek, J. T., Sheng, S., Rodriguez, A. L., Katzenellenbogen, J. A., Katzenellenbogen, B. S., and Greene, G. L. (2004). Allosteric control of ligand selectivity between estrogen receptors α and β: Implications for other nuclear receptors. *Mol Cell* 13, 317–327.

485. Shang, Y. and Brown, M. (2002). Molecular determinants for the tissue specificity of SERMs. *Science* 295, 2465–2468.

CHAPTER 6

ANDROGEN RECEPTOR

JAMES T. DALTON[1] AND WENQING GAO[2]

[1] *GTx Inc., Memphis, TN 38163, USA*

[2] *Department of Pharmaceutical Sciences, School of Pharmacy and Pharmaceutical Sciences, University at Buffalo SUNY, Buffalo, NY 14260, USA*

Abstract: Androgen receptor (NR3C4, or AR) is another important steroid hormone receptor that is activated by endogenous androgens, mainly testosterone and 5α-dihydrotestosterone (5α-DHT). AR function can be regulated by the binding of its ligands, either agonists or antagonists, which initiate sequential conformational changes of the receptor that will affect receptor function. At the cellular level, AR mainly acts as a transcription factor to regulate downstream target genes expression, while emerging research also revealed that nongenomic actions contribute to AR signaling and function. AR mediates the physiological actions of androgen, which is responsible for male sexual differentiation and pubertal changes. Historically, various steroidal androgens were used to treat androgen-related endocrine disorders, but the advent of aromatase inhibitors and recombinant erythropoietin supplanted the use of androgens in many of these conditions. With the rapidly increasing knowledge of AR protein structure and molecular mechanism of action in the last few decades, a large variety of nonsteroidal ligands, have been developed to selectively modulate AR action with much improved tissue and/or function specificity and safety profiles as compared to steroids, which could greatly expand the therapeutic uses of androgens. This chapter will focus on the chemistry and structural biology of AR, and its role and potential as drug target in disease treatment, with emphasis on the recent development of selective androgen receptor modulators (SARMs). We will provide a brief introduction of AR structure and function, followed by a detailed discussion of well-characterized synthetic AR ligands (steroidal and nonsteroidal), with integrated discussions regarding the molecular mechanism of action and potential therapeutic applications for both existing and emerging classes of AR ligands.

6.1. INTRODUCTION

Androgen receptor (NR3C4, or AR) is another important steroid hormone receptor that is activated by endogenous androgens, mainly testosterone and 5α-dihydrotestosterone (5α-DHT). As described in earlier chapters of this book, only five vertebrate steroid receptors have been identified among this large family of proteins: glucocorticoid (NR3C1, or GR), mineralocorticoid (NR3C2, or MR), progesterone (NR3C3, or PR), androgen (NR3C4, or AR), and estrogen receptors (ER),

C.M. Bunce, M.J. Campbell (eds.), *Nuclear Receptors*, Proteins and Cell Regulation 8, 143–182. DOI 10.1007/978-90-481-3303-1_6,

including ERα (NR3A1) and ERβ (NR3A2) [1]. Similar to other steroid receptors, AR function is regulated by the binding of its ligands, either agonists or antagonists, which initiate sequential conformational changes of the receptor that will affect receptor function. At the cellular level, AR mainly acts as a transcription factor to regulate downstream target genes expression. However, emerging research in the past few years has also revealed that nongenomic actions contribute to AR signaling and function.

As a sex hormone receptor, AR mediates the physiological actions of androgen, which is responsible for male sexual differentiation and pubertal changes. Historically (starting in the 1950s), various steroidal androgens were used to treat androgen-related endocrine disorders, such as hopogonadism, anemia and/or breast cancer. However, the advent of aromatase inhibitors and recombinant erythropoietin supplanted the use of androgens in many of these conditions. With the rapidly increasing knowledge of AR protein structure and molecular mechanism of action in the last few decades, a large variety of AR ligands, largely nonsteroidal ligands, have been developed to selectively modulate AR action. The newer generations of AR ligands demonstrate much improved tissue and/or function specificity and safety profiles as compared to steroids, which could greatly expand the therapeutic uses of androgens.

Considering its importance as a drug target, the search for more functionally specific AR ligands will continue to expand. This chapter will focus on the chemistry and structural biology of AR, and its role and potential as drug target in disease treatment, with emphasis on the recent development of selective androgen receptor modulators (SARMs). We will provide a brief introduction of AR structure and function, as well as the chemistry of endogenous AR ligands. This will be followed by a detailed discussion of well-characterized synthetic AR ligands (steroidal and nonsteroidal) that bind to AR with high affinity and specificity, with integrated discussions regarding the molecular mechanism of action and potential therapeutic applications for both existing and emerging classes of AR ligands.

6.1.1. Physiologic Roles and Clinical Application of Androgens

Physiologically, functional AR is responsible for male sexual differentiation in utero and for male pubertal changes. In adult males, androgen is mainly responsible for maintaining libido, spermatogenesis, muscle mass and strength, bone mineral density, and stimulating erythropoiesis [2–4]. Accordingly, AR is mainly expressed in androgen target tissues, such as the reproductive tissues (i.e., prostate, seminal vesicle, epididymis, testes), skeletal muscle, liver, and central nervous system (CNS), with the highest expression level observed in the prostate, adrenal gland, and epididymis as determined by real-time PCR [5]. In general, the actions of androgen in the reproductive tissues are known as the androgenic effects, while the nitrogen-retaining effects of androgen in muscle and bone are referred to as the anabolic effects.

Classically, testosterone is primarily used to treat endocrine disorders related to androgen deficiency, like male hypogonadism and Klinefelter's syndrome. In recent years, hormone replacement therapy has also been proposed as an approach to overcome age-related androgen insufficiency in aging males, in hope of improving body composition, bone and cartilage metabolism, certain domains of brain function, and even decrease cardiovascular risk [6]. Besides treating the endocrine disorders, androgens have also been used to stimulate erythropoiesis in the treatment of anemia (e.g., aplastic anemia, and anemia secondary to chronic renal failure), and improve nitrogen balance and muscle development in patients with catabolic states due to cancer, burns, traumas, or Acquired Immunodeficiency Syndrome (AIDS). In contrast to its therapeutic uses, androgens are sometimes abused by athletes in the belief that its anabolic effects would help improve their athletic performance. Although androgens do promote muscle growth in individuals with androgen deficiency, it is rather controversial whether androgens, at high pharmacological doses, would have any beneficial effects on muscle development in sexually mature men. Occasionally, androgens are also used to treat short stature, breast cancer (as anti-estrogen), and hereditary angioedema [2].

As the major reproductive organ, prostate function is particularly sensitive to androgen action. Prostate diseases like BPH (Benign Prostatic Hyperplasia) and prostate cancer represent some of the greatest threats to men's heath. The etiology and treatment of both diseases are linked to excess androgen stimulation, particularly by DHT. Various therapeutic strategies [7] have been developed to block androgen action by either reducing tissue DHT concentrations (e.g., 5α-reductase inhibitor) or blocking DHT binding to AR (i.e., antiandrogens). Therefore, antiandrogens are widely used for the treatment of BPH and prostate cancer by directly blocking DHT binding to the AR.

Another potential clinical application for androgen is as a component of hormonal male contraception. Since testosterone also regulates gonadotropin releases via feedback regulation of the hypothalamus-pituitary-testis axis, supra-physiological doses of testosterone can be given alone or in combination with gonadotropin-suppression agents (i.e., progestin, GnRH antagonists) to suppress spermatogenesis while replacing testicular androgen at the same time.

Although there are many potential clinical applications for androgens, the therapeutic uses of steroidal androgens are mainly limited by their poor pharmacokinetic profiles (i.e., low bioavailability) and side effects due to a lack of tissue selectivity. Over the past thirty years, tremendous research effort has been devoted to the discovery of nonsteroidal ligands with improved tissue selectivity in the hope of expanding the therapeutic use of androgens and antiandrogens.

Additionally, disorders related to various AR mutations have been identified (The Androgen Receptor Gene Mutations Database World Wide Web Server, http://www.androgendb.mcgill.ca/). The majority of these mutations are associated with diseases, like Androgen Insensitivity Syndrome and prostate cancer. The antiandrogen withdrawal syndrome (AWS) observed in prostate cancer therapy also appears to be related to certain AR mutations, such as T877A and W741C mutations, which convert some AR antagonists into agonists (see more discussion in Section 6.2.2.2). Besides the site mutations documented, AR gene polymorphism has also been identified, particularly, the poly-Q (CAG)n at exon I. The polymorphic $(CAG)_{10–35}$ triplet repeat sequence, starting from codon 58, codes for polyglutamine. The length of the repeat is inversely correlated with the transactivation activity of AR [6]. Although it is difficult to manage the genetic disorders with AR ligands, attempt to identify better antiandrogens that may overcome AWS for the treatment of prostate cancer are underway in many laboratories.

6.1.2. Gene and Protein Structure and Function

6.1.2.1. AR gene and protein structure

The AR gene was localized to the human X chromosome in 1981 by Migeon et al. [8], with its genomic DNA being first cloned in 1998 by several research groups [9, 10, 11]. To date, only one AR gene has been identified in humans.

The AR gene is more than 90 kb long and codes for a protein of 919 amino acids that contains the major functional domains of NRs, as illustrated in Figure 6.1. The N-terminal domain (NTD), which serves a modulatory function, is encoded by exon 1 (1,586 bp). The DNA-binding domain (DBD) is encoded by exons 2 and 3 (152 and 117 bp, respectively) [12]. The ligand-binding domain (LBD) is encoded by five exons which vary from 131 to 288 bp in size. There is also a small hinge region between the DBD and LBD. Two transactivation functions have been identified. The N-terminal activation function 1 (AF1) is constitutively active in truncated receptor that does not contain LBD, and is not conserved in sequence compared to other steroid receptors (Figure 6.2); whereas the C-terminal activation function 2 (AF2) functions in a ligand-dependent manner and is relatively more conserved in

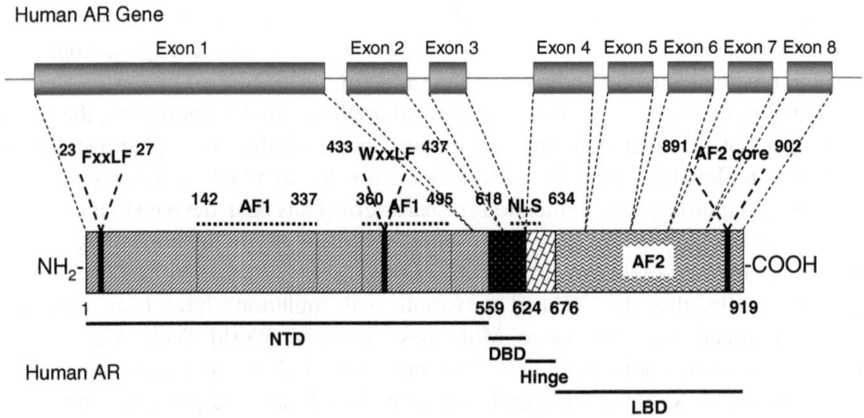

Figure 6.1. Structural organization of AR gene and protein

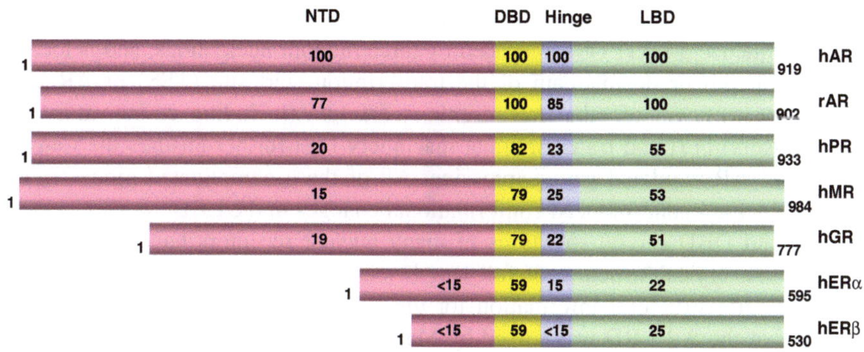

Figure 6.2. Amino acid sequence identity among members of the steroid receptor family (adapted from reference [160]). Sequence alignment was performed using William Pearson's LALIGN program [161]

sequence as compared to other steroid hormone receptors, particularly with regard to the charge-clamp residues [13]. A nuclear localization signal (NLS) spans the region between the DBD and the hinge region.

Human and rat AR share high similarity in amino acid sequence, with 85% overall identity and entirely identical sequences in the DBD and LBD [10] (Figure 6.2). AR also shares similar organization with other steroid receptors, an individual NTD, conserved DBD, and C-terminal LBD. The NTD shows the least conservation of sequence as compared to other steroid receptors (less than 25% identity), while the central DBD is well conserved (59–82%), reflecting the common need to bind to DNA, while the variation is responsible for the selection of different target sequences. The LBD also shows low sequence identity ranging from 22 to 55%, reflecting receptor specificity for individual hormones. Among all the steroid receptors, human AR LBD shares more sequence identity with human PR, GR, and MR LBDs (all around 50%) [14], with an overall sequence homology of 88% to the PR LBD when including conservative mutations. Despite the relatively low sequence identity, all of the steroid receptor LBD assume a similar three-dimensional structure with certain highly conserved structural features, including a "charge clamp" and helical features as revealed by crystal structures. These similarities in LBD conformation provide the structural basis for the cross reactivity that is often observed with synthetic steroids.

6.1.2.2. *AR protein conformation and function*

Similar to the other steroid receptors, unbound AR is mainly located in the cytoplasm and associated with a complex of heat shock proteins (HSPs) through interactions with LBD [15]. Upon agonist binding [16], AR goes through a series of conformational changes: the heat shock proteins dissociate from AR, and the transformed AR undergoes dimerization, phosphorylation, and translocation to the nucleus, which is mediated by the NLS. Translocated receptor then binds to the androgen response element (ARE), which is characterized by the 6-nucleotide half-site consensus sequence 5′-TGTTCT-3′ spaced by three random nucleotides, and is located either upstream or downstream of the transcription start site (TSS) of AR gene targets. Recruitment of other transcription cofactors (including co-activators and co-repressors) [17] and transcriptional machinery [18] further ensures the transactivation of AR-regulated gene expression. All of these complicated processes are initiated by the ligand-induced conformational changes in the LBD.

LBD and DBD crystal structures of many nuclear receptors have been solved, and the first crystal structure of full-length receptor was recently solved in 2008 [19], but little structural information for the NTD was revealed. The first crystal structure of the AR LBD was solved by Matias et al. [20]. AR LBD shares similar three-dimensional structure with other agonist-bound steroid receptors (Figure 6.3), as it contains eleven α-helices (H) and two short β-turns that are arranged in three layers to form an anti-parallel "α-helical sandwich". Different from other steroid receptors, H2 is not present in AR LBD, but the same numbering of the other helices was retained for easy comparison. Helices H1 and H3 form one face, while helices H4, H5, the first β-turn, and helices H8 and H9 form the central layer of the structure, and helices H6, H7, H10 and H11 constitute the second face. H5, the N-terminal of H3 and the C-terminal region of H10 and H11 form the main part of the hydrophobic ligand binding pocket (LBP). Upon agonist binding, H12 is repositioned and serves as the "lid" of the LBP to stabilize the ligand, and the very end of the C-terminal of the LBD forms the second β-turn (next to H8

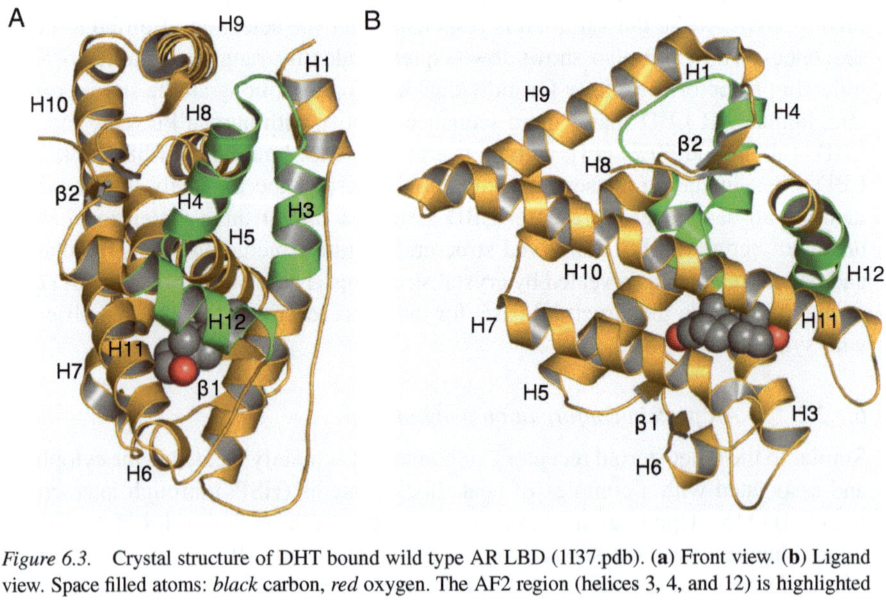

Figure 6.3. Crystal structure of DHT bound wild type AR LBD (1I37.pdb). (**a**) Front view. (**b**) Ligand view. Space filled atoms: *black* carbon, *red* oxygen. The AF2 region (helices 3, 4, and 12) is highlighted in *green*

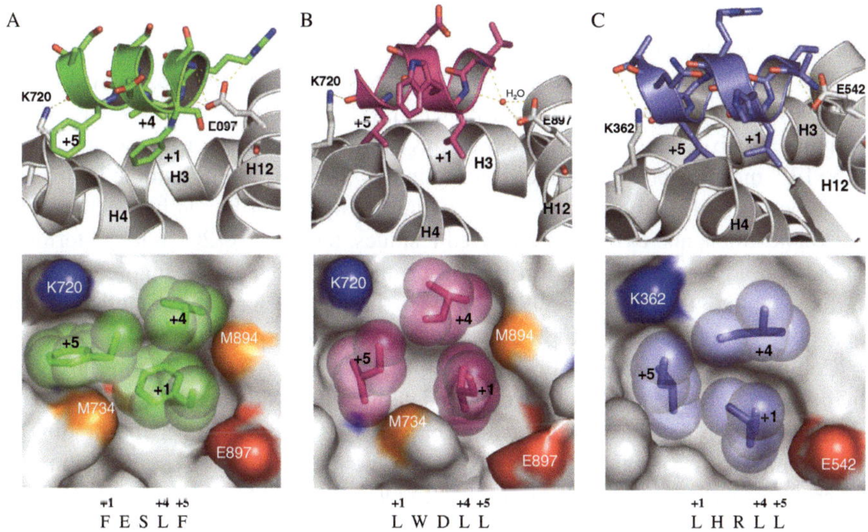

Figure 6.4. Interactions of FxxLF and/or LxxLL motifs with the AR and/or ER LBD. *Upper panels*, hydrogen-bonding interactions (shown as *yellow dotted line*) between peptide and AF2 region residues. *Lower panels*, surface view of the AF2 interface, side chains of the hydrophobic residues of FxxLF and LxxLL motifs are shown as spheres. (**a**) FxxLF bound to DHT-AR AF2 interface (1T7R.pdb). (**b**) LxxLL bound to DHT-AR AF2 interface (1T7F.pdb). (**c**) LxxLL bound to raloxifene-ERα AF2 interface (1GWQ.pdb)

and H10), which works as a "lock" to further stabilize the "lid" (H12) conformation. The agonist-induced conformational change in the LBD allows the formation of a functional AF2 region on the surface of LBD (Figure 6.3, highlighted in green), which is crucial for both cofactor recruitment during transcriptional activation (Figure 6.4a,b) and the intra-receptor N-terminal and C-terminal interaction (N/C interaction).

The AF2 region is a surface hydrophobic groove (Figure 6.4a,b) formed by the C-terminal region of H3, loop 3–4, H4, and H12, a region that covers the highly conserved NR LBD signature motif [21] and the AF2 core, which is similar to the structure that is also observed in agonist bound ER [22] (Figure 6.4C). A functional AF2 region is believed to be crucial for co-activator recruitment, as the so-called nuclear receptor box "LxxLL" motif [23] from the nuclear-receptor-interacting domain (NID) of co-activators specifically binds to this surface [24]. Unlike other steroid receptors, the intra-receptor N/C interaction [25] that occurs upon agonist binding is considered the most significant AF2 interaction in AR, and is responsible for stabilization of the agonist bound LBD. Similar motifs from AR NTD, "^{23}FxxLF^{27}" and/or "^{433}WxxLF^{437}" (Figure 6.1), can also interact with the AF2 region [25]. Therefore, both interactions may compete for the AF2 region upon agonist binding. The detailed topology change induced by different ligands within this region might provide the structural basis for recognition of specific binding motifs and cofactor interactions [26, 27].

The steroidal androgen bound AR LBD has been co-crystallized with short peptides that contain the NR box, either "LxxLL" motif or the "FxxLF" motif (Figure 6.4a,b). Hydrophobic interactions between the leucine residues in the "LxxLL" motif or phenylalanine residues in the "FxxLF" motif, and the hydrophobic groove, hold the peptide in place; while the hydrogen bonds between peptide backbone atoms and two well conserved residues, a lysine (K720) at the C-terminal of H3 and a glutamate (E897) in H12, form the "charge clamp" (Figure 6.4a,b) to further stabilize the interaction [26, 28]. Very similar mechanism of interaction was also observed in LxxLL motif bound ERα LBD AF2 region (Figure 6.4c) [29].

Compared to ERα-LBD (Figure 6.4c), DHT-bound AR LBD forms a deeper hydrophobic groove that could accommodate the bulky side chain of phenylalanine residues, which provides the structural basis for its preference for the binding of FxxLF motifs than LxxLL motifs, suggesting the idea that the N/C interaction may be preferred over co-activator recruitment in DHT-bound AR. In contrast, ERα-LBD prefers the binding of LxxLL motif to FxxLF motif [28]. On the other hand, when LxxLL motif binds to AR LBD (Figure 6.4b) [26], the peptide backbone only forms hydrogen bond with K720, a shift in peptide position precludes the direct hydrogen bonding with E897, which could also explain the relatively lower affinity of AR for this LxxLL motif.

Various steroid receptor ligands have been shown to regulate receptor function by inducing specific conformational changes of the AF2 region. The well-known "AF2 antagonist model", as illustrated in Figure 6.5, has been established for both GR [30] and ER [22]: H12 (shown in red) adopts a conformation that helps form a functional AF2 in estradiol (agonist) bound LBD, but it is displaced from agonist conformation and folds over the AF2 region in raloxifene (antagonist) bound LBD. The "^{540}LLEML^{544}" motif in H12 binds to the AF2 region in a similar way as the LxxLL motif and blocks the binding of co-activators.

The distinct preferences of AR for N/C interaction could also become targets for new drug discovery [31]. Studies [32, 33] have shown that ligand binding induced AR N/C interaction correlates with its ability to activate transcription, where disruption of the N/C interaction might become an effective strategy to develop antagonists. It has been hypothesized that the "AF2-antagonist" model might apply to AR as well, and H12 is repositioned to bind to the AF2 region upon antagonist binding via the "^{895}MAEII^{899}" motif, which is believed to block the binding of other motifs [24]. However, this hypothesis has not been proved by crystallography studies.

Besides the "AF2 antagonist model", other mechanisms have been proposed to explain AR antagonist activity. Although the agonist-induced conformational changes in the LBD result in the dissociation of the chaperone protein complex from AR [34], the dissociation does not seem to happen upon antagonist binding in AR [35], which might also account for the antagonist activity by simply blocking access to the AF2 region. Additionally, some evidence [36] suggested that AR antagonist, bicalutamide, could stimulate AR translocation and DNA binding, but it could not

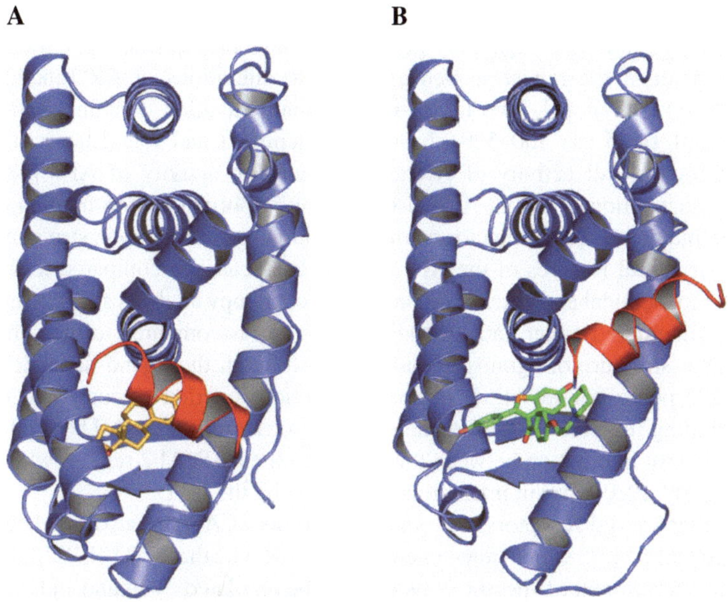

Figure 6.5. "AF2 antagonist model", as illustrated by crystal structures of wild type ERα LBD bound with (**a**) estradiol (1ERE.pdb) and (**b**) raloxifene (1ERR.pdb). Estradiol is shown in *yellow*, raloxifene is shown in *green*. Helix 12 is highlighted in *red*. Helix 12 folds over the AF2 region when antagonist (raloxifene) binds to ERα LBD

mediate co-activator recruitment, instead, bicalutamide binding mediated the recruitment of co-repressor, NCoR [37]. Therefore, differential recruitment of cofactors is also considered as a possible mechanism of action for AR antagonists. Despite all the theories proposed above, the precise mechanism of action for AR antagonists and structure of antagonist-bound AR LBD remain unclear.

Ligand-induced AR conformational changes provide the structural basis for the recruitment of cofactor proteins and transcriptional machinery, which is also required for the assembly of AR-mediated transcription complexes [18]. Shang et al. [18] showed that the formation of an activation complex involves AR, co-activators, and RNA polymerase II recruitment to both the enhancer and promoter regions of the prostate specific antigen (PSA) gene, whereas the formation of a repression complex involves factors bound only at the promoter, but not the enhancer. Since the formation of a functional AF2 region provides a structural basis for ligand-induced protein-protein interaction, ligand-specific recruitment of cofactors might be crucial for the agonist or antagonist activity of AR ligands [37]. On the other hand, as has been demonstrated with bicalutamide, possible ligand-specific interactions could be directly regulated by the surface topology of the AF2 region as shown by Sathya et al. [27]. Due to the subtle change in surface topology, formation of a functional AF2 does not guarantee effective N/C interaction. In other words, lack of N/C interaction does not correlate with the loss of AF2 functionality.

The AR NTD is characterized by a number of amino acid repeat sequences, including the poly-glutamine (Q), poly-glycine (G) and poly-proline (P) tracts, with polymorphism of the repeat sequence related to alterations in AR function or even disease conditions [6]. Two ligand-independent activation domains, Tau-1 (aa 100–370) and Tau-5 (aa 360–529), have been identified and found necessary and sufficient for the full activity of AR and the intrinsic activity of AF1 [38]. Besides its transactivation function, AF1 also plays an important role in mediating protein-protein interactions and cofactor recruitment due to its flexibility in structure. However, the structural features of the AF1 are less characterized compared to the AF2, and both biochemical and circular dichroism spectroscopy analysis suggest that NTD is highly disordered under native condition [39–41] as compared to the compact and globular structure of agonist-bound LBD. Although the ligand-dependent AF2 in the LBD possesses very little intrinsic activation function as compared to AF1, ligand binding may indirectly modulate the conformation of the NTD via the unique intra-receptor interaction between the NTD and the LBD AF2 region, which could consequently affect protein interactions mediated by the NTD.

Similar to other steroid receptors, the genomic actions of AR are also regulated by various cofactors [42], which may interact with AR via the NTD, LBD, and even DBD. The recruitment of specific cofactors can be regulated by ligand-induced conformational changes, as well as post-translational modifications [12] of different functional domains, including LBD and NTD. Additionally, tissue and/or cell type-specific expression of cofactors also contribute to the interactions between AR and cofactors. Although a large number of proteins [42] have been identified as AR cofactors, knowledge related to the structural basis of the interactions is rather limited to the AF2-dependent interactions (Figure 6.4).

DNA binding is also required for AR-regulated gene expression, which is known as the classic genomic function of AR and has been characterized by DNA microarray studies [43–45]. The ARE half-site sequence can be arranged as either inverted repeats or direct repeats [46, 47], and AR recognizes and binds to the ARE site through two zinc fingers located in the DBD. Ligand-bound AR forms homodimers, and appears to form “head-to-head” dimers [48] even when it is bound to the direct repeats of ARE. Selective recognition of specific ARE sequences could be regulated by ligand binding [49] and/or the presence of other transcriptional factors, which bind to their own DNA binding sites as well (combinatorial regulation) [50].

6.1.2.3. Nongenomic actions of androgens

The nongenomic actions of steroid receptors are well recognized, and have been discussed in previous chapters of this book. As compared to the genomic actions of the steroid receptors, the nongenomic actions are characterized by the rapidity of action, which varies from seconds to an hour or so; and interaction with cytoplasmic and/or plasma membrane-associated signaling pathways [51]. The nongenomic actions of AR have also been observed in various tissues and cell types, including Sertoli cells [52], oocytes [53], skeletal muscle cells [54], osteoblasts [55, 56], and prostate cancer cells [57, 58]. Functionally, the nongenomic actions of androgen were shown to involve either rapid activation of certain kinase-signaling cascades or modulation of intra-cellular calcium levels, which could be related to stimulation of gap junction communication, neuronal plasticity, and aortic relaxation etc. [59]. Structurally, the nongenomic actions are likely mediated by direct interactions between AR and various membrane or cytosolic proteins [60], however, the detailed structural basis for these interactions is unclear. Even so, separation of the genomic and nongenomic functions of steroid receptors using specific ligands has been proposed as a new strategy to achieve tissue selectivity [51, 61], although the structural features that are essential for achieving the separation need to be further investigated, and the therapeutic benefits of nongenomic action specific AR ligands are yet to be defined.

6.1.3. Androgen Biochemistry

6.1.3.1. Testosterone synthesis, metabolism, and tissue disposition

Testosterone and its active metabolite, 5α-DHT, are the major endogenous androgens. Testosterone is primarily synthesized (Figure 6.6) in Leydig cells in the testes, with smaller amount of testosterone being synthesized in adrenal cortex, liver, and ovary. The rate-limiting step in testosterone synthesis, cholesterol side chain cleavage by P450scc, is regulated by luteinizing hormone (LH), which is under the feedback regulation by testosterone at both the pituitary and hypothalamic levels. Dehydroepiandrosterone (DHEA) also has weak agonist activity. Testosterone cross-reacts at low affinity with other steroid receptors, such as PR and ER. In comparison, 5α-DHT binds to AR with more specificity and higher affinity.

Testosterone is the major circulating androgen (~95%), and is highly bound to plasma proteins, with 40% sequestered with high affinity to sex hormone-binding globulin (SHBG) and 60% bound with low affinity to albumin, leaving only about 2% as free, unbound hormone. In comparison, 5α-DHT has greater binding affinity to SHBG, accounts for only 5% of circulating androgen, and is largely derived from peripheral metabolism of testosterone. Since the metabolic clearance of testosterone is dependent on free testosterone concentration, plasma SHBG concentration may directly affect the half life of circulating testosterone. Endogenous testosterone secretion is pulsatile and diurnal in healthy adult men, who typically produce approximately 3–10 mg of testosterone per day with the highest serum concentration occurring about 8:00AM and the lowest at about 8:00PM. Average serum concentrations and diurnal variation diminish as men age [62], while circulating levels of SHBG increase to cause further decreases in free testosterone levels [63].

The incidence of prostate cancer and benign prostate hyperplasia (BPH) increases with age, despite the decrease in free testosterone concentrations. Although there is no direct evidence to suggest that testosterone causes the disease, early-stage prostate cancer is clearly dependent on androgen. Evidence is also accumulating to suggest that residual and/or adrenal androgens and the AR play a role in "hormone-refractory" prostate cancer [64].

Circulating testosterone is mainly inactivated in the liver [65, 66, 67] through reduction, oxidation, glucuronidation and eventual renal excretion (Figure 6.7). It

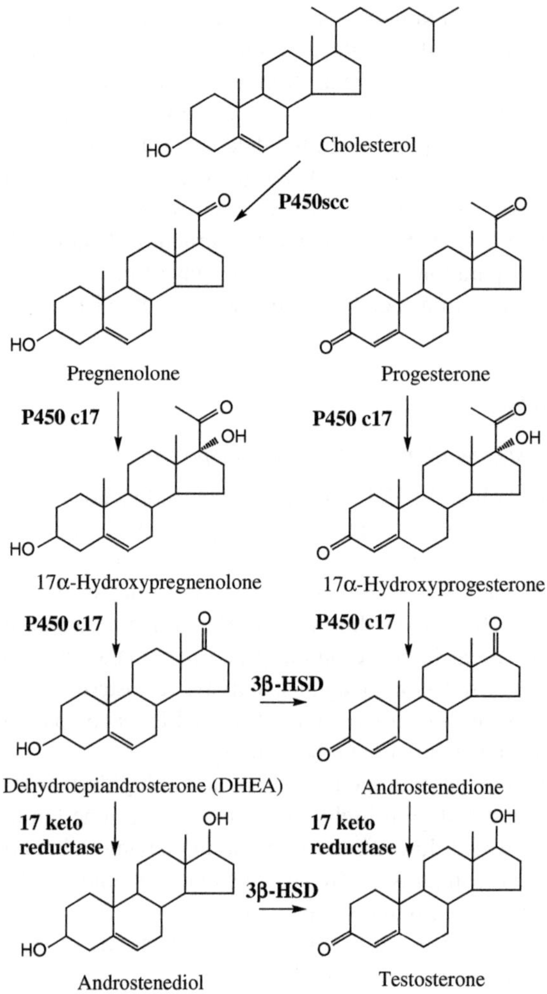

Figure 6.6. Testosterone synthesis (adapted from reference [162]). Abbreviations: P450scc, cholesterol side-chain cleavage enzyme; HSD, hydroxy steroid dehydrogenase

can be metabolized to androstenedione through oxidation of the 17β-OH group, and androstanedione with 5α-reduction of ring A. Androstanedione can be further converted to androsterone after 3-keto group reduction, and androstenedione can also be converted to etiocholanolone through 5β- and 3-keto reduction. Similarly, 5α-DHT can be converted to androstanedione, androsterone, and androstanediol [68]. Most of the metabolites mentioned above undergo extensive glucuronidation of the 3α- or 17β-OH groups as well, either in the target tissues or the liver [69], and are further excreted in the urine. Androsterone and etiocholanolone are major urinary

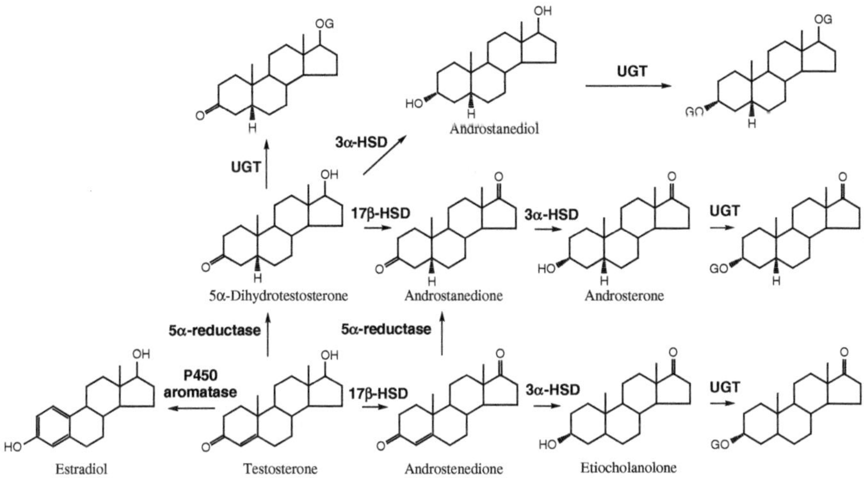

Figure 6.7. Testosterone metabolism (adapted from reference [69]). Abbreviations: G, glucuronide; HSD, hydroxy steroid dehydrogenase; UGT, UDP-glucuronosyltransferase

metabolites. Both are inactive metabolites, and are excreted mainly as glucuronide conjugates, or to a lesser extent as sulfate conjugates [65].

Due to extensive hepatic metabolism, testosterone has a short plasma half life of less than 30 min. It is not orally available. Around 90% of the oral dose of testosterone is metabolized before it reaches the systemic circulation. To improve the bioavailability, most of the testosterone preparations are delivered through topical gel, transdermal patch or intramuscular injections. A variety of testosterone esters (Figure 6.8) are commonly administered parenterally for prolonged activity. Esterification enhances the lipid solubility of the steroid and permits the formation

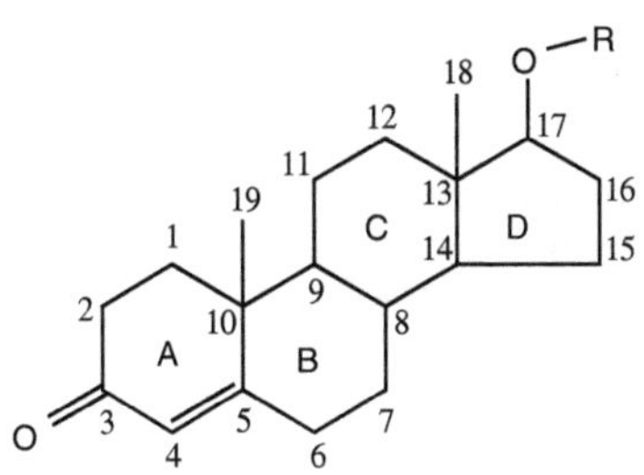

Propionate: R = CH_3CH_2CO
Enanthanate: R = $CH_3(CH_2)_5CO$

Figure 6.8. Testosterone esters

of a local depot after intramuscular injection, with ester hydrolysis releasing testosterone as the active species in vivo.

6.1.3.2. Testosterone modes of action

Testosterone can be metabolized in either its target tissues or the liver (Figure 6.9). It can be reduced to 5α-DHT by 5α-reductase (type 1 and/or 2) [70] in the prostate gland, skin, and liver [71]. A small amount of testosterone (0.2%) can also be converted to estradiol by aromatase through the cleavage of the C_{19} methyl group, and aromatization of ring A, which mainly occurs in the adipose tissue, ovary, and testes to a much lesser extend [3, 72]. Both 5α-reduction and aromatization are irreversible processes, which generate 5α-DHT and estradiol as the major active metabolites of testosterone.

Due to tissue-specific metabolism, endogenous testosterone has three modes of action. It may directly act through AR in target tissues where 5α-reductase is not expressed, be converted to 5α-DHT (5–10%) by 5α-reductase before binding to AR, or be aromatized to estrogen (0.2%) and act through the estrogen receptor [2]. The formation of 5α-DHT is a natural way for the "DHT-dependent" tissues, such as prostate and seminal vesicle, to amplify the androgenic activity of testosterone since 5α-DHT is a more potent AR agonist than testosterone. It binds to AR with higher affinity, and has two to ten fold higher potency than testosterone in androgen-responsive tissues [6]. On the other hand, estrogen appears to play an important role in regulating metabolic processes [73, 74], mood and cognition [75], cardiovascular disease [76, 77], sexual function including libido [78], and bone turnover in men [79, 80]. Besides these active metabolites, testosterone is the major androgen that acts in the "DHT-independent" tissues, such as skeletal muscle, where 5α-reductase

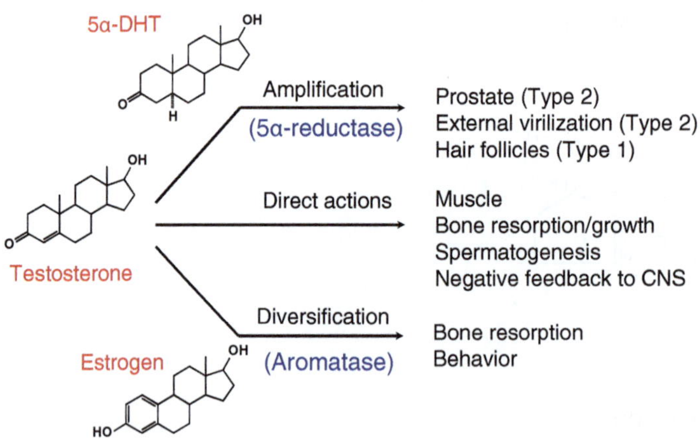

Figure 6.9. Three modes of the physiological actions of testosterone (T). T may act directly through AR, be converted to 5α-DHT by 5α-reductase, and be aromatized to estrogen and act through the ER

is not expressed or expressed at a very low level [71]. It directly regulates skeletal muscle growth, bone formation, fat distribution, and sexual function.

6.2. CHEMISTRY OF AR LIGANDS

AR ligands can be classified into two structural classes (steroidal or nonsteroidal) and two pharmacologic classes (agonist or antagonist). Novel AR ligands are usually first identified by in vitro receptor binding assay, using either rat prostate cytosolic AR, recombinant AR protein, or cells that express AR [81]. Although binding affinity are often determined in different research groups using slightly different methods, the AR binding affinity of all ligands is commonly presented as the relative binding affinity (RBA) compared to the synthetic ligand R1881, which has a K_d value of 0.53 nM as determined [82, 83] using rat prostate cytosolic AR. The agonist or antagonist activity of the ligand is often examined in vitro using reporter gene assay. Sometimes, the antagonist activity of the ligand is also tested as its ability to suppress androgen-dependent prostate cancer cell growth. Further in vivo testing of the androgenic and anabolic activities, or antagonist activity, is mainly done by measuring the prostate (measure of androgenic activity) and levator ani muscle (measure of anabolic activity) sizes in castrated rat models, after treatment [84] with the compound alone or in combination with testosterone preparations. The antiandrogen activity can also be tested in androgen sensitive prostate cancer xenograft model, such as LNCaP xenografts.

6.2.1. Steroidal AR Ligands

6.2.1.1. Chemistry, structure-activity relationship

Synthetic AR ligands (Table 6.1) were developed as early as the 1950s, by modifying the steroidal structure of testosterone. The structure-activity relationship of these steroidal AR ligands is well documented [7, 65, 85]. However, low bioavailability, poor pharmacokinetic properties, and undesirable side effects due to the lack of tissue-specific activities have limited the use of most steroidal ligands. For a long period of time, majority of the research efforts to find new androgens were devoted to the development of orally available anabolic steroids.

For steroidal androgens, the presence of 3-keto or 3α-OH groups appear to enhance the androgenic activity of the ligand, and removal of the 19-methyl group (e.g., nandrolone) seems to be favorable for the separation of androgenic and anabolic activity. However, complete separation of androgenic and anabolic activity has not been accomplished with synthetic steroids. 17α-Alkylation (e.g., oxandrolone) significantly slows the hepatic metabolism of the 17β-OH group, increases the oral bioavailability, and prolongs the elimination half life of these ligands. However, 17α-alkylated androgens are more likely to cause hepatotoxicity after long term use [86, 87], which is the most serious side effect of the synthetic steroids.

Table 6.1. Steroidal AR ligands

Compounds	Compound structure	RBA (%)	References
R1881	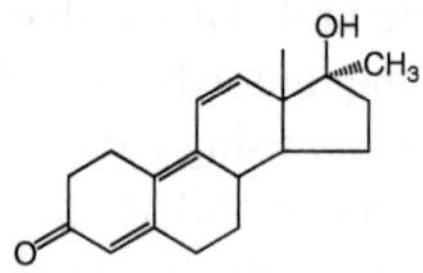	100	
Testosterone	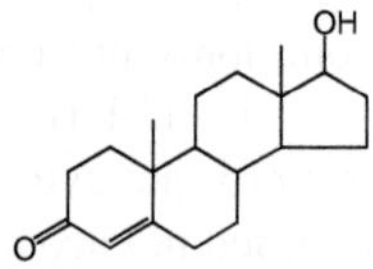	25	[163]
5α-DHT	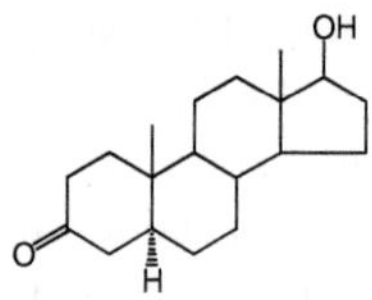	88	[163]
Nandrolone	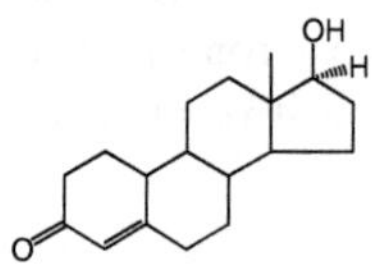	31	[164]
Oxandrolone	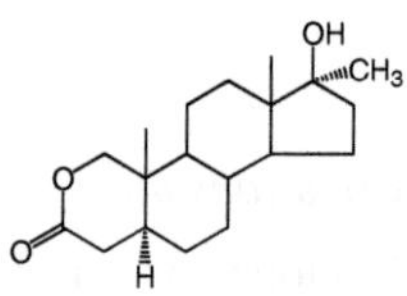	0.3	[33]
Oxendolone	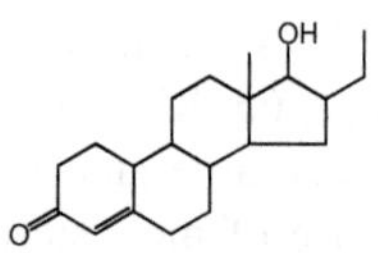	0.8	[165]
Cyproterone acetate (CPA)	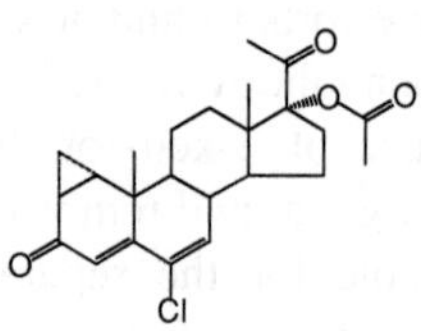	1.7	[163]
Spironolactone	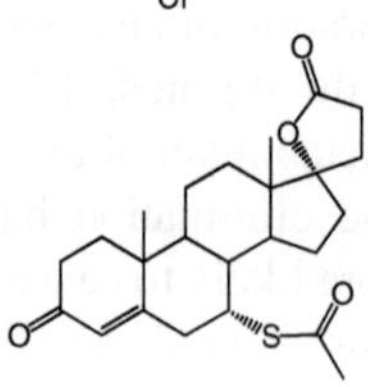	0.4	[81]

Similar to endogenous androgens, reduction of the 3-keto and 4,5 double bond are common routes of metabolism. Both metabolites and parent drug are eliminated either directly or as glucuronidate or sulfate conjugate.

A few steroidal ligands have been used clinically as antiandrogens, including cyproterone acetate (CPA), oxendolone, and spironolactone. CPA is moderately effective in reducing hirsutism alone or in combination with an oral contraceptive [88], but is not approved for use in the United States. Spironolactone was initially developed as an aldosterone inhibitor for the treatment of fluid retention and hypertension. It is approved by Food and Drug Administration (FDA) to treat hirsutism in women [89], although it may cause irregular menses. Gynecomastia is a common side effect in men [90], which limits its application in this gender.

6.2.1.2. Clinical applications

Steroidal androgens are mainly used for the treatment of hypogonadism. However, the intrinsic androgenic activity of the steroidal androgens may suppress gonadotropin secretion and can often cause undesirable side effects during therapy, such as the virilizing and masculinizing actions in women and children. Additionally, due to the structural similarity in the steroid skeleton, steroidal androgens also tend to cross react with other steroid receptors and cause undesirable side effects like salt and water retention, or even edema [91].

Despite these side effects, anabolic steroids are clinically used to treat diseases related to muscle wasting [92] (i.e., oxandrolone), and until recently to stimulate production of erythropoietin in the treatment of anemia resulting from bone marrow failure [93] (i.e., oxymetholone). As oral anabolic steroids, oxandrolone, stanozolol, and oxymetholone show lesser androgenic activity, but the shared 17α-methyl group of these agents is often associated with potential hepatotoxicity.

Antiandrogens have therapeutic potential in the treatment of prostate cancer, BPH, acne, virilization in women, and male contraception. However, none of the steroidal antiandrogens were able to achieve complete androgen blockade.

6.2.2. Nonsteroidal AR Ligands

As discussed above, the clinical application of steroidal AR ligands has been limited by poor oral bioavailability, potential hepatotoxicity, lack of tissue selectivity, and occasionally, cross reaction with other steroid receptors. Further structural modification of the steroidal ligands is somewhat limited by the steroid skeleton. Nonsteroidal ligands were developed to achieve high AR specificity and improve oral bioavailability, which also allow more flexible structural modifications if necessary.

Several structural classes of nonsteroidal AR ligands, either agonist or antagonist, have been discovered. The well-characterized pharmacophores include aryl propionamide analogs, hydantoin analogs, and quinoline analogs, most of which possess high binding affinity to AR. Further structural modifications of these pharmacophores also led to the discovery of a novel class of AR ligand, selective androgen receptor modulators (SARMs).

6.2.2.1. Chemistry, structure-activity relationship

Aryl propionamide analogs were first developed as nonsteroidal antiandrogens, with bicalutamide, flutamide, and nilutamide (Table 6.2) being the most successful examples of this class. Unlike the steroidal antiandrogens, they are considered pure antiandrogens since they possess little if any intrinsic androgenic activity when bound to wild type AR and do not cross-react with any of the other steroid receptors. These ligands are orally available with in vivo half lives ranging from 8 h to 6 days in humans. The nonsteroidal antiandrogens are used almost exclusively to treat prostate cancer.

Hydrolysis of the amide bond represents the major metabolic pathway for flutamide and its active metabolite hydroxyl-flutamide [94]. To achieve complete AR blockage, flutamide is generally used at doses of 750 mg/day. Extensive hepatic metabolism of the drug generates a large amount of the hydrolysis product, 3-trifluoromethyl-4-nitroaniline, which might be related to the hepatotoxicity of flutamide [95]. Clinical use of flutamide has been limited by its hepatotoxicity after long term administration and short biological half life.

As a hydantoin analog of flutamide, nilutamide is also eliminated exclusively by metabolism [96], mainly reduction of the aromatic nitro group. Although the carbonyl functions of the imidazolinedione are also labile, it is less susceptible to hepatic metabolism at the amide bond than hydroxyflutamide, which results in a much longer half life of two days in humans. Even so, the nitro anion-free radical formed during nitro reduction might still be associated with hepatotoxicity [97, 98] in humans, especially when using the relatively high dosage (150–300 mg/day) employed for androgen blockage. Further structural modification of nilutamide led to the development of a series of hydantoin analogs that act as AR antagonists [99, 100, 101, 102] that bind to the wild type AR with high affinity. SAR studies demonstrated that the 4-nitro-naphthyl structure is favorable for high affinity binding of these ligands (e.g., BMS-5 as shown in Table 6.3) [100].

Bicalutamide replaced flutamide and nilutamide as the antiandrogen of choice for prostate cancer treatment, since it has less hepatotoxicity and a longer half life (6 days in humans) [103] that allow once a day administration at relatively lower dosage (50 mg/day). As a structural analog, bicalutamide shares the amide bond structure with flutamide. However, the amide bond hydrolysis was observed in rat, but not in humans [104, 105], which could explain the prolonged half life of bicalutamide in humans. Bicalutamide is mainly metabolized by hydroxylation and glucuronidation. With the presence of the chiral carbon in structure, bicalutamide is administered as racemate. However, the in vivo antiandrogenic activity of bicalutamide arises almost entirely from its R-isomer, which has approximately 30-fold greater binding affinity

Table 6.2. Nonsteroidal AR ligands

Chemotype	General Chemical Structure	Lead Compound (Antagonist)		Lead Compound (SARM)	
Aryl Propionamide Analogs	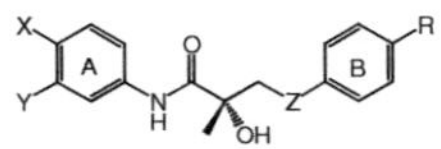	Bicalutamide	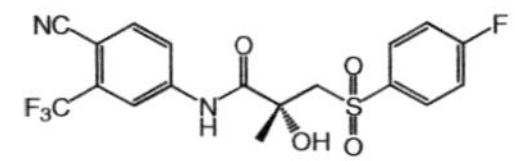	S4	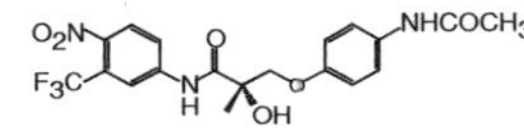
		Hydroxy-Flutamide	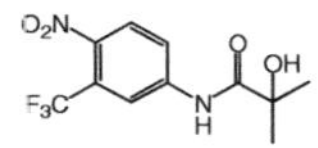		
Hydantoin Analogs	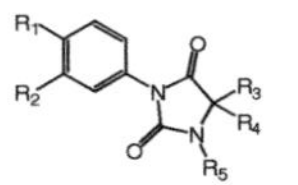	Nilutamide	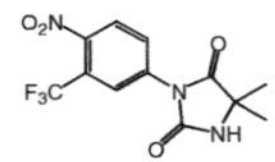	BMS 564929	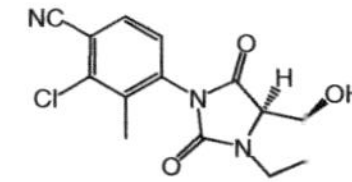
Quinoline Analogs (Tri-cyclic)	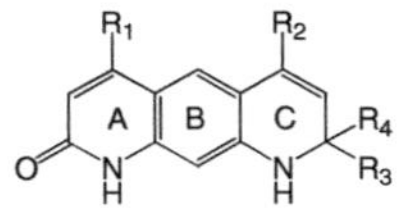	LG120907	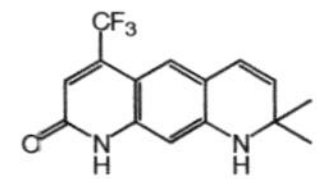		
(Bi-cyclic)	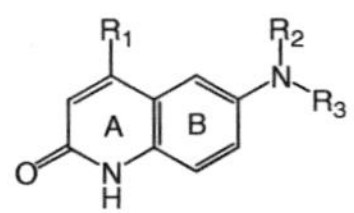			LGD2226	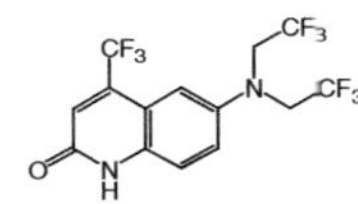
Tetrahydro-quinoline Analogs	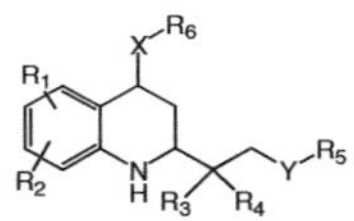			S-40503	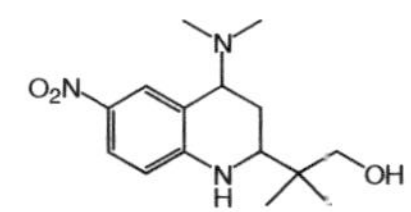

Table 6.3. Crystal structures of various ligand bound AR LBD

	Structural classes		Ligand	Ligand structure	LBD	RBA (%) (Reference)	PDB code	References
Steroidal		Agonist	R1881	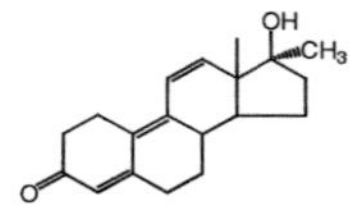	Human AR WT	100 [82, 83]	1E3G	[20]
		Agonist	Testosterone	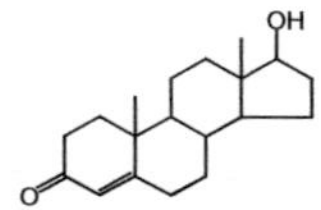	Human AR WT	25 [163]	2AM9	[151]
		Agonist	5α-DHT	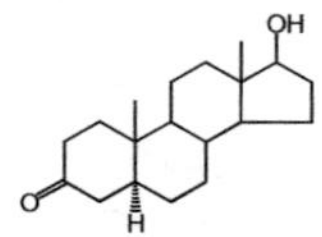	Rat AR WT	88 [163]	1I37	[158]
		Antagonist	Cyproterone Acetate (CPA)	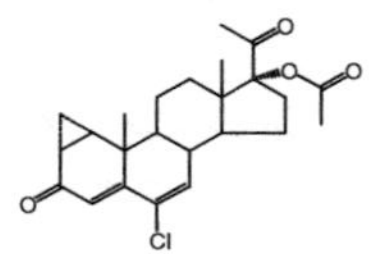	Human AR T877A	1.7 [163]	2OZ7	[153]
Nonsteroidal	Aryl Propionamide	Antagonist	Hydroxy-flutamide	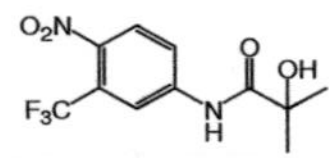	Human AR T877A	0.1 [163]	2AX6	[145]
		Antagonist	R-Bicalutamide	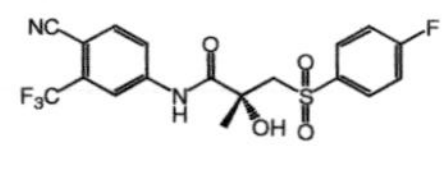	Human AR W741L	0.4 [163]	1Z95	[155]

Table 6.3. (Continued)

Structural classes		Ligand	Ligand structure	LBD	RBA (%) (Reference)	PDB code	References
	Agonist	S-1	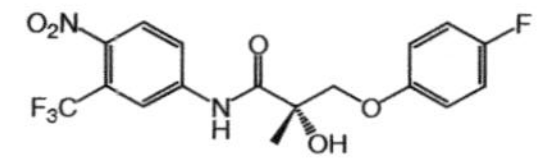	Human AR WT	4 [122]	2AXA	[145]
Hydantoin Analogs	Antagonist	BMS-5	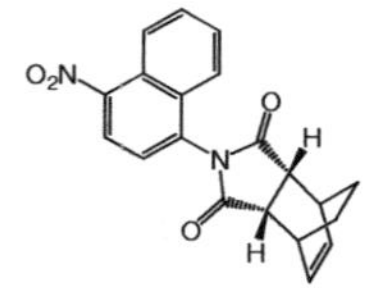	Rat AR T877A	NA[a]	1XNN	[99]
	Agonist	BMS-10b	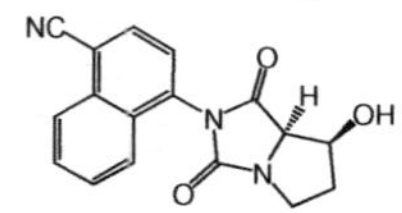	Rat AR WT	NA[a]	2IHQ	[159]
	Agonist	BMS 564929	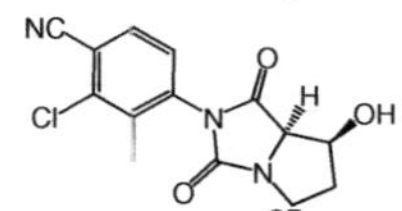	Rat AR WT	3 [136]	2NW4	[126]
Quinoline	Agonist	LGD2226	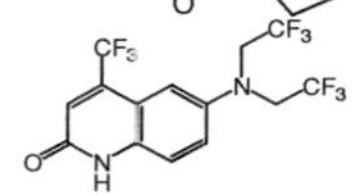	Human AR WT	0.2 [102]	2HVC	[157]

[a]NA: Not available

and is cleared at a rate 1/100th of the S-isomer. Although bicalutamide appeared to be peripherally selective in rats [106] with less antiandrogen activity in the pituitary, which could be related to its low tissue distribution in central nervous system, similar tissue selectivity was not observed in humans.

The greatly improved specificity and favorable pharmacokinetic profile of nonsteroidal antiandrogens, as compared to steroidal antiandrogens, affords much more efficient androgen blockage for prostate cancer treatment. At therapeutic doses, due to the competitive blockage of AR in both prostate and pituitary, these drugs often trigger significant increases in LH release, which further stimulates higher serum testosterone concentrations. Therefore, they are used primarily in combination with a gonadotropin-releasing hormone (GnRH) agonist, which shuts down the testicular but not adrenal testosterone production, in therapy commonly referred to as total androgen blockade. The treatments are similarly effective as surgical castration [107]. However, this so-called "chemical castration" also abolishes libido and the anabolic activity of androgens in the muscle and bone, causing undesirable side effects that may include muscle wasting and osteoporosis. In addition to their use for prostate disease, antiandrogens are also effective for the treatment for alopecia [108], hirsutism [88, 109], and acne [110]. Hydantoin derivatives, like RU58841, were developed in Europe for topical treatment of acne and alopecia [111, 112].

6.2.2.2. *Antiandrogen withdrawal syndrome*

Prostate cancer often advances to a "hormone-refractory" state in which the disease progresses in the presence of continued androgen ablation or antiandrogen therapy, suggesting the development of androgen-independent prostate cancer cells or the ability of adrenal androgens to support tumor growth. Instances of antiandrogen withdrawal syndrome (AWS) have also been reported after prolonged treatment with antiandrogens [113]. AWS is defined in terms of the tumor regression or symptomatic relief observed upon cessation of antiandrogen therapy. AR mutations that result in receptor promiscuity and the ability of these antiandrogens to exhibit agonist activity might at least partially account for this phenomenon. For example, hydroxyflutamide and bicalutamide act as AR agonists in T877A and W741L/W741C AR mutants [114, 115], respectively. Research efforts devoted to the development of new generation of "pan-antagonists" that would inhibit both wild-type and mutant AR are underway.

6.2.3. Selective Androgen Receptor Modulators (SARMs)

The successful development and marketing of selective estrogen receptor modulators (SERMs) kindled the interest in the development of selective ligands for other nuclear hormone receptors. The concept of selective androgen receptor modulators (SARMs) [116, 117] was proposed in 1999: a compound that has (1) high specificity for the AR; (2) desirable oral bioavailability and pharmacokinetic profiles; and most importantly, (3) desirable tissue-selective pharmacological activities. The major discriminating criterion is the tissue selectivity of the ligand in vivo. For an ideal SARM, the weak agonist or antagonist activity in the prostate would reduce or eliminate concern for the potential to stimulate nascent or undetected prostate cancer; while the strong agonist activity in the anabolic tissues like muscle and bone will help expand the therapeutic use of SARMs to the treatment of hypogonadism, osteoporosis, or muscle-wasting conditions associated with aging, severe burn injury, and cancer [118].

6.2.3.1. Chemistry

Early development of nonsteroidal antiandrogens identified several pharmacophores with high oral bioavailability and much more flexibility for further structural modification. As such, several diverse classes of nonsteroidal SARMs have been identified [119], including but not limited to aryl propionamide analogs, quinoline analogs (bicyclic or tricyclic), and hydantoin analogs (Table 6.2).

The discovery and development of SARMs is at an early but rapidly advancing stage, with most of the compounds under preclinical development and a handful now completing either phase I or phase II clinical trials [119]. Most of the SARMs identified to date are nonsteroidal anabolic agents, with the first generation aryl propionamide [120] and quinoline [121] analogs reported in 1998. The aryl propionamide SARMs were the first to demonstrate in vivo tissue selectivity for skeletal muscle in 2003 [122, 123], followed by a tetrahydroquinoline (THQ) SARM [124] later that year, quinoline [125] SARM in 2006, and hydantoin [126] SARM in 2007. All of these anabolic SARMs demonstrate some degree of tissue selectivity in castrated animal models, invariably with stronger agonist activities in anabolic tissues (e.g., levator ani muscle, bone) than in androgenic tissues (e.g., prostate).

6.2.3.2. Tissue selectivity of SARMs

The aryl propionamide SARMs are structurally similar to bicalutamide, but demonstrate distinguishing pharmacologic properties including tissue selective anabolic activity in animal models. The earliest lead compound in this class, known as S4, prevented tissue weight loss associated with androgen deficiency (castration) after short term treatment (two weeks), and behaved as partial agonists in the prostate ($ED_{50} = 2$ mg/kg/day), but full agonists in the levator ani muscle ($ED_{50} =$ 0.6 mg/kg/day) [122]. Similar tissue selectivity was also observed after prolonged treatment (8 weeks) [127]. This SARM (3 mg/kg/day) also demonstrated strong anabolic effects in restoring skeletal muscle (i.e., soleus muscle) strength, total body bone mineral density and lean mass, with suppression of LH and FSH only at highest doses [128]. Furthermore, S4 also prevented orchidectomy-induced bone loss in a female rat model for osteoporosis [129].

In intact male rats [123], S4 worked as antagonists in the prostate while maintaining anabolic effects in the levator ani muscle. It did not affect gonadotropin release or plasma testosterone concentrations, suggesting that SARMs with low intrinsic activity in the prostate might serve as an alternative therapy for benign prostate hyperplasia (BPH) or even prostate cancer. Furthermore, the ability of this class of SARMs to replace or supplement the anabolic effects of testosterone in muscle and bone of both intact [123] and castrated rats [130] suggest that such compounds might also be used in combination with progestins to achieve oral, hormonal male contraception.

Furthermore, these nonsteroidal ligands are orally available [131, 132], with in vivo half lives of about four hours in rats. Metabolism studies [133, 134] have shown that, similar to hydroxyflutamide and nilutamide, amide bond hydrolysis and nitro reduction are the major metabolic pathway observed in rats [135]. However, their structural similarity to bicalutamide suggests that amide bond hydrolysis is not likely to occur in humans. These derivatives are eliminated exclusively through hepatic metabolism. This is the first class of nonsteroidal AR ligands that demonstrated the desired tissue selectivity of SARMs.

Another important class of SARMs that have shown tissue selectivity is the hydantoin derivatives (Table 6.2) [136]. Lead compound BMS-564929 binds AR with high affinity (RBA 3%) and high specificity. In castrated rats, BMS-564929 also demonstrated tissue selectivity, with $ED_{50} = 141$ μg/kg/day in the prostate, $ED_{50} = 0.9$ μg/kg/day in the levator ani muscle, and $ID_{50} = 8$ μg/kg/day in suppressing LH release. The compound is orally available in humans, with an in vivo half life of 8–14 h. Compared to the bicalutamide derivatives, the prolonged in vivo half life of these ligands could explain the lower dose needed to achieve its pharmacological activities in animal models, since the in vivo activities seem to be more related to the tissue exposure of the ligands when they share similar binding affinity and intrinsic activity. However, the more potent suppression of LH observed with these compounds may have implications for their use in androgen replacement therapy. Studies regarding the effects of BMS-564929 on bone or other androgenic and anabolic tissues have not been reported to date.

The quinolone derivatives (Table 6.2), either bicyclic or tricyclic, bind to AR with relatively high affinity (nanomolar range). Bicyclic derivative LGD2226 was shown to be tissue selective after two weeks treatment in castrated rat model [137]. The anabolic effects of LGD2226 in bone were also reported [138]. During a four-month treatment period, LGD2226 prevented castration-induced bone loss and maintained bone quality by stimulating bone formation and inhibiting bone turnover.

In summary, for all three classes of SARMs described above, slight structural modifications change the ligand from AR antagonist to agonist. Despite the detailed structure-activity relationships that have been elucidated for these compounds, it remains unclear what kind of ligand receptor interaction determines the agonist or antagonist activity of the ligand, largely due to the limited knowledge about the receptor conformation upon antagonist binding.

6.2.3.3. *Mechanisms of tissue selectivity*

Various mechanisms of action have been proposed to explain the tissue selectivity of SARMs [139–141], including differential tissue distribution of the ligands or receptor cofactors, potential interactions with 5α-reductase and/or aromatase at the tissue level, ligand-specific regulation of gene expression (genomic action), and/or nongenomic actions at molecular level.

Most of these concepts have been "adopted" from more advanced research in the area of SERMs, with the idea that the mechanisms of action identified for SERMs should apply to other selective nuclear receptor modulators like the SARMs. However, there are fundamental differences between androgen and estrogen physiology and signaling that are commonly overlooked. First of all, there are two forms of ER (ERα and ERβ) but only one endogenous ligand (estrogen). ERα and ERβ exhibit differences in their structure, ligand discrimination, tissue distribution patterns, transcriptional properties, and biological roles. The presence of two forms of ER provides a natural way to regulate estrogen action in different tissues. In contrast, there is only one AR but two endogenous ligands (testosterone and DHT). The presence of two endogenous androgens provides a different mechanism to regulate androgen action in different tissues.

As summarized in Figure 6.9, testosterone action in the prostate is naturally "amplified" through conversion to DHT by 5α-reducatse (Type 2). Inhibition of 5α-reducatse with finasteride abolishes the amplification process and leaves testosterone the dominant androgen in the prostate (>80%), and reveals the intrinsic actions of testosterone itself as a partial agonist in the prostate. As demonstrated by Wright et al. [142], testosterone is much less potent than DHT in stimulating prostate growth, and the absence of the amplification process led to a significant right-shift of the dose response curve (Figure 6.10a) [142]

Studies [123, 136] have shown that currently identified nonsteroidal SARMs are not substrates for 5α-reductase and cannot be converted to 5α-reduced metabolites, indicating that the action of these ligands in androgenic tissues is not amplified in a similar way as testosterone. Tissue-selective SARMs are not as potent as 5α-DHT in the androgenic tissues. As such, they are more likely to behave as partial agonist. As the only active androgen in the prostate, the dose-response curve of SARMs (e.g., BMS-564929) in the prostate demonstrated a significant right shift (Figure 6.10b) similar to that observed with testosterone alone (with the presence of finasteride, Figure 6.10a). However, in the anabolic tissues wherein androgen action is DHT-independent, SARMs perfectly mimic the direct action of testosterone and work as full agonists. Therefore, although more intriguing explanations suggestive of novel AR conformations and tissue-specific cofactor recruitment are often sought, existing evidence with BMS-564929 and other SARMs strongly suggest that tissue-specific expression of 5α-reductase plays an overwhelming role in determining the tissue selectivity of SARMs.

The role of aromatase and tissue distribution of the ligand appear to be small compared to other mechanisms [17]. Tissue distribution studies with bicalutamide and hydantoin derivatives [136, 143] showed that preferential accumulation of these drugs in anabolic tissues does not occur. Furthermore, although the conversion of testosterone to estrogen does play some role in the development of prostate disease, its contribution to the overall growth of the normal prostate is rather limited [144].

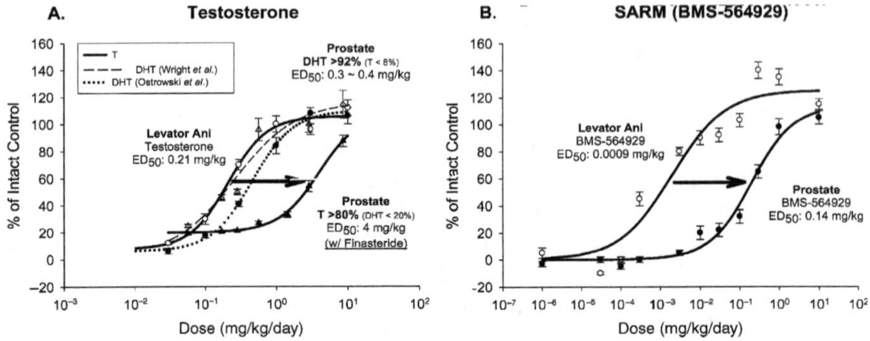

Figure 6.10. Tissue-specific expression of 5α-reductase contributes to the tissue selectivity of SARMs. (**a**) Partial agonist activity of testosterone in the prostate was revealed when co-administered with type 2 5α-reductase inhibitor finasteride. (Dose-response curves of T in the levator ani muscle (•) and T in the prostate (○ and *dotted line*) were redrawn with permission from Ostrowski et al. [126]; and dose-response curves of T in the prostate with (▲) or without (Δ and *dash line*) co-administration of finasteride were redrawn with permission from Wright et al. [142].) T data from two different studies (Wright et al. *dotted line*; and Ostrowski et al. *dash line*) largely overlap, with similar ED_{50} values between 0.3 and 0.4 mg/kg/day, showing that these two studies are comparable. (**b**) Partial agonist activity of SARM (hydantoin analog BMS-564929) in the prostate when administered alone. (Dose-response curves redrawn with permission from Ostrowski et al. [126]) BMS-564929 was orally administered since it is orally available

Ligand-specific regulation of genomic and/or nongenomic actions is frequently proposed as the major mechanism responsible for the tissue selectivity of SARMs, and has been the focus of the vast majority of the research effort. Ligand binding induces specific conformational changes in the LBD, which could further modulate the surface topology of the protein and subsequent protein-protein interactions between the receptor and other cellular proteins: either cytosolic proteins involved in different signaling pathways (nongenomic pathway) or cofactors involved in transcriptional activation (genomic pathway). Furthermore, ligand-specific receptor conformation and protein-protein interactions could also result in ligand-specific gene regulation, due to potential changes in recognition of ARE, or interactions with cofactors and/or other transcription factors.

Although changes in AR function are initiated by ligand-induced conformational changes of the LBD, crystal structures of SARM-bound AR LBD (see following discussion) solved so far have not revealed conformational changes as significant as those observed in SERM-bound ER LBD [126, 145], particularly the AF2 region [22, 146]. Furthermore, testosterone and DHT-bound AR LBD structures were considered “essentially identical” [126] despite the considerable differences in their androgenic actions in the prostate (Figure 6.10a). Although crystal structures are very useful for determining ligand binding mechanisms, it is difficult to predict potential changes in receptor function that would occur under native conditions. Subtle changes in protein conformation, surface topology, or both that might affect protein-protein interaction and receptor function need to be assessed more carefully under native conditions to identify meaningful consequences in function, if any. Furthermore, although peptide mapping and protein-protein interaction assays [147, 148] have identified ligand-specific interaction profiles of the receptor, the significance of these differential profiles in function has not been fully evaluated at the tissue level or in vivo. As such, it is unclear if these differences play a significant role in ligand-stimulated AR action in different tissues. Further research is necessary to prove that certain molecular mechanisms contribute equally or at all to the tissue selectivity of SARMs.

Ligand-induced protein-protein interaction is known to contribute to N/C interaction [33] or co-activator recruitment with the AR [149]. Both interactions seem to be mediated by the interaction between the AF2 region and the "FxxLF" or "LxxLL" binding motifs [150]. The hydrophobic groove present in the AF2 region of AR LBD appears to be more favorable for phenylalanine side chain binding, which suggests that the N/C interaction is preferred over co-activator recruitment upon agonist binding. In comparison, antagonist-binding does not initiate N/C interaction in AR [27], and it recruits corepressor [37], NCoR, instead of co-activators, demonstrating dramatic differences in receptor conformation. Although nonsteroidal SARM-bound AR LBD conformation has not been well characterized, Sathya et al. [27] investigated some proposed steroidal SARMs in vitro. It is obvious that a synthetic AR agonist can induce an "activating" conformational change in AR without facilitating N/C interactions of the receptor, which suggests that ligand-specific conformational change is achievable with synthetic ligands, and provides the structural basis for possible differences in protein-protein interactions.

In conclusion, there has been considerable progress in the discovery and development of SARMs in the past decade. SARMs with much improved specificity for AR, in vivo pharmacokinetic profiles, and higher degree of tissue selectivity will undoubtedly expand the clinical applications of androgens. However, unlike SERMs, the tissue selectivity of SARMs may depend at least partially on the tissue-specific expression of an enzyme, 5α-reductase, and lack of interaction between SARM and this activating enzyme. Further research is necessary to determine if the molecular mechanisms identified in SERM research apply to SARM as well. It is generally believed that ligand-specific conformation changes and tissue-specific differences in cofactor expression and interaction play a role in SARM pharmacology, but the question remains: to what extent do they contribute to the observed differences in tissue response (e.g., prostate and muscle) during SARM treatment? Although 5α-reductase expression and activity may not fully explain the tissue-selective pharmacologic actions of SARMs, it may indeed play the predominant but less beguiling role in the observed tissue-selective pharmacologic activity of SARMs.

6.3. AR STRUCTURAL BIOLOGY

Crystallography remains the best way to examine the molecular binding mechanism of different ligands. A large number of x-ray crystal structures of various AR ligand-bound AR LBD were solved (Table 6.3). Although these crystal structures help determine the binding mechanisms for ligands of various structural classes (steroidal or nonsteroidal) and functional classes (agonist or antagonist), all of these structures are solved with the "agonist" conformation of the LBD. For example, although antiandrogen-bound LBD structures are available, these structures were solved with mutant forms of the LBD, which convert these antagonists to agonists. Since the "antagonist" conformation of AR LBD is not yet available, there is not enough evidence to support the hypothesis that "AF2-antagonist" model (Figure 6.5) also applies to AR. Therefore, despite the diversity of the chemical structure of the ligands, the AR LBD crystal structures available to date are very similar and highly super-imposable.

6.3.1. Crystal Structures of Steroid-Bound AR LBD

The crystal structures of various steroidal ligands bound LBD (Table 6.3) revealed that the steroid skeleton interacts with the ligand-binding pocket largely through hydrophobic interactions, while hydrogen bonding at both extremities of the structure strongly tether the ligand to the binding site. For steroidal androgens, including testosterone, 5α-DHT, and R1881, the binding mechanisms are nearly identical (Figure 6.11). Mainly, the 3-keto group on the ring A forms H bonds with the side chains of residues Q711 and R752 directly or indirectly (through H_2O molecule); ring C is in close contact with the main chain of L704 and side chain of N705; and the 17β-OH group H bonds with the side chains of N705 and T877. The interactions between the ligand and residues N705 (N-terminal of H3) and T877 (C-terminal of H11) appear to be crucial to the conformational changes that pull the N-terminal of H3 and C-terminal of H11 towards the LBP, which serves as part of the mechanism to close the hydrophobic LBP upon ligand binding [24].

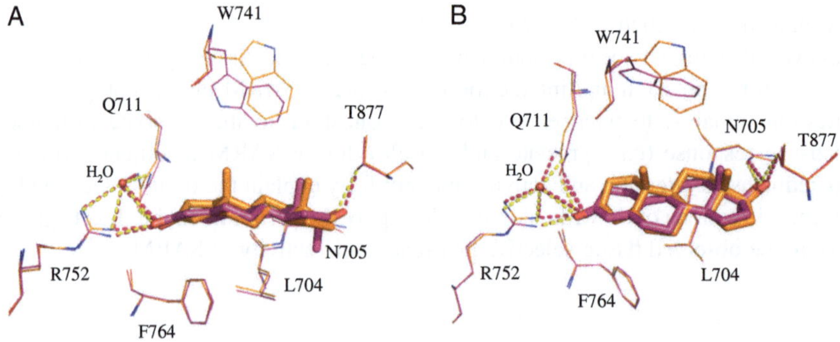

Figure 6.11. Superimposition of x-ray crystal structures of R1881-bound LBD and DHT-bound LBD. R1881 (*magenta*) WT complex (1E3G.pdb) and DHT (*orange*) WT complex (1I37.pdb), shown as side view (**a**) and top views (**b**) of the steroidal plane. H bonds are labeled as *dotted lines*

Within the LBP, the side chain conformation of various residues is highly super-imposable between the steroidal agonists, and is particularly similar between testosterone- and DHT-bound LBD. Therefore, the difference in their binding affinity (Table 6.3) is very likely due to the slight difference in structure and the consequent difference in H-bond strength [151]. In comparison to DHT-bound LBD, significant conformational change of the W741 and M895 side chain was observed in R1881-bound LBD (Figure 6.11), which could have significant functional consequence considering that both residues are in close proximity to the AF2 region (W741) and H12 (M895).

In the mutant form of AR LBD (T877A), the mutation introduces more space around the D-ring to accommodate a larger substituent at the 17-position, which allows ligands with greater steric bulk at this position, like CPA [152], to be accommodated into the binding pocket as well, and allows the LBD adopt the "agonist" conformation [153]. On the other hand, absence of the threonine hydroxyl group abolishes the hydrogen bond between the 17α-OH group of steroids and the receptor, significantly reducing the binding affinity of DHT to the mutant AR.

In general, most of the synthetic steroids, no matter agonists or antagonists, contain an A-ring 3-keto group that forms H bonds with Q711 and R752, which helps maintain the high binding affinity of the ligand. On the other hand, modification of the D ring 17β-OH group interferes with the H bonding to N705 and T877, which not only decreases the binding affinity of ligand but may also contribute to the antagonist activity of the ligand, considering the close proximity of these residues to H12 and the opening side of the binding pocket.

6.3.2. Crystal Structures of Nonsteroidal Ligand-Bound AR LBD

6.3.2.1. Bicalutamide bound AR LBD (W741L)

Several structural classes of nonsteroidal compounds demonstrate high binding affinity [154] (nanomolar range) to AR, and the binding mechanism for these compounds has been revealed by the crystal structures (Table 6.3). The first nonsteroidal ligand bound AR LBD crystal structure was R-bicalutamide-bound mutant LBD (W741L), with which bicalutamide functions as an agonist [155].

As shown in the crystal structure (Figure 6.12), bicalutamide adopts a greatly bent conformation upon binding to the LBD, with the majority of the molecule (A-ring and the amide bond) overlapping the steroidal plane, while the B-ring of the molecule folds away from the plane, pointing to the top of the ligand-binding pocket, which represents a unique structural feature of this class of ligands. On one side of the molecule, the A-ring cyano group forms H bonds with Q711 and R752, similar to the 3-keto group in 5α-DHT. On the other side of the molecule, the chiral hydroxyl group forms H bonds with L704 and N705, mimicking ring C and the 17β-OH group in 5α-DHT. These H bonding interactions are believed to be critical for high binding affinity of the ligand. Furthermore, intra-molecular hydrogen bonding in bicalutamide was also observed between the amide bond hydrogen, the chiral

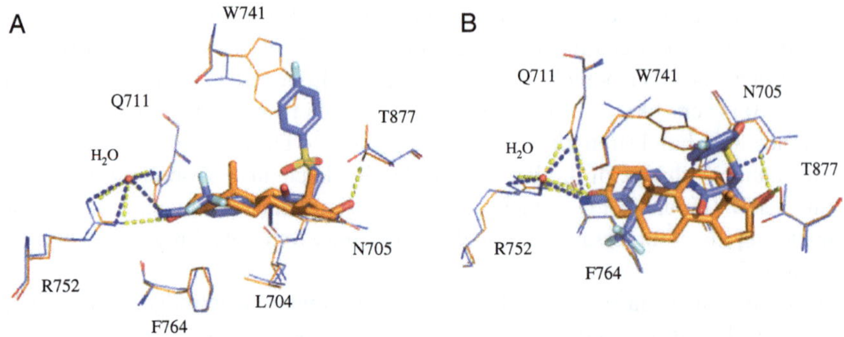

Figure 6.12. Superimposition of x-ray crystal structures of R-bicalutamide (aryl propionamide antagonist)-bound LBD and DHT-bound LBD. R-bicalutamide (*blue*) W741L complex (1Z95.pdb) and DHT (*orange*) WT complex (1I37.pdb), shown as side view (**a**) and top views (**b**) of the steroidal plane. H bonds are labeled as *dotted lines*

hydroxyl group, and the sulfonyl linkage oxygen, likely helping achieve the ligand geometry shown in Figure 6.12.

There is no H bonding between bicalutamide and the T877 side chain, and the B-ring folds into the position of the W741 side chain and makes direct contact with H12. The absence of a favorable H bond with T877 might partially explain the relatively lower binding affinity of bicalutamide to wild type AR, since the interaction between DHT 17β-OH group and T877 side chain is believed to be crucial for the bending of the C-terminus of H11 (Figure 6.3a). In comparison, the absence of W741 side chain in the W741L mutant exposes more space to accommodate the B-ring and allows the formation of a functional AF2 region, which might explain the agonist activity of bicalutamide in this mutant.

Assuming similar ligand geometry in wild type AR, the B-ring of bicalutamide would displace the W741 side chain from the ligand-binding pocket (Figure 6.12); sterically disrupting the formation of the AF2 region and providing a plausible mechanism for the antagonist activity of bicalutamide. Additionally, lack of the H bond with T877 and disruption of W741 side chain could result in a less folded overall conformation of the LBD and interrupt efficient dissociation of HSP90 complex [35].

6.3.2.2. *Aryl propionamide bound AR LBD (WT)*

The binding mechanism of the aryl propionamide analogs was further revealed by the crystal structure of S1-bound wild type AR LBD [145]. As an AR partial agonist, S1 was identified as one of the lead compounds of the aryl propionamide SARMs. The overall structure of the LBD adopts the typical "agonist" conformation. Within the binding pocket, S1 adopts similar ligand geometry and general hydrogen bonding pattern as that observed with bicalutamide due to their similarities in structure (Figure 6.13). However, with the flexibility of the ether linkage in these ligands, the S1 molecule adopt an even more bent conformation as compared to bicalutamide,

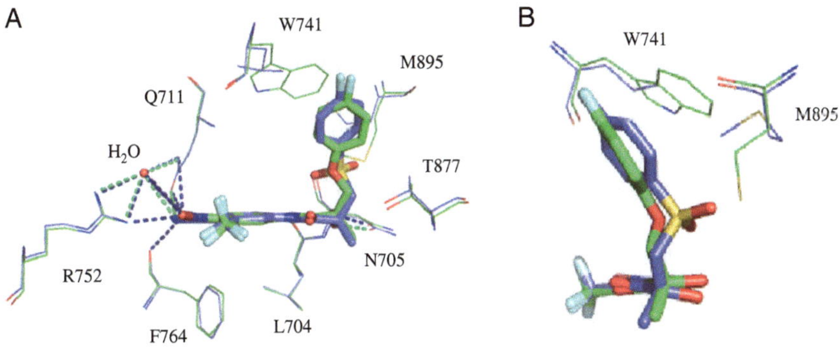

Figure 6.13. Superimposition of x-ray crystal structures of R-bicalutamide (aryl propionamide antagonist)-bound LBD and S1 (aryl propionamide SARM)-bound LBD. R-bicalutamide (*blue*) W741L complex (1Z95.pdb) and S1 (*green*) WT complex (2AXA.pdb), shown as side views (A. face to B-ring; B. side to B-ring) of the steroidal plane. H bonds are labeled as *dotted lines*

which allows the accommodation of the B-ring without displacing the bulky side chain of W741 in wild type LBD, and potentially explains the agonist activity observed in compounds incorporating ether linkages.

Also, when superimposed with DHT-bound LBD (Figure 6.14), besides the lack of H bond with T877, the most significant change within the binding pocket is the movement of W741 side chain, which is "pushed" away from its "original" position in DHT-bound LBD in order to accommodate the B-ring. SAR studies of this class of ligands has shown that the B-ring para-position substitution is critical for the agonist activity of the bicalutamide derivatives [156], which is supported by the fact that the B-ring in S1 is in close contact with W741 and is oriented nearly parallel to H12. Structural modifications of the B-ring may indirectly affect the AF2 region by changing the positioning of W741 side chain and H12.

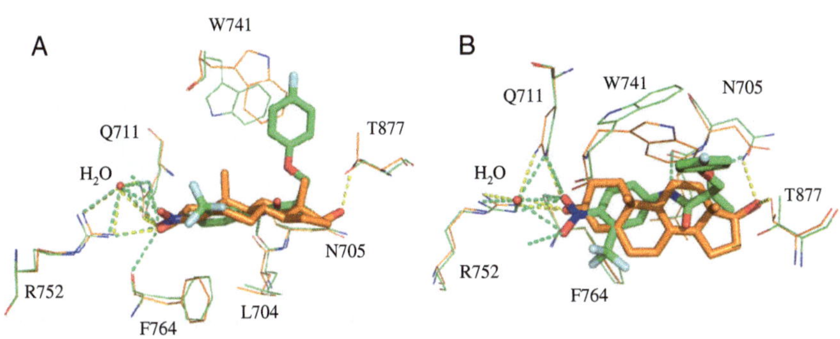

Figure 6.14. Superimposition of x-ray crystal structures of S1 (aryl propionamide SARM)-bound LBD and DHT-bound LBD. S1 (*green*) WT complex (2AXA.pdb) and DHT (*orange*) WT complex (1I37.pdb), shown as side view (**a**) and top views (**b**) of the steroidal plane. H bonds are labeled as *dotted lines*

6.3.2.3. *Hydantoin derivatives bound AR LBD (WT)*

The hydantoin derivatives share similar A-ring structure as bicalutamide (Table 6.2). As such, the cyano or nitro group interacts with Q711 and R752 [126] in a similar way as bicalutamide (Figure 6.15). In general, the benzene ring or the naphthyl group, together with the hydantoin ring, overlaps the steroid plane as well. On the other side of the ligand, the hydroxyl group forms H bond with N705, while H bond with T877 is absent.

As the lead compound of a series of hydantoin SARMs, BMS-564929 is also a partial agonist both in vitro and in vivo [126]. Structurally, BMS-564929 shares similar a binding mechanism to other hydantoin derivatives, including the interaction of the side chain of F764 with the phenyl ring of BMS-564929 via a π-edge to face interaction (5.20Å, ring centroids). Compared to DHT-bound LBD, BMS-56429 also displaces the W741 side chain away from the binding pocket, which could be due to the slight difference in orientation between the hydantoin ring in BMS-654929 and the C ring of DHT.

6.3.2.4. *Quinolone derivatives bound AR LBD (WT)*

The quinolone derivatives (Table 6.2), either bicyclic or tricyclic, can bind to AR with relatively high affinity (nanomolar range). The crystal structure of LGD2226 (SARM) bound wt AR LBD [157] (Figure 6.16) also revealed the binding mechanism for the bicyclic quinolone pharmacophore. The A-ring keto group in LGC2226 mimics the A-ring keto group in DHT and forms hydrogen bonds with Q711 and R752, which anchor the ligand to the receptor; while the additional hydrogen bond between the A-ring secondary amine group and Q711 helps to strengthen the interaction. On the other side of the binding pocket, although the N-linked trifluoroethyl

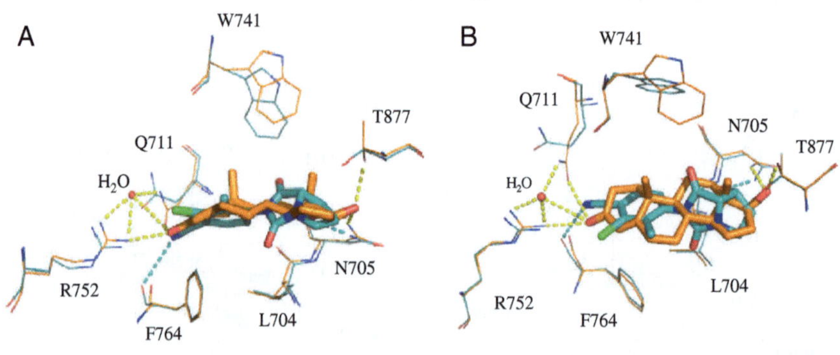

Figure 6.15. Superimposition of x-ray crystal structures of BMS564929 (hydantoin SARM)-bound LBD and DHT-bound LBD. BMS564929 (*teal*) WT complex (2NW4.pdb) and DHT (*orange*) WT complex (1I37.pdb), shown as side view (**a**) and top views (**b**) of the steroidal plane. H bonds are labeled as *dotted lines*

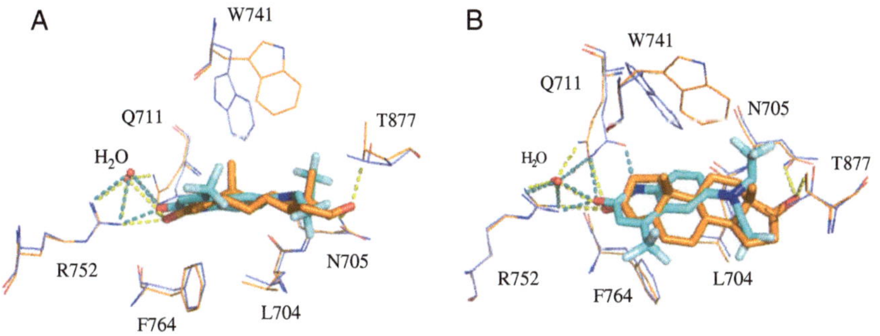

Figure 6.16. Superimposition of x-ray crystal structures of LGD2226 (quinoline SARM)-bound LBD and DHT-bound LBD. LGD2226 (*cyan*) WT complex (2HVC.pdb) and DHT (*orange*) WT complex (1I37.pdb), shown as side view (**a**) and top views (**b**) of the steroidal plane. H bonds are labeled as *dotted lines*

groups do not form any hydrogen bonds with the receptor, they do occupy a similar space as the C and D rings of steroids. Similar to other nonsteroidal agonists, LGD2226 binding also displaces the W741 side chain away from the top of the binding pocket (Figure 6.16), which may be related to the trifluoroethyl group that extrudes the binding plane.

In summary, despite of the structural diversity of AR ligands, there are significant similarities in the binding mechanisms of either steroidal or nonsteroidal ligands. In general: (1). the hydrogen bonding that mimics the interaction between the 3-keto group of DHT and Q711 and/or R752 is critical for "anchoring" the ligand to the binding pocket; (2). planar structure (see detailed discussion above) is usually present to mimic the steroidal plane and maintain the hydrophobic interactions; and lastly, (3) structural features in proximity to helix 12 and the AF2 region demonstrate more diversity in binding mechanism and appear to be more critical in determining the pharmacological activities of the ligand.

Compared to the steroidal agonists, nonsteroidal agonists demonstrate some noticeable differences in binding mechanism: none of the nonsteroidal ligands forms H bond with T877, and all of them more or less displace the side chain of W741, which may at least partially explain the relatively low binding affinity of these ligands. On the other hand, considering the unique position of W741 as related to the AF2 region, altering the conformation of the W741 side chain may have significant functional consequences via modulation of the AF2 surface topology.

Alignment of various crystal structures clearly revealed W741 and M895 as the main "moving" residues within the LBP (Figure 6.17). On one hand, M895 is part of the "^{895}MAEII^{899}" motif in H12, potential interaction between the ligand and M895 may affect the proper positioning of H12. On the other hand, W741 side chain lies on the top of the binding pocket and right underneath the AF2 surface, so significant conformational change of the side chain may also affect the depth and shape of the AF2 surface groove, which is critical for modulating the overall function of

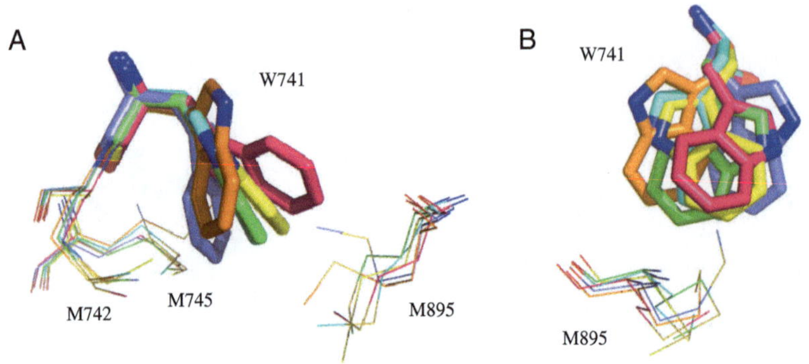

Figure 6.17. Superimposition of the moving residues of the LBP of AR LBD upon binding of various ligands, including DHT (*orange*), R1881 (*green*), S1 (*magenta*), BMS564929 (*yellow*), and LGD2226 (*blue*). W741 is shown as stick. M742, M745, and M895 are shown as lines. Both side view (**a**) and surface view (**b**) of W741 side chain are presented

the receptor and cofactor recruitment (see discussion in Section 6.1.2.2). Therefore, lack of H bonding with T877 and displacement of W741 side chain may account for the reduced binding affinity and partial agonist activities of the nonsteroidal SARMs identified so far.

Thus, all of the crystal structures of SARM-bound LBD adopt very similar overall conformations as compared to DHT-bound LBD. However, subtle differences in surface conformation, particularly the AF2 region, that are known to be important for AR structure and function may not be revealed by the crystal structure. Nevertheless, SARM binding do not appear to cause LBD conformational changes that are as significant as those observed with SERMs, which also raises the challenging question: what is the structural basis for the tissue selectivity of SARMs?

6.4. FUTURE PERSPECTIVES

Although the agonism and antagonism of ER ligands are well established [22], and appear to apply to some of the other NRs as well [30, 140], it is unclear if the molecular mechanisms of SERM tissue selectivity or the "AF2 antagonist model" (H12 displacement) for antiestrogen action apply to AR ligands. Nevertheless, even if the molecular mechanism underlying the antagonism of AR ligands is different from the ER, both appear to be related to the alteration or disruption of the formation of a functional AF2 region. Several new approaches targeting the AF2 region have been proposed [31] to block the agonist activity of AR for the treatment of prostate cancer, including small peptides that contain the "LxxLL" binding motif [148]; or small ligands that can introduce conformational changes to interrupt AF2 function [27]. Therefore, the LBD AF2 region has become a popular drug target site.

As opposed to pure agonists and antagonists, it seems probable that significant drug discovery and development efforts will remain devoted to SARMs. Due to their flexibility in structural modification, SARMs represent an interesting new class of AR ligands that could potentially regulate multiple aspects of AR activation: ligand-induced receptor conformation changes; protein-protein interactions; and protein-DNA interactions. "Selective ARE (androgen response elements)" [49] were proposed recently as a mechanism for ligand-specific regulation of AR function. SARMs with various intrinsic activities could provide a promising opportunity to better understand the role of receptor-protein and/or receptor-DNA interaction in determining the tissue selectivity of AR ligands.

More and more evidence has emerged to suggest that nongenomic pathways also play an important role in AR function. Tissue selectivity may also be achieved by developing ligands that specifically activate the nongenomic pathway, as demonstrated for the ER ligand, estren [55, 61]. The separation of the genomic and nongenomic pathways is likely dependent on ligand-specific regulation of receptor-protein interaction. However, better understanding of the molecular pharmacology and structural basis of the process will be necessary to identify potential targeting sites.

In summary, the development of AR ligands will continue, particularly focusing on the search for ligands that are AR specific, metabolically stable, safe, tissue selective, and discriminate between genomic and non-genomic pathways. A better understanding of the protein chemistry of the AR will help design ligands that could specifically target important processes during AR activation. This new generation of AR ligands appears poised to provide new drugs for the treatment of many diseases.

REFERENCES

1. Mosselman, S., Polman, J., and Dijkema, R. (1996). *FEBS Lett* 392, 49–53.
2. Goodman, L. S., Hardman, J. G., Limbird, L. E., and Gilman, A. G. (2001). *Goodman & Gilman's the pharmacological basis of therapeutics*, 10th ed. McGraw-Hill Medical Pub. Division, New York.
3. Johansen, K. L. (2004). *Semin Dial* 17, 202–208.
4. Mooradian, A. D., Morley, J. E., and Korenman, S. G. (1987). *Endocr Rev* 8, 1–28.
5. Keller, E. T., Ershler, W. B., and Chang, C. (1996). *Front Biosci* 1, d59–d71.
6. Oettel, M. (2003). *Aging Male* 6, 230–256.
7. Foye, W. O., Williams, D. A., and Lemke, T. L. (2002). *Foye's principles of medicinal chemistry*, 5th ed. Lippincott Williams & Wilkins, Philadelphia.
8. Migeon, B. R., Brown, T. R., Axelman, J., and Migeon, C. J. (1981). *Proc Natl Acad Sci USA* 78, 6339–6343.
9. Chang, C. S., Kokontis, J., and Liao, S. T. (1988). *Science* 240, 324–326.
10. Lubahn, D. B., Joseph, D. R., Sar, M., Tan, J., Higgs, H. N., Larson, R. E., French, F. S., and Wilson, E. M. (1988). *Mol Endocrinol* 2, 1265–1275.
11. Trapman, J., Klaassen, P., Kuiper, G. G., van der Korput, J. A., Faber, P. W., van Rooij, H. C., Geurts van Kessel, A., Voorhorst, M. M., Mulder, E., and Brinkmann, A. O. (1988). *Biochem Biophys Res Commun* 153, 241–248.
12. McEwan, I. J. (2004). *Endocr Relat Cancer* 11, 281–293.

13. Chawnshang, C. (2002). *Androgens and androgen receptor: Mechanisms, functions, and clinical applications*. Kluwer Academic Publishers, Boston.
14. Marhefka, C. A., Moore, B. M., 2nd, Bishop, T. C., Kirkovsky, L., Mukherjee, A., Dalton, J. T., and Miller, D. D. (2001). *J Med Chem* 44, 1729–1740.
15. Pratt, W. B. and Toft, D. O. (1997). *Endocr Rev* 18, 306–360.
16. Freedman, L. P. (1998). *Molecular biology of steroid and nuclear hormone receptors*. Birkhauser, Boston.
17. Heinlein, C. A. and Chang, C. (2002). *Endocr Rev* 23, 175–200.
18. Shang, Y., Myers, M., and Brown, M. (2002). *Mol Cell* 9, 601–610.
19. Chandra, V., Huang, P., Hamuro, Y., Raghuram, S., Wang, Y., Burris, T. P., and Rastinejad, F. (2008). *Nature* 456, 350–356.
20. Matias, P. M., Donner, P., Coelho, R., Thomaz, M., Peixoto, C., Macedo, S., Otto, N., Joschko, S., Scholz, P., Wegg, A., Basler, S., Schafer, M., Egner, U., and Carrondo, M. A. (2000). *J Biol Chem* 275, 26164–26171.
21. Wurtz, J. M., Bourguet, W., Renaud, J. P., Vivat, V., Chambon, P., Moras, D., and Gronemeyer, H. (1996). *Nat Struct Biol* 3, 87–94.
22. Brzozowski, A. M., Pike, A. C., Dauter, Z., Hubbard, R. E., Bonn, T., Engstrom, O., Ohman, L., Greene, G. L., Gustafsson, J. A., and Carlquist, M. (1997). *Nature* 389, 753–758.
23. Heery, D. M., Kalkhoven, E., Hoare, S., and Parker, M. G. (1997). *Nature* 387, 733–736.
24. Bourguet, W., Germain, P., and Gronemeyer, H. (2000). *Trends Pharmacol Sci* 21, 381–388.
25. He, B., Minges, J. T., Lee, L. W., and Wilson, E. M. (2002). *J Biol Chem* 277, 10226–10235.
26. Hur, E., Pfaff, S. J., Payne, E. S., Gron, H., Buehrer, B. M., and Fletterick, R. J. (2004). *PLoS Biol* 2, E274.
27. Sathya, G., Chang, C. Y., Kazmin, D., Cook, C. E., and McDonnell, D. P. (2003). *Cancer Res* 63, 8029–8036.
28. He, B., Gampe, R. T., Jr., Kole, A. J., Hnat, A. T., Stanley, T. B., An, G., Stewart, E. L., Kalman, R. I., Minges, J. T., and Wilson, E. M. (2004). *Mol Cell* 16, 425–438.
29. Warnmark, A., Treuter, E., Gustafsson, J. A., Hubbard, R. E., Brzozowski, A. M., and Pike, A. C. (2002). *J Biol Chem* 277, 21862–21868.
30. Kauppi, B., Jakob, C., Farnegardh, M., Yang, J., Ahola, H., Alarcon, M., Calles, K., Engstrom, O., Harlan, J., Muchmore, S., Ramqvist, A. K., Thorell, S., Ohman, L., Greer, J., Gustafsson, J. A., Carlstedt-Duke, J., and Carlquist, M. (2003). *J Biol Chem* 278, 22748–22754.
31. Chang, C. Y. and McDonnell, D. P. (2005). *Trends Pharmacol Sci* 26, 225–228.
32. Langley, E., Kemppainen, J. A., and Wilson, E. M. (1998). *J Biol Chem* 273, 92–101.
33. Kemppainen, J. A., Langley, E., Wong, C. I., Bobseine, K., Kelce, W. R., and Wilson, E. M. (1999). *Mol Endocrinol* 13, 440–454.
34. Veldscholte, J., Berrevoets, C. A., Zegers, N. D., van der Kwast, T. H., Grootegoed, J. A., and Mulder, E. (1992). *Biochemistry* 31, 7422–7430.
35. Kuil, C. W., Berrevoets, C. A., and Mulder, E. (1995). *J Biol Chem* 270, 27569–27576.
36. Masiello, D., Cheng, S., Bubley, G. J., Lu, M. L., and Balk, S. P. (2002). *J Biol Chem* 277, 26321–26326.
37. Hodgson, M. C., Astapova, I., Cheng, S., Lee, L. J., Verhoeven, M. C., Choi, E., Balk, S. P., and Hollenberg, A. N. (2005). *J Biol Chem* 280, 6511–6519.
38. Callewaert, L., Van Tilborgh, N., and Claessens, F. (2006). *Cancer Res* 66, 543–553.
39. Reid, J., Kelly, S. M., Watt, K., Price, N. C., and McEwan, I. J. (2002). *J Biol Chem* 277, 20079–20086.
40. Davies, P., Watt, K., Kelly, S. M., Clark, C., Price, N. C., and McEwan, I. J. (2008). *J Mol Endocrinol* 41, 301–314.
41. Lavery, D. N. and McEwan, I. J. (2008). *Biochemistry* 47, 3360–3369.
42. Heemers, H. V. and Tindall, D. J. (2007). *Endocr Rev* 28, 778–808.
43. Nantermet, P. V., Xu, J., Yu, Y., Hodor, P., Holder, D., Adamski, S., Gentile, M. A., Kimmel, D. B., Harada, S., Gerhold, D., Freedman, L. P., and Ray, W. J. (2004). *J Biol Chem* 279, 1310–1322.

44. Nelson, P. S., Clegg, N., Arnold, H., Ferguson, C., Bonham, M., White, J., Hood, L., and Lin, B. (2002). *Proc Natl Acad Sci U S A* 99, 11890–11895.
45. DePrimo, S. E., Diehn, M., Nelson, J. B., Reiter, R. E., Matese, J., Fero, M., Tibshirani, R., Brown, P. O., and Brooks, J. D. (2002). *Genome Biol* 3, RESEARCH0032.
46. Claessens, F., Verrijdt, G., Schoenmakers, E., Haelens, A., Peeters, B., Verhoeven, G., and Rombauts, W. (2001). *J Steroid Biochem Mol Biol* 76, 23–30.
47. Haelens, A., Verrijdt, G., Callewaert, L., Christiaens, V., Schauwaers, K., Peeters, B., Rombauts, W., and Claessens, F. (2003). *Biochem J* 369, 141–151.
48. Shaffer, P. L., Jivan, A., Dollins, D. E., Claessens, F., and Gewirth, D. T. (2004). *Proc Natl Acad Sci U S A* 101, 4758–4763.
49. Hsiao, P. W., Thin, T. H., Lin, D. L., and Chang, C. (2000). *Mol Cell Biochem* 206, 169–175.
50. Remenyi, A., Scholer, H. R., and Wilmanns, M. (2004). *Nat Struct Mol Biol* 11, 812–815.
51. Norman, A. W., Mizwicki, M. T., and Norman, D. P. (2004). *Nat Rev Drug Discov* 3, 27–41.
52. Rahman, F. and Christian, H. C. (2007). *Trends Endocrinol Metab* 18, 371–378.
53. Lutz, L. B., Jamnongjit, M., Yang, W. H., Jahani, D., Gill, A., and Hammes, S. R. (2003). *Mol Endocrinol* 17, 1106–1116.
54. Estrada, M., Espinosa, A., Muller, M., and Jaimovich, E. (2003). *Endocrinology* 144, 3586–3597.
55. Kousteni, S., Bellido, T., Plotkin, L. I., O'Brien, C. A., Bodenner, D. L., Han, L., Han, K., DiGregorio, G. B., Katzenellenbogen, J. A., Katzenellenbogen, B. S., Roberson, P. K., Weinstein, R. S., Jilka, R. L., and Manolagas, S. C. (2001). *Cell* 104, 719–730.
56. Zagar, Y., Chaumaz, G., and Lieberherr, M. (2004). *J Biol Chem* 279, 2403–2413.
57. Kampa, M., Papakonstanti, E. A., Hatzoglou, A., Stathopoulos, E. N., Stournaras, C., and Castanas, E. (2002). *Faseb J* 16, 1429–1431.
58. Unni, E., Sun, S., Nan, B., McPhaul, M. J., Cheskis, B., Mancini, M. A., and Marcelli, M. (2004). *Cancer Res* 64, 7156–7168.
59. Simoncini, T. and Genazzani, A. R. (2003). *Eur J Endocrinol* 148, 281–292.
60. Migliaccio, A., Castoria, G., Di Domenico, M., de Falco, A., Bilancio, A., Lombardi, M., Barone, M. V., Ametrano, D., Zannini, M. S., Abbondanza, C., and Auricchio, F. (2000). *Embo J* 19, 5406–5417.
61. Kousteni, S., Chen, J. R., Bellido, T., Han, L., Ali, A. A., O'Brien, C. A., Plotkin, L., Fu, Q., Mancino, A. T., Wen, Y., Vertino, A. M., Powers, C. C., Stewart, S. A., Ebert, R., Parfitt, A. M., Weinstein, R. S., Jilka, R. L., and Manolagas, S. C. (2002). *Science* 298, 843–846.
62. Bremner, W. J., Vitiello, M. V., and Prinz, P. N. (1983). *J Clin Endocrinol Metab* 56, 1278–1281.
63. Schulman, C. and Lunenfeld, B. (2002). *World J Urol* 20, 4–10.
64. Hakimi, J. M., Rondinelli, R. H., Schoenberg, M. P., and Barrack, E. R. (1996). *World J Urol* 14, 329–337.
65. Burger, A. and Abraham, D. J. (2003). *Burger's medicinal chemistry and drug discovery*, 6th ed. Wiley, Hoboken, NJ.
66. Dorfman, R. I. and Ungar, F. (1965). *Metabolism of steroid hormones*, 2nd ed. Academic Press, New York.
67. Verderame, M. (1986). *CRC handbook of hormones, vitamins, and radiopaques*. CRC Press, Boca Raton, FL.
68. Fotherby, K., James, F., Atkinson, L., and Dumasia, M. (1972). *J Endocrinol* 52, v–vi.
69. Belanger, A., Pelletier, G., Labrie, F., Barbier, O., and Chouinard, S. (2003). *Trends Endocrinol Metab* 14, 473–479.
70. Russell, D. W., Berman, D. M., Bryant, J. T., Cala, K. M., Davis, D. L., Landrum, C. P., Prihoda, J. S., Silver, R. I., Thigpen, A. E., and Wigley, W. C. (1994). *Recent Prog Horm Res* 49, 275–284.
71. Thigpen, A. E., Silver, R. I., Guileyardo, J. M., Casey, M. L., McConnell, J. D., and Russell, D. W. (1993). *J Clin Invest* 92, 903–910.
72. Vermeulen, A., Kaufman, J. M., Goemaere, S., and van Pottelberg, I. (2002). *Aging Male* 5, 98–102.
73. Oettel, M. (2002). *Aging Male* 5, 248–257.

74. de Ronde, W., Pols, H. A., van Leeuwen, J. P., and de Jong, F. H. (2003). *Clin Endocrinol (Oxf)* 58, 529–542.
75. Barrett-Connor, E., Goodman-Gruen, D., and Patay, B. (1999). *J Clin Endocrinol Metab* 84, 3681–3685.
76. Van Pottelbergh, I., Braeckman, L., De Bacquer, D., De Backer, G., and Kaufman, J. M. (2003). *Atherosclerosis* 166, 95–102.
77. Mukherjee, T. K., Dinh, H., Chaudhuri, G., and Nathan, L. (2002). *Proc Natl Acad Sci USA* 99, 4055–4060.
78. Toda, K., Okada, T., Takeda, K., Akira, S., Saibara, T., Shiraishi, M., Onishi, S., and Shizuta, Y. (2001). *J Endocrinol* 168, 455–463.
79. Falahati-Nini, A., Riggs, B. L., Atkinson, E. J., O'Fallon, W. M., Eastell, R., and Khosla, S. (2000). *J Clin Invest* 106, 1553–1560.
80. Orwoll, E. S. (2003). *Osteoporos Int* 14, 93–98.
81. Fang, H., Tong, W., Branham, W. S., Moland, C. L., Dial, S. L., Hong, H., Xie, Q., Perkins, R., Owens, W., and Sheehan, D. M. (2003). *Chem Res Toxicol* 16, 1338–1358.
82. Kelce, W. R., Monosson, E., Gamcsik, M. P., Laws, S. C., and Gray, L. E., Jr. (1994). *Toxicol Appl Pharmacol* 126, 276–285.
83. Waller, C. L., Juma, B. W., Gray, L. E., Jr., and Kelce, W. R. (1996). *Toxicol Appl Pharmacol* 137, 219–227.
84. Hershberger, L. G., Shipley, E. G., and Meyer, R. K. (1953). *Proc Soc Exp Biol Med* 83, 175–180.
85. Singh, S. M., Gauthier, S., and Labrie, F. (2000). *Curr Med Chem* 7, 211–247.
86. Cabasso, A. (1994). *Med Sci Sports Exerc* 26, 2–4.
87. Petera, V., Bobek, K., and Lahn, V. (1962). *Clin Chim Acta* 7, 604–606.
88. Venturoli, S., Marescalchi, O., Colombo, F. M., Macrelli, S., Ravaioli, B., Bagnoli, A., Paradisi, R., and Flamigni, C. (1999). *J Clin Endocrinol Metab* 84, 1304–1310.
89. Cumming, D. C., Yang, J. C., Rebar, R. W., and Yen, S. S. (1982). *Jama* 247, 1295–1298.
90. Caminos-Torres, R., Ma, L., and Snyder, P. J. (1977). *J Clin Endocrinol Metab* 45, 255–260.
91. Mayer, M. and Rosen, F. (1975). *Am J Physiol* 229, 1381–1386.
92. Langer, C. J., Hoffman, J. P., and Ottery, F. D. (2001). *Nutrition* 17, S1–S20.
93. Pavlatos, A. M., Fultz, O., Monberg, M. J., and Vootkur, A. (2001). *Pharmd Clin Ther* 23, 789–801, discussion 771.
94. Katchen, B. and Buxbaum, S. (1975). *J Clin Endocrinol Metab* 41, 373–379.
95. Fau, D., Eugene, D., Berson, A., Letteron, P., Fromenty, B., Fisch, C., and Pessayre, D. (1994). *J Pharmacol Exp Ther* 269, 954–962.
96. Creaven, P. J., Pendyala, L., and Tremblay, D. (1991). *Urology* 37, 13–19.
97. Berson, A., Wolf, C., Berger, V., Fau, D., Chachaty, C., Fromenty, B., and Pessayre, D. (1991). *J Pharmacol Exp Ther* 257, 714–719.
98. Fau, D., Berson, A., Eugene, D., Fromenty, B., Fisch, C., and Pessayre, D. (1992). *J Pharmacol Exp Ther* 263, 69–77.
99. Salvati, M. E., Balog, A., Shan, W., Wei, D. D., Pickering, D., Attar, R. M., Geng, J., Rizzo, C. A., Gottardis, M. M., Weinmann, R., Krystek, S. R., Sack, J., An, Y., and Kish, K. (2005). *Bioorg Med Chem Lett* 15, 271–276.
100. Balog, A., Salvati, M. E., Shan, W., Mathur, A., Leith, L. W., Wei, D. D., Attar, R. M., Geng, J., Rizzo, C. A., Wang, C., Krystek, S. R., Tokarski, J. S., Hunt, J. T., Gottardis, M., and Weinmann, R. (2004). *Bioorg Med Chem Lett* 14, 6107–6111.
101. Salvati, M. E., Balog, A., Wei, D. D., Pickering, D., Attar, R. M., Geng, J., Rizzo, C. A., Hunt, J. T., Gottardis, M. M., Weinmann, R., and Martinez, R. (2005). *Bioorg Med Chem Lett* 15, 389–393.
102. Salvati, M. E., Attar, R. M., Krystek, S. R., Lee, F., Hunt, J. T., Gottardis, M., Jure-Kunkeld, M., Sack, J., Giese, S., Martinez, R., Mitt, T., Pickering, D., Shan, W., Wei, D. D., Zhu, H., Cullinan, C., Geng, J., Rizzo, C. A., Wang, C., Mathur, A., Leith, L. W., Kish, K., Vite, G., Dell, J., Spires, T., An, Y., and Weinmann, R. (2004). *AACR* abstract # 1948.
103. Cockshott, I. D. (2004). *Clin Pharmacokinet* 43, 855–878.

104. Boyle, G. W., McKillop, D., Phillips, P. J., Harding, J. R., Pickford, R., and McCormick, A. D. (1993). *Xenobiotica* 23, 781–798.
105. McKillop, D., Boyle, G. W., Cockshott, I. D., Jones, D. C., Phillips, P. J., and Yates, R. A. (1993). *Xenobiotica* 23, 1241–1253.
106. Furr, B. J. and Tucker, H. (1996). *Urology* 47, 13–25, discussion 29–32.
107. Schellhammer, P., Sharifi, R., Block, N., Soloway, M., Venner, P., Patterson, A. L., Sarosdy, M., Vogelzang, N., Jones, J., and Kolvenbag, G. (1995). *Urology* 45, 745–752.
108. Diamanti-Kandarakis, E. (1999). *Curr Pharm Des* 5, 707–723.
109. Muderris, I. I., Bayram, F., Ozcelik, B., and Guven, M. (2002). *Gynecol Endocrinol* 16, 63–66.
110. Carmina, E. and Lobo, R. A. (2002). *Clin Endocrinol (Oxf)* 57, 231–234.
111. Van Dort, M. E. and Jung, Y. W. (2001). *Bioorg Med Chem Lett* 11, 1045–1047.
112. Battmann, T., Bonfils, A., Branche, C., Humbert, J., Goubet, F., Teutsch, G., and Philibert, D. (1994). *J Steroid Biochem Mol Biol* 48, 55–60.
113. Taplin, M. E., Rajeshkumar, B., Halabi, S., Werner, C. P., Woda, B. A., Picus, J., Stadler, W., Hayes, D. F., Kantoff, P. W., Vogelzang, N. J., and Small, E. J. (2003). *J Clin Oncol* 21, 2673–2678.
114. Suzuki, H., Akakura, K., Komiya, A., Aida, S., Akimoto, S., and Shimazaki, J. (1996). *Prostate* 29, 153–158.
115. Hara, T., Miyazaki, J., Araki, H., Yamaoka, M., Kanzaki, N., Kusaka, M., and Miyamoto, M. (2003). *Cancer Res* 63, 149–153.
116. Zhi, L. and Martinborough, E. (2001). *Annu Rep Med Chem* 36, 169–180.
117. Negro-Vilar, A. (1999). *J Clin Endocrinol Metab 84*, 3459–3462.
118. Gao, W. and Dalton, J. T. (2007). *Drug Discov Today* 12, 241–248.
119. Gao, W., Kim, J., and Dalton, J. T. (2006). *Pharm Res* 23, 1641–1658.
120. Dalton, J. T., Mukherjee, A., Zhu, Z., Kirkovsky, L., and Miller, D. D. (1998). *Biochem Biophys Res Commun* 244, 1–4.
121. Edwards, J. P., West, S. J., Pooley, C. L., Marschke, K. B., Farmer, L. J., and Jones, T. K. (1998). *Bioorg Med Chem Lett* 8, 745–750.
122. Yin, D., Gao, W., Kearbey, J. D., Xu, H., Chung, K., He, Y., Marhefka, C. A., Veverka, K. A., Miller, D. D., and Dalton, J. T. (2003). *J Pharmacol Exp Ther* 304, 1334–1340.
123. Gao, W., Kearbey, J. D., Nair, V. A., Chung, K., Parlow, A. F., Miller, D. D., and Dalton, J. T. (2004). *Endocrinology* 145, 5420–5428.
124. Hanada, K., Furuya, K., Yamamoto, N., Nejishima, H., Ichikawa, K., Nakamura, T., Miyakawa, M., Amano, S., Sumita, Y., and Oguro, N. (2003). *Biol Pharm Bull* 26, 1563–1569.
125. Miner, J. N., Chang, W., Chapman, M. S., Finn, P. D., Hong, M. H., Lopez, F. J., Marschke, K. B., Rosen, J., Schrader, W., Turner, R., van Oeveren, A., Viveros, H., Zhi, L., and Negro-Vilar, A. (2006). *Endocrinology* doi:10.1210/en.2006–0793.
126. Ostrowski, J., Kuhns, J. E., Lupisella, J. A., Manfredi, M. C., Beehler, B. C., Krystek, S. R., Jr., Bi, Y., Sun, C., Seethala, R., Golla, R., Sleph, P. G., Fura, A., An, Y., Kish, K. F., Sack, J. S., Mookhtiar, K. A., Grover, G. J., and Hamann, L. G. (2007). *Endocrinology* 148, 4–12.
127. Gao, W., Reiser, P. J., Coss, C. C., Phelps, M. A., Kearbey, J. D., Miller, D. D., and Dalton, J. T. (2005). *Endocrinology* 146(11): 4887–4897.
128. Gao, W., Reiser, P. J., Kearbey, J. D., Phelps, M. A., Coss, C. C., Miller, D. D., and Dalton, J. T. (2004). *The Endocrine Society, New Orleans.*
129. Kearbey, J. D., Gao, W., Miller, D. D., and Dalton, J. T. (2003). *AAPS PharmSci* 5.
130. Chen, J., Hwang, D. J., Bohl, C. E., Miller, D. D., and Dalton, J. T. (2005). *J Pharmacol Exp Ther* 312(2): 546–553.
131. Kearbey, J. D., Wu, D., Gao, W., Miller, D. D., and Dalton, J. T. (2004). *Xenobiotica* 34, 273–280.
132. Wu, D., Gao, W., Kearbey, J. D., Chung, K., Miller, D. D., and Dalton, J. T. (2003). *AAPS PharmSci* 5.
133. Gao, W., Johnston, J. S., Miller, D. D., and Dalton, J. T. (2006). *Drug Metab Dispos* 34, 254–260.
134. Gao, W., Wu, Z., Bohl, C. E., Yang, J., Miller, D. D., and Dalton, J. T. (2006). *Drug Metab Dispos* 34, 243–253.
135. Gao, W., Wu, Z., Chung, K., Miller, D. D., and Dalton, J. T. (2003). *AAPS PharmSci.*

136. Hamann, L. G. *227th National Meeting of the American Chemical Society Medicinal Chemistry Division* 2004, MEDI–175.
137. Miner, J. N., Chang, W., Chapman, M. S., Finn, P. D., Hong, M. H., Lopez, F. J., Marschke, K. B., Rosen, J., Schrader, W., Turner, R., van Oeveren, A., Viveros, H., Zhi, L., and Negro-Vilar, A. (2007). *Endocrinology* 148, 363–373.
138. Rosen, J. and Negro-Vilar, A. (2002). *J Musculoskelet Neuronal Interact* 2, 222–224.
139. Gronemeyer, H., Gustafsson, J. A., and Laudet, V. (2004). *Nat Rev Drug Discov* 3, 950–964.
140. Smith, C. L. and O'Malley, B. W. (2004). *Endocr Rev* 25, 45–71.
141. Katzenellenbogen, B. S. and Katzenellenbogen, J. A. (2002). *Science* 295, 2380–2381.
142. Wright, A. S., Douglas, R. C., Thomas, L. N., Lazier, C. B., and Rittmaster, R. S. (1999). *Endocrinology* 140, 4509–4515.
143. Parent, E. E., Jenks, C., Sharp, T., Welch, M. J., and Katzenellenbogen, J. A. (2006). *Nucl Med Biol* 33, 705–713.
144. Brodie, A. M., Son, C., King, D. A., Meyer, K. M., and Inkster, S. E. (1989). *Cancer Res* 49, 6551–6555.
145. Bohl, C. E., Miller, D. D., Chen, J., Bell, C. E., and Dalton, J. T. (2005). *J Biol Chem* 280, 37747–37754.
146. Wu, Y. L., Yang, X., Ren, Z., McDonnell, D. P., Norris, J. D., Willson, T. M., and Greene, G. L. (2005). *Mol Cell* 18, 413–424.
147. Kazmin, D., Prytkova, T., Cook, C. E., Wolfinger, R., Chu, T. M., Beratan, D., Norris, J. D., Chang, C. Y., and McDonnell, D. P. (2006). *Mol Endocrinol* 20, 1201–1217.
148. Chang, C. Y. and McDonnell, D. P. (2002). *Mol Endocrinol* 16, 647–660.
149. Masiello, D., Chen, S. Y., Xu, Y., Verhoeven, M. C., Choi, E., Hollenberg, A. N., and Balk, S. P. (2004). *Mol Endocrinol* 18, 2388–2401.
150. Song, L. N., Herrell, R., Byers, S., Shah, S., Wilson, E. M., and Gelmann, E. P. (2003). *Mol Cell Biol* 23, 1674–1687.
151. Pereira de Jesus-Tran, K., Cote, P. L., Cantin, L., Blanchet, J., Labrie, F., and Breton, R. (2006). *Protein Sci* 15, 987–999.
152. Matias, P. M., Carrondo, M. A., Coelho, R., Thomaz, M., Zhao, X. Y., Wegg, A., Crusius, K., Egner, U., and Donner, P. (2002). *J Med Chem* 45, 1439–1446.
153. Bohl, C. E., Wu, Z., Miller, D. D., Bell, C. E., and Dalton, J. T. (2007). *J Biol Chem* 282, 13648–13655.
154. Allan, G. F. and Sui, Z. (2003). *NURSA e-J* 1, ID# 3.09172003.1.
155. Bohl, C. E., Gao, W., Miller, D. D., Bell, C. E., and Dalton, J. T. (2005). *Proc Natl Acad Sci USA* 102, 6201–6206.
156. Yin, D., He, Y., Perera, M. A., Hong, S. S., Marhefka, C., Stourman, N., Kirkovsky, L., Miller, D. D., and Dalton, J. T. (2003). *Mol Pharmacol* 63, 211–223.
157. Wang, F., Liu, X. Q., Li, H., Liang, K. N., Miner, J. N., Hong, M., Kallel, E. A., van Oeveren, A., Zhi, L., and Jiang, T. (2006). *Acta Crystallogr Sect F Struct Biol Cryst Commun* 62, 1067–1071.
158. Sack, J. S., Kish, K. F., Wang, C., Attar, R. M., Kiefer, S. E., An, Y., Wu, G. Y., Scheffler, J. E., Salvati, M. E., Krystek, S. R., Jr., Weinmann, R., and Einspahr, H. M. (2001). *Proc Natl Acad Sci USA* 98, 4904–4909.
159. Sun, C., Robl, J. A., Wang, T. C., Huang, Y., Kuhns, J. E., Lupisella, J. A., Beehler, B. C., Golla, R., Sleph, P. G., Seethala, R., Fura, A., Krystek, S. R., Jr., An, Y., Malley, M. F., Sack, J. S., Salvati, M. E., Grover, G. J., Ostrowski, J., and Hamann, L. G. (2006). *J Med Chem* 49, 7596–7599.
160. Zhou, Z. X., Wong, C. I., Sar, M., and Wilson, E. M. (1994). *Recent Prog Horm Res* 49, 249–274.
161. Myers, E. W. and Miller, W. (1989). *Bull Math Biol* 51, 5–37.
162. Miller, W. L. (1988). *Endocr Rev* 9, 295–318.
163. Christiansen, R. G., Bell, M. R., D'Ambra, T. E., Mallamo, J. P., Herrmann, J. L., Ackerman, J. H., Opalka, C. J., Kullnig, R. K., Winneker, R. C., Snyder, B. W. et al. (1990). *J Med Chem* 33, 2094–2100.
164. Litwack, G. (1970). *Biochemical actions of hormones*. Academic Press, New York.
165. Wakeling, A. E., Furr, B. J., Glen, A. T., and Hughes, L. R. (1981). *J Steroid Biochem* 15, 355–359.

CHAPTER 7

THYROID HORMONE RECEPTORS

BJÖRN VENNSTRÖM[1], HONG LIU[2], AND DOUGLAS FORREST[2]

[1] *Department of Cell and Molecular Biology, Karolinska Institute, S-171 77 Stockholm, Sweden*

[2] *NIDDK, Clinical Endocrinology Branch, National Institutes of Health, Bethesda, MD 20892-1772, USA*

Abstract: Thyroid hormone promotes a diverse range of developmental, neurological and metabolic functions in vertebrate species. Human thyroid disorders result in a correspondingly wide range of disease symptoms. The functions of thyroid hormone are mediated by a small group of thyroid hormone receptors encoded by two conserved genes. Thyroid hormone receptors were among the first nuclear receptors to be identified and act as ligand-regulated transcription factors. These receptors are particularly versatile since they also have the potential to mediate ligand-independent transcriptional control. Genetic analyses have revealed both specific and overlapping roles for each receptor, revealing how a small receptor family can mediate an extended range of biological functions.

7.1. HISTORICAL INTRODUCTION TO THYROID HORMONE RECEPTORS

Abnormalities of the thyroid gland have long been associated with developmental and metabolic disease in human populations. Towards the end of the nineteenth century, a functional basis for this association was provided when it was found that a substance extracted from the thyroid had potent biological activity. In 1891, Murray reported that injections of sheep thyroid extracts gave a marked recovery in a woman suffering from myxoedema, a condition of fatigue and reduced metabolic activity that is nowadays known to be caused by a lack of thyroid hormone [1]. Soon thereafter, dramatic improvements were also observed following similar treatments for childhood cretinism, a condition characterized by retarded growth and mental function that we know today is caused by a developmental lack of thyroid hormone [2]. These landmark studies in the nascent field of endocrinology demonstrated how a substance produced by a glandular tissue stimulated responses in distant target organs. The extractable substance was subsequently identified as thyroid hormone [3] (Figure 7.1).

These early findings raised two fundamental questions that still remain pertinent today: (i) How does thyroid hormone elicit its biological responses in target tissues?

C.M. Bunce, M.J. Campbell (eds.), *Nuclear Receptors*, Proteins and Cell Regulation 8, 183–201. DOI 10.1007/978-90-481-3303-1_7,

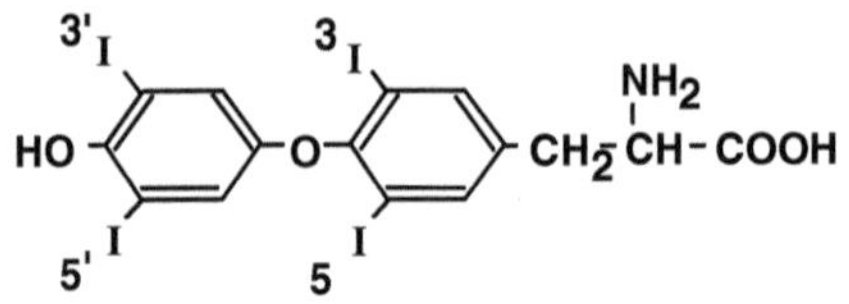

T4, thyroxine

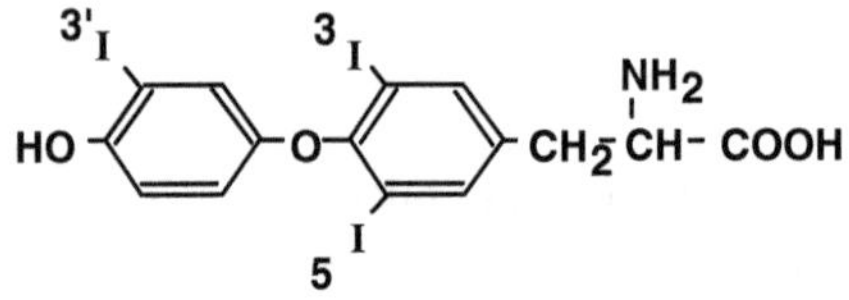

T3, tri-iodothyronine

Figure 7.1. The thyroid gland releases both thyroxine (T4) and tri-iodothyronine (T3) into the circulation. T4 is the main form of thyroid hormone released into the circulation. T3 is presnt at lower levels than T4 in serum but it is the primary biologically active form of thyroid hormone that binds the TR with highest affinity. Deiodinase enzymes can convert T4 into T3 in peripheral tissues

(ii) How does this hormone stimulate so many different functions in different tissues? Moreover, the same hormone has other actions in non-mammalian species, for example, in triggering metamorphosis, the radical change in body form and function that accompanies amphibian development [4–7]. Thus, thyroid hormone provides a highly adaptable signal in vertebrate species. Remarkably, these many functions of thyroid hormone are mediated by similar receptors found across all classes of vertebrate species.

The thyroid hormone receptor (TR), one of the first nuclear receptors to be identified, is critical for the cellular response to thyroid hormone. Early studies had shown that thyroid hormone elicited changes in mRNA expression that were sensitive to actinomycin D, thus indicating transcription as a key intermediary step in the hormonal response [8]. In 1986, cDNAs encoding the TR were cloned from chick and human tissue [9, 10]. Like steroid receptors, the TR was found to stimulate ligand-dependent transcription of reporter genes in transfection assays [11–14]. Subsequent studies identified a small family of TRα and TRβ isoforms encoded by two genes, *Thra* (*Nr1a1*) and *Thrb (Nr1a2)*, respectively, both of which are conserved across vertebrate species (the human gene symbols are *THRA and THRB*, respectively; these genes were previously also known as *ERBA*α and *ERBA*β) (Figure 7.2). Mutagenesis in mammalian models has revealed a wealth of information on how TR isoforms mediate diverse functions of thyroid hormone in vivo.

The purpose of this chapter is to provide an up-to-date overview of the TR genes and their biological functions that may be useful for students, research trainees or readers with a general interest in the subject. For further specialist information on the subject, readers may find helpful details in other articles [15–19].

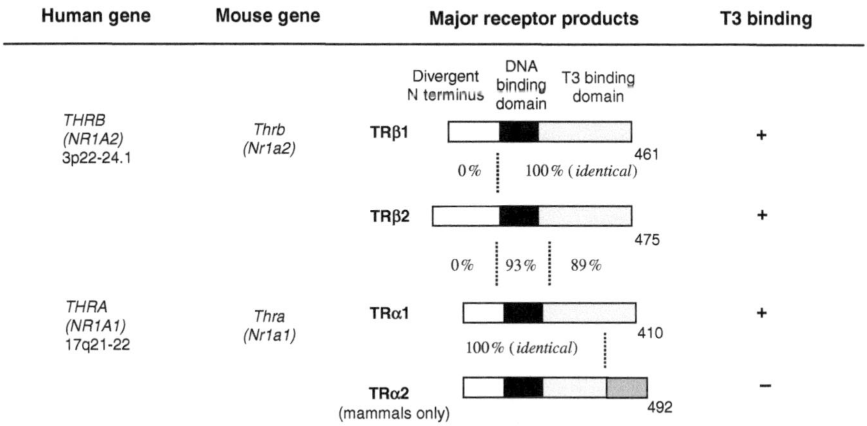

Figure 7.2. Thyroid hormone receptors. TRβ1, TRβ2 and TRα1 are canonical T3 receptors found across vertebrate species including amphibia, reptiles, fish, birds and mammals. The TR isoforms are expressed in different tissue-specific patterns. TRα2 is found only in mammals and is not a T3 receptor because of a C-terminal splice change in the ligand binding domain. TRα2 is widely co-expressed with TRα1 in mammalian tissues but its function remains enigmatic. Chromosomal locations of the human *THRB* and *THRA* genes are indicated

7.2. PRODUCTION OF THYROID HORMONE: SYSTEMIC AND LOCAL CONTROL

The major product released by the thyroid gland into the circulation is thyroxine (T4), a tetra-iodinated form of the hormone, which is derived from tyrosine residues in the thyroglobulin protein in the thyroid follicles (Figure 7.1). The thyroid also releases smaller amounts of tri-iodothyronine (T3), the most biologically active form of the hormone. The thyroid gland releases T4 and T3 in response to thyroid-stimulating hormone (TSH, or thyrotropin) which is produced by the anterior pituitary gland. TSH release is in turn stimulated by TSH-releasing hormone (TRH) produced by the hypothalamus. The hypothalamic-pituitary-thyroid axis is subject to negative feedback control by thyroid hormone, which auto-regulates the output of thyroid hormone into the circulation. Thus, elevated T4 and T3 levels would suppress TSH to reduce the stimulation of the thyroid gland. Conversely, low T4 and T3 levels would result in increased TSH production and increased stimulation of the thyroid gland as a means of maintaining an appropriate hormonal status in the circulation.

T3, the major biologically active form of thyroid hormone, binds to the TR with much higher affinity than T4. The concentration of T3 in the circulation is lower than that of T4 but levels of T3 can be boosted in peripheral tissues by deiodinase enzymes that convert T4 into T3 [20–22]. Type 2 deiodinase (encoded by *Dio2*) is an activating enzyme with 5′-deiodination activity that removes iodine from the outer ring of T4 to produce T3. In contrast, type 3 deiodinase (encoded by *Dio3*) is an inactivating enzyme with 5-deiodinase activity that removes iodine from the inner ring of T4 and T3, thus depleting both forms of the hormone. Type 1 deiodinase (encoded by *Dio1*) is enriched in liver and kidney and has both 5′- and 5-deiodination activities and is thought to contribute to production of T3 in the circulation. Recent evidence indicates the importance of these enzymes, particularly type 2 and type 3 deiodinases, in increasing or depleting local levels of T3 in specific tissues at key stages of development, thereby regulating TR activity in target tissues [23–28].

7.3. TRANSCRIPTIONAL ACTIVITY OF THE TR

Analysis of gene expression profiles in cultured cells or whole animal tissues show that T3 both induces and represses large numbers of genes [29–34]. Although in most cases, it is still unknown if most of these genes are directly or indirectly regulated by the TR, the data indicate that T3-stimulated repression and induction of different sets of genes are likely to be equally important for biological responses.

The majority of in vitro studies of TR transcriptional activity have concerned examples of T3-induced rather than repressed gene expression. The TR can bind stably to a T3 response element (T3RE) in a target gene as a homodimer, or heterodimer together with a retinoid X receptor (RXR), or in some cases, weakly as a monomer. A typical T3RE in a positively-induced response gene consists of a repeat DNA sequence based on the motif 5′-AGGTCA-3′. In natural target genes, T3RE configurations vary and include motifs arranged as direct, palindromic or inverted palindromic repeats [12, 14, 35].

It is important to bear in mind that negatively-regulated genes are probably as important as positively-regulated genes in vivo, although few negative target genes have been studied in detail. The mechanism by which the T3-bound TR represses transcription remains unclear. It is unclear if there is a specific type of T3RE that mediates negative rather than positive regulation. In some cases, evidence suggests that other types of transcription factors bind to the target gene while the TR when bound by T3 interacts with these other transcription factors to modulate the transcriptional response. An example of these complex interactions concerns the negative regulation of the TSH β subunit gene by T3 which involves interaction of TRβ with the GATA-2 transcription factor [36, 37].

The TR, like other nuclear receptors, can interact with a variety of other classes of transcription factors and cofactors. These interactions can provide specificity in the type of target gene recognized and the net outcome on the level of transcriptional activity of the target gene. Such interactions can potentially confer great specificity in the function of the TR in different cell types at different developmental stages, or in homeostatic responses of adult tissues. A number of corepressors and coactivators that influence TR function have been reported [38, 39] but these are not the subject of this article, which makes only brief mention of cofactors (see below). Further details on the DNA binding and transcriptional activities of the TR may be found elsewhere [16, 40, 18].

7.4. T3-DEPENDENT AND T3-INDEPENDENT FUNCTIONS OF THE TR

The TR associates with nuclear chromatin in the absence or presence of T3 and is considered to possess both T3-independent and T3-dependent activities. This has been shown for some target genes that are positively induced by T3, in which T3 binding switches the TR from being a T3-independent repressor to a T3-dependent activator. In the absence of T3, the TR may bind to the T3RE and repress basal transcription in association with corepressors such as NCoR or SMRT [41]. The TR/corepressor complex interacts with chromatin modifying enzymes to inhibit transcriptional activation. Binding of T3 promotes a conformational change in the TR that dissociates corepressors and allows the recruitment of coactivators. The TR/coactivator complex recruits other histone modifying enzymes that promote the opening of chromatin, to allow transcriptional induction of the target gene [42–46]. The significance of T3-dependent functions of the TR is clearly evident from numerous studies of the response to T3 in whole organisms or in cells in culture.

However, the potential significance of T3-independent functions of the TR is only partly explored and the elucidation of such T3-independent functions requires a somewhat more complicated experimental design. The T3-independent activity of the TR may potentially be important in both normal biological processes and in disease pathogenesis. An example of a normal situation in which a TR may act with minimal hormonal stimulation could be in early embryogenesis when levels of T3 may be potentially very low in certain tissues [47–49].

There is growing evidence that the T3-independent activity of the TR is important in the pathogenesis of hypothyroid disease. Thus, studies in mice have suggested that the lack of thyroid hormone causes more severe developmental retardation than does the loss of TRs. This has been explained by a role for the unoccupied TR in actively exacerbating the symptoms of hypothyroidism. For example, postnatal growth in mice is retarded by the deletion of all TRs but to a lesser extent than by hypothyroidism [50]. Also, the absence of TRs is generally compatible with normal lifespan whereas severe congenital hypothyroidism causes a high incidence of post-weaning lethality in mice [51, 50, 52].

This point is illustrated by the well known example of the retarded cerebellar development caused by hypothyroidism [53]. Hypothyroidism produces severe cerebellar retardation in mice only if TRs are present and not in TRα1-deficient mice [54]. These findings together with genetic studies of dominant negative TRs that do not bind T3 provide evidence that the unoccupied TR actively impairs neurodevelopment [55–57]. The cerebellar retardation may be explained by the TR being locked in an unoccupied state which produces worse transcriptional dysregulation of target genes than does the absence of the TR. The details of which target genes are involved and how their regulation is altered are unknown. Nonetheless, such findings have potentially wider implications for human disease states both in development and in the adult. For example, adult-onset hypothyroidism is relatively common and it is possible that some of the consequences, include aberrant metabolic and cardiac function, may involve chronic, unregulated activity by TRs.

7.5. THE TR FAMILY

The *Thra* and *Thrb* genes are conserved in all classes of vertebrate species including mammals, birds, amphibia, fish and reptiles. Although cDNAs for a number of variant gene products have been cloned in diverse species, there are only three widely conserved, or canonical, TR isoforms: (i) the TRα1-like isoform encoded by *Thra*; (ii) the TRβ1-like isoform encoded by *Thrb*; (iii) the TRβ2 isoform, also encoded by *Thrb*. Of these isoforms, TRβ2 is the most highly conserved with its unique N-terminus being encoded by a single, highly conserved exon that is located proximally to the common exons of the *Thrb* gene (Figure 7.3). The N-terminus of the TRβ1-like isoform is encoded by exons that are more scattered and variable in number between species than the single TRβ2-specific exon [58–60]. In humans, the TRβ1-specific exons have been shown to include several short non-coding exons resulting in variable 5′ untranslated sequences in TRβ1 mRNA [61, 62].

An assortment of other splice variants and truncated products of both receptor genes have been described but these are not consistently present across vertebrate

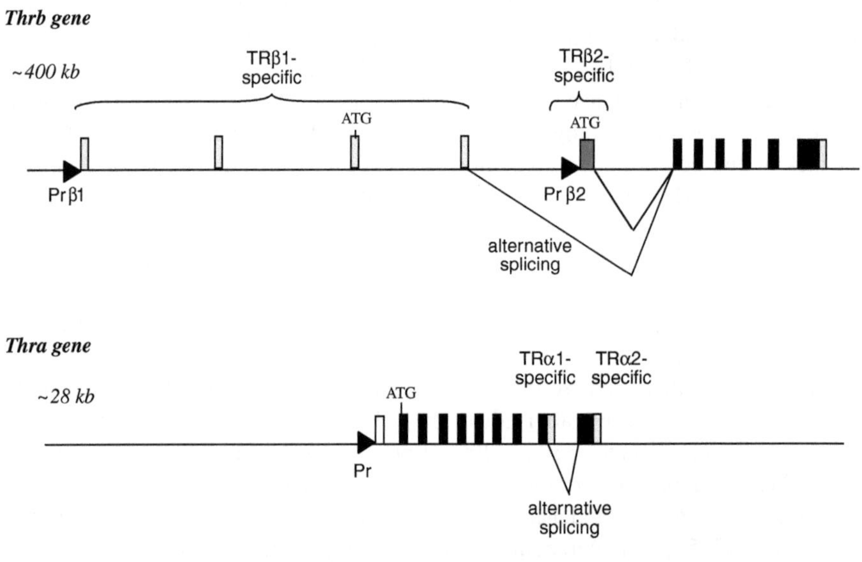

Figure 7.3. Thyroid hormone receptor genes in mammals. The mouse *Thrb* and *Thra* genes are shown and are representative of conserved genes in other mammalian species including humans. The *Thrb* gene is complex and large, spanning ~400 kb of genomic DNA. *Thrb* employs alternative promoters (Pr-β1 and Pr-β2) and splicing to give differential expression of TRβ1 and TRβ2 isoforms. The *Thra* gene is smaller and in mammals, it employs alternative splicing of 3′ exons to produce the T3 receptor TRα1 and the non-receptor product TRα2 of unknown function

species and the biological function of these products is open to debate [63, 64]. Of these, mention is briefly made here of the TRα2 product which is found consistently in mammals but not non-mammalian species, and may represent an unusual mammalian adaptation. TRα2 is a C-terminal splice variant product of the *Thra* gene that does not bind T3 nor function as a receptor and it may instead act as a weak repressor [65–67]. TRα2 is widely co-expressed with TRα1 but its biological role remains enigmatic. It has been suggested that TRα2 might counter-balance normal TR activity although it is unexplained why this would be necessary in mammals but not non-mammalian species. Another possibility is that TRα2 has no intrinsic function but that in tissues where its expression is high, it functionally reduces the amount of TRα1 present since TRα1 and TRα2 are mutually exclusive splice products of the same primary mRNA [68]. For example, the high ratio of TRα2: TRα1 in brain tissues might make neuronal tissues somewhat less prone to damage from over-stimulation by T3.

7.6. DIFFERENTIAL EXPRESSION OF TRs

Early studies showed that the *Thra* and *Thrb* genes are expressed in different developmental and tissue-specific patterns, giving a clue that the two genes mediate individual functions in vivo [69, 47, 70–72]. TRα1 mRNA is expressed widely in many tissues from early stages of embryogenesis with some variations in levels in different tissues. In contrast, TRβ1 and TRβ2 mRNAs are expressed with specific timecourses in more restricted tissues than TRα1. For example, TRβ1 is expressed in liver, kidney, brain, pituitary and cochlea but shows little if any expression in spleen or haemopoietic tissues. TRβ2 shows the most restricted expression pattern of any TR isoform, being found in the hypothalamus, anterior pituitary gland, cochlea and cone photoreceptors in the retina [73–75].

As development progresses, both TRα1 and TRβ1 isoforms tend to be co-expressed in several tissues suggesting that in some situations at certain times, TR isoforms may serve overlapping functions. More detailed expression mapping studies have been performed in a number of tissues including brain regions, pituitary, cochlea [70, 75], retina [74] and heart [76]. The predicted individual and overlapping functions of the TR gene have been largely substantiated by functional analyses based on targeted mutagenesis in mice (see below).

Given the complexity of TR functions on different cells even in a defined tissue (see next section), there is a need in contemporary research for more refined mapping of expression patterns of specific TR isoforms at the level of single cells.

7.7. BIOLOGICAL FUNCTIONS OF TRs

The *Thra* and *Thrb* genes have been subjected to intensive mutagenesis in mouse models giving many knockout (null) and knock-in (change of function) alleles. Details of these strains may be found in other review articles [17, 77–80]. The study

Table 7.1. Some phenotypes caused by mutations in the *Thrb and Thra genes* in mice

	Receptor gene		
Phenotype	*Thra*	*Thrb*	*Combined Thra/Thrb*
Sensory deficits (auditory, visual)	–	++	+++
Hypothalamic-pituitary-thyroid axis, hyperactive	–	+	+++
Goitre	–	+	+++
Liver, defective response to T3	+/–	++	+++
Dwarfism	+/–	–	++
Infertility (Female)	–	–	++
Low body temperature	+	–	+
Bradycardia, prolonged QTend duration	+	–	+
Intestinal defects	+	–	++
Defective bone development, osteosclerosis	+	–	++
Sertoli cell defects	+	–	n.a.
Haemopoietic defects (erythroid, lymphoid)	+	–	n.a.

Phenotypic severity: –, no phenotype; +, least, +++ most severe;
n.a.: data not available.
The information in the table is based primarily on knockout (not knockin) mutations.
For references, see the text.

of "knockout" strains deficient for one or more TR isoforms has yielded a wealth of knowledge concerning tissue-specific TR functions (Table 7.1).

The "knockin" strains carrying change-of-function mutations in the ligand-binding domain of the receptors provide another perspective on TR functions in vivo. These types of mutations, originally found in the *THRB* gene of human patients with resistance to thyroid hormone [15] typically reduce the affinity of the TR for T3 causing it to became a T3-independent, dominant negative inhibitor, a so-called aporeceptor. Such mutations introduced into the mouse *Thrb* or *Thra* genes have resulted in a multitude of phenotypes caused by "receptor-mediated hypothyroidism" in specific target tissues. These phenotypes are often more pronounced than those observed in knockout strains, thereby highlighting the strong aporeceptor activity of the unoccupied TR.

Table 7.1 shows that phenotypes in mouse strains with TR gene mutations arise in multiple tissues and organs, reflecting the extensive range of functions that are subject to control by the TRs. Phenotypes include defects in endocrine, metabolic, thermogenic and cardiac functions as well as developmental defects, for example, in bone and intestine. In accord with the well known requirement for thyroid hormone in the nervous system, several phenotypes have also been identified in neurological functions and behavior (Table 7.2).

Providing a full catalogue of known phenotypes is beyond the scope of this article. However, a useful classification of phenotypes may be made according to which TR gene is primarily responsible for a given function. Thus, *Thra*-specific functions include cardiac and thermogenic functions [81–86] intestinal maturation [87–89]

Table 7.2. Some nervous system phenotypes associated with mutations in *Thrb* and *Thra genes* in mice

System phenotype	Receptor gene	Primary reference
Sensory systems		
Deafness	Thrb	[112, 103]
Color blindness	Thrb	[106, 110, 114]
Whisker-barrel synaptic impairment (tactile)	Thra	[96]
Behavior		
Anxiety-like behavior	Thra	[56]
Learning defects	Thra	[113], [56], [129]
Hyperactivity	Thrb	[111]
Mating behavior defects	Thra, Thrb	[130]
Audiogenic seizures	Thrb	[106, 110, 114]
Cellular defects		
Cerebellar defects	Thra, Thrb	[54], [97], [55]
Impaired glucose utilization (synaptic impairment)	Thra	[95]
Oligodendrocyte defects	Thra	[101], [100]
GABAergic interneuron defects	Thra	[128], [99], [57]

haemopoietic functions [90, 91] and bone maturation [92–94]. A number of neural functions are influenced by *Thra* mutations. For example, glucose utilization is a pre-requisite for neuronal activity and reflects synaptic activity. *Thra* mutations impair glucose utilization as revealed by brain imaging studies [95, 96]. *Thra* mutations are also associated with anxiety-like behavior in mice [56, 57] and defects in the differentiation of some types of neurons [97, 98, 56, 99, 57] and oligodendrocytes [100, 101].

Thrb mutations result in a different spectrum of phenotypes than do *Thra* mutations and have revealed a major role for the *Thrb* gene in endocrine and sensory functions. The feedback control of the hypothalamic-pituitary-thyroid (HPT) axis is regulated by TRβ, probably involving both TRβ1 and TRβ2 isoforms [102–108]. An interplay of both TRb1 and TRb2 isoforms is likely to be involved at both the hypothalamic and pituitary levels of the axis. Liver responses to T3 are primarily regulated by TRβ1 [109, 31]. *Thrb* gene mutations also result in certain aberrant behaviours, including audiogenic seizures [110] and hyperactivity [111]. Sensory deficiencies cause by *Thrb* mutations include severe deafness [112] and a defect in cone photoreceptors resulting in a form of colour blindness [106].

These genetic analyses in mice, especially of the *Thrb* gene, have revealed sensory systems to be major targets of T3 action with novel phenotypes being identified not only in the auditory and colour visual systems but also in tactile systems [96]. Sensory systems have received relatively less attention than the brain as targets of T3 action but the evidence reveals that the promotion of sensory development is amongst the most important functions of thyroid hormone.

7.8. COOPERATIVE FUNCTIONS OF TR ISOFORMS

The above genetic analyses of the two TR genes indicate that the tissue-specific or cell-specific expression of variety of TR isoforms allows regulation of a diverse range of functions. Perhaps the clearest example of a cell-specific function for a TR isoform is that of TRβ2 in the retinal cones, the photoreceptors that mediate colour vision [106]. TRβ2 is specifically induced in the small population of newly-generated cones during embryogenesis but not in other retinal cell types. In TRβ2-deficient mice, cones remain intact but they lack medium wavelength-sensitive (M) opsin resulting in a form of colour blindness. Other examples of TR isoform-specific functions include the role of TRα1 in hippocampal neuron populations [113] and TRβ1 as the major isoform that regulates liver responses to T3 [31].

However, in many organs and tissues, both TRα1 and TRβ isforms are co-expressed in varying proportions. Potentially, both TRα1 and TRβ isoforms may contribute to some extent to the control of the same functions. One of the first demonstrations of this higher level of complexity was revealed in the HPT axis. Thus, deletion of TRα1 has a minimal impact on the levels of T4, T3 and TSH in serum (WIkstrom 98), whereas deletion of TRβ isoforms results in almost 3-fold elevation of T4, T3 and TSH levels (112, 103). However, the combined deletion of TRβ1, TRβ2 and TRα1 causes a much exacerbated phenotype compared to deletion of TRβ isoforms alone, with >10-fold elevated T4 and T3 levels and >100-fold elevated TSH levels [104, 50]. The potential of TRα1 to regulate the HPT axis was further evident from another targeted mutation that overexpressed TRα1, which could largely rescue the defects in the HPT axis caused by loss of TRβ [114, 68].

Such interchangeability or overlap would explain why mice with single gene mutations show relatively limited phenotypes and why mice with combined deletions have much more severe phenotypes [104, 50]. Thus, substitution between isoforms masks some of the phenotypic consequences of deletion of any given isoform. However, in combined mutant mice lacking all known T3 receptors, substitution is precluded with the result that a range of novel and exacerbated phenotypes arise, including retarded growth and bone development, a large goitre, hyperactivity of the pituitary-thyroid axis with extremely high levels of serum TSH, T4 and T3, abnormalities in other hormone systems such as growth hormone and IGF1 and impaired female fertility [50].

The individual tissue-specific functions of given TR isoform may therefore be accounted for in many cases by the expression level of the isoform rather than intrinsically different properties. However, it is possible that specific TR isoforms possess certain individual properties that are determined by structural characteristics. For example, variations in the N-terminus or subtle amino acid differences in the DNA binding domain or ligand binding domain may influence the conformation adopted by a TRα1 or TRβ complex. Although subtle, such properties may influence the specificity of binding to target gene response elements or the types of transcriptional cofactors that bind with the TR complex to determine the transcriptional outcome [102, 115–117].

7.9. DISEASE

Human resistance to thyroid hormone is a rare syndrome that is associated with mutations in the *THRB* gene [15, 118]. The syndrome shows autosomal dominant inheritance in almost all cases which is explained by the dominant negative activity acquired by the mutant TRb isoforms. In vitro, these mutant receptors can inhibit normal TR activity in transfection assays. Presumably, in human patients, the mutant receptors interfere with normal TRβ or TRα1 activity in different tissues to produce varying degrees of resistance to T3 action. The symptoms can vary widely but often include tachycardia and goitre and in many cases, some form of mental retardation or attention deficit-hyperactivity disorders. The syndrome is characterized by increased serum levels of thyroid hormones and inappropriately normal or elevated TSH, reflecting resistance of the pituitary to feedback regulation by thyroid hormone. A single atypical kindred showed autosomal recessive inheritance [119] and a small proportion of cases do not carry identified mutations in the *THRB* coding exons. It is thought that mutations in other genes may produce similar symptoms.

To date, no inherited mutations in the *THRA* gene are known in human disease. In mice, *Thra* mutations can produce pronounced phenotypes with growth retardation and behavioral abnormalities [120, 92, 121]. Interestingly, serum thyroid hormone levels are relatively little altered in most mouse strains with *Thra* mutations in contrast to the marked changes in thyroid hormone levels and TSH caused by *Thrb* mutations in mice or *THRB* mutations in humans. It is possible therefore that human patients with *THRA* mutations exist but will not be identified on the basis of thyroid hormone abnormalities.

Other analyses have screened human population groups for polymorphic changes in *THRA* and *THRB* genes that may be associated with subtler symptoms, for example in serum hormone parameters [122].

Somatic mutations in *THRB* and *THRA* genes have been described in different types of human cancers, including liver, lung and breast cancers, suggesting that thyroid hormone may influence the progression of tumorigenesis [123, 79]. A recent screen has shown that human retinoblastomas possess characteristics of cone photoreceptor precursors including the expression of TRβ2, suggesting that TRβ2 may contribute to the tumourigenic properties of retinoblastoma cells [124]. The chicken retroviral oncogene v-erbA carried by avian erythroleukemia virus ES4 encodes a mutant form of TRα1 that does not bind T3 and which acts as a dominant negative repressor [125]. This oncogene contributes to the induction of erythroleukemias in the chick.

7.10. CONCLUDING REMARKS

The approach to elucidating thyroid hormone action at the level of the receptors has been particularly fruitful, yielding a wealth of new biological knowledge relevant to many areas of mammalian development and homeostasis. A picture has emerged in which distinct TR isoforms mediate a wide variety of cell- and tissue-specific functions thus explaining in part how thyroid hormone is able to elicit its a diverse range of biological responses. Detailed investigation of the receptor genes has also revealed new functions of thyroid hormone that were previously unrecognized. An example is the unexpectedly critical role of the TRβ2 isoform in the development of the colour visual system, a finding that was unpredicted from the fields of research into either endocrinology or retinal biology. It is likely that further exploratory investigations of receptors in other systems may reveal other novel functions.

The new understanding of receptor functions in vivo has created a need to understand the networks of downstream genes that underlie the specific functions identified for TR isoforms. There is great scope for investigating these underlying target gene networks. Both primary and secondary response genes are likely to be important in the net biological outcome of a given system.

Understanding the functions of TRs at the tissue level may have practical implications as reflected by the growing interest in therapeutic agents that could modify thyroid hormone action in tissue-specific manner. For example, there is growing interest in the development of agents that act as TR agonists and antagonists [126, 127].

ACKNOWLEDGEMENTS

This work was supported in part by the intramural research program at NIDDK at the National Institutes of Health.

REFERENCES

1. Murray, G. (1891). Note on the treatment of myxoedema by hypodermic injections of an extract of the thyroid gland of a sheep. *Br Med J* 2, 796–797.
2. Osler, W. (1897). Sporadic cretinism in America. *Transactions Congress of American Physicians and Surgeons*, 4 169–206.
3. Harington, C. R. (1935). Biochemical basis of thyroid function. *The Lancet* 225, 1199–1204.
4. Gudernatsch, J. (1912). Feeding experiments on tadpoles. I. The influence of specific organs given as food on growth and differentiation. *Roux Arch Entwicklungsmechanik der Organismen* 35, 457–483.
5. Inui, Y., and Miwa, S. (1985). Thyroid hormone induces metamorphosis of flounder larvae. *Gen Comp Endocrinol* 60, 450–454.
6. Brown, D. D. (1997). The role of thyroid hormone in zebrafish and axolotl development. *Proc Natl Acad Sci U S A* 94, 13011–13016.
7. Tata, J. R. (2006). Amphibian metamorphosis as a model for the developmental actions of thyroid hormone. *Mol Cell Endocrinol* 246, 10–20.
8. Tata, J. R. (1963). Inhibition of the biological action of thyroid hormones by actinomycin D and puromycin. *Nature* 197, 1167–1168.

9. Sap, J., Muñoz, A., Damm, K., Goldberg, Y., Ghysdael, J., Leutz, A., Beug, H., and Vennström, B. (1986). The c-*erb*A protein is a high affinity receptor for thyroid hormone. *Nature* 324, 635–640.
10. Weinberger, C., Thompson, C. C., Ong, E. S., Lebo, R., Gruol, D. J., and Evans, R. M. (1986). The *c-erb*-A gene encodes a thyroid hormone receptor. *Nature* 324, 641–646.
11. Glass, C. K., Franco, R., Weinberger, C., Albert, V. R., Evans, R. M., and Rosenfeld, M. G. (1987). A c-erb-A binding site in rat growth hormone gene mediates trans-activation by thyroid hormone. *Nature* 329, 738–741.
12. Umesono, K., and Evans, R. M. (1989). Determinants of target gene specificity for steroid/thyroid hormone receptors. *Cell* 57, 1139–1146.
13. Sap, J., de Magistris, L., Stunnenberg, H., and Vennström, B. (1990). A major thyroid hormone response element in the third intron of the rat growth hormone gene. *EMBO J* 9, 887–896.
14. Brent, G., Moore, D., and Larsen, P. (1991). Thyroid hormone regulation of gene expression. *Ann Rev Physiol* 53, 17–35.
15. Refetoff, S., Weiss, R. E., and Usala, S. J. (1993). The syndromes of resistance to thyroid hormone. *Endocrine Rev* 14, 348–399.
16. Koenig, R. J. (1998). Thyroid hormone receptor coactivators and corepressors. *Thyroid* 8, 703–713.
17. Forrest, D., and Vennström, B. (2000). Functions of thyroid hormone receptors in mice. *Thyroid* 10, 41–52.
18. Zhang, J., and Lazar, M. A. (2000). The mechanism of action of thyroid hormones. *Annu Rev Physiol* 62, 439–466.
19. Tata, J. R. (2007). A hormone for all seasons. *Perspect Biol Med* 50, 89–103.
20. Köhrle, J. (1999). Local activation and inactivation of thyroid hormones: the deiodinase family. *Mol Cell Endocrinol* 151, 103–119.
21. Bianco, A. C., Salvatore, D., Gereben, B., Berry, M. J., and Larsen, P. R. (2002). Biochemistry, cellular and molecular biology, and physiological roles of the iodothyronine selenodeiodinases. *Endocr Rev* 23, 38–89.
22. St Germain, D. L., Hernandez, A., Schneider, M. J., and Galton, V. A. (2005). Insights into the role of deiodinases from studies of genetically modified animals. *Thyroid* 15, 905–916.
23. Ng, L., Goodyear, R. J., Woods, C. A., Schneider, M. J., Diamond, E., Richardson, G. P., Kelley, M. W., Germain, D. L., Galton, V. A., and Forrest, D. (2004). Hearing loss and retarded cochlear development in mice lacking type 2 iodothyronine deiodinase. *Proc Natl Acad Sci U S A* 101, 3474–3479.
24. Brown, D. D. (2005). The role of deiodinases in amphibian metamorphosis. *Thyroid* 15, 815–821.
25. Bianco, A. C., and Kim, B. W. (2006). Deiodinases: implications of the local control of thyroid hormone action. *J Clin Invest* 116, 2571–2579.
26. Hernandez, A., Martinez, M. E., Fiering, S., Galton, V. A., and St Germain, D. (2006). Type 3 deiodinase is critical for the maturation and function of the thyroid axis. *J Clin Invest* 116, 476–484.
27. Galton, V. A., Wood, E. T., St Germain, E. A., Withrow, C. A., Aldrich, G., St Germain, G. M., Clark, A. S., and St Germain, D. L. (2007). Thyroid hormone homeostasis and action in the type 2 deiodinase-deficient rodent brain during development. *Endocrinology* 148, 3080–3088.
28. Ng, L., Hernandez, A., He, W., Ren, T., Srinivas, M., Ma, M., Galton, V. A., St Germain, D. L., and Forrest, D. (2009). A protective role for type 3 deiodinase, a thyroid hormone-inactivating enzyme, in cochlear development and auditory function. *Endocrinology* 150, 1952–1960.
29. Iglesias, T., Caubin, J., Stunnenberg, H. G., Zaballos, A., Bernal, J., and Munoz, A. (1996). Thyroid hormone-dependent transcriptional repression of neural cell adhesion molecule during brain maturation. *EMBO J* 15, 4307–4316.
30. Miller, L. D., Park, K. S., Guo, Q. M., Alkharouf, N. W., Malek, R. L., Lee, N. H., Liu, E. T., and Cheng, S. Y. (2001). Silencing of Wnt signaling and activation of multiple metabolic pathways in response to thyroid hormone-stimulated cell proliferation. *Mol Cell Biol* 21, 6626–6639.
31. Flores-Morales, A., Gullberg, H., Fernandez, L., Stahlberg, N., Lee, N. H., Vennström, B., and Norstedt, G. (2002). Patterns of liver gene expression governed by TRβ. *Mol Endocrinol* 16, 1257–1268.

32. Poguet, A. L., Legrand, C., Feng, X., Yen, P. M., Meltzer, P., Samarut, J., and Flamant, F. (2003). Microarray analysis of knockout mice identifies cyclin D2 as a possible mediator for the action of thyroid hormone during the postnatal development of the cerebellum. *Dev Biol* 254, 188–199.
33. Yen, P. M., Feng, X., Flamant, F., Chen, Y., Walker, R. L., Weiss, R. E., Chassande, O., Samarut, J., Refetoff, S., and Meltzer, P. S. (2003). Effects of ligand and thyroid hormone receptor isoforms on hepatic gene expression profiles of thyroid hormone receptor knockout mice. *EMBO Rep* 4, 581–587.
34. Bernal, J. (2005). Thyroid hormones and brain development. *Vitam Horm* 71, 95–122.
35. Malo, M. S., Zhang, W., Alkhoury, F., Pushpakaran, P., Abedrapo, M. A., Mozumder, M., Fleming, E., Siddique, A., Henderson, J. W., and Hodin, R. A. (2004). Thyroid hormone positively regulates the enterocyte differentiation marker intestinal alkaline phosphatase gene via an atypical response element. *Mol Endocrinol* 18, 1941–1962.
36. Gordon, D. F., Tucker, E. A., Tundwal, K., Hall, H., Wood, W. M., and Ridgway, E. C. (2006). MED220/thyroid receptor-associated protein 220 functions as a transcriptional coactivator with Pit-1 and GATA-2 on the thyrotropin-beta promoter in thyrotropes. *Mol Endocrinol* 20, 1073–1089.
37. Matsushita, A., Sasaki, S., Kashiwabara, Y., Nagayama, K., Ohba, K., Iwaki, H., Misawa, H., Ishizuka, K., and Nakamura, H. (2007). Essential role of GATA2 in the negative regulation of thyrotropin beta gene by thyroid hormone and its receptors. *Mol Endocrinol* 21, 865–884.
38. Potter, G. B., Zarach, J. M., Sisk, J. M., and Thompson, C. C. (2002). The thyroid hormone-regulated corepressor hairless associates with histone deacetylases in neonatal rat brain. *Mol Endocrinol* 16, 2547–2560.
39. Tenbaum, S. P., Juenemann, S., Schlitt, T., Bernal, J., Renkawitz, R., Munoz, A., and Baniahmad, A. (2003). Alien/CSN2 gene expression is regulated by thyroid hormone in rat brain. *Dev Biol* 254, 149–160.
40. Cheng, S. Y. (2000). Multiple mechanisms for regulation of the transcriptional activity of thyroid hormone receptors. *Rev Endocr Metab Disord* 1, 9–18.
41. Chen, J. D., and Evans, R. M. (1995). A transcriptional co-repressor that interacts with nuclear hormone receptors. *Nature* 377, 454–458.
42. Lemon, B. D., and Freedman, L. P. (1999). Nuclear receptor cofactors as chromatin remodelers. *Curr Opin Genet Dev* 9, 499–504.
43. Ito, M., and Roeder, R. G. (2001). The TRAP/SMCC/Mediator complex and thyroid hormone receptor function. *Trends Endocrinol Metab* 12, 127–134.
44. Sachs, L. M., Jones, P. L., Havis, E., Rouse, N., Demeneix, B. A., and Shi, Y. B. (2002). Nuclear receptor corepressor recruitment by unliganded thyroid hormone receptor in gene repression during Xenopus laevis development. *Mol Cell Biol* 22, 8527–8538.
45. Paul, B. D., Fu, L., Buchholz, D. R., and Shi, Y. B. (2005). Coactivator recruitment is essential for liganded thyroid hormone receptor to initiate amphibian metamorphosis. *Mol Cell Biol* 25, 5712–5724.
46. Sato, Y., Buchholz, D. R., Paul, B. D., and Shi, Y. B. (2007). A role of unliganded thyroid hormone receptor in postembryonic development in Xenopus laevis. *Mech Dev* 124, 476–488.
47. Forrest, D., Sjöberg, M., and Vennström, B. (1990). Contrasting developmental and tissue-specific expression of α and β thyroid hormone receptor genes. *EMBO J* 9, 1519–1528.
48. Banker, D. E., Bigler, J., and Eisenman, R. N. (1991). The thyroid hormone receptor gene (c-erbA alpha) is expressed in advance of thyroid gland maturation during the early embryonic development of Xenopus laevis. *Mol Cell Biol* 11, 5079–5089.
49. Havis, E., Le Mevel, S., Morvan Dubois, G., Shi, D. L., Scanlan, T. S., Demeneix, B. A., and Sachs, L. M. (2006). Unliganded thyroid hormone receptor is essential for Xenopus laevis eye development. *EMBO J* 25, 4943–4951.
50. Göthe, S., Wang, Z., Ng, L., Nilsson, J., Campos-Barros, A., Ohlsson, C., Vennström, B., and Forrest, D. (1999). Mice devoid of all known thyroid hormone receptors are viable but exhibit disorders of the pituitary-thyroid axis, growth and bone maturation. *Genes Dev* 13, 1329–1341.

51. Mansouri, A., Chowdhury, K., and Gruss, P. (1998). Follicular cells of the thyroid gland require *Pax8* gene function. *Nat Genet* 19, 87–90.
52. Marians, R. C., Ng, L., Blair, H. C., Unger, P., Graves, P. N., and Davies, T. F. (2002). Defining thyrotropin-dependent and -independent steps of thyroid hormone synthesis by using thyrotropin receptor-null mice. *Proc Natl Acad Sci U S A* 99, 15776–15781.
53. Legrand, J. (1979). Morphogenetic actions of thyroid hormones. *TINS* 2, 234–236.
54. Morte, B., Manzano, J., Scanlan, T., Vennström, B., and Bernal, J. (2002). Deletion of the thyroid hormone receptor alpha 1 prevents the structural alterations of the cerebellum induced by hypothyroidism. *Proc Natl Acad Sci U S A* 99, 3985–3989.
55. Hashimoto, K., Curty, F. H., Borges, P. P., Lee, C. E., Abel, E. D., Elmquist, J. K., Cohen, R. N., and Wondisford, F. E. (2001). An unliganded thyroid hormone receptor causes severe neurological dysfunction. *Proc Natl Acad Sci U S A* 98, 3998–4003.
56. Venero, C., Guadano-Ferraz, A., Herrero, A. I., Nordstrom, K., Manzano, J., de Escobar, G. M., Bernal, J., and Vennstrom, B. (2005). Anxiety, memory impairment, and locomotor dysfunction caused by a mutant thyroid hormone receptor alpha1 can be ameliorated by T3 treatment. *Genes Dev* 19(18), 2152–2163.
57. Wallis, K., Sjogren, M., van Hogerlinden, M., Silberberg, G., Fisahn, A., Nordstrom, K., Larsson, L., Westerblad, H., Morreale de Escobar, G., Shupliakov, O., and Vennstrom, B. (2008). Locomotor deficiencies and aberrant development of subtype-specific GABAergic interneurons caused by an unliganded thyroid hormone receptor alpha1. *J Neurosci* 28, 1904–1915.
58. Yaoita, Y., Shi, Y.-B., and Brown, D. (1990). *Xenopus laevis* a and b thyroid hormone receptors. *Proc Natl Acad Sci USA* 87, 7090–7094.
59. Shi, Y. B., Yaoita, Y., and Brown, D. D. (1992). Genomic organization and alternative promoter usage of the two thyroid hormone receptor β genes in Xenopus laevis. *J Biol Chem* 267, 733–738.
60. Jones, I., Srinivas, M., Ng, L., and Forrest, D. (2003). The thyroid hormone receptor β gene: structure and functions in the brain and sensory systems. *Thyroid* 13, 1057–1068.
61. Sakurai, A., Miyamoto, T., and DeGroot, L. J. (1992). Cloning and characterization of the human thyroid hormone receptor β 1 gene promoter. *Biochem Biophys Res Commun* 185, 78–84.
62. Frankton, S., Harvey, C. B., Gleason, L. M., Fadel, A., and Williams, G. R. (2004). Multiple messenger ribonucleic acid variants regulate cell-specific expression of human thyroid hormone receptor β1. *Mol Endocrinol* 18, 1631–1642.
63. Chassande, O., Fraichard, A., Gauthier, K., Flamant, F., Legrand, C., Savatier, P., Laudet, V., and Samarut, J. (1997). Identification of transcripts initiated from an internal promoter in the c-erbA alpha locus that encode inhibitors of retinoic acid receptor- alpha and triiodothyronine receptor activities. *Mol Endocrinol* 11, 1278–1290.
64. Williams, G. R. (2000). Cloning and characterization of two novel thyroid hormone receptor β isoforms. *Mol Cell Biol* 20, 8329–8342.
65. Izumo, S., and Mahdavi, V. (1988). Thyroid hormone receptor alpha isoforms generated by alternative splicing differentially activate myosin HC gene transcription. *Nature* 334, 539–542.
66. Lazar, M., Hodin, R., Darling, D., and Chin, W. (1988). Identification of a rat c-erbA α-related protein which binds deoxyribonucleic acid but does not bind thyroid hormone. *Mol Endocrinol* 2, 893–901.
67. Koenig, R. J., Lazar, M. A., Hodin, R. A., Brent, G. A., Larsen, P. R., Chin, W. W., and Moore, D. D. (1989). Inhibition of thyroid hormone action by a non-hormone binding c-erbA protein generated by alternative mRNA splicing. *Nature* 337, 659–661.
68. Saltó, C., Kindblom, J. M., Johansson, C., Wang, Z., Gullberg, H., Nordström, K., Mansén, A., Ohlsson, C., Thorén, P., Forrest, D., and Vennström, B. (2001). Ablation of TRα2 and a concomitant overexpression of α1 yields a mixed hypo- and hyperthyroid phenotype in mice. *Mol Endocrinol* 15, 2115–2128.
69. Murray, M. B., Zilz, N. D., McCreary, N. L., Macdonald, M. J., and Towle, H. C. (1988). Isolation and characterization of rat cDNA clones for two distinct thyroid hormone receptors. *J Biol Chem* 263, 12770–12777.

70. Bradley, D. J., Towle, H. C., and Young, W. S., 3rd (1992). Spatial and temporal expression of α- and β-thyroid hormone receptor mRNAs, including the β 2-subtype, in the developing mammalian nervous system. *J Neurosci* 12, 2288–2302.
71. Schwartz, H. L., Strait, K. A., Ling, N. C., and Oppenheimer, J. H. (1992). Quantitation of rat tissue thyroid hormone binding receptor isoforms by immunoprecipitation of nuclear triiodothyronine binding capacity. *J Biol Chem* 267, 11794–11799.
72. Thompson, C. C. (1996). Thyroid hormone-responsive genes in developing cerebellum include a novel synaptotagmin and a hairless homolog. *J Neurosci* 16, 7832–7840.
73. Hodin, R. A., Lazar, M. A., Wintman, B. I., Darling, D. S., Koenig, R. J., Larsen, P. R., Moore, D. D., and Chin, W. W. (1989). Identification of a thyroid hormone receptor that is pituitary-specific. *Science* 244, 76–78.
74. Sjöberg, M., Vennström, B., and Forrest, D. (1992). Thyroid hormone receptors in chick retinal development: differential expression of mRNAs for α and N-terminal variant β receptors. *Development* 114, 39–47.
75. Bradley, D. J., Towle, H. C., and Young, W. S., 3rd (1994). α and β thyroid hormone receptor (TR) gene expression during auditory neurogenesis: evidence for TR isoform-specific transcriptional regulation in vivo. *Proc Natl Acad Sci U S A* 91, 439–443.
76. Stoykov, I., Zandieh-Doulabi, B., Moorman, A. F., Christoffels, V., Wiersinga, W. M., and Bakker, O. (2006). Expression pattern and ontogenesis of thyroid hormone receptor isoforms in the mouse heart. *J Endocrinol* 189, 231–245.
77. Flamant, F., and Samarut, J. (2003). Thyroid hormone receptors: lessons from knockout and knock-in mutant mice. *Trends Endocrinol Metab* 14, 85–90.
78. Wondisford, F. E. (2003). Thyroid hormone action: insight from transgenic mouse models. *J Investig Med* 51, 215–220.
79. Cheng, S. Y. (2005). Thyroid hormone receptor mutations and disease: beyond thyroid hormone resistance. *Trends Endocrinol Metab* 16, 176–182.
80. Ng, L., and Forrest, D. (2006). Developmental roles of the thyroid hormone receptor α and β genes. *Adv Dev Biol* 16, 1–31.
81. Wikström, L., Johansson, C., Saltó, C., Barlow, C., Campos Barros, A., Baas, F., Forrest, D., Thorén, P., and Vennström, B. (1998). Abnormal heart rate and body temperature in mice lacking thyroid hormone receptor α1. *EMBO J* 17, 455–461.
82. Johansson, C., Göthe, S., Forrest, D., Vennström, B., and Thorén, P. (1999). Cardiovascular phenotype and temperature control in mice lacking thyroid hormone receptor β or both α1 and β. *Am J Physiol (Heart Circ Physiol)* 276, H2006–H2012.
83. Gloss, B., Trost, S., Bluhm, W., Swanson, E., Clark, R., Winkfein, R., Janzen, K., Giles, W., Chassande, O., Samarut, J., and Dillmann, W. (2001). Cardiac ion channel expression and contractile function in mice with deletion of thyroid hormone receptor α or β. *Endocrinology* 142, 544–550.
84. Johansson, C., Koopmann, R., Vennström, B., and Benndorf, K. (2002). Accelerated inactivation of voltage-dependent K+ outward current in cardiomyocytes from thyroid hormone receptor alpha1-deficient mice. *J Cardiovasc Electrophysiol* 13, 44–50.
85. Marrif, H., Schifman, A., Stepanyan, Z., Gillis, M. A., Calderone, A., Weiss, R. E., Samarut, J., and Silva, J. E. (2005). Temperature homeostasis in transgenic mice lacking thyroid hormone receptor-alpha gene products. *Endocrinology* 146, 2872–2884.
86. Pelletier, P., Gauthier, K., Sideleva, O., Samarut, J., and Silva, J. E. (2008). Mice lacking the thyroid hormone receptor-alpha gene spend more energy in thermogenesis, burn more fat, and are less sensitive to high-fat diet-induced obesity. *Endocrinology* 149, 6471–6486.
87. Plateroti, M., Chassande, O., Fraichard, A., Gauthier, K., Freund, J. N., Samarut, J., and Kedinger, M. (1999). Involvement of T3Ralpha- and beta-receptor subtypes in mediation of T3 functions during postnatal murine intestinal development. *Gastroenterology* 116, 1367–1378.
88. Plateroti, M., Gauthier, K., Domon-Dell, C., Freund, J. N., Samarut, J., and Chassande, O. (2001). Functional interference between thyroid hormone receptor α and natural truncated TR delta α isoforms in the control of intestine development. *Mol Cell Biol* 21, 4761–4772.

89. Plateroti, M., Kress, E., Mori, J. I., and Samarut, J. (2006). Thyroid hormone receptor alpha1 directly controls transcription of the beta-catenin gene in intestinal epithelial cells. *Mol Cell Biol* 26, 3204–3214.
90. Arpin, C., Pihlgren, M., Fraichard, A., Aubert, D., Samarut, J., Chassande, O., and Marvel, J. (2000). Effects of T3Rα1 and T3Rα2 gene deletion on T and B lymphocyte development. *J Immunol* 164, 152–160.
91. Angelin-Duclos, C., Domenget, C., Kolbus, A., Beug, H., Jurdic, P., and Samarut, J. (2005). Thyroid hormone T3 acting through the thyroid hormone {alpha} receptor is necessary for implementation of erythropoiesis in the neonatal spleen environment in the mouse. *Development* 132, 925–934.
92. Tinnikov, A., Nordström, K., Thorén, P., Kindblom, J. M., Malin, S., Rozell, B., Adams, M., Rajanayagam, O., Pettersson, S., Ohlsson, C., et al. (2002). Retardation of post-natal development caused by a negatively acting thyroid hormone receptor α1. *EMBO J* 21, 5079–5087.
93. O'Shea, P. J., Harvey, C. B., Suzuki, H., Kaneshige, M., Kaneshige, K., Cheng, S. Y., and Williams, G. R. (2003). A thyrotoxic skeletal phenotype of advanced bone formation in mice with resistance to thyroid hormone. *Mol Endocrinol* 17, 1410–1424.
94. O'Shea, P. J., Bassett, J. H., Sriskantharajah, S., Ying, H., Cheng, S. Y., and Williams, G. R. (2005). Contrasting skeletal phenotypes in mice with an identical mutation targeted to thyroid hormone receptor {alpha}1 or {beta}. *Mol Endocrinol* 19, 3045–3059.
95. Itoh, Y., Esaki, T., Kaneshige, M., Suzuki, H., Cook, M., Sokoloff, L., Cheng, S. Y., and Nunez, J. (2001). Brain glucose utilization in mice with a targeted mutation in the thyroid hormone α or β receptor gene. *Proc Natl Acad Sci U S A* 98, 9913–9918.
96. Esaki, T., Suzuki, H., Cook, M., Shimoji, K., Cheng, S. Y., Sokoloff, L., and Nunez, J. (2003). Functional activation of cerebral metabolism in mice with mutated thyroid hormone nuclear receptors. *Endocrinology* 144, 4117–4122.
97. Heuer, H., and Mason, C. A. (2003). Thyroid hormone induces cerebellar Purkinje cell dendritic development via the thyroid hormone receptor alpha1. *J Neurosci* 23, 10604–10612.
98. Lemkine, G. F., Raj, A., Alfama, G., Turque, N., Hassani, Z., Alegria-Prevot, O., Samarut, J., Levi, G., and Demeneix, B. A. (2005). Adult neural stem cell cycling in vivo requires thyroid hormone and its alpha receptor. *FASEB J* 19, 863–865.
99. Gilbert, M. E., Sui, L., Walker, M. J., Anderson, W., Thomas, S., Smoller, S. N., Schon, J. P., Phani, S., and Goodman, J. H. (2007). Thyroid hormone insufficiency during brain development reduces parvalbumin immunoreactivity and inhibitory function in the hippocampus. *Endocrinology* 148, 92–102.
100. Baas, D., Legrand, C., Samarut, J., and Flamant, F. (2002). Persistence of oligodendrocyte precursor cells and altered myelination in optic nerve associated to retina degeneration in mice devoid of all thyroid hormone receptors. *Proc Natl Acad Sci U S A* 99, 2907–2911.
101. Billon, N., Jolicoeur, C., Tokumoto, Y., Vennström, B., and Raff, M. (2002). Normal timing of oligodendrocyte development depends on thyroid hormone receptor α 1. *EMBO J* 21, 6452–6460.
102. Lezoualc'h, F., Hassan, A. H. S., Giraud, P., Loeffler, J.-P., Lee, S. L., and Demeneix, B. A. (1992). Assignment of the β-thyroid hormone receptor to 3,5,3′-triiodothyronine-dependent inhibition of transcription from the thyrotropin-releasing hormone promoter in chick hypothalamic neurons. *Mol Endocrinol* 6, 1797–1804.
103. Forrest, D., Hanebuth, E., Smeyne, R. J., Everds, N., Stewart, C. L., Wehner, J. M., and Curran, T. (1996b). Recessive resistance to thyroid hormone in mice lacking thyroid hormone receptor β: evidence for tissue-specific modulation of receptor function. *EMBO J* 15, 3006–3015.
104. Gauthier, K., Chassande, O., Plateroti, M., Roux, J. P., Legrand, C., Pain, B., Rousset, B., Weiss, R., Trouillas, J., and Samarut, J. (1999). Different functions for the thyroid hormone receptors TRα and TRβ in the control of thyroid hormone production and post-natal development. *EMBO J* 18, 623–631.
105. Abel, E. D., Ahima, R. S., Boers, M. E., Elmquist, J. K., and Wondisford, F. E. (2001). Critical role for thyroid hormone receptor β2 in the regulation of paraventricular thyrotropin-releasing hormone neurons. *J Clin Invest* 107, 1017–1023.

106. Ng, L., Hurley, J. B., Dierks, B., Srinivas, M., Saltó, C., Vennström, B., Reh, T. A., and Forrest, D. (2001a). A thyroid hormone receptor that is required for the development of green cone photoreceptors. *Nat Genet* 27, 94–98.
107. Abel, E. D., Moura, E. G., Ahima, R. S., Campos-Barros, A., Pazos-Moura, C. C., Boers, M. E., Kaulbach, H. C., Forrest, D., and Wondisford, F. E. (2003). Dominant inhibition of thyroid hormone action selectively in the pituitary of thyroid hormone receptor-beta null mice abolishes the regulation of thyrotropin by thyroid hormone. *Mol Endocrinol* 17, 1767–1776.
108. Dupre, S. M., Guissouma, H., Flamant, F., Seugnet, I., Scanlan, T. S., Baxter, J. D., Samarut, J., Demeneix, B. A., and Becker, N. (2004). Both thyroid hormone receptor (TR)beta 1 and TR beta 2 isoforms contribute to the regulation of hypothalamic thyrotropin-releasing hormone. *Endocrinology* 145, 2337–2345.
109. Amma, L. L., Campos-Barros, A., Wang, Z., Vennström, B., and Forrest, D. (2001). Distinct tissue-specific roles for thyroid hormone receptors β and α1 in regulation of type 1 deiodinase expression. *Mol Endocrinol* 15, 467–475.
110. Ng, L., Pedraza, P. E., Faris, J. S., Vennstrom, B., Curran, T., Morreale de Escobar, G., and Forrest, D. (2001b). Audiogenic seizure susceptibility in thyroid hormone receptor beta-deficient mice. *Neuroreport* 12, 2359–2362.
111. Siesser, W. B., Cheng, S. Y., and McDonald, M. P. (2005). Hyperactivity, impaired learning on a vigilance task, and a differential response to methylphenidate in the TRbetaPV knock-in mouse. *Psychopharmacology (Berl)* 181, 653–663.
112. Forrest, D., Erway, L. C., Ng, L., Altschuler, R., and Curran, T. (1996a). Thyroid hormone receptor β is essential for development of auditory function. *Nature Genet* 13, 354–357.
113. Guadaño-Ferraz, A., Benavides-Piccione, R., Venero, C., Lancha, C., Vennström, B., Sandi, C., DeFelipe, J., and Bernal, J. (2003). Lack of thyroid hormone receptor alpha1 is associated with selective alterations in behavior and hippocampal circuits. *Mol Psychiatry* 8, 30–38.
114. Ng, L., Rusch, A., Amma, L. L., Nordstrom, K., Erway, L. C., Vennstrom, B., and Forrest, D. (2001c). Suppression of the deafness and thyroid dysfunction in Thrb-null mice by an independent mutation in the Thra thyroid hormone receptor alpha gene. *Hum Mol Genet* 10, 2701–2708.
115. Sjöberg, M., and Vennström, B. (1995). Ligand-dependent and -independent transactivation by thyroid hormone receptor β2 is determined by the structure of the hormone response element. *Mol Cell Biol* 15, 4718–4726.
116. Zhu, X.-G., McPhie, P., Lin, K.-H., and Cheng, S.-Y. (1997). The differential hormone-dependent transcriptional activation of thyroid hormone receptor isoforms is mediated by interplay of their domains. *J Biol Chem* 272, 9048–9054.
117. Yang, Z., and Privalsky, M. L. (2001). Isoform-specific transcriptional regulation by thyroid hormone receptors: hormone-independent activation operates through a steroid receptor mode of co-activator interaction. *Mol Endocrinol* 15, 1170–1185.
118. Chatterjee, V., and Beck-Peccoz, P. (2001). Resistance to thyroid hormone, In L. de Groot and J. Jameson, eds., Endocrinology. WB Saunders Company, Philadelphia, PA, pp. 1609–1615.
119. Refetoff, S., DeWind, L. T., and DeGroot, L. J. (1967). Familial syndrome combining deaf-mutism, stippled epiphyses, goiter, and abnormally high PBI: possible target organ refractoriness to thyroid hormone. *J Clin Endocrinol Metab* 27, 279–294.
120. Kaneshige, M., Kaneshige, K., Zhu, X., Dace, A., Garrett, L., Carter, T. A., Kazlauskaite, R., Pankratz, D. G., Wynshaw-Boris, A., Refetoff, S., et al. (2000). Mice with a targeted mutation in the thyroid hormone β receptor gene exhibit impaired growth and resistance to thyroid hormone. *Proc Natl Acad Sci U S A* 97, 13209–13214.
121. Liu, Y. Y., Schultz, J. J., and Brent, G. A. (2003). A thyroid hormone receptor α gene mutation (P398H) is associated with visceral adiposity and impaired catecholamine-stimulated lipolysis in mice. *J Biol Chem* 278, 38913–38920.
122. Sorensen H. G., van der Deure, W. M., Hansen, P. S., Peeters, R. P., Breteler, M. M., Kyvik, K. O., Sorensen, T. I., Hegedus. L., and Visser, T. J. (2008). Identification and consequences of polymorphisms in the thyroid hormone receptor alpha and beta genes. *Thyroid* 18:1087–1094.

123. Gonzalez-Sancho, J. M., Garcia, V., Bonilla, F., and Munoz, A. (2003). Thyroid hormone receptors/THR genes in human cancer. *Cancer Lett* 192, 121–132.
124. Lee, T. C., Almeida, D., Claros, N., Abramson, D. H., and Cobrinik, D. (2006). Cell cycle-specific and cell type-specific expression of rb in the developing human retina. *Invest Ophthalmol Vis Sci* 47, 5590–5598.
125. Sap, J., Muñoz, A., Schmitt, J., Stunnenberg, H., and Vennström, B. (1989). Repression of transcription mediated at a thyroid hormone response element by the v-erbA oncogene product. *Nature* 340, 242–244.
126. Baxter, J. D., Webb, P., Grover, G., and Scanlan, T. S. (2004). Selective activation of thyroid hormone signaling pathways by GC-1: a new approach to controlling cholesterol and body weight. *Trends Endocrinol Metab* 15, 154–157.
127. Berkenstam, A., Kristensen, J., Mellstrom, K., Carlsson, B., Malm, J., Rehnmark, S., Garg, N., Andersson, C. M., Rudling, M., Sjoberg, F., et al. (2008). The thyroid hormone mimetic compound KB2115 lowers plasma LDL cholesterol and stimulates bile acid synthesis without cardiac effects in humans. *Proc Natl Acad Sci U S A* 105, 663–667.
128. Cuevas, E., Ausó, E., Telefont, M., Morreale de Escobar, G., Sotelo, C., and Berbel, P. (2005 Aug). Transient maternal hypothyroxinemia at onset of corticogenesis alters tangential migration of medial ganglionic eminence-derived neurons. *Eur J Neurosci* 22(3), 541–551.
129. Wilcoxon, J. S., Nadolski, G. J., Samarut, J., Chassande, O., and Redei, E. E. (2007) Behavioral inhibition and impaired spatial learning and memory in hypothyroid mice lacking thyroid hormone receptor alpha. *Behav Brain Res* 177, 109–116.
130. Dellovade, T. L., Chan, J., Vennstrom, B., Forrest, D., and Pfaff, D. W. (2000). The two thyroid hormone receptor genes have opposite effects on estrogen-stimulated sex behaviors. *Nat Neurosci* 3, 472–475.

CHAPTER 8

THE VITAMIN D RECEPTOR (NR1I1)

ORLA MAGUIRE AND MORAY J. CAMPBELL

Department of Pharmacology and Therapeutics, Roswell Park Cancer Institute, Elm & Carlton Streets, Buffalo, NY 14263, USA

Abstract: The vitamin D receptor (VDR) is designated NR1I1 using the nuclear receptor superfamily nomenclature. An appreciation of the VDR endocrine system precedes the isolation of the receptor by approximately 300 years. The VDR plays a significant role in regulating serum calcium levels through its actions on the cellular machinery to absorb calcium in the gut, to re-absorb it in the kidneys and deposit it in the bones. Consequently, insufficiency of the VDR ligand 1α,25dihydroxyvitaminD_3 (1α25$(OH)_2D_3$) in a child leads to the classic symptoms of rickets.

Rickets was first described in the seventeenth century by Daniel Whistler in the Netherlands, and subsequently by other physicians in London and elsewhere. The molecular etiology for rickets remained unresolved until the beginning of the twentieth century when the central deficiency of an active vitamin D hormone was revealed. Sir Edward Mellanby in 1919 discovered that the dietary deficiency that caused rickets could be ameliorated by fish oil extracts and that the active ingredient was identified as vitamin D_2 (Ergocalciferol). Around the same time Huldschinsky, Hess and Unger found rickets could be cured by exposure to UV radiation or vegetable oil. This led to the identification of 1α,25$(OH)_2D_3$ as the active hormone.

The increase in understanding of rickets and its prevention by dietary and environmental factors (UV) were entirely co-incident with the rapid rise of the syndrome in the urbanized and industrialized living conditions of the cities of northern Europe where poor access to sunlight and a restricted diet made the effects of under-active VDR all too common. Further studies revealed that vitamin D_2 and vitamin D_3 are actually not vitamins but are secosteroids derived from Ergosterol and 7-dehydrocholesterol respectively. Indeed such was the vigor and significance of the increase in understanding the role of 1α25$(OH)_2D_3$ that its chemical identification and syntheses by Adolf Windaus was sufficient to receive the Nobel Prize in Chemistry (1928). Continued work led in the 1960s to analyses of 1α25$(OH)_2D_3$ metabolism and the identification of 25(OH)D followed by the identification in 1970s of vitamin D binding proteins in target cells and the final cloning of the VDR in 1988 by Bert O'Malley and colleagues lead to a functional understanding of the VDR endocrine system.

C.M. Bunce, M.J. Campbell (eds.), *Nuclear Receptors*, Proteins and Cell Regulation 8, 203–236. DOI 10.1007/978-90-481-3303-1_8,

In the subsequent decades remarkable strides have been made in describing the diverse biology that the VDR participates in. Researchers accommodated this diversity of biological actions by separating functions into the so-called Classical (the regulation of serum calcium levels) and Non-Classical (everything else). Perhaps now these views are beginning to be consolidated into more unified views of the actions of this receptor.

8.1. THE VITAMIN D RECEPTOR

8.1.1. $1,25(OH)_2D_3$ Synthesis Is Initiated Extremely Effectively in the Skin and Forms Part of an Endocrine Signaling Loop

Systemic monitoring and regulation of serum calcium levels are fundamentally important processes owing to the vital function that calcium plays in a wide range of cellular functions. The VDR plays a well-established endocrine role in the regulation of calcium homeostasis by regulating calcium absorption in the gut and kidney, and bone mineralization. In turn, $1\alpha,25(OH)_2D_3$ status is dependent upon cutaneous synthesis initiated by solar radiation and also on dietary intake – a reduction of either one or both sources leads to insufficiency. The contribution from the UV-initiated cutaneous conversion of 7-dehydrocholesterol to vitamin D_3 is the greater, contributing over 90% towards final $1\alpha,25(OH)_2D_3$ synthesis in a vitamin D sufficient individual. The importance of the relationships between solar exposure and the ability to capture UV-mediated energy is underscored by the inverse correlation between human skin pigmentation and latitude. That is, the individual capacity to generate vitamin D_3 in response to solar UV-exposure is intimately associated with forebear environmental adaptation. The correct and sufficient level of solar exposure and serum vitamin D_3 are matters of considerable debate. Current recommendations for daily vitamin D_3 intake are in the range of 400–800 IU/day [1]. More recently, reassessment of the $1\alpha,25(OH)_2D_3$ impact on the prevention of osteoporosis has suggested that the correct level may be as high as 2–3,000 IU/day, which may reflect more accurately "ancestral" serum levels [2].

The central relationship between UV exposure and calcium homeostasis has underpinned the development of the endocrine view of $1\alpha,25(OH)_2D_3$ signaling, with spatially distinct sites within the body of incremental vitamin D activation. Thus, vitamin D_3 produced in the skin is converted in the liver to 25-hydroxyvitamin D_3, (25(OH)-D), and circulating levels of this metabolite serve as a useful index of vitamin D status. A further hydroxylation occurs in the kidney at the carbon 1 position by 25-hydroxyvitamin D-1α-hydroxylase (encoded by *CYP27b1*) to produce the biologically active hormone, $1\alpha,25(OH)_2D_3$. A second mitochondrial cytochrome P450 enzyme, the 24-hydroxylase (encoded by *CYP24*), can utilize both 25OH-D and $1\alpha,25(OH)_2D_3$ as substrates and is the first step in the inactivation pathway for these metabolites.

Another role of $1\alpha25(OH)_2D_3$ in the kidney is regulation of its own production, mediated by the VDR via a negative feedback mechanism. The presence of

$1\alpha 25(OH)_2D_3$ inhibits renal expression of 1α-hydroxlase and enhances the production of 24-hydroxylase thus catabolising $1\alpha 25(OH)_2D_3$ into less active metabolites and allowing their excretion. Thus, elevated levels of $1\alpha,25(OH)_2D_3$ appear to block its synthesis and induce its own inactivation [3], in a classical negative feedback loop. These effects are in concert with parathyroid hormone (PTH), also via a feedback mechanism. Lowering calcium levels stimulates an increase in PTH expression which leads to increased renal calcium resorption and osteoblast formation. These effects, in part by altering serum phosphate levels, stimulate 1α-hydroxlase and activate $1\alpha 25(OH)_2D_3$ thereby increasing calcium absorption in the intestines [4]. $1\alpha,25(OH)_2D_3$ activation of the VDR in the parathyroid gland suppresses PTH expression [5] and further maintains correct serum calcium. $1\alpha,25(OH)_2D_3$ has also been shown to inhibit the proliferation of parathyroid gland cells thus suggesting a role for VDR in suppression of hyperparathyroidism independent of its role in calcium regulation.

More recently, the expression of 1α-hydroxlase has been identified in keratinocytes and a wide range of other cell types suggesting an autocrine/paracrine role for the local synthesis and signaling of $1\alpha,25(OH)_2D_3$ [6–11]. Thus, not only the liver and kidney endocrine loop, but in multiple target tissues 25OH-D may enter into an intracellular VDR signalling axis that co-ordinates the local autocrine synthesis, metabolism and signal transduction of $1\alpha,25(OH)_2D_3$. The components of this axis are regulated dynamically, similarly to the kidney, with *CYP27b1* being repressed by $1\alpha,25(OH)_2D_3$ and correspondingly *CYP24* positively regulated by $1\alpha,25(OH)_2D_3$. The biological significance of these autocrine actions have been the subject of intense investigation, and support the concept that the VDR has two, perhaps distinct, broad biological roles. Namely, the endocrine regulation of serum calcium and the autocrine/paracrine regulation of biological functions associated with the regulation of cell proliferation and differentiation, and with the modulation of immune responses.

8.1.2. Homology Within Nuclear Receptor Superfamily

A highly conserved VDR has been found in animals with a calcified skeleton and is undetectable in non-chordate species, although is present in certain non-calcified chordates such as the lamprey (reviewed in [12]). Within prokaryotes there appears to be the capacity to undertake UV-catalyzed metabolism of cholesterol compounds and suggests that the evolution of vitamin D biochemistry is very ancient. These findings suggest that the VDR system has been adapted to regulate calcium function and retains other functions that are calcium independent. Some of these functions, in common with other nuclear receptors, include the capacity to be regulated by ligand-independent events and may reflect a pre-ligand-activated evolutionary state [13].

Phylogenetic classification has defined seven nuclear receptor subfamilies, and within these the VDR is in the group 1 sub-family, sharing homology with the

LXRs and FXR, and more distantly the PPARs. The receptors within this sub-family preferentially form homodimers or heterodimeric complexes with RXR acting as a common central partner for VDR, PPARs, LXRs and FXR. Thus the receptors in the group appear to be all responsive to either bile acid or xenobiotics and therefore widely integrated with bile acid homeostasis and detoxification. In keeping with this capacity the bile acid lithocholic acid (LCA) has recently been shown to be a potent ligand for the VDR all be it with lower mM affinity [14].

$1\alpha,25(OH)_2D_3$ and its pre-cursor $25(OH)D_3$, in common with most other nuclear receptor ligands, are highly hydrophobic and therefore transported in the aqueous blood stream associated with a specific binding protein (DBP) [15, 16]. At the cell membrane they are free to diffuse across the lipid membrane, although the identification of Megalin as an active transport protein for $25(OH)D_3$ suggests that transport into the cell of vitamin D_3 metabolites maybe more tightly regulated than by merely by passive diffusion alone [17]. Once in target cells, $1\alpha,25(OH)_2D_3$ associates with the VDR. It is therefore reasonable to speculate that the VDR originated during early chordate evolution as a transcription factor, probably involved in detoxification, and subsequently developed a role in calcium homeostasis in response to UV-catalysed synthesis of $1\alpha,25(OH)_2D_3$.

8.1.3. The Choreography of Transcriptional Regulation

8.1.3.1. Generic VDR transcriptional regulation

The VDR appears to have the capacity to be distributed between the cell membrane, the cytoplasm and the nucleus. In the absence of ligand, the VDR is predominantly located in the nucleus, although there is evidence of cytoplasmic expression and cell-membrane associated VDR, which may mediate rapid non-genomic responses [4]. This is a feature of several nuclear receptors, such as the ERα, where the nuclear receptor is cycled through caveolae at the cell membrane to initiate signal transduction pathways (see Chapter 6). The contribution of these actions to the overall functions of $1\alpha,25(OH)_2D_3$ remains to be clarified fully. Equally, there is evidence for cytoplasmic VDR that is trafficked into the nucleus upon ligand activation. Ligand activation induces nuclear localisation, in tandem with RXRs [18], each in association with specific importins [19].

Structurally, the VDR is uncommon, compared to other NRs, as it does not contain an activation domain at its amino terminus (AF1). In most other receptors this is an important domain for activation, for example for autonomous ligand-independent AF function domain. The VDR relies on the activation domain in the carboxy terminus (AF-2) for activation, whereas other domains in the VDR that are involved in heterodimerisation with RXR [20]. The VDR ligand binding pocket contains hydrophobic residues such as His-305 and -397 that are important in the binding of $1\alpha25(OH)_2D_3$. Ligand binding specifically requires interaction of the hydroxyl group of the A ring at carbon 1 of $1\alpha,25(OH)_2D_3$, which is added by the action of the 1α hydroxylase enzyme. The binding of ligand causes an LBD conformational change, which allows the C-terminal helix 12 of the AF2 domain to reposition into an active conformation exposing a docking surface for transcriptional co-regulators [21, 22]. Both the un-liganded and liganded VDR associates with diverse proteins involved with both transcriptional suppression and activation.

In the absence of ligand, the VDR exist in an "*apo*" state, associated with co-repressors (e.g., NCOR1 and NCOR2/SMRT) as part of large complexes (~2.0 MDa) [23–26] and bound to response element (RE) sequences. These complexes in turn actively recruit a range of enzymes that post-translationally modify histone tails, for example histone deacetylases (HDACs) and methyltransferases, and thereby maintain a locally condensed chromatin structure around response element sequences [26–33].

Ligand binding induces a so-called *holo* state, facilitating the association of the VDR-RXR dimer with co-activator complexes. A large number of interacting co-activator proteins have been described, which can be divided into multiple families including the p160 family, the non-p160 members, and members of the large "bridging" DRIP/TRAP/ARC complex, which links the receptor complex to the co-integrators CBP/p300 and basal transcriptional machinery [34, 35].

These receptor co-activator complexes induce different functions that are required to choreograph the recruitment of basal transcriptional machinary and initiate transcription. The DRIP/TRAP complex directly binds to the VDR, and is thought to be the anchor between the VDR and the other co-regulatory proteins in the transcriptional machinery. Although the DRIPs were first thought to be VDR specific they were also identified in the regulation of the other nuclear receptors including the GR and TR [36, 37]. Indeed the exact specificity/selectivity of many of the co-regulatory factors remains to be established fully, although there is some suggestion that certain co-activators are specific for the VDR, for example NCoA-62 [38]. Binding of DRIP205 and SRC to VDR target gene promoters are $1\alpha,25(OH)_2D_3$ dependent. Following the DRIP complex binding the co-activator binding leads to the co-ordinate activation of an antagonistic battery of chromatin remodeling enzymes, such as histone acetyltransferases, and thereby induce the reorganization of local chromatin regions at the RE of the target gene promoter.

The complex choreography of these events has recently emerged from the study of the VDR [27, 39–44] and other nuclear receptors [45–48] and involves cyclical rounds of promoter-specific complex assembly, gene transactivation, complex disassembly and proteosome-mediated receptor degradation co-incident with co-repressor binding and silencing of transcription. This gives rise to the characteristic periodicity of nuclear receptor transcriptional activation and pulsatile mRNA and protein accumulation. The periodicity of VDR induced mRNA accumulation of target genes is not common, but rather tends towards patterns that are specific for individual target genes and suggests that promoter-specific complexes combine to determine the precise periodicity [49, 50].

8.1.3.2. Signal specificity

The diversity of VDR expression sites, being expressed in virtually all cells of the human, and the disparate phenotypic effects, from regulating calcium transport to sensing redox potential and DNA damage beg the question of cell specificity of actions. That is, what governs the spatial-temporal regulation of VDR-dependent transcriptomes in different cell types. These apparent conundrums in understanding are increasingly common for a number of nuclear receptors and other transcription factors.

The specificity of VDR signaling may arise due to integration with other perhaps more dominant transcription factors. Again, for other nuclear receptors (e.g. AR and ERα), so-called pioneer factors appear to be highly influential in determining choice and magnitude of transcriptional actions [51]. It remains to be established to what extent the VDR interacts similarly with other transcription factors.

Aside from the established co-regulators, some chaperone proteins have been reported to be regulators of VDR mediated transcription. HSP70 has been shown to interact with the VDR and thereby repress transcription [52], whereas BAG1L, an HSP70 binding protein, has been shown to bind to the VDR, and enhances VDR mediated transcription [53]. Very recently, p23 and HSP90 have been shown to release the VDR/co-activator complex from the promoter of target genes in the presence of $1\alpha,25(OH)_2D_3$ [54].

It is becoming clear that co-repressors such as NCOR1 and NCOR2/SMRT, which are broadly expressed, are tailored further for specific receptor and tissue functions by variation in expression, localisation, and isoform composition [55–57]. Equally the list continues to grow of novel co-repressor proteins that the VDR interacts with and includes TRIP15/Alien [58], Hairless [59] and DREAM [60]. Hairless has been shown to block VDR mediated differentiation of keratinocytes, whereas addition of $1\alpha,25(OH)_2D_3$ displaces Hairless from the promoter of target genes and recruits co-activators to promote differentiation. DREAM (downstream regulatory element antagonist modulator), usually binds to direct repeat response elements in the promoters of target genes to enhance transcription in VDR and RAR target genes in a calcium dependent manner, and suggests that specificity arises from the interactions of VDR with tissue-specific co-factors. Williams syndrome transcription factor (WSTF) containing WINAC complex, identified by Kato and colleagues, directly interacts with unliganded VDR and mediates binding to promoter sequences where ligand can then bind and recruit other co-regulatory proteins. WINAC has ATP-dependent chromatin-remodeling activity and contains both SWI/SNF components and DNA replication-related factors. WINAC mediates the recruitment of unliganded VDR to its target sites in promoters, and may organize local nucleosomal positioning to allow promoters access to co-regulators. This suggests a novel mechanism in transcriptional regulation, in which VDR binds to gene promoters before ligand is present [61, 62]. More recently, the same group has identified a role for rapid changes in DNA methylation patterns to govern further the actions of VDR [63]. Collectively these findings suggest that the individual tissue-specific responses are tailored by specific expression of co-factors.

Post-translational modifications (PTM) possibly confer further VDR specificity of function. PTMs resulting from signal transduction processes for example bring about phosphorylation, acetylation and ubiquitinylation events on the AR. The VDR has been less extensively studied, but crucial roles have emerged for the phosphorylation of serine and threonine residues. VDR was identified as a phospho-protein over 15 years ago with the N-terminal being identified as a substrate for kinase activity. Since then several residues have been identified which appear to regulate DNA binding and co-factor recruitment.

The zinc finger DNA binding domain is located at the N terminal of the VDR and adjacent to this domain is the Serine 51 residue. This residue has been identified as crucial for ligand induced and phosphorylation dependent transcriptional activation by the VDR. When Ser51 is mutated phosphorylation of the VDR, by PKC at least, is all but completely abolished and its transcriptional activity as measured by reporter assay in response to $1\alpha,25(OH)_2D_3$ is markedly reduced [64]. It is intriguing that the crucial site of PKC activity is located so close to the DNA binding domain, but whether there are allosteric or biochemical changes which alter the ability of the VDR to bind DNA remain to be elucidated.

The common NR partner RXR can also be phosphorylated and as a result alters recruitment of co-factors to its holocomplexes. Ser260 is located within the ligand binding domain of the RXR and appears crucial for mediating co-factor binding and ligand induced transcriptional responses. When phosphorylated, Ser260 allows binding between the RXR and VDR, but presumably through allosteric changes to the complex, limits the recruitment of co-factors to the complex [65].

The recruitment of co-factors to the VDR holocomplex also appears to be regulated by the presence of PTMs for example, kinase CK-II. The phospho-mimic mutant VDRS208D does not increase or decrease VDR-DNA, VDR-RXR or VDR-SRC interactions but it does increase the levels of VDR-DRIP205 complexes present. CK-II which specifically phosphorylates Ser208 enhances $1,25(OH)_2D_3$ induced transactivation of VDR targets [66, 67]. In addition, phosphatase inhibitors (okadoic acid) in combination with $1,25(OH)_2D_3$ shifts the co-factor preference from GRIP-1 to DRIP205 [68]. Taken together these data suggest that the DRIP205 co-activator complex enhances the transcriptional response by VDR and is recruited by CK-II dependent phosphorylation of the VDR at Ser208.

Establishing the specificity of function and selectivity of VDR interactions has to an extent been limited by technical approaches. What is required now is isolation of VDR (membrane, cyctoplasm, nucleus) in either individual or very pure populations of cells and using unbiased approaches to dissect VDR interactions rather than candidate driven procedures.

8.1.4. Vitamin D Response Elements

Simple REs are formed by two recognition motives and their relative distance and orientation contributes to receptor binding specificity. More recently larger, composite and integrated elements have been identified, suggesting a more intricate control involving integration with other transcription factors (TFs), for example p53 and C/EBPα as demonstrated on the promoter/enhancer regions of *CDKN1A* and

SULT2A1 respectively [49, 69]. Thus, the combinatorial actions of the VDR with other TFs most likely go someway towards explaining the apparent diversity of VDR biological actions.

The most abundant VDRE is the DR3 – an imperfect hexameric direct repeat sequence AGTTCA with a spacer of three nucleotides The first type of vitamin D responsive genes identified contained a direct repeat (DR) three element in the promoter region of the gene. In the DR3 configuration, RXR, the heterodimer partner is believed to occupy the upstream half-site and VDR the downstream motif with two half-sites spaced by three nucleotides. Other types of VDREs have been since been identified. One such VDRE is a palindromic sequence with a nine base-pair nucleotide spacer (IR9) [70]. This sequence was identified in the human calbindin D9K gene and like most VDREs the VDR/RXR binds this sequence in a 5′-RXR-VDR-3′ polarity. More recently a novel everted repeat sequence with a 6 base-pair nucleotide spacer (ER6) has been identified, for example in the gene for CYP3A4 (an enzyme important in xenobiotic metabolism) in addition to the DR3 already known to be present in this gene. An inverted repeat with no spacer (IR0) has also been identified in the SULT2A1 gene. To date, the majority of VDR/RXR target genes are regulated by the heterodimer formation through a DR3 element in the genes regulatory sequences.

The ability of VDR to display transrepression, that is ligand-dependent transcriptional repression has received significant interest. Although the classical DR3 element binds the VDR to elicit a transcriptional response, analysis of the avian PTH gene has revealed a ligand-dependent repression of this gene by VDR [71]. The element mediating this effect was identified as a DR3, and since it resulted in transcriptional repression, the motif was referred to as a negative nVDRE. A nVDRE has been identified in the human kidney in the 1-α-hydroxylase gene 530 bp upstream of the transcription start site. Interestingly, the VDR does not bind directly to this sequence, binding has been shown to be mediated by an intermediary factor known as bHLH-type transcription factor VDR interacting repressor (VDIR). It has since been shown that liganded VDR binds to the VDIR and indirectly causes repression through HDAC mechanisms [71–73]. Furhter, the ability of the VDR to play roles in both transactivation and transrepression reflects emerging themes for other nuclear receptors, e.g. PPAR's [74, 75] and suggests a hitherto unsuspected flexibility of the VDR to associate with a diverse array of protein factors to adapt function [73, 76].

8.2. VDR ACTIONS IN NORMAL TISSUES

8.2.1. Lessons from Murine Models

Key insights have been gained into VDR actions through the use of knockout mice. Independently, three groups have generated *Vdr*-deficient mice with a range of targeting strategies [70, 77, 78]. The Vdr is expressed widely during murine embryonic development in tissues involved in calcium homeostasis and bone development.

Vdr disruption results in a profound phenotype in these models, which is principally observed post weaning and is associated with the alteration of duodenal calcium absorption and bone mineralization, resulting in hypocalcemia, secondary hyperparathyroidism, osteomalacia, rickets, impaired bone formation and elevated serum levels of $1\alpha,25(OH)_2D_3$. In parallel, a range of more subtle effects are seen more clearly when the animals are rescued with dietary calcium supplementation. The animals became growth retarded, display alopecia, uterine hypoplasia, impaired ovarian folliculogenesis, reproductive dysfunction, cardiac hypertrophy, and enhanced thrombogenicity.

These phenotypes reflect the reduced ability of these mice to absorb calcium in the intestine and reabsorb it in the kidney. Both processes involve both passive and active transport and it appears a component of the latter is regulated by $1\alpha25(OH)_2D_3$ activation of the Vdr, resulting in increased serum calcium concentration. This has been the subject of intensive investigation and reflects the combination of three calcium transport processes; namely into the cell from the extra-luminal space (e.g. either nephron or GI tract), transport across the cell cytoplasm and transport out of the cell into the serum. The lack of a pre-weaning phenotype demonstrates that other non-Vdr dependent processes combined with the passive diffusion can satiate the calcaemic requirements of the embryo in utero and of neo-nates. Furthermore intestinal calcium absorption is decreased significantly but not lost completely in Vdr –/– in the post-weaning animals, and the Vdr regulated gene targets reflect this. Similarly, urinary calcium content is increased reflecting loss of kidney reabsorbtion. In both tissues, calcium channels, for example of the transient receptor potential vanilloid (TRPV) family members, have been found to be Vdr targets and shown to be functionally responsible for calcium influx at the apical membrane of enterocytes. TRPV5 and 6 function in the kidney and duodenum respectively. Calcium binding proteins (CaBPs) of differing sizes are responsible of cytoplasmic transport, again with CaBP-9k operating in the intestine whereas it works in co-operation CaBP-28k in the kidney. Calcium efflux is governed by the plasma membrane calcium ATPase (PMCA1b) in the intestine, whereas the sodium/calcium exchanger (NCX1) and PMCA1b are involved in the kidney. VDR influences the expression of these components, but neither exclusively nor absolutely. That is serum calcium levels appears to trigger the regulated expression of these factors in non-Vdr dependent mechanisms also; an absence of Vdr does not reduce expression of these factors completely and also calcium maybe actively transported by as yet unknown mechanisms. Nonetheless the reduced ability to absorb and retain calcium is sufficient to give rise to a range of phenotypes in the adult animal (reviewed in [79]).

8.2.1.1. Calcified tissues

Embryonic bone development is normal in the Vdr –/– animals with normal morphology, mineral content and gene expression patterns in pups compared to wildtype litter mates. After weaning the longitudinal growth of the long bones display features of advanced rickets, specifically widening and disorganization of the epiphyseal growth plate. This is partly the result of impaired activation of caspase 9 in chondrocytes leading to a loss of apoptosis and resultant hypertrophy. This may in part also reflect the fact that the Vdr –/– animals also display elevated serum phosphate levels.

Normally the actions of VDR and PTH signals combine with serum phosphate levels in an endocrine loop to regulate serum calcium. The signaling between bone and the parathyroid glands is mediated in part by the secretion of PTH, which is ultimately responsive to serum calcium levels in a feedback loop involving the regulation of PTH levels and $25(OH)D_3$ activation in the kidney (via *CYP27b1*). In bone osteoblasts, $1\alpha,25(OH)_2D_3$ interacts with PTH to enhance the expression of receptor activator of nuclear factor (NF)-κB (RANK) ligand (RANKL). Once RANKL is on the osteoblast surface it interacts with RANK in the pre-osteoclast and triggers maturation to an osteoclast. The mature osteoclast can then release collagenases to destroy the matrix and release calcium and phosphorous into the metabolism. More recently, an active VDRE has been found on the RANKL promoter suggesting that the formation of mature osteoclasts could be mediated directly via the VDR. $1\alpha25(OH)_2D_3$ has also been shown to stimulate the differentiation of osteoblasts, and cause them to develop a calcified matrix in vitro. In osteoblasts VDR can also induce the phosphaturic factor FGF23 in osteoblastic cells. Parallel studies have identified roles for VDR to induce NFκB and VEGF signaling resulting in increased bone vascularization and increased osteoclast numbers. Consequently, the low serum calcium in the Vdr –/– mice leads to hyperparathyroidism, increased PTH mRNA levels in the parathyroids, and elevated serum PTH levels that is normalized by feeding these animals a rescue diet of high calcium.

These findings suggest that the VDR is involved, both positively and negatively, in the direct and indiect regulation of serum calcium availability by facilitating both deposition and release of this vital mineral from bone. Again, this highly integrated biology is reflected by the fact that VDR is not obligative for bone formation but rather facilitative of bone maturation.

Teeth are another tissue where calcium deposition is important, notably in the enamel. The process of tooth formation, eruption and calcification requires a range of support cells that are responsive to the VDR to regulate genes which handle calcium transport. Consequently, reflecting the bone maturation phenotype for the vdr –/– there is a loss of mineralization in the teeth of vdr –/– and impaired dentin maturation.

8.2.1.2. Skin and hair

The Vdr is readily detected in keratinocytes and co-treatment of calcium and $1\alpha,25(OH)_2D_3$ decreases proliferation and promotes differentiation of cultured keratinocytes. The Vdr is also detected in outer root sheath and hair follicle bulb, as well as in the sebaceous glands. Vdr –/– mice develop hair loss and ultimately alopecia totalis, associated with large dermal cysts [80, 81]. The alopecia arises is complete from a failure to initiate anagen, which is the first post-natal hair growth phase. Subsequently the hair follicles convert into epidermal cysts. In mice the alopecia phenotype is not prevented by the high calcium rescue diet.

Hair follicle formation requires highly co-ordinated signaling between different cell types including contributions from the stem cells components and therefore the alopecia phenotype has attracted significant research interest as it may represent a role for the VDR in stem cell maintenance [82]. Interestingly these effects appear independent of ligand binding, in that they can be rescued even with Vdr is mutated in the LBD, but not completely if the AF2 domain is interrupted, suggesting that the association with co-factors is required. Notably the co-repressor, Hairless plays a clear role in hair formation with the knockout resulting in alopecia strikingly similar to that observed in the Vdr null mice [59, 83, 84]. Hairless binds to the VDR and represses VDR-mediated transactivation and also other nuclear receptors including the Thyroid receptor and the retinoic acid receptor. Subsequent studies have demonstrated that a failure to maintain hair follicles in Vdr –/– animals does not actually reflect a loss of follicle stem cells but rather an inability of the primitive progenitor cells to migrate along the follicle at the onset of anagen [85].

Wnt singalling is one of the major processes regulating postmorphogenic hair follicle development. Interestingly, the development of dermal cysts and increase in sebaceous glands observed in the Vdr and Hairless –/– mice are also similar to mice expressing a keratinocyte-specific disruption to β-catenin [86–88]. These findings have raised the possibility that one function of the Vdr may be to co-regulate aspects of Wnt singalling. This is supported further by the physical association of VDR in a complex with β-catenin and other Wnt components [85].

In summary alopecia appears one of the most exclusive phenotypes of the Vdr as a result of a critical although not wholly resolved role that the receptor plays in regulating post-morphogenic cycles of hair growth. This role appears to connect VDR function to migration of primitive progenitor cells from the bulb in the hair follicle and the development of new hair. Equally the biology of hair regeneration reflects the choreographed actions of multiple nuclear receptors and other regulatory processes including Wnt signaling.

8.2.1.3. The reproductive organs

One unexpected finding of the Vdr –/– animals was the uterine hypoplasia and impaired ovarian function in the females which lead to dramatically reduced fertility. Similarly to the hair phenotype, this was not restored by the rescue diet of high calcium [89]. Estradiol supplementation, however, of the female mice restored uterine function and fertility and suggests the fault lies with an inability to generate estrogen. A possible explanation for this lies in the ability of Vdr to regulate aromatase.

The mammary gland has been studied in a comprehensive series of experiments by Welsh and co-workers [90, 91]. In mammals, the requirement for calcium increases during pregnancy and lactation, to meet the increased needs of the fetal skeleton for mineralisation and to provide calcium for milk. Analysis of the mammary gland in Vdr deficient mice supports important anti-mitotic and pro-differentiative roles for the receptor. The glands in the knockout mice display enhanced growth compared to their wild-type littermates and are heavier, with an elevated number of terminal end buds, greater ductal outgrowth and enhanced secondary branch points. This accelerated growth is exacerbated further during the pregnancy-associated proliferative burst, and the post-lactation apoptosis associated with involution is delayed. Similarly, in cells derived from these mice and in *ex vivo* organ cultures, the response to exogenous estrogen and progesterone appears enhanced. Thus, the mammary gland represents an intriguing tissue where endocrine (calcaemic) and autocrine (anti-mitotic, pro-differentiative, pro-apoptotic) effects of the VDR appear to converge.

8.2.1.4. The cardiovascular system

The VDR also appears to have effects on the cardiovascular system. Multiple cell types in this system express the VDR and respond to $1\alpha,25(OH)_2D_3$. One component of this system is the Renin-Angiotensin system. Renin is a protease that cleaves angiotensinogen to release, ultimately, angiotensinII that in turn activates specific receptors and regulates electrolyte, volume, and blood pressure homeostasis. Sustained activation in this manner can lead to hypertension and cardiac hypertrophy. A VDRE has been identified in the promoter of the renin gene, and direct $1\alpha,25(OH)_2D_3$ treatment appears to regulate renin expression negatively in vitro and in vivo. Consequently some reports have demonstrated elevated renin in Vdr –/– animals and systemic hypertension [92, 93].

8.3. THE PATHOBIOLOGY OF VDR

8.3.1. Bone Phenotpyes

Disruption to VDR signaling is evident in a number of pathological settings. Bone pathologies are perhaps the most readily anticipated pathologies that arise from an understanding of the etiology of human rickets and the phenotype of the Vdr –/– animals. In a very small pool of individuals (perhaps only 70 kindreds world-wide) VDR point mutations, for example in the LBD (Arg274Leu), have been identified that give rise to classic symptoms of rickets, and yet are essentially resistant to supplementation with exogenous vitamin D3, so-called type II rickets.

Other workers have investigated the possibility that widespread SNPs within the VDR gene locus are predicative of bone strength. Bone is a highly dynamic tissue, and its strength varies by gender, age, ethinicity, combined with environmental factors such as geography and diet. Therefore it is highly challenging to establish causative relationships between a given polymorphism and a pathophysiological outcome, such as osteoperosis. Despite these difficulties a large number of investigators have investigated such relationships and have both claimed and failed to establish the validity of such a hypothesis. Thus it seems likely that VDR function in a human is central to calcium homeostasis and, dependent upon the status of other factors, may penetrate through to bone pathologies.

8.3.2. VDR and Cancer

A parallel and highly significant area of investigation has emerged within the last thirty years and that is the ability of the VDR to govern cell growth and differentiation. This has had the greatest realization within the arena of cancer cell biology. The VDR is expressed in a wide variety of human cell types, and in vitro $1\alpha,25(OH)_2D_3$ is able to regulate proliferation of a range of normal cells, including keratinocytes, cells from the epithelial linings of the prostate, breast, colon, and myeloid CD34-positive precursors [94]. Whilst these tissues display divergent phenotypes they all share in common the capacity to undergo self-renewal and contain tissue-specific stem cell components. It is likely therefore that the common actions of the VDR relate to its ability to influence the cell cycle progression. In this regard perhaps the largest area of study for the VDR is in the control of proliferation and differentiation of cancer cell types.

8.3.2.1. Evidence of VDR involvement in cancer

In 1981, $1\alpha,25(OH)_2D_3$ was shown to inhibit human melanoma cell proliferation significantly in vitro at nanomolar concentrations [95], and was subsequently found to induce differentiation in cultured mouse and human myeloid leukemia cells [96, 97]. Following these studies, anti-proliferative effects have been demonstrated in a wide variety of cancer cell lines, including those from prostate, breast, colon [98–105].

To identify critical target genes that mediate these actions, comprehensive genome-wide *in silico* and transcriptomic screens have analysed the anti-proliferative VDR transcriptome and revealed broad consensus on certain targets, but has also highlighted variability [98, 106–108]. This heterogeneity may in part reflect experimental conditions, cell line differences, and genuine tissue-specific differences of co-factor expression that alter the amplitude and periodicity of VDR transcriptional actions. The common antiproliferative VDR functions are associated with arrest at G_0/G_1 of the cell cycle, coupled with up-regulation of a number of cell cycle inhibitors including $p21^{(waf1/cip1)}$ and $p27^{(kip1)}$. Promoter characterisation studies have demonstrated a series of VDREs in the promoter/enhancer region *CDKN1A* (encodes $p21^{(waf1/cip1)}$) [39, 109]. By contrast, $p27^{(kip1)}$ protein levels appear to be regulated by a range of post-transcriptional mechanisms, such as enhanced mRNA translation, and attenuating degradative mechanisms, often in a cell type specific manner [110–112]. The up-regulation of $p21^{(waf1/cip1)}$ and $p27^{(kip1)}$ principally mediate G_1 cell cycle arrest, but $1\alpha,25(OH)_2D_3$ has been shown to mediate a G_2/M cell cycle arrest in a number of cancer cell lines via direct induction of *GADD45α* [25, 107, 113]. Again, this regulation appears to combine direct gene transcription and a range of post-transcriptional mechanisms. These studies highlight the difficulty of establishing strict transcriptional effects of the VDR, as a range of post-transcriptional effects act in concert to regulate target protein levels. Concomitant with changes in the cell cycle there is some evidence that $1\alpha,25(OH)_2D_3$ also induces differentiation. This is most clearly evidenced in myeloid cell lines, but also supported by other cell types and most likely reflects the intimate links that exist between the regulation of the G1 transition, the expression of CDKIs such as $p21^{(waf1/cip1)}$ and the induction of cellular differentiation. Another VDR effect is associated with elevated expression of a number of brush-border-associated enzymes such as alkaline phosphatase, as well as intermediate filaments, vinculin, ZO-1, ZO-2, desmosomes and E-cadherin, which collectively enhance adhesion and suppress migration [114], actions that appear to be intimately governed by the direct VDR regulation of cystatin D (*CST5*) [115].

VDR actions, notably in MCF-7 breast cancer cells, are associated with a profound and rapid induction of apoptosis, irrespective of p53 content. This may reflect the VDR role in the involution of the post-lactating mammary gland. The direct transcriptional targets which regulate these apoptotic actions remain elusive, although there is evidence of an involvement of the *BAX* family of proteins [116, 117]. Induction of programmed cell death following $1\alpha,25(OH)_2D_3$ treatment is also associated with increased generation of reactive oxygen species (ROS). $1\alpha,25(OH)_2D_3$ treatment up-regulates *VDUP1* encoding vitamin D up-regulated protein 1, which binds to the disulfide reducing protein thioredoxin and inhibits its ability to neutralize ROS, thereby potentiating stress-induced apoptosis [118, 119]. In other cells, the apoptotic response is delayed and not so pronounced, probably reflecting less direct effects. Taken together, these data suggest that the extent and timing of apoptotic events depend on the integration of VDR actions with other cell signaling systems. This regulation of apoptosis in human cancer cell lines reflects, of course, the absence of apoptosis in chondrocytes in the Vdr –/– animals.

In support of these in vitro findings a large number of epidemiological studies have identified an association between environments of reduced serum 25(OH)D and cancer rates. Initially Garland and co-workers demonstrated that intensity of local sunlight is inversely correlated with risk of certain cancers including breast, prostatic and colorectal carcinoma [120–125]. Supportively, levels of 25OH-D, the major circulating metabolite of vitamin D, are significantly lower in breast cancer patients than in age-matched controls [126].

Epidemiological studies have also linked the incidence of prostate cancer to vitamin D insufficiency as a result of either diet or environment. In 1990 Schwartz and colleagues suggested a role for vitamin D in decreasing the risk for prostate cancer based on the observation that mortality rates in the US are inversely related to incident solar radiation [124]. Recently a study of men in the San Francisco Bay area reported a reduced risk of advanced prostate cancer associated with high sun exposure and similar relationships have been established in UK populations [122, 127].

As with breast and prostate cancer, some epidemiological studies have noted that colon cancer risk and mortality increase with increasing latitude, for example, adjusted death rates from colon cancer in Caucasian males in the US were nearly three times higher in north eastern than sunnier more southerly states [128].

These relationships do not appear universal as negative epidemiological findings also exist and probably reflect the difficulty in establishing direct causative relationships between UV exposure, serum levels 25(OH)D over the lifetime of an individual, and the rates of cancer initiation and progression in the individual and general population.

8.3.2.2. In vivo studies

The efficacy of $1\alpha,25(OH)_2D_3$ and its analogs has also been extensively tested in carcinogen-induced models in vivo, indicating a range of protective effects against both tumour initiation, progression and invasion, and supporting VDR chemoprevention and chemotherapy applications. A large series of studies have examined the ability of dietary or pharmacological addition of vitamin D compounds either to prevent tumour formation or to inhibit growth of transplanted tumour xenografts [129, 130].

A number of workers have utilized the *Vdr*-deficient mice as tools to elucidate more clearly the actions of the VDR in cancer settings. A series of mice have been generated in which the VDR ablated background has been crossed into different tumor disposition phenotypes. Thus, crossing the *vdr*-deficient and heterozygote mice with mouse mammary tumor virus (MMTV)-*neu* transgenic mice, has generated animals which show a degree of VDR haplo-sufficiency. The mammary tumour burden in the crossed mice is reduced by the presence of one wild-type *vdr* allele, and further by two wild-type *vdr* alleles [131]. In addition, *vdr* –/– mice demonstrate greater susceptibility to carcinogen challenge. For example, treatment of these mice with DMBA induced more pre-neoplasic lesions in the mammary glands than in wild-type mice [132]. Similarly others have used the same models to establish UV-induced carcinogenesis in the skin and demonstrated a clear protective role for the VDR to limit carcinogenesis in this tissue [133].

Focusing on dietary regimes which demonstrate tumour pre-disposition, long-term studies on mice fed with a Western-style diet (for example high fat and phosphate and low 25(OH)D and calcium content) increased colonic epithelial cell hyperproliferation. Acute exposure to these diets, for example over 12 weeks, proved sufficient to induce colon-crypt hyperplasia; effects which could be ameliorated by addition of calcium and $1\alpha,25(OH)_2D_3$ [134].

Another important model to test chemoprevention and chemotherapy is the Apc_{min} mouse. APC is a key negative regulator of β-catenin actions and is commonly disrupted in humans developing colon cancer. The rate of polyp formation in Apc_{min} mice was significantly increased in mice fed a Western diet compared to animals on standard chow. Only moderate effects of $1\alpha,25(OH)_2D_3$ were found to suppress polyp formation in this model, associated with marked hypercalcaemia. However, the effects were more pronounced and significant when a potent analog of $1\alpha,25(OH)_2D_3$ was used, which also displayed reduced toxicity [135].

A complementary approach to these studies has examined the capacity of $1\alpha,25(OH)_2D_3$ to interact with other dietary components, which are known to be chemoprotective. One such strategy has focused on the ability to enhance local autocrine synthesis and signalling of $1\alpha,25(OH)_2D_3$. For example, phytoestrogens, such as soy or genestein, are known to be protective, and in vivo feeding these substances appears to increase *CYP27B1* and reduce *CYP24* expression in the mouse colon, resulting in locally elevated levels of $1\alpha,25(OH)_2D_3$ [136]. These results would support the concept that Asian diets, rich in phytoestrogens and vitamin D, may, in part, explain the traditionally low rates of breast, prostate and colon cancer in this region.

8.3.2.3. The VDR in DNA damage and repair

An important and emergent area, both in terms of physiology and therapeutic exploitation, is the role the liganded VDR appears to play in maintaining genomic integrity and facilitating DNA repair. There appears to be close co-operation between VDR actions and the p53 tumour suppressor pathway. The maintenance of genomic fidelity against a back drop of self-renewal is central to the normal development and adult function of many tissues including the mammary and prostate glands, and the colon. For example, in the mammary gland p53 family members play a role in gland development and maintenance. *P63 –/–* animals have an absence of mammary and other epithelial structures, associated with a failure of lineage commitment (reviewed in [137]), whereas *p53 –/–* animals have delayed mammary gland involution, reflecting the *Vdr –/–* animals, and wider tumour suspectability (reviewed in [138]).

The overlap between p53 and VDR appears to extend beyond cellular phenotypes. The *VDR* is a common transcriptional target of both p53 and p63 [2, 3] and VDR and p53 share a cohort of direct target genes associated with cell cycle arrest, signal transduction and programmed cell death including *CDKN1A GADD45A, RB1, PCNA, Bax, IGFBP3, TGFB1/2* and *EGFR* [5–8]. At the transcriptional level both VDR heterodimers and p53 tetramers associate, for example, with chromatin remodelling factors CBP/p300 and the SWI/SNF to initiate transactivation [10–13]. By contrast, in the gene repressive state VDR and p53 appear to associate with distinct repressor proteins, for example p53 with mSin3A [14, 15] and SnoN [16], and VDR with NCoR1 [17], suggesting the possibly association with distinct sets of histone deacetylases.

Indeed, *CDKN1A* promoter-dissection studies revealed adjacent p53 and VDR binding sites, suggesting composite responsive regions [39] and reflect composite VDR responsive regions identified also in other gene targets. Together, these findings suggest that $1\alpha,25(OH)_2D_3$-replete environments enhance p53 to regulate mitosis negatively.

Similarly the role of $1\alpha,25(OH)_2D_3$ in the skin is also suggestive of its chemopreventative effects. UV light from sun exposure has several effects in the skin; UVA light induces DNA damage through increasing the level of reactive oxygen species (ROS), but importantly UVB light also catalyzes the conversion of 7-dehydroxycholesterol to 25(OH)-D and induces the expression of VDR.

In addition, antimicrobial and anti-inflammatory genes are another subset of VDR targets that are induced by UV radiation. Suppression of the adaptive inflammatory response is thought to be protective for several reasons, inflamed tissues contain more ROS, which in turn can damage DNA and prevent proper function of DNA repair machinery. Also the induction of cytokines and growth factors associated with inflammation act to increase the proliferative potential of the cells. NF-κB is a key mediator of inflammation and the VDR attenuates this process by negatively regulating NF-κB signalling [139]. This control by VDR is underscored by studies showing *Vdr–/–* mice are more sensitive to chemicals that induce inflammation than their wild-types counterparts [140]. The normally protective effect of inflammation that occurs under other conditions is lost through this VDR mediated suppression but is compensated for by the induction of a cohort of antimicrobial and antifungal genes via the innate immune response [141–143]. The induction of antimicrobials not only prevents infection in damaged tissue but can be cytotoxic for cells with increased levels of anion phospholipids within their membranes, a common feature of transformed cells [144], experimental results are however conflicting. Antimicrobials like DCDMNQ show potent antiproliferative effects in prostate cancer cells lines such as PC-3 and Du-145 [145] and derivatives of 1,2,4-trizole are cytotoxic against some colon and breast cancer cell lines [146].

The key question, and central to exploiting any therapeutic potential of this receptor, is why should the VDR exert such pleiotropic actions? One possible explanation for this pleiotropism is that it represents an adaptation of the skin to UV exposure, coupling the paramount importance of initiating $1\alpha,25(OH)_2D_3$ synthesis with protection of cell and tissue integrity. Thus, VDR actions are able to maximize UV-initiated synthesis of $1\alpha,25(OH)_2D_3$ production, whilst controlling the extent of local inflammation that can result from sun exposure. To compensate for the potential loss of protection associated with immunosuppression, the VDR mediates a range of antimicrobial actions [78, 79]. Equally, local genomic protection is ensured through the up regulation of target genes which induce G_0/G_1 arrest, co-operation with p53 and the induction of cell differentiation. It remains a tantalising possibility that the functional convergence between p53 family and VDR signalling arose in the dermis as an evolutionary adaptation to counterbalance the conflicting physiological requirements of vitamin D synthesis and genome protection, are sustained in epithelial systems, such as the lining of the mammary gland, to protects against geno-toxic insults derived from either the environment or local inflammation.

8.3.2.4. Therapeutic exploitation

One major hindrance to this is the hypercalcaemic effects of vitamin D that can lead to calcium toxicity. Various non-calcaemic analogues of vitamin D have been developed and shown promising results in vitro, in mouse tumor studies and in small pilot human intervention studies and support the concept that the calcemic effects of the VDR can be dissociated from the non-calcemic ones (reviewed in [147].) Attempting to capture the desirable the anti-mitotic and pro-differentiation actions of the VDR has been investigated with arguably the greatest detail in both chemotherapy and chemoprevention settings in prostate cancer, and has highlighted a number therapeutic questions and development issues.

The prostate is the site of both significant benign and malignant diseases. Benign prostate hyperplasia (BPH) is a complex, frequent, syndrome characterized by a static component related to disease conditions benign prostate overgrowth, a dynamic component responsible for urinary irritative symptoms, and an inflammatory component. Preclinical data demonstrate that a selective, potent VDR agonist reduces the static component of BPH by inhibiting the activity of intra-prostatic growth factors downstream of the androgen receptor, and the dynamic component by targeting bladder cells [148, 149]. These data led to a proof-of-concept clinical study that successfully demonstrated arrest of prostate growth in BPH patients treated with elocalcitol and a favourable safety profile. In addition, calcaemia and calciuria, the most sensitive parameters to VDR agonist administration, remained within normal values, indicating separation of calcaemic from non-calcaemic effects.

The etiology of prostate cancer potentially also affords an opportunity to target VDR effects in a chemoprevention setting by studying it's actions in men on active surveillance. This approach, for the management of early prostate cancer, offers the hope of avoiding "unnecessary" treatment. It also provides a window for the identification of biomarkers of prostate cancer behavior by monitoring disease progression in terms of PSA kinetics and repeat biopsies [150]. Workers at the Institute of Cancer Research/Royal Marsden Hospital, London, UK, lead by Dr. Chris C. Parker are currently conducting a double-blind, placebo-controlled, randomised phase II trial of vitamin D supplementation in a unique cohort of men with localised prostate cancer (personal communication). The expression and responsiveness of the VDR signaling axis is being compared against established biomarkers of prostate cancer progression, for example PSA kinetics and histological measurements from repeat biopsies, to establish prognostic significance.

By contrast other workers have aimed to target the VDR with the natural hormone as a chemotherapeutic strategy and demonstrated potency and efficacy utilizing well tolerated, high pulse doses of $1\alpha,25(OH)_2D_3$, defined by optimizing dosing protocols, and have enhanced therapeutic potency further by combinations with established and novel chemotherapy regimens [151–154]. For example a Phase II trial using high-dose, intermittent treatment of androgen-independent prostate cancer with $1\alpha,25(OH)_2D_3$ and dexamethasone showed relatively little toxicity and anti-tumour activity. These in vitro, in vivo, epidemiological and small scale phase 2 clinical trials encouraged the commission by the pharmaceutical industry of a large-scale phase 3 multicenter, randomized, double-blind trial of patients with androgen independent prostate cancer in the US, Canada and Europe. The trial, known as ASCENT (AIPC Study of Calcitriol Enhancing Taxotere), was to examine the benefit of combining a vitamin D3 compound with the chemotherapeutic agent Taxotere [151]. Unfortunately this trial was terminated due to higher death rate in the Vitamin D3 plus Taxotere treatment group with about 900 of the planned total of 1,200 patients enrolled. At the time of writing the reason for the unexpected deaths has not been fully established. These findings suggest the precise manner remains elusive in which to exploit the VDR in the regulation of cell cycle and anti-neoplastic capacities.

8.3.3. Autoimmune Diseases and Graft Rejection

Many cells of the immune system express VDR and respond to 1α,25$(OH)_2D_3$, notably antigen-presenting cells (APCs), such as macrophages and dendritic cells (DCs). In particular, VDR agonists markedly modulate DC phenotype and function. Studies performed either on monocyte-derived DCs from human peripheral blood or on bone-marrow derived mouse DCs, have consistently shown with the capacity of VDR agonists to down regulate expression of the co-stimulatory molecules CD40, CD80, CD86, and to decrease IL-12 levels while enhancing IL-10 production, resulting in impaired T-cell activation. The combination of these effects can explain the capacity of VDR agonists to induce DCs with tolerogenic properties that favour regulatory T cell enhancement. DCs are able to synthesize 1α,25$(OH)_2D_3$ in vitro as a consequence of increased 1α-hydroxylase activity, which could also promote regulatory T cell induction. 1α,25$(OH)_2D_3$ may also participate in the physiological control of immune responses, and be involved in maintaining tolerance to self antigens, as suggested by the enlarged lymph nodes in VDR-deficient mice containing a higher frequency of mature DCs (reviewed in [155]).

Autoimmune diseases occur when there is a failure of tolerance mechanisms. The vitamin D endocrine system is involved in a variety of biological processes able to modulate immune responses, and the tolerogenic properties of VDR agonists render this class of compounds particularly suitable for the treatment of autoimmune disease. Beneficial effects of VDR agonists have been observed in several autoimmune disease models, including experimental allergic encephalomyelitis, collagen-induced arthritis, inflammatory bowel disease, and autoimmune diabetes.

The clinically proven capacity of VDR agonists to treat psoriasis reflects again the combined VDR actions in the skin. Psoriasis is a relatively common chronic inflammatory skin disease, which involves a Th1-mediated hyperproliferation of keratinocytes. Given the capacity of VDR agonists to modulate both cell types, their success in treating psoriasis is perhaps not surprising. 1α,25$(OH)_2D_3$ and other VDR agonists are currently the mainstay treatment in mild and moderate psoriasis, accounting for about 50% of all drugs used to treat this disease. At present, they are used only topically, because a safe analog for systemic use has not yet been developed. Mechanistically, the beneficial effects include inhibition of proliferation and inflammatory cytokine production by skin-infiltrating T cells. Pro-apoptotic actions of VDR agonists on keratinocytes in psoriatic lesions are also involved [156–158].

Another important potential area of application, as a result of the ability of VDR agonists to modulate both APCs and T cells, is the inhibition of graft rejection. The beneficial immunoregulatory properties of 1α,25$(OH)_2D_3$ and its analogues have indeed been demonstrated in different models of experimental organ transplantation. In many cases, the inhibition of acute allograft rejection has been found comparable to that induced by treatment with optimal doses of cyclosporine A (CsA), indicating the potent properties of VDR agonists in transplantation. In most experimental models, acute rejection has been further delayed by combining VDR agonists with a suboptimal dose of CsA or other immunosuppressive agents, suggesting the possibility of combined treatment in clinical settings. Treatment with VDR agonists has consistently shown efficacy in delaying acute allograft rejection, but the effects on chronic rejection are probably most interesting in terms of potential clinical application. DCs and T cells are direct targets of the hormone, leading to the inhibition of pathogenic effector T cells and enhancing the frequency of T cells with suppressive properties, largely via induction of tolerogenic DCs. In addition, inhibition of chemokine production and smooth muscle cell proliferation could represent very important properties of VDR agonists in the control of chronic allograft rejection [159, 160].

8.3.4. Antimicrobial Actions

For over 100 years it has been well-established that sunlight exposure can ameliorate tuberculosis, hence the popularity of sanatoriums in the late nineteenth and early twentieth century. At the time the mechanism was not apparent, but recently a role for VDR signaling has emerged. $1\alpha,25(OH)_2D_3$ is a key link between toll-like receptor activation and antibacterial response in innate immunity against tuberculosis, via induction of anti-microbial peptides like cathelicidin [124–126]. A clinical correlate is provided by the observation that sera from African-American individuals, known to have increased susceptibility to tuberculosis, have reduced levels of $25(OH)D_3$ and are inefficient in cathelicidin mRNA induction, suggesting that vitamin D sufficiency contributes to decreased susceptibility to microbial infections. Similar mechanisms probably operate in individuals affected by skin tuberculosis, a condition known as lupus vulgaris where exposure to high-intensity light, produced from an electric arc lamp, initiates tissue self-renewal thereby generating uninfected normal skin.

8.3.5. Mechanisms of Disruption

A major limitation in the therapeutic exploitation of VDR in cancer therapies and potentially other disease settings is the resistance of cells towards $1\alpha25(OH)_2D_3$. Although the development of potent analogs of $1\alpha25(OH)_2D_3$ have demonstrated some enhanced potency but resistance remains an issue.

8.3.5.1. Reduced environmental availability of $1\alpha,25(oh)_2D_3$

There are data to support disrupted local generation of active $1\alpha25(OH)_2D_3$ in tumors. There are reduced *CYP27b1* mRNA and protein levels in breast cancer cell lines and primary tumours. Comparative genome hybridization studies have found that *CYP24* is amplified in human breast cancer and *CYP24*, associated with altered patterns of $1\alpha25(OH)_2D_3$ metabolism [11, 161]. Therefore over-expression of 24-hydroxylase may further abrogate growth control mediated by $1\alpha25(OH)_2D_3$, via target cell inactivation of the hormone. Thus for example, in breast cancer it has been proposed that low circulating concentrations of 25OH-D, arising as a result of reduced exposure to sunlight, altered dietary patterns and impaired generation of $1\alpha25(OH)_2D_3$ within breast tissue all contribute to the incidence of disease [11, 162–166].

As with breast cancer, the proposed mechanism for the protective effects of sunlight on prostate risk involves the local generation of $1\alpha25(OH)_2D_3$ from circulating 25OH-D in prostate epithelial cells. Cancerous prostate cells express reduced 1α-hydroxylase activity. Prediagnostic serum levels of 25OH-D have been assessed in several prospective studies with some reporting increased risk among men with low circulating levels of the vitamin D metabolite, and a suggestion of an inverse relationship with advanced disease [125, 163–165].

8.3.5.2. Cellular resistance

Reflecting the co-operative an integrated nature of the VDR to function as a transcription factor a number of workers have identified mechanisms by which more dominant signaling process are able either to ablate or attenuate VDR signaling. For example, Munoz and co-workers have dissected the inter-relationships between the VDR, E-Cadherin and the Wnt signalling pathway in colon cancer cell lines and primary tumours. In these studies the induction of *CDH1* (encodes E-Cadherin) was seen in subpopulations of SW480 colon cancer cells, which express the VDR and respond to $1\alpha,25(OH)_2D_3$. The VDR thereby limits the transcriptional effects of β-catenin by physically and directly binding it in the nucleus, and by up regulating E-cadherin to sequestrate β-catenin in the cytoplasm. In malignancy, these actions are corrupted through down regulation of *VDR* mRNA, which appears to be a direct consequence of binding by the transcriptional repressor SNAIL; a key regulator of the epithelial-mesenchyme transition, which is over-expressed in colon cancer. Equally underscoring the central importance of β-catenin, it has recently been shown to be post-translationally modified and act as VDR co-activator and supports a model of checks and balances between these two signalling processes [167–169].

8.3.5.3. Genetic resistance

Outside of the very limited pool of mutations reported in the VDR in type II rickets, the receptor, generally, is neither mutated nor does it appear to the subject of cytogenetic abnormalities [170]. By contrast the polymorphic variations of the *VDR* have been widely reported. Polymorphisms in the 3′ and 5′ regions of the gene have been described and variously associated with risk of breast, prostate and colon cancer, although the functional consequences remain to be established clearly. For example, a start codon polymorphism in exon II at the 5′ end of the gene, determined using the *fok*-I restriction enzyme, result in a truncated protein. At the 3′ end of the gene, three polymorphisms have been identified that do not lead to any change in either the transcribed mRNA or the translated protein. The first two sequences generate *Bsm*I and *Apa*I restriction sites and are intronic, lying between exons 8 and 9. The third polymorphism, which generates a *Taq*I restriction site, lies in exon 9 and leads to a silent codon change (from ATT to ATC) which both encode an isoleucine residue at position 352. These three polymorphisms are linked to a further gene variation, a variable length adenosine sequence within the 3′ untranslated region (3′UTR). The poly(A) sequence varies in length and can be segregated into two groups; long sequences of 18–24 adenosines or short ones [125, 171–173]. The length of the poly(A) tail can determine mRNA stability [174–176] so the polymorphisms resulting in long poly(A) tails may increase the local levels of the VDR protein.

Multiple studies have addressed the association between *VDR* genotype and cancer risk and progression. In breast cancer the *Apa*I polymorphism shows a significant association with breast cancer risk, as indeed have *Bsm*I and the "L" poly(A) variant. Similarly the *Apa*I polymorphism is associated with metastases to bone [177, 178]. The functional consequences of the *Bsm*I, *Apa*I and *Taq*I polymorphisms are unclear but because of genetic linkage may act as a marker for the poly(A) sequence within the 3′UTR, which in turn determine transcript stability. Interestingly combined polymorphisms and serum 25OH-D levels have been shown to compound breast cancer risk and disease severity further [179].

Earlier studies suggested that polymorphisms in the VDR gene might also be associated with risk factor of prostate cancer. Ntais and co-workers performed a meta-analysis of 14 published studies with four common gene polymorphisms (*Taq*1, poly A repeat, *Bsm*1 and *Fok*1) in individuals of European, Asian and African descent. They concluded that these polymorphisms are unlikely to be major determinants of susceptibility to prostate cancer on a wide population basis [180]. Equally studies in colon cancer have yet to reveal conclusive relationships and may be dependent upon ethnicity of the population studied.

8.3.5.4. Epigenetic resistance

In cancer cells the lack of an anti-proliferative response is reflected by a suppression of the transcriptional responsiveness of anti-proliferative target genes such as *CDKN1A*, *CDKN1B*, *GADD45A* and *IGFBPs*, *BRCA1* [25, 100, 181, 182]. Paradoxically, VDR transactivation is sustained or even enhanced, as measured by induction of the highly $1\alpha25(OH)_2D_3$-inducible *CYP24* gene [183, 184]. Together these data suggest that the lack of functional VDR alone cannot explain resistance and instead the VDR transcriptome is skewed in cancer cells to disfavour anti-proliferative target genes. Various workers have proposed that this apparent $1\alpha25(OH)_2D_3$-insensitivity is the result of epigenetic events that selectively suppress the ability of the VDR to transactivate target genes.

Supportively, in malignant prostate primary cultures and cell lines, with reduced $1\alpha25(OH)_2D_3$ anti-proliferative responses, elevated co-repressor mRNA expression has been frequently identified, most commonly involving *NCoR2/SMRT*. These data indicated that the ratio of *VDR* to co-repressor maybe critical to determine $1\alpha25(OH)_2D_3$ responsiveness in cancer cells. A siRNA approach towards *NCoR2/SMRT* demonstrated the significant role this co-repressor plays in regulating this response, with its repression resulting in profound enhancement of the induction of *GADD45* α in response to $1\alpha25(OH)_2D_3$ [185]. These data support a central role for elevated *NCoR2/SMRT* levels to suppress the induction of key target genes resulting in loss of sensitivity to the anti-proliferative action of $1\alpha25(OH)_2D_3$. Parallel studies have demonstrated a similar spectrum of reduced $1\alpha25(OH)_2D_3$-responsiveness between non-malignant breast epithelial cells and breast cancer cell lines. Again, this was not determined solely by a linear relationship between the levels of $1\alpha25(OH)_2D_3$ and *VDR* mRNA expression. Rather, elevated co-repressors mRNA levels, notably *NCoR1*, in ERα negative breast cancer cell lines and primary cultures, was associated with $1\alpha25(OH)_2D_3$ insensitivity; parallel events have been identified in bladder cancer too [186].

Equally this molecular lesion can be targeted by co-treatment of ligand ($1\alpha25(OH)_2D_3$) plus the HDAC inhibitors such as trichostatin A (TSA). These approaches restored the $1\alpha25(OH)_2D_3$-response of androgen-independent PC-3 cells to levels indistinguishable from control normal prostate epithelial cells. This reversal of $1\alpha25(OH)_2D_3$ insensitivity was associated with re-expression of gene targets associated with the control of proliferation and induction of apoptosis, notably *GADD45A* [100, 182, 185]. Targeting this molecular lesion in breast cancer cells through co-treatments of $1\alpha25(OH)_2D_3$ with HDAC inhibitors co-ordinately regulated VDR targets such as *CDKN1A* and *GADD45A* and restored antiproliferative responsiveness [24, 187]. Similarly other workers have used combinatorial chemistry to combine aspects of the structure of $1\alpha,25(OH)_2D_3$ and HDAC inhibitors into single molecule which demonstrates very significant potency [188].

Together these data support the concept that altered patterns of co-repressors inappropriately sustains histone deacetylation around the VDRE of target gene promoter/enhancer regions, and shifts the dynamic equilibrium between *apo* and *holo* receptor conformations, to favour transcriptional repression of key target genes. Thus VDR gene targets are less responsive in $1\alpha25(OH)_2D_3$-insensitive cancer cells compared to non-malignant counterparts. Furthermore targeting this molecular lesion with co-treatments of vitamin D_3 compounds plus HDAC inhibitors generates a temporal window where the equilibrium point between *apo* and *holo* complexes is shifted to favour a more transcriptionally permissive environment.

These findings compliment a number of parallel studies undertaken by others, which have established cooperativity between $1\alpha25(OH)_2D_3$ and butyrate compounds, such as sodium butyrate (NaB) [189–194]. These compounds are short-chain fatty acids produced during fermentation by endogenous intestinal bacteria and have the capacity to act as HDAC inhibitors. Stein and co-workers have identified the effects in colon cancer cells of $1\alpha25(OH)_2D_3$ plus NaB co-treatments to include the co-ordinate regulation of the VDR itself. Together these studies underscore further the importance of the dietary-derived milieu to regulate epithelial proliferation and differentiation beyond sites of action in the gut.

8.4. TOWARDS A UNIFIED UNDERSTANDING OF THE VDR

VDR biology participates in at least three fundamental areas of biology required for human health, and which are disrupted in human disease. It participates in the regulation of serum calcium, and by implication the maintenance of bone integrity; the control of cell proliferation and differentiation, and by implication the disruption of these actions in malignancy; and as a modifier of immune responses and by implication contributes towards auto-immune diseases. The divergence of these actions may make the VDR a particularly challenging receptor to understand in terms of biology and to exploit therapeutically.

For "next generation" developments to occur it will be necessary to adopt a broader view of VDR signaling. Historically, researchers have studied the abilities of single nuclear receptors such as the VDR to regulate a discrete group of gene targets and influence cell function. This has led to substantial knowledge concerning many of these receptors, individually. Cell and organism function however depends on the dynamic interactions of a collection of receptors, through the networks that link them, and against the backdrop of intrinsic cellular programs, such as those governing development and differentiation. The current lack of an integral view of how these interactions bring about function and dysfunction, for example in the aging human individual, can be attributed to the limitations of previously available techniques and tools to undertake such studies. The implementation of post-genomic techniques together with bioinformatics and systems biology methodology is expected to generate an integral view, thereby revealing and quantifying the mechanisms by which cells, tissues and organisms interact with environmental factors such as diet [195, 196].

Nuclear receptors act as an adaptive homeostatic network in several tissues to sense environmental dietary and xenobiotic lipophilic compounds and sustain the cell, for example, through the diurnal patterns of fast and feeding (reviewed in [197, 198, 160]). The VDR was originally described for a central endocrine role in maintenance of serum calcium levels. Similarly the FXR and LXRs were described for their central role in cholesterol metabolism and bile acid synthesis in the entero-hepatic system. However their broad and integrated expression in multiple target tissues suggest a broader role. Examination of the known target genes for VDR, RARs, PPARs, FXR and LXRs reveals that they share in common the regulation of cell cycle, programmed cell death, differentiation and xenobiotic and metabolic clearance.

Specifically dietary-derived fatty acids, and bile acids cycle rapidly in response to dietary intake and work hormonally to co-ordinate multiple aspects of tissue function in response to changing energetic status. Thus, it is probably naive to assume that the VDR alone plays a key and dominant role in cell and tissue function by acting singularly, but instead is intimately linked to the actions of related nuclear receptors (e.g. PPARs, FXR and LXR) and co-factors.

Viewed in this manner, the actions of the VDR to regulate cell growth and differentiation, as part of a network of environmental and dietary sensing receptors, may be the central and common function for this receptor. The differentiated phenotype of these cells then participate in diverse biology that range from calcium transport to dermis formation and mammary gland function.

The challenge is to model the spatio-temporal actions of the nuclear receptor network and, in particular, the extent to which the VDR exerts critical control over transcription and translation. Such an understanding requires a clear awareness of the chromatin architecture and context of the promoter regions (e.g. histone modifications, DNA methylation), genomic organization, gene regulation hierarchies and $1\alpha,25(OH)_2D_3$-based metabolomic cascades, all within the context of specific cell backgrounds. The ultimate research goal will be to translate this understanding to strategies that can predict the capacity of subsets of VDR actions to be regulated in targeted in distinct cell types and exploited in discrete disease settings.

REFERENCES

1. Roux, C. et al. (2008). New insights into the role of vitamin D and calcium in osteoporosis management: An expert roundtable discussion. *Curr Med Res Opin* 24, 1363–1370.
2. wson-Hughes, B. et al. (2005). Estimates of optimal vitamin D status. *Osteoporos Int* 16, 713–716.
3. Takeyama, K. *et al.* (1997). 25-Hydroxyvitamin D3 1alpha-hydroxylase and vitamin D synthesis. *Science* 277, 1827–1830.
4. Vertino, A. M. et al. (2005). Nongenotropic, anti-apoptotic signaling of 1alpha,25(OH)2-vitamin D3 and analogs through the ligand binding domain of the vitamin D receptor in osteoblasts and osteocytes. Mediation by Src, phosphatidylinositol 3-, and JNK kinases. *J Biol Chem* 280, 14130–14137.
5. Demay, M. B., Kiernan, M. S., DeLuca, H. F., and Kronenberg, H. M. (1992). Sequences in the human parathyroid hormone gene that bind the 1,25-dihydroxyvitamin D3 receptor and mediate transcriptional repression in response to 1,25-dihydroxyvitamin D3. *Proc Natl Acad Sci USA* 89, 8097–8101.
6. Zehnder, D. *et al.* (2001). Extrarenal expression of 25-hydroxyvitamin d(3)-1 alpha-hydroxylase. *J Clin Endocrinol Metab* 86, 888–894.
7. Schwartz, G. G., Whitlatch, L. W., Chen, T. C., Lokeshwar, B. L., and Holick, M. F. (1998). Human prostate cells synthesize 1,25-dihydroxyvitamin D3 from 25-hydroxyvitamin D3. *Cancer Epidemiol Biomarkers Prev* 7, 391–395.
8. Schwartz, G. G. *et al.* (2004). Pancreatic cancer cells express 25-hydroxyvitamin D-1 alpha-hydroxylase and their proliferation is inhibited by the prohormone 25-hydroxyvitamin D3. *Carcinogenesis* 25, 1015–1026.
9. Diaz, L. *et al.* (2000). Identification of a 25-hydroxyvitamin D3 1alpha-hydroxylase gene transcription product in cultures of human syncytiotrophoblast cells. *J Clin Endocrinol Metab* 85, 2543–2549.
10. Friedrich, M. *et al.* (2002). Analysis of 25-hydroxyvitamin D3-1alpha-hydroxylase in cervical tissue. *Anticancer Res* 22, 183–186.
11. Townsend, K. *et al.* (2005). Autocrine metabolism of vitamin D in normal and malignant breast tissue. *Clin Cancer Res* 11, 3579–3586.
12. Krasowski, M. D., Yasuda, K., Hagey, L. R., and Schuetz, E. G. (2005). Evolutionary selection across the nuclear hormone receptor superfamily with a focus on the NR1I subfamily (vitamin D, pregnane X, and constitutive androstane receptors). *Nucl Recept* 3, 2.
13. Ellison, T. I., Eckert, R. L., and MacDonald, P. N. (2007). Evidence for 1,25-dihydroxyvitamin D3-independent transactivation by the vitamin D receptor: Uncoupling the receptor and ligand in keratinocytes. *J Biol Chem* 282, 10953–10962.
14. Makishima, M. *et al.* (2002). Vitamin D receptor as an intestinal bile acid sensor. *Science* 296, 1313–1316.

15. Imawari, M., Kida, K., and Goodman, D. S. (1976). The transport of vitamin D and its 25-hydroxy metabolite in human plasma. Isolation and partial characterization of vitamin D and 25-hydroxyvitamin D binding protein. *J Clin Invest* 58, 514–523.
16. Bouillon, R., Van Assche, F. A., Van Baelen, H., Heyns, W., and De Moor, P. (1981). Influence of the vitamin D-binding protein on the serum concentration of 1,25-dihydroxyvitamin D3. Significance of the free 1,25-dihydroxyvitamin D3 concentration. *J Clin Invest* 67, 589–596.
17. Nykjaer, A. *et al.* (1999). An endocytic pathway essential for renal uptake and activation of the steroid 25-(OH) vitamin D3. *Cell* 96, 507–515.
18. Prufer, K., Racz, A., Lin, G. C., and Barsony, J. (2000). Dimerization with retinoid X receptors promotes nuclear localization and subnuclear targeting of vitamin D receptors. *J Biol Chem* 275, 41114–41123.
19. Yasmin, R., Williams, R. M., Xu, M., and Noy, N. (2005). Nuclear import of the retinoid X receptor, the vitamin D receptor, and their mutual heterodimer. *J Biol Chem* 280, 40152–40160.
20. Quack, M. and Carlberg, C. (2000). The impact of functional vitamin D(3) receptor conformations on DNA-dependent vitamin D(3) signaling. *Mol Pharmacol* 57, 375–384.
21. Renaud, J. P. *et al.* (1995). Crystal structure of the RAR-gamma ligand-binding domain bound to all-trans retinoic acid. *Nature* 378, 681–689.
22. Nakabayashi, M. *et al.* (2008). Crystal structures of rat vitamin D receptor bound to adamantyl vitamin D analogs: Structural basis for vitamin D receptor antagonism and partial agonism. *J Med Chem* 51, 5320–5329.
23. Saramaki, A. et al. (2009). Cyclical chromatin looping and transcription factor association on the regulatory regions of the p21 (CDKN1A) gene in response to 1alpha,25-dihydroxyvitamin D3. *J Biol Chem* 284(12), 8073-8082.
24. Banwell, C. M. *et al.* (2006). Altered nuclear receptor corepressor expression attenuates vitamin D receptor signaling in breast cancer cells. *Clin Cancer Res* 12, 2004–2013.
25. Khanim, F. L. *et al.* (2004). Altered SMRT levels disrupt vitamin D3 receptor signalling in prostate cancer cells. *Oncogene* 23, 6712–6725.
26. Li, J. *et al.* (2000). Both corepressor proteins SMRT and N-CoR exist in large protein complexes containing HDAC3. *EMBO J* 19, 4342–4350.
27. Malinen, M. *et al* (2008). Distinct HDACs regulate the transcriptional response of human cyclin-dependent kinase inhibitor genes to trichostatin A and 1{alpha},25-dihydroxyvitamin D3. *Nucleic Acids Res* 36(1), 121-132.
28. Yoon, H. G. et al. (2003). Purification and functional characterization of the human N-CoR complex: The roles of HDAC3, TBL1 and TBLR1. *EMBO J* 22, 1336–1346.
29. Alenghat, T., Yu, J., and Lazar, M. A. (2006). The N-CoR complex enables chromatin remodeler SNF2H to enhance repression by thyroid hormone receptor. *EMBO J* 25(17), 3966-3974.
30. Picard, F. *et al.* (2004). Sirtl promotes fat mobilization in white adipocytes by repressing PPAR-gamma. *Nature* 429, 771–776.
31. Yu, C. *et al* (2005). The nuclear receptor corepressors NCoR and SMRT decrease PPARgamma transcriptional activity and repress 3T3-L1 adipogenesis. *J Biol Chem* 280(14), 13600–13605.
32. Takada, I. *et al.* (2007). A histone lysine methyltransferase activated by non-canonical Wnt signalling suppresses PPAR-gamma transactivation. *Nat Cell Biol* 9, 1273–1285.
33. Picard, F. *et al.* (2004). Sirtl promotes fat mobilization in white adipocytes by repressing PPAR-gamma. *Nature* 429, 771–776.
34. Oda, Y. et al. (2003). Two distinct coactivators, DRIP/mediator and SRC/p160, are differentially involved in vitamin D receptor transactivation during keratinocyte differentiation. *Mol Endocrinol* 17, 2329–2339.
35. Rachez, C. *et al.* (2000). The DRIP complex and SRC-1/p160 coactivators share similar nuclear receptor binding determinants but constitute functionally distinct complexes. *Mol Cell Biol* 20, 2718–2726.
36. Ito, M. and Roeder, R. G. (2001). The TRAP/SMCC/Mediator complex and thyroid hormone receptor function. *Trends Endocrinol Metab* 12, 127–134.

37. Ding, X. F. *et al.* (1998). Nuclear receptor-binding sites of coactivators glucocorticoid receptor interacting protein 1 (GRIP1) and steroid receptor coactivator 1 (SRC-1): Multiple motifs with different binding specificities. *Mol Endocrinol* 12, 302–313.
38. Zhang, C. *et al.* (2003). Nuclear coactivator-62 kDa/Ski-interacting protein is a nuclear matrix-associated coactivator that may couple vitamin D receptor-mediated transcription and RNA splicing. *J Biol Chem* 278, 35325–35336.
39. Saramaki, A., Banwell, C. M., Campbell, M. J., and Carlberg, C. (2006). Regulation of the human p21(waf1/cip1) gene promoter via multiple binding sites for p53 and the vitamin D3 receptor. *Nucleic Acids Res* 34, 543–554.
40. Vaisanen, S., Dunlop, T. W., Sinkkonen, L., Frank, C., and Carlberg, C. (2005). Spatio-temporal activation of chromatin on the human CYP24 gene promoter in the presence of 1alpha,25-dihydroxyvitamin D(3). *J Mol Biol* 350(1), 65-77.
41. Zella, L. A., Kim, S., Shevde, N. K., and Pike, J. W. (2006). Enhancers located within two introns of the vitamin D receptor gene mediate transcriptional autoregulation by 1,25-dihydroxyvitamin D3. *Mol Endocrinol* 20(6), 1231–1247.
42. Kim, S., Shevde, N. K., and Pike, J. W. (2005). 1,25-Dihydroxyvitamin D3 stimulates cyclic vitamin D receptor/retinoid X receptor DNA-binding, co-activator recruitment, and histone acetylation in intact osteoblasts. *J Bone Miner Res* 20, 305–317.
43. Seo, Y. K. *et al.* (2007). Xenobiotic- and vitamin D-responsive induction of the steroid/bile acid-sulfotransferase Sult2A1 in young and old mice: The role of a gene enhancer in the liver chromatin. *Gene* 386, 218–223.
44. Meyer, M. B., Watanuki, M., Kim, S., Shevde, N. K., and Pike, J. W. (2006). The human TRPV6 distal promoter contains multiple vitamin D receptor binding sites that mediate activation by 1,25-dihydroxyvitamin D3 in intestinal cells. *Mol Endocrinol* 20(6), 1447–1461.
45. Reid, G. et al. (2003). Cyclic, proteasome-mediated turnover of unliganded and liganded ERalpha on responsive promoters is an integral feature of estrogen signaling. *Mol Cell* 11, 695–707.
46. Metivier, R. et al. (2003). Estrogen receptor-alpha directs ordered, cyclical, and combinatorial recruitment of cofactors on a natural target promoter. *Cell* 115, 751–763.
47. Yang, X. *et al.* (2006). Nuclear receptor expression links the circadian clock to metabolism. *Cell* 126, 801–810.
48. Carroll, J. S. *et al.* (2006). Genome-wide analysis of estrogen receptor binding sites. *Nat Genet* 38, 1289–1297.
49. Saramaki, A., Banwell, C. M., Campbell, M. J., and Carlberg, C. (2006). Regulation of the human p21(waf1/cip1) gene promoter via multiple binding sites for p53 and the vitamin D3 receptor. *Nucleic Acids Res* 34, 543–554.
50. Vaisanen, S., Dunlop, T. W., Sinkkonen, L., Frank, C., and Carlberg, C. (2005). Spatio-temporal activation of chromatin on the human CYP24 gene promoter in the presence of 1alpha,25-Dihydroxyvitamin D3. *J Mol Biol* 350, 65–77.
51. Eeckhoute, J., Carroll, J. S., Geistlinger, T. R., Torres-Arzayus, M. I., and Brown, M. (2006). A cell-type-specific transcriptional network required for estrogen regulation of cyclin D1 and cell cycle progression in breast cancer. *Genes Dev* 20, 2513–2526.
52. Lutz, W., Kohno, K., and Kumar, R. (2001). The role of heat shock protein 70 in vitamin D receptor function. *Biochem Biophys Res Commun* 282, 1211–1219.
53. Guzey, M., Takayama, S., and Reed, J. C. (2000). BAG1L enhances trans-activation function of the vitamin D receptor. *J Biol Chem* 275, 40749–40756.
54. Bikle, D., Teichert, A., Hawker, N., Xie, Z., and Oda, Y. (2007). Sequential regulation of keratinocyte differentiation by 1,25(OH)2D3, VDR, and its coregulators. *J Steroid Biochem Mol Biol* 103, 396–404.
55. Hollenberg, A. N., Monden, T., Madura, J. P., Lee, K., and Wondisford, F. E. (1996). Function of nuclear co-repressor protein on thyroid hormone response elements is regulated by the receptor A/B domain. *J Biol Chem* 271, 28516–28520.

56. Jonas, B. A., Varlakhanova, N., Hayakawa, F., Goodson, M., and Privalsky, M. L. (2007). Response of SMRT and N-CoR corepressors to MAP Kinase Kinase Kinase cascades is determined by alternative mRNA splicing. *Mol Endocrinol* 21, 1924-1929.
57. Goodson, M. L., Jonas, B. A., and Privalsky, M. L. (2005). Alternative mRNA splicing of SMRT creates functional diversity by generating corepressor isoforms with different affinities for different nuclear receptors. *J Biol Chem* 280, 7493–7503.
58. Polly, P. et al (2000). VDR-Alien: A novel, DNA-selective vitamin D(3) receptor-corepressor partnership. *FASEB J* 14, 1455–1463.
59. Xie, Z., Chang, S., Oda, Y., and Bikle, D. D. (2005). Hairless suppresses vitamin D receptor transactivation in human keratinocytes. *Endocrinology* 147(1), 314-323.
60. Scsucova, S. *et al.* (2005). The repressor DREAM acts as a transcriptional activator on Vitamin D and retinoic acid response elements. *Nucleic Acids Res* 33, 2269–2279.
61. Fujiki, R. *et al.* (2005). Ligand-induced transrepression by VDR through association of WSTF with acetylated histones. *EMBO J* 24, 3881–3894.
62. Kitagawa, H. *et al.* (2003). The chromatin-remodeling complex WINAC targets a nuclear receptor to promoters and is impaired in Williams syndrome. *Cell* 113, 905–917.
63. Kim, M. S. *et al.* (2009). DNA demethylation in hormone-induced transcriptional derepression. *Nature* 461, 1007–1012.
64. Hsieh, J. C. et al. (1991). Human vitamin D receptor is selectively phosphorylated by protein kinase C on serine 51, a residue crucial to its trans-activation function. *Proc Natl Acad Sci USA* 88, 9315–9319.
65. Macoritto, M. *et al.* (2008). Phosphorylation of the human retinoid X receptor {alpha} at serine 260 impairs coactivator(s) recruitment and induces hormone resistance to multiple ligands. *J Biol Chem* 283, 4943–4956.
66. Arriagada, G. et al. (2007). Phosphorylation at serine 208 of the 1[alpha],25-dihydroxy Vitamin D3 receptor modulates the interaction with transcriptional coactivators. *J Steroid Biochem Mol Biol* 103, 425–429.
67. Jurutka, P. W. *et al.* (1996). Human vitamin D receptor phosphorylation by casein kinase II at Ser-208 potentiates transcriptional activation. *Proc Natl Acad Sci USA* 93, 3519–3524.
68. Barletta, F., Freedman, L. P., and Christakos, S. (2002). Enhancement of VDR-mediated transcription by phosphorylation: Correlation with increased interaction between the VDR and DRIP205, a subunit of the VDR-interacting protein coactivator complex. *Mol Endocrinol* 16, 301–314.
69. Song, C. S. *et al* (2005). An essential role of the CAAT/enhancer binding protein-{alpha} in the vitamin D induced expression of the human steroid/bile acid-sulfotransferase (SULT2A1). *Mol Endocrinol* 20(4), 795-808.
70. Yoshizawa, T. et al. (1997). Mice lacking the vitamin D receptor exhibit impaired bone formation, uterine hypoplasia and growth retardation after weaning. *Nat Genet* 16, 391–396.
71. Kim, M. S. et al. (2007). 1Alpha,25(OH)2D3-induced transrepression by vitamin D receptor through E-box-type elements in the human parathyroid hormone gene promoter. *Mol Endocrinol* 21, 334–342.
72. Turunen, M. M., Dunlop, T. W., Carlberg, C., and Vaisanen, S. (2007). Selective use of multiple vitamin D response elements underlies the 1 alpha,25-dihydroxyvitamin D3-mediated negative regulation of the human CYP27B1 gene. *Nucleic Acids Res* 35, 2734–2747.
73. Murayama, A., Kim, M. S., Yanagisawa, J., Takeyama, K., and Kato, S. (2004). Transrepression by a liganded nuclear receptor via a bHLH activator through co-regulator switching. *EMBO J* 23, 1598–1608.
74. Rosenfeld, M. G., Lunyak, V. V., and Glass, C. K. (2006). Sensors and signals: A coactivator/corepressor/epigenetic code for integrating signal-dependent programs of transcriptional response. *Genes Dev* 20, 1405–1428.
75. Chen, C. D. *et al.* (2004). Molecular determinants of resistance to antiandrogen therapy. *Nat Med* 10, 33–39.

76. Fujiki, R. *et al.* (2005). Ligand-induced transrepression by VDR through association of WSTF with acetylated histones. *Embo J* 24, 3881–3894.
77. Li, Y. C. *et al.* (1997). Targeted ablation of the vitamin D receptor: An animal model of vitamin D-dependent rickets type II with alopecia. *Proc Natl Acad Sci USA* 94, 9831–9835.
78. Van Cromphaut, S. J. *et al.* (2001). Duodenal calcium absorption in vitamin D receptor-knockout mice: Functional and molecular aspects. *Proc Natl Acad Sci USA* 98, 13324–13329.
79. Bouillon, R. et al. (2008). Vitamin D and human health: Lessons from vitamin D receptor null mice. *Endocr Rev* 29, 726–776.
80. Hawker, N. P., Pennypacker, S. D., Chang, S. M., and Bikle, D. D. (2007). Regulation of human epidermal keratinocyte differentiation by the vitamin D receptor and its coactivators DRIP205, SRC2, and SRC3. *J Invest Dermatol* 127, 874–880.
81. Hawker, N. P., Pennypacker, S. D., Chang, S. M., and Bikle, D. D. (2006). Regulation of human epidermal keratinocyte differentiation by the vitamin D receptor and its coactivators DRIP205, SRC2, and SRC3. *J Invest Dermatol* 127, 874–880.
82. Cianferotti, L., Cox, M., Skorija, K., and Demay, M. B. (2007). Vitamin D receptor is essential for normal keratinocyte stem cell function. *Proc Natl Acad Sci USA* 104(22), 9428-9433.
83. Wang, J., Malloy, P. J., and Feldman, D. (2007). Interactions of the vitamin D receptor with the corepressor hairless: Analysis of hairless mutants in atrichia with papular lesions. *J Biol Chem* 282, 25231–25239
84. Beaudoin, G. M., 3rd, Sisk, J. M., Coulombe, P. A., and Thompson, C. C. (2005). Hairless triggers reactivation of hair growth by promoting Wnt signaling. *Proc Natl Acad Sci USA* 102, 14653–14658.
85. Palmer, H. G., Martinez, D., Carmeliet, G., and Watt, F. M. (2008). The vitamin D receptor is required for mouse hair cycle progression but not for maintenance of the epidermal stem cell compartment. *J Invest Dermatol* 128, 2113–2117.
86. Yang, S. H. *et al.* (2008). Pathological responses to oncogenic Hedgehog signaling in skin are dependent on canonical Wnt/beta3-catenin signaling. *Nat Genet* 40, 1130–1135.
87. Andl, T., Reddy, S. T., Gaddapara, T., and Millar, S. E. (2002). WNT signals are required for the initiation of hair follicle development. *Dev Cell* 2, 643–653.
88. Millar, S. E. *et al.* (1999). WNT signaling in the control of hair growth and structure. *Dev Biol* 207, 133–149.
89. Yoshizawa, T. et al. (1997). Mice lacking the vitamin D receptor exhibit impaired bone formation, uterine hypoplasia and growth retardation after weaning. *Nat Genet* 16, 391–396.
90. Zinser, G., Packman, K., and Welsh, J. (2002). Vitamin D(3) receptor ablation alters mammary gland morphogenesis. *Development* 129, 3067–3076.
91. Zinser, G. M. and Welsh, J. (2004). Accelerated mammary gland development during pregnancy and delayed postlactational involution in vitamin D3 receptor null mice. *Mol Endocrinol* 18, 2208–2223.
92. Wu-Wong, J. R. (2007). Vitamin D receptor: A highly versatile nuclear receptor. *Kidney Int* 72, 237–239.
93. Xiang, W. et al. (2005). Cardiac hypertrophy in vitamin D receptor knockout mice: Role of the systemic and cardiac renin-angiotensin systems. *Am J Physiol Endocrinol Metab* 288, E125–E132.
94. Bikle, D. D., Xie, Z., Ng, D., Tu, C. L., and Oda, Y. (2003). Squamous cell carcinomas fail to respond to the prodifferentiating actions of 1,25(OH)2D: Why?. *Recent Results Cancer Res* 164, 111–122.
95. Colston, K., Colston, M. J., and Feldman, D. (1981). 1,25-dihydroxyvitamin D3 and malignant melanoma: The presence of receptors and inhibition of cell growth in culture. *Endocrinology* 108, 1083–1086.
96. Miyaura, C. et al. (1981). alpha,25-Dihydroxyvitamin D3 induces differentiation of human myeloid leukemia cells. *Biochem Biophys Res Commun* 102, 937–943.
97. Abe, E. et al. (1981). Differentiation of mouse myeloid leukemia cells induced by 1 alpha,25-dihydroxyvitamin D3. *Proc Natl Acad Sci USA* 78, 4990–4994.

98. Palmer, H. G. et al. (2003). Genetic signatures of differentiation induced by 1alpha,25-dihydroxyvitamin D3 in human colon cancer cells. *Cancer Res* 63, 7799–7806.
99. Koike, M. *et al.* (1997). 19-nor-hexafluoride analogue of vitamin D3: A novel class of potent inhibitors of proliferation of human breast cell lines. *Cancer Res* 57, 4545–4550.
100. Campbell, M. J., Elstner, E., Holden, S., Uskokovic, M., and Koeffler, H. P. (1997). Inhibition of proliferation of prostate cancer cells by a 19-nor-hexafluoride vitamin D3 analogue involves the induction of p21waf1, p27kip1 and E-cadherin. *J Mol Endocrinol* 19, 15–27.
101. Elstner, E. *et al.* (1999). Novel 20-epi-vitamin D3 analog combined with 9-cis-retinoic acid markedly inhibits colony growth of prostate cancer cells. *Prostate* 40, 141–149.
102. Peehl, D. M. et al. (1994). Antiproliferative effects of 1,25-dihydroxyvitamin D3 on primary cultures of human prostatic cells. *Cancer Res* 54, 805–810.
103. Welsh, J. *et al.* (2002). Impact of the Vitamin D3 receptor on growth-regulatory pathways in mammary gland and breast cancer. *J Steroid Biochem Mol Biol* 83, 85–92.
104. Colston, K. W., Berger, U., and Coombes, R. C. (1989). Possible role for vitamin D in controlling breast cancer cell proliferation. *Lancet* 1, 188–191.
105. Colston, K., Colston, M. J., Fieldsteel, A. H., and Feldman, D. (1982). 1,25-dihydroxyvitamin D3 receptors in human epithelial cancer cell lines. *Cancer Res* 42, 856–859.
106. Eelen, G. *et al* (2004). Microarray analysis of 1alpha,25-dihydroxyvitamin D3-treated MC3T3-E1 cells. *J Steroid Biochem Mol Biol* 89–90, 405–407.
107. Akutsu, N. et al. (2001). Regulation of gene expression by 1alpha,25-dihydroxyvitamin D3 and Its analog EB1089 under growth-inhibitory conditions in squamous carcinoma Cells. *Mol Endocrinol* 15, 1127–1139.
108. Wang, T. T. *et al* (2005). Large-scale *in silico* and microarray-based identification of direct 1,25-dihydroxyvitamin D3 target genes. *Mol Endocrinol* 19(11), 2685–2695.
109. Liu, M., Lee, M. H., Cohen, M., Bommakanti, M., and Freedman, L. P. (1996). Transcriptional activation of the Cdk inhibitor p21 by vitamin D3 leads to the induced differentiation of the myelomonocytic cell line U937. *Genes Dev* 10, 142–153.
110. Li, P. et al. (2004). p27(Kip1) stabilization and G(1) arrest by 1,25-dihydroxyvitamin D(3) in ovarian cancer cells mediated through down-regulation of cyclin E/cyclin-dependent kinase 2 and Skp1-Cullin-F-box protein/Skp2 ubiquitin ligase. *J Biol Chem* 279, 25260–25267.
111. Huang, Y. C., Chen, J. Y., and Hung, W. C. (2004). Vitamin D(3) receptor/Sp1 complex is required for the induction of p27(Kip1) expression by vitamin D(3). *Oncogene* 23, 4856–4861.
112. Hengst, L. and Reed, S. I. (1996). Translational control of p27Kip1 accumulation during the cell cycle. *Science* 271, 1861–1864.
113. Jiang, F., Li, P., Fornace, A. J., Jr., Nicosia, S. V., and Bai, W. (2003). G2/M arrest by 1,25-dihydroxyvitamin D3 in ovarian cancer cells mediated through the induction of GADD45 via an exonic enhancer. *J Biol Chem* 278, 48030–48040.
114. Palmer, H. G. *et al.* (2001). Vitamin D(3) promotes the differentiation of colon carcinoma cells by the induction of E-cadherin and the inhibition of beta-catenin signaling. *J Cell Biol* 154, 369–387.
115. Alvarez-Diaz, S. *et al.* (2009). Cystatin D is a candidate tumor suppressor gene induced by vitamin D in human colon cancer cells. *J Clin Invest* 119, 2343–2358.
116. Blutt, S. E., McDonnell, T. J., Polek, T. C., and Weigel, N. L. (2000). Calcitriol-induced apoptosis in LNCaP cells is blocked by overexpression of Bcl-2. *Endocrinology* 141, 10–17.
117. Mathiasen, I. S., Lademann, U., and Jaattela, M. (1999). Apoptosis induced by vitamin D compounds in breast cancer cells is inhibited by Bcl-2 but does not involve known caspases or p53. *Cancer Res* 59, 4848–4856.
118. Song, H. *et al.* (2003). Vitamin D(3) up-regulating protein 1 (VDUP1) antisense DNA regulates tumorigenicity and melanogenesis of murine melanoma cells via regulating the expression of fas ligand and reactive oxygen species. *Immunol Lett* 86, 235–247.
119. Han, S. H. et al. (2003). VDUP1 upregulated by TGF-beta1 and 1,25-dihydorxyvitamin D3 inhibits tumor cell growth by blocking cell-cycle progression. *Oncogene* 22, 4035–4046.

120. Garland, C. F. and Garland, F. C. (1980). Do sunlight and vitamin D reduce the likelihood of colon cancer?. *Int J Epidemiol* 9, 227–231.
121. Garland, F. C., Garland, C. F., Gorham, E. D., and Young, J. F. (1990). Geographic variation in breast cancer mortality in the United States: A hypothesis involving exposure to solar radiation. *Prev Med* 19, 614–622.
122. Luscombe, C. J. et al. (2001). Prostate cancer risk: Associations with ultraviolet radiation, tyrosinase and melanocortin-1 receptor genotypes. *Br J Cancer* 85, 1504–1509.
123. Giovannucci, E. (2005). The epidemiology of vitamin D and cancer incidence and mortality: A review (United States). *Cancer Causes Control* 16, 83–95.
124. Schwartz, G. G. and Hulka, B. S. (1990). Is vitamin D deficiency a risk factor for prostate cancer? (Hypothesis). *Anticancer Res* 10, 1307–1311.
125. John, E. M., Schwartz, G. G., Koo, J., Van Den, B. D., and Ingles, S. A. (2005). Sun exposure, vitamin D receptor gene polymorphisms, and risk of advanced prostate cancer. *Cancer Res* 65, 5470–5479.
126. Lowe, L. C. et al. (2005). Plasma 25-hydroxy vitamin D concentrations, vitamin D receptor genotype and breast cancer risk in a UK Caucasian population. *Eur J Cancer* 41, 1164–1169.
127. Luscombe, C. J. *et al.* (2001). Outcome in prostate cancer associations with skin type and polymorphism in pigmentation-related genes. *Carcinogenesis* 22, 1343–1347.
128. Slattery, M. L. et al. (2004). Dietary calcium, vitamin D, VDR genotypes and colorectal cancer. *Int J Cancer* 111, 750–756.
129. Anzano, M. A. et al. (1994). 1 alpha,25-Dihydroxy-16-ene-23-yne-26,27-hexafluorocholecalciferol (Ro24–5531), a new deltanoid (vitamin D analogue) for prevention of breast cancer in the rat. *Cancer Res* 54, 1653–1656.
130. Belleli, A., Shany, S., Levy, J., Guberman, R., and Lamprecht, S. A. (1992). A protective role of 1,25-dihydroxyvitamin D3 in chemically induced rat colon carcinogenesis. *Carcinogenesis* 13, 2293–2298.
131. Zinser, G. M. and Welsh, J. (2004). Vitamin D receptor status alters mammary gland morphology and tumorigenesis in MMTV-neu mice. *Carcinogenesis* 25(12), 2361-2372.
132. Zinser, G. M. and Welsh, J. (2004). Effect of Vitamin D(3) receptor ablation on murine mammary gland development and tumorigenesis. *J Steroid Biochem Mol Biol* 89–90, 433–436.
133. Ellison, T. I., Smith, M. K., Gilliam, A. C., and MacDonald, P. N. (2008). Inactivation of the vitamin D receptor enhances susceptibility of murine skin to UV-induced tumorigenesis. *J Invest Dermatol* 128, 2508–2517.
134. Xue, L., Lipkin, M., Newmark, H., and Wang, J. (1999). Influence of dietary calcium and vitamin D on diet-induced epithelial cell hyperproliferation in mice. *J Natl Cancer Inst* 91, 176–181.
135. Huerta, S. et al. (2002). 1alpha,25-(OH)(2)-D(3) and its synthetic analogue decrease tumor load in the Apc(min) mouse. *Cancer Res* 62, 741–746.
136. Cross, H. S. *et al* (2004). Phytoestrogens and vitamin D metabolism: A new concept for the prevention and therapy of colorectal, prostate, and mammary carcinomas. *J Nutr* 134, 1207S–1212S.
137. Tsai, H. C. and Norman, A. W. (1973). Studies on calciferol metabolism. 8. Evidence for a cytoplasmic receptor for 1,25-dihydroxy-vitamin D3 in the intestinal mucosa. *J Biol Chem* 248, 5967–5975.
138. Brumbaugh, P. F., Hughes, M. R., and Haussler, M. R. (1975). Cytoplasmic and nuclear binding components for 1alpha25-dihydroxyvitamin D3 in chick parathyroid glands. *Proc Natl Acad Sci USA* 72, 4871–4875.
139. Szeto, F. L. *et al.* (2007). Involvement of the vitamin D receptor in the regulation of NF-[kappa]B activity in fibroblasts. *J Steroid Biochem Mol Biol* 103, 563–566.
140. Froicu, M. and Cantorna, M. (2007). Vitamin D and the vitamin D receptor are critical for control of the innate immune response to colonic injury. *BMC Immunol* 8, 5.
141. Gombart, A. F., Borregaard, N., and Koeffler, H. P. (2005). Human cathelicidin antimicrobial peptide (CAMP) gene is a direct target of the vitamin D receptor and is strongly up-regulated in myeloid cells by 1,25-dihydroxyvitamin D3. *FASEB J* 19, 1067–1077.

142. Wang, T. -T. et al. (2004). Cutting edge: 1,25-dihydroxyvitamin D3 is a direct inducer of antimicrobial peptide gene expression. *J Immunol* 173, 2909–2912.
143. Mallbris, L., Edström, D. W., Sundblad, L., Granath, F., and Stahle, M. (2005). UVB upregulates the antimicrobial protein hCAP18 mRNA in human skin. *J Investig Dermatol* 125, 1072–1074.
144. Zasloff, M. (2005). Sunlight, Vitamin D, and the innate immune defenses of the human skin. *J Investig Dermatol* 125, xvi–xvii.
145. Copeland, R. L., Das, J., Bakare, O., Enwerem, N. M., Berhe, S., Hillaire, K., White, D., Beyene, D., Kassim, O. O., and Kanaan, Y. M. (2007). Cytotoxicity of 2,3-dichloro-5,8-dimethoxy-1,4-naphthoquinone in androgen-dependent and -independent prostate cancer cell lines. *Anticancer Res* 27, 1537–1546.
146. Sztanke, K., Tuzimski, T., Rzymowska, J., Pasternak, K., and Kandefer-Szerszen, M. (2008). Synthesis, determination of the lipophilicity, anticancer and antimicrobial properties of some fused 1,2,4-triazole derivatives. *Eur J Med Chem* 43(2), 404–419.
147. Campbell, M. J. and Adorini, L. (2006). The vitamin D receptor as a therapeutic target. *Expert Opin Ther Targets* 10, 735–748.
148. Morelli, A. *et al.* (2008). The vitamin D receptor agonist elocalcitol upregulates L-type calcium channel activity in human and rat bladder. *Am J Physiol Cell Physiol* 294, C1206–C1214.
149. Morelli, A. et al. (2007). BXL-628, a vitamin D receptor agonist effective in benign prostatic hyperplasia treatment, prevents RhoA activation and inhibits RhoA/Rho kinase signaling in rat and human bladder. *Prostate* 67, 234–247.
150. Jhavar, S. *et al* (2008). Biopsy tissue microarray study of Ki-67 expression in untreated, localized prostate cancer managed by active surveillance. *Prostate Cancer Prostatic Dis* 12(2), 143-147.
151. Beer, T. M. (2005). ASCENT: The androgen-independent prostate cancer study of calcitriol enhancing taxotere. *BJU Int* 96, 508–513.
152. Beer, T. M. et al. (2005). Pharmacokinetics and tolerability of a single dose of DN-101, a new formulation of calcitriol, in patients with cancer. *Clin Cancer Res* 11, 7794–7799.
153. Trump, D. L., Potter, D. M., Muindi, J., Brufsky, A., and Johnson, C. S. (2006). Phase II trial of high-dose, intermittent calcitriol (1,25 dihydroxyvitamin D3) and dexamethasone in androgen-independent prostate cancer. *Cancer* 106, 2136–2142.
154. Beer, T. M., Myrthue, A., and Eilers, K. M. (2005). Rationale for the development and current status of calcitriol in androgen-independent prostate cancer. *World J Urol* 23, 28–32.
155. Adorini, L. and Penna, G. (2008). Control of autoimmune diseases by the vitamin D endocrine system. *Nat Clin Pract Rheumatol* 4, 404–412.
156. Lu, J. *et al.* (2005). Transcriptional profiling of keratinocytes reveals a vitamin D-regulated epidermal differentiation network. *J Invest Dermatol* 124, 778–785.
157. van de Kerkhof, P. C. and Franssen, M. E. (2001). Psoriasis of the scalp. Diagnosis and management. *Am J Clin Dermatol* 2, 159–165.
158. Ashcroft, D. M., Po, A. L., Williams, H. C., and Griffiths, C. E. (2000). Systematic review of comparative efficacy and tolerability of calcipotriol in treating chronic plaque psoriasis. *Br Med J* 320, 963–967.
159. Sottili, M. et al. (2009). Immunomodulatory effects of BXL-01-0029, a less hypercalcemic vitamin D analogue, in human cardiomyocytes and T cells. *Exp Cell Res* 315, 264–273.
160. Mathieu, C. and Adorini, L. (2002). The coming of age of 1,25-dihydroxyvitamin D(3) analogs as immunomodulatory agents. *Trends Mol Med* 8, 174–179.
161. Albertson, D. G. *et al.* (2000). Quantitative mapping of amplicon structure by array CGH identifies CYP24 as a candidate oncogene. *Nat Genet* 25, 144–146.
162. Feskanich, D. *et al.* (2004). Plasma vitamin D metabolites and risk of colorectal cancer in women. *Cancer Epidemiol Biomarkers Prev* 13, 1502–1508.
163. Ahonen, M. H., Tenkanen, L., Teppo, L., Hakama, M., and Tuohimaa, P. (2000). Prostate cancer risk and prediagnostic serum 25-hydroxyvitamin D levels (Finland). *Cancer Causes Control* 11, 847–852.

164. Hsu, J. Y., Feldman, D., McNeal, J. E., and Peehl, D. M. (2001). Reduced 1alpha-hydroxylase activity in human prostate cancer cells correlates with decreased susceptibility to 25-hydroxyvitamin D3-induced growth inhibition. *Cancer Res* 61, 2852–2856.
165. Chen, T. C., Wang, L., Whitlatch, L. W., Flanagan, J. N., and Holick, M. F. (2003). Prostatic 25-hydroxyvitamin D-1alpha-hydroxylase and its implication in prostate cancer. *J Cell Biochem* 88, 315–322.
166. Garland, C. *et al.* (1985). Dietary vitamin D and calcium and risk of colorectal cancer: A 19-year prospective study in men. *Lancet* 1, 307–309.
167. Pendas-Franco, N. et al. (2008). DICKKOPF-4 is induced by TCF/beta-catenin and upregulated in human colon cancer, promotes tumour cell invasion and angiogenesis and is repressed by 1alpha,25-dihydroxyvitamin D3. *Oncogene* 27, 4467–4477.
168. Aguilera, O. et al. (2007). The Wnt antagonist DICKKOPF-1 gene is induced by 1alpha,25-dihydroxyvitamin D3 associated to the differentiation of human colon cancer cells. *Carcinogenesis* 28, 1877–1884.
169. Palmer, H. G. *et al.* (2004). The transcription factor SNAIL represses vitamin D receptor expression and responsiveness in human colon cancer. *Nat Med* 10, 917–919.
170. Miller, C. W., Morosetti, R., Campbell, M. J., Mendoza, S., and Koeffler, H. P. (1997). Integrity of the 1,25-dihydroxyvitamin D3 receptor in bone, lung, and other cancers. *Mol Carcinog* 19, 254–257.
171. Guy, M., Lowe, L. C., Bretherton-Watt, D., Mansi, J. L., and Colston, K. W. (2003). Approaches to evaluating the association of vitamin D receptor gene polymorphisms with breast cancer risk. *Recent Results Cancer Res* 164, 43–54.
172. Ingles, S. A. *et al.* (1998). Association of prostate cancer with vitamin D receptor haplotypes in African-Americans. *Cancer Res* 58, 1620–1623.
173. Ma, J. et al. (1998). Vitamin D receptor polymorphisms, circulating vitamin D metabolites, and risk of prostate cancer in United States physicians. *Cancer Epidemiol Biomarkers Prev* 7, 385–390.
174. Gorlach, M., Burd, C. G., and Dreyfuss, G. (1994). The mRNA poly(A)-binding protein: Localization, abundance, and RNA-binding specificity. *Exp Cell Res* 211, 400–407.
175. Kim, J. G. *et al.* (2003). Association between vitamin D receptor gene haplotypes and bone mass in postmenopausal Korean women. *Am J Obstet Gynecol* 189, 1234–1240.
176. Kuraishi, T., Sun, Y., Aoki, F., Imakawa, K., and Sakai, S. (2000). The poly(A) tail length of casein mRNA in the lactating mammary gland changes depending upon the accumulation and removal of milk. *Biochem J* 347, 579–583.
177. Schondorf, T. *et al.* (2003). Association of the vitamin D receptor genotype with bone metastases in breast cancer patients. *Oncology* 64, 154–159.
178. Lundin, A. C., Soderkvist, P., Eriksson, B., Bergman-Jungestrom, M., and Wingren, S. (1999). Association of breast cancer progression with a vitamin D receptor gene polymorphism. South-East Sweden Breast Cancer Group. *Cancer Res* 59, 2332–2334.
179. Guy, M. *et al.* (2004). Vitamin d receptor gene polymorphisms and breast cancer risk. *Clin Cancer Res* 10, 5472–5481.
180. Ntais, C., Polycarpou, A., and Ioannidis, J. P. (2003). Vitamin D receptor gene polymorphisms and risk of prostate cancer: A meta-analysis. *Cancer Epidemiol Biomarkers Prev* 12, 1395–1402.
181. Rashid, S. F. et al. (2001). Synergistic growth inhibition of prostate cancer cells by 1 alpha,25 Dihydroxyvitamin D(3) and its 19-nor-hexafluoride analogs in combination with either sodium butyrate or trichostatin A. *Oncogene* 20, 1860–1872.
182. Campbell, M. J., Gombart, A. F., Kwok, S. H., Park, S., and Koeffler, H. P. (2000). The anti-proliferative effects of 1alpha,25(OH)2D3 on breast and prostate cancer cells are associated with induction of BRCA1 gene expression. *Oncogene* 19, 5091–5097.
183. Miller, G. J., Stapleton, G. E., Hedlund, T. E., and Moffat, K. A. (1995). Vitamin D receptor expression, 24-hydroxylase activity, and inhibition of growth by 1alpha,25-dihydroxyvitamin D3 in seven human prostatic carcinoma cell lines. *Clin Cancer Res* 1, 997–1003.

184. Rashid, S. F., Mountford, J. C., Gombart, A. F., and Campbell, M. J. (2001). 1alpha,25-dihydroxyvitamin D(3) displays divergent growth effects in both normal and malignant cells. *Steroids* 66, 433–440.
185. Khanim, F. L. *et al.* (2004). Altered SMRT levels disrupt vitamin D(3) receptor signalling in prostate cancer cells. *Oncogene* 23, 6712–6725.
186. Abedin, S. A. *et al.* (2009). Elevated NCOR1 disrupts a network of dietary-sensing nuclear receptors in bladder cancer cells. *Carcinogenesis* 30, 449–456.
187. Banwell, C. M., O'Neill, L. P., Uskokovic, M. R., and Campbell, M. J. (2004). Targeting 1alpha,25-dihydroxyvitamin D3 antiproliferative insensitivity in breast cancer cells by co-treatment with histone deacetylation inhibitors. *J Steroid Biochem Mol Biol* 89–90, 245–249.
188. Tavera-Mendoza, L. E. *et al.* (2008). Incorporation of histone deacetylase inhibition into the structure of a nuclear receptor agonist. *Proc Natl Acad Sci USA* 105, 8250–8255.
189. Costa, E. M. and Feldman, D. (1987). Modulation of 1,25-dihydroxyvitamin D3 receptor binding and action by sodium butyrate in cultured pig kidney cells (LLC-PK1). *J Bone Miner Res* 2, 151–159.
190. Gaschott, T. and Stein, J. (2003). Short-chain fatty acids and colon cancer cells: The vitamin D receptor–butyrate connection. *Recent Results Cancer Res* 164, 247–257.
191. Daniel, C., Schroder, O., Zahn, N., Gaschott, T., and Stein, J. (2004). p38 MAPK signaling pathway is involved in butyrate-induced vitamin D receptor expression. *Biochem Biophys Res Commun* 324, 1220–1226.
192. Chen, J. S., Faller, D. V., and Spanjaard, R. A. (2003). Short-chain fatty acid inhibitors of histone deacetylases: Promising anticancer therapeutics?. *Curr Cancer Drug Targets* 3, 219–236.
193. Gaschott, T., Werz, O., Steinmeyer, A., Steinhilber, D., and Stein, J. (2001). Butyrate-induced differentiation of Caco-2 cells is mediated by vitamin D receptor. *Biochem Biophys Res Commun* 288, 690–696.
194. Tanaka, Y., Bush, K. K., Klauck, T. M., and Higgins, P. J. (1989). Enhancement of butyrate-induced differentiation of HT-29 human colon carcinoma cells by 1,25-dihydroxyvitamin D3. *Biochem Pharmacol* 38, 3859–3865.
195. Westerhoff, H. V. and Palsson, B. O. (2004). The evolution of molecular biology into systems biology. *Nat Biotechnol* 22, 1249–1252.
196. Muller, M. and Kersten, S. (2003). Nutrigenomics: Goals and strategies. *Nat Rev Genet* 4, 315–322.
197. Evans, R. M. (2005). The nuclear receptor superfamily: A rosetta stone for physiology. *Mol Endocrinol* 19, 1429–1438.
198. Carlberg, C. and Dunlop, T. W. (2006). An integrated biological approach to nuclear receptor signaling in physiological control and disease. *Crit Rev Eukaryot Gene Expr* 16, 1–22.

CHAPTER 9

RETINOIC ACID RECEPTORS

AUDREY CRAS[1], FABIEN GUIDEZ[2], AND CHRISTINE CHOMIENNE[1]

[1] *Inserm UMR-S-718, Hôpital Saint Louis, Institut Universitaire d'Hématologie, Avenue Claude Vellefaux, 75010, Paris, France; Université Paris- Diderot UMR-S-718, 75010, Paris, France*

[2] *Department of Medical and Molecular Genetics, King's College London, Guy's Hospital, London SE1 9RT, UK*

Abstract: Retinoids, a group of structural and functional derivatives of vitamin A are known to regulate a large number of essential biological processes such as cell growth, differentiation and death. The retinoic acid (RA) signalling pathway involves the precise regulation of retinoid levels and the control of RA-dependent gene expression in target cells. The effects are mainly mediated by two types of nuclear receptors – retinoic acid receptors and retinoid X receptors – which act as ligand-dependent transcription factors. Different alterations of retinoid receptors or in the RA signalling pathway have been found to be associated with tumorigenesis. The success of retinoid-based differentiation therapy in acute promyelocytic leukemia largely contributes to the understanding of the molecular and pharmacological actions of retinoids. The development of selective receptor retinoids offers a great promising class of compounds for cancer therapy and prevention.

9.1. RETINOID RECEPTORS

9.1.1. Structure of Retinoic Receptors

The retinoid signal is transduced by two families of nuclear receptors, the family of retinoic acid receptors (RARs) and that of the retinoid X receptors (RXRs). Each family consists of three subtypes: RARα (NR1B1), RARβ (NR1B2), RARγ (NR1B3) and RXRα (NR2B1), RXRβ (NR2B2), RXRγ (NR2B3) [1]. These subtypes are encoded by separate genes (localised to chromosome 17q21, 3p24, 12q13 for *RARA*, *RARB*, *RARG* and to 9q34.3, 6q21.3, 1q22–q23 for *RXRA*, *RXRB*, *RXRG*, respectively) giving rise to multiple isoforms after alternative splicing and promoter usage [2]. There are two major isoforms for RARα (α1 and α2) and RARγ (γ1 and γ2) and four major isoforms for RARβ (β1 and β3 initiated at the P1 promoter and β2 and β4 initiated at the P2 promoter). There are two major isoforms each for RXRα (α1 and α2), RXRβ (β1 and β2), and RXRγ (γ1 and γ2). Retinoid receptors work as RXR/RAR heterodimers or RXR/RXR homodimers [3]. RXR were

C.M. Bunce, M.J. Campbell (eds.), *Nuclear Receptors*, Proteins and Cell Regulation 8, 237–258. DOI 10.1007/978-90-481-3303-1_9,

also shown to form heterodimers with other members of the nuclear receptors family [4] among which the thyroid receptors (TRs), vitamin D receptor (VDR), peroxisome proliferator-activated receptors (PPARs), and more recently identified liver X receptors (LXRs), farnesoid X receptor (FXR), pregnane X receptor (PXR), and constitutively activated receptors (CARs) (Figure 9.1). RXR heterodimers can be functionally classified into distinct permissive or nonpermissive RXR subordination groups [5]. When associated with permissive partners (PPARs, LXRs, FXR, and PXR), heterodimers can be activated by either the RXR agonist or the agonist of the partner receptor. When both RXR and the partner receptor are bound to their respective agonist a synergic activation is observed. In contrast, heterodimers formed by RXR and a nonpermissive partner (RARs, TRs, and VDR) cannot be activated by an RXR agonist but only by the agonist of the partner receptor.

Retinoid receptors have a modular structure common to all nuclear receptors composed of six conserved regions designated A–F [4] (Figure 9.1). The highly conserved C region harbors the DNA-binding domain (DBD) which consists of two cysteine-rich zinc-finger motifs and confers sequence specific DNA recognition.

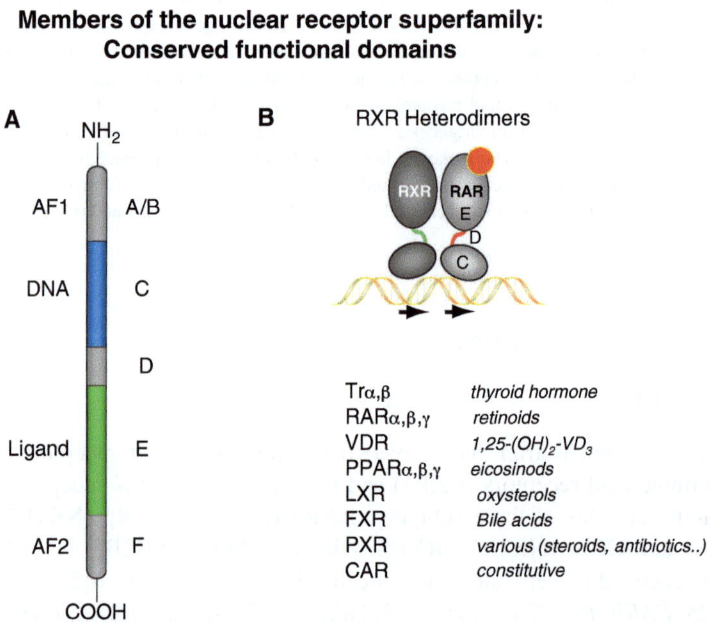

Figure 9.1. Members of the nuclear receptor superfamily: conserved functional domains. (**a**) Schematic representation of a typical nuclear receptor (denoted A–F regions). Regions A/B and F contain activation function (AF) and regions C and D contain the conserved DNA binding domain (DNA) and Ligand binding domain (ligand) respectively, see text for details. (**b**) Schematic representation of RXR/RAR heterodimer bound on DNA and a list of the nuclear receptor superfamily members functioning as RXR heterodimers with their respective ligands

Region E is the second most conserved region and corresponds to the ligand binding domain (LBD). This multifunctional complex contains the ligand-binding pocket, the main dimerization domain and the ligand-dependent transactivation function (AF-2). The LBD is formed by 12 conserved α-helices and a β-turn which are folded into a three-layered, antiparallel helical sandwich with H4, H5, H8, H9 and H11 sandwiched between H1, H2 and H3 on one side and H6, H7 and H10 on the other. In this structure, the C-terminal helix H12, which encompasses the AF-2 activation domain, points away from the LBD core. In this structure, H9 and H10 constitute the core of the dimerization interface, whereas hydrophobic residues mainly from H3, H5, H7 and the β-sheet contribute to the ligand-binding pocket. Importantly, the LBD contains a conserved serine residue which is phosphorylated upon PKA signalling activation. The N-terminal A/B region exhibits a ligand-independent transcriptional activation function (AF-1) which contains also several consensus phosphorylation sites for proline-dependent kinases which include cyclin-dependent kinases (CDKs) and MAP kinases [6]. The D region harbors nuclear localization signals and serves as a hinge between the DBD and the LBD, allowing rotation of the DBD. A F region is present in RARs, but absent in RXRs.

9.1.2. Transcription of Retinoid-Target Genes

In the absence of ligand, retinoid receptors bind as oriented RXR/RAR heterodimers to specific DNA sequences, named RA response elements (RAREs), found in the promoters of a large number of retinoid-target genes (Figure 9.2). The P2 promoters of RAR are induced by retinoids owing to the presence of a RARE. RAREs are typically composed of two direct repeats of a core hexameric motif, PuG(G/T)TCA [2]. The classical RARE is a 5-bp-spaced direct repeat (referred to as DR5). However, RAR/RXR heterodimers also bind to direct repeats separated by 1 bp (DR1) or 2 bp (DR2). RXRs also bind to DR1 as RXR/RXR homodimers. On DR2 and DR5 elements, RXR occupies the 5′hexameric motif, whereas the RAR partner occupies the 3′motif [7] (Figure 9.2). In contrast, on DR1 elements, the polarity is reversed (5′RAR/RXR-3′), switching the activity of the heterodimer from an activator to a repressor of transcription [8, 9].

In the classical model of gene regulation by retinoids, ligand-free RXR/RAR heterodimer bound to RARE repress transcription through the recruitment of the nuclear receptor corepressor protein (NCoR) and the silencing mediator of retinoid and thyroid hormone receptors (SMRT) [10]. Corepressors recruit complexes with histone deacetylase activity (HDACs) which increase the interaction of the N-terminal histone tails with the nucleosomal DNA thus compacting the chromatin (Figure 9.2). The ligand binding induces conformational changes in the LBD involving helix H12 which makes direct contacts with the ligand and seals the lid of the ligand-binding pocket, further stabilizing ligand binding. They also cause corepressor release and create a new hydrophobic cleft formed between H3, H4 and H12 which constitutes a surface where coactivators can bind. Two families of coactivators have

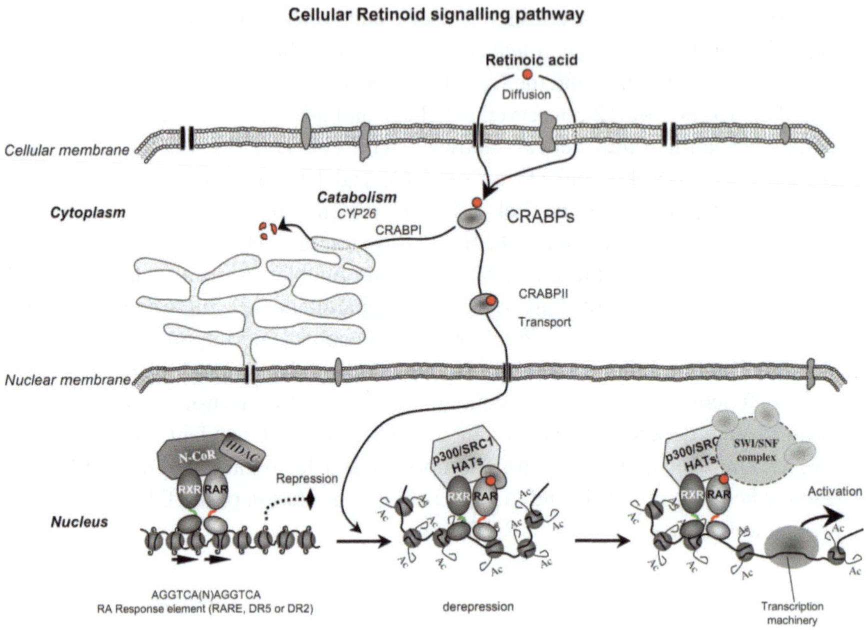

Figure 9.2. Cellular retinoid signalling pathway. Retinoic acid enters the cell via diffusion through the cellular membrane and binds to the cellular retinoic acid-binding proteins (CRABPs). CRABPI protein is involved in the CYP26-associated catabolism of RA while CRABPII protein acts as a coactivator of nuclear receptor. After binding RA, CRABPII protein translocates to the nucleus and docks to the retinoic acid receptor (RAR). In the absence of ligand, RAR heterodimerizes with RXR and acts as a transcriptional repressor by recruiting nuclear receptor corepressors, including N-CoR and histone deacetylase (HDAC). The presence of RA induces an allosteric change in the receptor leading to the release of the co-repressor complex and the recruitment of the co-activator complex (including p300 and SRC1) inducing acetylation of histone through their histone acetyltransferase activity (HAT, and Ac represents the acetyl group added to an histone tail). Locally modified chromatin structure then permits the recruitment of chromatin remodeler (SWI/SNF) allowing the association of the transcription machinery to the promoter region thus leading to the activation of transcription of genes required for cellular differentiation and growth inhibition

been identified: the SRC/p160 family which includes SRC-1/NCoA-1, TIF-2/GRIP-1/NCoA-2/SRC-2 and pCIP/ACTR/AIB1/TRAM1/RAC3/SRC-3, and the p300/CBP and CARM-1 family (Figure 9.2). We showed that cellular retinoic acid-binding protein II (CRABPII) also acts as a coactivator of nuclear retinoid receptors [11]. In contrast to most coactivator interactions, CRABPII does not interact with the AF-2 domain of RAR and RXR. This feature is correlated with the absence of an LXXLL motif in CRABPII and therefore stresses that CRABPII defines another level of regulation of nuclear receptor activity. We also characterized two domains of CRABPII (NRID1 and NRID2) that directly interact with RAR and RXR. These domains contain key structures of the ligand pocket entrance of CRABPII but not the LBD itself [12]. After binding RA, CRABPII translocates into the nucleus and docks to the apo-receptors bound to their promoters (Figure 9.2). This establishes a channel that allows the release of RA from holo-CRABPII to apo-receptors. Cyclin D3 forms a ternary complex and allows the stability of the CRABPII-RAR-RA interaction [13]. Because the CRABPII-RAR interaction does not require the presence of RA, apo-CRABPII could remain bound to the holo-receptor, preventing dissociation of RA from the nuclear receptor. Thus, CRABPII and cyclin D3 would increase the stability of the DNA-bound RXR/RAR complex, further contributing to the enhancement of RA-mediated transcription.

Coactivators locally modify chromatin structure through their histone acetyltransferase (HAT) or their histone methyltransferase (HMT) activity. Altogether, these histone modifications create binding sites that form an «histone code» read by a specialized domain present in the chromatin modifiers [14]. This code would allow the recruitment of ATP-dependent chromatin remodelers (SWI/SNF) allowing the formation of nucleosome-free or nucleosome-spaced regions [15]. Once repressive chromatin has been decondensed, the retinoid receptors become able to recruit the transcription machinery via their association with the SMCC (Srb and Mediator protein containing complex) [16]. After these last steps, the mediator facilitates recruitment of the transcriptional machinery to the promoter through its interaction with the RNA Pol II and GTFs (general transcription factors) (Figure 9.2).

9.1.3. Regulation of RXR/RAR-Mediated Transcription

RXR/RAR-mediated transcription regulation, as well as others nuclear receptors and transcription factors, involves phosphorylation processes by a cross-talk with several signalling pathways through the activation of kinase cascades [6]. Indeed, upon activation of PKA signalling, the LBD of RARα is phosphorylated at a conserved serine residue [17]. The RXRα LBD can also be phosphorylated by MAPKs [18]. It has been shown that the AF-1 domain of RARγ is a target for phosphorylation by the kinase CDK7, the activity of which depends on its association with cyclin H and MAT1 to form the ternary cyclin-dependent kinase (CDK)-activating kinase (CAK) complex of the GTF TFIIH [19]. It has been equally demonstrated that phosphorylation of the AF-1 domain of the RARβ isotype induces the dissociation of vinexin β, a repressor of RARβ-mediated transcription [20]. Finally, the activity of RARs is negatively modulated in response to stress upon phosphorylation of the LBD of their RXR heterodimerization partner [21].

In response to RA, RARs, their coactivators, corepressors and components of the GTF machinery are degraded by the ubiquitin-proteasome pathway [22]. This degradation might provide an efficient way for regulating the cyclic interaction of RARs with the promoter. Components of the ubiquitin-proteasome pathway may function as modulators of transcriptional activity through their capacity in regulating the transcriptional complex. For example, SUG-1, which is one of the six ATPases of the 19S regulatory complex of the proteasome, interacts with nuclear receptors including RARs, RXRs, TRs, VDR and estrogen receptors, with the general transcription factor TFIIH, and RNA Pol II and is recruited to transcriptionally active genes [23, 24].

9.2. ORIGIN AND ROLES OF RETINOIDS

9.2.1. Origin, Synthesis and Metabolism of Retinoids

There are three main natural retinoids, all derived from retinol (vitamin A): all-*trans* (ATRA), 9-*cis* (9cRA) and 13-*cis* retinoic acid (13cRA). Retinol is present in animal products, including milk products, eggs and fish oils, and can also be obtain by processing of the provitamin β-carotene present in various vegetables. After esterification to retinyl-ester in the gut, vitamin A is stored in the liver. Synthesis of RA from retinol is a two step process starting by cleavage of retinyl ester. Oxidation of vitamin A to all-*trans* retinaldehyde is achieved by alcohol deshydrogenase then followed by a second oxidation by retinaldehyde deshydrogenase leading to ATRA [25]. This last step is the limiting stage of ATRA biosynthesis. ATRA which is the most biologically active form is metabolized to hydroxylated metabolites by the cytochrome P450 (CYP26). In the blood, retinoids are associated with retinol-binding protein and transthyretin. In cells, retinoids are associated as non-covalent complexes with cellular retinol-binding (CRBP and CRBPII) and retinoic acid-binding proteins (CRABPI and CRABPII). These latter proteins appear to play a central role in biological activities of retinoids, by modulating their degradation and their nuclear biodisponibility (Figure 9.2) [26, 27]. 9cRA an isomerisation product of ATRA can bind both RARs and RXRs whereas ATRA can bind only RARs.

9.2.2. Natural and Synthetic Retinoids

Natural retinoids, are largely used as therapeutic agents for the treatment and prevention of cancer and hyperproliferative diseases [28]. ATRA is a central therapeutic agent in acute promyelocytic leukaemia (APL) (see below) and 13cRA is currently been tested in clinical trials alone or in association with interferon-α or chemotherapy in solid tumors. However, teratogenicity, severe headaches and mucocutaneous toxicity associated with RA use is a limiting factor to their wide use. Moreover, intrinsic or acquired retinoid resistance has limited the clinical activity of ATRA. In this context, researches were directed to synthesise selective RAR or RXR agonists and antagonists, to improve their efficiency while limiting side effects [29].

Retinoid structure is composed of three regions: the hydrophobic region, the polar region -usually a carboxylic acid to establish an ionic bridge with the LBD of receptor- and the central polyene linker. Structural modifications in the hydrophobic and central regions, most notably the replacement of the polyene chain with aromatic groups, lead to an increased chemical stability and prevent metabolic oxidative degradation. These derivatives, named arotinoids, present a great binding capacity to the receptor as shown by the high activity of the pan-agonist 4-[E-2-(5,6,7,8-Tetrahydro-5,5,8,8-tetramethyl-2-naphthalenyl)-1propenyl] 4 benzoic acid (TTNPB) [30]. A detailed analysis of the LBD of RARα, β and γ showed that the three isotypes differ by only three amino acids, which are sufficient to confer the selectivity of the RAR-specific compounds. Aromatic retinoids containing an amide in the linker region, such as AM580, AM80, AGN193836 and BMS-194753, were found to be RARα selective, whereas retinoids with a larger hydrophobic region, such as BMS185411 and AGN193639, have been reported to be RARβ specific agonists. Rexinoids, ligands for RXRs, are becoming increasingly appreciated not simply as silent heterodimerization partners of other NRs, but also as active transducers of tumour-suppressive signals independant of retinoids. In contrast to RAR, the amino acid sequences of the RXRα, β and γ LBD are identical, thus complicating the search for RXR isotype-specific ligands. Bexarotene or LG1069 (Targretin®) has been approved for the treatment of cutaneous T-cell lymphoma [31] and is being tested in clinical trials for the treatment of breast and lung cancer [32, 33, 34, 35]. AGN194204 and LG100268 have been reported to present some anticancer activity in breast and pancreatic cancer cells but their therapeutic potential in clinical trials has yet not been established [36, 37].

Other approaches include the use of atypical retinoids, also named retinoid-related molecules (RRMs) that have retinoid receptor-independent properties. The two main compounds in this class are 4-hydroxyphenylretinamide (4-HPR; fenretinide) and 6-[3-(1-adamantyl)- 4-hydroxyphenyl]-2-naphthalene carboxylic acid (AHPN; CD437). These synthetic compounds bind to and transactivate RARs mainly RARγ and, more weakly, RARβ but this activity does not explain all their growth-inhibitory and apoptogenic effects. Indeed they are active in retinoid-resistant cells and retinoid antagonists do not completely block their activity. 4-HPR is currently tested in numerous solid tumor phase I trials and demonstrated a protective effect in preventing second breast cancer [38]. To overcome the limitation associated with exogenous ATRA therapy, RA metabolism-blocking agents (RAMBA) such as liarozole have been developed [39]. This strategy consists to modulate and/or increase the levels of endogenous ATRA by inhibiting the cytochrome P450-dependent ATRA-4-hydroxylase enzymes responsible for ATRA metabolism.

9.2.3. Roles of Retinoids and Their Receptors

Both experimental and clinical studies have revealed that retinoids regulate a wide variety of essential biological processes, such as vertebrate embryonic morphogenesis and organogenesis [40], cell growth arrest, differentiation and apoptosis [41]. In adults, retinoids and their receptors are required for the proper functioning of a number of organs including the skin, lung, liver and neuronal and immune systems. In situ hybridization revealed that whereas RARα is present in most tissues, both RARβ and RARγ expressions are more selective [42]. The expression pattern of the RXR subtypes is rather different: RXRα and RXRβ are expressed widely and can be found in almost every tissue whereas RXRγ expression is mainly restricted to the muscle and brain [43]. These differences in tissue distribution suggest that retinoid receptors have distinct physiological functions. Interestingly large similarities were noted between RAR and thyroid hormone receptor (TR) whereas no such observation were made with RXR, suggesting involvement of each type of receptor in distinct regulation pathways, and thus a double retinoid signalling pathway [44].

Low levels of vitamin A causes severe malformation, whereas high concentrations of retinoids are teratogenic, supporting the idea that RA plays a central role during early embryogenesis. Gene-ablation experiments have shown that nearly all the congenital malformations and increased susceptibility to infections that are caused by vitamin A deficiency (VAD) are due to the absence of RAR or RXR functions. Pierre Chambon's group developed different models of deficient mice involving RAR and RXR subtypes/isoforms aiming to precise their own implication in physiological functions during homeostasis [45]. Mice deficient for one type of RAR are viable, but present different types of foetal and postnatal patterns observed in the VAD syndrome. On the other hand, mice lacking more than one subtypes/isoforms of RAR are associated with a dramatic decrease in viability and with all the known manifestations of the VAD syndrome in surviving mice. Inactivation of the RXRα gene has more severe consequences than that of RXRβ or RXRγ. The loss of RXRα is lethal during fetal life due to hypoplasia and failure of the myocardium. The ablation of RXRβ led to defects of spermatozoa and RXRγ-null mutants have an increased metabolic rate, thyroxine and thyroid-stimulating hormone [45]. Because of their partnering with many others NRs, RXRs play unique modulatory and integratives roles across multiple metabolic systems in adult tissues either during homeostasis or during pathological processes such as diabetes, obesity, inflammatory diseases and atherosclerosis [46].

9.3. RETINOID RECEPTOR ALTERATIONS IN HUMAN CANCERS

9.3.1. RARα: Molecular Genetics of APL

In most cases, the origin of APL is a t(15;17)(22; q11.2–12) chromosomal translocation that fuses the promyelocytic leukaemia gene *PML* and the *RARA* gene [47]. Within the *RARA* gene, breakpoints map to genomic sequences lying upstream of exon 4 which encodes the B-region of the RAR and N-terminal part of the DBD. Consequently, the *PML-RARA* fusion gene product always contains the same regions of RARα (B–F). In myeloid cell lines responsive to RA, PML-RARα blocks RA-mediated transactivation and totally abrogates the RA mediated granulocytic differentiation [48]. The fact that in presence of high dose of ATRA, PML-RARα can activate RA-inducible reporter genes is in agreement with the structural conservation of the LBD in the PML-RARα protein and a lower binding affinity of ATRA to endogenous RA-receptors in APL NB4 cell line compared to HL60 cell line as reference [49, 50, 51]. PML-RARα binds to RAREs in the promoter region of RA target genes and recruits corepressor proteins (Sin3A/NcoR/HDAC) on both PML and RARα moieties. PML-RARα has a higher affinity than wildtype RARα for HDACs, so physiological doses of ATRA can not dissociate the HDAC-containing corepressor complex [52]. In addition, PML-RARα is able to recruit DNA methyl transferase (DNMT1 and DNMT3a) leading to the hypermethylation of the RA downstream

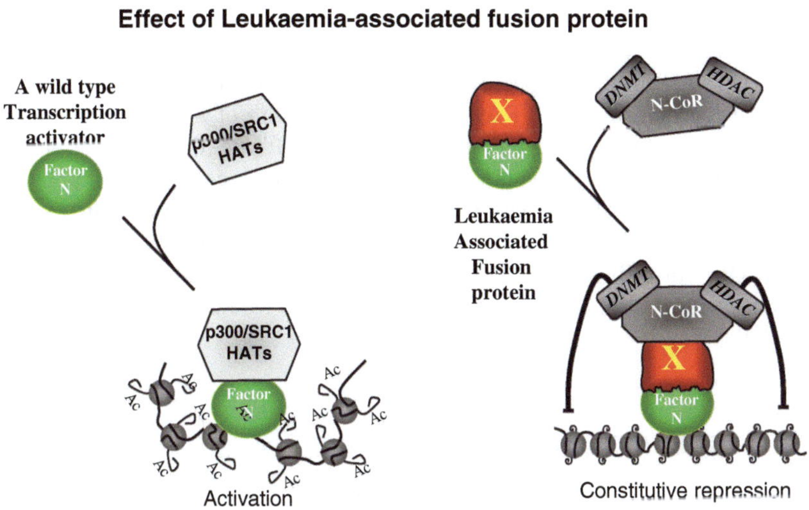

Figure 9.3. Effect of leukaemia-associated fusion protein. Wild type transcription factors (Factor N, for example RAR, AML1) act as transcriptional activators by recruiting coactivator complex (including p300 and SRC1) inducing acetylation (Ac) of the local chromatin (*left*); while leukaemia associated proteins (X being the fusion partner, for example PML-RAR or AML1-ETO) act as repressors by recruiting corepressor complex (including HDAC and/or DNA methyltransferase (DNMT), see text for details)

gene promoter enhancing the transcriptional repression [53] (Figure 9.3). The coiled-coil region of PML is responsible for oligomerization of the fusion protein resulting in enhanced repression [54]. Unlike the RAR-RXR heterodimer, PML-RARα can also form homodimers that recruit two corepressor molecules [55]. As PML-RARα binds to RAREs, they act as dominant silencing transcription factors that repress transcription activation mediated by the RXR-RAR heterodimer still produced from the intact *RARA* allele. The consequence is a block of differentiation at the promyelocytic stage during myelopoiesis. In addition, a specific sumoylation site in PML is absolutely required for APL transformation and allows the recruitment of a transcriptionnal repressor, Daxx [56]. Recently, it has been shown that RXR is an essential component of the oncogenic PML-RARα complex [57]. Indeed PML-RARα enhances posttranslational modifications of RXRα, including its sumoylation, suggesting that PML-bound sumoylation enzymes target RXRα and possibly other PML-RARα-bound chromatin proteins, further contributing to deregulated transcription. Finally, the altered functionality of PML in the fusion protein, such as loss of its pro-apoptotic activity, or abrogation of corepressor function, adds to the growth potential or survival capacity of APL blasts [58].

In addition to the PML-RARα rearrangement, four other APL associated translocations of the *RARA* gene have been characterized to date at the molecular level. The t(5;17)(q35;q21), t(11;17)(q23;q21), t(11;17)(q13;q21) and t(17;17)(q11;q21) fuse

RARA to the nucleophosmin (*NPM*), promyelocytic leukemia zinc finger (*PLZF*), nuclear mitotic apparatus (*NUMA*) and signal transducer and activator of transcription 5B (*STAT5B*) genes, respectively [59]. As in the case of t(15;17) translocation, the above rearrangements of RARα lead to expression of a fusion protein which acts as a dominant negative inhibitor of wild-type RARα through aberrant interaction with corepressor molecules. APL fusion proteins could form homodimers, thus exposing the RARα moiety in such a way as to offer extra binding sites for corepressors. In addition, the PLZF-RARα fusion protein behaves distinctly from PML-RARα since PLZF itself recruits corepressors with high affinity prohibiting their release at pharmacological doses of ATRA and leading to a resistant form of APL [60].

9.3.2. RARβ a Tumour Suppressor Gene

When compared to adjacent normal tissues, RARβ levels have been demonstrated to be reduced or absent in a variety of solid tumours including non-small cell lung cancer, squamous cell carcinomas of the head and neck, and breast cancer [61–64]. The suggestion that RARβ acts as a tumor suppressor is also consistent with the fact that the reactivation of silenced RARβ expression upon RA treatment leads to growth inhibition and apoptosis [65, 66]. The expression of RARβ is selectively lost in premalignant oral lesions, and restoration of the RARβ expression by 13cRA is associated with a clinical response [67]. Although the RARβ2 and RARβ4 isoforms are transcribed from the same RA-responsive promoter P2, there is evidence that RARβ2 functions as a tumor suppressor gene whereas RARβ4 acts as a dominant negative receptor [68, 69]. A number of reports suggest that in other cell contexts tumor growth and proliferation is not under RARβ expression, but depends on other retinoid receptors for retinoid response. For example, RARγ expression was strongly correlated with retinoid-induced growth inhibition of squamous cell carcinomas from the head and neck [70]. Interestingly, it has been reported in retinoid-resistant ovarian tumor cells that overexpression of any RAR subtype could restore retinoid-induced growth inhibition [71].

Transfection of tumor cells with a RARE-dependent reporter gene demonstrates an induction in transcriptional activity even in RA-resistant cells suggesting a functionnal RA signalling pathway. Thus lack of RA efficacy and restoration of RARβ expression may be due to abnormalities located at the endogenous *RARB* promoter region. Contrary to APL and RARα translocation, there is no evidence for genetic alteration in the *RARB* gene. Likewise it has not been possible to demonstrate a correlation beetwen lack of RARβ expression and the loss of heterozygosity on chromosome 3p24, the locus of *RARB* gene. Another possible mechanism that has been explored is the role of various other transcription factors in the regulation of RARβ expression, as it was shown for the two orphan receptors, nurr77 and COUP-TF [72, 73]. In human lung cancer cell lines, a high COUP-TF and nurr77 expression is correlated with RA sensitivity and resistance respectively. Moreover overexpression of COUP-TF results in the restoration of RARβ expression and growth inhibition. It has been suggested that COUP-TF binds to a DR-8 response element in the *RARB* P2 promoter in a RARα-dependent manner, leading to the recruitment of CBP to the *RARB* promoter by RARα. The activation of RXR-Nurr77 heterodimer by RXR agonist induces RARβ expression and apoptosis. The subcellular localization of RXRα has been linked to retinoid responsiveness; in retinoid sensitive cells, RXRα is localized in the nucleoplasm, while in resistant cells, it was located in the splicing factor compartment.

Epigenetic changes of the *RARB* promoter have also been demonstrated to contribute to the transcriptional silencing of *RARB*. Cote and Momparler were the first to suggest that DNA methylation is responsible for the lack of expression of RARβ in colon tumors [74]. Since many studies have confirmed an association between hypermethylation of the *RARB* P2 promoter and either loss of RARβ expression in a large variety of cancer cell lines and primary tumors. Treatment of tumor cells with the demethylating agent 5-aza-2′-deoxycytidine (5-Aza-dC) can restore either the expression of RARβ or the inducibility of RARβ expression by RA [75–78]. Finally, several reports have demonstrated the reactivation of the expression of RARβ upon treatment of tumor cells with an HDAC inhibitor either alone or in combination with demethylating agent. Aberrant histone H3 acetylation due to a DNA hypermethylation consistently correlates with RA resistance in lung cancer cells lines and loss of RARβ expression [79]. In thyroid cancer cells, we have shown that the *RARB* P2 promoter is in a non permissive status due to absence of histone H3 and H4 acetylation resulting in lack of RA transcriptional activity and differentiation which may be restored upon combination with an HDAC inhibitor [80]. Combination of RA and an HDAC inhibitor leads to a reduction of human renal cell carcinoma proliferation in a xenograft tumor model [81].

9.4. ANTIMOUR ACTIVITY OF RA

RA therapy in malignant cells is based on functional alternative RA signalling pathways that, upon pharmacological concentrations of a given retinoid, restore control of cell death, differentiation and proliferation. Various mechanisms are involved, including the transcriptional control of the tissue-specific RAR gene via endogenous receptors. A number of target genes have been identified that are either directly or indirectly regulated by RAR and RXR [82]. Microarray data revealed novel as well as previously recognized candidate retinoid target genes, and confirm the expected cell context differences [83–85, 76, 77, 86, 87].

9.4.1. RA and Differentiation Therapy

The ATRA treatment of APL constitutes the first model of molecular target-based induction of differentiation [88–90]. Optimization of the ATRA-based regimens combining ATRA and chemotherapy has further raised the complete remission rate up to 90–95%. Primary cultures of APL blasts reflect the APL sensitivity to different retinoids in relation to their structure-dose [91]. ATRA treatment is associated with RARα reexpression and differentiation of blasts [92]. A correlation between ATRA-induced differentiation, intracellular concentrations of ATRA and differential gene expression in APL cells was observed and allows to predict ATRA response both in vitro and in vivo [49, 50, 93, 94]. Differentiation of myeloid leukemic cell lines by retinoids is enhanced by the addition of certain cytokines. In APL, a strong correlation between the expression and secretion of tumor necrosis factor α (TNFα), interleukin-6 (IL-6), IL-8, and IL-1β and differentiation with ATRA was found [95, 96].

ATRA, at pharmacological doses (10^{-7} to 10^{-6} M), binds to the PML-RARα LBD, resulting in dissociation of the corepressor complex from the receptor. This relieves the HDAC-dependent block of differentiation and, through association of co-activator complexes, triggers the transcriptional programmes that are usually controlled by RXR-RARα heterodimers. Another effect of ATRA in modulating PML-RARα is to induce its degradation. It was reported that ATRA could trigger caspase-mediated cleavage of the PML-RARα [97] and induces proteasome-dependent degradation of RARα and PML-RARα. [98]. The use of arsenic trioxide (ATO) in combination with ATRA further improved the clinical outcome of refractory or relapsed as well as newly diagnosed APL. Under high concentrations ATO induces apoptosis, mainly through activating the mitochondria-mediated intrinsic apoptotic pathway, whereas low concentrations and a longer treatment course tends to promote differentiation of APL cells [99, 100]. In contrast to transcriptional remodeling of ATRA-induced differentiation, the effects of ATO reside mainly at the proteome level notably *via* sumoylation and subsequent proteasomal degradation of PML and PML-RARα, creating a molecular foundation for the synergistic effects between ATRA and ATO [101].

The important connection between the formation of RAR fusion proteins and the aberrant recruitment of HDAC and DNMT complexes to RAR target genes in APL underlines the importance of these enzyme systems for anticancer therapy. Association therapy including retinoids and epigenetic drugs such as 5-Aza-dC (DNMT inhibitor) and MS275, vorinostat or valproic acid (HDAC inhibitors) in APL treatment are supported by several observations. These effects are of particular interest in t(15;17) RA-resistant blasts, as it was shown in mice models [60] and clinical reports [102]. Many observations confirm the notion of an aberrant recruitment of HDACs on retinoid target genes in AML others than APL. AML2 subtype with the t(8;21) chromosomal rearrangement giving rise to AML1-ETO protein fusion which is not responsive to ATRA differentiation induction but to G-CSF [103] is one example where HDAC inhibition is associated with a recovery of retinoid-dependent transactivation and terminal differentiation [104] (Figure 9.3). In addition to HDAC inhibition, reversal of DNA hypermethylation by demethylating agents has been shown to restore ATRA-mediated differentiation [105]. Other AMLs patients have now been shown to achieve complete remission when ATRA is combined with epigenetic drugs [106–109].

9.4.2. RA and Death Signalling Pathway

In APL cells, ATRA has been shown to induce not only differentiation but also apoptosis leading to complete remission in patients. These two pathways are distinct and temporally orchestrated by the ability of ATRA to induce both proapoptotic and antiapoptotic programs. Apoptosis in this context is under the dependency of tumour-necrosis factor (TNF)-related apoptosis-inducing ligand (TRAIL; also called Apo2L) pathway, through an increase in expression of DR4 and DR5 which are TRAIL receptors 1 and 2, and increasing expression of TRAIL [110]. In addition, retinoids are able to induce and activate key components of interferon-signalling pathways such as interferon-regulatory factor (IRF) [111]. The recognition of IRF as a central convergent point between interferon and TRAIL pathways was shown to be the result of the presence of an IRF1 responsive element in the TRAIL promoter [112].

RXRs are unable to induce transcription in the absence of an RAR ligand, even in association with their own ligand rexinoid. Rexinoids however are not destitute of activity, as illustrated by their pro-apoptotic properties in APL cells [113]. Cross-talk between rexinoids and protein kinase A (PKA) signalling pathway can experimentally induce differentiation and apoptosis, and it is of particular interest that such observations were made even in ATRA resistant APL blasts. Activation of PKA by cyclic AMP leads to releasing of corepressive complex from heterodimers thus allowing rexinoid induced recruitment of the coactivator complex [114]. The apoptotic process triggered by atypical retinoids seems to proceed through TRAIL and a mitochondrial pathway. Indeed AHPN induces the translocation of Nurr77 from the nucleus to mitochondria leading to the cytochrome c release and apoptosis [115]. In human lung cancer it has been shown that AHPN-induced apoptosis via induction of Egr-1 and Nur77 which requires active ERK1/2 independently of p38 mitogen-activated protein kinase [116]. Involvement of TRAIL pathway by CD437 due to increased expression of DR4 is mediated by activation of NF-kappaB [117].

9.4.3. RA in Chemoprevention

Retinoids have been tested as chemopreventive agents through their antiproliferative differentiation-inducing and proapoptotic effect and could target multiple steps of carcinogenesis. Thus, blocking pre-neoplastic conversion might be more feasible than reversing cancer. Much evidence supports the idea that retinoids can prevent cancer by inhibiting progression from premalignant to malignant stages [118].

Classical retinoids had been shown to be effective for the treatment of preneoplasic diseases such as leukoplakia, actinic keratosis and cervical dysplasia and to delay the development of skin cancer in individuals with xeroderma pigmentosum. Incidence of basal cell carcinoma in mouse models can be reduced by tazarotene an RARβ and RARγ agonist [119]. Skin chemical carcinogenesis is a multistep process that involves initiation, promotion and progression. AP1, a nuclear factor involved in cell proliferation is known to cross talk with the RAR pathway. These interactions can lead to positive and negative transcriptional effects depending on promoter structure and cell-context. Retinoids can act either by inhibiting phorbol-ester-induced AP1 activity, preventing the promotion step of chemical skin carcinogenesis [120] or by reversing established tumor promotion [121].

RA chemopreventive effect is not limited to skin tumors. For examples, primary hepatocarcinomas seem preventable by atypical retinoids [122] and controlateral breast cancer in premenauposal women could be reduced by 4-HPR in combination with tamoxifen [38]. Bexarotene had proven useful for the prevention of the development of premalignant mammary lesions in experimental animal models. The authors found that several rexinoid genes involved in growth regulation were modulated in the mammary glands from bexarotene-treated mice [123]. In the same way bexarotene induces dosage-dependent repression of growth in responsive, but not resistant human bronchial epithelial cells. Bexarotene activity was linked to a G1 arrest due to cyclin D1 and epidermal growth factor receptor (EGFR) repression. Similar biomarker changes occur in patients with lung tumors treated by bexarotene suggesting that these genes can be used as biomarkers to demonstrate a biological effect on the target tissue [124].

9.5. CONCLUSION

Since initial studies in 1925 identifying the important role of vitamin A in maintaining the homeostasis of tissues, researches have been dedicated to elucidating the molecular and cellular networks that are induced by retinoids in order to exploit their anti-proliferative and differentiative properties for cancer therapy and prevention. Future development in drug discovery will continue to explore the efficacy of retinoids and rexinoids, either alone or in combination with other differentiation-inducing, cytotoxic or chromatin remodelling agents, as well as the use of receptor-selective and non classical retinoids.

REFERENCES

1. Nuclear Receptors Nomenclature Committee (1999). A unified nomenclature system for the nuclear receptor superfamily. *Cell* 97, 161–163.
2. Leid, M., Kastner, P., and Chambon, P. (1992). Multiplicity generates diversity in the retinoic acid signalling pathways. *Trends Biochem Sci* 17, 427–433.
3. Lehmann, J. M., Jong, L., Fanjul, A., Cameron, J. F., Lu, X. P., Haefner, P., Dawson, M. I., and Pfahl, M. (1992). Retinoids selective for retinoid X receptor response pathways. *Science* 258, 1944–1946.
4. Chambon, P. (2005). The nuclear receptor superfamily: A personal retrospect on the first two decades. *Mol Endocrinol* 19, 1418–1428.
5. Shulman, A. I., Larson, C., Mangelsdorf, D. J., and Ranganathan, R. (2004). Structural determinants of allosteric ligand activation in RXR heterodimers. *Cell* 116, 417–429.
6. Rochette-Egly, C. (2003). Nuclear receptors: Integration of multiple signalling pathways through phosphorylation. *Cell Signal* 15, 355–366.

7. Predki, P. F., Zamble, D., Sarkar, B., and Giguere, V. (1994). Ordered binding of retinoic acid and retinoid-X receptors to asymmetric response elements involves determinants adjacent to the DNA-binding domain. *Mol Endocrinol* 8, 31–39.
8. Kurokawa, R., Soderstrom, M., Horlein, A., Halachmi, S., Brown, M., Rosenfeld, M. G., and Glass, C. K. (1995). Polarity-specific activities of retinoic acid receptors determined by a co-repressor. *Nature* 377, 451–454.
9. Rastinejad, F., Wagner, T., Zhao, Q., and Khorasanizadeh, S. (2000). Structure of the RXR-RAR DNA-binding complex on the retinoic acid response element DR1. *EMBO J* 19, 1045–1054.
10. Dilworth, F. J. and Chambon, P. (2001). Nuclear receptors coordinate the activities of chromatin remodeling complexes and coactivators to facilitate initiation of transcription. *Oncogene* 20, 3047–3054.
11. Delva, L., Bastie, J. N., Rochette-Egly, C., Kraiba, R., Balitrand, N., Despouy, G., Chambon, P., and Chomienne, C. (1999). Physical and functional interactions between cellular retinoic acid binding protein II and the retinoic acid-dependent nuclear complex. *Mol Cell Biol* 19, 7158–7167.
12. Bastie, J. N., Despouy, G., Balitrand, N., Rochette-Egly, C., Chomienne, C., and Delva, L. (2001). The novel co-activator CRABPII binds to RARalpha and RXRalpha via two nuclear receptor interacting domains and does not require the AF-2 'core'. *FEBS Lett* 507, 67–73.
13. Despouy, G., Bastie, J. N., Deshaies, S., Balitrand, N., Mazharian, A., Rochette-Egly, C., Chomienne, C., and Delva, L. (2003). Cyclin D3 is a cofactor of retinoic acid receptors, modulating their activity in the presence of cellular retinoic acid-binding protein II. *J Biol Chem* 278, 6355–6362.
14. Strahl, B. D. and Allis, C. D. (2000). The language of covalent histone modifications. *Nature* 403, 41–45.
15. Gangaraju, V. K. and Bartholomew, B. (2007). Mechanisms of ATP dependent chromatin remodeling. *Mutat Res* 618, 3–17.
16. Lewis, B. A. and Reinberg, D. (2003). The mediator coactivator complex: Functional and physical roles in transcriptional regulation. *J Cell Sci* 116, 3667–3675.
17. Rochette-Egly, C., Oulad-Abdelghani, M., Staub, A., Pfister, V., Scheuer, I., Chambon, P., and Gaub, M. P. (1995). Phosphorylation of the retinoic acid receptor-alpha by protein kinase A. *Mol Endocrinol* 9, 860–871.
18. Adam-Stitah, S., Penna, L., Chambon, P., and Rochette-Egly, C. (1999). Hyperphosphorylation of the retinoid X receptor alpha by activated c-Jun NH2-terminal kinases. *J Biol Chem* 274, 18932–18941.
19. Bastien, J., Adam-Stitah, S., Riedl, T., Egly, J. M., Chambon, P., and Rochette-Egly, C. (2000). TFIIH interacts with the retinoic acid receptor gamma and phosphorylates its AF-1-activating domain through cdk7. *J Biol Chem* 275, 21896–21904.
20. Bour, G., Plassat, J. L., Bauer, A., Lalevee, S., and Rochette-Egly, C. (2005). Vinexin beta interacts with the non-phosphorylated AF-1 domain of retinoid receptor gamma (RARgamma) and represses RARgamma-mediated transcription. *J Biol Chem* 280, 17027–17037.
21. Bruck, N., Bastien, J., Bour, G., Tarrade, A., Plassat, J. L., Bauer, A., Adam-Stitah, S., and Rochette-Egly, C. (2005). Phosphorylation of the retinoid x receptor at the omega loop, modulates the expression of retinoic-acid-target genes with a promoter context specificity. *Cell Signal* 17, 1229–1239.
22. Bour, G., Lalevee, S., and Rochette-Egly, C. (2007). Protein kinases and the proteasome join in the combinatorial control of transcription by nuclear retinoic acid receptors. *Trends Cell Biol* 17, 302–309.
23. Gianni, M., Bauer, A., Garattini, E., Chambon, P., and Rochette-Egly, C. (2002). Phosphorylation by p38MAPK and recruitment of SUG-1 are required for RA-induced RAR gamma degradation and transactivation. *EMBO J* 21, 3760–3769.
24. vom Baur, E., Zechel, C., Heery, D., Heine, M. J., Garnier, J. M., Vivat, V., Le Douarin, B., Gronemeyer, H., Chambon, P., and Losson, R. (1996). Differential ligand-dependent interactions

between the AF-2 activating domain of nuclear receptors and the putative transcriptional intermediary factors mSUG1 and TIF1. *EMBO J* 15, 110–124.
25. Napoli, J. L. (1996). Retinoic acid biosynthesis and metabolism. *FASEB J* 10, 993–1001.
26. Guidez, F., Parks, S., Wong, H., Jovanovic, J. V., Mays, A., Gilkes, A. F., Mills, K. I., Guillemin, M. C., Hobbs, R. M., Pandolfi, P. P., de The, H., Solomon, E., and Grimwade, D. (2007). RARalpha-PLZF overcomes PLZF-mediated repression of CRABPI, contributing to retinoid resistance in t(11;17) acute promyelocytic leukemia. *Proc Natl Acad Sci U S A* 104, 18694–18699.
27. Won, J. Y., Nam, E. C., Yoo, S. J., Kwon, H. J., Um, S. J., Han, H. S., Kim, S. H., Byun, Y., and Kim, S. Y. (2004). The effect of cellular retinoic acid binding protein-I expression on the CYP26-mediated catabolism of all-trans retinoic acid and cell proliferation in head and neck squamous cell carcinoma. *Metabolism* 53, 1007–1012.
28. Altucci, L., Leibowitz, M. D., Ogilvie, K. M., de Lera, A. R., and Gronemeyer, H. (2007). RAR and RXR modulation in cancer and metabolic disease. *Nat Rev Drug Discov* 6, 793–810.
29. de Lera, A. R., Bourguet, W., Altucci, L., and Gronemeyer, H. (2007). Design of selective nuclear receptor modulators: RAR and RXR as a case study. *Nat Rev Drug Discov* 6, 811–820.
30. Dawson, M. I. and Zhang, X. K. (2002). Discovery and design of retinoic acid receptor and retinoid X receptor class- and subtype-selective synthetic analogs of all-trans-retinoic acid and 9-cis-retinoic acid. *Curr Med Chem* 9, 623–637.
31. Duvic, M., Hymes, K., Heald, P., Breneman, D., Martin, A. G., Myskowski, P., Crowley, C., and Yocum, R. C. (2001). Bexarotene is effective and safe for treatment of refractory advanced-stage cutaneous T-cell lymphoma: Multinational phase II-III trial results. *J Clin Oncol* 19, 2456–2471.
32. Edelman, M. J., Smith, R., Hausner, P., Doyle, L. A., Kalra, K., Kendall, J., Bedor, M., and Bisaccia, S. (2005). Phase II trial of the novel retinoid, bexarotene, and gemcitabine plus carboplatin in advanced non-small-cell lung cancer. *J Clin Oncol* 23, 5774–5778.
33. Esteva, F. J., Glaspy, J., Baidas, S., Laufman, L., Hutchins, L., Dickler, M., Tripathy, D., Cohen, R., DeMichele, A., Yocum, R. C., Osborne, C. K., Hayes, D. F., Hortobagyi, G. N., Winer, E., and Demetri, G. D. (2003). Multicenter phase II study of oral bexarotene for patients with metastatic breast cancer. *J Clin Oncol* 21, 999–1006.
34. Govindan, R., Crowley, J., Schwartzberg, L., Kennedy, P., Williams, C., Ekstrand, B., Sandler, A., Jaunakais, D., Bolejack, V., and Ghalie, R. (2006). Phase II trial of bexarotene capsules in patients with advanced non-small-cell lung cancer after failure of two or more previous therapies. *J Clin Oncol* 24, 4848–4854.
35. Khuri, F. R., Rigas, J. R., Figlin, R. A., Gralla, R. J., Shin, D. M., Munden, R., Fox, N., Huyghe, M. R., Kean, Y., Reich, S. D., and Hong, W. K. (2001). Multi-institutional phase I/II trial of oral bexarotene in combination with cisplatin and vinorelbine in previously untreated patients with advanced non-small-cell lung cancer. *J Clin Oncol* 19, 2626–2637.
36. Balasubramanian, S., Chandraratna, R. A., and Eckert, R. L. (2004). Suppression of human pancreatic cancer cell proliferation by AGN194204, an RXR-selective retinoid. *Carcinogenesis* 25, 1377–1385.
37. Crowe, D. L. and Chandraratna, R. A. (2004). A retinoid X receptor (RXR)-selective retinoid reveals that RXR-alpha is potentially a therapeutic target in breast cancer cell lines, and that it potentiates antiproliferative and apoptotic responses to peroxisome proliferator-activated receptor ligands. *Breast Cancer Res* 6, R546–R555.
38. Veronesi, U., Mariani, L., Decensi, A., Formelli, F., Camerini, T., Miceli, R., Di Mauro, M. G., Costa, A., Marubini, E., Sporn, M. B., and De Palo, G. (2006). Fifteen-year results of a randomized phase III trial of fenretinide to prevent second breast cancer. *Ann Oncol* 17, 1065–1071.
39. Njar, V. C., Gediya, L., Purushottamachar, P., Chopra, P., Vasaitis, T. S., Khandelwal, A., Mehta, J., Huynh, C., Belosay, A., and Patel, J. (2006). Retinoic acid metabolism blocking agents (RAMBAs) for treatment of cancer and dermatological diseases. *Bioorg Med Chem* 14, 4323–4340.
40. Niederreither, K., Subbarayan, V., Dolle, P., and Chambon, P. (1999). Embryonic retinoic acid synthesis is essential for early mouse post-implantation development. *Nat Genet* 21, 444–448.

41. Love, J. M. and Gudas, L. J. (1994). Vitamin A, differentiation and cancer. *Curr Opin Cell Biol* 6, 825–831.
42. Dolle, P., Ruberte, E., Leroy, P., Morriss-Kay, G., and Chambon, P. (1990). Retinoic acid receptors and cellular retinoid binding proteins. I. A systematic study of their differential pattern of transcription during mouse organogenesis. *Development* 110, 1133–1151.
43. Mangelsdorf, D. J., Borgmeyer, U., Heyman, R. A., Zhou, J. Y., Ong, E. S., Oro, A. E., Kakizuka, A., and Evans, R. M. (1992). Characterization of three RXR genes that mediate the action of 9-cis retinoic acid. *Genes Dev* 6, 329–344.
44. Heyman, R. A., Mangelsdorf, D. J., Dyck, J. A., Stein, R. B., Eichele, G., Evans, R. M., and Thaller, C. (1992). 9-cis retinoic acid is a high affinity ligand for the retinoid X receptor. *Cell* 68, 397–406.
45. Mark, M., Ghyselinck, N. B., and Chambon, P. (2006). Function of retinoid nuclear receptors: Lessons from genetic and pharmacological dissections of the retinoic acid signaling pathway during mouse embryogenesis. *Annu Rev Pharmacol Toxicol* 46, 451–480.
46. Szanto, A., Narkar, V., Shen, Q., Uray, I. P., Davies, P. J., and Nagy, L. (2004). Retinoid X receptors: X-ploring their (patho)physiological functions. *Cell Death Differ* 11(Suppl 2), S126–S143.
47. de The, H., Vivanco-Ruiz, M. M., Tiollais, P., Stunnenberg, H., and Dejean, A. (1990). Identification of a retinoic acid responsive element in the retinoic acid receptor beta gene. *Nature* 343, 177–180.
48. Rousselot, P., Hardas, B., Patel, A., Guidez, F., Gaken, J., Castaigne, S., Dejean, A., de The, H., Degos, L., Farzaneh, F. et al. (1994). The PML-RAR alpha gene product of the t(15;17) translocation inhibits retinoic acid-induced granulocytic differentiation and mediated transactivation in human myeloid cells. *Oncogene* 9, 545–551.
49. Agadir, A., Cornic, M., Jerome, M., Menot, M. L., Cambier, N., Gaub, M. P., Gourmel, B., Lefebvre, P., Degos, L., and Chomienne, C. (1995). Characterization of nuclear retinoic acid binding activity in sensitive leukemic cell lines: Cell specific uptake of ATRA and RAR alpha protein modulation. *Biochem Biophys Res Commun* 213, 112–122.
50. Agadir, A., Cornic, M., Lefebvre, P., Gourmel, B., Balitrand, N., Degos, L., and Chomienne, C. (1995). Differential uptake of all-trans retinoic acid by acute promyelocytic leukemic cells: Evidence for its role in retinoic acid efficacy. *Leukemia* 9, 139–145.
51. Agadir, A., Cornic, M., Lefebvre, P., Gourmel, B., Jerome, M., Degos, L., Fenaux, P., and Chomienne, C. (1995). All-trans retinoic acid pharmacokinetics and bioavailability in acute promyelocytic leukemia: Intracellular concentrations and biologic response relationship. *J Clin Oncol* 13, 2517–2523.
52. Grignani, F., De Matteis, S., Nervi, C., Tomassoni, L., Gelmetti, V., Cioce, M., Fanelli, M., Ruthardt, M., Ferrara, F. F., Zamir, I., Seiser, C., Grignani, F., Lazar, M. A., Minucci, S., and Pelicci, P. G. (1998). Fusion proteins of the retinoic acid receptor-alpha recruit histone deacetylase in promyelocytic leukaemia. *Nature* 391, 815–818.
53. Di Croce, L., Raker, V. A., Corsaro, M., Fazi, F., Fanelli, M., Faretta, M., Fuks, F., Lo Coco, F., Kouzarides, T., Nervi, C., Minucci, S., and Pelicci, P. G. (2002). Methyltransferase recruitment and DNA hypermethylation of target promoters by an oncogenic transcription factor. *Science* 295, 1079–1082.
54. Minucci, S., Maccarana, M., Cioce, M., De Luca, P., Gelmetti, V., Segalla, S., Di Croce, L., Giavara, S., Matteucci, C., Gobbi, A., Bianchini, A., Colombo, E., Schiavoni, I., Badaracco, G., Hu, X., Lazar, M. A., Landsberger, N., Nervi, C., and Pelicci, P. G. (2000). Oligomerization of RAR and AML1 transcription factors as a novel mechanism of oncogenic activation. *Mol Cell* 5, 811–820.
55. Lin, R. J. and Evans, R. M. (2000). Acquisition of oncogenic potential by RAR chimeras in acute promyelocytic leukemia through formation of homodimers. *Mol Cell* 5, 821–830.
56. Zhu, J., Zhou, J., Peres, L., Riaucoux, F., Honore, N., Kogan, S., and de The, H. (2005). A sumoylation site in PML/RARA is essential for leukemic transformation. *Cancer Cell* 7, 143–153.
57. Zhu, J., Nasr, R., Peres, L., Riaucoux-Lormiere, F., Honore, N., Berthier, C., Kamashev, D., Zhou, J., Vitoux, D., Lavau, C., and de The, H. (2007). RXR is an essential component of the oncogenic PML/RARA complex in vivo. *Cancer Cell* 12, 23–35.

58. Khan, M. M., Nomura, T., Kim, H., Kaul, S. C., Wadhwa, R., Shinagawa, T., Ichikawa-Iwata, E., Zhong, S., Pandolfi, P. P., and Ishii, S. (2001). Role of PML and PML-RARalpha in Mad-mediated transcriptional repression. *Mol Cell* 7, 1233–1243.
59. Zelent, A., Guidez, F., Melnick, A., Waxman, S., and Licht, J. D. (2001). Translocations of the RARalpha gene in acute promyelocytic leukemia. *Oncogene* 20, 7186–7203.
60. He, L. Z., Guidez, F., Tribioli, C., Peruzzi, D., Ruthardt, M., Zelent, A., and Pandolfi, P. P. (1998). Distinct interactions of PML-RARalpha and PLZF-RARalpha with co-repressors determine differential responses to RA in APL. *Nat Genet* 18, 126–135.
61. Castillo, L., Milano, G., Santini, J., Demard, F., and Pierrefite, V. (1997). Analysis of retinoic acid receptor beta expression in normal and malignant laryngeal mucosa by a sensitive and routine applicable reverse transcription-polymerase chain reaction enzyme-linked immunosorbent assay method. *Clin Cancer Res* 3, 2137–2142.
62. Widschwendter, M., Berger, J., Daxenbichler, G., Muller-Holzner, E., Widschwendter, A., Mayr, A., Marth, C., and Zeimet, A. G. (1997). Loss of retinoic acid receptor beta expression in breast cancer and morphologically normal adjacent tissue but not in the normal breast tissue distant from the cancer. *Cancer Res* 57, 4158–4161.
63. Xu, X. C., Sneige, N., Liu, X., Nandagiri, R., Lee, J. J., Lukmanji, F., Hortobagyi, G., Lippman, S. M., Dhingra, K., and Lotan, R. (1997). Progressive decrease in nuclear retinoic acid receptor beta messenger RNA level during breast carcinogenesis. *Cancer Res* 57, 4992–4996.
64. Xu, X. C., Sozzi, G., Lee, J. S., Lee, J. J., Pastorino, U., Pilotti, S., Kurie, J. M., Hong, W. K., and Lotan, R. (1997). Suppression of retinoic acid receptor beta in non-small-cell lung cancer in vivo: Implications for lung cancer development. *J Natl Cancer Inst* 89, 624–629.
65. Li, Y., Dawson, M. I., Agadir, A., Lee, M. O., Jong, L., Hobbs, P. D., and Zhang, X. K. (1998). Regulation of RAR beta expression by RAR- and RXR-selective retinoids in human lung cancer cell lines: Effect on growth inhibition and apoptosis induction. *Int J Cancer* 75, 88–95.
66. Liu, Y., Lee, M. O., Wang, H. G., Li, Y., Hashimoto, Y., Klaus, M., Reed, J. C., and Zhang, X. (1996). Retinoic acid receptor beta mediates the growth-inhibitory effect of retinoic acid by promoting apoptosis in human breast cancer cells. *Mol Cell Biol* 16, 1138–1149.
67. Lotan, R., Xu, X. C., Lippman, S. M., Ro, J. Y., Lee, J. S., Lee, J. J., and Hong, W. K. (1995). Suppression of retinoic acid receptor-beta in premalignant oral lesions and its up-regulation by isotretinoin. *N Engl J Med* 332, 1405–1410.
68. Berard, J., Gaboury, L., Landers, M., De Repentigny, Y., Houle, B., Kothary, R., and Bradley, W. E. (1994). Hyperplasia and tumours in lung, breast and other tissues in mice carrying a RAR beta 4-like transgene. *EMBO J* 13, 5570–5580.
69. Berard, J., Laboune, F., Mukuna, M., Masse, S., Kothary, R., and Bradley, W. E. (1996). Lung tumors in mice expressing an antisense RARbeta2 transgene. *FASEB J* 10, 1091–1097.
70. Klaassen, I., Brakenhoff, R. H., Smeets, S. J., Snow, G. B., and Braakhuis, B. J. (2001). Expression of retinoic acid receptor gamma correlates with retinoic acid sensitivity and metabolism in head and neck squamous cell carcinoma cell lines. *Int J Cancer* 92, 661–665.
71. Wu, S., Zhang, D., Zhang, Z. P., Soprano, D. R., and Soprano, K. J. (1998). Critical role of both retinoid nuclear receptors and retinoid-X-receptors in mediating growth inhibition of ovarian cancer cells by all-trans retinoic acid. *Oncogene* 17, 2839–2849.
72. Lin, B., Chen, G. Q., Xiao, D., Kolluri, S. K., Cao, X., Su, H., and Zhang, X. K. (2000). Orphan receptor COUP-TF is required for induction of retinoic acid receptor beta, growth inhibition, and apoptosis by retinoic acid in cancer cells. *Mol Cell Biol* 20, 957–970.
73. Wu, Q., Li, Y., Liu, R., Agadir, A., Lee, M. O., Liu, Y., and Zhang, X. (1997). Modulation of retinoic acid sensitivity in lung cancer cells through dynamic balance of orphan receptors nur77 and COUP-TF and their heterodimerization. *EMBO J* 16, 1656–1669.
74. Cote, S. and Momparler, R. L. (1997). Activation of the retinoic acid receptor beta gene by 5-aza-2'-deoxycytidine in human DLD-1 colon carcinoma cells. *Anticancer Drugs* 8, 56–61.

75. Cote, S., Sinnett, D., and Momparler, R. L. (1998). Demethylation by 5-aza-2'-deoxycytidine of specific 5-methylcytosine sites in the promoter region of the retinoic acid receptor beta gene in human colon carcinoma cells. *Anticancer Drugs* 9, 743–750.
76. Liu, T., Bohlken, A., Kuljaca, S., Lee, M., Nguyen, T., Smith, S., Cheung, B., Norris, M. D., Haber, M., Holloway, A. J., Bowtell, D. D., and Marshall, G. M. (2005). The retinoid anticancer signal: Mechanisms of target gene regulation. *Br J Cancer* 93, 310–318.
77. Liu, Z., Zhang, L., Ding, F., Li, J., Guo, M., Li, W., Wang, Y., Yu, Z., Zhan, Q., Wu, M., and Liu, Z. (2005). 5-Aza-2'-deoxycytidine induces retinoic acid receptor-beta(2) demethylation and growth inhibition in esophageal squamous carcinoma cells. *Cancer Lett* 230, 271–283.
78. Sirchia, S. M., Ferguson, A. T., Sironi, E., Subramanyan, S., Orlandi, R., Sukumar, S., and Sacchi, N. (2000). Evidence of epigenetic changes affecting the chromatin state of the retinoic acid receptor beta2 promoter in breast cancer cells. *Oncogene* 19, 1556–1563.
79. Suh, Y. A., Lee, H. Y., Virmani, A., Wong, J., Mann, K. K., Miller, W. H., Jr., Gazdar, A., and Kurie, J. M. (2002). Loss of retinoic acid receptor beta gene expression is linked to aberrant histone H3 acetylation in lung cancer cell lines. *Cancer Res* 62, 3945–3949.
80. Cras, A., Darsin-Bettinger, D., Balitrand, N., Cassinat, B., Soulie, A., Toubert, M. E., Delva, L., and Chomienne, C. (2007). Epigenetic patterns of the retinoic acid receptor beta2 promoter in retinoic acid-resistant thyroid cancer cells. *Oncogene* 26, 4018–4024.
81. Touma, S. E., Goldberg, J. S., Moench, P., Guo, X., Tickoo, S. K., Gudas, L. J., and Nanus, D. M. (2005). Retinoic acid and the histone deacetylase inhibitor trichostatin a inhibit the proliferation of human renal cell carcinoma in a xenograft tumor model. *Clin Cancer Res* 11, 3558–3566.
82. Balmer, J. E. and Blomhoff, R. (2002). Gene expression regulation by retinoic acid. *J Lipid Res* 43, 1773–1808.
83. Freemantle, S. J., Kerley, J. S., Olsen, S. L., Gross, R. H., and Spinella, M. J. (2002). Developmentally-related candidate retinoic acid target genes regulated early during neuronal differentiation of human embryonal carcinoma. *Oncogene* 21, 2880–2889.
84. Kim, H. T., Kong, G., Denardo, D., Li, Y., Uray, I., Pal, S., Mohsin, S., Hilsenbeck, S. G., Bissonnette, R., Lamph, W. W., Johnson, K., and Brown, P. H. (2006). Identification of biomarkers modulated by the rexinoid LGD1069 (bexarotene) in human breast cells using oligonucleotide arrays. *Cancer Res* 66, 12009–12018.
85. Lian, Z., Kluger, Y., Greenbaum, D. S., Tuck, D., Gerstein, M., Berliner, N., Weissman, S. M., and Newburger, P. E. (2002). Genomic and proteomic analysis of the myeloid differentiation program: Global analysis of gene expression during induced differentiation in the MPRO cell line. *Blood* 100, 3209–3220.
86. Meani, N., Minardi, S., Licciulli, S., Gelmetti, V., Coco, F. L., Nervi, C., Pelicci, P. G., Muller, H., and Alcalay, M. (2005). Molecular signature of retinoic acid treatment in acute promyelocytic leukemia. *Oncogene* 24, 3358–3368.
87. Tagliafico, E., Tenedini, E., Manfredini, R., Grande, A., Ferrari, F., Roncaglia, E., Bicciato, S., Zini, R., Salati, S., Bianchi, E., Gemelli, C., Montanari, M., Vignudelli, T., Zanocco-Marani, T., Parenti, S., Paolucci, P., Martinelli, G., Piccaluga, P. P., Baccarani, M., Specchia, G., Torelli, U., and Ferrari, S. (2006). Identification of a molecular signature predictive of sensitivity to differentiation induction in acute myeloid leukemia. *Leukemia* 20, 1751–1758.
88. Castaigne, S., Chomienne, C., Daniel, M. T., Ballerini, P., Berger, R., Fenaux, P., and Degos, L. (1990). All-trans retinoic acid as a differentiation therapy for acute promyelocytic leukemia. I. Clinical results. *Blood* 76, 1704–1709.
89. Chomienne, C., Ballerini, P., Balitrand, N., Amar, M., Bernard, J. F., Boivin, P., Daniel, M. T., Berger, R., Castaigne, S., and Degos, L. (1989). Retinoic acid therapy for promyelocytic leukaemia. *Lancet* 2, 746–747.
90. Degos, L., Chomienne, C., Daniel, M. T., Berger, R., Dombret, H., Fenaux, P., and Castaigne, S. (1990). Treatment of first relapse in acute promyelocytic leukaemia with all-trans retinoic acid. *Lancet* 336, 1440–1441.

91. Chomienne, C., Ballerini, P., Balitrand, N., Daniel, M. T., Fenaux, P., Castaigne, S., and Degos, L. (1990). All-trans retinoic acid in acute promyelocytic leukemias. II. In vitro studies: Structure-function relationship. *Blood* 76, 1710–1717.
92. Chomienne, C., Balitrand, N., Ballerini, P., Castaigne, S., de The, H., and Degos, L. (1991). All-trans retinoic acid modulates the retinoic acid receptor-alpha in promyelocytic cells. *J Clin Invest* 88, 2150–2154.
93. Cassinat, B., Chevret, S., Zassadowski, F., Balitrand, N., Guillemot, I., Menot, M. L., Degos, L., Fenaux, P., and Chomienne, C. (2001). In vitro all-trans retinoic acid sensitivity of acute promyelocytic leukemia blasts: A novel indicator of poor patient outcome. *Blood* 98, 2862–2864.
94. Quere, R., Baudet, A., Cassinat, B., Bertrand, G., Marti, J., Manchon, L., Piquemal, D., Chomienne, C., and Commes, T. (2007). Pharmacogenomic analysis of acute promyelocytic leukemia cells highlights CYP26 cytochrome metabolism in differential all-trans retinoic acid sensitivity. *Blood* 109, 4450–4460.
95. Dubois, C., Schlageter, M. H., de Gentile, A., Balitrand, N., Toubert, M. E., Krawice, I., Fenaux, P., Castaigne, S., Najean, Y., Degos, L. et al. (1994). Modulation of IL-8, IL-1 beta, and G-CSF secretion by all-trans retinoic acid in acute promyelocytic leukemia. *Leukemia* 8, 1750–1757.
96. Dubois, C., Schlageter, M. H., de Gentile, A., Guidez, F., Balitrand, N., Toubert, M. E., Krawice, I., Fenaux, P., Castaigne, S., Najean, Y. et al. (1994). Hematopoietic growth factor expression and ATRA sensitivity in acute promyelocytic blast cells. *Blood* 83, 3264–3270.
97. Nervi, C., Ferrara, F. F., Fanelli, M., Rippo, M. R., Tomassini, B., Ferrucci, P. F., Ruthardt, M., Gelmetti, V., Gambacorti-Passerini, C., Diverio, D., Grignani, F., Pelicci, P. G., and Testi, R. (1998). Caspases mediate retinoic acid-induced degradation of the acute promyelocytic leukemia PML/RARalpha fusion protein. *Blood* 92, 2244–2251.
98. Zhu, J., Gianni, M., Kopf, E., Honore, N., Chelbi-Alix, M., Koken, M., Quignon, F., Rochette-Egly, C., and de The, H. (1999). Retinoic acid induces proteasome-dependent degradation of retinoic acid receptor alpha (RARalpha) and oncogenic RARalpha fusion proteins. *Proc Natl Acad Sci U S A* 96, 14807–14812.
99. Chen, G. Q., Shi, X. G., Tang, W., Xiong, S. M., Zhu, J., Cai, X., Han, Z. G., Ni, J. H., Shi, G. Y., Jia, P. M., Liu, M. M., He, K. L., Niu, C., Ma, J., Zhang, P., Zhang, T. D., Paul, P., Naoe, T., Kitamura, K., Miller, W., Waxman, S., Wang, Z. Y., de The, H., Chen, S. J., and Chen, Z. (1997). Use of arsenic trioxide (As2O3) in the treatment of acute promyelocytic leukemia (APL): I. As2O3 exerts dose-dependent dual effects on APL cells. *Blood* 89, 3345–3353.
100. Chen, G. Q., Zhu, J., Shi, X. G., Ni, J. H., Zhong, H. J., Si, G. Y., Jin, X. L., Tang, W., Li, X. S., Xong, S. M., Shen, Z. X., Sun, G. L., Ma, J., Zhang, P., Zhang, T. D., Gazin, C., Naoe, T., Chen, S. J., Wang, Z. Y., and Chen, Z. (1996). In vitro studies on cellular and molecular mechanisms of arsenic trioxide (As2O3) in the treatment of acute promyelocytic leukemia: As2O3 induces NB4 cell apoptosis with downregulation of Bcl-2 expression and modulation of PML-RAR alpha/PML proteins. *Blood* 88, 1052–1061.
101. Zheng, P. Z., Wang, K. K., Zhang, Q. Y., Huang, Q. H., Du, Y. Z., Zhang, Q. H., Xiao, D. K., Shen, S. H., Imbeaud, S., Eveno, E., Zhao, C. J., Chen, Y. L., Fan, H. Y., Waxman, S., Auffray, C., Jin, G., Chen, S. J., Chen, Z., and Zhang, J. (2005). Systems analysis of transcriptome and proteome in retinoic acid/arsenic trioxide-induced cell differentiation/apoptosis of promyelocytic leukemia. *Proc Natl Acad Sci U S A* 102, 7653–7658.
102. Warrell, R. P., Jr., He, L. Z., Richon, V., Calleja, E., and Pandolfi, P. P. (1998). Therapeutic targeting of transcription in acute promyelocytic leukemia by use of an inhibitor of histone deacetylase. *J Natl Cancer Inst* 90, 1621–1625.
103. Da Silva, N., Meyer-Monard, S., Menot, M. L., Parrado, A., Lebel, A., Balitrand, N., Fenaux, P., Miclea, J. M., Rousselot, P., Degos, L., Dombret, H., and Chomienne, C. (2000). Functional G-CSF pathways in t(8;21) leukemic cells allow for differentiation induction and degradation of AML1-ETO. *Hematol J* 1, 316–328.
104. Ferrara, F. F., Fazi, F., Bianchini, A., Padula, F., Gelmetti, V., Minucci, S., Mancini, M., Pelicci, P. G., Lo Coco, F., and Nervi, C. (2001). Histone deacetylase-targeted treatment restores retinoic acid signaling and differentiation in acute myeloid leukemia. *Cancer Res* 61, 2–7.

105. Fazi, F., Zardo, G., Gelmetti, V., Travaglini, L., Ciolfi, A., Di Croce, L., Rosa, A., Bozzoni, I., Grignani, F., Lo-Coco, F., Pelicci, P. G., and Nervi, C. (2007). Heterochromatic gene repression of the retinoic acid pathway in acute myeloid leukemia. *Blood* 109, 4432–4440.
106. Bug, G., Ritter, M., Wassmann, B., Schoch, C., Heinzel, T., Schwarz, K., Romanski, A., Kramer, O. H., Kampfmann, M., Hoelzer, D., Neubauer, A., Ruthardt, M., and Ottmann, O. G. (2005). Clinical trial of valproic acid and all-trans retinoic acid in patients with poor-risk acute myeloid leukemia. *Cancer* 104, 2717–2725.
107. Cimino, G., Lo-Coco, F., Fenu, S., Travaglini, L., Finolezzi, E., Mancini, M., Nanni, M., Careddu, A., Fazi, F., Padula, F., Fiorini, R., Spiriti, M. A., Petti, M. C., Venditti, A., Amadori, S., Mandelli, F., Pelicci, P. G., and Nervi, C. (2006). Sequential valproic acid/all-trans retinoic acid treatment reprograms differentiation in refractory and high-risk acute myeloid leukemia. *Cancer Res* 66, 8903–8911.
108. Kuendgen, A., Schmid, M., Schlenk, R., Knipp, S., Hildebrandt, B., Steidl, C., Germing, U., Haas, R., Dohner, H., and Gattermann, N. (2006). The histone deacetylase (HDAC) inhibitor valproic acid as monotherapy or in combination with all-trans retinoic acid in patients with acute myeloid leukemia. *Cancer* 106, 112–119.
109. Soriano, A. O., Yang, H., Faderl, S., Estrov, Z., Giles, F., Ravandi, F., Cortes, J., Wierda, W. G., Ouzounian, S., Quezada, A., Pierce, S., Estey, E. H., Issa, J. P., Kantarjian, H. M., and Garcia-Manero, G. (2007). Safety and clinical activity of the combination of 5-azacytidine, valproic acid, and all-trans retinoic acid in acute myeloid leukemia and myelodysplastic syndrome. *Blood* 110, 2302–2308.
110. Altucci, L., Rossin, A., Raffelsberger, W., Reitmair, A., Chomienne, C., and Gronemeyer, H. (2001). Retinoic acid-induced apoptosis in leukemia cells is mediated by paracrine action of tumor-selective death ligand TRAIL. *Nat Med* 7, 680–686.
111. Matikainen, S., Ronni, T., Hurme, M., Pine, R., and Julkunen, I. (1996). Retinoic acid activates interferon regulatory factor-1 gene expression in myeloid cells. *Blood* 88, 114–123.
112. Clarke, N., Jimenez-Lara, A. M., Voltz, E., and Gronemeyer, H. (2004). Tumor suppressor IRF-1 mediates retinoid and interferon anticancer signaling to death ligand TRAIL. *EMBO J* 23, 3051–3060.
113. Benoit, G. R., Flexor, M., Besancon, F., Altucci, L., Rossin, A., Hillion, J., Balajthy, Z., Legres, L., Segal-Bendirdjian, E., Gronemeyer, H., and Lanotte, M. (2001). Autonomous rexinoid death signaling is suppressed by converging signaling pathways in immature leukemia cells. *Mol Endocrinol* 15, 1154–1169.
114. Altucci, L., Rossin, A., Hirsch, O., Nebbioso, A., Vitoux, D., Wilhelm, E., Guidez, F., De Simone, M., Schiavone, E. M., Grimwade, D., Zelent, A., de The, H., and Gronemeyer, H. (2005). Rexinoid-triggered differentiation and tumor-selective apoptosis of acute myeloid leukemia by protein kinase A-mediated desubordination of retinoid X receptor. *Cancer Res* 65, 8754–8765.
115. Li, H., Kolluri, S. K., Gu, J., Dawson, M. I., Cao, X., Hobbs, P. D., Lin, B., Chen, G., Lu, J., Lin, F., Xie, Z., Fontana, J. A., Reed, J. C., and Zhang, X. (2000). Cytochrome c release and apoptosis induced by mitochondrial targeting of nuclear orphan receptor TR3. *Science* 289, 1159–1164.
116. Sakaue, M., Adachi, H., Dawson, M., and Jetten, A. M. (2001). Induction of Egr-1 expression by the retinoid AHPN in human lung carcinoma cells is dependent on activated ERK1/2. *Cell Death Differ* 8, 411–424.
117. Jin, F., Liu, X., Zhou, Z., Yue, P., Lotan, R., Khuri, F. R., Chung, L. W., and Sun, S. Y. (2005). Activation of nuclear factor-kappaB contributes to induction of death receptors and apoptosis by the synthetic retinoid CD437 in DU145 human prostate cancer cells. *Cancer Res* 65, 6354–6363.
118. Sun, S. Y. and Lotan, R. (2002). Retinoids and their receptors in cancer development and chemoprevention. *Crit Rev Oncol Hematol* 41, 41–55.
119. So, P. L., Lee, K., Hebert, J., Walker, P., Lu, Y., Hwang, J., Kopelovich, L., Athar, M., Bickers, D., Aszterbaum, M., and Epstein, E. H., Jr. (2004). Topical tazarotene chemoprevention reduces

Basal cell carcinoma number and size in Ptch1+/- mice exposed to ultraviolet or ionizing radiation. *Cancer Res* 64, 4385–4389.

120. Dong, Z., Birrer, M. J., Watts, R. G., Matrisian, L. M., and Colburn, N. H. (1994). Blocking of tumor promoter-induced AP-1 activity inhibits induced transformation in JB6 mouse epidermal cells. *Proc Natl Acad Sci U S A* 91, 609–613.
121. Chen, J. Y., Penco, S., Ostrowski, J., Balaguer, P., Pons, M., Starrett, J. E., Reczek, P., Chambon, P., and Gronemeyer, H. (1995). RAR-specific agonist/antagonists which dissociate transactivation and AP1 transrepression inhibit anchorage-independent cell proliferation. *EMBO J* 14, 1187–1197.
122. Kojima, S., Okuno, M., Matsushima-Nishiwaki, R., Friedman, S. L., and Moriwaki, H. (2004). Acyclic retinoid in the chemoprevention of hepatocellular carcinoma (review). *Int J Oncol* 24, 797–805.
123. Li, Y., Zhang, Y., Hill, J., Kim, H. T., Shen, Q., Bissonnette, R. P., Lamph, W. W., and Brown, P. H. (2008). The rexinoid, bexarotene, prevents the development of premalignant lesions in MMTV-erbB2 mice. *Br J Cancer* 98, 1380–1388.
124. Dragnev, K. H., Petty, W. J., Shah, S. J., Lewis, L. D., Black, C. C., Memoli, V., Nugent, W. C., Hermann, T., Negro-Vilar, A., Rigas, J. R., and Dmitrovsky, E. (2007). A proof-of-principle clinical trial of bexarotene in patients with non-small cell lung cancer. *Clin Cancer Res* 13, 1794–1800.

CHAPTER 10

PPARs: IMPORTANT REGULATORS IN METABOLISM AND INFLAMMATION

LINDA M. SANDERSON AND SANDER KERSTEN

Nutrigenomics Consortium, TI Food and Nutrition, Nieuwe Kanaal 9A, 6709 PA Wageningen, The Netherlands

Nutrition, Metabolism and Genomics Group, Division of Human Nutrition, Wageningen University, Bomenweg 2, 6703 HD Wageningen, The Netherlands

Abstract: The ligand-activated family of peroxisome proliferator activated receptors (PPARs) consists of three members named PPARα, PPARδ and PPARγ. Each PPAR subtype is characterized by a specific tissue expression pattern, partially accounting for distinct biological functions. Analogous to many other nuclear receptors, PPARs form heterodimers with the retinoid X receptor and regulate DNA transcription by binding to specific response elements present in target genes.

PPARα (NR1C1) is highly expressed in liver, heart, intestine, skeletal muscle, and various immune cells. Agonists for PPARα include the lipid-lowering fibrate drugs as well as numerous fatty acids and eicosanoids. PPARα is known to play an important role in many different metabolic processes, especially under conditions of fasting, and has proven to be an important regulator of inflammation via inhibition of gene expression.

PPARδ (NR1C2) is ubiquitously expressed but its function has mainly been studied in skin, heart, and skeletal muscle. In the past few years, it has become evident that PPARδ is involved in numerous biological processes including lipid metabolism, wound healing and inflammation.

The most studied PPAR subtype is PPARγ (NR1C3), which is expressed at high levels in adipose tissue, macrophages and vascular cells. PPARγ drives adipocyte differentiation, has important regulatory roles during fat storage and glucose metabolism, and is an important suppressor of inflammation. Importantly, it serves as the molecular target for the thiazolidionedione drugs.

In this chapter we provide an overview of the major functions of the three PPAR subtypes, and focus on their role in metabolic and inflammatory processes.

10.1. INTRODUCTION

The peroxisome proliferator activated receptors (PPARs) are ligand-activated transcription factors that belong to the superfamily of nuclear hormone receptors, and more specifically to the class II nuclear receptors. Three subtypes encoded by separate genes can be distinguished: PPARα, PPARβ/δ and PPARγ [1, 2]. Until now,

C.M. Bunce, M.J. Campbell (eds.), *Nuclear Receptors*, Proteins and Cell Regulation 8, 259–285. DOI 10.1007/978-90-481-3303-1_10,

PPARs have been cloned from several species, including human, mouse, rat, chicken, fish, guinea pig, hamster and amphibian [2, 3].

PPARs, like many other nuclear receptors, bind to DNA and regulate transcription in the form of a heterodimer with the nuclear receptor retinoid X receptor (RXR) (Figure 1a). Recent evidence suggests that to a large extent PPARs and retinoid X receptors are associated even in the absence of ligand [4]. Activation of target genes occurs through direct binding of the PPAR:RXR heterodimer to specific nucleotide sequences called peroxisome proliferator response elements (PPREs) [5]. These response elements are of the direct repeat 1 (DR-1) type, which is defined by a direct repetition of the consensus sequence AGGTCA with a single nucleotide insertion between the two repeats [5]. PPAR will bind to the $5'$ part of the response element, whereas RXR binds to the $3'$ half-site [5]. A PPRE is commonly present as one or multiple copies located in or close to the promoter region of a target gene [5].

Members of the nuclear hormone receptor family are related to each other with respect to amino acid sequence and their molecular mode of function within cells. They share a highly conserved DNA binding domain (DBD) responsible for binding to the response element sequence in target genes. A reasonably well conserved ligand-binding domain (LBD) is present at the C-terminal end of the receptors, which binds ligands and interacts with coactivators. Crystallographic structures of PPARs reveal an exceptionally spacious ligand-binding pocket compared to other nuclear receptors, thereby explaining the promiscuity in ligand binding [6–8]. Ligands of PPARs include numerous fatty acids and their derivates, as well as a large number of industrially synthesized compounds.

The ability of PPARs to regulate transcription is controlled by complex interactions involving coactivators and corepressors (Figure 1a). PPARs are present in the nucleus of the cell, both in the presence and absence of ligand. In the latter case, compression of the chromatin is caused by nuclear corepressors (NCoRs), which connect PPARs with enzymes expressing histone deacetylase activity [9–11]. Once ligand binding occurs, these corepressor protein complexes dissociate, and subsequent recruitment of several coactivator proteins leads to a conformational change within the ligand-binding domain of the receptor [10, 11]. Several nuclear receptor coactivators have been identified in recent years. Coactivators contain a so-called LXXLL (L: leucine; X: any amino acid) motif, where direct binding with the LBD of the nuclear receptor takes place. Coactivators often contain several LXXLL motifs, suggesting that they are able to bind to several different nuclear receptors simultaneously.

Some coactivators possess histone acetyltransferase (HAT) activity (Figure 1a). This group of coactivators is able to remodel chromatin structure and in this way facilitate gene transcription [11, 12]. Examples of this type of coactivators are CBP (CREB binding protein) and p300 (adenovirus E1A-associated protein). SRC-1 (steroid receptor coactivator) is a coactivator that binds to the LBD of the receptor in a ligand-dependent manner and forms a complex with CBP/p300 proteins, thereby recruiting HAT activity.

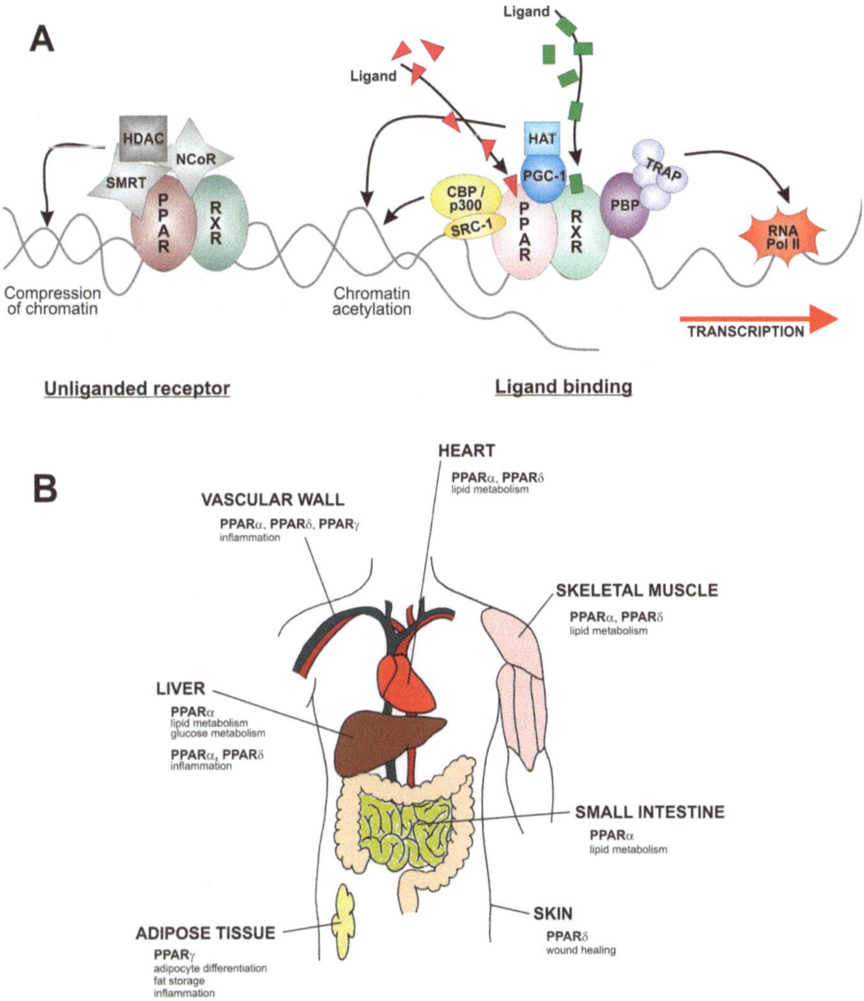

Figure 10.1. Schematic depiction of the mechanism of gene regulation by PPARs, and their regulatory role in different tissues. (**a**) Transcriptional regulation and the interplay between PPARs and cofactors. PPARs bind to DNA and regulate transcription as a heterodimer with the nuclear receptor retinoid X receptor (RXR). In the absence of ligand, corepressor proteins connect the PPAR:RXR complex with enzymes expressing histone deacetylase activity (HDAC). This causes compression of the chromatin structure and subsequent repression of gene transcription. Ligand binding is followed by a dissociation of the corepressor complexes and recruitment of coactivator proteins. CREB binding protein (CBP) and adenovirus E1A-associated protein (p300) are coactivators that possess histone acetyltransferase (HAT) activity, leading to remodeling of the chromatin structure and a facilitation of gene transcription. Steroid receptor coactivator 1 (SRC-1) forms a complex with CBP/p300 proteins, thereby recruiting HAT activity. PPAR gamma coactivator 1 (PGC-1) has the ability to recruit enzymes that posses HAT activity. PPAR binding protein (PBP) serves as an anchor within multi-subunit coactivator complexes, called TRAP complexes, which may function as docking platforms for RNA polymerase II during the transcriptional process. (**b**) Overview of tissue distribution and different processes where PPARs have proven to play an important regulatory role

A different group of coactivators consists of proteins that form multi-subunit coactivator complexes. These coactivators are recruited to the nuclear receptor in a ligand-dependent manner, and may function as docking platforms for RNA polymerase II during the transcriptional process. The most important protein in this context is the coactivator PBP (PPAR binding protein) which serves as an anchor inside the complex. Experiments involving liver specific knock-outs of PBP showed a dysfunctional PPARα-mediated transcriptional response [13]. Another important coactivator is PGC-1 (PPAR gamma coactivator 1). PGC-1 interacts with PPARγ to induce mitochondrial biogenesis and induction of a brown adipocyte-specific gene expression program [14]. Additionally, PGC-1 is able to coactivate PPARα-dependent gene regulation, especially towards fatty acid catabolic processes [15]. This protein thus plays an important role during fasting in the liver by upregulating genes involved in gluconeogenesis and fatty acid oxidation/ketogenesis [16, 17].

Each PPAR subtype is characterized by a specific tissue expression pattern, thereby accounting for their distinct biological functions. Within a tissue, PPARs exhibit differential activity towards target genes, which is partly due to differential availability of receptor-specific ligands and coactivators [5, 18, 19]. Nevertheless, studies using PPAR agonists have shown considerable overlap in gene regulation between PPARs, even within a certain tissue. To what extent this is an artifact of pharmacological activation of PPARs or reflects the limited receptor specificity of synthetic agonists remains to be determined.

Here we provide an overview of the major functions of PPARs, separated by PPAR subtype, and focus on their role in metabolic and inflammatory processes.

10.2. PPARα (NR1C1)

Of the three PPAR subtypes, PPARα was discovered first [1]. Since then it has been cloned in several different species, including human [20, 21], rat [22], frog [23] and rabbit [24]. Human PPARα is situated on chromosome 22, and its mouse namesake has been mapped to chromosome 15.

In rodent as well as in human, PPARα is expressed in many tissues that actively metabolize fatty acids (Figure 1b). It is highly expressed in liver, with expression levels in mouse reportedly exceeding those in human [25, 26]. In addition, PPARα is relatively well expressed in heart, kidney, intestine, skeletal muscle and brown adipose tissue [1, 19, 27–31]. PPARα has also been found in different types of immune cells, such as macrophages and T and B cells [32–40]. Finally, PPARα has been detected in vascular endothelial, vascular smooth muscle cells, and atherosclerotic lesions [33, 41–44].

PPARα serves as a receptor for a structurally diverse set of compounds, both natural and synthetic. A major group of synthetic PPARα agonists are the fibrates, which comprises a cluster of lipid-lowering drugs used for treatment of dyslipidemia, including gemfibrozil, bezafibrate, clofibrate, fenofibrate and WY14643 [2, 5, 45–48]. In addition, PPARα is activated by a variety of plasticizers, insecticides, and other rodent hepatic carcinogens. Endogenous ligands for PPARα include a variety of (long-chain polyunsaturated) fatty acids and eicosanoids [22, 49, 50]. Recently, it was shown that the effects of dietary unsaturated fatty acids on hepatic gene expression are almost entirely mediated by PPARα and mimic those of synthetic PPARα agonists [51].

10.2.1. Metabolism

PPARα plays an important regulatory role in many different metabolic processes, especially under conditions of fasting [52–54] (Figure 1b). In addition, it governs the metabolic response in liver to acute and chronic dietary fat feeding [51, 55]. Here, we summarize the involvement of PPARα in nutrient metabolism, emphasizing pathways in liver.

10.2.1.1. Lipid metabolism

The first connection between PPARα and fatty acid catabolism was made in 1992 when it was shown that the peroxisomal acyl-coenzyme A (CoA) oxidase gene was a direct target gene of PPARα [23, 56]. This enzyme carries out the first step of the oxidation of very long chain polyunsaturated fatty acids in peroxisomes. Since then, numerous genes involved in mitochondrial and peroxisomal fatty acid oxidation have been shown to be under control of PPARα, especially in the fasted state.

During fasting, free fatty acids (FFA) are released from adipose tissue storage and are transported to the liver. The liver, being a central player in maintaining metabolic homeostasis, responds by increasing the rate of fatty acid beta-oxidation and, after prolonged fasting, ketogenesis. Fasting PPARα null mice exhibit a severe impairment in hepatic mitochondrial beta-oxidation leading to hypoketonemia, hepatic steatosis, myocardial lipid accumulation and hypoglycemia [52–54]. It is now evident that PPARα not merely governs expression of a few key genes such as *Cpt1*, *Acadm*, and *Hmgcs2* [2, 57–59], but actually regulates entire pathways involved in different aspects of hepatic fatty acid metabolism, including fatty acid uptake across cell membranes (*Cd36, Fatp*) [60–63], intracellular fatty acid binding and transport (*Fabp*) [64–66], fatty acid activation (*Acsl*), microsomal fatty acid oxidation (*Cyp4a*), peroxisomal fatty acid oxidation (*Acaa, Acot, Ehhadh, Decr2, Acox1*), mitochondrial fatty acid oxidation (*Acad, Cpt, Hadh, Acot*), and ketogenesis (*Hmgcs2, Hmgcl*) [67]. Furthermore, PPARα upregulates expression of several lipases (*Mgll, Lipg, Pnpla2*), as well as many genes involved in fatty acid synthesis/elongation/desaturation (*Fads, Agpat, Scd1, Dgat1*). Thus, PPARα can be considered a master regulator of hepatic lipid metabolism. Interestingly, although fasting increases delivery of free fatty acids to the liver, recent studies suggest that free fatty acids are unable to ligand-activate hepatic PPARα, in contrast to lipoprotein-derived fatty acids [68, 69].

PPARα has a similar, though less comprehensive role in regulation of lipid metabolism in cardiac and skeletal muscle [70–72]. Moreover, recent studies also reveal a major role for PPARα in lipid metabolism in the small intestine [73] (Figure 1b).

10.2.1.2. Glucose metabolism

A role of PPARα in glucose metabolism is supported by the severe hypoglycemia in fasted PPARα null mice [53, 54, 74–76]. While it is tempting to relate the reduced plasma glucose to defective fatty acid oxidation, which in many fatty acid oxidation disorders gives rise to hypoglycemia, there is compelling evidence for direct regulation of glucose synthesis by PPARα. Several genes involved in synthesis of glucose from gluconeogenic precursors have been identified as direct PPARα targets, including *PCK1*, *Pcx*, and *Gpd*, although the former gene only in human [77]. Specifically the conversion of glycerol into gluconeogenic intermediates is under direct control of PPARα [77]. The effect of PPARα deletion on gluconeogenic fluxes reveals a more complex picture [74]. It has been reported that PPARα null mice exhibit an increased gluconeogenic flux towards glycogen, thereby diminishing hepatic glucose output [78]. Synthesis of glycogen is also affected in PPARα null mice, which is likely mediated via defective regulation of *Gys2* [79].

The hypoglycemia witnessed in PPARα null mice may also partially be due to an increased rate of glucose utilization [80]. Decreased expression of *Pdk4*, which is a PPARα target gene in liver, heart, kidney and skeletal muscle, may relieve the block on pyruvate oxidation and thus glucose utilization [81–85]. Furthermore, regulation of glucose utilization may occur via PPARα in the brain [80].

10.2.1.3. Amino acid metabolism

In addition to lipid and glucose metabolism, PPARα also governs metabolism of amino acids [86, 87]. The expression of numerous genes involved in the ammonia detoxification pathway and urea synthesis including *Cps1*, *Otc*, *Ass1*, and *Asl* are downregulated by PPARα. Consequently, plasma ammonia levels are increased after WY14643 treatment, while plasma urea levels are increased in PPARα null mice [87, 88]. Currently, the molecular mechanisms behind this regulation remain elusive and require more detailed investigation [89].

10.2.2. Inflammation

A delayed inflammatory response to topical administration of leukotriene B4 and arachidonic acid provided the first evidence for a link between PPARα and inflammation [90]. Follow-up studies have shown that PPARα is an extremely important regulator of inflammation, mainly by inhibiting inflammatory gene expression (Figure 1b).

10.2.2.1. *Hepatic inflammation*

Numerous studies have shown a reduction of hepatic cytokine-stimulated inflammation and production of acute phase proteins upon PPARα activation. Several molecular mechanisms behind the anti-inflammatory effects of PPARα have been suggested. These include interference with several proinflammatory transcription factors including signal transducer and activator of transcription (STAT), activator protein-1 (AP-1), and NF-kB [91]. Further studies have revealed that PPARα diminishes the activity of the proinflammatory transcription factor CAATT/enhancer binding proteins (C/EBP) via sequestration of the coactivator glucocorticoid receptor-interacting protein-1/transcriptional intermediary factor-2 (GRIP1/TIF2) [92]. Finally, PPARα can also inhibit cytokine signaling pathways via downregulation of the IL-6 receptor [93, 94] and upregulation of sIL-1 receptor antagonist [95], leading to diminished inflammatory responses. Its potent anti-inflammatory activity in liver may confer a protective role for PPARα against steatohepatitis. Indeed, several studies in mice have shown that activation of PPARα can slow down or even reverse progression of steatohepatitis [96–99]. Part of the effect of PPARα may be linked to preventing upregulation of the *Cox2* gene, which has been directly linked to the progression of steatosis to steatohepatitis [100]. Conversely, PPARα ablation accelerates development of steatohepatitis in mice rendered obese by chronic high-fat feeding. PPARα may protect against steatohepatitis by a combination of reducing hepatic lipid storage and direct suppression of pro-inflammatory gene expression [101].

10.2.2.2. *Inflammation in vascular wall*

Inflammation in the arterial wall is an important contributor to atherogenesis [102]. In addition to suppressing inflammatory responses in liver, PPARα also modulates inflammatory reactions in the arterial wall. As PPARα is expressed in various cell types present in atherosclerotic lesions, including smooth muscle cells, endothelial cells, and macrophages, the effect of PPARα on lesion development is rather complex. Immune-modulating effects of specific PPARα activation have been reported in these various cell types. However, some controversy still exists about the exact role of PPARα in the vascular wall as both pro- and antiatherogenic effects of PPARα have been demonstrated.

PPARα has been shown to suppress expression of several proinflammatory genes in the vascular wall of animals with extensive atherosclerosis, including monocyte chemotactic protein-1 (*Mcp-1*), tumor necrosis factor alpha (*Tnfα*), vascular cell adhesion molecule-I (*Vcam I*), intercellular adhesion molecule-I (*Icam I*), and interferon-γ (*Ifnγ*) [52]. Other studies have shown that the anti-inflammatory role of PPARα in the vascular wall depends on the severity of inflammation or vascular lesion. In the absence of inflammation or in early lesions, the effects of PPARα are mainly proatherogenic [53, 54], whereas the development of severe lesions accompanied by inflammation is strongly reduced by PPARα activation.

Overall, it is clear that PPARα has a major impact on metabolic and inflammatory gene expression, especially in the liver and vascular wall. In general, these effects are positive in the context of specific metabolic diseases, including dyslipidemia and atherosclerosis.

10.3. PPARβ/δ (NR1C2)

The PPARβ/δ subtype was first identified in *Xenopus laevis* under the name PPARβ [23]. Shortly thereafter the receptor was cloned in mouse [18, 103] and human as NUC1 or PPARδ [20]. Throughout the remainder of this chapter we will refer to the receptor as PPARδ, which represents the official gene name. The human *PPARδ* gene has been mapped to chromosome 6, while its mouse counterpart is present on chromosome 17. Human and rodent PPARδ protein are highly homologous, sharing ~90% sequence identity in the LBD [47].

Similar to PPARα, PPARδ is ubiquitously expressed. High levels of mRNA have been found in the skin [104–107], heart [108, 109], skeletal muscle [110], adipose tissue [111, 112], small intestine [19, 28], and brain [19, 28, 113]. A recent study indicated that PPARδ protein is especially abundant in mouse small intestine, followed by keratinocytes, liver, and at much lower levels in heart and skeletal muscle [114]. In comparison to PPARα and PPARγ, the function of PPARδ is generally less well understood. However, in the past few years, using specific PPARδ agonists and/or PPARδ null mice, significant progress in the characterization of PPARδ has been made. It is now evident that PPARδ is involved in numerous biological processes including lipid metabolism [47, 113, 115–119], wound healing [120, 121], inflammation [122], placental development [47, 123–125], brain function and development [113, 126] and colon cancer [112, 127–130]. Here the focus will be on the role of PPARδ in inflammatory and metabolic processes (Figure 1b).

PPARδ can bind both endogenous and synthetic agonists. Endogenous PPARδ ligands include naturally occurring fatty acids [8, 131] as well as various eicosanoids such as prostaglandin A1 (PGA_1), prostaglandin D2 (PGD_2) and prostaglandin I2 (PGI_2) [50, 132, 133]. Recently, evidence was provided that PPARδ also binds retinoic acid, which may be selectively delivered to PPARδ via FABP5 [134].

Synthetic ligands of PPARδ are currently explored for their potential to improve plasma lipoprotein levels and include GW501516 and GW0742 [135]. In a primate model for type 2 diabetes, GW501516 increased serum HDL cholesterol, improved insulin sensitivity and reduced adiposity in diet-induced obese mice [118]. In addition, GW501516 was shown to reverse multiple abnormalities of the metabolic syndrome in mice and humans, causing significant reductions in plasma triglycerides, apolipoprotein B, LDL cholesterol, insulin, and glucose tolerance [119, 136–138]. So far none of the PPARδ agonists have been launched onto the market, though GW501516 and MBX-8025 have entered phase 2 clinical trails.

10.3.1. Lipid Metabolism

Analogous to PPARα and PPARγ, evidence is accumulating that PPARδ plays a role in the regulation of lipid metabolism (Figure 1b). Consequently, PPARδ has become an interesting target for the treatment of metabolic syndrome. Effects of PPARδ have been demonstrated in several tissues, including skeletal muscle, heart, adipose tissue, and liver.

10.3.1.1. Lipid metabolism in skeletal muscle

Numerous in vitro studies have shown a stimulatory effect of PPARδ overexpression or activation on expression of genes involved in fatty acid catabolism, including mitochondrial- (*Lcad*, *Hadha*, *Decr*) and peroxisomal fatty acid oxidation (*Ech*), fatty acid transport (*Fatp*, *Lipe*, *Acsl*, *Cpt1*) and energy uncoupling (*Ucp1*, *-2*, *-3*) [75, 119, 139–141] (Figure 1b). In line with these data, administration of GW501516 was found to induce fatty acid oxidation in skeletal muscle of C57BL/6J mice [119]. Similarly, the PPARδ agonist GW610742X decreased muscle lipid content and shifted fuel use towards fatty acids, while inducing expression of specific genes (*Pdk4*, *Cpt1b*, *Ucp3*).

Besides governing fatty acid oxidation, PPARδ also determines muscle fiber type characteristics. Forced overexpression of PPARδ in skeletal muscle is associated with a selective increase in type 2a fast-oxidative fibers or, when expressed at supraphysiological levels, causes fiber type transformation towards type I fibers concurrent with an increase in endurance exercise performance [142, 143]. Endurance exercise performance is also increased by GW501516, at least when combined with exercise training [144]. Conversely, selective deletion of PPARδ in skeletal muscle myocytes is associated with a reduced muscle oxidative capacity and a switch in muscle fiber-type characteristics toward less oxidative fibers [110]. Surprisingly, the role of PPARδ in skeletal muscle fatty acid oxidation is not supported by studies using whole body PPARδ null mice [145].

10.3.1.2. Lipid metabolism in heart

Similar to the situation in skeletal muscle, PPARδ appears to stimulate fatty acid oxidation in heart, although the picture is far from clear. Treatment of neonatal rat cardiomyocytes with PPARδ ligands L-165041 and GW501516 increases fatty acid oxidation rate and expression of selected genes involved in fatty acid catabolism [109]. In vivo, the absence of PPARδ in heart leads to myocardial lipid accumulation and cardiomyopathy, as well as downregulation of several key fatty acid oxidation genes such as carnitine palmitoyltransferase 1 (*Cpt1*) and acyl-coenzyme A oxidase 1 (*Acox1*) [146]. Conversely, treatment of rats with the PPARδ specific agonist GW610742X stimulated fatty acid oxidation rate. The metabolic changes were associated with increased expression of genes involved in lipid catabolism (*Cd36*, *Cpt1*, *Ucp3*) [147]. Also, transgenic overexpression of PPARδ in heart increased expression of a subset of PPARδ target genes involved in fatty acid catabolism, which did not translate into increased rate of fatty acid oxidation. Instead, PPARδ stimulated myocardial glucose utilization, possibly via induction of *Slc2a4* (*Glut4*) and phosphofructokinase [108]. Thus, while the importance of PPARδ in normal heart functioning is evident, its specific impact on cardiac fatty acid oxidation remains somewhat ambiguous.

10.3.1.3. Lipid metabolism in adipose tissue

A gain of function study has shown that PPARδ promotes fatty acid oxidation in adipocytes [137]. On the other hand, PPARδ seems to have a facilitative, yet important role in lipo- and adipogenesis [148]. Accordingly, the role of PPARδ in adipose tissue remains somewhat ambiguous.

10.3.1.4. Lipoprotein metabolism

Several studies support a role for PPARδ in lipoprotein metabolism. PPARδ agonists were shown to increase plasma HDL levels in mice [117, 149], rhesus monkeys [118], and human subjects [150]. Although the precise mechanism is unknown, the effect may be mediated by the cholesterol transporter *Abca1*, which is a target gene of PPARδ [118].

Apart from elevated levels of plasma HDL, primates also show a decrease in plasma triglycerides upon PPARδ activation [118, 150]. Consistent with these data, plasma TG levels are increased in PPARδ null mice [151]. Due to their beneficial effect on plasma lipoproteins, PPARδ agonists are currently explored for the treatment of dyslipidemia.

10.3.2. Wound Healing

PPARδ is the dominant PPAR subtype in the skin and has been shown to be involved in different phases of the healing process of epidermal wounds [30, 121, 152] (Figure 1b). PPARδ becomes induced in keratinocytes at the wound edge of damaged skin and, in contrast to PPARα, which is expressed during the early inflammatory phase of the healing, PPARδ remains active until the wound healing process has been completed [121]. Induction of PPARδ expression is mediated by inflammatory cytokines, which via induction of stress-associated kinase pathway target a AP-1 site in the PPARδ promoter [153]. The increase in PPARδ activity promotes keratinocyte differentiation and protects against apoptosis, thereby stimulating wound closure. Suppression of apoptosis is mediated by PPARδ-dependent upregulation of integrin-linked kinase and 3-phosphoinositide-dependent kinase-1 (*Pdpk1*), which phosphorylates protein kinase B-alpha (*Akt1*) [154]. In addition, PPARδ stimulates wound healing by altering actin cytoskeleton plasticity and integrin function, resulting in increased cell migration [155]. At later stages in wound healing, normal PPARδ expression is restored by a TGFβ and SMAD3-mediated suppression of c-JUN binding to the PPARδ promoter [156].

10.3.3. Inflammation

The role of PPARδ in inflammation has primarily been studied in the context of atherosclerosis (Figure 1b). Treatment with synthetic PPARδ agonist has been repeatedly shown to suppress inflammation in atherosclerotic lesions and lipid loaded macrophages, possibly by down-regulating expression of the chemoattractant *Ccl2* (*Mcp-1*) [157–160]. Recent studies suggest that the inflammatory properties of PPARδ extend to other cell types present in the vascular wall, including endothelial cells and vascular smooth muscle cells [161, 162]. Inhibition of inflammation likely accounts for the marked reduction in atherosclerotic lesion size upon PPARδ activation as observed in several but not all studies [151, 152, 160, 163]. Remarkably, macrophage-specific deletion of PPARδ is also associated with a significant reduction in atherosclerotic lesion size [157]. Together, these data suggest a complex role for PPARδ in atherosclerosis via its presence in macrophages and other cells that may involve both pro- and anti-inflammatory activities, as well as specific effects on plasma lipoproteins [157]. Possibly, the effects of PPARδ on these pathways are compounded by its ability to promote lipid accumulation in macrophages via induction of genes involved in lipid uptake and storage (*CD36*, *MSR1*) [164].

The general anti-inflammatory properties of PPARδ are further substantiated by recent data showing that PPARδ promotes alternative activation of macrophages resident in liver and adipose tissue, which confers a protection against insulin resistance and hepatic steatosis [165, 166].

10.4. PPARγ (NR1C3)

PPARγ is clearly the most widely studied PPAR subtype, which is explained by it serving as the molecular target for the insulin-sensitizing thiazolidinedione drugs (TZDs). PPARγ also plays a key role in adipogenesis and consequently has been extensively studied for its involvement in obesity development.

Four distinct transcript variants of PPARγ are known, labeled PPARγ1 through PPARγ4, that yield two protein variants differing at their N-terminus by the addition of 28 (human PPARγ2) or 30 (mouse PPARγ2) amino acids [23, 167–171]. Whereas PPARγ2 is expressed selectively in adipose tissue, [23, 167, 172–176], PPARγ1 has a broader expression pattern and is found in gut, brain, vascular cells and macrophages [167, 173, 174, 177].

Both common and rare sequence variants are known for the human *PPARγ* gene. The common Pro12Ala variant has been shown to be associated with a lower BMI, improved insulin sensitivity and reduced incidence of type 2 diabetes [178–182]. Rare sequence variants of PPARγ lead to the formation of a dysfunctional protein that via dominant negative action interferes with transcriptional activation, possibly by sequestering coactivator proteins [183]. Afflicted patients suffer from a form of lipodystrophy characterized by loss of fat from the gluteal region, dyslipidemia, hepatic steatosis, and severe insulin resistance [184].

Similar to other PPARs, PPARγ is able to bind both endogenous and synthetic ligands. The endogenous ligands for PPARγ remain poorly characterized. In contrast to PPARα and PPARδ, dietary (poly-unsaturated) fatty acids appear to be relatively weak ligands for PPARγ [47, 49, 185, 186]. Ligand-activation of PPARγ in specific cell types may occur by fatty acid nitration products [187], as well as by oxidized fatty acids such as the linoleic acid metabolites 9-HODE and 13-HODE [188]. It has been shown that prostaglandin 15d-PGJ2 efficiently binds and activates PPARγ, yet due to its low concentration, its relevance as a physiological PPARγ agonist can be questioned. As mentioned above, PPARγ is activated by synthetic ligands belonging to the antidiabetic thiazolidinedione drugs, which include troglitazone, rosiglitazone and pioglitazone [189–194]. These drugs improve insulin sensitivity and are used in the treatment of type 2 diabetes [190–196]. In addition, tyrosine derivative drugs like glitazars [197], as well as NSAIDs like ibuprofen and fenoprofen [198] have been identified as PPARγ ligands.

10.4.1. Metabolism

PPARγ is best known for its ability to stimulate adipocyte differentiation, fat storage, and glucose metabolism [199–204] (Figure 1b). Moreover, PPARγ suppresses inflammation. Although it is becoming more evident that metabolism and inflammation are intertwined, for the sake of simplicity the impact of PPARγ on the two processes will be discussed separately.

10.4.1.1. Adipose tissue

Studies with PPARγ null mice as well as PPARγ null stem cells have shown that PPARγ is absolutely required for adipocyte differentiation [205–208]. For detailed coverage of the role of PPARγ in adipogenesis the reader is referred to several excellent reviews [199–204]. Here the focus will be on the impact of PPARγ on the fully developed adipose tissue.

In the mature adipocyte, PPARγ stimulates the expression of numerous genes involved in fatty acid uptake (*Cd36*, *Slc27a1*, *Slc27a3*), fatty acid synthesis (*Elovls*, *Mogat, Acly*), lipolysis (*Lipe*, *Pnpla2*, *Mgll*, *Lpl*), lipid droplet proteins (*Cidea*, *Cidec*, *Adfp*, *Plin*, *S3-12*), glucose metabolism (*Pck1*, *Pdk4*, *Gys2*, *Slc2a4*), and glycerol metabolism (*Aqp7*, *Aqp3*, *Gpd1*, *Gyk, Pck1*). Additionally, a set of miscellaneous genes is regulated by PPARγ (*G0s2*, *Ucp1*, *Ucp2*, *Abca1*). The overall effect is enhanced extraction of fatty acids from circulating triglycerides, enhanced fatty acid uptake, enhanced fatty acid re-esterification, and enhanced storage as triglycerides. In addition, uptake of glucose as well as its conversion to fatty acid and glycerol phosphate is stimulated, contributing to increased energy storage. From a physiological perspective, PPARγ is thus particularly important in the fed state to drive storage of consumed nutrients in the adipose tissue.

From a clinical perspective, removal of fatty acids and triglycerides from the circulating pool towards storage in the adipose tissue improves dyslipidemia and minimizes ectopic fat storage. Indirectly, these changes may also promote insulin sensitivity. This is exemplified by the phenotype of hypomorphic PPARγ mice [209] and more distinctly adipose tissue specific PPARγ null mice, which show inborn and progressive lipodystrophy characterized by accumulation of TG in non-adipose tissue such as liver and skeletal muscle, leading to insulin resistance [210, 211]. The effect of PPARγ on plasma FFA and ectopic fat storage may provide a mechanistic basis for the insulin sensitizing effect of TZDs.

In addition to its primary function as an energy storage organ, adipose tissue produces various hormones such as leptin, adiponectin, resistin, and TNFα that play an active role in the regulation of energy metabolism. The insulin-sensitizing effect of TZDs may be partially accounted for by altered production of these hormones. Indeed, adiponectin, production of which is elevated by PPARγ, promotes insulin responsiveness and glucose uptake. Other so called adipokines that are under control of PPARγ and that may mediate effects of PPARγ activation include RBP4 and ANGPTL4 [212].

In addition to the mechanisms eluded to above, alternative explanations for the insulin sensitizing effect of PPARγ agonists include an increase in the number of small, insulin sensitive adipocytes [213], as well as a direct effect on macrophages, as will be further discussed below.

10.4.1.2. Non-adipose tissue

PPARγ has been shown to induce macrophage expression of scavenger receptor CD36, which is involved in uptake of oxidized LDL into the macrophage [188, 214]. Overall, much overlap is observed between the effect of PPARγ on gene expression in macrophages and adipocytes. In addition, macrophage PPARγ governs cholesterol esterification and intracellular cholesterol distribution, and stimulates cholesterol removal from the macrophage via the target genes *ABCA1*, *ABCG1*, caveolin and *APOE1* [34, 215–219]. Collectively, these effects of PPARγ beneficially impact macrophage foam cell formation [158].

Although PPARγ is only weakly expressed in skeletal muscle and liver, muscle or liver specific PPARγ null mice show a major and complex metabolic phenotype. Presently, the relative importance of skeletal muscle PPARγ in TZD-induced muscle glucose disposal remains undecided [220, 221]. In liver, PPARγ expression increases during over- and high fat feeding, concurrent with development of hepatic steatosis, which is aggravated in PPARα null mice [55, 208, 222]. Hepatic PPARγ is also upregulated in models of lipoatrophy and leptin deficiency, and studies employing PPARγ overexpression or deletion indicate that PPARγ is necessary and sufficient for inducing fatty liver [223–226].

10.4.2. Inflammation

Inflammation has become a prime area of interest as a candidate process linking obesity to many of its comorbidities. Numerous studies suggest that obesity is associated with a state of low grade inflammation, which likely originates from white adipose tissue and has been suggested to impact insulin sensitivity. In addition, inflammation importantly contributes to the process of atherosclerosis.

10.4.2.1. Atherosclerosis

Macrophages play an important role in both innate and adaptive immune responses, including phagocytosis of pathogens and defective or dying cells. In addition, by taking up oxidized LDL and converting into foam cells, macrophages are key contributors to development of atherosclerosis.

Although PPARγ may stimulate uptake of oxidized LDL into macrophages, which is pro-atherogenic, the anti-atherogenic effects of PPARγ via suppression of inflammation seem to dominate. PPARγ activation reduces arteriosclerotic lesions at least partially by inhibiting inflammatory gene expression in macrophages, including *MCP-1*, *VCAM-1*, *ICAM-1*, *IFNγ* and *TNFα* [158]. Furthermore, reduced amounts of cytokines [227], nitric oxide and macrophage-scavenger receptor class A (SRA) [228] have been observed upon PPARγ activation.

Anti-inflammatory effects resulting from PPARγ activation has been shown in human and mouse macrophages. Treatment with PPARγ agonist 15d-PGJ2 was shown to have an anti-inflammatory effect by decreasing production of inflammatory cytokines such as interleukin IL1β, IL6 and TNFα in human peripheral blood mononuclear cells (PBMCs) [227]. Furthermore, treatment of activated peritoneal macrophages with 15d-PGJ2 reduced expression of inducible nitric oxide synthase, gelatinase B and scavenger receptor A genes, partially by inhibiting the transcription factors AP-1, STAT and NFκB [229]. PPARγ activation was also shown to suppress expression of *COX2*, mainly via preventing activation and translocation of NFκB [230–232]. However, because of its limited specificity for PPARγ, many experimental outcomes following 15d-PGJ2 treatment may be only partially dependent on PPARγ, which complicates interpretation of the data.

Nevertheless, studies with synthetic PPARγ agonists support a general anti-inflammatory effect of PPARγ [233–235], which plays a role in the anti-atherogenic effects of PPARγ, as assessed by measurement of carotid arterial intimal and medial complex thickness [236–238].

Several molecular mechanisms have been proposed to underlie the anti-inflammatory effects of PPARγ. A major mechanism involves transrepression, which describes the DNA-binding independent protein-protein interaction between PPARγ and other (pro-inflammatory) transcription factors such as NFκB, STAT and AP-1, causing a change in transcriptional activity [229, 239]. In addition, PPARγ may compete with pro-inflammatory transcription factors for limited amounts of coactivators such as SRC-1, TIF2, AIB-1, CBP, p300, TRAP220, and DRIP205 in the cell [240]. Another possibility involves binding of PPARγ to nuclear receptor corepressor (NCoR)- histone deacetylase-3 (HDAC3) complexes, and preventing the removal of these corepressor complexes from promoter regions of inflammatory genes and causing a suppression of gene transcription [241]. Binding of PPARγ to NCoR is initiated by ligand-dependent SUMOylation of the PPARγ ligand binding domain [242, 243].

10.4.2.2. Adipose tissue

Macrophages are abundant in adipose tissue and together with adipocytes contribute to the secretion of a variety of pro- and anti-inflammatory cytokines. It is now evident that in obese individuals adipocyte hypertrophy leads to the recruitment of macrophages in adipose tissue, thereby altering its secretory profile [244–249]. Recent evidence suggests that these macrophages are primarily classically activated and mainly secrete pro-inflammatory cytokines [250]. Indeed, compared with lean individuals, obese persons have been observed to have a higher expression of TNFα, IL-6, MCP-1, INOS, and TGFß1 [251–261]. Presently, the trigger leading to the infiltration of macrophages is unclear but may involve local hypoxia as well as adipose cell death.

Recent studies support a major role for PPARγ in regulating not only the amount of macrophages present in adipose tissue but also their phenotype and secretory profile. Treatment of mice with PPARγ agonist stimulated infiltration of alternatively activated macrophages into adipose tissue, thereby reducing pro-inflammatory gene expression [262]. Conversely, macrophage-specific deletion of PPARγ decreased expression of markers of alternatively activated macrophages in adipose tissue, and increased inflammatory gene expression [263, 264]. The increased abundance of classically activated macrophages led to worsening of insulin resistance, especially after high fat feeding [263, 264]. These data suggest that macrophage PPARγ plays a major role in determining macrophage polarization in adipose tissue, and may mediate the effect of TZDs on insulin sensitivity.

Overall, it is evident that PPARγ has a major influence on metabolic and inflammatory gene expression, especially in adipocytes and macrophages. In general, these effects are positive in the context of specific metabolic diseases, including insulin sensitivity and atherosclerosis.

10.5. CONCLUDING REMARKS

It is now evident that the primary functions of PPARs are governing metabolic processes and inflammation. Presently, PPARα and PPARγ serve as therapeutic targets for dyslipidemia and insulin resistance, respectively, and synthetic ligands of PPARδ are currently being explored for their potential to improve plasma lipoprotein levels. As our understanding of the link between inflammation and metabolism advances, better insight will be obtained into the mechanism underlying the therapeutic actions of PPAR agonists.

REFERENCES

1. Issemann, I. and Green, S. (1990). Activation of a member of the steroid hormone receptor superfamily by peroxisome proliferators. *Nature* 347(6294), 645–650.
2. Mandard, S., Muller, M., and Kersten, S. (2004). Peroxisome proliferator-activated receptor alpha target genes. *Cell Mol Life Sci* 61(4), 393–416.
3. Huang, J. C. (2008). The role of peroxisome proliferator-activated receptors in the development and physiology of gametes and preimplantation embryos. *PPAR Res* 2008, 732303.
4. Feige, J. N. et al. (2005). Fluorescence imaging reveals the nuclear behavior of peroxisome proliferator-activated receptor/retinoid X receptor heterodimers in the absence and presence of ligand. *J Biol Chem* 280(18), 17880–17890.
5. Desvergne, B. and Wahli, W. (1999). Peroxisome proliferator-activated receptors: Nuclear control of metabolism. *Endocr Rev* 20(5), 649–688.
6. Nolte, R. T. et al. (1998). Ligand binding and co-activator assembly of the peroxisome proliferator-activated receptor-gamma. *Nature* 395(6698), 137–143.
7. Xu, H. E. et al. (2001). Structural determinants of ligand binding selectivity between the peroxisome proliferator-activated receptors. *Proc Natl Acad Sci U S A* 98(24), 13919–13924.
8. Xu, H. E. et al. (1999). Molecular recognition of fatty acids by peroxisome proliferator-activated receptors. *Mol Cell* 3(3), 397–403.
9. Krogsdam, A. M. et al. (2002). Nuclear receptor corepressor-dependent repression of peroxisome-proliferator-activated receptor delta-mediated transactivation. *Biochem J* 363(Pt 1), 157–165.
10. Degenhardt, T. et al. (2006). The insulin-like growth factor-binding protein 1 gene is a primary target of peroxisome proliferator-activated receptors. *J Biol Chem* 281(51), 39607–39619.
11. Yu, S. and Reddy, J. K. (2007). Transcription coactivators for peroxisome proliferator-activated receptors. *Biochim Biophys Acta* 1771(8), 936–951.
12. Rosenfeld, M. G. and Glass, C. K. (2001). Coregulator codes of transcriptional regulation by nuclear receptors. *J Biol Chem* 276(40), 36865–36868.
13. Jia, Y. et al. (2004). Transcription coactivator PBP, the peroxisome proliferator-activated receptor (PPAR)-binding protein, is required for PPARalpha-regulated gene expression in liver. *J Biol Chem* 279(23), 24427–24434.
14. Puigserver, P. et al. (1998). A cold-inducible coactivator of nuclear receptors linked to adaptive thermogenesis. *Cell* 92(6), 829–839.
15. Vega, R. B., Huss, J. M., and Kelly, D. P. (2000). The coactivator PGC-1 cooperates with peroxisome proliferator-activated receptor alpha in transcriptional control of nuclear genes encoding mitochondrial fatty acid oxidation enzymes. *Mol Cell Biol* 20(5), 1868–1876.
16. Yoon, J. C. et al. (2001). Control of hepatic gluconeogenesis through the transcriptional coactivator PGC-1. *Nature* 413(6852), 131–138.
17. Rhee, J. et al. (2003). Regulation of hepatic fasting response by PPARgamma coactivator-1alpha (PGC-1): Requirement for hepatocyte nuclear factor 4alpha in gluconeogenesis. *Proc Natl Acad Sci U S A* 100(7), 4012–4017.
18. Kliewer, S. A. et al. (1994). Differential expression and activation of a family of murine peroxisome proliferator-activated receptors. *Proc Natl Acad Sci U S A* 91(15), 7355–7359.
19. Braissant, O. et al. (1996). Differential expression of peroxisome proliferator-activated receptors (PPARs): Tissue distribution of PPAR-alpha, -beta, and -gamma in the adult rat. *Endocrinology* 137(1), 354–366.
20. Schmidt, A. et al. (1992). Identification of a new member of the steroid hormone receptor superfamily that is activated by a peroxisome proliferator and fatty acids. *Mol Endocrinol* 6(10), 1634–1641.
21. Sher, T. et al. (1993). cDNA cloning, chromosomal mapping, and functional characterization of the human peroxisome proliferator activated receptor. *Biochemistry* 32(21), 5598–5604.
22. Gottlicher, M. et al. (1992). Fatty acids activate a chimera of the clofibric acid-activated receptor and the glucocorticoid receptor. *Proc Natl Acad Sci USA* 89(10), 4653–4657.

23. Dreyer, C. et al. (1992). Control of the peroxisomal beta-oxidation pathway by a novel family of nuclear hormone receptors. *Cell* 68(5), 879–887.
24. Guan, Y. et al. (1997). Expression of peroxisome proliferator-activated receptors in urinary tract of rabbits and humans. *Am J Physiol* 273(6 Pt 2), F1013–F1022.
25. Hoekstra, M. et al. (2003). Specific gene expression of ATP binding cassette transporters and nuclear hormone receptors in rat liver parenchymal, endothelial, and Kupffer cells. *J Biol Chem* 278(28), 25448–25453.
26. Palmer, C. N. et al. (1998). Peroxisome proliferator activated receptor-alpha expression in human liver. *Mol Pharmacol* 53(1), 14–22.
27. Jones, P. S. et al. (1995). Chromosomal localisation, inducibility, tissue-specific expression and strain differences in three murine peroxisome-proliferator-activated-receptor genes. *Eur J Biochem* 233(1), 219–226.
28. Escher, P. et al. (2001). Rat PPARs: Quantitative analysis in adult rat tissues and regulation in fasting and refeeding. *Endocrinology* 142(10), 4195–4202.
29. Auboeuf, D. et al. (1997). Tissue distribution and quantification of the expression of mRNAs of peroxisome proliferator-activated receptors and liver X receptor-alpha in humans: No alteration in adipose tissue of obese and NIDDM patients. *Diabetes* 46(8), 1319–1327.
30. Braissant, O. and Wahli, W. (1998). Differential expression of peroxisome proliferator-activated receptor-alpha, -beta, and -gamma during rat embryonic development. *Endocrinology* 139(6), 2748–2754.
31. Loviscach, M. et al. (2000). Distribution of peroxisome proliferator-activated receptors (PPARs) in human skeletal muscle and adipose tissue: Relation to insulin action. *Diabetologia* 43(3), 304–311.
32. Chinetti, G. et al. (1998). Activation of proliferator-activated receptors alpha and gamma induces apoptosis of human monocyte-derived macrophages. *J Biol Chem* 273(40), 25573–25580.
33. Chinetti, G. et al. (2000). CLA-1/SR-BI is expressed in atherosclerotic lesion macrophages and regulated by activators of peroxisome proliferator-activated receptors. *Circulation* 101(20), 2411–2417.
34. Chinetti, G. et al. (2001). PPAR-alpha and PPAR-gamma activators induce cholesterol removal from human macrophage foam cells through stimulation of the ABCA1 pathway. *Nat Med* 7(1), 53–58.
35. Gbaguidi, F. G. et al. (2002). Peroxisome proliferator-activated receptor (PPAR) agonists decrease lipoprotein lipase secretion and glycated LDL uptake by human macrophages. *FEBS Lett* 512(1-3), 85–90.
36. Marx, N. et al. (2001). PPARalpha activators inhibit tissue factor expression and activity in human monocytes. *Circulation* 103(2), 213–219.
37. Neve, B. P. et al. (2001). PPARalpha agonists inhibit tissue factor expression in human monocytes and macrophages. *Circulation* 103(2), 207–212.
38. Cunard, R. et al. (2002). Regulation of cytokine expression by ligands of peroxisome proliferator activated receptors. *J Immunol* 168(6), 2795–2802.
39. Jones, D. C., Ding, X., and Daynes, R. A. (2002). Nuclear receptor peroxisome proliferator-activated receptor alpha (PPARalpha) is expressed in resting murine lymphocytes. The PPARalpha in T and B lymphocytes is both transactivation and transrepression competent. *J Biol Chem* 277(9), 6838–6845.
40. Marx, N. et al. (2002). PPAR activators as antiinflammatory mediators in human T lymphocytes: Implications for atherosclerosis and transplantation-associated arteriosclerosis. *Circ Res* 90(6), 703–710.
41. Staels, B. et al. (1998). Activation of human aortic smooth-muscle cells is inhibited by PPARalpha but not by PPARgamma activators. *Nature* 393(6687), 790–793.
42. Marx, N. et al. (1999). PPARalpha activators inhibit cytokine-induced vascular cell adhesion molecule-1 expression in human endothelial cells. *Circulation* 99(24), 3125–3131.

43. Delerive, P. et al. (1999). Peroxisome proliferator-activated receptor activators inhibit thrombin-induced endothelin-1 production in human vascular endothelial cells by inhibiting the activator protein-1 signaling pathway. *Circ Res* 85(5), 394–402.
44. Inoue, I. et al. (1998). Expression of peroxisome proliferator-activated receptor alpha (PPAR alpha) in primary cultures of human vascular endothelial cells. *Biochem Biophys Res Commun* 246(2), 370–374.
45. Kersten, S., Desvergne, B., and Wahli, W. (2000). Roles of PPARs in health and disease. *Nature* 405(6785), 421–424.
46. Thorp, J. M. and Waring, W. S. (1962). Modification of metabolism and distribution of lipids by ethyl chlorophenoxyisobutyrate. *Nature* 194, 948–949.
47. Willson, T. M. et al. (2000). The PPARs: From orphan receptors to drug discovery. *J Med Chem* 43(4), 527–550.
48. Berger, J. and Moller, D. E. (2002). The mechanisms of action of PPARs. *Annu Rev Med* 53, 409–435.
49. Kliewer, S. A. et al. (1997). Fatty acids and eicosanoids regulate gene expression through direct interactions with peroxisome proliferator-activated receptors alpha and gamma. *Proc Natl Acad Sci U S A* 94(9), 4318–4323.
50. Forman, B. M., Chen, J., and Evans, R. M. (1997). Hypolipidemic drugs, polyunsaturated fatty acids, and eicosanoids are ligands for peroxisome proliferator-activated receptors alpha and delta. *Proc Natl Acad Sci U S A* 94(9), 4312–4317.
51. Sanderson, L. M. et al. (2008). Effect of synthetic dietary triglycerides: A novel research paradigm for nutrigenomics. *PLoS ONE* 3(2), e1681.
52. Hashimoto, T. et al. (2000). Defect in peroxisome proliferator-activated receptor alpha-inducible fatty acid oxidation determines the severity of hepatic steatosis in response to fasting. *J Biol Chem* 275(37), 28918–28928.
53. Leone, T. C., Weinheimer, C. J., and Kelly, D. P. (1999). A critical role for the peroxisome proliferator-activated receptor alpha (PPARalpha) in the cellular fasting response: The PPARalpha-null mouse as a model of fatty acid oxidation disorders. *Proc Natl Acad Sci U S A* 96(13), 7473–7478.
54. Kersten, S. et al. (1999). Peroxisome proliferator-activated receptor alpha mediates the adaptive response to fasting. *J Clin Invest* 103(11), 1489–1498.
55. Patsouris, D. et al. (2006). Peroxisome proliferator-activated receptor alpha mediates the effects of high-fat diet on hepatic gene expression. *Endocrinology* 147(3), 1508–1516.
56. Tugwood, J. D. et al. (1992). The mouse peroxisome proliferator activated receptor recognizes a response element in the 5′ flanking sequence of the rat acyl CoA oxidase gene. *EMBO J* 11(2), 433–439.
57. Rodriguez, J. C. et al. (1994). Peroxisome proliferator-activated receptor mediates induction of the mitochondrial 3-hydroxy-3-methylglutaryl-CoA synthase gene by fatty acids. *J Biol Chem* 269(29), 18767–18772.
58. Hegardt, F. G. (1998). Transcriptional regulation of mitochondrial HMG-CoA synthase in the control of ketogenesis. *Biochimie* 80(10), 803–806.
59. Le May, C. et al. (2000). Reduced hepatic fatty acid oxidation in fasting PPARalpha null mice is due to impaired mitochondrial hydroxymethylglutaryl-CoA synthase gene expression. *FEBS Lett* 475(3), 163–166.
60. Martin, G. et al. (1997). Coordinate regulation of the expression of the fatty acid transport protein and acyl-CoA synthetase genes by PPARalpha and PPARgamma activators. *J Biol Chem* 272(45), 28210–28217.
61. Frohnert, B. I., Hui, T. Y., and Bernlohr, D. A. (1999). Identification of a functional peroxisome proliferator-responsive element in the murine fatty acid transport protein gene. *J Biol Chem* 274(7), 3970–3977.

62. Motojima, K. et al. (1998). Expression of putative fatty acid transporter genes are regulated by peroxisome proliferator-activated receptor alpha and gamma activators in a tissue- and inducer-specific manner. *J Biol Chem* 273(27), 16710–16714.
63. Sato, O. et al. (2002). Dual promoter structure of mouse and human fatty acid translocase/CD36 genes and unique transcriptional activation by peroxisome proliferator-activated receptor alpha and gamma ligands. *J Biol Chem* 277(18), 15703–15711.
64. Wolfrum, C. et al. (2001). Fatty acids and hypolipidemic drugs regulate peroxisome proliferator-activated receptors alpha - and gamma-mediated gene expression via liver fatty acid binding protein: A signaling path to the nucleus. *Proc Natl Acad Sci U S A* 98(5), 2323–2328.
65. Tan, N. S. et al. (2002). Selective cooperation between fatty acid binding proteins and peroxisome proliferator-activated receptors in regulating transcription. *Mol Cell Biol* 22(14), 5114–5127.
66. Poirier, H. et al. (2001). Differential involvement of peroxisome-proliferator-activated receptors alpha and delta in fibrate and fatty-acid-mediated inductions of the gene encoding liver fatty-acid-binding protein in the liver and the small intestine. *Biochem J* 355(Pt 2), 481–488.
67. Rakhshandehroo, M. et al. (2007). Comprehensive analysis of PPARalpha-dependent regulation of hepatic lipid metabolism by expression profiling. *PPAR Res* 2007, 26839.
68. Chakravarthy, M. V. et al. (2005). "New" hepatic fat activates PPARalpha to maintain glucose, lipid, and cholesterol homeostasis. *Cell Metab* 1(5), 309–322.
69. Sanderson, L. M. et al., (2009) PPARβ/δ but not PPARα serves as plasma free fatty acid sensor in liver. *Mol Cell Biol*, 29(23), 6257–6267.
70. Brandt, J. M., Djouadi, F., and Kelly, D. P. (1998). Fatty acids activate transcription of the muscle carnitine palmitoyltransferase I gene in cardiac myocytes via the peroxisome proliferator-activated receptor alpha. *J Biol Chem* 273(37), 23786–23792.
71. Finck, B. N. et al. (2002). The cardiac phenotype induced by PPARalpha overexpression mimics that caused by diabetes mellitus. *J Clin Invest* 109(1), 121–130.
72. Finck, B. N. et al. (2005). A potential link between muscle peroxisome proliferator- activated receptor-alpha signaling and obesity-related diabetes. *Cell Metab* 1(2), 133–144.
73. Bunger, M. et al. (2007). Genome-wide analysis of PPARalpha activation in murine small intestine. *Physiol Genomics* 30(2), 192–204.
74. Xu, J. et al. (2002). Peroxisome proliferator-activated receptor alpha (PPARalpha) influences substrate utilization for hepatic glucose production. *J Biol Chem* 277(52), 50237–50244.
75. Muoio, D. M. et al. (2002). Fatty acid homeostasis and induction of lipid regulatory genes in skeletal muscles of peroxisome proliferator-activated receptor (PPAR) alpha knock-out mice. Evidence for compensatory regulation by PPAR delta. *J Biol Chem* 277(29), 26089–26097.
76. Sugden, M. C. et al. (2002). Peroxisome-proliferator-activated receptor-alpha (PPARalpha) deficiency leads to dysregulation of hepatic lipid and carbohydrate metabolism by fatty acids and insulin. *Biochem J* 364(Pt 2), 361–368.
77. Patsouris, D. et al. (2004). PPARalpha governs glycerol metabolism. *J Clin Invest* 114(1), 94–103.
78. Bandsma, R. H. et al. (2004). Hepatic de novo synthesis of glucose 6-phosphate is not affected in peroxisome proliferator-activated receptor alpha-deficient mice but is preferentially directed toward hepatic glycogen stores after a short term fast. *J Biol Chem* 279(10), 8930–8937.
79. Mandard, S. et al. (2007). Glycogen synthase 2 is a novel target gene of peroxisome proliferator-activated receptors. *Cell Mol Life Sci* 64(9), 1145–1157.
80. Knauf, C. et al. (2006). Peroxisome proliferator-activated receptor-alpha-null mice have increased white adipose tissue glucose utilization, GLUT4, and fat mass: Role in liver and brain. *Endocrinology* 147(9), 4067–4078.
81. Sugden, M. C. et al. (2001). Role of peroxisome proliferator-activated receptor-alpha in the mechanism underlying changes in renal pyruvate dehydrogenase kinase isoform 4 protein expression in starvation and after refeeding. *Arch Biochem Biophys* 395(2), 246–252.
82. Wu, P., Peters, J. M., and Harris, R. A. (2001). Adaptive increase in pyruvate dehydrogenase kinase 4 during starvation is mediated by peroxisome proliferator-activated receptor alpha. *Biochem Biophys Res Commun* 287(2), 391–396.

83. Wu, P. et al. (1999). Mechanism responsible for inactivation of skeletal muscle pyruvate dehydrogenase complex in starvation and diabetes. *Diabetes* 48(8), 1593–1599.
84. Holness, M. J. et al. (2002). Up-regulation of pyruvate dehydrogenase kinase isoform 4 (PDK4) protein expression in oxidative skeletal muscle does not require the obligatory participation of peroxisome-proliferator-activated receptor alpha (PPARalpha). *Biochem J* 366(Pt 3), 839–846.
85. Holness, M. J. et al. (2002). Evaluation of the role of peroxisome-proliferator-activated receptor alpha in the regulation of cardiac pyruvate dehydrogenase kinase 4 protein expression in response to starvation, high-fat feeding and hyperthyroidism. *Biochem J* 364(Pt 3), 687–694.
86. Edgar, A. D. et al. (1998). Fenofibrate modifies transaminase gene expression via a peroxisome proliferator activated receptor alpha-dependent pathway. *Toxicol Lett* 98(1-2), 13–23.
87. Kersten, S. et al. (2001). The peroxisome proliferator-activated receptor alpha regulates amino acid metabolism. *FASEB J* 15(11), 1971–1978.
88. Sheikh, K. et al. (2007). Beyond lipids, pharmacological PPARalpha activation has important effects on amino acid metabolism as studied in the rat. *Am J Physiol Endocrinol Metab* 292(4), E1157–E1165.
89. Desvergne, B., Michalik, L., and Wahli, W. (2006). Transcriptional regulation of metabolism. *Physiol Rev* 86(2), 465–514.
90. Devchand, P. R. et al. (1996). The PPARalpha-leukotriene B4 pathway to inflammation control. *Nature* 384(6604), 39–43.
91. Delerive, P. et al. (1999). Peroxisome proliferator-activated receptor alpha negatively regulates the vascular inflammatory gene response by negative cross-talk with transcription factors NF-kappaB and AP-1. *J Biol Chem* 274(45), 32048–32054.
92. Gervois, P. et al. (2001). Negative regulation of human fibrinogen gene expression by peroxisome proliferator-activated receptor alpha agonists via inhibition of CCAAT box/enhancer-binding protein beta. *J Biol Chem* 276(36), 33471–33477.
93. Delerive, P., Fruchart, J. C., and Staels, B. (2001). Peroxisome proliferator-activated receptors in inflammation control. *J Endocrinol* 169(3), 453–459.
94. Gervois, P. et al. (2004). Global suppression of IL-6-induced acute phase response gene expression after chronic in vivo treatment with the peroxisome proliferator-activated receptor-alpha activator fenofibrate. *J Biol Chem* 279(16), 16154–16160.
95. Stienstra, R. et al. (2007). The Interleukin-1 receptor antagonist is a direct target gene of PPARalpha in liver. *J Hepatol* 46(5), 869–877.
96. Ip, E. et al. (2003). Central role of PPARalpha-dependent hepatic lipid turnover in dietary steatohepatitis in mice. *Hepatology* 38(1), 123–132.
97. Ip, E. et al. (2004). Administration of the potent PPARalpha agonist, Wy-14,643, reverses nutritional fibrosis and steatohepatitis in mice. *Hepatology* 39(5), 1286–1296.
98. Kashireddy, P. V. and Rao, M. S. (2004). Lack of peroxisome proliferator-activated receptor alpha in mice enhances methionine and choline deficient diet-induced steatohepatitis. *Hepatol Res* 30(2), 104–110.
99. Shiri-Sverdlov, R. et al. (2006). Early diet-induced non-alcoholic steatohepatitis in APOE2 knock-in mice and its prevention by fibrates. *J Hepatol* 44(4), 732–741.
100. Yu, J. et al. (2006). COX-2 induction in mice with experimental nutritional steatohepatitis: Role as pro-inflammatory mediator. *Hepatology* 43(4), 826–836.
101. Stienstra, R. et al. (2007). Peroxisome proliferator-activated receptor alpha protects against obesity-induced hepatic inflammation. *Endocrinology* 148(6), 2753–2763.
102. Lusis, A. J. (2000). Atherosclerosis. *Nature* 407(6801), 233–241.
103. Chen, F., Law, S. W., and O'Malley, B. W. (1993). Identification of two mPPAR related receptors and evidence for the existence of five subfamily members. *Biochem Biophys Res Commun* 196(2), 671–677.
104. Di-Poi, N. et al. (2004). Functions of peroxisome proliferator-activated receptors (PPAR) in skin homeostasis. *Lipids* 39(11), 1093–1099.
105. Burdick, A. D. et al. (2006). The role of peroxisome proliferator-activated receptor-beta/delta in epithelial cell growth and differentiation. *Cell Signal* 18(1), 9–20.

106. Kuenzli, S. and Saurat, J. H. (2003). Peroxisome proliferator-activated receptors in cutaneous biology. *Br J Dermatol* 149(2), 229–236.
107. Michalik, L. et al. (2000). Nuclear hormone receptors and mouse skin homeostasis: Implication of PPARbeta. *Horm Res* 54(5-6), 263–268.
108. Burkart, E. M. et al. (2007). Nuclear receptors PPARbeta/delta and PPARalpha direct distinct metabolic regulatory programs in the mouse heart. *J Clin Invest* 117(12), 3930–3939.
109. Gilde, A. J. et al. (2003). Peroxisome proliferator-activated receptor (PPAR) alpha and PPARbeta/delta, but not PPARgamma, modulate the expression of genes involved in cardiac lipid metabolism. *Circ Res* 92(5), 518–524.
110. Schuler, M. et al. (2006). PGC1alpha expression is controlled in skeletal muscles by PPARbeta, whose ablation results in fiber-type switching, obesity, and type 2 diabetes. *Cell Metab* 4(5), 407–414.
111. Amri, E. Z. et al. (1995). Cloning of a protein that mediates transcriptional effects of fatty acids in preadipocytes. Homology to peroxisome proliferator-activated receptors. *J Biol Chem* 270(5), 2367–2371.
112. Barak, Y. et al. (2002). Effects of peroxisome proliferator-activated receptor delta on placentation, adiposity, and colorectal cancer. *Proc Natl Acad Sci U S A* 99(1), 303–308.
113. Basu-Modak, S. et al. (1999). Peroxisome proliferator-activated receptor beta regulates acyl-CoA synthetase 2 in reaggregated rat brain cell cultures. *J Biol Chem* 274(50), 35881–35888.
114. Girroir, E. E. et al. (2008). Quantitative expression patterns of peroxisome proliferator-activated receptor-beta/delta (PPARbeta/delta) protein in mice. *Biochem Biophys Res Commun* 371(3), 456–461.
115. Kota, B. P., Huang, T. H., and Roufogalis, B. D. (2005). An overview on biological mechanisms of PPARs. *Pharmacol Res* 51(2), 85–94.
116. Peters, J. M. et al. (2000). Growth, adipose, brain, and skin alterations resulting from targeted disruption of the mouse peroxisome proliferator-activated receptor beta(delta). *Mol Cell Biol* 20(14), 5119–5128.
117. Leibowitz, M. D. et al. (2000). Activation of PPARdelta alters lipid metabolism in db/db mice. *FEBS Lett* 473(3), 333–336.
118. Oliver, W. R., Jr. et al. (2001). A selective peroxisome proliferator-activated receptor delta agonist promotes reverse cholesterol transport. *Proc Natl Acad Sci U S A* 98(9), 5306–5311.
119. Tanaka, T. et al. (2003). Activation of peroxisome proliferator-activated receptor delta induces fatty acid beta-oxidation in skeletal muscle and attenuates metabolic syndrome. *Proc Natl Acad Sci USA* 100(26), 15924–15929.
120. Barish, G. D., Narkar, V. A., and Evans, R. M. (2006). PPAR delta: A dagger in the heart of the metabolic syndrome. *J Clin Invest* 116(3), 590–597.
121. Michalik, L. et al. (2001). Impaired skin wound healing in peroxisome proliferator-activated receptor (PPAR)alpha and PPARbeta mutant mice. *J Cell Biol* 154(4), 799–814.
122. Schmuth, M. et al. (2004). Peroxisome proliferator-activated receptor (PPAR)-beta/delta stimulates differentiation and lipid accumulation in keratinocytes. *J Invest Dermatol* 122(4), 971–983.
123. Lim, H. et al. (1999). Cyclo-oxygenase-2-derived prostacyclin mediates embryo implantation in the mouse via PPARdelta. *Genes Dev* 13(12), 1561–1574.
124. Ding, N. Z. et al. (2003). Differential expression of peroxisome proliferator-activated receptor delta at implantation sites and in decidual cells of rat uterus. *Reproduction* 125(6), 817–825.
125. Rees, W. D., McNeil, C. J., and Maloney, C. A. (2008). The Roles of PPARs in the Fetal Origins of Metabolic Health and Disease. *PPAR Res* 2008, 459030.
126. Granneman, J., Skoff, R., and Yang, X. (1998). Member of the peroxisome proliferator-activated receptor family of transcription factors is differentially expressed by oligodendrocytes. *J Neurosci Res* 51(5), 563–573.
127. Wang, D. et al. (2004). Prostaglandin E(2) promotes colorectal adenoma growth via transactivation of the nuclear peroxisome proliferator-activated receptor delta. *Cancer Cell* 6(3), 285–295.
128. He, T. C. et al. (1999). PPARdelta is an APC-regulated target of nonsteroidal anti-inflammatory drugs. *Cell* 99(3), 335–345.

129. Gupta, R. A. et al. (2000). Prostacyclin-mediated activation of peroxisome proliferator-activated receptor delta in colorectal cancer. *Proc Natl Acad Sci U S A* 97(24), 13275–13280.
130. Park, B. H., Vogelstein, B., and Kinzler, K. W. (2001). Genetic disruption of PPARdelta decreases the tumorigenicity of human colon cancer cells. *Proc Natl Acad Sci U S A* 98(5), 2598–2603.
131. Fyffe, S. A. et al. (2006). Recombinant human PPAR-beta/delta ligand-binding domain is locked in an activated conformation by endogenous fatty acids. *J Mol Biol* 356(4), 1005–1013.
132. Yu, K. et al. (1995). Differential activation of peroxisome proliferator-activated receptors by eicosanoids. *J Biol Chem* 270(41), 23975–23983.
133. Lim, H. and Dey, S. K. (2002). A novel pathway of prostacyclin signaling-hanging out with nuclear receptors. *Endocrinology* 143(9), 3207–3210.
134. Schug, T. T. et al. (2007). Opposing effects of retinoic acid on cell growth result from alternate activation of two different nuclear receptors. *Cell* 129(4), 723–733.
135. Sznaidman, M. L. et al. (2003). Novel selective small molecule agonists for peroxisome proliferator-activated receptor delta (PPARdelta)–synthesis and biological activity. *Bioorg Med Chem Lett* 13(9), 1517–1521.
136. Riserus, U. et al. (2008). Activation of peroxisome proliferator-activated receptor (PPAR)delta promotes reversal of multiple metabolic abnormalities, reduces oxidative stress, and increases fatty acid oxidation in moderately obese men. *Diabetes* 57(2), 332–339.
137. Wang, Y. X. et al. (2003). Peroxisome-proliferator-activated receptor delta activates fat metabolism to prevent obesity. *Cell* 113(2), 159–170.
138. Lee, C. H. et al. (2006). PPARdelta regulates glucose metabolism and insulin sensitivity. *Proc Natl Acad Sci U S A* 103(9), 3444–3449.
139. Dressel, U. et al. (2003). The peroxisome proliferator-activated receptor beta/delta agonist, GW501516, regulates the expression of genes involved in lipid catabolism and energy uncoupling in skeletal muscle cells. *Mol Endocrinol* 17(12), 2477–2493.
140. Chevillotte, E. et al. (2001). The regulation of uncoupling protein-2 gene expression by omega-6 polyunsaturated fatty acids in human skeletal muscle cells involves multiple pathways, including the nuclear receptor peroxisome proliferator-activated receptor beta. *J Biol Chem* 276(14), 10853–10860.
141. Holst, D. et al. (2003). Nutritional regulation and role of peroxisome proliferator-activated receptor delta in fatty acid catabolism in skeletal muscle. *Biochim Biophys Acta* 1633(1), 43–50.
142. Luquet, S. et al. (2003). Peroxisome proliferator-activated receptor delta controls muscle development and oxidative capability. *FASEB J* 17(15), 2299–2301.
143. Wang, Y. X. et al. (2004). Regulation of muscle fiber type and running endurance by PPARdelta. *PLoS Biol* 2(10), e294.
144. Narkar, V. A. et al. (2008). AMPK and PPARdelta agonists are exercise mimetics. *Cell* 134(3), 405–415.
145. Bedu, E. et al. (2007). Double gene deletion reveals the lack of cooperation between PPARalpha and PPARbeta in skeletal muscle. *Biochem Biophys Res Commun* 357(4), 877–881.
146. Cheng, L. et al. (2004). Cardiomyocyte-restricted peroxisome proliferator-activated receptor-delta deletion perturbs myocardial fatty acid oxidation and leads to cardiomyopathy. *Nat Med* 10(11), 1245–1250.
147. Jucker, B. M. et al. (2007). PPARdelta activation normalizes cardiac substrate metabolism and reduces right ventricular hypertrophy in congestive heart failure. *J Cardiovasc Pharmacol* 50(1), 25–34.
148. Matsusue, K., Peters, J. M., and Gonzalez, F. J. (2004). PPARbeta/delta potentiates PPARgamma-stimulated adipocyte differentiation. *FASEB J* 18(12), 1477–1479.
149. van der Veen, J. N. et al. (2005). Reduced cholesterol absorption upon PPARdelta activation coincides with decreased intestinal expression of NPC1L1. *J Lipid Res* 46(3), 526–534.
150. Sprecher, D. L. et al. (2007). Triglyceride:high-density lipoprotein cholesterol effects in healthy subjects administered a peroxisome proliferator activated receptor delta agonist. *Arterioscler Thromb Vasc Biol* 27(2), 359–365.

151. Akiyama, T. E. et al. (2004). Peroxisome proliferator-activated receptor beta/delta regulates very low density lipoprotein production and catabolism in mice on a Western diet. *J Biol Chem* 279(20), 20874–20881.
152. Westergaard, M. et al. (2001). Modulation of keratinocyte gene expression and differentiation by PPAR-selective ligands and tetradecylthioacetic acid. *J Invest Dermatol* 116(5), 702–712.
153. Tan, N. S. et al. (2001). Critical roles of PPAR beta/delta in keratinocyte response to inflammation. *Genes Dev* 15(24), 3263–3277.
154. Di-Poi, N. et al. (2002). Antiapoptotic role of PPARbeta in keratinocytes via transcriptional control of the Akt1 signaling pathway. *Mol Cell* 10(4), 721–733.
155. Tan, N. S. et al. (2007). The nuclear hormone receptor peroxisome proliferator-activated receptor beta/delta potentiates cell chemotactism, polarization, and migration. *Mol Cell Biol* 27(20), 7161–7175.
156. Tan, N. S. et al. (2004). Essential role of Smad3 in the inhibition of inflammation-induced PPARbeta/delta expression. *EMBO J* 23(21), 4211–4221.
157. Lee, C. H. et al. (2003). Transcriptional repression of atherogenic inflammation: Modulation by PPARdelta. *Science* 302(5644), 453–457.
158. Li, A. C. et al. (2004). Differential inhibition of macrophage foam-cell formation and atherosclerosis in mice by PPARalpha, beta/delta, and gamma. *J Clin Invest* 114(11), 1564–1576.
159. Graham, T. L. et al. (2005). The PPARdelta agonist GW0742X reduces atherosclerosis in LDLR(-/-) mice. *Atherosclerosis* 181(1), 29–37.
160. Barish, G. D. et al. (2008). PPARdelta regulates multiple proinflammatory pathways to suppress atherosclerosis. *Proc Natl Acad Sci U S A* 105(11), 4271–4276.
161. Fan, Y. et al. (2008). Suppression of pro-inflammatory adhesion molecules by PPAR-delta in human vascular endothelial cells. *Arterioscler Thromb Vasc Biol* 28(2), 315–321.
162. Kim, H. J. et al. (2008). Transforming growth factor-beta1 is a molecular target for the peroxisome proliferator-activated receptor delta. *Circ Res* 102(2), 193–200.
163. Takata, Y. et al. (2008). PPARdelta-mediated antiinflammatory mechanisms inhibit angiotensin II-accelerated atherosclerosis. *Proc Natl Acad Sci U S A* 105(11), 4277–4282.
164. Vosper, H. et al. (2001). The peroxisome proliferator-activated receptor delta promotes lipid accumulation in human macrophages. *J Biol Chem* 276(47), 44258–44265.
165. Kang, K. et al. (2008). Adipocyte-derived Th2 cytokines and myeloid PPARdelta regulate macrophage polarization and insulin sensitivity. *Cell Metab* 7(6), 485–495.
166. Odegaard, J. I. et al. (2008). Alternative M2 activation of Kupffer cells by PPARdelta ameliorates obesity-induced insulin resistance. *Cell Metab* 7(6), 496–507.
167. Zhu, Y. et al. (1995). Structural organization of mouse peroxisome proliferator-activated receptor gamma (mPPAR gamma) gene: Alternative promoter use and different splicing yield two mPPAR gamma isoforms. *Proc Natl Acad Sci U S A* 92(17), 7921–7925.
168. Zhu, Y. et al. (1993). Cloning of a new member of the peroxisome proliferator-activated receptor gene family from mouse liver. *J Biol Chem* 268(36), 26817–26820.
169. Tontonoz, P. et al. (1994). mPPAR gamma 2: Tissue-specific regulator of an adipocyte enhancer. *Genes Dev* 8(10), 1224–1234.
170. Greene, M. E. et al. (1995). Isolation of the human peroxisome proliferator activated receptor gamma cDNA: Expression in hematopoietic cells and chromosomal mapping. *Gene Expr* 4(4-5), 281–299.
171. Fajas, L. et al. (1997). The organization, promoter analysis, and expression of the human PPARgamma gene. *J Biol Chem* 272(30), 18779–18789.
172. Chawla, A. et al. (1994). Peroxisome proliferator-activated receptor (PPAR) gamma: Adipose-predominant expression and induction early in adipocyte differentiation. *Endocrinology* 135(2), 798–800.
173. Tontonoz, P., Hu, E., and Spiegelman, B. M. (1994). Stimulation of adipogenesis in fibroblasts by PPAR gamma 2, a lipid-activated transcription factor. *Cell* 79(7), 1147–1156.

174. Vidal-Puig, A. J. et al. (1997). Peroxisome proliferator-activated receptor gene expression in human tissues. Effects of obesity, weight loss, and regulation by insulin and glucocorticoids. *J Clin Invest* 99(10), 2416–2422.
175. Lazar, M. A. (2005). PPAR gamma, 10 years later. *Biochimie* 87(1), 9–13.
176. Lambe, K. G. and Tugwood, J. D. (1996). A human peroxisome-proliferator-activated receptor-gamma is activated by inducers of adipogenesis, including thiazolidinedione drugs. *Eur J Biochem* 239(1), 1–7.
177. Ricote, M. et al. (1998). Expression of the peroxisome proliferator-activated receptor gamma (PPARgamma) in human atherosclerosis and regulation in macrophages by colony stimulating factors and oxidized low density lipoprotein. *Proc Natl Acad Sci U S A* 95(13), 7614–7619.
178. Yen, C. J. et al. (1997). Molecular scanning of the human peroxisome proliferator activated receptor gamma (hPPAR gamma) gene in diabetic Caucasians: Identification of a Pro12Ala PPAR gamma 2 missense mutation. *Biochem Biophys Res Commun* 241(2), 270–274.
179. Beamer, B. A. et al. (1998). Association of the Pro12Ala variant in the peroxisome proliferator-activated receptor-gamma2 gene with obesity in two Caucasian populations. *Diabetes* 47(11), 1806–1808.
180. Deeb, S. S. et al. (1998). A Pro12Ala substitution in PPARgamma2 associated with decreased receptor activity, lower body mass index and improved insulin sensitivity. *Nat Genet* 20(3), 284–287.
181. Altshuler, D. et al. (2000). The common PPARgamma Pro12Ala polymorphism is associated with decreased risk of type 2 diabetes. *Nat Genet* 26(1), 76–80.
182. Tonjes, A. et al. (2006). Association of Pro12Ala polymorphism in peroxisome proliferator-activated receptor gamma with Pre-diabetic phenotypes: Meta-analysis of 57 studies on nondiabetic individuals. *Diabetes Care* 29(11), 2489–2497.
183. Agostini, M. et al. (2006). Non-DNA binding, dominant-negative, human PPARgamma mutations cause lipodystrophic insulin resistance. *Cell Metab* 4(4), 303–311.
184. Semple, R. K., Chatterjee, V. K., and O'Rahilly, S. (2006). PPAR gamma and human metabolic disease. *J Clin Invest* 116(3), 581–589.
185. Shiraki, T. et al. (2005). Alpha,beta-unsaturated ketone is a core moiety of natural ligands for covalent binding to peroxisome proliferator-activated receptor gamma. *J Biol Chem* 280(14), 14145–14153.
186. Krey, G. et al. (1997). Fatty acids, eicosanoids, and hypolipidemic agents identified as ligands of peroxisome proliferator-activated receptors by coactivator-dependent receptor ligand assay. *Mol Endocrinol* 11(6), 779–791.
187. Schopfer, F. J. et al. (2005). Nitrolinoleic acid: An endogenous peroxisome proliferator-activated receptor gamma ligand. *Proc Natl Acad Sci U S A* 102(7), 2340–2345.
188. Nagy, L. et al. (1998). Oxidized LDL regulates macrophage gene expression through ligand activation of PPARgamma. *Cell* 93(2), 229–240.
189. Lehmann, J. M. et al. (1995). An antidiabetic thiazolidinedione is a high affinity ligand for peroxisome proliferator-activated receptor gamma (PPAR gamma). *J Biol Chem* 270(22), 12953–12956.
190. Yu, J. G. et al. (2002). The effect of thiazolidinediones on plasma adiponectin levels in normal, obese, and type 2 diabetic subjects. *Diabetes* 51(10), 2968–2974.
191. Maeda, N. et al. (2001). PPARgamma ligands increase expression and plasma concentrations of adiponectin, an adipose-derived protein. *Diabetes* 50(9), 2094–2099.
192. Yamauchi, T. et al. (2001). The mechanisms by which both heterozygous peroxisome proliferator-activated receptor gamma (PPARgamma) deficiency and PPARgamma agonist improve insulin resistance. *J Biol Chem* 276(44), 41245–41254.
193. Miyazaki, Y. et al. (2004). Effect of pioglitazone on circulating adipocytokine levels and insulin sensitivity in type 2 diabetic patients. *J Clin Endocrinol Metab* 89(9), 4312–4319.
194. Bajaj, M. et al. (2004). Plasma resistin concentration, hepatic fat content, and hepatic and peripheral insulin resistance in pioglitazone-treated type II diabetic patients. *Int J Obes Relat Metab Disord* 28(6), 783–789.

195. Lazar, M. A. (2005). How obesity causes diabetes: Not a tall tale. *Science* 307(5708), 373–375.
196. Patsouris, D., Muller, M., and Kersten, S. (2004). Peroxisome proliferator activated receptor ligands for the treatment of insulin resistance. *Curr Opin Investig Drugs* 5(10), 1045–1050.
197. Henke, B. R. et al. (1998). N-(2-Benzoylphenyl)-L-tyrosine PPARgamma agonists. 1. Discovery of a novel series of potent antihyperglycemic and antihyperlipidemic agents. *J Med Chem* 41(25), 5020–5036.
198. Lehmann, J. M. et al. (1997). Peroxisome proliferator-activated receptors alpha and gamma are activated by indomethacin and other non-steroidal anti-inflammatory drugs. *J Biol Chem* 272(6), 3406–3410.
199. Rosen, E. D. and Spiegelman, B. M. (2001). PPARgamma : A nuclear regulator of metabolism, differentiation, and cell growth. *J Biol Chem* 276(41), 37731–37734.
200. Knouff, C. and Auwerx, J. (2004). Peroxisome proliferator-activated receptor-gamma calls for activation in moderation: Lessons from genetics and pharmacology. *Endocr Rev* 25(6), 899–918.
201. Lehrke, M. and Lazar, M. A. (2005). The many faces of PPARgamma. *Cell* 123(6), 993–999.
202. Cock, T. A., Houten, S. M., and Auwerx, J. (2004). Peroxisome proliferator-activated receptor-gamma: Too much of a good thing causes harm. *EMBO Rep* 5(2), 142–147.
203. Rosen, E. D. and Spiegelman, B. M. (2000). Molecular regulation of adipogenesis. *Annu Rev Cell Dev Biol* 16, 145–171.
204. Schoonjans, K. et al. (1996). PPARalpha and PPARgamma activators direct a distinct tissue-specific transcriptional response via a PPRE in the lipoprotein lipase gene. *EMBO J* 15(19), 5336–5348.
205. Barak, Y. et al. (1999). PPAR gamma is required for placental, cardiac, and adipose tissue development. *Mol Cell* 4(4), 585–595.
206. Rosen, E. D. et al. (1999). PPAR gamma is required for the differentiation of adipose tissue in vivo and in vitro. *Mol Cell* 4(4), 611–617.
207. Ren, D. et al. (2002). PPARgamma knockdown by engineered transcription factors: Exogenous PPARgamma2 but not PPARgamma1 reactivates adipogenesis. *Genes Dev* 16(1), 27–32.
208. Zhang, J. et al. (2004). Selective disruption of PPARgamma 2 impairs the development of adipose tissue and insulin sensitivity. *Proc Natl Acad Sci U S A* 101(29), 10703–10708.
209. Koutnikova, H. et al. (2003). Compensation by the muscle limits the metabolic consequences of lipodystrophy in PPAR gamma hypomorphic mice. *Proc Natl Acad Sci U S A* 100(24), 14457–14462.
210. He, W. et al. (2003). Adipose-specific peroxisome proliferator-activated receptor gamma knockout causes insulin resistance in fat and liver but not in muscle. *Proc Natl Acad Sci U S A* 100(26), 15712–15717.
211. Jones, J. R. et al. (2005). Deletion of PPARgamma in adipose tissues of mice protects against high fat diet-induced obesity and insulin resistance. *Proc Natl Acad Sci U S A* 102(17), 6207–6212.
212. Mandard, S. et al. (2004). The direct peroxisome proliferator-activated receptor target fasting-induced adipose factor (FIAF/PGAR/ANGPTL4) is present in blood plasma as a truncated protein that is increased by fenofibrate treatment. *J Biol Chem* 279(33), 34411–34420.
213. Hauner, H. (2002). The mode of action of thiazolidinediones. *Diabetes Metab Res Rev* 18(Suppl 2), S10–S15.
214. Tontonoz, P. et al. (1998). PPARgamma promotes monocyte/macrophage differentiation and uptake of oxidized LDL. *Cell* 93(2), 241–252.
215. Hirakata, M. et al. (2004). Comparison of the effects of pioglitazone and rosiglitazone on macrophage foam cell formation. *Biochem Biophys Res Commun* 323(3), 782–788.
216. Argmann, C. A. et al. (2003). Activation of peroxisome proliferator-activated receptor gamma and retinoid X receptor results in net depletion of cellular cholesteryl esters in macrophages exposed to oxidized lipoproteins. *Arterioscler Thromb Vasc Biol* 23(3), 475–482.
217. Galetto, R. et al. (2001). Identification of a peroxisome-proliferator-activated-receptor response element in the apolipoprotein E gene control region. *Biochem J* 357(Pt 2), 521–527.
218. Llaverias, G. et al. (2004). Rosiglitazone upregulates caveolin-1 expression in THP-1 cells through a PPAR-dependent mechanism. *J Lipid Res* 45(11), 2015–2024.

219. Chawla, A. et al. (2001). A PPAR gamma-LXR-ABCA1 pathway in macrophages is involved in cholesterol efflux and atherogenesis. *Mol Cell* 7(1), 161–171.
220. Hevener, A. L. et al. (2003). Muscle-specific Pparg deletion causes insulin resistance. *Nat Med* 9(12), 1491–1497.
221. Norris, A. W. et al. (2003). Muscle-specific PPARgamma-deficient mice develop increased adiposity and insulin resistance but respond to thiazolidinediones. *J Clin Invest* 112(4), 608–618.
222. Medina-Gomez, G. et al. (2005). The link between nutritional status and insulin sensitivity is dependent on the adipocyte-specific peroxisome proliferator-activated receptor-gamma2 isoform. *Diabetes* 54(6), 1706–1716.
223. Yu, S. et al. (2003). Adipocyte-specific gene expression and adipogenic steatosis in the mouse liver due to peroxisome proliferator-activated receptor gamma1 (PPARgamma1) overexpression. *J Biol Chem* 278(1), 498–505.
224. Memon, R. A. et al. (2000). Up-regulation of peroxisome proliferator-activated receptors (PPAR-alpha) and PPAR-gamma messenger ribonucleic acid expression in the liver in murine obesity: Troglitazone induces expression of PPAR-gamma-responsive adipose tissue-specific genes in the liver of obese diabetic mice. *Endocrinology* 141(11), 4021–4031.
225. Gavrilova, O. et al. (2003). Liver peroxisome proliferator-activated receptor gamma contributes to hepatic steatosis, triglyceride clearance, and regulation of body fat mass. *J Biol Chem* 278(36), 34268–34276.
226. Matsusue, K. et al. (2003). Liver-specific disruption of PPARgamma in leptin-deficient mice improves fatty liver but aggravates diabetic phenotypes. *J Clin Invest* 111(5), 737–747.
227. Jiang, C., Ting, A. T., and Seed, B. (1998). PPAR-gamma agonists inhibit production of monocyte inflammatory cytokines. *Nature* 391(6662), 82–86.
228. Moore, K. J. et al. (2001). The role of PPAR-gamma in macrophage differentiation and cholesterol uptake. *Nat Med* 7(1), 41–47.
229. Ricote, M. et al. (1998). The peroxisome proliferator-activated receptor-gamma is a negative regulator of macrophage activation. *Nature* 391(6662), 79–82.
230. Inoue, H., Tanabe, T., and Umesono, K. (2000). Feedback control of cyclooxygenase-2 expression through PPARgamma. *J Biol Chem* 275(36), 28028–28032.
231. Abdelrahman, M., Sivarajah, A., and Thiemermann, C. (2005). Beneficial effects of PPAR-gamma ligands in ischemia-reperfusion injury, inflammation and shock. *Cardiovasc Res* 65(4), 772–781.
232. Maggi, L. B., Jr. et al. (2000). Anti-inflammatory actions of 15-deoxy-delta 12,14-prostaglandin J2 and troglitazone: Evidence for heat shock-dependent and -independent inhibition of cytokine-induced inducible nitric oxide synthase expression. *Diabetes* 49(3), 346–355.
233. Cuzzocrea, S. et al. (2004). Rosiglitazone, a ligand of the peroxisome proliferator-activated receptor-gamma, reduces acute inflammation. *Eur J Pharmacol* 483(1), 79–93.
234. Cuzzocrea, S. et al. (2004). Rosiglitazone, a ligand of the peroxisome proliferator-activated receptor-gamma, reduces acute pancreatitis induced by cerulein. *Intensive Care Med* 30(5), 951–956.
235. Ellis, C. N. et al. (2000). Troglitazone improves psoriasis and normalizes models of proliferative skin disease: Ligands for peroxisome proliferator-activated receptor-gamma inhibit keratinocyte proliferation. *Arch Dermatol* 136(5), 609–616.
236. Minamikawa, J. et al. (1998). Potent inhibitory effect of troglitazone on carotid arterial wall thickness in type 2 diabetes. *J Clin Endocrinol Metab* 83(5), 1818–1820.
237. Koshiyama, H. et al. (2001). Rapid communication: Inhibitory effect of pioglitazone on carotid arterial wall thickness in type 2 diabetes. *J Clin Endocrinol Metab* 86(7), 3452–3456.
238. Dormandy, J. A. et al. (2005). Secondary prevention of macrovascular events in patients with type 2 diabetes in the PROactive Study (PROspective pioglitAzone Clinical Trial In macroVascular Events): A randomised controlled trial. *Lancet* 366(9493), 1279–1289.
239. Ricote, M. and Glass, C. K. (2007). PPARs and molecular mechanisms of transrepression. *Biochim Biophys Acta* 1771(8), 926–935.
240. Kodera, Y. et al. (2000). Ligand type-specific interactions of peroxisome proliferator-activated receptor gamma with transcriptional coactivators. *J Biol Chem* 275(43), 33201–33204.

241. Pascual, G. et al. (2005). A SUMOylation-dependent pathway mediates transrepression of inflammatory response genes by PPAR-gamma. *Nature* 437(7059), 759–763.
242. Ohshima, T., Koga, H., and Shimotohno, K. (2004). Transcriptional activity of peroxisome proliferator-activated receptor gamma is modulated by SUMO-1 modification. *J Biol Chem* 279(28), 29551–29557.
243. Floyd, Z. E. and Stephens, J. M. (2004). Control of peroxisome proliferator-activated receptor gamma2 stability and activity by SUMOylation. *Obes Res* 12(6), 921–928.
244. Weisberg, S. P. et al. (2003). Obesity is associated with macrophage accumulation in adipose tissue. *J Clin Invest* 112(12), 1796–1808.
245. Fain, J. N. (2006). Release of interleukins and other inflammatory cytokines by human adipose tissue is enhanced in obesity and primarily due to the nonfat cells. *Vitam Horm* 74, 443–477.
246. Kahn, B. B. and Flier, J. S. (2000). Obesity and insulin resistance. *J Clin Invest* 106(4), 473–481.
247. Greenberg, A. S. and Obin, M. S. (2006). Obesity and the role of adipose tissue in inflammation and metabolism. *Am J Clin Nutr* 83(2), 461S–465S.
248. Permana, P. A., Menge, C., and Reaven, P. D. (2006). Macrophage-secreted factors induce adipocyte inflammation and insulin resistance. *Biochem Biophys Res Commun* 341(2), 507–514.
249. Suganami, T., Nishida, J., and Ogawa, Y. (2005). A paracrine loop between adipocytes and macrophages aggravates inflammatory changes: Role of free fatty acids and tumor necrosis factor alpha. *Arterioscler Thromb Vasc Biol* 25(10), 2062–2068.
250. Lumeng, C. N., Bodzin, J. L., and Saltiel, A. R. (2007). Obesity induces a phenotypic switch in adipose tissue macrophage polarization. *J Clin Invest* 117(1), 175–184.
251. Hotamisligil, G. S., Shargill, N. S., and Spiegelman, B. M. (1993). Adipose expression of tumor necrosis factor-alpha: Direct role in obesity-linked insulin resistance. *Science* 259(5091), 87–91.
252. Vgontzas, A. N. et al. (1997). Elevation of plasma cytokines in disorders of excessive daytime sleepiness: Role of sleep disturbance and obesity. *J Clin Endocrinol Metab* 82(5), 1313–1316.
253. Perreault, M. and Marette, A. (2001). Targeted disruption of inducible nitric oxide synthase protects against obesity-linked insulin resistance in muscle. *Nat Med* 7(10), 1138–1143.
254. Samad, F. et al. (1997). Elevated expression of transforming growth factor-beta in adipose tissue from obese mice. *Mol Med* 3(1), 37–48.
255. Visser, M. et al. (1999). Elevated C-reactive protein levels in overweight and obese adults. *J Am Med Assoc* 282(22), 2131–2135.
256. Weyer, C. et al. (2002). Humoral markers of inflammation and endothelial dysfunction in relation to adiposity and in vivo insulin action in Pima Indians. *Atherosclerosis* 161(1), 233–242.
257. Sartipy, P. and Loskutoff, D. J. (2003). Monocyte chemoattractant protein 1 in obesity and insulin resistance. *Proc Natl Acad Sci U S A* 100(12), 7265–7270.
258. Samad, F., Yamamoto, K., and Loskutoff, D. J. (1996). Distribution and regulation of plasminogen activator inhibitor-1 in murine adipose tissue in vivo. Induction by tumor necrosis factor-alpha and lipopolysaccharide. *J Clin Invest* 97(1), 37–46.
259. Samad, F., Pandey, M., and Loskutoff, D. J. (1998). Tissue factor gene expression in the adipose tissues of obese mice. *Proc Natl Acad Sci U S A* 95(13), 7591–7596.
260. De Pergola, G. and Pannacciulli, N. (2002). Coagulation and fibrinolysis abnormalities in obesity. *J Endocrinol Invest* 25(10), 899–904.
261. Fried, S. K., Bunkin, D. A., and Greenberg, A. S. (1998). Omental and subcutaneous adipose tissues of obese subjects release interleukin-6: Depot difference and regulation by glucocorticoid. *J Clin Endocrinol Metab* 83(3), 847–850.
262. Stienstra, R. et al. (2008). PPARgamma activation promotes infiltration of alternatively activated macrophages into adipose tissue. *J Biol Chem* 283(33), 22620-22627.
263. Odegaard, J. I. et al. (2007). Macrophage-specific PPARgamma controls alternative activation and improves insulin resistance. *Nature* 447(7148), 1116–1120.
264. Hevener, A. L. et al. (2007). Macrophage PPAR gamma is required for normal skeletal muscle and hepatic insulin sensitivity and full antidiabetic effects of thiazolidinediones. *J Clin Invest* 117(6), 1658–1669.

CHAPTER 11

XENOBIOTIC RECEPTORS CAR AND PXR

CURTIS KLAASSEN AND HONG LU

Department of Pharmacology, Toxicology and Therapeutics, University of Kansas Medical Center, Kansas City, KS, USA

Abstract: The absorption, distribution, metabolism, and excretion (ADME) of chemicals are coordinately fulfilled by drug transporters as well as phase-I and phase-II drug-metabolizing enzymes. The two so-called "xenobiotic receptors", namely constitutive active/androstane receptor (CAR, NR1I3) and pregnane X receptor (PXR, NR1I2), are predominantly expressed in liver and gastrointestinal tract, where CAR and PXR control the expression of a large battery of ADME genes. Although CAR and PXR overlap largely on their target genes, CAR and PXR differ considerably in the regulation of their transcriptional activities, namely gene expression of nuclear receptors, ligand recognition, and activation/suppression by phosphorylation. Unlike PXR, a xenobiotic sensor activated by diverse agonist ligands, ligand-binding is unlikely the major mechanism of CAR activation. In contrast, the transactivation activities of CAR are modulated by phosphorylation as well as various coactivators and corepressors. By binding to multiple DNA response elements that CAR and PXR share with other nuclear receptors, physically interacting with other nuclear receptors, and/or competing with other nuclear receptors for the limited intracellular supply of coactivators, CAR and PXR are not only essential in regulating ADME genes, but also important in metabolic homeostasis of cholesterol, steroid hormones, bile acids, lipids, and glucose. Thus, CAR and PXR appear to be critical links between pharmacokinetics and pharmacodynamics of therapeutic drugs. The species difference between humans and mice in CAR and PXR ligands is a major problem in toxicopharmacological and preclinical studies. However, the availability of CAR and PXR "humanized" mice may help to solve this problem.

11.1. INTRODUCTION

The absorption, distribution, metabolism, and excretion (ADME) of chemicals are coordinately fulfilled by drug transporters as well as phase-I and phase-II drug-metabolizing enzymes. The constitutive and altered expression of these ADME genes are largely under the transcriptional control of certain nuclear receptor family members [1–3]. The two so-called 'xenobiotic receptors', namely constitutive active/androstane receptor (CAR, NR1I3) and pregnane X receptor (PXR, NR1I2), are predominantly expressed in liver and gastrointestinal tract, where CAR and

C.M. Bunce, M.J. Campbell (eds.), *Nuclear Receptors*, Proteins and Cell Regulation 8, 287–305. DOI 10.1007/978-90-481-3303-1_11,

PXR control the expression of a large battery of ADME genes. In the last decade, remarkable progress has been made on studies of CAR and PXR regarding their target genes and regulatory mechanisms of their transcriptional activities, resulting in a much better understanding of the xenobiotic-mediated induction process and their potential clinical relevance to drug therapy. In addition to regulating xenobiotic metabolism, CAR and PXR both play physiological roles in regulating metabolic pathways important for the elimination and/or detoxification of cholesterol, bile acids, and hormones. The regulatory roles of CAR and PXR in the metabolism of glucose and lipids are also starting to emerge.

11.2. REGULATION OF XENOBIOTIC AND ENDOBIOTIC METABOLISM BY CAR AND PXR

11.2.1. CAR and PXR in Xenobiotic Metabolism and Liver Pathophysiology

11.2.1.1. Induction of ADME genes by CAR

First cloned in 1994 [4], CAR has emerged as a key nuclear receptor regulating ADME genes. Unlike the classical ligand-dependent nuclear receptors, CAR is constitutively active [5], but is inhibited by the inverse agonist ligands androstanol and androstenol [6]. The detoxification/elimination of potentially toxic chemicals depends on the concerted action of metabolic enzymes and transporters. CAR is highly expressed in liver and small intestine, two key tissues expressing xenobiotic-metabolizing enzymes and transporters, and mediates the induction of these ADME genes by the antiepileptic drug phenobarbital (PB) and a synthetic inducer 1,4-bis[2-(3,5-dichloropyridyloxy)]benzene (TCPOBOP) in mice [7]. TCPOBOP is an agonist ligand for CAR [8], whereas PB induces nuclear translocation of CAR [9] via dephosphorylation of CAR [10], resulting in increased expression of CAR-target genes, such as phase-I enzymes cytochrome P450 1A (CYP1A), CYP2B, CYP3A, phase-II enzymes UDP-glucuronosyltransferase (UGT) 1A1, sulfotransferases (SULT) 1E and 2A, as well as efflux transporters multidrug resistance-associated proteins (MRPs) [11, 7].

The induction of xenobiotic-metabolizing enzymes by CAR is generally protective, but can be harmful if toxic metabolites are produced. For example, CAR activators induce hepatic expression of 3 acetaminophen (APAP)-metabolizing enzymes in wild-type but not CAR-null mice, and CAR-null mice are resistant to APAP toxicity [12]. Similarly, CAR-null mice are also resistant to hepatotoxicity induced by carbon tetrachloride [13]. Additionally, CAR activation protects against a stress response elicited by hyperbilirubinemia [14, 15]. Studies show that PB treatment disrupts thyroid hormone homeostasis [16–19]; a key role of CAR in this process was confirmed using CAR-null mice [20, 21]. The role of CAR in endobiotic and xenobiotic metabolism has important clinical implications in disease prevention, drug-drug interactions, and development of better drug treatments [7, 22].

11.2.1.2. Induction of ADME genes by PXR

First cloned in 1998 in mice [23] and humans [24], PXR is also termed steroid and xenobiotic receptor (SXR), because PXR/SXR is also activated by naturally occurring steroids, such as pregnenolone and progesterone, as well as synthetic glucocorticoids and antiglucocorticoids [23]. As two closely related nuclear receptors, PXR and CAR overlap regarding DNA-binding sites and their target genes. The classical genes induced by PXR include CYP3A4, CYP2B, CYP2C, glutathione-*S*-transferase (GST), SULT, and UGT [25–28]. PXR also induces the expression of some drug transporter genes, such as organic anion transporting polypeptide 1a4 (Oatp1a4), multidrug resistance protein 1 (Mdr1), and Mrp2 [29–31].

11.2.1.3. CAR and PXR in drug–drug and food–drug interactions

Adverse drug reactions (ADR) are major clinical problems contributing to patient mortality and morbidity. CAR activation plays an important role in drug-drug interactions. Enzyme induction can increase reactive metabolites of drugs; for example, PB activation of CAR potentiates acetaminophen-induced hepatotoxicity. More often, enzyme induction results in accelerated clearance and loss of drug efficacy, a classical example being coadministration of PB and contraceptives (e.g. ethinylestradiol) leads to a 25-fold higher risk of pill failure, due to increased metabolic clearance of estrogens [32]. Beyond drug-drug interactions, food-drug and disease-drug interactions are major factors responsible for ADR and loss of efficacy [33, 34]. In addition to various therapeutic drugs [35], PXR is also activated by many natural products, such as St. John's Wort (SJW), gugulipid, kava kava, *Coleus forskolii*, *Hypoxis*, *Sutherlandia*, qing hao (Sweet Wormwood), wu wei zi (*Schisandra chinensis)*, and gan cao (Glycyrrhiza) [36]. SJW is an herbal remedy widely used for the treatment of depression. Although SJW inhibits CYP3A and P-glycoprotein in vitro [37], long-term consumption of SJW induces CYP3A via activation of PXR by hyperforin, an active constituent within SJW [38], resulting in decreased blood concentrations of many therapeutic drugs, such as amitriptyline, cyclosporine, digoxin, indinavir, irinotecan, warfarin, phenprocoumon, alprazolam, dextrometorphane, simvastatin, and oral contraceptives [39, 40]. Additionally, activation of PXR increases the hepatotoxicity of acetaminophen in mice, probably due to increased bioactivation of acetaminophen by Cyp3a11 [41]. Yin Zhi Huang, a traditional herbal medicine used widely in Asia to treat neonatal jaundice, enhances bilirubin clearance by activating CAR [42]. The use of food supplements is increasing in the United States, raising concern of food-drug interactions. Induction of drug metabolism due to activation of CAR and/or PXR is an important source of adverse drug-drug or food-drug interactions.

11.2.2. CAR and PXR in Endobiotic Metabolism and Liver Pathophysiology

In addition to its pivotal role in ADME of xenobiotics, CAR is important in regulating thyroid hormone homeostasis through inducing key phase-II enzymes responsible for the inactivation of thyroid hormones [43, 44, 20, 21]. CAR activation promotes hepatocyte proliferation and blocks apoptosis, and is essential for rodent hepatocarcinogenesis promoted by PB and TCPOBOP [45, 46, 7]. However, there is no evidence that PB promotes hepatocarcinogenesis in humans. Recently, CAR is implicated important in the pathogenesis of non-alcoholic steatohepatitis [47]. Activation of PXR induces CD36, a fatty acid transporter, causing hepatic steatosis in wild-type mice and 'PXR-humanized' mice [48]. Activation of PXR also increases the proliferation of hepatocytes [49], and PXR has been shown essential for normal progression of liver regeneration after partial hepatectomy [50]. It is now known that CAR and PXR crosstalk with other nuclear receptors to regulate endobiotic metabolism and liver pathophysiology [51–53].

11.2.2.1. Role of CAR and PXR in protection against bile acid toxicity

Accumulation of bile acids during cholestasis causes hepatotoxicity. Activation of CAR and PXR protect mice from bile acid induced hepatotoxicity, as indicated by the increased liver injury in CAR- and PXR-null mice after bile-duct ligation (for extrahepatic cholestasis) [54], as well as the protection of mice by CAR or PXR activators against hepatotoxicity induced by the hydrophobic lithocholic acid (for intrahepatic cholestasis) [55, 56]. Interestingly, CAR-null mice, but not PXR-null mice, have increased hepatotoxicity after treatment with lithocholic acid; the induction of CAR in PXR-null mice may compensate for the loss of PXR in inducing hepatic expression of detoxifying enzymes such as Cyp3a11 and Sult2a1, which catalyze the detoxification of lithocholic acid to more water-soluble bile acids for urinary excretion [56, 57]. Additionally, the decrease of bile-acid synthesis through suppressing hepatic expression of the rate-limiting enzyme Cyp7a1 by activated CAR and/or PXR also contributes to the detoxification process during cholestasis [55, 56].

11.2.2.2. Negative crosstalk of CAR with other nuclear receptors

As a master regulator of liver development, hepatocyte nuclear factor 4α (HNF4α) plays a key role in hepatic expression of CAR [58]. Interestingly, a functional inhibitory cross-talk between CAR and HNF4α in hepatic lipid/glucose metabolism was reported recently. CAR inhibits HNF4α-target genes through competing for common coactivators and/or competing with HNF4α for binding to the direct repeat 1 (DR1) motif of the Cyp7a1 promoter [59]. Accordingly, the CAR activator TCPOBOP decreased hepatic expression of Cyp7a1 and Cyp8b1 in mice. Phosphoenolpyruvate carboxykinase (PEPCK) and carnitine palmitoyltransferase (CPT) are key enzymes in gluconeogenesis and oxidation of long-chain fatty acids, respectively. The forkhead transcription factor FoxO1 positively controls the expression of genes involved in gluconeogenesis, and is the target of insulin suppressive action on the gluconeogenesis pathway [52]. FoxO1 is a coactivator for CAR- and PXR-mediated transcription; in contrast, CAR, acting as corepressor, down-regulates FoxO1-mediated transcription in the presence of TCPOBOP [60]. Interestingly, TCPOBOP inhibited fasting-induction of CPT and the FoxO1 target gene PEPCK [61], but had no effect on their expression in fed mice [59]. Similarly, PB suppressed hepatic expression of PEPCK and CPT in wild-type, but not in CAR-null mice [62]. Such a negative cross-talk between CAR and FoxO1 may partially explain the decreased glucose levels and increased insulin sensitivity observed in PB-treated diabetic patients [63]. Thus, CAR not only induces enzymes and transporters that metabolize or transport endobiotics and xenobiotics, but also suppresses enzymes involved in bile-acid production, gluconeogenesis, and fatty acid β-oxidation in liver. Interestingly, PB induces Cyp4a10 and Cyp4a14, two target genes of PPARα, only in CAR-null mice [62], supporting the role of CAR as a transcriptional blocker for certain genes. Additionally, CAR can inhibit estrogen receptor signaling through competing for the limited amount of coactivators [64]. As two closely related nuclear receptors, CAR and PXR overlap largely regarding DNA-binding sites and target genes; However, CAR can compete with HNF4α on binding to a DR1 site, to which PXR does not bind [59]. In summary, various endogenous and exogenous stimuli converge in altering gene expression and function of CAR, and activation of CAR results in induction of xenobiotic-metabolizing enzymes and transporters, but inhibition of enzymes involved in bile-acid production, gluconeogenesis, and fatty acid β-oxidation in liver.

11.2.2.3. Crosstalk of PXR with other nuclear receptors

Activation of PXR induces hepatic expression of PPARγ [65]. PXR and FoxA2 physically interact through their ligand and DNA-binding domains, respectively. This interaction prevents the binding of FoxA2 to its DNA response elements in the CPT1 promoter, resulting in down-regulation of CPT1 and decreased energy metabolism [66]. Additionally, PXR inhibits the expression of CYP7A1, the rate-limiting enzyme in bile-acid biosynthesis, likely due to the competition of PXR with HNF4α for the common coactivator PPARγ coactivator 1α (PGC-1α) [67]. Similar to CAR, PXR also suppresses transcriptional activity of FoxO1 [60], implicating PXR as a negative transcriptional regulator of genes involved in glucose metabolism.

11.2.3. CAR and PXR in Regulation of ADME Gene Expression and Pathophysiology in Small Intestine

The CAR activator phenobarbital induces CYP2B6, CYP3A4, and MDR1 mRNA in intestinal precision-cut slices of human jejunum [68, 69]. Activation of PXR increases intestinal expression of Cyp3a11 and Mdr1a but decreases concentrative nucleoside transporter 2 in wild-type mice, whereas PXR-null mice have altered intestinal basal expression of certain nucleoside transporters [70]. Activation of PXR also increases the expression of PPARα in mouse small intestine [71]. In the human colon carcinoma cell line LS174T, the activated PXR induces MDR1 through binding to a DR4 motif in the upstream enhancer of MDR1 gene [29]. In human intestinal precision-cut slices, the PXR ligand rifampicin induces CYP2B6, CYP3A4, UGT1A6, and MDR1 mRNA in proximal jejunum, as well as UGT1A6 and MDR1 mRNA in colon [69]. PXR activation regulates cholesterol metabolism in the intestine through inducing CYP27A1 and ABCA1 [70, 72].

Down-regulation of PXR and polymorphisms in the PXR gene locus have been associated with inflammatory bowel diseases [73, 74]. PXR activation ameliorates inflammatory bowel disease in mice via inhibition of NF-kappaB target gene expression [75]. Rifaximin, a rifamycin analog approved for the treatment of travelers' diarrhea, is also beneficial in the treatment of multiple chronic gastrointestinal disorders. Interestingly, rifaximin was identified as a gut-specific human PXR ligand [76]. A few studies demonstrate that PXR differentially modulates hepatic and intestinal gene expression [77, 70, 71], which may be due to the differential interaction of PXR with other tissue-specific transcription factors. Therefore, the biological significances of CAR and PXR in the small intestine warrant further investigation.

11.3. REGULATORY MECHANISMS OF THE TRANSCRIPTIONAL ACTIVITIES AND GENE EXPRESSION OF CAR AND PXR

11.3.1. Regulation of the Transcriptional Activities of CAR and PXR

11.3.1.1. Regulation of the transcriptional activity of CAR

CAR is constitutively active, and can be further activated by different stimuli. Without stimulus, most CAR protein is sequestered in the cytosol by CAR retention protein (CCRP) and heat shock protein 90 [78], whereas a small portion of CAR is in the nucleus, which maintains the basal expression of certain CAR-target genes, such as CYP2B. The human CAR ligand pocket is smaller than that of human vitamin D receptor and PXR (870 $Å^3$ and 1290–1540 $Å^3$, respectively, versus 675 $Å^3$ for CAR), which accounts for the relatively small size of CAR ligands compared to those for human PXR (i.e. taxol and rifampicin) [11]. Due to the small and rigid nature of the ligand-binding pocket of CAR, only a small number of CAR ligands have been identified.

In addition to xenobiotic activators, such as PB and TCPOBOP, CAR is activated by high levels of estrogens, bilirubin, and bile acids [7]. Upon activation, CAR translocates into the nucleus, binding to a response element on the DNA sequence, heterodimerizes with retinoid X receptor alpha (RXRα), and increases gene transcription. Phosphorylation/dephosphorylation plays a key role in regulating the activation and nuclear translocation of CAR. Phosphorylation of CAR by extracellular signal-regulated kinase (ERK) retains CAR in the cytosol [79], whereas dephosphorylation of CAR by protein phosphatase 2A results in its activation and nuclear translocation [10]. The transactivation activity of CAR is regulated by various coactivators and corepressors. Coactivators PGC-1α [80], steroid receptor coactivator-1 [81], glucocorticoid receptor-interacting protein, PPAR binding protein (PBP) [82], and activating signal cointegrator-2 [83] are able to activate CAR *ligand-independently*. Interaction of CAR with coactivators increases nuclear accumulation of CAR [84].

11.3.1.2. Regulation of the transcriptional activity of PXR

PXR is an 'authentic' xenobiotic sensor, because its large and expandable ligand-binding pocket allows it to accept a large variety of xenobiotics as ligands [85–87, 22]. It is noteworthy that T0901317, a high-affinity ligand for liver X receptor (LXR) that has been widely used in animals to study LXR function, is also a high-affinity ligand for PXR; the in vivo effects of this compound ascribed to LXR activation should be re-examined [88]. Although the endogenous ligand for PXR remains unclear, PXR-null mice have a marked decrease in hepatic expression of Oatp1a4 [70], a classical PXR target gene, suggesting that PXR is constitutively active and responsible for maintaining the basal expression of certain genes. Reports on the cellular localization of PXR are controversial. Two in vitro studies showed that human PXR is localized exclusively in the nucleus, regardless of the presence or absence of ligands [89, 90]. However, two in vivo studies demonstrated that mouse PXR is sequestered in the cytosol by CCRP in untreated mouse liver, and translocated to the nucleus upon activation by ligands, where PXR heterodimerizes with RXRα and increases gene transcription [89, 78]. Currently, it remains unclear whether such conflicting results are due to species differences or differences in experimental systems (in vitro for human PXR versus in vivo for mouse PXR). PXR is activated after phosphorylation by protein kinase A (PKA) [91]. In contrast, protein kinase C (PKC) was shown to repress the signaling of mouse and human PXR [92]. PXR shares similar coactivators with CAR, such as PGC-1α, SRC-1, and PBP. Additionally, the transcriptional activity of PXR (and CAR) is suppressed by LXR and a corepressor small heterodimer partner (SHP) [53].

11.3.1.3. Importance of expression levels of CAR in determining hepatic basal expression and induction of CAR-target genes

Female Wistar-Kyoto rat livers have much lower CAR than males, which is associated with a much lower induction of CYP2B by PB [93]. Consistently, more prominent induction of CAR-target genes CYP2B1/2 and UGT2B1 by trans-stilbene oxide, a CAR activator, was found in male versus female Wistar-Kyoto rats [94]. In contrast, in female liver of CD-1 mice, higher CAR mRNA was associated with higher basal expression and TCPOBOP-mediated induction of Cyp2b10 [95]. Interestingly, in humans, women also have higher amounts of CYP2B6 mRNA (3.9-fold) and protein (1.7-fold) than do men [96].

Neonates undergo postnatal development and have marked individual variations in the development of drug metabolizing systems [97]. Fetal and neonatal hepatic expression of PXR and CAR is low and highly variable in humans [98]. The developmental expression of CYP3A isofroms has a major impact on the efficacy and safety of drug treatment in children [99]. Acetaminophen overdose is a major cause of acute liver failure; in pediatric populations, the overwhelming majority of acetaminophen overdoses are due to unintentional overdoses [100]. CAR is essential in mediating acetaminophen hepatotoxicity [12]. The potential contribution of variable expression of CAR to acetaminophen hepatotoxicity in children warrants investigation. Hyperbilirubinaemia is one of the most common and important pathological conditions in the newborn. In human livers, expression of CAR correlates highly with expression of CYPs, UGT1A1, and MRP2 [98, 101], key factors in bilirubin catabolism. It is proposed that a functional deficit of CAR activity, due to low expression of CAR, may contribute to neonatal jaundice [15]. Interestingly, although maternal administration of the CAR activator phenobarbital for weeks is successful in preventing neonatal jaundice [102], two randomized, double-blinded, and placebo controlled trials demonstrate that phenobarbital treatment is ineffective in neonatal jaundice [103, 104]. Thus, certain neonates appear to have poor response to CAR- and PXR-mediated induction of drug metabolizing genes, probably due to very low hepatic expression of CAR and PXR in these neonates.

11.3.1.4. Interaction between CAR and PXR

Although the induction of Cyp3a11 by pregnenolone-16α-carbonitrile (PCN) is lost in PXR-null mouse liver, hepatic basal expression of Cyp3a11 increases in PXR-null mouse liver compared to wild-type [49], probably due to the overlapping roles of PXR and CAR in the induction of Cyp3a11 [27]. As two closely related nuclear receptors, CAR and PXR overlap largely regarding DNA-binding sites and target genes; both of them can bind to diverse DNA sequences, such as DR3-5, inverted repeat-6, everted repeat-6 and -8 response elements [11, 105, 85, 106, 7]. However, CAR can compete with HNF4α on binding to a DR1 site, to which PXR does not bind [59]. Additionally, CAR can induce enoyl-CoA hydratase/3-hydroxyacyl-CoA dehydrogenase, a peroxisomal beta-oxidation enzyme, by binding to a DR2 site in its promoter, which PXR has not been shown to bind to [107].

There is less overlapping of ligands between CAR and PXR. A human CAR activator 6-(4-chlorophenyl)imidazo[2,1-b][1,3]thiazole-5-carbaldehyde O-(3,4-dichlorobenzyl) oxime (CITCO) selectively activates CAR at low concentration (100 nM), but strongly activates both CAR and PXR at high concentrations (> 1 μM) [108]. Phenobarbital and 5β-pregnane-3,20 dione are activators of both CAR and PXR, whereas clotrimazole and androstanol are activators of PXR but inverse agonists of CAR. Similarly, bile acids, such as cholic acid, 12-ketolithocholic acid, or 7-ketodeoxycholic acid methyl ester, which are primary ligands of farnesoid X receptor (FXR) and/or LXR, are activators of PXR but suppressors of CAR [109, 53]. Moreover, guggulsterone, an active ingredient of guggulipids, is an FXR antagonist and appears to be a PXR activator [110] but a CAR inverse agonist [111].

11.3.2. Regulation of CAR and PXR Gene Expression

Both CAR and PXR transcripts are predominantly expressed in liver and small intestine [112]. There are multiple polymorphic or splicing variants of CAR and PXR transcripts with differential tissue distribution patterns and transcriptional activities, which may influence inter-tissue and interindividual differences in target gene induction and drug-drug interactions [113–115]. PXR mRNA expression has also been reported in the stomach, heart, brain, and breast tissue [116, 117]. Human CAR mRNA is also detected in kidney, testis, adrenal, and brain caudate nucleus, whereas spleen, heart, and prostate express only splicing variants of CAR [118].

In mice, hepatic CAR mRNA is very low before birth, but rapidly increases to a high level immediately after birth. In humans, hepatic CAR mRNA is also much lower in prenatals [98]. In both pediatric [98] and adult humans [119], hepatic CAR mRNA correlates highly with HNF4α mRNA, consistent with a key role of HNF4α1 in basal expression of CAR in mouse [58] and human liver [120]. The expression of human CAR and PXR is also controlled by glucocorticoids through the glucocorticoid receptor [121, 122]; there is no evidence that mouse CAR or PXR expression is also controlled by the glucocorticoid receptor [123]. Hepatic expression of CAR is markedly less in mice null for the ubiquitous coactivator PBP [82]. Thus, HNF4α1 and PBP play key roles in maintaining the liver-predominant basal expression of CAR. Nevertheless, both HNF4α1 and PBP are expressed at similar levels in liver and kidney [124, 58]; however, renal expression of CAR is very low [5, 125]. Apparently, HNF4α1 and PBP are required, but not sufficient for the liver-predominant expression of CAR. Hepatic expression of CAR is induced by cAMP signaling during starvation in mice [123].

In contrast to CAR, whose expression is dependent on HNF4α in both fetal and adult liver, expression of PXR is dependent on HNF4α only in fetal liver, but not adult liver [126]. Currently, the mechanism of such difference in hepatic expression of CAR and PXR remains unknown. A major difference between fetal and adult liver is the absence and presence of bile acid signaling mediated by the bile acid receptor FXR, which induces PXR gene expression [127]. In contrast, hepatic expression of CAR is induced in FXR-null mice [128]. Thus, difference in FXR signaling may play a major role in determining the differential expression of CAR and PXR in HNF4α-null fetal and adult mouse liver. Additionally, hepatic protein expression of human PXR is regulated by microRNA [129].

11.4. SPECIES DIFFERENCE IN GENE REGULATION AND FUNCTION OF CAR AND PXR BETWEEN HUMANS AND MICE

Most nuclear hormone receptors share high homology between humans and rodents. In contrast, human and rodent CAR differ largely in their ligand-binding domain, with only 71% sequence identity between human and mouse CAR. Similarly, human and rodent PXR share approximately 95% identity in their DNA-binding domains, but only 75–80% identity in their amino acid sequences in the ligand-binding domain [130, 131]. PXR and CAR genes may have adapted to species differences in toxic compound exposure. Although PB, a non-ligand CAR activator, activates both human and mouse CAR, most ligand agonists of CAR have divergent effects on human and mouse CAR. For example, although TCPOBOP is a potent agonist of mouse CAR (EC_{50} ~20 nM), it does not activate human or rat CAR [8]. In contrast, CITCO is an effective agonist of human CAR but not mouse CAR [108]. Moreover, a histamine receptor-blocking drug meclizine acts as an agonist activator for mouse CAR, but an inverse agonist (inactivator) for human CAR [132]. Furthermore, human and mouse CAR differ in their responses to endogenous hormones: pharmacological levels of estrogens activate mouse CAR but not human CAR [133], whereas androstane metabolites are efficacious deactivators of mouse CAR but not human CAR [8]. In summary, there are considerable species differences between humans and mice in the response of CAR to xenobiotics and endogenous chemicals. Considerable species differences in ligand selectivity also exist between human and rodent PXR. The potent rodent PXR activator PCN is a poor ligand for human PXR, whereas the potent human PXR activators rifampin and SR12813 do not appreciably activate mouse PXR [134, 135].

To overcome species differences in ligand recognition between human and mouse CAR and PXR, CAR and PXR 'humanized' mice have been generated. The 'CAR-humanized' mice were produced by crossing an albumin promoter/human CAR transgene into the CAR null background [12]. These 'CAR-humanized' mice respond to CITCO but not TCPOBOP [42], and are thus an improved model for investigating in vivo effects of activation of human CAR in mouse livers. Nevertheless, because these 'CAR-humanized' mice lack intestinal expression of CAR, and the expression levels of human CAR in mouse liver are likely different from the endogenous CAR gene, caution is needed when using these 'CAR-humanized' mice.

Similarly, the 'PXR-humanized' mice have been produced by crossing an albumin promoter/human PXR transgene into the PXR null background [136]. These 'PXR-humanized' mice displayed hepatic induction of Cyp3a11 by a human PXR ligand rifampicin but not a mouse PXR ligand PCN. Recently, another PXR-humanized mouse model was generated by bacterial artificial chromosome (BAC) transgenesis in PXR-null mice using a BAC clone containing the complete human PXR gene and 5′- and 3′-flanking sequences [137]. In this PXR-humanized mouse model, human PXR is selectively expressed in the liver and intestine, and both hepatic and intestinal CYP3As were strongly induced by rifampicin but not PCN. This PXR-humanized mouse model would be a more appropriate in vivo tool for evaluating the overall pharmacokinetic consequences of human PXR activation by xenobiotics [137].

11.5. SUMMARY

CAR and PXR have emerged as key nuclear receptors regulating ADME of xenobiotics and endobiotics. Although CAR and PXR overlap largely on their target genes, CAR and PXR differ considerably in the regulation of their transcriptional activities, namely gene expression of nuclear receptors, ligand recognition, and activation/suppression by phosphorylation. Unlike PXR, a xenobiotic sensor activated by diverse agonist ligands, ligand-binding is unlikely the major mechanism of CAR activation. In contrast, the transactivation activities of CAR are modulated by phosphorylation as well as various coactivators and corepressors, which may play a key role in the ligand-independent constitutive activity of CAR. By binding to multiple DNA response elements that CAR and PXR share with other nuclear receptors, physically interacting with other nuclear receptors, and/or competing with other nuclear receptors for the limited intracellular supply of coactivators, CAR and PXR are not only essential in regulating ADME genes, but also important in metabolic homeostasis of cholesterol, steroid hormones, bile acids, lipids, and glucose. Thus, CAR and PXR appear to be critical links between pharmacokinetics and pharmacodynamics of therapeutic drugs. The species difference between humans and mice in CAR and PXR ligands is a major problem in toxicopharmacological and preclinical studies. However, the availability of CAR and PXR 'humanized' mice may help to solve this problem. A thorough understanding of the regulation and function of CAR and PXR will greatly improve the efficacy and safety of therapeutic drugs, and better protect us from environmental chemicals.

REFERENCES

1. Dixit, S. G., Tirona, R. G., and Kim, R. B. (2005). Beyond CAR and PXR. *Curr Drug Metab* 6, 385–397.
2. Eloranta, J. J., Meier, P. J., and Kullak-Ublick, G. A. (2005). Coordinate transcriptional regulation of transport and metabolism. *Methods Enzymol* 400, 511–530.
3. Klaassen, C. D. and Slitt, A. L. (2005). Regulation of hepatic transporters by xenobiotic receptors. *Curr Drug Metab* 6, 309–328.
4. Baes, M., Gulick, T., Choi, H. S., Martinoli, M. G., Simha, D., and Moore, D. D. (1994). A new orphan member of the nuclear hormone receptor superfamily that interacts with a subset of retinoic acid response elements. *Mol Cell Biol* 14, 1544–1552.
5. Choi, H. S., Chung, M., Tzameli, I., Simha, D., Lee, Y. K., Seol, W., and Moore, D. D. (1997). Differential transactivation by two isoforms of the orphan nuclear hormone receptor CAR. *J Biol Chem* 272, 23565–23571.
6. Forman, B. M., Tzameli, I., Choi, H. S., Chen, J., Simha, D., Seol, W., Evans, R. M., and Moore, D. D. (1998). Androstane metabolites bind to and deactivate the nuclear receptor CAR-beta. *Nature* 395, 612–615.
7. Qatanani, M. and Moore, D. D. (2005). CAR, the continuously advancing receptor, in drug metabolism and disease. *Curr Drug Metab* 6, 329–339.
8. Tzameli, I., Pissios, P., Schuetz, E. G., and Moore, D. D. (2000). The xenobiotic compound 1,4-bis[2-(3,5-dichloropyridyloxy)]benzene is an agonist ligand for the nuclear receptor CAR. *Mol Cell Biol* 20, 2951–2958.
9. Kawamoto, T., Sueyoshi, T., Zelko, I., Moore, R., Washburn, K., and Negishi, M. (1999). Phenobarbital-responsive nuclear translocation of the receptor CAR in induction of the CYP2B gene. *Mol Cell Biol* 19, 6318–6322.
10. Hosseinpour, F., Moore, R., Negishi, M., and Sueyoshi, T. (2006). Serine 202 regulates the nuclear translocation of constitutive active/androstane receptor. *Mol Pharmacol* 69, 1095–1102.
11. Willson, T. M. and Kliewer, S. A. (2002). PXR, CAR and drug metabolism. *Nat Rev Drug Discov* 1, 259–266.

12. Zhang, J., Huang, W., Chua, S. S., Wei, P., and Moore, D. D. (2002). Modulation of acetaminophen-induced hepatotoxicity by the xenobiotic receptor CAR. *Science* 298, 422–424.
13. Yamazaki, Y., Kakizaki, S., Horiguchi, N., Takagi, H., Mori, M., and Negishi, M. (2005). Role of nuclear receptor CAR in carbon tetrachloride-induced hepatotoxicity. *World J Gastroenterol* 11, 5966–5972.
14. Sugatani, J., Kojima, H., Ueda, A., Kakizaki, S., Yoshinari, K., Gong, Q. H., Owens, I. S., Negishi, M., and Sueyoshi, T. (2001). The phenobarbital response enhancer module in the human bilirubin UDP-glucuronosyltransferase UGT1A1 gene and regulation by the nuclear receptor CAR. *Hepatology* 33, 1232–1238.
15. Huang, W., Zhang, J., Chua, S. S., Qatanani, M., Han, Y., Granata, R., and Moore, D. D. (2003). Induction of bilirubin clearance by the constitutive androstane receptor (CAR). *Proc Natl Acad Sci USA* 100, 4156–4161.
16. Barter, R. A. and Klaassen, C. D. (1992). UDP-glucuronosyltransferase inducers reduce thyroid hormone levels in rats by an extrathyroidal mechanism. *Toxicol Appl Pharmacol* 113, 36–42.
17. Liu, J., Liu, Y., Barter, R. A., and Klaassen, C. D. (1995). Alteration of thyroid homeostasis by UDP-glucuronosyltransferase inducers in rats: A dose-response study. *J Pharmacol Exp Ther* 273, 977–985.
18. Klaassen, C. D. and Hood, A. M. (2001). Effects of microsomal enzyme inducers on thyroid follicular cell proliferation and thyroid hormone metabolism. *Toxicol Pathol* 29, 34–40.
19. Hood, A., Allen, M. L., Liu, Y., Liu, J., and Klaassen, C. D. (2003). Induction of T(4) UDP-GT activity, serum thyroid stimulating hormone, and thyroid follicular cell proliferation in mice treated with microsomal enzyme inducers. *Toxicol Appl Pharmacol* 188, 6–13.
20. Maglich, J. M., Watson, J., McMillen, P. J., Goodwin, B., Willson, T. M., and Moore, J. T. (2004). The nuclear receptor CAR is a regulator of thyroid hormone metabolism during caloric restriction. *J Biol Chem* 279, 19832–19838.
21. Qatanani, M., Zhang, J., and Moore, D. D. (2005). Role of the constitutive androstane receptor in xenobiotic-induced thyroid hormone metabolism. *Endocrinology* 146, 995–1002.
22. Timsit, Y. E. and Negishi, M. (2007). CAR and PXR: The xenobiotic-sensing receptors. *Steroids* 72, 231–246.
23. Kliewer, S. A., Moore, J. T., Wade, L., Staudinger, J. L., Watson, M. A., Jones, S. A., McKee, D. D., Oliver, B. B., Willson, T. M., Zetterstrom, R. H., Perlmann, T., and Lehmann, J. M. (1998). An orphan nuclear receptor activated by pregnanes defines a novel steroid signaling pathway. *Cell* 92, 73–82.
24. Bertilsson, G., Heidrich, J., Svensson, K., Asman, M., Jendeberg, L., Sydow-Backman, M., Ohlsson, R., Postlind, H., Blomquist, P., and Berkenstam, A. (1998). Identification of a human nuclear receptor defines a new signaling pathway for CYP3A induction. *Proc Natl Acad Sci USA* 95, 12208–12213.
25. Maglich, J. M., Stoltz, C. M., Goodwin, B., Hawkins-Brown, D., Moore, J. T., and Kliewer, S. A. (2002). Nuclear pregnane x receptor and constitutive androstane receptor regulate overlapping but distinct sets of genes involved in xenobiotic detoxification. *Mol Pharmacol* 62, 638–646.
26. Sonoda, J., Xie, W., Rosenfeld, J. M., Barwick, J. L., Guzelian, P. S., and Evans, R. M. (2002). Regulation of a xenobiotic sulfonation cascade by nuclear pregnane X receptor (PXR). *Proc Natl Acad Sci U S A* 99, 13801–13806.
27. Wei, P., Zhang, J., Dowhan, D. H., Han, Y., and Moore, D. D. (2002). Specific and overlapping functions of the nuclear hormone receptors CAR and PXR in xenobiotic response. *Pharmacogenomics J* 2, 117–126.
28. Chen, C., Staudinger, J. L., and Klaassen, C. D. (2003). Nuclear receptor, pregname X receptor, is required for induction of UDP-glucuronosyltranferases in mouse liver by pregnenolone-16 alpha-carbonitrile. *Drug Metab Dispos* 31, 908–915.
29. Geick, A., Eichelbaum, M., and Burk, O. (2001). Nuclear receptor response elements mediate induction of intestinal MDR1 by rifampin. *J Biol Chem* 276, 14581–14587.

30. Kast, H. R., Goodwin, B., Tarr, P. T., Jones, S. A., Anisfeld, A. M., Stoltz, C. M., Tontonoz, P., Kliewer, S., Willson, T. M., and Edwards, P. A. (2002). Regulation of multidrug resistance-associated protein 2 (ABCC2) by the nuclear receptors pregnane X receptor, farnesoid X-activated receptor, and constitutive androstane receptor. *J Biol Chem* 277, 2908–2915.
31. Staudinger, J. L., Madan, A., Carol, K. M., and Parkinson, A. (2003). Regulation of drug transporter gene expression by nuclear receptors. *Drug Metab Dispos* 31, 523–527.
32. Lewis, J. R. (1979). Oral contraceptive and antiepileptic drug interaction resulting in contraceptive failure. *J Am Med Assoc* 241, 2743.
33. Santos, C. A. and Boullata, J. I. (2005). An approach to evaluating drug-nutrient interactions. *Pharmacotherapy* 25, 1789–1800.
34. Carlberg, C. and Dunlop, T. W. (2006). An integrated biological approach to nuclear receptor signaling in physiological control and disease. *Crit Rev Eukaryot Gene Expr* 16, 1–22.
35. Moore, J. T. and Kliewer, S. A. (2000). Use of the nuclear receptor PXR to predict drug interactions. *Toxicology* 153, 1–10.
36. Staudinger, J. L., Ding, X., and Lichti, K. (2006). Pregnane X receptor and natural products: Beyond drug-drug interactions. *Expert Opin Drug Metab Toxicol* 2, 847–857.
37. Hellum, B. H. and Nilsen, O. G. (2008). In vitro inhibition of CYP3A4 metabolism and P-glycoprotein-mediated transport by trade herbal products. *Basic Clin Pharmacol Toxicol* 102, 466–475.
38. Moore, L. B., Goodwin, B., Jones, S. A., Wisely, G. B., Serabjit-Singh, C. J., Willson, T. M., Collins, J. L., and Kliewer, S. A. (2000). St. John's wort induces hepatic drug metabolism through activation of the pregnane X receptor. *Proc Natl Acad Sci U S A* 97, 7500–7502.
39. Madabushi, R., Frank, B., Drewelow, B., Derendorf, H., and Butterweck, V. (2006). Hyperforin in St. John's wort drug interactions. *Eur J Clin Pharmacol* 62, 225–233.
40. Pal, D. and Mitra, A. K. (2006). MDR- and CYP3A4-mediated drug-herbal interactions. *Life Sci* 78, 2131–2145.
41. Guo, G. L., Moffit, J. S., Nicol, C. J., Ward, J. M., Aleksunes, L. A., Slitt, A. L., Kliewer, S. A., Manautou, J. E., and Gonzalez, F. J. (2004). Enhanced acetaminophen toxicity by activation of the pregnane X receptor. *Toxicol Sci* 82, 374–380.
42. Huang, W., Zhang, J., and Moore, D. D. (2004a). A traditional herbal medicine enhances bilirubin clearance by activating the nuclear receptor CAR. *J Clin Invest* 113, 137–143.
43. Vansell, N. R. and Klaassen, C. D. (2002a). Effect of microsomal enzyme inducers on the biliary excretion of triiodothyronine (T(3)) and its metabolites. *Toxicol Sci* 65, 184–191.
44. Vansell, N. R. and Klaassen, C. D. (2002b). Increase in rat liver UDP-glucuronosyltransferase mRNA by microsomal enzyme inducers that enhance thyroid hormone glucuronidation. *Drug Metab Dispos* 30, 240–246.
45. Yamamoto, Y., Moore, R., Goldsworthy, T. L., Negishi, M., and Maronpot, R. R. (2004). The orphan nuclear receptor constitutive active/androstane receptor is essential for liver tumor promotion by phenobarbital in mice. *Cancer Res* 64, 7197–7200.
46. Huang, W., Zhang, J., Washington, M., Liu, J., Parant, J. M., Lozano, G., and Moore, D. D. (2005). Xenobiotic stress induces hepatomegaly and liver tumors via the nuclear receptor constitutive androstane receptor. *Mol Endocrinol* 19, 1646–1653.
47. Yamazaki, Y., Kakizaki, S., Horiguchi, N., Sohara, N., Sato, K., Takagi, H., Mori, M., and Negishi, M. (2007). The role of the nuclear receptor constitutive androstane receptor in the pathogenesis of non-alcoholic steatohepatitis. *Gut* 56, 565–574.
48. Zhou, J., Zhai, Y., Mu, Y., Gong, H., Uppal, H., Toma, D., Ren, S., Evans, R. M., and Xie, W. (2006). A novel pregnane X receptor-mediated and sterol regulatory element-binding protein-independent lipogenic pathway. *J Biol Chem* 281, 15013–15020.
49. Staudinger, J., Liu, Y., Madan, A., Habeebu, S., and Klaassen, C. D. (2001a). Coordinate regulation of xenobiotic and bile acid homeostasis by pregnane X receptor. *Drug Metab Dispos* 29, 1467–1472.

50. Dai, G., He, L., Bu, P., and Wan, Y. J. (2008). Pregnane X receptor is essential for normal progression of liver regeneration. *Hepatology* 47, 1277–1287.
51. Kakizaki, S., Yamazaki, Y., Takizawa, D., and Negishi, M. (2008). New insights on the xenobiotic-sensing nuclear receptors in liver diseases- CAR and PXR. *Curr Drug Metab* 9, 614–621.
52. Konno, Y., Negishi, M., and Kodama, S. (2008). The roles of nuclear receptors CAR and PXR in hepatic energy metabolism. *Drug Metab Pharmacokinet* 23, 8–13.
53. Pascussi, J. M., Gerbal-Chaloin, S., Duret, C., Daujat-Chavanieu, M., Vilarem, M. J., and Maurel, P. (2008). The tangle of nuclear receptors that controls xenobiotic metabolism and transport: Crosstalk and consequences. *Annu Rev Pharmacol Toxicol* 48, 1–32.
54. Stedman, C. A., Liddle, C., Coulter, S. A., Sonoda, J., Alvarez, J. G., Moore, D. D., Evans, R. M., and Downes, M. (2005). Nuclear receptors constitutive androstane receptor and pregnane X receptor ameliorate cholestatic liver injury. *Proc Natl Acad Sci U S A* 102, 2063–2068.
55. Staudinger, J. L., Goodwin, B., Jones, S. A., Hawkins-Brown, D., MacKenzie, K. I., LaTour, A., Liu, Y., Klaassen, C. D., Brown, K. K., Reinhard, J., Willson, T. M., Koller, B. H., and Kliewer, S. A. (2001b). The nuclear receptor PXR is a lithocholic acid sensor that protects against liver toxicity. *Proc Natl Acad Sci U S A* 98, 3369–3374.
56. Zhang, J., Huang, W., Qatanani, M., Evans, R. M., and Moore, D. D. (2004). The constitutive androstane receptor and pregnane X receptor function coordinately to prevent bile acid-induced hepatotoxicity. *J Biol Chem* 279, 49517–49522.
57. Wagner, M., Halilbasic, E., Marschall, H. U., Zollner, G., Fickert, P., Langner, C., Zatloukal, K., Denk, H., and Trauner, M. (2005). CAR and PXR agonists stimulate hepatic bile acid and bilirubin detoxification and elimination pathways in mice. *Hepatology* 42, 420–430.
58. Briancon, N. and Weiss, M. C. (2006). In vivo role of the HNF4alpha AF-1 activation domain revealed by exon swapping. *EMBO J* 25, 1253–1262.
59. Miao, J., Fang, S., Bae, Y., and Kemper, J. K. (2006). Functional inhibitory cross-talk between car and HNF-4 in hepatic lipid/glucose metabolism is mediated by competition for binding to the DR1 motif and to the common coactivators, GRIP-1 and PGC-1alpha. *J Biol Chem* 281(21), 14537-14546.
60. Kodama, S., Koike, C., Negishi, M., and Yamamoto, Y. (2004). Nuclear receptors CAR and PXR cross talk with FOXO1 to regulate genes that encode drug-metabolizing and gluconeogenic enzymes. *Mol Cell Biol* 24, 7931–7940.
61. Hall, R. K., Yamasaki, T., Kucera, T., Waltner-Law, M., O'Brien, R., and Granner, D. K. (2000). Regulation of phosphoenolpyruvate carboxykinase and insulin-like growth factor-binding protein-1 gene expression by insulin. The role of winged helix/forkhead proteins. *J Biol Chem* 275, 30169–30175.
62. Ueda, A., Hamadeh, H. K., Webb, H. K., Yamamoto, Y., Sueyoshi, T., Afshari, C. A., Lehmann, J. M., and Negishi, M. (2002). Diverse roles of the nuclear orphan receptor CAR in regulating hepatic genes in response to phenobarbital. *Mol Pharmacol* 61, 1–6.
63. Lahtela, J. T., Arranto, A. J., and Sotaniemi, E. A. (1985). Enzyme inducers improve insulin sensitivity in non-insulin-dependent diabetic subjects. *Diabetes* 34, 911–916.
64. Min, G., Kim, H., Bae, Y., Petz, L., and Kemper, J. K. (2002b). Inhibitory cross-talk between estrogen receptor (ER) and constitutively activated androstane receptor (CAR). CAR inhibits ER-mediated signaling pathway by squelching p160 coactivators. *J Biol Chem* 277, 34626–34633.
65. Lee, J. H., Zhou, J., and Xie, W. (2008). PXR and LXR in hepatic steatosis: A new dog and an old dog with new tricks. *Mol Pharm* 5, 60–66.
66. Nakamura, K., Moore, R., Negishi, M., and Sueyoshi, T. (2007). Nuclear pregnane X receptor cross-talk with FoxA2 to mediate drug-induced regulation of lipid metabolism in fasting mouse liver. *J Biol Chem* 282, 9768–9776.
67. Li, T. and Chiang, J. Y. (2005). Mechanism of rifampicin and pregnane X receptor inhibition of human cholesterol 7 alpha-hydroxylase gene transcription. *Am J Physiol Gastrointest Liver Physiol* 288, G74–G84.

68. Burk, O., Arnold, K. A., Geick, A., Tegude, H., and Eichelbaum, M. (2005). A role for constitutive androstane receptor in the regulation of human intestinal MDR1 expression. *Biol Chem* 386, 503–513.
69. van de Kerkhof, E. G., de Graaf, I. A., Ungell, A. L., and Groothuis, G. M. (2008). Induction of metabolism and transport in human intestine: Validation of precision-cut slices as a tool to study induction of drug metabolism in human intestine in vitro. *Drug Metab Dispos* 36, 604–613.
70. Cheng, X. and Klaassen, C. D. (2006). Regulation of mRNA expression of xenobiotic transporters by the pregnane x receptor in mouse liver, kidney, and intestine. *Drug Metab Dispos* 34, 1863–1867.
71. Inoue, J., Satoh, S. I., Kita, M., Nakahara, M., Hachimura, S., Miyata, M., Nishimaki-Mogami, T., and Sato, R. (2008). PPARalpha gene expression is up-regulated by LXR and PXR activators in the small intestine. *Biochem Biophys Res Commun* 371(4), 675-678.
72. Li, T., Chen, W., and Chiang, J. Y. (2007). PXR induces CYP27A1 and regulates cholesterol metabolism in the intestine. *J Lipid Res* 48, 373–384.
73. Langmann, T., Moehle, C., Mauerer, R., Scharl, M., Liebisch, G., Zahn, A., Stremmel, W., and Schmitz, G. (2004). Loss of detoxification in inflammatory bowel disease: Dysregulation of pregnane X receptor target genes. *Gastroenterology* 127, 26–40.
74. Martinez, A., Marquez, A., Mendoza, J., Taxonera, C., Fernandez-Arquero, M., Diaz-Rubio, M., de la Concha, E. G., and Urcelay, E. (2007). Role of the PXR gene locus in inflammatory bowel diseases. *Inflamm Bowel Dis* 13, 1484–1487.
75. Shah, Y. M., Ma, X., Morimura, K., Kim, I., and Gonzalez, F. J. (2007). Pregnane X receptor activation ameliorates DSS-induced inflammatory bowel disease via inhibition of NF-kappaB target gene expression. *Am J Physiol Gastrointest Liver Physiol* 292, G1114–G1122.
76. Ma, X., Shah, Y. M., Guo, G. L., Wang, T., Krausz, K. W., Idle, J. R., and Gonzalez, F. J. (2007b). Rifaximin is a gut-specific human pregnane X receptor activator. *J Pharmacol Exp Ther* 322, 391–398.
77. Hartley, D. P., Dai, X., He, Y. D., Carlini, E. J., Wang, B., Huskey, S. E., Ulrich, R. G., Rushmore, T. H., Evers, R., and Evans, D. C. (2004). Activators of the rat pregnane X receptor differentially modulate hepatic and intestinal gene expression. *Mol Pharmacol* 65, 1159–1171.
78. Squires, E. J., Sueyoshi, T., and Negishi, M. (2004). Cytoplasmic localization of pregnane X receptor and ligand-dependent nuclear translocation in mouse liver. *J Biol Chem* 279, 49307–49314.
79. Koike, C., Moore, R., and Negishi, M. (2007). Extracellular signal-regulated kinase is an endogenous signal retaining the nuclear constitutive active/androstane receptor (CAR) in the cytoplasm of mouse primary hepatocytes. *Mol Pharmacol* 71, 1217–1221.
80. Shiraki, T., Sakai, N., Kanaya, E., and Jingami, H. (2003). Activation of orphan nuclear constitutive androstane receptor requires subnuclear targeting by peroxisome proliferator-activated receptor gamma coactivator-1 alpha. A possible link between xenobiotic response and nutritional state. *J Biol Chem* 278, 11344–11350.
81. Muangmoonchai, R., Smirlis, D., Wong, S. C., Edwards, M., Phillips, I. R., and Shephard, E. A. (2001). Xenobiotic induction of cytochrome P450 2B1 (CYP2B1) is mediated by the orphan nuclear receptor constitutive androstane receptor (CAR) and requires steroid co-activator 1 (SRC-1) and the transcription factor Sp1. *Biochem J* 355, 71–78.
82. Jia, Y., Guo, G. L., Surapureddi, S., Sarkar, J., Qi, C., Guo, D., Xia, J., Kashireddi, P., Yu, S., Cho, Y. W., Rao, M. S., Kemper, B., Ge, K., Gonzalez, F. J., and Reddy, J. K. (2005). Transcription coactivator peroxisome proliferator-activated receptor-binding protein/mediator 1 deficiency abrogates acetaminophen hepatotoxicity. *Proc Natl Acad Sci U S A* 102, 12531–12536.
83. Min, G., Kemper, J. K., and Kemper, B. (2002a). Glucocorticoid receptor-interacting protein 1 mediates ligand-independent nuclear translocation and activation of constitutive androstane receptor in vivo. *J Biol Chem* 277, 26356–26363.
84. Xia, J. and Kemper, B. (2005). Structural determinants of constitutive androstane receptor required for its glucocorticoid receptor interacting protein-1-mediated nuclear accumulation. *J Biol Chem* 280, 7285–7293.

85. Moore, J. T., Moore, L. B., Maglich, J. M., and Kliewer, S. A. (2003). Functional and structural comparison of PXR and CAR. *Biochim Biophys Acta* 1619, 235–238.
86. Carnahan, V. E. and Redinbo, M. R. (2005). Structure and function of the human nuclear xenobiotic receptor PXR. *Curr Drug Metab* 6, 357–367.
87. Matic, M., Mahns, A., Tsoli, M., Corradin, A., Polly, P., and Robertson, G. R. (2007). Pregnane X receptor: Promiscuous regulator of detoxification pathways. *Int J Biochem Cell Biol* 39, 478–483.
88. Mitro, N., Vargas, L., Romeo, R., Koder, A., and Saez, E. (2007). T0901317 is a potent PXR ligand: Implications for the biology ascribed to LXR. *FEBS Lett* 581, 1721–1726.
89. Kawana, K., Ikuta, T., Kobayashi, Y., Gotoh, O., Takeda, K., and Kawajiri, K. (2003). Molecular mechanism of nuclear translocation of an orphan nuclear receptor, SXR. *Mol Pharmacol* 63, 524–531.
90. Koyano, S., Kurose, K., Saito, Y., Ozawa, S., Hasegawa, R., Komamura, K., Ueno, K., Kamakura, S., Kitakaze, M., Nakajima, T., Matsumoto, K., Akasawa, A., Saito, H., and Sawada, J. (2004). Functional characterization of four naturally occurring variants of human pregnane X receptor (PXR): One variant causes dramatic loss of both DNA binding activity and the transactivation of the CYP3A4 promoter/enhancer region. *Drug Metab Dispos* 32, 149–154.
91. Ding, X. and Staudinger, J. L. (2005a). Induction of drug metabolism by forskolin: The role of the pregnane X receptor and the protein kinase a signal transduction pathway. *J Pharmacol Exp Ther* 312, 849–856.
92. Ding, X. and Staudinger, J. L. (2005c). Repression of PXR-mediated induction of hepatic CYP3A gene expression by protein kinase C. *Biochem Pharmacol* 69, 867–873.
93. Yoshinari, K., Sueyoshi, T., Moore, R., and Negishi, M. (2001). Nuclear receptor CAR as a regulatory factor for the sexually dimorphic induction of CYB2B1 gene by phenobarbital in rat livers. *Mol Pharmacol* 59, 278–284.
94. Slitt, A. L., Cherrington, N., Fisher, C., Negishi, M., and Klaassen, C. D. (2006). Induction of genes for metabolism and transport by trans-stilbene oxide in livers of Sprague-Dawley and Wistar-Kyoto Rats. *Drug Metab Dispos* 34(7), 1190-1197.
95. Ledda-Columbano, G. M., Pibiri, M., Concas, D., Molotzu, F., Simbula, G., Cossu, C., and Columbano, A. (2003). Sex difference in the proliferative response of mouse hepatocytes to treatment with the CAR ligand, TCPOBOP. *Carcinogenesis* 24, 1059–1065.
96. Lamba, V., Lamba, J., Yasuda, K., Strom, S., Davila, J., Hancock, M. L., Fackenthal, J. D., Rogan, P. K., Ring, B., Wrighton, S. A., and Schuetz, E. G. (2003). Hepatic CYP2B6 expression: Gender and ethnic differences and relationship to CYP2B6 genotype and CAR (constitutive androstane receptor) expression. *J Pharmacol Exp Ther* 307, 906–922.
97. Blake, M. J., Castro, L., Leeder, J. S., and Kearns, G. L. (2005). Ontogeny of drug metabolizing enzymes in the neonate. *Semin Fetal Neonatal Med* 10, 123–138.
98. Vyhlidal, C. A., Gaedigk, R., and Leeder, J. S. (2006). Nuclear receptor expression in fetal and pediatric liver: Correlation with CYP3A expression. *Drug Metab Dispos* 34, 131–137.
99. Stevens, J. C. (2006). New perspectives on the impact of cytochrome P450 3A expression for pediatric pharmacology. *Drug Discov Today* 11, 440–445.
100. Amar, P. J. and Schiff, E. R. (2007). Acetaminophen safety and hepatotoxicity – where do we go from here?. *Expert Opin Drug Saf* 6, 341–355.
101. Wortham, M., Czerwinski, M., He, L., Parkinson, A., and Wan, Y. J. (2007a). Expression of CAR, HNF4{alpha}, and POR Genes Determine Interindividual Variability in Basal Expression and Activity of a Broad Scope of Xenobiotic Metabolism Genes in the Human Liver. *Drug Metab Dispos* 35, 1700–1710.
102. Valaes, T., Kipouros, K., Petmezaki, S., Solman, M., and Doxiadis, S. A. (1980). Effectiveness and safety of prenatal phenobarbital for the prevention of neonatal jaundice. *Pediatr Res* 14, 947–952.

103. Arya, V. B., Agarwal, R., Paul, V. K., and Deorari, A. K. (2004). Efficacy of oral phenobarbitone in term "at risk" neonates in decreasing neonatal hyperbilirubinemia: A randomized double-blinded, placebo controlled trial. *Indian Pediatr* 41, 327–332.
104. Murki, S., Dutta, S., Narang, A., Sarkar, U., and Garewal, G. (2005). A randomized, triple-blind, placebo-controlled trial of prophylactic oral phenobarbital to reduce the need for phototherapy in G6PD-deficient neonates. *J Perinatol* 25, 325–330.
105. Honkakoski, P., Sueyoshi, T., and Negishi, M. (2003). Drug-activated nuclear receptors CAR and PXR. *Ann Med* 35, 172–182.
106. Kretschmer, X. C. and Baldwin, W. S. (2005). CAR and PXR: Xenosensors of endocrine disrupters?. *Chem Biol Interact* 155, 111–128.
107. Kassam, A., Winrow, C. J., Fernandez-Rachubinski, F., Capone, J. P., and Rachubinski, R. A. (2000). The peroxisome proliferator response element of the gene encoding the peroxisomal beta-oxidation enzyme enoyl-CoA hydratase/3-hydroxyacyl-CoA dehydrogenase is a target for constitutive androstane receptor beta/9-cis-retinoic acid receptor-mediated transactivation. *J Biol Chem* 275, 4345–4350.
108. Maglich, J. M., Parks, D. J., Moore, L. B., Collins, J. L., Goodwin, B., Billin, A. N., Stoltz, C. A., Kliewer, S. A., Lambert, M. H., Willson, T. M., and Moore, J. T. (2003). Identification of a novel human constitutive androstane receptor (CAR) agonist and its use in the identification of CAR target genes. *J Biol Chem* 278, 17277–17283.
109. Ekins, S., Mirny, L., and Schuetz, E. G. (2002). A ligand-based approach to understanding selectivity of nuclear hormone receptors PXR, CAR, FXR, LXRalpha, and LXRbeta. *Pharm Res* 19, 1788–1800.
110. Owsley, E. and Chiang, J. Y. (2003). Guggulsterone antagonizes farnesoid X receptor induction of bile salt export pump but activates pregnane X receptor to inhibit cholesterol 7alpha-hydroxylase gene. *Biochem Biophys Res Commun* 304, 191–195.
111. Ding, X. and Staudinger, J. L. (2005b). The ratio of constitutive androstane receptor to pregnane X receptor determines the activity of guggulsterone against the Cyp2b10 promoter. *J Pharmacol Exp Ther* 314, 120–127.
112. Bookout, A. L., Jeong, Y., Downes, M., Yu, R. T., Evans, R. M., and Mangelsdorf, D. J. (2006). Anatomical profiling of nuclear receptor expression reveals a hierarchical transcriptional network. *Cell* 126, 789–799.
113. Kanno, Y., Aoki, S., Mochizuki, M., Mori, E., Nakahama, T., and Inouye, Y. (2005). Expression of constitutive androstane receptor splice variants in rat liver and lung and their functional properties. *Biol Pharm Bull* 28, 2058–2062.
114. Lamba, J., Lamba, V., and Schuetz, E. (2005). Genetic variants of PXR (NR1I2) and CAR (NR1I3) and their implications in drug metabolism and pharmacogenetics. *Curr Drug Metab* 6, 369–383.
115. Auerbach, S. S., Dekeyser, J. G., Stoner, M. A., and Omiecinski, C. J. (2007). CAR2 displays unique ligand binding and RXRalpha heterodimerization characteristics. *Drug Metab Dispos* 35, 428–439.
116. Dotzlaw, H., Leygue, E., Watson, P., and Murphy, L. C. (1999). The human orphan receptor PXR messenger RNA is expressed in both normal and neoplastic breast tissue. *Clin Cancer Res* 5, 2103–2107.
117. Lamba, V., Yasuda, K., Lamba, J. K., Assem, M., Davila, J., Strom, S., and Schuetz, E. G. (2004b). PXR (NR1I2): Splice variants in human tissues, including brain, and identification of neurosteroids and nicotine as PXR activators. *Toxicol Appl Pharmacol* 199, 251–265.
118. Lamba, J. K., Lamba, V., Yasuda, K., Lin, Y. S., Assem, M., Thompson, E., Strom, S., and Schuetz, E. (2004a). Expression of constitutive androstane receptor splice variants in human tissues and their functional consequences. *J Pharmacol Exp Ther* 311, 811–821.
119. Wortham, M., Czerwinski, M., He, L., Parkinson, A., and Wan, Y. J. (2007b). Expression of constitutive androstane receptor, hepatic nuclear factor 4 alpha, and P450 oxidoreductase genes

determines interindividual variability in basal expression and activity of a broad scope of xenobiotic metabolism genes in the human liver. *Drug Metab Dispos* 35, 1700–1710.

120. Pascussi, J. M., Robert, A., Moreau, A., Ramos, J., Bioulac-Sage, P., Navarro, F., Blanc, P., Assenat, E., Maurel, P., and Vilarem, M. J. (2007). Differential regulation of constitutive androstane receptor expression by hepatocyte nuclear factor4alpha isoforms. *Hepatology* 45, 1146–1153.
121. Pascussi, J. M., Drocourt, L., Fabre, J. M., Maurel, P., and Vilarem, M. J. (2000). Dexamethasone induces pregnane X receptor and retinoid X receptor-alpha expression in human hepatocytes: Synergistic increase of CYP3A4 induction by pregnane X receptor activators. *Mol Pharmacol* 58, 361–372.
122. Pascussi, J. M., Drocourt, L., Gerbal-Chaloin, S., Fabre, J. M., Maurel, P., and Vilarem, M. J. (2001). Dual effect of dexamethasone on CYP3A4 gene expression in human hepatocytes. Sequential role of glucocorticoid receptor and pregnane X receptor. *Eur J Biochem* 268, 6346–6358.
123. Ding, X., Lichti, K., Kim, I., Gonzalez, F. J., and Staudinger, J. L. (2006). Regulation of constitutive androstane receptor and its target genes by fasting, cAMP, hepatocyte nuclear factor alpha, and the coactivator peroxisome proliferator-activated receptor gamma coactivator-1alpha. *J Biol Chem* 281, 26540–26551.
124. Zhu, Y., Qi, C., Jain, S., Rao, M. S., and Reddy, J. K. (1997). Isolation and characterization of PBP, a protein that interacts with peroxisome proliferator-activated receptor. *J Biol Chem* 272, 25500–25506.
125. Petrick, J. S. and Klaassen, C. D. (2007). Importance of hepatic induction of constitutive androstane receptor and other transcription factors that regulate xenobiotic metabolism and transport. *Drug Metab Dispos* 35, 1806–1815.
126. Kyrmizi, I., Hatzis, P., Katrakili, N., Tronche, F., Gonzalez, F. J., and Talianidis, I. (2006). Plasticity and expanding complexity of the hepatic transcription factor network during liver development. *Genes Dev* 20, 2293–2305.
127. Jung, D., Mangelsdorf, D. J., and Meyer, U. A. (2006). Pregnane X receptor is a target of farnesoid X receptor. *J Biol Chem* 281, 19081–19091.
128. Guo, G. L., Lambert, G., Negishi, M., Ward, J. M., Brewer, H. B., Jr., Kliewer, S. A., Gonzalez, F. J., and Sinal, C. J. (2003). Complementary roles of farnesoid X receptor, pregnane X receptor, and constitutive androstane receptor in protection against bile acid toxicity. *J Biol Chem* 278, 45062–45071.
129. Takagi, S., Nakajima, M., Mohri, T., and Yokoi, T. (2008). Post-transcriptional regulation of human pregnane X receptor by microRNA affects the expression of cytochrome P450 3A4. *J Biol Chem* 283, 9674-9680.
130. Krasowski, M. D., Yasuda, K., Hagey, L. R., and Schuetz, E. G. (2005). Evolutionary selection across the nuclear hormone receptor superfamily with a focus on the NR1I subfamily (vitamin D, pregnane X, and constitutive androstane receptors). *Nucl Recept* 3, 2.
131. Poso, A. and Honkakoski, P. (2006). Ligand recognition by drug-activated nuclear receptors PXR and CAR: Structural, site-directed mutagenesis and molecular modeling studies. *Mini Rev Med Chem* 6, 937–947.
132. Huang, W., Zhang, J., Wei, P., Schrader, W. T., and Moore, D. D. (2004b). Meclizine is an agonist ligand for mouse constitutive androstane receptor (CAR) and an inverse agonist for human CAR. *Mol Endocrinol* 18, 2402–2408.
133. Kawamoto, T., Kakizaki, S., Yoshinari, K., and Negishi, M. (2000). Estrogen activation of the nuclear orphan receptor CAR (constitutive active receptor) in induction of the mouse Cyp2b10 gene. *Mol Endocrinol* 14, 1897–1905.
134. LeCluyse, E. L. (2001). Pregnane X receptor: Molecular basis for species differences in CYP3A induction by xenobiotics. *Chem Biol Interact* 134, 283–289.
135. Lemaire, G., Mnif, W., Pascussi, J. M., Pillon, A., Rabenoelina, F., Fenet, H., Gomez, E., Casellas, C., Nicolas, J. C., Cavailles, V., Duchesne, M. J., and Balaguer, P. (2006). Identification of new

human pregnane X receptor ligands among pesticides using a stable reporter cell system. *Toxicol Sci* 91, 501–509.

136. Xie, W., Barwick, J. L., Downes, M., Blumberg, B., Simon, C. M., Nelson, M. C., Neuschwander-Tetri, B. A., Brunt, E. M., Guzelian, P. S., and Evans, R. M. (2000). Humanized xenobiotic response in mice expressing nuclear receptor SXR. *Nature* 406, 435–439.
137. Ma, X., Shah, Y., Cheung, C., Guo, G. L., Feigenbaum, L., Krausz, K. W., Idle, J. R., and Gonzalez, F. J. (2007a). The PREgnane X receptor gene-humanized mouse: A model for investigating drug-drug interactions mediated by cytochromes P450 3A. *Drug Metab Dispos* 35, 194–200.

CHAPTER 12

FXR

YANDONG WANG, WEIDONG CHEN, XIAOSONG CHEN, AND WENDONG HUANG

Department of Gene Regulation and Drug Discovery, Beckman Research Institute, City of Hope National Medical Center, 1500 E. Duarte Road, Duarte, CA 91010, USA

Abstract: Great progress has been made in the understanding of the physiological roles of the nuclear receptor farnesoid X receptor (FXR) during the last several years. Roles for FXR were initially identified in the regulation of bile acid, cholesterol, triglyceride, and glucose metabolism. More recently, additional functions of FXR are identified in enteroprotection, liver regeneration, cancer and aging. These exciting findings suggest that FXR has a broader role than previously thought, and also highlight potential opportunities for using FXR as a drug target for different diseases.

12.1. INTRODUCTION

The farnesoid X receptor (FXR, NR1H4) was first isolated from a rat liver cDNA library in 1995 [1]. FXR was originally considered an 'orphan' nuclear receptor because its natural ligands were unknown. Subsequently, FXR was 'adopted' following the discovery that metabolites of bile acids binded to and activated this receptor at physiological concentrations [2–4]. FXR is highly expressed in the liver, intestine, kidney, and adrenals. It is now very clear that FXR is an important regulator for diverse metabolic pathways, including the cholesterol, bile acid, lipid and glucose metabolisms. FXR binds to DNA as heterodimers with its partner, retinoid X receptor (RXR, NR2B1) to regulate the mRNA levels of various gene transcripts [5]. There are four known FXR isoforms (FXRα1, FXRα2, FXR α3 and FXR α4) in humans and mice [5, 6]. Some FXR target genes, including intestinal bile acid binding protein (IBABP) and the three fibrinogen subunits (FBG, -alpha, -beta and -gamma), are regulated in an isoform-dependent manner [7]; but most FXR target genes, including bile salt export pump (BSEP, ABCb11) and small heterodimer partner (SHP), are regulated in an isoform-independent manner. The four isoforms of FXR are expressed in a tissue-dependent manner [5]. A second FXR gene, FXRβ (NR1H5) is identified in rodents, rabbits and dogs. However, it is a pseudogene in human and primates. The function of FXRβ is still unknown so far.

C.M. Bunce, M.J. Campbell (eds.), *Nuclear Receptors*, Proteins and Cell Regulation 8, 307–326. DOI 10.1007/978-90-481-3303-1_12,

FXR expression can be induced by prolonged fasting, probably through HNF4α and peroxisome proliferators-associated receptor coactivator 1α (PGC-1α) coactivation [8]. HNF1α is also shown to activate FXR promoter [9]. Furthermore, FXR mRNA levels are increased in response to high levels of glucose by an unknown mechanism [10]. On the other hand, hypoxia is shown to downregulate FXR expression [11]. In addition to transcriptional regulation, FXR is also shown to be posttranslationally modified by PKC phosphorylation, which enhances FXR transactivity [12, 13]. But more works are expected to understand the regulation of FXR expression and protein modifications.

Over the past decade, a number of studies on FXR have established FXR as a key regulator of metabolism. However, recent studies indicate that FXR has additional functions in enteroprotection, liver regeneration, cancer and aging. As such, FXR has become a promising target for the treatment of different diseases.

12.2. FXR LIGANDS

Ligand identification for nuclear receptors always greatly facilitates their study. FXR was originally proposed to be a receptor for an intermediary metabolite, farnesol [1]. However, the supraphysiological concentrations required to activate the FXR impede the use of farnesoid as a ligand. The major breakthrough in FXR biology was the discovery that bile acids are endogenous ligands for this nuclear receptor [2, 3, 4]. In fact, both conjugated and unconjugated bile salts are able to activate FXR at physiological concentrations. The hydrophobic BA chenodeoxycholic acid (CDCA) is the most effective activator of FXR. The secondary bile acids, deoxycholic acid (DCA) and lithocholic acid (LCA), can both activate FXR but to a much lesser extent than CDCA does, whereas hydrophilic ursodeoxycholic (UDCA) and muricholic acids can not activate FXR [2]. Recently, Deng et al. and Wang et al. reported that oxysterol 22(R)-hydroxycholesterol and androsterone are FXR natural ligands, respectively [14, 15]. However, whether they are real endogenous ligands of FXR and what are the physiological consequences of FXR activation by them still need to be established (Table 12.1). Some polyunsaturated fatty acids such as arachidonic acid and decosahexaenoic acid [16] and bile acid metabolites such as 26- or 25-hydroxylated bile alcohols were also identified as weak FXR ligands [17]. Recently, two Japanese groups identified several natural products as potential FXR ligands such as ginkgolic acids, coumestrol, marchantin A and marchantin E by reporter assay screening [18–20]. In addition, several synthetic FXR ligands have been generated. They include GW4064 [21], 6ECDCA [22], AGN29, AGN31 [23] and MFA-1 [24]. The most widely used FXR ligand is the non-steroidal isoxazole analog GW4064. But the potential cell-toxic effect and uncertain bioavailability restrict its further use. Several GW4064 analogs are synthesized but still do not solve the problem of the poor bioavailability [25]. Instead, 6 alpha-Ethyl-chenodeoxycholic acid (6-ECDCA), a novel compound derived from the natural FXR ligand CDCA,

Table 12.1. Summary of related FXR information

Gene	NR1H4 12q23.3
Expression	Liver Small intestine Kidney Adrenals Vascular smooth muscle Adipose tissue Heart
Natural agonists	Primary bile acid: CA, CDCA; Secondary bile acid: LCA and DCA; Polyunsaturated fatty acids: arachidonic acid; docosahexaenoic acid, and linolenic acid (Endogenous and selective bile acid receptor modulators that specifically regulate expression of certain FXR targets); Bile acid metabolites: 26- or 25-hydroxylated bile alcohols; Oxysterols: oxysterol 22(R)-hydroxysholesterol; Androsterone (very weak activity) Cafestol, Ginkgolic acids, Coumestrol, Marchantin
Synthetic agonists	GW4064 (high affinity agonist), 6ECDCA (semisynthetic bile acid), AGN29, AGN31, MFA-1
Antagonists	Guggulsterone, lithocholate, AGN34, Stigmasterol, Scalarane Sesterterpenes
Response elements	IR-1: GAGTTAaTGACCT; GGGTGAaTAACCT; GGGACAtTGATCCT; AGGTCAaGTGCCT; GGGTCAgTGACCC DR-1: AGAGCAnAGGGGA ER-8: TGAACTcttaaccaAGTTCA Monomer binding site: GATCCTTGAACTCT; TGAACT
Relevant diseases	Cholestasis Diabetes Atherosclerosis Cholesterol gallstone disease Liver fibrosis Liver cancer Breast cancer Colon cancer Aging

has become an alternative agonist ligand for FXR and is currently under clinical trials for the treatment of several liver diseases [26 , 22]. In contrast to FXR agonists, only a small number of FXR antagonists are identified. These include guggulsterone [27], Stigmasterol [28] and scalarane sesterterpenes [29, 30].

12.3. FXR TARGET GENES AND FXREs

FXR regulates the expression of a wide variety of target genes by binding either as a monomer or as a heterodimer with RXR to FXR response elements (FXREs) that consist of an inverted repeat (IR) of the canonical AGGTCA hexanucleotide core motif spaced by 0 bp (IR-0) [31] or 1 bp (IR-1) [32–34]. The highest level of transactivation by FXR/RXR was mediated by an IR-1 where one half-site had the sequence GGGTCA. In 1999, Grober et al. showed that BAs induce expression of the human IBABP gene through an interaction between an FXR/RXR heterodimer and an IR-1 element in the proximal IBABP promoter (Table 12.2). The physiological relevance of these data is supported by the observation that BA-mediated regulation of IBABP expression is also found in vivo in the mouse [35]. Goodwin et al. [36] report that FXR directly binds and activates the SHP-1 promoter by binding to an IR-1 element (GAGTTAaTGACCT). SHP-1, in turn, interacts with liver receptor homolog 1 (LRH-1) and represses the expression of cholesterol 7 alpha-hydroxylase (CYP7A1) and cytochrome P450 sterol 12α-hydroxylase (CYP8B1). Ananthanarayanan et al. [34] show that FXR directly activates the transcription of BSEP by binding to an IR-1 element in the BSEP promoter. FXR also induces bile acid-CoA:amino acid N-acetyltransferase (BAT) and phospholipid transfer protein (PLTP) via IR-1 elements in the promoters of these genes [37, 27]. In addition, FXR activates other genes that are critical for bile acid enterohepatic circulation, such as human organic anion transporting polypeptide 8 (OATP8) and organic solute transporters α and β (OSTα/β), via IR-1 elements [38, 39]. To summarize, IR-1 is the primary binding sequence for FXR.

Additionally, FXR has also been shown to activate an IR-0 in the dehydroepiandrosterone sulfotransferase (STD) gene, which encodes an enzyme with bile acid sulfo-conjugating activity [31]. Moreover, in addition to IR-1 and IR-0 elements, FXR/RXR heterodimers can also recognize other DNA motifs with varying affinity, such as direct repeats (DRs) of the hexanucleotide core sequence with different spacing [40]. For example, FXR uses a DR-1 element to bind to the promoter of the syndecan-1 (SDC-1) gene, a transmembrane heparan sulfate proteoglycan that participates in the binding and internalization of extracellular ligands [41]. Furthermore, an everted repeat (ER) of the core motif separated by eight nucleotides (ER-8) was shown to mediate the induction of the multidrug resistant-associated protein 2 (MRP2) by BAs [42]. Finally, Claudel et al. [43] showed that apolipoprotein A-I (apoA-I) is regulated by FXR via the DNA binding of a monomeric form. In summary, FXR can bind to a variety of FXREs with differing

Table 12.2. Genes regulated by FXR and the related FXREs

Metabolism	Gene	Regulation	FXREs	References
Cholesterol and bile acid	SHP	Induced	IR-1, GAGTTAaTGACCT (human)	Goodwin et al. [36]
	CYP7A1	Repressed	–	Goodwin et al. [36]
	CYP8B1	Repressed	–	Goodwin et al. [36]
	IBABP	Induced	IR-1, GGGTGAaTAACCT	Grober et al. [35]
	BSEP	Induced	IR-1, GGGACAtTGATCCT	Ananthanarayanan et al. [34]
	rMRP2	Induced	ER-8, TGAACTcttaacca AGTTCA	Kast et al. [42]
	BAT	Induced	IR-1, AGGTCAaGTGCCT	Pircher et al. [37]
	STD	Induced	IR-0, GGGTCA TGAACT	Song et al. [31]
	ASBT	Repressed	–	Li et al. [32, 33]
	Hepatic insig-2	Induced	IR-1, AGGTCAaCGACCT, AGGACAtTGCCCC	Hubbert et al. [44]
	OATP8	Induced	IR-1, AGGACAaTGACCT	Jung et al. [38]
	hOSTα/ hOSTβ	Induced	OSTα, IR-1, GGGTGAaTGACCT, AGGCCAgTGACCC, GGGTCAgGGCCCT, GGGTAAtTAAACC OSTβ, IR-1, AGGTCAgTCACCC, AGGTGAtACACCT	Lee et al. [39]

Table 12.2. (Continued)

Metabolism	Gene	Regulation	FXREs	References
	BACS	Induced	IR-1, GGGGCAaAGACCT	Pircher et al. [37]
Lipid	MDR3	Induced	IR-1, ATGTCAaTAACCT	Huang et al. [98]
	Human Complement C3	Induced	IR-1, AGGTTAcTCACCC	Li et al. [32]
	SDC1	DR-1, AGAGC AnAGGGGA	Anisfeld et al. [39]	
	VLDLR	Induced	–	Sirvent et al. [51]
	PLTP	Induced	IR-1, GGGTCAgTGACCC	Urizar et al. [27]
	Apo-I	Repressed	monomeric form, GATCCT TGAACTCT	Claudel et al. [43]
	Hepatic lipase	Repressed	–	Sirvent et al. [99]
Glucose	PEPCK	Induced or repressed	–	Stayrook et al. [16], Ma et al. [55], Zhang et al. [53, 54], and Cariou et al. [100]

affinities. This allows FXR to regulate the expression of different genes involved in different metabolic pathways (Table 12.2).

12.4. FXR REGULATES DIVERSE METABOLIC PATHWAYS AND CELL HOMEOSTASIS

As described above, the various FXREs enable FXR to regulate a battery of genes involved in a variety of metabolic pathways.

Bile acids are amphipathic molecules with detergent-like properties that are essential for nutritional absorption. Cells, especially those of organs that participate in enterohepatic circulation, must be able to sense intracellular bile acid levels and to elicit an adequate response in case of depletion or over-load. Physiological concentrations of both primary and secondary bile acids efficiently activate FXR. Elevated bile acid levels are toxic, and therefore their synthesis and enterohepatic circulation is tightly controlled. It is now well established that FXR functions as a primary bile acid sensor. In accordance with this role, many FXR-target genes have been identified that are involved in bile salt and cholesterol metabolism [44, 39, 31]. FXR not only down-regulates the expression of CYP7A1, the rate-limiting enzyme in the classic pathway of bile acid synthesis from cholesterol, but also represses the expression of another key enzyme in bile acid synthesis, CYP8B1 [45] (Figure 12.1). In addition, it is clear that bile acid activating FXR in the intestine leads to the induction of mouse Fgf 15 [46] or its human ortholog FGF19 [47], thereby suppressing CYP7A1 expression through a JNK-dependent signaling cascade. Recently, Hubbert et al. [44] reported that FXRα induces the expression of hepatic Insig-2, which represses lanosterol 14α-demethylase, and reduces HMG-CoA reductase protein levels. This FXRα-mediated regulation results in the repression of cholesterol synthesis (Figure 12.1). These findings indicate that FXR not only directly represses the synthesis of bile acids, but also inhibits the synthesis of cholesterol, the precursor for bile acids. In addition, FXR controls enterohepatic circulation of bile acids by regulating genes involved in bile acid secretion, such as BSEP, the major hepatic bile salt exporter in liver, and MRP2, which mediates the efflux of several conjugated compounds across the apical membrane of the hepatocyte into the bile canaliculi [34, 42], as well as proteins involved in bile acid transport, such as IBABP, an intestinal protein that binds bile salts with high affinity in the cytosol of enterocytes; the apical sodium-dependent bile acid cotransporter (ASBT/SLC10A2), which is the primary bile salt uptake protein in the intestine; and the sodium-dependent taurocholate cotransporting protein (NTCP), the major hepatic bile salt importer. FXR also regulates genes involved in bile acid detoxification, such as BAT, dehydroepiandrosterone–sulfotransferase (SULT2A1) and bile acid CoA synthetase (BACS) [48]. In conclusion, FXR is a master regulator of the homeostasis of bile acids.

Subsequent studies demonstrate that FXR also regulates a set of genes that participate in lipid and glucose metabolism. Sinal et al. originally proposed that FXR is involved in the control of plasma lipid levels [49]. Furthermore, the studies of

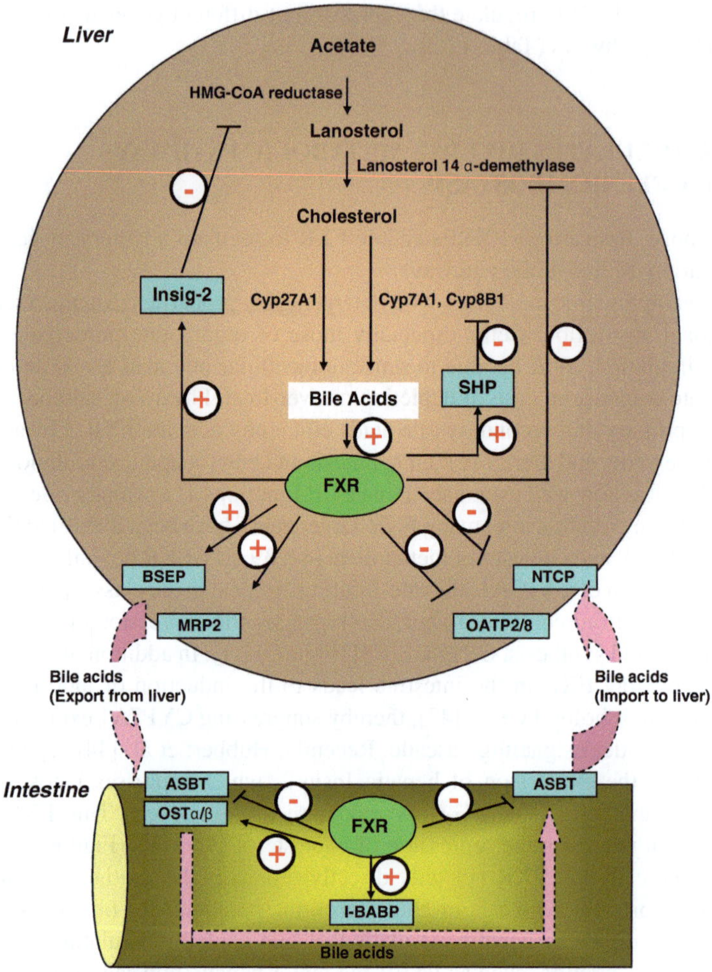

Figure 12.1. FXR as a master regulator of bile acid homeostasis in liver and intestine. In the liver, FXR negatively regulates bile acid production by repressing CYP7A1, the rate-limiting enzyme of the synthetic pathway of bile acids, and inhibiting HMG-CoA reductase and lanosterol 14 α-demethylase, which play key roles in the synthesis of cholesterol. FXR induces the expression of BSEP and MRP2, which are involved in bile acid export, and simultaneously represses bile acid import by downregulating NTCP and OATP2/8. In the intestine, FXR induces the expression of IBABP, which provides the enterocytes with a shuttling device for the delivery of bile acids from the apical to the basolateral membrane. FXR does not directly regulate the ASBT gene, but it may influence the network of transcription factors that are involved in a response to bile acids. FXR activates the expression of OSTα/β, which serves to transport bile acids from the gut to the enterohepatic circulation where they are transported back to the liver

Edwards et al. showed that FXR alters the transcription of several genes involved in fatty acid and triglyceride synthesis, as well as lipoprotein metabolism. These genes include PLTP, SDC-1 and the very low density lipoprotein receptor (VLDLR) [50, 41, 51]. These results, together with the findings that activation of FXR leads to repression of SREBP-1c, a transcription factor that controls genes involved in fatty acid and triglyceride synthesis, provide a mechanism to account for the triglyceride-lowering effects of bile acids and some synthetic FXR agonist ligands [52, 10].

Early studies showed that expression of FXR is reduced in animal models of diabetes and FXR is regulated by glucose, likely via the pentose phosphate pathway [10]. Recent observations of glucose levels in $FXR^{-/-}$ mice have produced conflicting data, suggesting that glucose levels are either unchanged [53, 54], increased [55] or repressed [56] compared to wild-type littermates, indicating that other crucial factors have yet to be identified, nonetheless, these reports have provided direct evidence that activation of FXR in wild-type or diabetic [db/db or KKA-(y)] mice promotes hypoglycemia and increases insulin sensitivity. Ma et al. [55] showed that loss of FXR function in the liver results in increased hepatic lipid accumulation and elevation of non-esterified fatty acids (FFAs) in the serum. The development of insulin resistance in the liver, which fails to suppress gluconeogenesis, and in the skeletal muscle, which reduces glucose uptake, contributes to the dysregulation of glucose homeostasis in $FXR^{-/-}$ mice. Therefore, FXR is essential for normal glucose homeostasis. Zhang et al. [53, 54] demonstrated that activation of FXR lowers plasma glucose levels by sensitizing insulin action, which provides further evidence that FXR may be involved in the regulation of glucose homeostasis.

FXR is not only a key metabolic regulator, but also an important cell protector. Bile acids are known to inhibit intestinal bacterial growth and the consequent damage to intestine mucosal [57]. This seems to be mediated by FXR activation, which induces the enteroprotection genes including angiogenin and IL-18 [58]. Therefore, FXR plays a role in keeping the intestine barrier intact. Similarly, in addition to its key role to reduce the toxic levels of bile acids, FXR activation further protects hepatocytes from death by stimulating MAKP pathway [59]. FXR also suppresses NF-κB-mediated hepatic inflammation [60]. In parallel, this anti-inflammation function is identified in vascular smooth muscle and endothelial cells [61, 62], which suggests a potential general role of FXR in antagonizing NF-κB-mediated inflammatory responses.

12.5. FXR AND DIFFERENT DISEASES

12.5.1. FXR and Cholestasis

Cholestatic injury is associated with the accumulation of bile acids and activation of pro-inflammatory cytokines in liver. Cholestasis causes systemic and intrahepatic retention of potentially toxic bile acids that results in liver injury, and ultimately leads to biliary fibrosis and cirrhosis [63]. Miyata and colleagues [64] explore the FXR-dependent protective mechanism for cholic acid-induced toxicity. The investigators found that BSEP mRNA and protein levels increased in the wild-type mice, but decreased in the FXR-null mice after being fed a diet that contained cholic acid. As the concentration of cholic acid in the diet increased, the wild-type mice had a compensatory increase in bile acid output rate. In FXR-deficient mice, the bile acid output significantly decreased in the face of rising cholic acid concentrations. This study demonstrates that FXR-mediated adaptive enhancement of canalicular bile acid excretion is a critical protective mechanism to prevent cholic acid-induced toxicity in cholestasis. Similarly, Liu et al. [65] find that a ligand for FXR, GW4064, protects the liver from cholestatic injury in the bile duct-ligation and alpha-naphthylisothiocyanate models of cholestasis.

Estrogens can cause intrahepatic cholestasis in susceptible women during pregnancy and when prescribed for oral contraception or postmenopausal hormone replacement therapy. Fiorucci and colleagues [66, 67] found that in a rat model of estrogen-induced cholestasis, administration of 6-ethyl chenodeoxycholic acid, a semi-synthetic bile acid and potent FXR ligand, protected against cholestasis by dramatically increasing the expression of basolateral and canalicular bile acid transporters (BSEP, MRP2 and MDR2) and repressing bile acid biosynthesis. The investigators conclude that the development of FXR ligands may provide a new approach for treatment of cholestatic disorders. Also, together with the xenobiotic receptors, pregnane X receptor (PXR, NR1H2) and constitutive androstane receptor (CAR, NR1H3), FXR prevents and ameliorates cholestasis through the activation of hepatic CYP450s, phase II enzymes that are able to decrease the cholestatic xenobiotic noxae and to detoxify the bile acid pool [68, 69].

12.5.2. FXR and Atherosclerosis

$FXR^{-/-}$ mice have a proathergenic profile of lipids [49]. However, decreased atherosclerotic lesions were observed in either $FXR^{-/-}Ldlr^{-/-}$ or $FXR^{-/-}ApoE^{-/-}$ double knockout mouse models of atherosclerosis [70, 53, 54]. Using the same $FXR^{-/-}ApoE^{-/-}$ mice, Hanniman et al. observed an opposite results that $FXR^{-/-}ApoE^{-/-}$ are more susceptible to atherosclerosis lesions following a western diet [71]. So far, the role of FXR in atherogenesis is still unclear. On the other hand, Bishop-Bailey et al. demonstrated that FXR is expressed in atherosclerotic lesions and vascular smooth muscle cells of human vessels but not in macrophages [72]. CDCA treatment was found to reduce the expression of a potent vasoconstrictive peptide, endothelin-1 [73], and to increase expression of adhesion molecules ICAM-1 and VCAM-1 [74], which may stimulate the recruitment of macrophages. But the role of FXR in regulating the expression of these genes remains to be determined.

12.5.3. FXR and Diabetes

Several key processes regulated by FXR, such as bile acid and triglyceride metabolism, are impaired in diabetic individuals [75]. This observation has led many investigators to assess the potential role of FXR in animal models of type 1 and 2 diabetes. Duran-Sandoval et al. observed that the expression of FXR was decreased in the livers of streptozotocin-induced diabetic rats as well as in diabetic Zucker rats, which suggested a link between FXR and diabetes [10]. Stayrook et al. [76] reported that phosphoenolpyruvate carboxykinase (PEPCK) expression and glucose production are regulated by FXR, which provides evidence for an additional link between carbohydrate metabolism and the well characterized lipid metabolism pathways regulated by FXR. The findings that FXR may regulate gluconeogenesis, coupled with the discovery of Duran-Sandoval et al. [10] that FXR expression is regulated by glucose levels, suggest that a feedback loop may be operating. In addition, Stayrook et al. [76] found that treatment of C57BL6 mice with GW4064 significantly increased hepatic PEPCK expression. Therefore, they suggested that activation of FXR may actually be unfavorable in diabetics. However, despite this observation, other reports suggest that activation of FXR may be useful for inhibiting hepatic gluconeogenesis in diabetics [77]. Zhang et al. [53, 54] report that FXR null mice exhibited glucose intolerance and insulin insensitivity. They propose that the development of FXR agonists may prove useful for the treatment of diabetes. Ma et al. [55] showed that, in contrast to the results in $FXR^{-/-}$ mice, bile acid-induced activation of FXR in wild-type mice repressed expression of gluconeogenic genes and decreased serum glucose. They demonstrate that FXR is required for the maintenance of glucose homeostasis in vivo. Based on these findings, FXR selective agonists are potential pharmaceutical candidates for the management of type 2 diabetes and hypertriglyceridemia, which are two major symptoms of metabolic syndrome.

12.5.4. FXR and Gallstone Disease

Compared to the wild-type mice, FXR knockout mice are more susceptible to cholesterol gallstone formation following administration of a lithogenic diet [78, 45]. FXR ligand, GW4064, significantly reduces gallstone formation probably by inducing the expression of bile acid transporters, Bsep and Mdr2 that help transport bile acids into bile and reduce cholesterol crystallization. These results suggest a potential use of FXR ligands in the treatment of cholesterol gallstone disease.

12.5.5. FXR and Aging

The long-lived Little mice (Ghrhr(lit/lit)) showed a concerted up-regulation of xenobiotic detoxification genes. Surprisingly, the classic xenobiotic receptors CAR (Constitutive Androstane Receptor) and PXR (Pregnane X Receptor) are not required for the up-regulation of xenobiotic genes in Little mice. Instead, FXR is likely a mediator of the up-regulation of xenobiotic detoxification genes in Little mice [79]. Moreover, BA levels are considerably elevated in the bile, serum, and liver of Little mice. Treatment of wild-type animals with a CA diet mimics in large part the up-regulation of xenobiotic detoxification genes observed in Little mice. Additionally, the loss of FXR reduces the expression of the xenobiotic detoxification genes up-regulated in Little mice. The authors conclude that alterations in xenobiotic metabolism in Little mice constitute a form of increased stress resistance and may contribute to the extended longevity of these mice [79].

12.5.6. FXR and Liver Regeneration

Liver regeneration after the loss of hepatic tissue is a fundamental parameter of liver response to injury. It is now defined as an orchestrated response induced by specific external stimuli and involving sequential changes in gene expression, growth factor production, and morphologic structure. Normal liver regeneration is important for restoring the liver mass following liver injury. However, irregular regeneration of hepatocytes, which develops as a result of repeated cycles of necrosis and regeneration of hepatocytes in chronic hepatitis, has been reported as an important factor in the hepatocarcinogenesis [80].

In addition to controlling the levels of BAs [81], FXR also helps promote normal liver regeneration in response to increased BA stress after 70% hepatectomy. An intriguing homeostatic mechanism for determination of liver size is proposed, in which FXR and possibly other nuclear receptors sense the levels of endogenous metabolites to determine the liver's functional capacity. When liver function is decreased as a result of injury, the resulting accumulation of bile acids activates FXR, which stimulates signaling pathways to protect the liver from bile acid toxicity and also promotes liver growth to handle the overload [82]. However, the signaling pathways involved in FXR function in liver regeneration remain to be identified.

12.5.7. FXR and Hepatocarcinogenesis

The hepatoprotective role of FXR is essential for the maintenance of normal liver physiology and prevention of the deleterious effects of bile acids. Indeed, FXR null mice spontaneously develop liver tumors due to chronic liver injury as they age [83]. In the absence of FXR, Yang et al. detected significant hepatocellular apoptosis, chronic liver injury and irregular liver regeneration, which resulted in spontaneous liver tumor formation [83]. The findings indicate that FXR is an important factor in an intriguing link between metabolic regulation and hepatocarcinogenesis. However, the mechanisms by which FXR suppresses liver cancer remain to be investigated.

BAs have been implicated in the induction of liver apoptosis and injury [84]. BAs can promote liver tumors in an HBV transgenic mouse model and are thought to induce inflammation and liver tumorigenesis in mdr-2 knockout mice [85–87]. This is now further confirmed in $FXR^{-/-}$ mice that feeding of a cholic acid diet promoted chemical-induced hepatocarcinogenesis [83].

In parallel, Kim et al. [88] reported very similar findings that, at 12 months of age, both male and female $FXR^{-/-}$ mice had a high incidence of degenerative hepatic lesions, altered cell foci and liver tumors including hepatocellular adenoma, carcinoma and hepatocholangiocellular carcinoma. The major findings between these two studies are very similar, especially the pathological changes of liver in $FXR^{-/-}$ mice including liver injury, irregular regeneration and strong inflammation. The results indicate that FXR may provide an intriguing link between metabolic regulation and hepatocarcinogenesis. However, the mechanism by which FXR suppresses liver cancer remains to be investigated. The fact that FXR is required for both liver regeneration and protection against hepatocarcingenesis suggests an intrinsic link between liver regeneration and hepatocarcinogenesis. FXR may have a dual-role in helping liver resume normal homeostasis: one is by controlling the bile acid level in liver, the other is by promoting liver repair by regeneration (Figure 12.2). However, the FXR-dependent liver regeneration is to prevent the further liver injury and proliferation. Therefore, FXR is rather working as a tumor suppressor. In the absence of FXR, this cycle of injury and compensatory liver regeneration (irregular liver regeneration) provides a tumor-prone environment. This is consistent to recent reports regarding the role of NF-κB in liver injury and hepatocarcinogenesis [89].

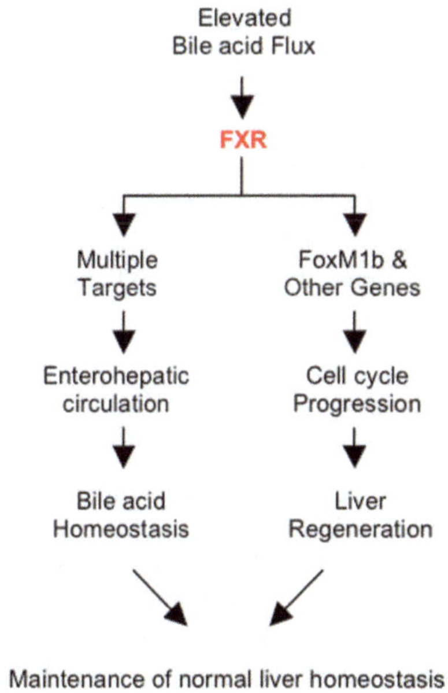

Figure 12.2. A model of dual effect of FXR function. In response to the increased bile acid flux, FXR regulates genes involved in both bile acid homeostasis and liver regeneration, which helps liver to resume homeostasis

The roles of FXR in hepatocarcinogenesis need to be further defined. However, a potential contribution of FXR in tumor suppression may be partially attributed to its anti-fibrosis function in liver. Chronic liver fibrosis has been linked to hepatocarcinogenesis when it finally develops into cirrhosis. Fiorucci et al. used a novel FXR ligand, 6-ECDCA, to demonstrate that activation of FXR in stellate cells inhibited pro-fibrosis gene expression in cooperation with additional two nuclear receptors, Shp and PPARα [90, 66, 67].

12.5.8. FXR and Other Cancers

The roles of FXR in carcinogenesis are not restricted to liver. FXR is expressed in nonenterohepatic tissues, including high levels in the kidneys and adrenal gland, which are 'nonclassic' bile acid targets [72], and low levels in the heart, vascular tissue, thymus, ovary, spleen, testes and adipose tissue [5, 53, 54]. The functions of FXR in these nonenterohepatic tissues are poorly understood. A potential link between BAs and colon cancer has been known for a long time. Recently, two separate reports suggest that FXR plays a protective role in intestinal cancer [91, 92]. Similarly, it is shown that FXR may also be involved in mediating the effects of bile acids on esophagus cancer [93, 94].

Recently, several reports suggest a potential link between FXR and breast cancer. It has been reported that FXR is expressed in human breast cancer tissue and cell lines [95, 96]. Breast cancer is epidemiologically linked to high-fat diets and high level of bile acids (BAs) in the body [97]. BAs are presented at high-concentration in the plasma of postmenopausal women with breast cancer and in breast cysts. FXR expression is detected in breast cancer tissues and cell lines [96]. Swales et al. indicated that FXR activation by CDCA and GW4064 induced breast cancer cell apoptosis [95]. In contrast, Journe et al. showed that FXR activator, farnesol, induced breast cancer cell growth. It still needs more studies to understand FXR function in breast cancer cell growth in order to better understand the roles of FXR in breast cancer development.

12.6. CONCLUDING REMARKS

FXR affects numerous signaling pathways via the genes it regulates directly and by its interference with other nuclear receptor signaling pathways. This creates a unique integrative mechanism to regulate the metabolism of bile acids, cholesterol, triglycerides, and glucose. Because FXR functions in diverse metabolic pathways, it is a promising therapeutic target for treating or preventing type 2 diabetes, hypertriglyceridemia, cholesterol gallstone disease, steato-hepatitis, and metabolic syndrome. The recent discovery that FXR is also involved in liver regeneration and liver carcinogenesis suggests that FXR is also a potential target for the liver transplantation and treatment of liver cancer.

ACKNOWLEDGMENTS

We apologize to colleagues whose work could not be cited due to space limitations. We thank Dr. Barry M. Forman for his suggestion. The authors are grateful to Keely Walker for proofreading. W.H. is supported by the Sidney Kimmel Foundation for Cancer Research, the Margaret E. Early Medical Research Trust and Concern Foundation.

REFERENCES

1. Forman, B. M., Goode, E., Chen, J., Oro, A. E., Bradley, D. J., Perlmann, T., Noonan, D. J., Burka, L. T., McMorris, T., Lamph, W. W., Evans, R. M., and Weinberger, C. (1995). Identification of a nuclear receptor that is activated by farnesol metabolites. *Cell* 81, 687–693.
2. Makishima, M., Okamoto, A. Y., Repa, J. J., Tu, H., Learned, R. M., Luk, A., Hull, M. V., Lustig, K. D., Mangelsdorf, D. J., and Shan, B. (1999). Identification of a nuclear receptor for bile acids. *Science* 284, 1362–1365.
3. Parks, D. J., Blanchard, S. G., Bledsoe, R. K., Chandra, G., Consler, T. G., Kliewer, S. A., Stimmel, J. B., Willson, T. M., Zavacki, A. M., Moore, D. D., and Lehmann, J. M. (1999). Bile acids: Natural ligands for an orphan nuclear receptor. *Science* 284, 1365–1368.
4. Wang, H., Chen, J., Hollister, K., Sowers, L. C., and Forman, B. M. (1999). Endogenous bile acids are ligands for the nuclear receptor FXR/BAR. *Mol Cell* 3, 543–553.
5. Zhang, Y., Kast-Woelbern, H. R., and Edwards, P. A. (2003). Natural structural variants of the nuclear receptor farnesoid X receptor affect transcriptional activation. *J Biol Chem* 278, 104–110.
6. Huber, R. M., Murphy, K., Miao, B., Link, J. R., Cunningham, M. R., Rupar, M. J., Gunyuzlu, P. L., Haws, T. F., Kassam, A., Powell, F., Hollis, G. F., Young, P. R., Mukherjee, R., and Burn, T. C. (2002). Generation of multiple farnesoid-X-receptor isoforms through the use of alternative promoters. *Gene* 290, 35–43.
7. Anisfeld, A. M., Kast-Woelbern, H. R., Lee, H., Zhang, Y., Lee, F. Y., and Edwards, P. A. (2005). Activation of the nuclear receptor FXR induces fibrinogen expression: A new role for bile acid signaling. *J Lipid Res* 46, 458–468.
8. Zhang, Y., Castellani, L. W., Sinal, C. J., Gonzalez, F. J., and Edwards, P. A. (2004). Peroxisome proliferator-activated receptor-g coactivator 1a (PGC-1a) regulates triglyceride metabolism by activation of the nuclear receptor FXR. *Genes Dev* 18, 157–169.
9. Lou, G., Li, Y., Chen, B., Chen, M., Chen, J., Liao, R., Zhang, Y., Wang, Y., and Zhou, D. (2007). Functional analysis on the 5′-flanking region of human FXR gene in HepG2 cells. *Gene* 396, 358–368.
10. Duran-Sandoval, D., Mautino, G., Martin, G., Percevault, F., Barbier, O., Fruchart, J. C., Kuipers, F., and Staels, B. (2004). Glucose regulates the expression of the farnesoid X receptor in liver. *Diabetes* 53, 890–898.
11. Fujino, T., Murakami, K., Ozawa, I., Minegishi, Y., Kashimura, R., Akita, T., Saitou, S., Atsumi, T., Sato, T., Ando, K., Hara, S., Kikugawa, K., and Hayakawa, M. (2009). Hypoxia downregulates farnesoid X receptor via a hypoxia-inducible factor independent but p38 mitogen-activated protein kinase-dependent pathway. *FEBS J* 276, 1319–1332.
12. Frankenberg, T., Miloh, T., Chen, F. Y., Ananthanarayanan, M., Sun, A. Q., Balasubramaniyan, N., Arias, I., Setchell, K. D., Suchy, F. J., and Shneider, B. L. (2008). The membrane protein ATPase class I type 8B member 1 signals through protein kinase C zeta to activate the farnesoid X receptor. *Hepatology* 48(6), 1896–1905, May 27 [Epub ahead of print].

13. Gineste, R., Sirvent, A., Paumelle, R., Helleboid, S., Aquilina, A., Darteil, R., Hum, D. W., Fruchart, J. C., and Staels, B. (2008). Phosphorylation of farnesoid X receptor by protein kinase C promotes its transcriptional activity. *Mol Endocrinol* 22(11), 2433–2447, Aug 28. [Epub ahead of print].
14. Deng, R., Yang, D., Yang, J., Yan, B. Oxysterol 22(R)-hydroxycholesterol induces the expression of the bile salt export pump through nuclear receptor farsenoid X receptor but not liver X receptor. *J Pharmacol Exp Ther* 2006;317: 317–325.
15. Wang, S., Lai, K., Moy, FJ., Bhat, A., Hartman, HB., Evans, MJ. The nuclear hormone receptor farnesoid X receptor (FXR) is activated by androsterone. *Endocrinology* 2006;147: 4025–4033.
16. Zhao, A., Yu, J., Lew, J. L., Huang, L., Wright, S. D., and Cui, J. (2004). Polyunsaturated fatty acids are FXR ligands and differentially regulate expression of FXR targets. *DNA Cell Biol* 23, 519–526.
17. Nishimaki-Mogami, T., Une, M., Fujino, T. et al. (2004). Identification of intermediates in the bile acid synthetic pathway as ligands for the farnesoid X receptor. *J Lipid Res* 45, 1538–1545.
18. Suzuki, T., Nishimaki-Mogami, T., Kawai, H., Kobayashi, T., Shinozaki, Y., Sato, Y., Hashimoto, T., Asakawa, Y., Inoue, K., Ohno, Y., Hayakawa, T., and Kawanishi, T. (2006). Screening of novel nuclear receptor agonists by a convenient reporter gene assay system using green fluorescent protein derivatives. *Phytomedicine* 13, 401–411.
19. Suzuki, T., Tamehiro, N., Sato, Y., Kobayashi, T., Ishii-Watabe, A., Shinozaki, Y., Nishimaki-Mogami, T., Hashimoto, T., Asakawa, Y., Inoue, K., Ohno, Y., Yamaguchi, T., and Kawanishi, T. (2008). The novel compounds that activate farnesoid X receptor: The diversity of their effects on gene expression. *J Pharmacol Sci* 107, 285–294.
20. Takahashi, M., Kanayama, T., Yashiro, T., Kondo, H., Murase, T., Hase, T., Tokimitsu, I., Nishikawa, J., and Sato, R. (2008). Effects of coumestrol on lipid and glucose metabolism as a farnesoid X receptor ligand. *Biochem Biophys Res Commun* 372, 395–399.
21. Maloney, P. R., Parks, D. J., Haffner, C. D. et al. (2000). Identification of a chemical tool for the orphan nuclear receptor FXR. *J Med Chem* 43, 2971–2974.
22. Pellicciari, R., Fiorucci, S., Camaioni, E. et al. (2002). 6alpha-ethyl-chenodeoxycholic acid (6-ECDCA), a potent and selective FXR agonist endowed with anticholestatic activity. *J Med Chem* 45, 3569–3572.
23. Dussault, I., Beard, R., Lin, M. et al. (2003). Identification of gene-selective modulators of the bile acid receptor FXR. *J Biol Chem* 278, 7027–7033.
24. Soisson, S. M., Parthasarathy, G., Adams, A. D., Sahoo, S., Sitlani, A., Sparrow, C., Cui, J., and Becker, J. W. (2008). Identification of a potent synthetic FXR agonist with an unexpected mode of binding and activation. *Proc Natl Acad Sci U S A* 105, 5337–5342.
25. Akwabi-Ameyaw, A., Bass, J. Y., Caldwell, R. D., Caravella, J. A., Chen, L., Creech, K. L., Deaton, D. N., Jones, S. A., Kaldor, I., Liu, Y., Madauss, K. P., Marr, H. B., McFadyen, R. B., Miller, A. B., Iii, F. N., Parks, D. J., Spearing, P. K., Todd, D., Williams, S. P., and Wisely, G. B. (2008). Conformationally constrained farnesoid X receptor (FXR) agonists: Naphthoic acid-based analogs of GW 4064. *Bioorg Med Chem Lett* 18, 4339–4343.
26. Costantino, G., Macchiarulo, A., Entrena-Guadix, A., Camaioni, E., and Pellicciari, R. (2003). Binding mode of 6ECDCA, a potent bile acid agonist of the farnesoid X receptor (FXR). *Bioorg Med Chem Lett* 13, 1865-1868.
27. Urizar, N. L., Dowhan, D. H., and Moore, D. D. (2000). The farnesoid X-activated receptor mediates bile acid activation of phospholipid transfer protein gene expression. *J Biol Chem* 275, 39313–39317.
28. Carter, B. A., Taylor, O. A., Prendergast, D. R., Zimmerman, T. L., Von Furstenberg, R., Moore, D. D., and Karpen, S. J. (2007). Stigmasterol, a soy lipid-derived phytosterol, is an antagonist of the bile acid nuclear receptor FXR. *Pediatr Res* 62, 301–306.
29. Nam, S. J., Ko, H., Shin, M., Ham, J., Chin, J., Kim, Y., Kim, H., Shin, K., Choi, H., and Kang, H. (2006). Farnesoid X-activated receptor antagonists from a marine sponge Spongia sp. *Bioorg Med Chem Lett* 16, 5398–5402.

30. Nam, S. J., Ko, H., Ju, M. K., Hwang, H., Chin, J., Ham, J., Lee, B., Lee, J., Won, D. H., Choi, H., Ko, J., Shin, K., Oh, T., Kim, S., Rho, J. R., and Kang, H. (2007). Scalarane sesterterpenes from a marine sponge of the genus Spongia and their FXR antagonistic activity. *J Nat Prod* 70, 1691–1695.
31. Song, C. S., Echchgadda, I., Baek, B. -S., Ahn, S. C., Oh, T., Roy, A. K., and Chatterjee, B. (2001). Dehydroepiandrosterone sulfotransferase gene induction by bile acid activated farnesoid X receptor. *J Biol Chem* 276, 42549–42556.
32. Li, H., Chen, F., Shang, Q., Pan, L., Shneider, B. L., Chiang, J. Y. L., Forman, B. M., Ananthanarayanan, M., Tint, G. S., Salen, G., and Xu, G. (2005). FXR-activating ligands inhibit rabbit ASBT expression via FXR-SHP-FTF cascade. *Am J Physiol Gastrointest Liver Physiol* 288, G60–G66.
33. Li, J., Pircher, P. C., Schulman, I. G., and Westin, S. K. (2005). Regulation of complement C3 expression by the bile acid receptor FXR. *J Biol Chem* 280, 7427–7434.
34. Ananthanarayanan, M., Balasubramanian, N., Makishima, M., Mangelsdorf, D. J., and Suchy, F. J. (2001). Human bile salt export pump promoter is transactivated by the farnesoid X receptor/bile acid receptor. *J Biol Chem* 276, 28857–28865.
35. Grober, J., Zaghini, I., Fujii, H., Jones, S. A., Kliewer, S. A., Willson, T. M., Ono, T., and Besnard, P. (1999). Identification of a bile acid-responsive element in the human ileal bile acid-binding protein gene. Involvement of the farnesoid X receptor/9-cis-retinoic acid receptor heterodimer. *J Biol Chem* 274, 29749–29754.
36. Goodwin, B., Jones, S. A., Price, R. R., Watson, M. A., McKee, D. D., Moore, L. B., Galardi, C., Wilson, J. G., Lewis, M. C., Roth, M. E., Maloney, P. R., Willson, T. M., and Kliewer, S. A. (2000). A regulatory cascade of the nuclear receptors FXR, SHP-1, and LRH-1 represses bile acid biosynthesis. *Mol Cell* 6, 517–526.
37. Pircher, P. C., Kitto, J. L., Petrowski, M. L., Tangirala, R. K., Bischoff, E. D., Schulman, I. G., and Westin, S. K. (2003). Farnesoid X Receptor regulates bile acid-amino acid conjugation. *J Biol Chem* 278, 27703–27711.
38. Jung, D., Podvinec, M., Meyer, U. A., Mangelsdorf, D. J., Fried, M., Meier, P. J., and Kullak-Ublick, G. A. (2002). Human organic anion transporting polypeptide 8 promoter is transactivated by the farnesoid X receptor/bile acid receptor. *Gastroenterology* 122, 1954–1966.
39. Lee, H., Zhang, Y., Lee, F. Y., Nelson, S. F., Gonzalez, F. J., and Edwards, P. A. (2006). FXR regulates organic solute transporters α and β in the adrenal gland, kidney, and intestine. *J Lipid Res* 47, 201–214.
40. Laffitte, B. A., Kast, H. R., Nguyen, C. M., Zavacki, A. M., Moore, D. D., and Edwards, P. A. (2000). Identification of the DNA binding specificity and potential target genes for the farnesoid X-activated receptor. *J Biol Chem* 275, 10638–10647.
41. Anisfeld, A. M., Kast-Woelbern, H. R., Meyer, M. E., Jones, S. A., Zhang, Y., Williams, K. J., Willson, T., and Edwards, P. A. (2003). Syndecan-1 expression is regulated in an isoform-specific manner by the farnesoid-X receptor. *J Biol Chem* 278, 20420–20428.
42. Kast, H. R., Goodwin, B., Tarr, P. T., Jones, S. A., Anisfeld, A. M., Stoltz, C. M., Tontonoz, P., Kliewer, S., Willson, T. M., and Edwards, P. A. (2002). Regulation of multidrug resistance-associated protein 2 (ABCC2) by the nuclear receptors pregnane X receptor, farnesoid X-activated receptor, and constitutive androstane receptor. *J Biol Chem* 277, 2908–2915.
43. Claudel, T., Sturm, E., Duez, H., Pineda Torra, I., Sirvent, A., Kosykh, V., Fruchart, J. C., Dallongeville, J., Hum, D. W., Kuipers, F., and Staels, B. (2002). Bile acid-activated nuclear receptor FXR suppresses apolipoprotein A-I transcription via a negative FXR response element. *J Clin Invest* 109, 961–971.
44. Hubbert, M. L., Zhang, Y., Lee, F. Y., and Edwards, P. A. (2007). Regulation of hepatic insig-2 by the farnesoid X-receptor. *Mol Endocrinol* 21, 1359–1369.
45. Moschetta, A., Bookout, A. L., and Mangelsdorf, D. J. (2004). Prevention of cholesterol gallstone disease by FXR agonists in a mouse model. *Nat Med* 10, 1352–1358.
46. Inagaki, T., Choi, M., Moschetta, A. et al. (2005). Fibroblast growth factor 15 functions as an enterohepatic signal to regulate bile acid homeostasis. *Cell Metab* 2, 217–225.

47. Holt, J. A., Luo, G., Billin, A. N. et al. (2003). Definition of a novel growth factor-dependent signal cascade for the suppression of bile acid biosynthesis. *Genes Dev* 17, 1581–1591.
48. Russell, D. W. (2003). The enzymes, regulation, and genetics of bile acid synthesis. *Annu Rev Biochem* 72, 137–174.
49. Sinal, C. J., Tohkin, M., Miyata, M., Ward, J. M., Lambert, G., and Gonzalez, F. J. (2000). Links targeted disruption of the nuclear receptor FXR/BAR impairs bile acid and lipid homeostasis. *Cell* 102, 731–744.
50. Edwards, P. A., Kast, H. R., and Anisfeld, A. M. (2002). BAREing it all: The adoption of LXR and FXR and their roles in lipid homeostasis. *J Lipid Res* 43, 2–12.
51. Sirvent, A., Claudel, T., Martin, G., Brozek, J., Kosykh, V., Darteil, R., Hum, D., Fruchart, J., and Staels, B. (2004a). The farnesoid X receptor induces very low density lipoprotein receptor gene expression. *FEBS Lett* 566, 173–177.
52. Watanabe, M., Houten, S. M., Wang, L., Moschetta, A., Mangelsdorf, D. J., Heyman, R. A., Moore, D. D., and Auwerx, J. (2004). Bile acids lower triglyceride levels via a pathway involving FXR, SHP, and SREBP-1c. *J Clin Invest* 113, 1408–1418.
53. Zhang, Y., Wang, X., Vales, C. et al. (2006). FXR deficiency causes reduced atherosclerosis in $Ldlr^{-/-}$ mice. *Arterioscler Thromb Vasc Biol* 26, 2316–2321.
54. Zhang, Y., Lee, F. Y., Barrera, G., Lee, H., Vales, C., Gonzalez, F. J., Willson, T. M., and Edwards, P. A. (2006). Activation of the nuclear receptor FXR improves hyperglycemia and hyperlipidemia in diabetic mice. *Proc Natl Acad Sci U S A* 103, 1006–1011.
55. Ma, K., Saha, P. K., Chan, L., and Moore, D. D. (2006). Farnesoid X receptor is essential for normal glucose homeostasis. *J Clin Invest* 116, 1102–1109.
56. Cariou, B., van Harmelen, K., Duran-Sandoval, D., van Dijk, T. H., Grefhorst, A., Abdelkarim, M., Caron, S., Torpier, G., Fruchart, J. C., Gonzalez, F. J., Kuipers, F., and Staels, B. (2006). The farnesoid X receptor modulates adiposity and peripheral insulin sensitivity in mice. *J Biol Chem* 281, 11039–11049.
57. Berg, R. D. (1995). Bacterial translocation from the gastrointestinal tract. *Trends Microbiol* 3, 149–154.
58. Inagaki, T., Moschetta, A., Lee, Y. K., Peng, L., Zhao, G., Downes, M., Yu, R. T., Shelton, J. M., Richardson, J. A., Repa, J. J., Mangelsdorf, D. J., and Kliewer, S. A. (2006). Regulation of antibacterial defense in the small intestine by the nuclear bile acid receptor. *Proc Natl Acad Sci U S A* 103, 3920–3925.
59. Wang, Y., Yang, F., Chen, W., Huang, X., Lai, L., Forman, M. B., and Huang, W. (2008a). Farnesoid X receptor (FXR) protects liver cells from apoptosis induced by serum deprivation in vitro and fasting in vivo. *Mol Endocrinol* 22, 1622–1632.
60. Wang, Y., Chen, W., Wang, M., Yu, D., Forman, B. M., and Huang, W. (2008b). Farnesoid X receptor antagonize NF-kB in hepatic inflammatory response. *Hepatology* 48, 1632–1643.
61. Li, Y. T., Swales, K. E., Thomas, G. J., Warner, T. D., and Bishop-Bailey, D. (2007). Farnesoid x receptor ligands inhibit vascular smooth muscle cell inflammation and migration. *Arterioscler Thromb Vasc Biol* 27, 2606–2611.
62. Li, J., Wilson, A., Kuruba, R., Zhang, Q., Gao, X., He, F., Zhang, L. M., Pitt, B. R., Xie, W., and Li, S. (2008). FXR-mediated regulation of eNOS expression in vascular endothelial cells. *Cardiovasc Res* 77, 169–177.
63. Trauner, M., Meier, P. J., and Boyer, J. L. (1998). Molecular pathogenesis of cholestasis. *N Engl J Med* 339, 1217–1227.
64. Miyata, M., Tozawa, A., Otsuka, H., Nakamura, T., Nagata, K., Gonzalez, F. J., and Yamazoe, Y. (2005). Role of Farnesoid X Receptor in the enhancement of canalicular bile acid output and excretion of unconjugated bile acids: A mechanism for protection against cholic acid-induced liver toxicity. *J Pharmacol Exp Ther* 312, 759–766.
65. Liu, Y., Binz, J., Numerick, M. J., Dennis, S., Luo, G., Desai, B., MacKenzie, K. I., Mansfield, T. A., Kliewer, S. A., Goodwin, B., and Jones, S. A. (2003). Hepatoprotection by the farnesoid X receptor agonist GW4064 in rat models of intra- and extrahepatic cholestasis. *J Clin Invest* 112, 1678–1687.

66. Fiorucci, S., Clerici, C., Antonelli, E., Orlandi, S., Goodwin, B., Sadeghpour, B. M., Sabatino, G., Russo, G., Castellani, D., Willson, T. M., Pruzanski, M., Pellicciari, R., and Morelli, A. (2005). Protective effects of 6-ethyl chenodeoxycholic acid, a farnesoid X receptor ligand, in estrogen-induced cholestasis. *J Pharmacol Exp Ther* 313, 604–612.
67. Fiorucci, S., Rizzo, G., Antonelli, E. et al. (2005). Cross-talk between farnesoid-X-receptor (FXR) and peroxisome proliferator-activated receptor gamma contributes to the antifibrotic activity of FXR ligands in rodent models of liver cirrhosis. *J Pharmacol Exp Ther* 315, 58–68.
68. Barbier, O., Torra, I. P., Sirvent, A., Claudel, T., Blanquart, C., Duran-Sandoval, D., Kuipers, F., Kosykh, V., Fruchart, J. C., and Staels, B. (2003). FXR induces the UGT2B4 enzyme in hepatocytes: A potential mechanism of negative feedback control of FXR activity. *Gastroenterology* 124, 1926–1940.
69. Jung, D., Mangelsdorf, D. J., and Meyer, U. A. (2006). PXR is a target of FXR. *J Biol Chem* 281, 19081–19091.
70. Guo, G. L., Santamarina-Fojo, S., Akiyama, T. E., Amar, M. J., Paigen, B. J., Brewer, B., Jr., and Gonzalez, F. J. (2006). Effects of FXR in foam-cell formation and atherosclerosis development. *Biochim Biophys Acta* 1761, 1401–1409.
71. Hanniman, E. A., Lambert, G., McCarthy, T. C., and Sinal, C. J. (2005). Loss of functional farnesoid X receptor increases atherosclerotic lesions in apolipoprotein E deficient mice. *J Lipid Res* 46, 2595–2604.
72. Bishop-Bailey, D., Walsh, D. T., and Warner, T. D. (2004). Expression and activation of the farnesoid X receptor in the vasculature. *Proc Natl Acad Sci U S A* 101, 3668–3673.
73. He, F., Li, J., Mu, Y., Kuruba, R., Ma, Z., Wilson, A., Alber, S., Jiang, Y., Stevens, T., Watkins, S., Pitt, B., Xie, W., and Li, S. (2006). Downregulation of endothelin-1 by farnesoid X receptor in vascular endothelial cells. *Circ Res* 98, 192–199.
74. Qin, P., Tang, X., Elloso, M. M., and Harnish, D. C. (2006). Bile acids induce adhesion molecule expression in endothelial cells through activation of reactive oxygen species, NF-kappaB, and p38. *Am J Physiol Heart Circ Physiol* 29, H741–H747.
75. Boland, L. L., Folsom, A. R., and Rosamond, W. D. (2002). Hyperinsulinemia, dyslipidemia, and obesity as risk factors for hospitalized gallbladder disease. A prospective study. *Ann Epidemiol* 12, 131–140.
76. Stayrook, K. R., Bramlett, K. S., Savkur, R. S., Ficorilli, J., Cook, T., Christe, M. E., Michael, L. F., and Burris, T. P. (2005). Regulation of carbohydrate metabolism by the farnesoid X receptor. *Endocrinology* 146, 984–991.
77. De Fabiani, E., Mitro, N., Gilardi, F., Caruso, D., Galli, G., and Crestani, M. (2003). Coordinated control of cholesterol catabolism to bile acids and of gluconeogenesis via a novel mechanism of transcription regulation linked to the fastedto- fed cycle. *J Biol Chem* 278, 39124–39132.
78. Wittenburg, H., Lyons, M. A., Li, R., Churchill, G. A., Carey, M. C., and Paigen, B. F. X. R. (2003). ABCG5/ABCG8 as determinants of cholesterol gallstone formation from quantitative trait locus mapping in mice. *Gastroenterology* 125, 868–881.
79. Amador-Noguez, D., Dean, A., Huang, W., Setchell, K., Moore, D., and Darlington, G. (2007). Alterations in xenobiotic metabolism in the long-lived Little mice. *Aging Cell* 6, 453–470.
80. Ueno, Y., Moriyama, M., Uchida, T., Arakawa, Y. Irregular regeneration of hepatocytes is an important factor in the hepatocarcinogenesis of liver disease. Hepatology 2001;33: 357–362.
81. Zhang, L., Huang, X., Meng, Z., Dong, B., Shiah, S., Moore, D. D., and Huang, W. (2009). Significance and mechanism of CYP7a1 gene regulation in acute phase of liver regeneration. *Mol Endocrinol* 23, 137–145.
82. Huang, W., Ma, K., Zhang, J., Qatanani, M., Cuvillier, J., Liu, J., Dong, B., Huang, X., and Moore, D. D. (2006). Nuclear receptor-dependent bile acid signaling is required for normal liver regeneration. *Science* 312, 233–236.
83. Yang, F., Huang, X., Yi, T., Yen, Y., Moore, D. D., and Huang, W. (2007). Spontaneous development of liver tumors in the absence of the bile acid receptor farnesoid X receptor. *Cancer Res* 67, 863–867.

84. Higuchi, H. and Gores, G. J. (2003). Bile acid regulation of hepatic physiology: IV. Bile acids and death receptors. *Am J Physiol Gastrointest Liver Physiol* 284, G734–G738.
85. Mauad, T. H., van Nieuwkerk, C. M., Dingemans, K. P. et al. (1994). Mice with homozygous disruption of the mdr2 P-glycoprotein gene. A novel animal model for studies of nonsuppurative inflammatory cholangitis and hepatocarcinogenesis. *Am J Pathol* 145, 1237–1245.
86. Barone, M., Maiorano, E., Ladisa, R. et al. (2003). Influence of ursodeoxycholate-enriched diet on liver tumor growth in HBV transgenic mice. *Hepatology* 37, 880–886.
87. Katzenellenbogen, M., Pappo, O., Barash, H. et al. (2006). Multiple adaptive mechanisms to chronic liver disease revealed at early stages of liver carcinogenesis in the Mdr2-knockout mice. *Cancer Res* 66, 4001–4010.
88. Kim, I., Morimura, K., Shah, Y., Yang, Q., Ward, J. M., and Gonzalez, F. J. (2007). Spontaneous hepatocarcinogenesis in farnesoid X receptor-null mice. *Carcinogenesis* 28, 940–946.
89. Karin, M. (2006). Nuclear factor-kappaB in cancer development and progression. *Nature* 441, 431–436.
90. Fiorucci, S., Antonelli, E., Rizzo, G. et al. (2004). The nuclear receptor SHP mediates inhibition of hepatic stellate cells by FXR and protects against liver fibrosis. *Gastroenterology* 127, 1497–1512.
91. Modica, S., Murzilli, S., Salvatore, L., Schmidt, D. R., and Moschetta, A. (2008). Nuclear bile acid receptor FXR protects against intestinal tumorigenesis. *Cancer Res* 68, 9589–9594.
92. Maran, R. R., Thomas, A., Roth, M., Sheng, Z., Esterly, N., Pinson, D., Gao, X., Zhang, Y., Ganapathy, V., Gonzalez, F. J., and Guo, G. L. (2009). Farnesoid X receptor deficiency in mice leads to increased intestinal epithelial cell proliferation and tumordevelopment. *J Pharmacol Exp Ther* 328, 469–477.
93. De Gottardi, A., Dumonceau, J. M., Bruttin, F., Vonlaufen, A., Morard, I., Spahr, L., Rubbia-Brandt, L., Frossard, J. L., Dinjens, W. N., Rabinovitch, P. S., and Hadengue, A. (2006). Expression of the bile acid receptor FXR in Barrett's esophagus and enhancement of apoptosis by guggulsterone in vitro. *Mol Cancer* 5, 48.
94. Capello, A., Moons, L. M., Van de Winkel, A., Siersema, P. D., van Dekken, H., Kuipers, E. J., and Kusters, J. G. (2008). Bile acid-stimulated expression of the farnesoid X receptor enhances the immune response in Barrett esophagus. *Am J Gastroenterol* 103, 1510–1516.
95. Swales, K. E., Korbonits, M., Carpenter, R., Walsh, D. T., Warner, T. D., and Bishop-Bailey, D. (2006). The farnesoid X receptor is expressed in breast cancer and regulates apoptosis and aromatase expression. *Cancer Res* 66, 10120–10126.
96. Journe, F., Laurent, G., Chaboteaux, C. et al. (2008). Farnesol, a mevalonate pathway intermediate, stimulates MCF-7 breast cancer cell growth through farnesoid-X-receptor-mediated estrogen receptor activation. *Breast Cancer Res Treat* 107, 49–61.
97. Costarelli, V. and Sanders, T. A. (2002). Plasma deoxycholic acid concentration is elevated in postmenopausal women with newly diagnosed breast cancer. *Eur J Clin Nutr* 56, 925–927.
98. Huang, L., Zhao, A., Lew, J. L., Zhang, T., Hrywna, Y., Thompson, J. R., de Pedro, N., Royo, I., Blevins, R. A., Peláez, F., Wright, S. D., and Cui, J. (2003). Farnesoid X receptor activates transcription of the phospholipid pump MDR3. *J Biol Chem* 278, 51085–51090.
99. Sirvent, A., Verhoeven, A. J. M., Jansen, H., Kosykh, V., Darteil, R. J., Hum, D. W., Fruchart, J. -C., and Staels, B. (2004b). Farnesoid X receptor represses hepatic lipase gene expression. *J Lipid Res* 45, 2110–2115.
100. Cariou, B., van Harmelen, K., Duran-Sandoval, D., van Dijk, T., Grefhorst, A., Bouchaert, E., Fruchart, J. C., Gonzalez, F. J., Kuipers, F., and Staels, B. (2005). Transient impairment of the adaptive response to fasting in FXR-deficient mice. *FEBS Lett* 579, 4076–4080.

CHAPTER 13

PHYSIOLOGICAL FUNCTIONS OF TR2 AND TR4 ORPHAN NUCLEAR RECEPTOR

SU LIU, SHAOZHEN XIE, YI-FEN LEE, AND CHAWNSHANG CHANG
George H. Whipple Lab for Cancer Research, Departments of Urology and Pathology, University of Rochester Medical Center, Rochester, NY 14642, USA

Abstract: The human testicular receptor 2 and 4 (TR2 and TR4) are two evolutionarily related orphan nuclear receptors belonging to the same nuclear receptor subfamily (Lee et al. 2002, *J Steroid Biochem Mol Biol* 81(4–5), 291–308). They regulate gene expression by binding to DNA as homodimers or a heterodimer with each other. TR4 may also cross-talk with other nuclear receptors, to control its target genes. In vitro and in vivo studies have identified several TR4 target genes, including ciliary neurotrophic factor alpha (CNTFRα) (Young et al. 1997, *J Biol Chem* 272(5), 3109–3116), apolipoprotein E (ApoE) (Kim et al. 2003, *J Biol Chem* 278(47), 46919–46926) and phosphenolpyruvate carboxykinase (PEPCK) (Liu et al. 2007, *Diabetes* 56(12), 2901–2909). Recent studies using TR4 knockout ($TR4^{-/-}$) mice suggested that TR4 may play essential roles in growth, development, and metabolism (Zhang et al. 2007, *Mol Endocrinol* 21(4), 908–920; Kim et al. 2005, *Biochem Biophys Res Commun* 328(1), 85–90; Chen et al. 2005, *Mol Cell Biol* 25(7), 2722–2732; Mu et al. 2004, *Mol Cell Biol* 24(13), 5887–5899; Collins et al. 2004, *Proc Natl Acad Sci U S A* 101(42), 15058–15063). Mice with a germline deletion of TR4 are viable but have high early postnatal mortality, growth retardation, and profound reduction in body weight. Further studies showed that TR4 plays essential roles in the development and functioning in the central nervous system (Chen et al. 2005, *Mol Cell Biol* 25(7), 2722–2732), such as proper myelination and oligodendrocyte differentiation (Zhang et al. 2007, *Mol Endocrinol* 21(4), 908–920). Studies also showed that TR4 is important for spermatogenesis in male mice (Mu et al. 2004, *Mol Cell Biol* 24(13), 5887–5899) and folliculogenesis in female mice (Chen et al. 2008, *Mol Endocrinol* 22, 858–867). In addition, TR4 might be involved in skeletal muscle function and bone remodeling. TR4 and TR2 also regulate embryonic and fetal globin gene transcription (Tanabe et al. 2002, *EMBO J* 21(13), 3434–3442; Tanabe et al. 2007, *EMBO J* 26(9), 2295–2306). Surprisingly, mice lacking TR2 are viable and have no serious developmental defects. Thus, TR2 may either not be important in spermatogenesis and testis development, or its roles may be compensated by other closely related proteins such as TR4. Therefore, this chapter will focus on the in vivo roles of TR4.

C.M. Bunce, M.J. Campbell (eds.), *Nuclear Receptors*, Proteins and Cell Regulation 8, 327–343. DOI 10.1007/978-90-481-3303-1_13,

13.1. INTRODUCTION

Nuclear receptors comprise a superfamily of ligand-activated transcription factors whose target genes are important for a variety of biological processes, including growth, differentiation, development, cancer, ageing, and metabolism [1, 2]. Nuclear receptors share a common structure consisting of four domains: a variable N-terminal A/B transactivation domain, a highly conserved DNA-binding domain (DBD), a hinge domain, and a ligand binding domain (LBD) [3]. Orphan nuclear receptors were identified by their structure homology to nuclear receptors before investigating their ligands and physiological functions, a process termed 'reverse endocrinology' [4, 5]. To date, more than half of orphan receptors were 'adopted' when their naturally-occurred ligands were identified, which then leads to significant understanding of their physiological functions [6]. In addition, generating orphan receptor knockout mice by gene targeting has been proven to be another powerful way to study their physiological roles, especially for those orphan receptors whose ligands remain unknown, such as testicular orphan receptor 2 and 4 (TR2 and TR4). This chapter covers recent progress on the in vivo functions of TR4 and TR2 revealed by the studies using gene knockout mice.

TR4 (NR2C2) and TR2 (NR2C1) were initially isolated from human prostate and testis cDNA libraries on the basis of sequence homology to known nuclear receptors [7]. Presently, their putative ligands remain unknown. TR4 and TR2 are closely related, and regulate gene expression by binding to DNA as homodimers or a heterodimer with each other. TR4 and TR2 response elements (TR4RE, TR2RE) consist of an imperfect direct repeat (DR) of two consensus sequences, AGGTCA, separated by variable spacer nucleotides [8]. In vitro and in vivo studies have identified several TR4 target genes, including apolipoprotein E (ApoE) [9], and phosphenolpyruvate carboxykinase (PEPCK) [10]. TR4 may cross-talk with other nuclear receptors, such as peroxisome proliferator-activated receptors (PPAR), thyroid hormone receptor (TR), retinoic acid receptor (RAR), vitamin D receptor (VDR), androgen receptor (AR), and estrogen receptor (ER) [8, 11–14] to control its target genes. Many TR4REs have similar binding sequences as other nuclear receptors, including Retinoid-X-receptor alpha (RXRα)/PPAR (DR1), RXRα /RAR (DR1), VDR (DR3), TR (DR4), and RARβ (DR5) [15–19]. TR4 may also compete with other nuclear receptors for coregulators, such as RIP140 [20], or directly interact with other nuclear receptors [8]. Therefore, while TR4 may directly regulate its target gene expression, it may coordinate or compete with other nuclear receptors to regulate target gene expression in different cellular environments.

Recent studies using TR4 knockout ($TR4^{-/-}$) mice suggested that TR4 may play essential roles in growth, development, and metabolism [21–25]. $TR4^{-/-}$ mice were generated by replacing the DNA-binding domain of TR4 with a neomycin resistant gene cassette along with the lacZ gene. Mice with a germline deletion of TR4 are viable but have high early postnatal mortality, growth retardation, and profound reduction in body weight. The number of $TR4^{-/-}$ pups generated by the mating of TR4 heterzygous ($TR4^{+/-}$) mice is well under that prediction by the normal

Mendelian ratio. $TR4^{-/-}$ females show defects in reproduction and maternal behavior, with pups of $TR4^{-/-}$ dams dying soon after birth with no indication of milk intake [25]. Further studies showed that TR4 plays essential roles in the development and function of the central nervous system [23], as well as spermatogenesis in male mice [24] and folliculogenesis in female mice [26]. Additionally, TR4 is involved in the development of bone, muscles, and spinal cord, cancer and ageing [8]. TR4 and TR2 also regulate embryonic and fetal globin gene transcription [27, 28].

Surprisingly, mice lacking TR2 are viable and have no serious developmental defects. Male mice lacking TR2 are fertile with functional testes, including normal sperm number and motility [29]. Thus, TR2 may either not be important in spermatogenesis and testis development, or its roles may be compensated by other closely related proteins such as TR4. Therefore, this chapter will focus on the in vivo roles of TR4.

13.2. TR4 AND FERTILITY

13.2.1. TR4 and Male Fertility

TR4 is ubiquitously expressed, with relatively highest expression in testis, kidney, and muscle [30]. In mouse testis development, TR4 expression increases significantly at day 16 and reaches the highest level at around day 21. Such an expression pattern is consistent with the first wave of spermatogenesis at the meiotic prophase. The discovery that TR4 is specifically and stage-dependently expressed in the primary spermatocytes raise the interest into the role of TR4 in male fertility and spermatogenesis. $TR4^{-/-}$ male mice have reduced fertility, compared to $TR4^{+/+}$ mice. Further study found that the sperm production in $TR4^{-/-}$ mice is reduced at various stages. In contrast, $TR4^{-/-}$ and $TR4^{+/+}$ mice show no differences in sperm morphology and cauda epididymis sperm motility. $TR4^{-/-}$ mice have delayed spermatogenesis at stages X to XII in the first wave of spermatogenesis. This delay can be due to delay and disruption of spermatogenesis at the end of late meiotic prophase and subsequent meiotic divisions. Histological examination of testis sections from $TR4^{-/-}$ mice shows degenerated primary spermatocytes and some necrotic tubules. Among testis-specific genes, the expression of sperm 1 and cyclin A1, which express at the end of the meiotic prophase, was delayed, and their expression levels were decreased in $TR4^{-/-}$ mice. Taken together, TR4 is essential for normal spermatogenesis in vivo [24].

13.2.2. TR4 and Female Fertility

TR4 also plays essential roles for normal folliculogenesis and ovarian functions in female mice [26]. Like $TR4^{-/-}$ male mice, $TR4^{-/-}$ female mice displayed subfertility, with prolonged and irregular estrous cycles. Histological analysis found that $TR4^{-/-}$ ovaries have reduced size and weight, as well as fewer or no preovulatory follicles and corpora lutea. In contrast, primary, preantral, and antral follicles in TR4$^{-/-}$ ovaries are normal. After superovulation, TR4$^{-/-}$ female mice produced fewer oocytes, preovulatory follicles, and corpora lutea, suggesting that ovarian function defects in TR4$^{-/-}$ female mice may contribute to the reduced fertility. Moreover, Serum luteinizing hormone (LH) and follicle-stimulating hormone (FSH) concentrations were similar in TR4$^{-/-}$ and TR4$^{+/+}$ female mice after superovulation, suggesting that the subfertility in TR4$^{-/-}$ mice is due to ovary defects rather than a deficiency in the pituitary gonadotropin axis. In support of this hypothesis, more intensive granulosa apoptosis was found in TR4$^{-/-}$ ovaries. Luteinizing hormone receptor (LHR) signal plays essential roles in ovarian sex hormone production via steroidogenesis. Mechanical dissection found that both protein and mRNA expression of LHR in ovaries are reduced in TR4$^{-/-}$ female mice. Further study found TR4 can regulate LHR gene expression via direct binding to its 5′ promoter. Thus, loss of TR4 will result in reduced LHR gene expression, which leads to reduced gonadal sex hormones due to reduced expression of enzymes involved in steroidogenesis. Together, TR4 might play essential roles in normal folliculogenesis by modulating LHR signals [26].

13.3. TR4 AND CENTRAL NERVOUS SYSTEM

TR4 has the highest expression in the central nervous system (CNS), including the hypothalamus, hippocampus, and cerebellum [31]. It has been demonstrated that the expression of TR4 correlates with the onset of neurogenesis, suggesting a role for TR4 in the nervous system during development. In addition, TR4 may cross-talk with other signalings that are involved in neurogenesis, such as RAR and RXR, to participate in related events [31].

TR4$^{-/-}$ mice display several phenotypes that typically indicate impaired motor coordination, including unsteady gait, failure to maintain balance on a ledge, as well as sudden jerks and tremors [31]. TR4$^{-/-}$ mice displayed increased spontaneous locomotor activity when introduced to an unfamiliar environment. Furthermore, TR4$^{-/-}$ mice have hyperkinetic response following physical stimulation, manipulation, or handling, as revealed by increased horizontal ambulation and stereotypic counts. In contrast, rearing and vertical movement counts were not different between TR4$^{-/-}$ mice and controls, indicating that exploratory behavior is intact in TR4$^{-/-}$ mice. One phenotype of TR4$^{-/-}$ mice is the lack of nest building ability, both in home and new cages. Nest building behavior is correlated with thermoregulation and with the function of the hippocampus. However, no difference was found either in body temperature or in hippocampal architecture in TR4$^{-/-}$ mice [31, 32].

13.3.1. TR4 and Cerebellar Development

13.3.1.1. Abnormal cerebellum in the adult TR4$^{-/-}$ brain

TR4$^{-/-}$ mice exhibited severe behavioral deficits, as mentioned above, suggesting impaired cerebellar function [23, 31, 32]. In the TR4$^{-/-}$ brain, cerebellar restricted hypoplasia is severe and cerebellar vermal lobules VI and VII are underdeveloped, while no structural alterations in the cerebral cortex are observed. Histological analysis of the $TR4^{-/-}$ cerebellar cortex reveals reductions in granule cell density, as well as a decreased number of parallel fiber boutons that are enlarged in size. Together, these data suggest that the excitatory stimulus from granule cells to Purkinje cells may be compromised in the $TR4^{-/-}$ cerebellum. Further analyses reveal that the levels of both gamma-aminobutyric acid (GABA), the major inhibitory neurotransmitter, and its synthetic enzyme, glutamic acid decarboxylase (GAD) are decreased in both Purkinje cells and interneurons of the $TR4^{-/-}$ cerebellum, suggesting that the inhibitory circuits signaling within and from the cerebellum may be perturbed. In the $TR4^{-/-}$ cerebellum, the immunoreactivity of GluR2/3, the AMPA type glutamate receptor, was reduced in Purkinje cells, but increased in the deep cerebellar nuclei. In addition, no progressive atrophy was observed at various adult stages in the $TR4^{-/-}$ brain. Together, these results suggest that the behavioral phenotype of $TR4^{-/-}$ mice may result from disrupted inhibitory pathways in the cerebellum [31, 32].

13.3.1.2. TR4 and cerebellum development

Throughout postnatal and adult stages, $TR4^{-/-}$ mice exhibited behavioral deficits in motor coordination, suggesting motor coordination is impaired earlier in postnatal $TR4^{-/-}$ mice [23]. Analysis of the postnatal and adult $TR4^{-/-}$ cerebellum revealed gross abnormalities in foliation. $TR4^{-/-}$ cerebellar cortex shows changes in the lamination, including reduction in the thickness of both the molecular layer (ML) and the internal granule layer (IGL) [23]. Analyses of the developing $TR4^{-/-}$ cerebellum indicate that the lamination irregularities observed may result from disrupted granule cell proliferation within the external granule cell layer (EGL), delayed inward migration of post-mitotic granule cells, and increased apoptosis during cerebellar development [23]. In addition, abnormal development of Purkinje cells was observed in the postnatal $TR4^{-/-}$ cerebellum, as indicated by aberrant dendritic arborization. In postnatal, neuronal-specific $TR4^{-/-}$ mice, architectural changes in the cerebellum were similar to those seen in $TR4^{-/-}$ mice, suggesting that TR4 function in neuronal lineages might be important for cerebellar morphogenesis. In contrast, no significant morphological alterations were observed in the cerebella of the Purkinje cell-specific knockout mice [23]. Therefore, the effect on Purkinje cell development is likely mediated by changes elsewhere, such as in granule cells, or is highly dependent on developmental stage. Together, current findings from various TR4 knockout mouse models suggest that TR4 is functional in neuronal lineages and is important for cerebellar morphogenesis [23].

13.3.2. TR4 and Myelination in Mouse Forebrain

The formation of myelin in the nervous system ensures rapid and efficient conduction of nerve impulses along myelination axons, which is essential for the proper functioning of the nervous system. In addition to the abnormal cerebellum development, the differentiation of glial cells, particular oligodendrocytes, was abnormal in TR4$^{-/-}$ mice [33]. A temporal and spatial pattern of reduced myelination was observed in TR4$^{-/-}$ mice. Myelination is reduced significantly in TR4$^{-/-}$ forebrains and in early developmental stages. Further analysis reveals that CC-1-positive (CC-1+) oligodendrocytes are decreased in TR4$^{-/-}$ mice forebrains [33]. The number and proliferation rate of platelet-derived growth factor receptor alpha-positive (PDGFalphaR+) oligodendrocyte precursor cells (OPCs) remain unaffected in these regions, whereas the later stage OPC marker O4 signals are reduced in TR4$^{-/-}$ forebrains when examined at postnatal day 7, suggesting that loss of TR4 interrupts oligodendrocyte differentiation [33]. In support of this observation, CC-1+ oligodendrocytes derived from 5-bromo-2′-deoxyuridine incorporating OPCs are significantly reduced in TR4$^{-/-}$ mice forebrains. Mechnical dissection showed higher Jagged1 expression levels in axon fiber-enriched regions in TR4$^{-/-}$ mice forebrains, suggesting a more activated Notch signaling in these regions that correlates with previous reports showing that Notch activation inhibits oligodendrocyte differentiation [33]. Together, TR4 is required for proper myelination in the CNS and is particularly important for oligodendrocyte differentiation and maturation in the forebrain regions. The altered Jagged1-Notch signaling in TR4$^{-/-}$ forebrain underlies a potential mechanism that contributes to the reduced myelination in the forebrain.

13.4. TR4 IN GLUCOSE AND LIPID METABOLISM AND INSULIN SENSITIVITY

TR4$^{-/-}$ mice display reduced fat tissue volume and smaller adipocytes size. TR4$^{-/-}$ mice also display increased insulin sensitivity and low glucose levels at birth and in fasting conditions, suggesting that TR4 plays critical roles in metabolic pathways [10, 34]. Further studies in TR4$^{-/-}$ mice suggested that TR4 plays important roles in insulin sensitivity, glucose metabolism [10], and lipid metabolism [34].

TR4 is involved in liver glucose metabolism by directly regulating the expression of phosphoenolpyruvate carboxykinase (PEPCK) [10], the key gene in gluconeogenesis to maintain glucose homeostasis [35]. TR4 expression is also strongly associated with PEPCK expression and hepatic glucose production. Fasted TR4$^{-/-}$ mice showed hypoglycemia with reduced PEPCK induction. TR4$^{-/-}$ mice also exhibited better glucose clearance and improved insulin sensitivity that could be due to decreased PEPCK expression [10].

TR4 is also involved in lipid metabolism by directly regulating expression of the stearoyl-CoA desaturase 1 (SCD-1) gene [34]. SCD-1 is the rate limiting enzyme in the biosynthesis of monounsaturated fatty acid. SCD-1 activity has been implicated in obesity, diabetes, lipogenesis, beta-oxidation, and insulin sensitivity [36, 37]. In the leptin-deficient ob/ob mice, lacking SCD1 resulted in markedly reduced adiposity and improved insulin sensitivity. Similar phenotypes were found in TR4$^{-/-}$ mice: TR4$^{-/-}$ mice also showed reduced SCD-1 gene expression, reduced fat mass and TG deposition, and improved insulin sensitivity [34]. It is unknown whether

TR4 has direct roles in adipogenesis or lipolysis. Tissue-specific actions of TR4 in liver, skeletal muscle, and fat on systemic lipid and glucose metabolism and insulin sensitivity remain to be revealed. Together, current findings found TR4 contributes positively to insulin resistance by control of its target genes, PEPCK and SCD-1.

13.5. TR2/4 IN EMBRYONIC AND FETAL β-GLOBIN GENE REPRESSION

The human β-globin locus consists of five functional beta-like genes, including embryonic ε-, fetal Gγ- and Aγ-, and adult β-globin genes, which are expressed in a developmental-specific manner [38]. The ε-globin gene is active in the primitive erythroid cells of the yolk sac during the first 6 weeks, while the γ- and β-globin genes are silent. Then the first switch occurs resulting in the silencing of ε-globin and activation of the γ-globin gene expression in the definitive hematopoietic cells of the fetal liver. The second switch occurs shortly after birth to silence the γ-globin gene and to activate the β-globin gene in the bone marrow and spleen [39]. Globin gene switching and erythropoiesis are modulated by *cis*-acting elements and transcription factors, such as α-hemoglobin stabilizing protein (AHSP) [40], direct repeat erythroid definitive (DRED) [27], and GATA-1.

DRED is a 540 kDa complex that binds with high affinity to DR1 sites in the human embryonic ε- and fetal γ-globin gene promoters, but not adult β-globin promoter, which has no DR1 element [27]. Both mRNAs of TR4 and TR2 are expressed at all stages of murine and human erythropoiesis. One mutation in a DR1 site which reduces TR2/TR4 binding in vitro causes elevated γ-globin transcription. Transgenic expression of TR4 or TR2 reduces endogenous embryonic ε-globin transcription [27]. These data suggest that TR2/TR4 forms the core of a larger DRED complex that represses embryonic ε- and fetal γ-globin transcription in definitive erythroid cells, and therefore therapeutic roles of TR4 and TR2 for treating sickle cell anemia.

Further study using TR4 and TR4 knockout mice found that silencing of both the embryonic and fetal β-type globin genes is delayed in definitive erythroid cells of TR2 and TR4 null mutant mice [28]. Moreover, in transgenic mice that express dominant-negative TR4 (dnTR4), human embryonic ε-globin is activated in primitive and definitive erythroid cells. Interestingly, human fetal γ-globin is activated by dnTR4 only in definitive, but not in primitive erythroid cells, implicating TR2/TR4 as a stage-selective repressor. Forced expression of wild-type TR2 and TR4 leads to precocious repression of ε-globin, but induction of γ-globin in definitive erythroid cells. These temporally specific, gene-selective alterations in ε-and γ-globin gene expression, by gain and loss of TR2/TR4 function, provide the first genetic evidence for a role for these nuclear receptors in sequential, gene-autonomous silencing of the ε- and γ-globin genes during development, and suggest that their differential utilization controls stage-specific repression of the human ε- and γ-globin genes [28].

A recent study also found that TR2/TR4 directly represses Gata1/GATA1 transcription in murine and human erythroid progenitor cells through an evolutionarily conserved binding site within a well-characterized, tissue-specific Gata1 enhancer. GATA-1 plays an essential role in erythroid differentiation. GATA-1 is required for both primitive and definitive erythropoiesis [41]. Thus, TR4 and TR2 may modulate Gata1 to determine terminal erythroid maturation [42].

13.6. TR4 AND SKELETAL MUSCLE

Previous researches indicated a role of TR4 in skeletal muscle function. Among all 49 nuclear receptors, TR4 is one of the seven 'cyclic' receptors which are rhythmically expressed in skeletal muscles of C57BL6 mice [43]. TR4 expression in skeletal muscle peaks at *Zeitgeber* Time 12, which is the time when mice start to be active, suggesting TR4 is crucial for muscle function.

We found $TR4^{-/-}$ mice exhibited a variety of behavioral abnormalities, such as mild trembling, unsteady gait, and reduced cage mobility. $TR4^{-/-}$ mice also displayed classical hind limb contractures upon tail suspension. Previous ledge and wire-hanging tests further confirmed impaired motor function in $TR4^{-/-}$ mice [44].

To further explore the motor function and muscle strength of $TR4^{-/-}$ mice, rotarod tests and quantified muscle strength and fatiguability assays [45] were performed. $TR4^{-/-}$ mice showed declined capacity in both assays, indicating they might suffer muscle weakness (Liu et al., manuscript in preparation). Treadmill test designed to determine the sustained exercise capacity [46, 47] also showed $TR4^{-/-}$ mice had worse performance compared with their $TR4^{+/+}$ littermates with respect to cumulative running distance and total aerobic workload (Liu et al., manuscript in preparation), which suggests impaired sustainable exercise capacity in $TR4^{-/-}$ mice.

From frozen cryostat sections of $TR4^{-/-}$ and $TR4^{+/+}$ skeletal muscle tissues, H&E staining showed basophilic bands at the periphery of muscle fibers, which might be due to the abnormal deposits of mitochondria, only in $TR4^{-/-}$ mice. Gomori Trichrome staining also showed the mitochondrial myopathy typical 'ragged red fibers', the red staining in the peripheral rim of muscle fibers, in $TR4^{-/-}$ skeletal muscle samples (Liu et al., manuscript in preparation). These ragged red fibers are believed to consist of abnormal subsarcolemmal aggregates of mitochondria. Two more specific staining assays for mitochondrial function assessment, succinate dehydrogenase (SDH) staining [48] and cytochrome c oxidase (COX) staining [49], were also applied in our studies. Consistent with trichrome staining pattern, SDH enzyme staining also revealed hyper-reactivity of SDH, suggesting abnormal mitochondrial proliferation in the muscle fibers of $TR4^{-/-}$ mice. The COX reaction is particularly useful in the evaluation of mitochondrial myopathies because COX contains subunits encoded for by both the mitochondrial and nuclear genome [50]. Interestingly, the COX staining for muscle fibers in $TR4^{-/-}$ mice showed no difference compared to $TR4^{+/+}$ mice, which could suggest the defects in mitochondria in $TR4^{-/-}$ mice might not be due to mitochondrial DNA defects (Liu et al., manuscript in preparation).

We also took a closer look at skeletal muscle tissues of $TR4^{-/-}$ and $TR4^{+/+}$ mice by electron microscopy (EM), which showed subsarcolemmal aggregates of variable sized mitochondria with a high content of cristae in $TR4^{-/-}$ mice. Electron-dense particles of Ca^{2+} were also seen in the membranes of mitochondria of $TR4^{-/-}$ mice, which might also indicate the loss of calcium homeostasis in $TR4^{-/-}$ mice. Besides abnormal mitochondrial accumulation, we found tubular aggregates (TAs) with a 'honeycomb' appearance in $TR4^{-/-}$ mice. TAs have been observed in skeletal muscle in various pathological conditions, especially in periodic paralysis, dyskalaemia, and hyperornithinaemia [51]. The cut section through the A-I junction in $TR4^{-/-}$ mice showed close association of mitochondria with small segments of TAs of varied dimensions, which might suggest that the TAs here are a compensatory reaction to mitochondria defects. Together, muscle biopsies and histology illustrated mitochondrial myopathies in $TR4^{-/-}$ mice (Liu et al., manuscript in preparation).

An abnormal accumulation of lactic acid in the blood, also known as hyperlactatemia or lactic acidemia, may occur in association with poor perfusion, several metabolic diseases, or mitochondrial damages [52]. We examined the lactate level in $TR4^{-/-}$ and $TR4^{+/+}$ mice, and $TR4^{-/-}$ mice displayed hyperlactatemia (Liu et al., manuscript in preparation).

One of the key roles of mitochondria is to produce ATP, the energy currency in cells, to maintain normal cell function. The capacity for ATP generation is thus an important index for mitochondrial function. Researchers found ATP synthesis was markedly decreased in myoblasts derived from patients with mitochondrial disorders [53]. ATP production by isolated skeletal muscle mitochondria from $TR4^{-/-}$ and $TR4^{+/+}$ mice was measured in the presence of pyruvate (50 mM) and malate (20 mM). Mitochondria from $TR4^{-/-}$ skeletal muscle have markedly lower ATP production rate per gram mitochondrial protein compared with mitochondria from their $TR4^{+/+}$ littermates (Liu et al., manuscript in preparation).

The inner mitochondrial membrane located OXPHOS system is the ATP generating machinery [54]. It consists of five multiprotein enzymatic complexes: complex I (nicotinamide adenine dinucleotide [NADH]-ubiquinone oxidoreductase); complex II (succinate-ubiquinone oxidoreductase); complex III (ubiquinol-cytochrome c oxidoreductase); complex IV (cytochrome c oxidase); and adenosine 5′-triphosphate (ATP) synthase (complex V) [55]. Studies reveal that decreased activity of any of the five complexes would cause impaired mitochondrial energy output [56–59]. The activities of the five complexes from isolated skeletal muscle mitochondria of $TR4^{-/-}$ and $TR4^{+/+}$ mice were spectrophotometrically measured by using a multiwavelength spectrophotometer and reduced complex I activity was found in $TR4^{-/-}$ skeletal muscle tissue. When we reintroduced TR4 into $TR4^{-/-}$ primary myoblast cells, we could partially rescue ATP generation capacity and complex I activity, which further supported TR4's role in mitochondrial function through regulation of complex I (Liu et al., manuscript in preparation). We also found reduction of complex V activity in $TR4^{-/-}$ mice, however, due to high variations of complex V activity among the mice, tests on more samples are needed (Liu et al., manuscript in preparation).

Isolated deficiency of complex I (NADH-ubiquinone reductase) is relatively frequent among mitochondrial disorders. In the past several years, many disease-associated mutations have been discovered in some of the 36 nuclear-encoded subunits of complex I. However, in many cases of complex I deficiency, most screened structural genes remain intact, suggesting that there are still unknown assembly factors for complex I, or other gene products involved in its formation and activity [60].

Complex I activity impairment in $TR4^{-/-}$ mice suggests TR4 could regulate genes that are involved in complex I structure/assembly. To screen out these TR4 targeted genes, Q-RTPCR was used to compare the complex I gene mRNA levels between $TR4^{-/-}$ and $TR4^{+/+}$ mice skeletal muscles and reduced expression of NUDFA1, a gene that is responsible for the complex I assembly [61], was found in $TR4^{-/-}$ mice (Liu et al., manuscript in preparation). Several TR4RE were identified in the NDUFAF1 5′ promoter. Luciferase assay showed TR4 could dose dependently activate NDUFAF1 and ChIP assay showed the direct binding of TR4 to NDUFAF1 promoter (Liu et al., manuscript in preparation). Thus we confirmed NDUFAF1 is a TR4 target gene. More powerful evidence came from the rescue experiment, in which NDUFAF1 was reintroduced into $TR4^{-/-}$ primary myoblast cells and ATP generation capacity and complex I activity could be partially rescued (Liu et al., manuscript in preparation).

Our studies showed TR4 could regulate mitochondrial function through its target gene, the complex I assembly factor NDUFAF1 and we hope this study could lead to a better understanding and treatment of mitochondrial disease in the future. Human mitochondrial disorders have an estimated incidence around 1:5,000∼1:10,000 [62]. Despite improved characterization of the genetic defects that lead to mitochondrial disorders in the past decade, the pathogenetic mechanisms of these diseases are not well understood. The clinic presentations and genetic etiologies of mitochondrial disorders are highly diverse due to its genetic and biochemical complexity, which adds to the difficulties for diagnosis and treatment [60]. Our studies suggest a potential application of TR4 in the development of new diagnostic and therapeutic approaches for mitochondrial disorders.

Other than the role in skeletal muscle function through regulating mitochondrial energy production, TR4 might also contribute to muscle differentiation. Both TR4 mRNA and protein level were induced in differentiation medium treated murine myoblast C2C12 cells. We also found TR4 can transcriptionally activate an important muscle differentiation factor, myogenin [63]. Therefore, it's possible that TR4 contributes to muscle differentiation through regulation of myogenin, although further studies are needed (Chang et al., manuscript in preparation).

13.7. TR4 AND BONE

Bone is a mineralized tissue that provides the body most of its mechanical support. Its basic function also includes locomotion, protection, and mineral homeostasis. Previous studies found TR4 ubiquitously expressed in all tissues including bone in 129×1/SvJ and C57BL6 mice [64], TR4 was also found as the most abundant expressed gene among all nuclear receptors in primary cultures of chicken bone marrow cells [65].

$TR4^{-/-}$ mice were born visibly indistinguishable from their wide-type littermates. However, by the age of 6 months, most $TR4^{-/-}$ mice developed severe spine curvature (Kyphosis), further indicating a role of TR4 in bone health (Chang et al., manuscript in preparation).

DEXA scanning showed significant reduced bone mineral density (BMD) in 6 month old female and male $TR4^{-/-}$ mice. We examined BMD in the total body, legs, tails, and skull in the five pairs of $TR4^{-/-}$ and $TR4^{+/+}$ mice from both genders and found a significant difference in all parts we compared, except skull (Chang et al., manuscript in preparation).

In agreement with the BMD analyses by the DEXA scanning, H&E staining showed trabecular and cortical bone volume in thoracic and lumbar spine were reduced significantly in 6 month old $TR4^{-/-}$ mice compared with their wild-type littermates. Furthermore, a large amount of fat was found accumulated in the bone marrow of trabecular bone in 6 month old $TR4^{-/-}$ mice by HE staining, while age matched wild-type mice show normal bone with little adipocyte accumulation (Chang et al., manuscript in preparation). These osteoporosis resembling phenotypes found in $TR4^{-/-}$ mice suggest TR4 might play a role in bone health.

Osteoporosis is a skeletal disease characterized by low bone mass, microarchitectural deterioration, low bone mass, and increased risk of fracture [66]. It is a major public health issue because of the potentially devastating outcomes and high cumulative rate of fractures [67]. Osteoporotic fractures are a frequent and important cause of disability and medical costs worldwide [68]. BMD is the best known predictor of osteoporosis and many of its determinants such as age, sex, race, body weight, state of hypoestrogenism, calcium, and vitamin D intake are known [69]. Besides BMD, many other skeletal characteristics also contribute to bone strength, such as bone macroarchitecture, bone microarchitecture, matrix and mineral composition, and the rate of bone turnover [67].

The cellular make up of bone includes osteoblasts, osteocytes, bone lining cells, and osteoclasts [70]. Bone is constantly resorbed by osteoclasts and then replaced by the osteoblasts in a physiologic process called bone remodeling [71]. The unbalanced function of osteoblasts and osteoclasts is the major cause of problems in bone mass, including osteoporosis or osteopenia. The genetic control of bone density suggests that the functions of the osteoblasts and the osteoclasts are tightly regulated, although they are also influenced by environmental factors. Thus, the regulation of bone density can be studied by unraveling the complex array of genes that influence the development and function of the osteoblasts and the osteoclasts [72].

Assays showed osteoclasts from $TR4^{-/-}$ mice functioned as well as $TR4^{+/+}$ osteoclasts (Chang et al., manuscript in preparation), suggesting the problems in $TR4^{-/-}$ bones might come from the osteoblasts. Primary mouse osteoblastic cell cultures were then established from calvarial cells isolated from 6- to 8-day-old neonatal $TR4^{-/-}$ and $TR4^{+/+}$ mice and von Kossa staining [73] and alizarin red S staining [74] were applied to visualize mineralized nodules in cell culture. $TR4^{-/-}$ cells showed significant reduced staining density in both assays, indicating problems in osteoblast differentiation in $TR4^{-/-}$ cells.

The differentiation of precommitted osteoprogenitor cells to mature osteoblasts in vivo is regulated by osteogenic factors that mediate the staged expression of bone phenotypic genes [75]. At each stage of osteoblast differentiation, a subset of genes is transcriptionally activated to accommodate developmental stage-specific physiological requirements for bone-specific gene products.

Among the known osteoblast differentiation genes, BMP-2, BMP-4, and collagen Type X can be activated by TR4 in CV-1 cells and reduced level of Type I collagen, parathyroid hormone receptors (PTHR) and core binding factor-1 (Cbfa-1)/runt-related transcription factor (Runx)-2 was found in TR4KO tissues, indicating a role of TR4 in osteoblast differentiation (Chang et al., manuscript in preparation). We also found up-regulation of TR4 expression in BMP-2-induced bone cell lineage in C2C12 myoblast cells. C2C12 cells were treated with BMP-2 (20 ng/ml) or TGF-β (3 ng/ml) for 48 h, and then the conditioned media were harvested for determination of alkaline phosphatase activity for an indication of bone cell lineage. TR4 expression levels were up-regulated in BMP-2 treated C2C12 cells, and the cells have detectable alkaline phosphatase activity. However, TR4 expression level does not change in TGFβ-treated C2C12 cells (Chang et al., manuscript in preparation).

Interestingly, previous studies showed TR4 negatively modulated the estrogen signaling pathway [76] and vitamin D signaling [77], and the repression of TR4 on ER-mediated transactivation is bone cell-specific. Estrogen is essential for maintaining bone mass in women and some selective ER modulators (SERMs) have been approved in many countries for the prevention and treatment of postmenopausal osteoporosis [78]. Vitamin D signaling is important for the regulation of calcium homeostasis and thus bone density. These findings seem opposite to what we found in $TR4^{-/-}$ mice. A possible explanation could be that TR4 might regulate bone cell function through a complex downstream network, and the activation of each specific downstream pathway of TR4 could be context and physiological condition dependent. Further assays are necessary to explore the role of TR4 in bone, and the molecular mechanism through which TR4 modulates bone cell function. It will be interesting to find out how the interactions between TR4 and estrogen or vitamin D signaling contribute to the regulation of bone remodeling.

Our findings suggest TR4 plays an essential role in bone remodeling. We hope further elucidation of the molecular mechanism through which TR4 mediates bone function will enhance the understanding of the pathogenesis of osteoporosis and lead to early identification of individuals at high osteoporotic risk and the development of preventive programs.

Other than osteoporosis, $TR4^{-/-}$ mice also display a progressive radiographic osteoarthritis disease. There are thin layers of articular cartilage lining in the bones which provide a smooth, wear-resistant structure that reduces friction and absorbs impact forces. Loss or damage of articular cartilage is a hallmark of osteoarthritic disease (OA). OA is the most common disabling condition of humans in the western world. Millions of people are afflicted with OA, and it ultimately affects more than half of the population over the age of 65. Six month old and 11 month old $TR4^{-/-}$ and $TR4^{+/+}$ mice were examined by radiography. A sign of knee joint space narrowing was found in a 6 month old $TR4^{-/-}$ mouse, and irregular joint surface and bone formation (osteophytes) was found in a 11 month old $TR4^{-/-}$ mouse, while normal knee joints and spine radiography were observed in 6 and 11 month old $TR4^{+/+}$ littermates. This was further proven by bone histomorphometry analysis of 6 month old $TR4^{-/-}$ mice in which an early stage of OA shows the superficial cartilage erosion, intermediate nucleus duplication, and a sign of infiltration in the meniscus from knee joints, while a normal articular cartilage was observed in their wild-type littermates (Chang et al., manuscript in preparation). Further studies are needed to elucidate the role of TR4 in OA.

13.8. CONCLUDING REMARKS AND FUTURE DIRECTIONS

Previous in vitro studies suggest that TR4 functions as a master regulator to modulate many signaling pathways, including neurogenesis, via ciliary neurotrophic factor alpha (CNTFRα) [79], interfering with retinoic acid/RAR/RXR [80, 81], thyroid hormone/T3R [82], vitamin D 3/VDR [83], AR and ER-mediated pathways. The powerful genetic manipulation techniques allowed us to further explore the in vivo physiological functions of TR4 in animals. $TR4^{-/-}$ mice displayed high rates of early postnatal mortality, significant growth retardation [25], reproductive defects with reduced fertility seen in both genders [25], and abnormalities in cerebella development [44]. This review clearly indicates that TR4 plays essential roles in postnatal growth, neural development, spermatogenesis, metabolism, skeletal muscle function, and bone remodeling [21–25]. The TR4 knockout mouse model, along with emerging phenotypes, has provided perspective information on the less known mechanisms involved in the effects of TR4 under physiological and pathological conditions. However, there are still many questions that need to be answered regarding more detailed mechanisms and the cross-talk between TR4's effects in different tissues. Studies to more deeply explore TR4 function in other tissues, as well as in current projects, are ongoing. The use of tissue-specific TR4 knockout animals will further clarify TR4's role in each tissue without the primary or secondary effect concerns. The future studies in TR4 knockout mice and TR4 tissue specific knockout mice will improve our understanding of TR4's physiological and pathological functions that might lead to the potential application of TR4 in the development of new diagnostic and therapeutic approaches for human diseases.

REFERENCES

1. Popov, V. M. et al. (2007). The functional significance of nuclear receptor acetylation. *Steroids* 72(2), 221–230.
2. Robinson-Rechavi, M., Escriva Garcia, H., and Laudet, V. (2003). The nuclear receptor superfamily. *J Cell Sci* 116(Pt 4), 585–586.
3. McKenna, N. J., Lanz, R. B., and O'Malley, B. W. (1999). Nuclear receptor coregulators: cellular and molecular biology. *Endocr Rev* 20(3), 321–344.

4. Willson, T. M. and Moore, J. T. (2002). Genomics versus orphan nuclear receptors–a half-time report. *Mol Endocrinol* 16(6), 1135–1144.
5. Kliewer, S. A., Lehmann, J. M., and Willson, T. M. (1999). Orphan nuclear receptors: shifting endocrinology into reverse. *Science* 284(5415), 757–760.
6. Giguere, V. (1999). Orphan nuclear receptors: from gene to function. *Endocr Rev* 20(5), 689–725.
7. Chang, C. et al. (1994). Human and rat TR4 orphan receptors specify a subclass of the steroid receptor superfamily. *Proc Natl Acad Sci U S A* 91(13), 6040–6044.
8. Lee, Y. F., Lee, H. J., and Chang, C. (2002). Recent advances in the TR2 and TR4 orphan receptors of the nuclear receptor superfamily. *J Steroid Biochem Mol Biol* 81(4–5), 291–308.
9. Kim, E. et al. (2003). Disruption of TR4 orphan nuclear receptor reduces the expression of liver apolipoprotein E/C-I/C-II gene cluster. *J Biol Chem* 278(47), 46919–46926.
10. Liu, N. C. et al. (2007). Loss of TR4 orphan nuclear receptor reduces phosphoenolpyruvate carboxykinase-mediated gluconeogenesis. *Diabetes* 56(12), 2901–2909.
11. Lee, Y. F. et al. (1999). Differential regulation of direct repeat 3 vitamin D3 and direct repeat 4 thyroid hormone signaling pathways by the human TR4 orphan receptor. *J Biol Chem* 274(23), 16198–16205.
12. Lee, C. H., Chinpaisal, C., and Wei, L. N. (1998). A novel nuclear receptor heterodimerization pathway mediated by orphan receptors TR2 and TR4. *J Biol Chem* 273(39), 25209–25215.
13. Young, W. J. et al. (1998). A bidirectional regulation between the TR2/TR4 orphan receptors (TR2/TR4) and the ciliary neurotrophic factor (CNTF) signaling pathway. *J Biol Chem* 273(33), 20877–20885.
14. Shyr, C. R. et al. (2002). Modulation of estrogen receptor-mediated transactivation by orphan receptor TR4 in MCF-7 cells. *J Biol Chem* 277(17), 14622–14628.
15. Lee, Y. F. et al. (1998). Negative feedback control of the retinoid-retinoic acid/retinoid X receptor pathway by the human TR4 orphan receptor, a member of the steroid receptor superfamily. *J Biol Chem* 273(22), 13437–13443.
16. Lee, Y. F. et al. (1997). Identification of direct repeat 4 as a positive regulatory element for the human TR4 orphan receptor. A modulator for the thyroid hormone target genes. *J Biol Chem* 272(18), 12215–12220.
17. Young, W. J., Smith, S. M., and Chang, C. (1997). Induction of the intronic enhancer of the human ciliary neurotrophic factor receptor (CNTFRalpha) gene by the TR4 orphan receptor. A member of steroid receptor superfamily. *J Biol Chem* 272(5), 3109–3116.
18. Yoshikawa, T. et al. (1996). Splice variants of rat TR4 orphan receptor: differential expression of novel sequences in the 5′-untranslated region and C-terminal domain. *Endocrinology* 137(5), 1562–1571.
19. Lee, H. J. et al. (1995). Suppression of gene expression on the simian virus 40 major late promoter by human TR4 orphan receptor. A member of the steroid receptor superfamily. *J Biol Chem* 270(50), 30129–30133.
20. Yan, Z. H. et al. (1998). Regulation of peroxisome proliferator-activated receptor alpha-induced transactivation by the nuclear orphan receptor TAK1/TR4. *J Biol Chem* 273(18), 10948–10957.
21. Zhang, Y. et al. (2007). Loss of testicular orphan receptor 4 impairs normal myelination in mouse forebrain. *Mol Endocrinol* 21(4), 908–920.
22. Kim, E. et al. (2005). Induction of apolipoprotein E expression by TR4 orphan nuclear receptor via 5′ proximal promoter region. *Biochem Biophys Res Commun* 328(1), 85–90.
23. Chen, Y. T. et al. (2005). Deficits in motor coordination with aberrant cerebellar development in mice lacking testicular orphan nuclear receptor 4. *Mol Cell Biol* 25(7), 2722–2732.
24. Mu, X. et al. (2004). Targeted inactivation of testicular nuclear orphan receptor 4 delays and disrupts late meiotic prophase and subsequent meiotic divisions of spermatogenesis. *Mol Cell Biol* 24(13), 5887–5899.
25. Collins, L. L. et al. (2004). Growth retardation and abnormal maternal behavior in mice lacking testicular orphan nuclear receptor 4. *Proc Natl Acad Sci U S A* 101(42), 15058–15063.

26. Chen, L. M. et al. (2008). Subfertility with defective folliculogenesis in female mice lacking TR4 orphan nuclear receptor. *Mol Endocrinol* 22, 858–867.
27. Tanabe, O. et al. (2002). An embryonic/fetal beta-type globin gene repressor contains a nuclear receptor TR2/TR4 heterodimer. *EMBO J* 21(13), 3434–3442.
28. Tanabe, O. et al. (2007). Embryonic and fetal beta-globin gene repression by the orphan nuclear receptors, TR2 and TR4. *EMBO J* 26(9), 2295–2306.
29. Shyr, C. R. et al. (2002). Spermatogenesis and testis development are normal in mice lacking testicular orphan nuclear receptor 2. *Mol Cell Biol* 22(13), 4661–4666.
30. Bookout, A. L. et al. (2006). Anatomical profiling of nuclear receptor expression reveals [no-break space]a hierarchical transcriptional network. *Cell* 126(4), 789–799.
31. Chen, Y. T. et al. (2007). The roles of testicular orphan nuclear receptor 4 (TR4) in cerebellar development. *Cerebellum* 7, 9–17.
32. Chen, Y. T. et al. (2007). Abnormal cerebellar cytoarchitecture and impaired inhibitory signaling in adult mice lacking TR4 orphan nuclear receptor. *Brain Res* 1168, 72–82.
33. Zhang, Y. et al. (2007). Loss of testicular orphan receptor 4 impairs normal myelination in mouse forebrain. *Mol Endocrinol* 21(4), 908–920.
34. Kim, E.L., Ning-Chun, and Chang, Chawnshang. Metformin inhibits TR4-mediated SCD1 gene expression via an SREBP-1c-independent mechanism that results in increased insulin sensitivity. (manuscript submitted).
35. Postic, C., Dentin, R., and Girard, J. (2004). Role of the liver in the control of carbohydrate and lipid homeostasis. *Diabetes Metab* 30(5), 398–408.
36. Cohen, P., Ntambi, J. M., and Friedman, J. M. (2003). Stearoyl-CoA desaturase-1 and the metabolic syndrome. *Curr Drug Targets Immune Endocr Metabol Disord* 3(4), 271–280.
37. Cohen, P. and Friedman, J. M. (2004). Leptin and the control of metabolism: Role for stearoyl-CoA desaturase-1 (SCD-1). *J Nutr* 134(9), 2455S–2463S.
38. Kiefer, C. M. et al. (2008). Epigenetics of [beta]-globin gene regulation. *Mutation Research/ Fundamental and Molecular Mechanisms of Mutagenesis* 647(1–2), 68–76.
39. Grosveld, F. (1999). Activation by locus control regions?. *Curr Opin Genet Dev* 9(2), 152–157.
40. Kihm, A. J. et al. (2002). An abundant erythroid protein that stabilizes free alpha-haemoglobin. *Nature* 417(6890), 758–763.
41. Cantor, A. and Orkin, S. (2002). Transcriptional regulation of erythropoiesis: an affair involving multiple partners. *Oncogene* 21(21), 3368–3376.
42. Tanabe, O. et al. (2007). The TR2 and TR4 orphan nuclear receptors repress Gata1 transcription. *Genes Dev* 21(21), 2832–2844.
43. Yang, X. et al. (2006). Nuclear receptor expression links the circadian clock to metabolism. *Cell* 126(4), 801–810.
44. Chen, Y. -T. et al. (2005). Deficits in motor coordination with aberrant cerebellar development in mice lacking testicular orphan nuclear receptor 4. *Mol Cell Biol* 25(7), 2722–2732.
45. Joya, J. E. et al. (2004). Muscle weakness in a mouse model of nemaline myopathy can be reversed with exercise and reveals a novel myofiber repair mechanism. *Hum Mol Genet* 13(21), 2633–2645.
46. Massett, M. P. and Berk, B. C. (2005). Strain-dependent differences in responses to exercise training in inbred and hybrid mice. *Am J Physiol Regul Integr Comp Physiol* 288(4), R1006–R1013.
47. Kinugawa, S. et al. (2005). Limited exercise capacity in heterozygous manganese superoxide dismutase gene-knockout mice: Roles of superoxide anion and nitric oxide. *Circulation* 111(12), 1480–1486.
48. Taylor, R. W. et al. (2004). The diagnosis of mitochondrial muscle disease. *Neuromuscular Disorders* 14(4), 237.
49. Arpa, J., Cruz-Martínez, A., Campos, Y., Gutiérrez-Molina, M., García-Rio, F., Pérez-Conde, C., Martín, M. A., Rubio, J. C., Del Hoyo, P., Arpa-Fernández, A., and Arenas, J. (2003). Prevalence and progression of mitochondrial diseases: A study of 50 patients. *Muscle Nerve* 28(6), 690–695.
50. Müller-Höcker, J., Pongratz, D., and Hübner, G. (1983). Focal deficiency of cytochrome-c-oxidase in skeletal muscle of patients with progressive external ophthalmoplegia. *Virchows Arch* 402(1), 61–71.

51. Chevessier, F. et al. (2005). The origin of tubular aggregates in human myopathies. *J Pathol* 207(3), 313–323.
52. Tarnopolsky, M. A. and Raha, S. (2005). Mitochondrial myopathies: diagnosis, exercise intolerance, and treatment options. *Med Sci Sports Exerc* 37(12), 2086–2093.
53. Shepherd, R. et al. (2006). Measurement of ATP production in mitochondrial disorders. *J Inherit Metab Dis* 29(1), 86.
54. Wallace, D. C. (2005). A mitochondrial paradigm of metabolic and degenerative diseases, aging, and cancer: A dawn for evolutionary medicine. *Ann Rev Genet* 39(1), 359–407.
55. DiMauro, S. (2004). Mitochondrial medicine. *Biochimica et Biophysica Acta (BBA) – Bioenergetics* 1659(2–3), 107.
56. Gellerich, F. et al. (2004). Energetic depression caused by mitochondrial dysfunction. *Mol Cell Biochem* 256–257(1–2), 391.
57. McKenzie, M., Liolitsa, D., and Hanna, M. G. (2004). Mitochondrial disease: Mutations and mechanisms. *Neurochem Res* 29(3), 589–600.
58. Thorburn, D. R. et al. (2004). Biochemical and molecular diagnosis of mitochondrial respiratory chain disorders. *Biochimica et Biophysica Acta (BBA) – Bioenergetics* 1659(2–3), 121.
59. Houstek, J. et al. (2004). Mitochondrial diseases and ATPase defects of nuclear origin. *Biochimica et Biophysica Acta (BBA) – Bioenergetics* 1658(1–2), 115.
60. DiMauro, S. and Hirano, M. (2005). Mitochondrial encephalomyopathies: An update. *Neuromuscular Disorders* 15(4), 276.
61. Lazarou, M. et al. (2009). Assembly of mitochondrial complex I and defects in disease. *Biochim Biophys Acta* 1793(1), 78–88, Molecular Cell Research, Corrected Proof.
62. Thorburn, D. R. (2004). Mitochondrial disorders: Prevalence, myths and advances. *J Inherit Metab Dis* 27(3), 349.
63. Pownall, M. E., Gustafsson, M. K., and Emerson, C. P. (2002). Myogenic regulatory factors and the specification of muscle progenitors in vertebrate embryos. *Ann Rev Cell Dev Biol* 18(1), 747–783.
64. Bookout, A. L. et al. (2006). Anatomical profiling of nuclear receptor expression reveals a hierarchical transcriptional network. *Cell* 126(4), 789–799.
65. Koritschoner, N. P. et al. (2001). The nuclear orphan receptor TR4 promotes proliferation of myeloid progenitor cells. *Cell Growth Differ* 12(11), 563–572.
66. Peacock, M. et al. (2002). Genetics of osteoporosis. *Endocr Rev* 23(3), 303–326.
67. Sambrook, P. and Cooper, C. (2006). Osteoporosis. *Lancet* 367(9527), 2010–2018.
68. Cummings, S. R. and Melton, L. J. (2002). Epidemiology and outcomes of osteoporotic fractures. *Lancet* 359(9319), 1761–1767.
69. Yves Giguère, F. R. (2000). The genetics of osteoporosis: 'complexities and difficulties'. *Clinl Genet* 57(3), 161–169.
70. Downey, P. A. and Siegel, M. I. (2006). Bone biology and the clinical implications for osteoporosis. *Phys Ther* 86(1), 77–91.
71. Ducy, P., Schinke, T., and Karsenty, G. (2000). The osteoblast: A sophisticated fibroblast under central surveillance. *Science* 289(5484), 1501–1504.
72. Langman, C. B. (2005). Genetic regulation of bone mass: From bone density to bone strength. *Pediatric Nephrol* 20(3), 352–355.
73. Wang, Y. H., Liu, Y., Maye, P., and Rowe, D. W. (2006). Examination of mineralized nodule formation in living osteoblastic cultures using fluorescent dyes. *Biotechnol Progr* 22(6), 1697–1701.
74. Gartland, A. et al. (2005). In vitro chondrocyte differentiation using costochondral chondrocytes as a source of primary rat chondrocyte cultures: An improved isolation and cryopreservation method. *Bone* 37(4), 530–544.
75. Seeman, E. (2003). Reduced bone formation and increased bone resorption: rational targets for the treatment of osteoporosis. *Osteoporos Int* 14(Suppl 3), S2–S8.
76. Harada, H. et al. (1998). Cloning of rabbit TR4 and its bone cell-specific activity to suppress estrogen receptor-mediated transactivation. *Endocrinology* 139(1), 204–212.

77. Lee, Y. -F. et al. (1999). Differential regulation of direct repeat 3 vitamin D3 and direct repeat 4 thyroid hormone signaling pathways by the human TR4 orphan receptor. *J Biol Chem* 274(23), 16198–16205.
78. Mulder, J. E., Kolatkar, N. S., and LeBoff, M. S. (2006). Drug insight: Existing and emerging therapies for osteoporosis. *Nat Clin Pract End Met* 2(12), 670–680.
79. Young, W. -J., Smith, S. M., and Chang, C. (1997). Induction of the intronic enhancer of the human ciliary neurotrophic factor receptor (CNTFRalpha) gene by the TR4 orphan receptor. A member of steroid receptor superfamily. *J Biol Chem* 272(5), 3109–3116.
80. Hirose, T. et al. (1995). The orphan receptor TAK1 acts as a repressor of RAR-, RXR- and T3R-mediated signaling pathways. *Biochem Biophys Res Comm* 211(1), 83–91.
81. Lee, Y. -F. et al. (1998). Negative feedback control of the retinoid-retinoic acid/retinoid X receptor pathway by the human TR4 orphan receptor, a member of the steroid receptor superfamily. *J Biol Chem* 273(22), 13437–13443.
82. Hwang, S. -B., Burbach, J., and Chang, C. (1998). TR4 orphan receptor crosstalks to chicken ovalbumin upstream protein-transcription factor and thyroid hormone receptor to induce the transcriptional activity of the human immunodeficiency virus type 1 long-terminal repeat. *Endocrine* 8(2), 169–175.
83. Lee, Y. -F. et al. (1999). Differential regulation of direct repeat 3 vitamin D3 and direct repeat 4 thyroid hormone signaling pathways by the human TR4 orphan receptor. *J Biol Chem* 274(23), 16198–16205.

CHAPTER 14

NUCLEAR RECEPTORS AND ATP DEPENDENT CHROMATIN REMODELING: A COMPLEX STORY

CRAIG J. BURD AND TREVOR K. ARCHER

Laboratory of Molecular Carcinogenesis, National Institute of Environmental Health Sciences, Research Triangle Park, NC, USA

Abstract: Nuclear receptors are a class of tightly regulated and highly inducible transcription factors and thus represent an excellent model for the study of transcription. The activity of these transcription factors is controlled by their interactions with co-regulatory proteins that act to remodel chromatin, modify histones, and initiate the transcriptional process. This review will focus on the use of nuclear receptors in understanding the role of ATP-dependent chromatin remodeling enzymes in transcription.

14.1. INTRODUCTION

Nuclear receptors are a superfamily of transcription factors that share a conserved structural organization [1]. These receptors are typified by domains utilized for DNA binding, ligand binding, and transcriptional regulation. The superfamily is typically segregated into two classes. The steroid receptors, which include the estrogen (ER), glucocorticoid (GR), androgen (AR), mineralocoticoid (MR), and progesterone (PR), are typically inactive without ligand and bound to heat shock proteins. In general, upon ligand activation, steroid receptors form homodimers and consequently bind to hormone response elements (HREs) within gene promoters. The subsequent recruitment of co-factors allow for transcriptional activation. The second class within the superfamily can be defined by their association with response elements and include the thyroid (TR), vitamin D (VDR), prostanoid (PPAR), retinoid (RAR), and rexinoid (RXR) receptors. These receptors are frequently constitutively associated with response elements within promoters as heterodimers and rely upon conformational changes for the recruitment of the necessary co-factors to illicit an active or repressive transcriptional state [2]. These co-factors are classified as either co-activators by their ability to enhance transcription typically through promoting an open chromatin state or co-repressors by initiating a closed chromatin state [3]. The activation of nuclear receptors involves a number of distinct events (ligand binding,

C.M. Bunce, M.J. Campbell (eds.), *Nuclear Receptors*, Proteins and Cell Regulation 8, 345–363. DOI 10.1007/978-90-481-3303-1_14,

promoter association, co-factor recruitment, chromatin re-organization), is tightly controlled, and results in a measureable readout. These facts make them an optimum model for studying transcription.

The identification of DNA as the carrier of inheritable genetic information has been followed by years of research into understanding the mechanisms by which this information is controlled, utilized, and finally transmitted. The complexity of the situation is magnified by the packaging of DNA with proteins into chromatin. This packaging allows for even the smallest chromosomes at metaphase to be compacted 7,000-fold, thereby enabling large chromosomes to fit into the comparably tiny nuclei of cells [4]. However, chromatin architecture can generate an obstruction to the fundamental processes of transcription, replication, and repair. In fact, it has become quite clear that the organization of chromatin structure is a dynamic entity that plays a large role in the tightly controlled regulation of these processes [5].

Chromatin consists of a repeating array of core subunit nucleosomes generated by the winding of 146 bp of DNA around histone octomers [6]. Individual nucleosomes are made up of two copies of each of the histones H2A, H2B, H3, and H4. In addition to these four core histones, one bound copy of the linker histone H1 directs the entrance and exit of the DNA strand from the core nucleosome and participates in compaction of the chromatin into higher order structures [7, 8]. Nucleosomes are separated by regions of linker DNA that vary in length to a maximum of about 80 bp and generate an average of 200 bp of DNA per nucleosome [9]. In addition to these histones, chromatin also contains non-histone proteins that together allow for a multi-layered structure that can dynamically change the accessibility of the DNA strand to regulatory proteins. Changes in the structure can lead to an active state, termed euchromatin, wherein the DNA and protein are less tightly associated and thus more accessible. In contrast, extremely condensed structures called heterochromatin decrease the accessibility to the DNA and lead to an inactive state [8].

The process of transcription requires the collaboration of a number of DNA interacting proteins including transcription factors, co-regulators, and transcriptional machinery. It is clear that the regulation of these processes is dependent upon much more than just the DNA sequence. The importance of chromatin regulation is gaining strength due to evidence of inheritable changes in gene regulation independent of underlying sequence that affect biological processes and are implicated in pathological phenotypes [10]. This study, termed epigenetics, has concentrated on chromatin modifications involved in gene regulation. It is apparent that chromatin structure is critical to control inducible expression of target genes and is achieved through a variety of mechanisms, complexes, and enzymes [11].

14.1.1. DNA Methylation

One of the most common modifications to chromatin is the addition of methyl groups to cytosine residues within the DNA sequence. In fact, it has been reported that up to 90% of CpG dinucleotides are methylated in somatic cells [12]. This modification plays a critical role in long-term transcriptional silencing and formation of heterochromatin. In general, regulation of transcription through DNA methylation involves the alteration of cytosines in CpG islands, stretches of GC rich DNA, within gene promoters. The mechanism by which hypermethylation of promoter DNA silences transcription is not fully understood, however, it is hypothesized that the methylation marks prevent transcription factor binding and/or alter chromatin structure through affecting chromatin modifications and nucleosome occupancy. It is has become apparent that DNA methylation is physiologically critical for normal development and is aberrantly regulated in pathological states such as cancer.

14.1.2. Covalent Histone Modifications

A distinct class of chromatin regulators initiate covalent modifications to the histone molecules through acetylation, phophorylation, methylation, ubiquitination, sumoylation, glycosylation, and ADP ribosylation [13]. Each histone can harbor several modifications, usually within the tail of the protein, and the differing combination of these modifications act to control gene regulation. These modifications are highly dynamic and enzymes exist that either deposit or remove these covalent links. The two most highly studied histone modifications are acetylation and methylation. Acetylation of histone tails is usually, but not always, associated with transcriptional activation. It is believed that acetylation of the lysine residues within the histones neutralizes the positive charge and thus weakens a normally tight interaction with the negatively charged DNA [14]. Supporting this hypothesis, recent studies using recombinant nucleosomal arrays demonstrate that acetylation of H4K16 restricts the formation of specific chromatin structures [15]. Methylation of histones can be associated with either heterochromatin and transcriptional repression or euchromatin and transcriptional activation depending upon the location of the specific modifications. For example, trimethylation of H3K9, H3K27, and H4K20 are found in heterochromatin while H3K4, H3K36, and H3K79 marks are seen in euchromatin [16]. Further complexity of histone methylation is found in the number of marks found at each residue. Each lysine may contain between one and three methyl groups. The variety of histone modifcations combined with the multiple sites of alteration allow for an extremely complex mechanism of regulation that can be utilized across the broad genomic landscape.

14.1.3. ATP-Dependent Chromatin Remodeling

In addition to the covalent modifications of both DNA and histones, the regulation of transcription is also mediated by protein complexes that utilize the energy derived from hydrolysis of ATP to actively alter chromatin structure[17]. These chromatin remodeling complexes are believed to either loosen the DNA/protein contacts and/or shift nucleosome positioning [18]. The four characterized families of chromatin remodeling ATPases are SWI/SNF, ISWI, NuRD and INO80 which are distinguished

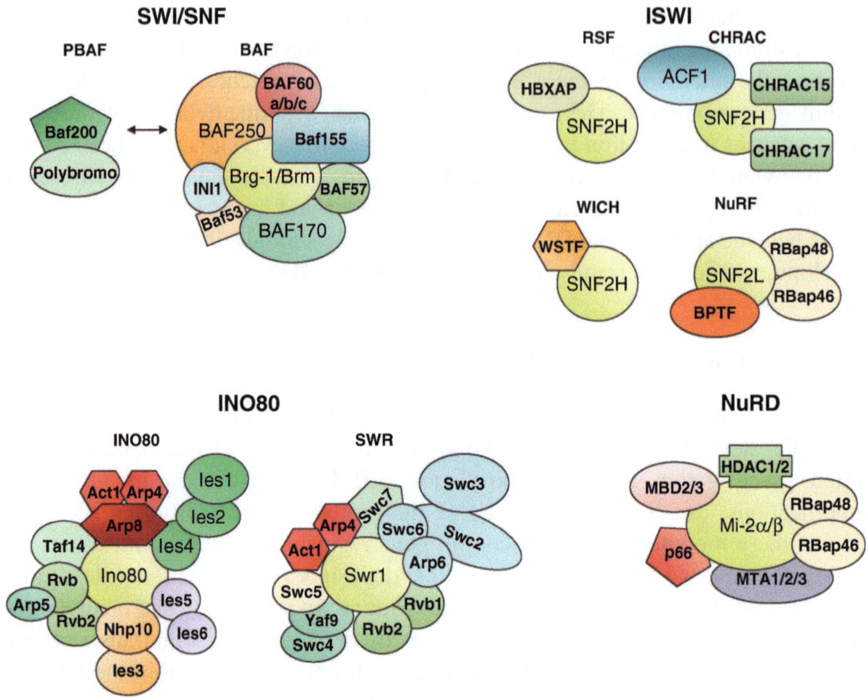

Figure 14.1. The ATPase dependent chromatin remodeling complexes are grouped into the four classes of **SWI/SNF** [46, 48–50], **ISWI** [72, 73, 75, 76], **NuRD** [82–85] , and **INO80** [95–98]. Each class may contain multiple complexes with alternate subunits

by their catalytic subunits, Brg1/Brm, ISWI, Mi-2 and INO80, respectively [19]. Although there is high homology between all of the ATPase domains within these proteins, each protein contains distinct chromatin-interacting regions such as bromo and SANT domains that may determine specificity of function between the families. In addition, each of these ATPases all exist in multi-subunit complexes with unique members that further add specificity (Figure 14.1). Many of these subunits are known to interact with transcription factors which may target activity to sites of activation or repression. This review will concentrate on the use of nuclear receptors in understanding the role of chromatin remodeling complexes as co-regulators of transcription.

14.2. CO-ACTIVATORS AND CO-REPRESSORS

A great deal of work has concentrated on the identification and specificity of novel co-regulatory molecules that are recruited to nuclear receptors at sites of gene transcription [20, 21]. While this list of co-regulators is ever increasing (http://www.nursa.org), it is clear that chromatin, and histones in particular, are a major target of these molecules, many of which contain intrinsic enzymatic activity. One of the first and most characterized group of nuclear receptor co-activators is the p160 class including SRC-1 (also NCoA-1), TIF2 (also GRIP1, SRC-2, NCoA-2), and AIB1 (also TRAM1, SRC-3) [22–24]. These molecules directly interact with nuclear receptors and either possesses intrinsic acetyltransferase activity (SRC-1, AIB1) and/or interacts with other histone acetyltransferases (HATS) such as CREB binding protein (CBP) or p300 CBP associated factor (PCAF) [22–27]. As previously mentioned, the ability to acetylate histones is typically associated with relaxation of the chromatin and transcriptional activity. In contrast, recruitment of co-repressors such as the NCOR and SMRT complexes that associate with histone deacetylases (HDACs) oppose the co-activator stimulation and repress transcription. The recruitment of these co-repressors to nuclear receptor occurs in either antagonist bound or unliganded receptors [28–32]. In addition to those molecules involved in regulating the acetylation of histones, there are a growing number of enzymatic co-regulators that affect all aspects of the histone code to carefully modulate transcription. The study of nuclear receptor regulation, however, has not been limited to the histone modifying class of co-regulators and has provided an excellent model for understanding the role of ATPase driven chromatin remodeling [33].

14.2.1. SWI/SNF as a Nuclear Receptor Co-regulator

The best characterized ATP-dependent chromatin remodeling complex is named SWI/SNF and refers to a yeast complex which was initially identified for its importance in mating type switching (SWI) and sucrose fermentation (SNF) [34–36]. Subsequently, this complex was shown to regulate nuclear receptor action in yeast studies of GR and ER [37–39]. The mammalian SWI/SNF complex is often referred to by the *Drosophila* homologue *Brahma*-related gene (Brg-1) or *Brahma* (Brm) and can contain either the Brg-1 or the Brm ATPase as its core [19, 40]. Despite the high homology between these proteins, differences in function have been described. The most notable distinction is the survival and relative normal phenotype of the Brm knockout animals as contrasted to the early lethality of the Brg-1 knockout [41, 42]. Furthermore, studies have demonstrated that each preferentially interacts with distinct classes of transcription factors [43]. In the context of nuclear receptor regulation differences between the ATPases have also been identified. For example, GR can utilize both Brm and Brg1 containing complexes for activation while AR appears to be selective for Brm containing complexes [37, 44, 45].

These ATPases act in complexes that contain a number of Brg-1 associating factors (BAFs) that are heterogeneous and add further complexity. The SWI/SNF complex is divided into two subgroups termed BAF and PBAF based upon complex makeup. Both complexes are made up of 10–12 factors (Figure 14.1) which include the core BAFs: BAF170, BAF155, BAF57, BAF53 and INI1 (also BAF47) [46]. Additionally, BAF contains either Brg1 or Brm and BAF250a/b while the PBAF complex only utilizes Brg1 and the two subunits, BAF200 and polybromo (also BAF180) [47–49].

The polybromo and BAF250 proteins are mutually exclusive in these complexes. These differences, along with the other BAFs, are believed to add specificity to the activity of SWI/SNF as well as enhance the intrinsic enzymatic function of the core ATPase. For example, in vitro studies have shown that the ability of Brg-1 to remodel nucleosomes is greatly enhanced by the addition of BAF170, BAF155, and INI1 [50]. The importance of these subunits is highlighted by targeted deletion of several subunits in mice. Deletions of polybromo, BAF250a, and INI1 all result in embryonic lethality [51–53]. Interestingly, lethality occurs during different stages of development further strengthening the belief that these subunits define a level of specificity. It is not surprising that a great deal of current research is concentrating on SWI/SNF function in development and nuclear receptor mediated effects may play a role in these studies.

The role of SWI/SNF in regulating nuclear receptor function has been studied using a variety of nuclear receptor systems. In fact, the steroid receptors GR, PR, ER, and AR as well as other receptors PPARγ, TR, VDR, and RAR have all been shown to interact with members of the SWI/SNF complex and regulate transcription [44, 54–60]. In most systems, full transcriptional activation of nuclear receptors depends upon the successful recruitment of SWI/SNF to target gene promoters in chromatin. As previously stated, much of the specificity of SWI/SNF depends upon the BAF subunits and consequently different BAFs have been associated with differential function. BAF60, BAF57 and BAF250 seem to be the clear links for nuclear receptor regulation and have been shown to directly interact with various receptors and be required for transcriptional activation [47, 54, 56, 60, 61]. Recently, it has been shown that the HSA domain located within Brg1 is critical for mediating interaction with BAF250 and consequently regulating GR activity [62]. This domain has also been shown to be very important in recruitment of actin-related proteins that are contained in ATP-ase remodeling complexes [63].

Receptor mediated SWI/SNF complex recruitment at a target promoter can elicit various effects [64]. The ability to reorganize nucleosomes allows for both activation and repression and thus, SWI/SNF can act as both a co-activator and co-repressor. While the role of SWI/SNF as a co-activator is much more characterized, the recruitment of the complex in nuclear receptor mediated repression has also been described [65]. Interestingly, on the same ER target, SWI/SNF may act as either a co-activator or –repressor depending upon which molecules recruit it to the promoter and the presence or absence of BAF170 within the complex [65]. This may be explained by the fact that many of these BAFs have also been shown to interact with alternate co-activators. For example, BAF57 interacts with SRC1 and thus connects the ATP dependent chromatin remodeling activity to the covalent modification of histones [54]. Recent work examining the global effect of SWI/SNF activation on GR regulation demonstrated that the complex is much more critical to regulating activation rather than repression [66]. Taken together, the complexity of SWI/SNF brought about by the various BAF subunits allows for the specificity needed for a broad acting transcriptional co-regulator.

Regardless of the transcriptional effect of SWI/SNF promoter occupancy, it has been demonstrated that recruitment results in chromatin remodeling and nucleosome repositioning. A great deal of effort has been spent in understanding the mechanistic dynamics of SWI/SNF mediated co-regulation of transcription. To this end, the mouse mammary tumor virus (MMTV) promoter has proven to be an excellent model system [67]. MMTV contains several HREs that are tightly controlled by various steroid receptors. Upon integration within the mammalian genome, the MMTV promoter arranges into a highly organized and well characterized array over six positioned nucleosomes [67]. Using this system, it has been shown that GR transactivation requires the specific recruitment of SWI/SNF which can not be rescued by other chromatin remodeling ATPases such as Mi-2 or ISWI [68]. In fact, chimera proteins utilizing the ATPase region of SNF2h in place of the Brg-1 region was also unable to rescue nuclear receptor activity indicating that some specificity is intrinsic to the ATPase domain of the given complex [69]. Furthermore, enzyme accessibility assays have specifically demonstrated nucleosome repositioning within the promoter upon hormone stimulation [44]. While SWI/SNF recruitment to the promoter is dependent upon GR, it has been shown that inhibition of ATPase and thus chromatin remodeling activity results in faster turnover of the receptor at the response element [70, 71]. Furthermore, mutations within ATPase domain of Brg-1 result in ablation of transcription [68]. Thus a dynamic balance between the nuclear receptor and chromatin remodeling is needed and consequently controlled by the chromatin architecture.

14.2.2. ISWI as a Nuclear Receptor Co-regulator

The ISWI family of chromatin modifiers was originally identified in *drosophila* and named after its homology to SWI2/SNF2 (Imitation of SWI) [72]. In mammals two orthologs of the ATPase have been identified and termed SNF2H and SNF2L [73]. Despite the homology between the two proteins, each has evolved to carry out distinct functions and is expressed differentially during development [73]. In progenitor lines, SNF2H seem to be the dominant ATPase in ISWI complexes, while SNF2L is more highly expressed in terminally differentiated cells [73, 74]. In general, these complexes usually possess much fewer subunits (between 2 and 4) than SWI/SNF. SNF2H is known to act as the core ATPase for complexes such as remodeling and spacing factor (RSF), Nucleosome remodeling complex (NoRC), and Williams syndrome transcription factor-ISWI chromatin remodeling (WICH) (Figure 14.1) [72, 73, 75, 76]. These complexes are involved in nucleosome assembly and chromosome replication. However, SNF2H has been implicated in TR regulation. SNF2H expression was required to repress an integrated reporter as well as an endogenous target of TR [77]. Interestingly, transient reporters which fail to develop an organized higher order chromatin structure were unaffected by SNF2H depletion. The SNF2L protein assembles into the nucleosome remodeling factor (NuRF) and cat eye critical region-2 containing factor (CERF) complexes believed to regulate transcription

(Figure 14.1) [78]. These complexes have also been linked to nuclear receptor function. In cell free systems, ISWI complexes enhance the transcription of RAR/RXR heterodimers and PR [74, 79, 80]. The interaction of ISWI and PR also involves the recruitment of the NuRF38 co-factor [80]. It has been further demonstrated in vivo, that NuRF interacts with and regulates the ecdysone receptor in *drosophila* [81]. In mammals, SNF2L was shown to interact with PR in mouse ovaries but the consequences of this interaction on target gene transcription have yet to be identified [74].

14.2.3. NuRD as a Nuclear Receptor Co-repressor

Another class of ATPase dependent chromatin remodeling complexes involves the nucleosome remodeling and deacetylation complex (NuRD). These complexes also utilize one of two possible core subunits, Mi-2α (CHD3) or Mi-2β (CHD4). The NuRD complex also contains methyl DNA binding proteins (MBD2 or MBD3), metastasis associated proteins (MTA1, MTA2, or MTA3), and retinoblastoma binding proteins (RbapP46 and Rbap48) [82–85]. The NuRD complex also associates with either HDAC1 or HDAC2 to link ATPase chromatin remodeling with histone deacytelation [82–85] (Figure 14.1). As nuclear receptor regulation has been shown to depend greatly upon both histone acetylation status and chromatin remodeling, it is not surprising that NuRD has been identified as a nuclear receptor co-repressor. Much of the work involving NuRD and nuclear receptors has investigated ER due to the link of MTA proteins in breast cancer progression [86]. MTA proteins as well as Rbap46 and Rbap48 have been shown to interact with ERα [87–90]. Additionally, the Mi-2β ATPase interacts with RORγ and is able to repress its transactivation domain [91]. The recruitment of NuRD to ER target promoters in the presence of antagonist also results in transcriptional repression [88].

The NuRD complex has also been linked to DNA methylation regulation. Recently, it has been shown in acute promyelocytic leukemia that NuRD is critical for the oncongenic fusion protein PML-RARα to mediate epigenetic DNA methylation [92]. It was demonstrated that PML-RARα recruited NuRD to target gene promoters and enhanced DNA methylation resulting in epigenetic silencing. Furthermore, treatment with retinoic acid, a common therapy for this disease, was able to reduce NuRD occupancy at these promoters [92]. The ability of NuRD to possibly direct DNA methylation has broad implications towards long term repression of gene transcription. The exact role NuRD plays in DNA methylation is unknown, but it along with numerous reports of methylation linked to nuclear receptor activity are being explored [93, 94].

14.2.4. INO80

The final family of chromatin remodeling complexes are represented by two distinct subgroups, INO80 and Swr1, which each contain a respective ATPase protein by the same name [95, 96]. These are larger complexes containing 14 (Swr1) or 15 (INO80) subunits [95–98] (Figure 14.1). The majority of work involving these complexes has been performed in yeast and nothing has been shown to link either complex with nuclear receptor regulation. While the majority of work implicates these complexes in DNA damage repair, null mutants in yeast suggest an important role in transcriptional regulation [99]. It is difficult to determine at this point whether INO80 complexes generally act as repressors or activators of transcription.

14.3. CURRENT APPROACHES TO STUDYING NUCLEAR RECEPTOR COUPLED CHROMATIN REMODELING

The challenge that lies ahead is to understand the mechanistic detail in both a spatial and temporal manner of how these complexes are recruited to sites of regulation, perform their enzymatic function, and the consequences of deregulation on the global epigenetic program. These questions have already begun to be addressed with a number of classical experimental approaches and a growing number of novel biochemical and genomic techniques.

14.3.1. Biochemical Approaches

The use of classical biochemical assays in nuclear receptor research has taught us a great deal about transcription and in particular, chromatin regulation. A number of approaches have been utilized to examine transcription in a controlled manner in the context of chromatin [100]. In vitro transcription assays involve the reconstitution of DNA templates into chromatin and the measurement of RNA transcriptions following the addition of crude cell extracts or purified proteins. These assays allow for the strict control of factors either through the complementation or factor depletion. These experiments have been utilized to identify the importance of each group of molecules and the sequential order of the necessary steps for transcription. In vitro model systems have demonstrated that RAR/ROR heterodimers are able to interact with the promoter but require the addition of ATPase remodeling activity to achieve strong interactions [79]. It was further shown that the preceding remodeling activity is required for HATs to exert full transcriptional activation. A similar experimental approach to in vitro chromatin assembly is the initiation of transcription reactions in Xenopus oocytes allowing for systems to be studied in an in vivo model. Upon injection of a single stranded DNA template into a Xenopus oocyte, the DNA is replicated and assembled into chromatin. The expression of specific transcriptional co-factors can be induced by the co-injection of mRNAs for the proteins of interest. Using this technique, it has been demonstrated that MMTV activation by GR results in a highly order positioning of the nucleosomes brought about by chromatin remodeling and is completely reversible [101]. The advantages of these systems lie in the ability to tightly control the expression of the co-factors being studied during the experiments which allows for close dissection of transcriptional events.

The ability to gauge the interaction between nuclear receptors and its target promoter has been managed in the past by the use of electromobility gel shift assay (EMSA). In fact, there is no substitute for measuring direct interactions of protein with a specific DNA sequence. In recent years, however, the chromatin immunoprecipitation assay (ChIP) has been more widely used to link proteins to specific promoter regions [102, 103]. This assay has the primary benefit of analyzing these interactions in a native chromatin conformation. ChIP assays involve cross-linking the protein and DNA complexes followed by dissection of the DNA into small fragments either through physical shearing or enzymatic cleavage [103]. The resulting small pieces of DNA/protein complex are then purified by immunoprecipitation using antibodies specific for the proteins of interest. Purified DNA fragments can then analyzed by PCR. Depending upon the DNA fragment sizes obtained, rather specific binding patterns can be obtained via this assay. In addition, antibodies specific for histone modifications can be utilized to study covalent modifications at sites of activation and/or repression. This technique has been critical in our understanding of covalent histone modifications at gene promoters [31, 104]. It is also important to realize that entire protein complexes are crosslinked and thus direct interaction with the DNA is not required to identify factor recruitment. While ChIP assays have become a very important tool in the field, they are limited by the availability of quality antibody reagents and the availability of open epitopes in cross-linked native conformation complexes. Due to this latter issue, appropriate controls are critical and negative results make it difficult to prove a lack of recruitment or occupancy. In all, the use of ChIP assays has been critical to demonstrate that nuclear receptor recruitment and subsequent co-regulator recruitment occurs rapidly following hormone activation.

When investigating chromatin remodeling, many assays rely on the principle that DNA tightly associated with a core histone is inaccessible to other protein interactions, most notably endonucleases [105]. DNase footprinting involves the treatment of samples with DNase which will cleave DNA that is not protected by protein association. For nucleosome mapping, MNase assays can be employed which utilize a similar concept. Micrococcal nuclease (MN) is able do induce double strand breaks in the linker regions between nucleosomes and treatment of genomic DNA with MN results in a 200 bp ladder indicative of varying multiples of intact nucleosomes. Using probes against DNA regions of interest, one can determine nucleosome positioning. Restriction enzymes can be also be utilized in this manner to determine if a specific site of DNA is contained within a nucleosome. Cleavage of the DNA can be determined by either southern blot or PCR. These assays have been extremely useful in understanding the MMTV array and nuclear receptor activation [67]. For instance, a restriction site for the Sst1 endonuclease is protected from digestion in the inactive conformation, but upon ligand stimulation and SWI/SNF mediated chromatin remodeling the site becomes accessible to the enzyme [106]. These experimental approaches are extremely powerful tools for investigating chromatin remodeling, but are limited by necessity of strong DNA/protein interactions. In addition, the technical aspects of the assays have many caveats. For example, MNase does have some sequence specificity for digestion and for many of these enzymes the duration of cleavage can initiate additional digestion of the DNA that is beyond the designed activity [107].

Most of the aforementioned techniques are excellent indicators of a transcriptional process taken at a single point in time. While time courses have been utilized in ChIP assays to demonstrate the duration of events, each individual assay is still a snapshot of the transcriptional event and does not measure dynamic processes. The use of fluorescent protein conjugates has allowed researchers to investigate these processes in living cells as a dynamic process [100]. Further advances in GFP derivatives that have distinct and specific absorption and emission spectra allow for new experimental techniques to address extremely difficult questions. Fluorescent Recovery After Photobleaching (FRAP) is a technique wherein proteins are bleached with a laser in a specific region of the cell, and the recovery of fluorescence by the migration of molecules from other cellular regions is monitored. Using this method, both nuclear receptors and their co-regulators have been shown to be highly dynamic on promoters with occupancies measured in the seconds [108–110]. This result was in stark contrast to ChIP data revealing nuclear receptor occupancy of promoters hours after ligand activation and residency times measured between 20 and 40 min [111, 112]. The real time imaging reveals hit-and-run model of nuclear receptor transcriptional regulation in which recruitment is followed by the necessary steps for transcription and subsequent disassociation. This event is then repeated over the course of several hours. Further analysis has been performed with GFP fused GR using an MMTV array integrated into cells. Due to the large number of repeats and thus HREs within the array, recruitment of GR upon ligand activation results in distinct fluorescent foci which can be used to study promoter dynamics. This system has been used to show that SWI/SNF is recruited to the MMTV following ligand activation and consequently results in an ATP dependent chromatin decondensation of the promoter [113]. The system has several built inherent caveats that deal with physiological relevance. The array utilizes a large number of response elements and the nuclear receptors are usually over expressed. Furthermore, the area of study using a microscope is not a compact and defined region of chromatin, so the data generated represents the average of regulated and unregulated promoter regions. Nonetheless, the growing sophistication of this technology is helping to address ever more intricate questions.

14.3.2. Genomic Approaches

Traditional biochemical experiments with individual promoters allow us to model transcriptional events that occur globally. However, newer technologies have developed that permit studies of transcriptional events on a genome-wide level [114]. The development of microarrays has revolutionized high-throughput study of the genome [115]. The technology involves hybridizing fluorescently labeled sample to an array consisting of thousands of oligonucleotide probes. The initial microarray experiments analyzed changes in transcription by globally monitoring levels of mRNA. These experiments were particularly useful indentifying the effects of ligands on nuclear receptor action and then later by either overexpression or knockdown of co-factors in cell culture systems. This technology was eventually combined with ChIP assays to create a very powerful approach to study transcription, referred to as ChIP-chip (or ChIP on Chip) [116]. The purification of DNA/protein complexes can then be mapped with tiled arrays of genomic sequence. This technique has been utilized to chart promoter binding elements, histone modifications, and nucleosome positioning [104, 117–121]. ChIP-chip analysis of nuclear receptor recruitment has yielded a vast array of data. Particularly interesting is the occupancy of receptors such as ER to regions of the genome that are far from the traditional or canonical regulated gene promoters [122]. The use of these elements to regulate transcription may involve higher order chromatin structures and chromosomal arrangements [8]. Enhanced mapping of the entire genome has since revealed that only 4% of ER binding is mapped to the near vicinity of transcriptional start sites [123]. The ability of the nuclear receptors to initiate intra- or inter-chromosomal loops to regulate transcription has broad implications towards the necessary chromatin remodeling machinery needed to initiate such an event. It has recently been utilized to investigate the role of SWI/SNF on GR promoter binding [66]. These studies demonstrated the identification of Brg-1 dependent and independent sites of GR regulation and highlight the importance of genomic studies as compared to a single promoter model system. However, genomic approaches may not be well suited to assess specific mechanistic questions at which the non-genomic systems excel, and it is clear that both systems will have to be utilized in combination. Despite the power of this genomic system, questions have also arisen due to the heterogeneity of the results obtained so far. Indeed, while several studies have investigated the recruitment of ERα in the breast cancer cell line MCF7, two seemingly identical studies had less then 20% overlap of ER binding sites between them [123, 124].

Recent advances in sequencing have changed the paradigm of studying genome wide events [125, 126]. These advances allow for rapid sequencing of multiple reactions simultaneously to achieve inexpensive expanses of sequence. The ability to generate this vast amount of sequence data quickly and inexpensively has made it possible to perform genome wide experiments at the nucleotide level [114]. Once again, ChIP assays have now been combined with this new technology to generate a new approach called ChIP-seq [127]. This technology addresses the same questions as ChIP-chip but has added advantages of requiring less replicates, lower cost, and eliminates biases associated with probe design [125].

14.4. SUMMARY

An increased understanding of transcriptional regulation will make it possible to address critical biological questions and the resulting progress of the field will undoubtedly lead to advances for human medicine. The growing interest in the field of epigenetics has broad implications towards key pathological states such as cancer [128]. While the histone code has begun to be deciphered and the mechanisms of chromatin remodeling uncovered, the vital question of how this vast array of co-regulators fit together and control such a highly complex yet conserved transcriptional program will be key. Even in the specific context of the remodeling complexes, the interplay and specificity are largely undefined at the nucleosome level. While certain complexes are clearly required for nuclear receptor function, each is able to reorder nucleosomes that have been repeatedly shown to be specific for each complex, adding further complexity is the tissue and cell type specificity of nuclear action [59]. It has yet to be assessed how some of the current dogmas on transcriptional mechanisms will be conserved as model systems grow. It appears that most of the players have been cast and the work is just beginning on how they will interact [129].

ACKNOWLEDGMENTS

This research was supported by the Intramural Research Program of the National Institute of Environmental Health Sciences, NIH; project number Z01 ES071006-09.

REFERENCES

1. Bain, D. L., Heneghan, A. F., Connaghan-Jones, K. D., and Miura, M. T. (2007). Nuclear receptor structure: Implications for function. *Annu Rev Physiol* 69, 201–220.
2. Giguere, V. (1999). Orphan nuclear receptors: From gene to function. *Endocr Rev* 20, 689–725.
3. McKenna, N. J., Lanz, R. B., and O'Malley, B. W. (1999). Nuclear receptor coregulators: Cellular and molecular biology. *Endocr Rev* 20, 321–344.
4. Uhlmann, F. (2001). Chromosome condensation: Packaging the genome. *Curr Biol* 11, R384–R387.
5. Bernstein, E. and Hake, S. B. (2006). The nucleosome: A little variation goes a long way. *Biochem Cell Biol* 84, 505–517.
6. Luger, K., Mader, A. W., Richmond, R. K., Sargent, D. F., and Richmond, T. J. (1997). Crystal structure of the nucleosome core particle at 2.8 A resolution. *Nature* 389, 251–260.
7. Aoyagi, S., Trotter, K. W., and Archer, T. K. (2005). ATP-dependent chromatin remodeling complexes and their role in nuclear receptor-dependent transcription in vivo. *Vitam Horm* 70, 281–307.
8. Hayes, J. J. and Hansen, J. C. (2001). Nucleosomes and the chromatin fiber. *Curr Opin Genet Dev* 11, 124–129.
9. Felsenfeld, G. and Groudine, M. (2003). Controlling the double helix. *Nature* 421, 448–453.
10. Archer, T. K., Lefebvre, P., Wolford, R. G., and Hager, G. L. (1992). Transcription factor loading on the MMTV promoter: A bimodal mechanism for promoter activation. *Science (New York, N.Y)* 255, 1573–1576.
11. Workman, J. L. and Kingston, R. E. (1998). Alteration of nucleosome structure as a mechanism of transcriptional regulation. *Annu Rev Biochem* 67, 545–579.
12. Miranda, T. B. and Jones, P. A. (2007). DNA methylation: The nuts and bolts of repression. *J Cell Physiol* 213, 384–390.
13. Kouzarides, T. (2007). Chromatin modifications and their function. *Cell* 128, 693–705.
14. Grunstein, M. (1997). Histone acetylation in chromatin structure and transcription. *Nature* 389, 349–352.

15. Shogren-Knaak, M., Ishii, H., Sun, J. M., Pazin, M. J., Davie, J. R., and Peterson, C. L. (2006). Histone H4-K16 acetylation controls chromatin structure and protein interactions. *Science (New York, N.Y)* 311, 844–847.
16. Wang, G. G., Allis, C. D., and Chi, P. (2007). Chromatin remodeling and cancer, Part I: Covalent histone modifications. *Trends Mol Med* 13, 363–372.
17. Kingston, R. E. and Narlikar, G. J. (1999). ATP-dependent remodeling and acetylation as regulators of chromatin fluidity. *Genes Dev* 13, 2339–2352.
18. Martens, J. A. and Winston, F. (2003). Recent advances in understanding chromatin remodeling by Swi/Snf complexes. *Curr Opin Genet Dev* 13, 136–142.
19. Trotter, K. W. and Archer, T. K. (2008). The BRG1 transcriptional coregulator. *Nucl Recept Signal* 6, e004.
20. Lonard, D. M. and O'Malley, B. W. (2007). Nuclear receptor coregulators: Judges, juries, and executioners of cellular regulation. *Mol Cell* 27, 691–700.
21. Collingwood, T. N., Urnov, F. D., and Wolffe, A. P. (1999). Nuclear receptors: Coactivators, corepressors and chromatin remodeling in the control of transcription. *J Mol Endocrinol* 23, 255–275.
22. Spencer, T. E., Jenster, G., Burcin, M. M., Allis, C. D., Zhou, J., Mizzen, C. A., McKenna, N. J., Onate, S. A., Tsai, S. Y., Tsai, M. J., and O'Malley, B. W. (1997). Steroid receptor coactivator-1 is a histone acetyltransferase. *Nature* 389, 194–198.
23. Voegel, J. J., Heine, M. J., Zechel, C., Chambon, P., and Gronemeyer, H. (1996). TIF2, a 160 kDa transcriptional mediator for the ligand-dependent activation function AF-2 of nuclear receptors. *EMBO J* 15, 3667–3675.
24. Anzick, S. L., Kononen, J., Walker, R. L., Azorsa, D. O., Tanner, M. M., Guan, X. Y., Sauter, G., Kallioniemi, O. P., Trent, J. M., and Meltzer, P. S. (1997). AIB1, a steroid receptor coactivator amplified in breast and ovarian cancer. *Science (New York, N.Y)* 277, 965–968.
25. Onate, S. A., Tsai, S. Y., Tsai, M. J., and O'Malley, B. W. (1995). Sequence and characterization of a coactivator for the steroid hormone receptor superfamily. *Science (New York, N.Y)* 270, 1354–1357.
26. Hong, H., Kohli, K., Trivedi, A., Johnson, D. L., and Stallcup, M. R. (1996). GRIP1, a novel mouse protein that serves as a transcriptional coactivator in yeast for the hormone binding domains of steroid receptors. *Proc Natl Acad Sci USA* 93, 4948–4952.
27. Hanstein, B., Eckner, R., DiRenzo, J., Halachmi, S., Liu, H., Searcy, B., Kurokawa, R., and Brown, M. (1996). p300 is a component of an estrogen receptor coactivator complex. *Proc Natl Acad Sci U S A* 93, 11540–11545.
28. Chen, J. D. and Evans, R. M. (1995). A transcriptional co-repressor that interacts with nuclear hormone receptors. *Nature* 377, 454–457.
29. Horlein, A. J., Naar, A. M., Heinzel, T., Torchia, J., Gloss, B., Kurokawa, R., Ryan, A., Kamei, Y., Soderstrom, M., Glass, C. K. et al. (1995). Ligand-independent repression by the thyroid hormone receptor mediated by a nuclear receptor co-repressor. *Nature* 377, 397–404.
30. Jackson, T. A., Richer, J. K., Bain, D. L., Takimoto, G. S., Tung, L., and Horwitz, K. B. (1997). The partial agonist activity of antagonist-occupied steroid receptors is controlled by a novel hinge domain-binding coactivator L7/SPA and the corepressors N-CoR or SMRT. *Mol Endocrinol (Baltimore, MD)* 11, 693–705.
31. Shang, Y. and Brown, M. (2002). Molecular determinants for the tissue specificity of SERMs. *Science (New York, N.Y)* 295, 2465–2468.
32. Dotzlaw, H., Moehren, U., Mink, S., Cato, A. C., Iniguez Lluhi, J. A., and Baniahmad, A. (2002). The amino terminus of the human AR is target for corepressor action and antihormone agonism. *Mol Endocrinol (Baltimore, Md)* 16, 661–673.
33. Chen, J., Kinyamu, H. K., and Archer, T. K. (2006). Changes in attitude, changes in latitude: Nuclear receptors remodeling chromatin to regulate transcription. *Mol Endocrinol (Baltimore, Md)* 20, 1–13.
34. Neigeborn, L. and Carlson, M. (1984). Genes affecting the regulation of SUC2 gene expression by glucose repression in *Saccharomyces cerevisiae*. *Genetics* 108, 845–858.

35. Winston, F. and Carlson, M. (1992). Yeast SNF/SWI transcriptional activators and the SPT/SIN chromatin connection. *Trends Genet* 8, 387–391.
36. Stern, M., Jensen, R., and Herskowitz, I. (1984). Five SWI genes are required for expression of the HO gene in yeast. *J Mol Biol* 178, 853–868.
37. Muchardt, C. and Yaniv, M. (1993). A human homologue of Saccharomyces cerevisiae SNF2/SWI2 and *Drosophila* brm genes potentiates transcriptional activation by the glucocorticoid receptor. *EMBO J* 12, 4279–4290.
38. Yoshinaga, S. K., Peterson, C. L., Herskowitz, I., and Yamamoto, K. R. (1992). Roles of SWI1, SWI2, and SWI3 proteins for transcriptional enhancement by steroid receptors. *Science (New York, NY)* 258, 1598–1604.
39. Ichinose, H., Garnier, J. M., Chambon, P., and Losson, R. (1997). Ligand-dependent interaction between the estrogen receptor and the human homologues of SWI2/SNF2. *Gene* 188, 95–100.
40. Tamkun, J. W., Deuring, R., Scott, M. P., Kissinger, M., Pattatucci, A. M., Kaufman, T. C., and Kennison, J. A. (1992). Brahma: A regulator of *Drosophila* homeotic genes structurally related to the yeast transcriptional activator SNF2/SWI2. *Cell* 68, 561–572.
41. Reyes, J. C., Barra, J., Muchardt, C., Camus, A., Babinet, C., and Yaniv, M. (1998). Altered control of cellular proliferation in the absence of mammalian brahma (SNF2alpha). *EMBO J* 17, 6979–6991.
42. Bultman, S., Gebuhr, T., Yee, D., La Mantia, C., Nicholson, J., Gilliam, A., Randazzo, F., Metzger, D., Chambon, P., Crabtree, G., and Magnuson, T. (2000). A Brg1 null mutation in the mouse reveals functional differences among mammalian SWI/SNF complexes. *Mol Cell* 6, 1287–1295.
43. Kadam, S. and Emerson, B. M. (2003). Transcriptional specificity of human SWI/SNF BRG1 and BRM chromatin remodeling complexes. *Mol Cell* 11, 377–389.
44. Fryer, C. J. and Archer, T. K. (1998). Chromatin remodelling by the glucocorticoid receptor requires the BRG1 complex. *Nature* 393, 88–91.
45. Marshall, T. W., Link, K. A., Petre-Draviam, C. E., and Knudsen, K. E. (2003). Differential requirement of SWI/SNF for androgen receptor activity. *J Biol Chem* 278, 30605–30613.
46. Wang, W., Cote, J., Xue, Y., Zhou, S., Khavari, P. A., Biggar, S. R., Muchardt, C., Kalpana, G. V., Goff, S. P., Yaniv, M., Workman, J. L., and Crabtree, G. R. (1996). Purification and biochemical heterogeneity of the mammalian SWI-SNF complex. *EMBO J* 15, 5370–5382.
47. Nie, Z., Xue, Y., Yang, D., Zhou, S., Deroo, B. J., Archer, T. K., and Wang, W. (2000). A specificity and targeting subunit of a human SWI/SNF family-related chromatin-remodeling complex. *Mol Cell Biol* 20, 8879–8888.
48. Lemon, B., Inouye, C., King, D. S., and Tjian, R. (2001). Selectivity of chromatin-remodelling cofactors for ligand-activated transcription. *Nature* 414, 924–928.
49. Yan, Z., Cui, K., Murray, D. M., Ling, C., Xue, Y., Gerstein, A., Parsons, R., Zhao, K., and Wang, W. (2005). PBAF chromatin-remodeling complex requires a novel specificity subunit, BAF200, to regulate expression of selective interferon-responsive genes. *Genes Dev* 19, 1662–1667.
50. Phelan, M. L., Sif, S., Narlikar, G. J., and Kingston, R. E. (1999). Reconstitution of a core chromatin remodeling complex from SWI/SNF subunits. *Mol Cell* 3, 247–253.
51. Guidi, C. J., Sands, A. T., Zambrowicz, B. P., Turner, T. K., Demers, D. A., Webster, W., Smith, T. W., Imbalzano, A. N., and Jones, S. N. (2001). Disruption of Ini1 leads to peri-implantation lethality and tumorigenesis in mice. *Mol Cell Biol* 21, 3598–3603.
52. Wang, Z., Zhai, W., Richardson, J. A., Olson, E. N., Meneses, J. J., Firpo, M. T., Kang, C., Skarnes, W. C., and Tjian, R. (2004). Polybromo protein BAF180 functions in mammalian cardiac chamber maturation. *Genes Dev* 18, 3106–3116.
53. Gao, X., Tate, P., Hu, P., Tjian, R., Skarnes, W. C., and Wang, Z. (2008). ES cell pluripotency and germ-layer formation require the SWI/SNF chromatin remodeling component BAF250a. *Proc Natl Acad Sci U S A* 105, 6656–6661.
54. Belandia, B., Orford, R. L., Hurst, H. C., and Parker, M. G. (2002). Targeting of SWI/SNF chromatin remodelling complexes to estrogen-responsive genes. *EMBO J* 21, 4094–4103.

55. Inoue, H., Furukawa, T., Giannakopoulos, S., Zhou, S., King, D. S., and Tanese, N. (2002). Largest subunits of the human SWI/SNF chromatin-remodeling complex promote transcriptional activation by steroifd hormone receptors. *J Biol Chem* 277, 41674–41685.
56. Link, K. A., Burd, C. J., Williams, E., Marshall, T., Rosson, G., Henry, E., Weissman, B., and Knudsen, K. E. (2005). BAF57 governs androgen receptor action and androgen-dependent proliferation through SWI/SNF. *Mol Cell Biol* 25, 2200–2215.
57. Debril, M. B., Gelman, L., Fayard, E., Annicotte, J. S., Rocchi, S., and Auwerx, J. (2004). Transcription factors and nuclear receptors interact with the SWI/SNF complex through the BAF60c subunit. *J Biol Chem* 279, 16677–16686.
58. Koszewski, N. J., Henry, K. W., Lubert, E. J., Gravatte, H., and Noonan, D. J. (2003). Use of a modified yeast one-hybrid screen to identify BAF60a interactions with the Vitamin D receptor heterodimer. *J Steroid Biochem Mol Biol* 87, 223–231.
59. Trotter, K. W. and Archer, T. K. (2007). Nuclear receptors and chromatin remodeling machinery. *Mol Cell Endocrinol* 265–266, 162–167.
60. Flajollet, S., Lefebvre, B., Cudejko, C., Staels, B., and Lefebvre, P. (2007). The core component of the mammalian SWI/SNF complex SMARCD3/BAF60c is a coactivator for the nuclear retinoic acid receptor. *Mol Cell Endocrinol* 270, 23–32.
61. Hsiao, P. W., Fryer, C. J., Trotter, K. W., Wang, W., and Archer, T. K. (2003). BAF60a mediates critical interactions between nuclear receptors and the BRG1 chromatin-remodeling complex for transactivation. *Mol Cell Biol* 23, 6210–6220.
62. Trotter, K. W., Fan, H. Y., Ivey, M. L., Kingston, R. E., and Archer, T. K. (2008). The HSA domain of BRG1 mediates critical interactions required for glucocorticoid receptor-dependent transcriptional activation in vivo. *Mol Cell Biol* 28, 1413–1426.
63. Szerlong, H., Hinata, K., Viswanathan, R., Erdjument-Bromage, H., Tempst, P., and Cairns, B. R. (2008). The HSA domain binds nuclear actin-related proteins to regulate chromatin-remodeling ATPases. *Nat Struct Mol Biol* 15, 469–476.
64. Hebbar, P. B. and Archer, T. K. (2003). Chromatin remodeling by nuclear receptors. *Chromosoma* 111, 495–504.
65. Zhang, B., Chambers, K. J., Faller, D. V., and Wang, S. (2007). Reprogramming of the SWI/SNF complex for co-activation or co-repression in prohibitin-mediated estrogen receptor regulation. *Oncogene* 26, 7153–7157.
66. John, S., Sabo, P. J., Johnson, T. A., Sung, M. H., Biddie, S. C., Lightman, S. L., Voss, T. C., Davis, S. R., Meltzer, P. S., Stamatoyannopoulos, J. A., and Hager, G. L. (2008). Interaction of the glucocorticoid receptor with the chromatin landscape. *Mol Cell* 29, 611–624.
67. Archer, T. K., Cordingley, M. G., Wolford, R. G., and Hager, G. L. (1991). Transcription factor access is mediated by accurately positioned nucleosomes on the mouse mammary tumor virus promoter. *Mol Cell Biol* 11, 688–698.
68. Trotter, K. W. and Archer, T. K. (2004). Reconstitution of glucocorticoid receptor-dependent transcription in vivo. *Mol Cell Biol* 24, 3347–3358.
69. Fan, H. Y., Trotter, K. W., Archer, T. K., and Kingston, R. E. (2005). Swapping function of two chromatin remodeling complexes. *Mol Cell* 17, 805–815.
70. Nagaich, A. K., Walker, D. A., Wolford, R., and Hager, G. L. (2004). Rapid periodic binding and displacement of the glucocorticoid receptor during chromatin remodeling. *Mol Cell* 14, 163–174.
71. Voss, T. C., John, S., and Hager, G. L. (2006). Single-cell analysis of glucocorticoid receptor action reveals that stochastic post-chromatin association mechanisms regulate ligand-specific transcription. *Mol Endocrinol (Baltimore, MD)* 20, 2641–2655.
72. Tsukiyama, T., Daniel, C., Tamkun, J., and Wu, C. (1995). ISWI, a member of the SWI2/SNF2 ATPase family, encodes the 140 kDa subunit of the nucleosome remodeling factor. *Cell* 83, 1021–1026.
73. Lazzaro, M. A. and Picketts, D. J. (2001). Cloning and characterization of the murine Imitation Switch (ISWI) genes: Differential expression patterns suggest distinct developmental roles for Snf2h and Snf2l. *J Neurochem* 77, 1145–1156.

74. Lazzaro, M. A., Pepin, D., Pescador, N., Murphy, B. D., Vanderhyden, B. C., and Picketts, D. J. (2006). The imitation switch protein SNF2L regulates steroidogenic acute regulatory protein expression during terminal differentiation of ovarian granulosa cells. *Mol Endocrinol (Baltimore, MD)* 20, 2406–2417.
75. Hakimi, M. A., Bochar, D. A., Schmiesing, J. A., Dong, Y., Barak, O. G., Speicher, D. W., Yokomori, K., and Shiekhattar, R. (2002). A chromatin remodelling complex that loads cohesin onto human chromosomes. *Nature* 418, 994–998.
76. LeRoy, G., Orphanides, G., Lane, W. S., and Reinberg, D. (1998). Requirement of RSF and FACT for transcription of chromatin templates in vitro. *Science (New York, N.Y)* 282, 1900–1904.
77. Alenghat, T., Yu, J., and Lazar, M. A. (2006). The N-CoR complex enables chromatin remodeler SNF2H to enhance repression by thyroid hormone receptor. *EMBO J* 25, 3966–3974.
78. Pepin, D., Vanderhyden, B. C., Picketts, D. J., and Murphy, B. D. (2007). ISWI chromatin remodeling in ovarian somatic and germ cells: Revenge of the NURFs. *Trends Endocrinol Metab* 18, 215–224.
79. Dilworth, F. J., Fromental-Ramain, C., Remboutsika, E., Benecke, A., and Chambon, P. (1999). Ligand-dependent activation of transcription in vitro by retinoic acid receptor alpha/retinoid X receptor alpha heterodimers that mimics transactivation by retinoids in vivo. *Proc Natl Acad Sci U S A* 96, 1995–2000.
80. Di Croce, L., Koop, R., Venditti, P., Westphal, H. M., Nightingale, K. P., Corona, D. F., Becker, P. B., and Beato, M. (1999). Two-step synergism between the progesterone receptor and the DNA-binding domain of nuclear factor 1 on MMTV minichromosomes. *Mol Cell* 4, 45–54.
81. Badenhorst, P., Xiao, H., Cherbas, L., Kwon, S. Y., Voas, M., Rebay, I., Cherbas, P., and Wu, C. (2005). The *Drosophila* nucleosome remodeling factor NURF is required for Ecdysteroid signaling and metamorphosis. *Genes Dev* 19, 2540–2545.
82. Tong, J. K., Hassig, C. A., Schnitzler, G. R., Kingston, R. E., and Schreiber, S. L. (1998). Chromatin deacetylation by an ATP-dependent nucleosome remodelling complex. *Nature* 395, 917–921.
83. Wade, P. A., Jones, P. L., Vermaak, D., and Wolffe, A. P. (1998). A multiple subunit Mi-2 histone deacetylase from Xenopus laevis cofractionates with an associated Snf2 superfamily ATPase. *Curr Biol* 8, 843–846.
84. Xue, Y., Wong, J., Moreno, G. T., Young, M. K., Cote, J., and Wang, W. (1998). NURD, a novel complex with both ATP-dependent chromatin-remodeling and histone deacetylase activities. *Mol Cell* 2, 851–861.
85. Zhang, Y., LeRoy, G., Seelig, H. P., Lane, W. S., and Reinberg, D. (1998). The dermatomyositis-specific autoantigen Mi2 is a component of a complex containing histone deacetylase and nucleosome remodeling activities. *Cell* 95, 279–289.
86. Kumar, R., Wang, R. A., and Bagheri-Yarmand, R. (2003). Emerging roles of MTA family members in human cancers. *Semin Oncol* 30, 30–37.
87. Mazumdar, A., Wang, R. A., Mishra, S. K., Adam, L., Bagheri-Yarmand, R., Mandal, M., Vadlamudi, R. K., and Kumar, R. (2001). Transcriptional repression of oestrogen receptor by metastasis-associated protein 1 corepressor. *Nat Cell Biol* 3, 30–37.
88. Fujita, N., Jaye, D. L., Kajita, M., Geigerman, C., Moreno, C. S., and Wade, P. A. (2003). MTA3, a Mi-2/NuRD complex subunit, regulates an invasive growth pathway in breast cancer. *Cell* 113, 207–219.
89. Dhasarathy, A., Kajita, M., and Wade, P. A. (2007). The transcription factor snail mediates epithelial to mesenchymal transitions by repression of estrogen receptor-alpha. *Mol Endocrinol (Baltimore, MD)* 21, 2907–2918.
90. Creekmore, A. L., Walt, K. A., Schultz-Norton, J. R., Ziegler, Y. S., McLeod, I. X., Yates, J. R., and Nardulli, A. M. (2008). The role of retinoblastoma-associated proteins 46 and 48 in estrogen receptor alpha mediated gene expression. *Mol Cell Endocrinol* 291, 79–86.
91. Johnson, D. R., Lovett, J. M., Hirsch, M., Xia, F., and Chen, J. D. (2004). NuRD complex component Mi-2beta binds to and represses RORgamma-mediated transcriptional activation. *Biochem Biophys Res Commun* 318, 714–718.

92. Morey, L., Brenner, C., Fazi, F., Villa, R., Gutierrez, A., Buschbeck, M., Nervi, C., Minucci, S., Fuks, F., and Di Croce, L. (2008). MBD3, a component of the NuRD complex, facilitates chromatin alteration and deposition of epigenetic marks. *Mol Cell Biol* 28, 5912–5923.
93. Darbre, P. D. and King, R. J. (1988). Role of receptor occupancy in the transition from responsive to unresponsive states in cultured breast tumor cells. *J Cell Biochem* 36, 83–89.
94. Cheng, A. S., Culhane, A. C., Chan, M. W., Venkataramu, C. R., Ehrich, M., Nasir, A., Rodriguez, B. A., Liu, J., Yan, P. S., Quackenbush, J., Nephew, K. P., Yeatman, T. J., and Huang, T. H. (2008). Epithelial progeny of estrogen-exposed breast progenitor cells display a cancer-like methylome. *Cancer Res* 68, 1786–1796.
95. Shen, X., Mizuguchi, G., Hamiche, A., and Wu, C. (2000). A chromatin remodelling complex involved in transcription and DNA processing. *Nature* 406, 541–544.
96. Mizuguchi, G., Shen, X., Landry, J., Wu, W. H., Sen, S., and Wu, C. (2004). ATP-driven exchange of histone H2AZ variant catalyzed by SWR1 chromatin remodeling complex. *Science (New York, NY)* 303, 343–348.
97. Shen, X., Ranallo, R., Choi, E., and Wu, C. (2003). Involvement of actin-related proteins in ATP-dependent chromatin remodeling. *Mol Cell* 12, 147–155.
98. Wu, W. H., Alami, S., Luk, E., Wu, C. H., Sen, S., Mizuguchi, G., Wei, D., and Wu, C. (2005). Swc2 is a widely conserved H2AZ-binding module essential for ATP-dependent histone exchange. *Nat Struct Mol Biol* 12, 1064–1071.
99. Ebbert, R., Birkmann, A., and Schuller, H. J. (1999). The product of the SNF2/SWI2 paralogue INO80 of Saccharomyces cerevisiae required for efficient expression of various yeast structural genes is part of a high-molecular-weight protein complex. *Mol Microbiol* 32, 741–751.
100. Lee, K. C. and Lee Kraus, W. (2001). Nuclear receptors, coactivators and chromatin: New approaches, new insights. *Trends Endocrinol Metab* 12, 191–197.
101. Belikov, S., Gelius, B., and Wrange, O. (2001). Hormone-induced nucleosome positioning in the MMTV promoter is reversible. *EMBO J* 20, 2802–2811.
102. Orlando, V., Strutt, H., and Paro, R. (1997). Analysis of chromatin structure by in vivo formaldehyde cross-linking. *Methods (San Diego, Calif)* 11, 205–214.
103. Solomon, M. J., Larsen, P. L., and Varshavsky, A. (1988). Mapping protein-DNA interactions in vivo with formaldehyde: Evidence that histone H4 is retained on a highly transcribed gene. *Cell* 53, 937–947.
104. Bernstein, B. E., Humphrey, E. L., Liu, C. L., and Schreiber, S. L. (2004). The use of chromatin immunoprecipitation assays in genome-wide analyses of histone modifications. *Methods Enzymol* 376, 349–360.
105. Fryer, C. J. and Archer, T. K. (2001). Analyzing the contributions of chromatin structure in nuclear hormone receptor activated transcription in vivo. *Methods Mol Biol* 176, 283–296.
106. Kinyamu, H. K., Fryer, C. J., Horwitz, K. B., and Archer, T. K. (2000). The mouse mammary tumor virus promoter adopts distinct chromatin structures in human breast cancer cells with and without glucocorticoid receptor. *J Biol Chem* 275, 20061–20068.
107. Flick, J. T., Eissenberg, J. C., and Elgin, S. C. (1986). Micrococcal nuclease as a DNA structural probe: Its recognition sequences, their genomic distribution and correlation with DNA structure determinants. *J Mol Biol* 190, 619–633.
108. Stenoien, D. L., Patel, K., Mancini, M. G., Dutertre, M., Smith, C. L., O'Malley, B. W., and Mancini, M. A. (2001). FRAP reveals that mobility of oestrogen receptor-alpha is ligand- and proteasome-dependent. *Nat Cell Biol* 3, 15–23.
109. Schaaf, M. J. and Cidlowski, J. A. (2003). Molecular determinants of glucocorticoid receptor mobility in living cells: The importance of ligand affinity. *Mol Cell Biol* 23, 1922–1934.
110. Rayasam, G. V., Elbi, C., Walker, D. A., Wolford, R., Fletcher, T. M., Edwards, D. P., and Hager, G. L. (2005). Ligand-specific dynamics of the progesterone receptor in living cells and during chromatin remodeling in vitro. *Mol Cell Biol* 25, 2406–2418.
111. Metivier, R., Penot, G., Hubner, M. R., Reid, G., Brand, H., Kos, M., and Gannon, F. (2003). Estrogen receptor-alpha directs ordered, cyclical, and combinatorial recruitment of cofactors on a natural target promoter. *Cell* 115, 751–763.

112. Reid, G., Hubner, M. R., Metivier, R., Brand, H., Denger, S., Manu, D., Beaudouin, J., Ellenberg, J., and Gannon, F. (2003). Cyclic, proteasome-mediated turnover of unliganded and liganded ERalpha on responsive promoters is an integral feature of estrogen signaling. *Mol Cell* 11, 695–707.
113. Johnson, T. A., Elbi, C., Parekh, B. S., Hager, G. L., and John, S. (2008). Chromatin remodeling complexes interact dynamically with a glucocorticoid receptor-regulated promoter. *Mol Biol Cell* 19, 3308–3322.
114. Kininis, M. and Kraus, W. L. (2008). A global view of transcriptional regulation by nuclear receptors: Gene expression, factor localization, and DNA sequence analysis. *Nucl Recept Signal* 6, e005.
115. Granjeaud, S., Bertucci, F., and Jordan, B. R. (1999). Expression profiling: DNA arrays in many guises. *Bioessays* 21, 781–790.
116. Buck, M. J. and Lieb, J. D. (2004). ChIP-chip: Considerations for the design, analysis, and application of genome-wide chromatin immunoprecipitation experiments. *Genomics* 83, 349–360.
117. Ren, B., Robert, F., Wyrick, J. J., Aparicio, O., Jennings, E. G., Simon, I., Zeitlinger, J., Schreiber, J., Hannett, N., Kanin, E., Volkert, T. L., Wilson, C. J., Bell, S. P., and Young, R. A. (2000). Genome-wide location and function of DNA binding proteins. *Science (New York, NY)* 290, 2306–2309.
118. Robyr, D. and Grunstein, M. (2003). Genomewide histone acetylation microarrays. *Methods (San Diego, Calif)* 31, 83–89.
119. Lee, W., Tillo, D., Bray, N., Morse, R. H., Davis, R. W., Hughes, T. R., and Nislow, C. (2007). A high-resolution atlas of nucleosome occupancy in yeast. *Nat Genet* 39, 1235–1244.
120. Ozsolak, F., Song, J. S., Liu, X. S., and Fisher, D. E. (2007). High-throughput mapping of the chromatin structure of human promoters. *Nat Biotechnol* 25, 244–248.
121. Bernstein, B. E., Humphrey, E. L., Erlich, R. L., Schneider, R., Bouman, P., Liu, J. S., Kouzarides, T., and Schreiber, S. L. (2002). Methylation of histone H3 Lys 4 in coding regions of active genes. *Proc Natl Acad Sci U S A* 99, 8695–8700.
122. Carroll, J. S., Liu, X. S., Brodsky, A. S., Li, W., Meyer, C. A., Szary, A. J., Eeckhoute, J., Shao, W., Hestermann, E. V., Geistlinger, T. R., Fox, E. A., Silver, P. A., and Brown, M. (2005). Chromosome-wide mapping of estrogen receptor binding reveals long-range regulation requiring the forkhead protein FoxA1. *Cell* 122, 33–43.
123. Carroll, J. S., Meyer, C. A., Song, J., Li, W., Geistlinger, T. R., Eeckhoute, J., Brodsky, A. S., Keeton, E. K., Fertuck, K. C., Hall, G. F., Wang, Q., Bekiranov, S., Sementchenko, V., Fox, E. A., Silver, P. A., Gingeras, T. R., Liu, X. S., and Brown, M. (2006). Genome-wide analysis of estrogen receptor binding sites. *Nat Genet* 38, 1289–1297.
124. Lin, C. Y., Vega, V. B., Thomsen, J. S., Zhang, T., Kong, S. L., Xie, M., Chiu, K. P., Lipovich, L., Barnett, D. H., Stossi, F., Yeo, A., George, J., Kuznetsov, V. A., Lee, Y. K., Charn, T. H., Palanisamy, N., Miller, L. D., Cheung, E., Katzenellenbogen, B. S., Ruan, Y., Bourque, G., Wei, C. L., and Liu, E. T. (2007). Whole-genome cartography of estrogen receptor alpha binding sites. *PLoS Genet* 3, e87.
125. Bentley, D. R. (2006). Whole-genome re-sequencing. *Curr Opin Genet Dev* 16, 545–552.
126. Robertson, G., Hirst, M., Bainbridge, M., Bilenky, M., Zhao, Y., Zeng, T., Euskirchen, G., Bernier, B., Varhol, R., Delaney, A., Thiessen, N., Griffith, O. L., He, A., Marra, M., Snyder, M., and Jones, S. (2007). Genome-wide profiles of STAT1 DNA association using chromatin immunoprecipitation and massively parallel sequencing. *Nature Methods* 4, 651–657.
127. Mardis, E. R. (2007). ChIP-seq: Welcome to the new frontier. *Nature Methods* 4, 613–614.
128. Wade, P. A. and Archer, T. K. (2006). Epigenetics: Environmental instructions for the genome. *Environ Health Perspect* 114, A.
129. O'Malley, B. W., Qin, J., and Lanz, R. B. (2008). Cracking the coregulator codes. *Curr Opin Cell Biol* 20, 310–315.

CHAPTER 15

NON-GENOMIC ACTION OF SEX STEROID HORMONES

ANTIMO MIGLIACCIO, GABRIELLA CASTORIA, AND FERDINANDO AURICCHIO

Department of General Pathology, II University of Naples, Via L. De Crecchio 7, 80138 Naples, Italy

Abstract: This chapter aims at giving a concise, up-to-date view of the non-genomic action of sex-steroid hormones and some related aspects. It includes an overview of the identity of the receptors mediating non-genomic action. To date, classical steroid receptors are known to be the principal mediators of this action, although there is now evidence that G-protein coupled receptors have a similar role. Non-genomic action has been separately analyzed in classical hormone-responsive cells (reproductive cells) and non-classical hormone-responsive cells (non-reproductive cells). With this distinction we intend to underline the different biological responses induced by the non-genomic action of steroids in these two cell types depending on their different functions, as well as the low level of receptor expression in some non-classical hormone-responsive cells, which might explain the absence of receptor transcriptional activity in such cells. One section focuses on the reversible integration between extra-nuclear and nuclear hormone action. Activation of kinase cascades by steroid hormones produces different effects in the cell nuclei. In addition to receptor or co-activator modifications, chromatin remodeling, receptor nuclear export, and increased expression of signaling effectors have been observed. The final section deals with the cross talk between growth factor and steroid signaling, a process which links hormonal and growth factor receptors, and regulates important cell functions. We apologize to scientists working in this area whose reports are not mentioned or extensively discussed owing to space limitations.

15.1. INTRODUCTION

Steroid hormone action is mediated by ligand-regulated transcriptional factors whose activity requires either direct interaction with target DNA sequences or association with proteins that in turn bind discrete nucleotide sequences. Thereafter, gene expression modification occurs, protein expression changes, and finally hormonal effects become evident. The entire process requires a relatively long time, from several minutes to hours or days. In contrast, non-genomic hormonal actions are rapid. They activate pathways that transduce signals from the cell membrane to the cell interior and are insensitive to RNA and protein synthesis inhibitors. Although the rapid, non-genomic actions of steroid hormones have only recently been intensively

C.M. Bunce, M.J. Campbell (eds.), *Nuclear Receptors*, Proteins and Cell Regulation 8, 365–379. DOI 10.1007/978-90-481-3303-1_15,

investigated, their existence was acknowledged a long time ago. The interest in non-classical action of steroids arises from the awareness that steroid hormone action cannot be entirely explained by steroid receptor transcriptional activity. Particularly, hormonal regulation of growth, immune system and behavior is not fully understood on that basis [1].

An important result deriving from the recent analysis of non-genomic hormone action is the accumulating evidence that steroid hormones activate classic signal transducing pathways which are the target of growth factors, cytokines and extracellular matrix. This activation is frequently triggered through engagement by steroid receptors of signaling effectors, such as Src and the p85 regulatory subunit of PI3-K. The non-genomic action of steroid receptors is independent of receptor transcriptional activity. This point is significantly illustrated in the present chapter by cells that respond to hormones in the absence of transcriptional action of the receptor. Non-genomic actions depend on different factors, such as cell type and hormone concentration. Estradiol stimulation of PI3-K/Akt pathway is a part of hormonal control of cell cycle in breast cancer MCF-7 cells [2], while estradiol activation of the same pathway triggers vasorelaxation in endothelial cells [3]. In NIH3T3 fibroblasts, a very low androgen concentration (1 pM) triggers S-phase entry through rapid action, whereas a much higher concentration (10 nM) induces membrane ruffling and cell migration ([4] and unpublished data). Non-genomic actions of steroid hormones regulate different processes such as proliferation, apoptosis, cell migration, and vasorelaxation under certain physiological conditions as well as in different diseases, including cancer.

Therefore, there is a growing expectation for knowledge of this hormonal action to translate from the bench to the treatment of patients.

15.2. IDENTITY OF EXTRA-NUCLEAR STEROID RECEPTORS AND RELATED ASPECTS

An important question related to non-genomic action of steroids is the identity of receptors mediating such an action. Candidates are classical steroid receptors, their truncated forms and G-protein coupled receptors (GPRs). In the more thoroughly analyzed systems, non-genomic actions of steroid hormones are mediated by classical steroid receptors, and much functional and immunochemical evidence supports this point. Examples of this evidence are here reported:

- Some non-hormone-responsive cells acquire non-genomic responsiveness to hormones after ectopic expression of classical steroid receptors. Cos-7 cells become estradiol-responsive in terms of rapid Src and Erk-2 activation after introduction of the wild-type human ER-alpha cDNA. The same cells respond to progestin with Erk-2 activation after co-transfection with hPR-B and hER-alpha cDNAs. Transfection of Cos-7 cells with hER-beta and/or hAR cDNAs confers estradiol and/or androgen responsiveness in terms of rapid Src activation [5–7].
- Knock down of ER-alpha by siRNA abrogates estradiol activation of MAPK in hormone-responsive cells [8]. ER-alpha- and beta-knockout mice, unlike wild type

animals, do not respond to estradiol with MAPK stimulation in the medial preoptic nucleus [9].

- Immunochemical approaches have revealed that a panel of antibodies directed against ER-alpha identified the full-length receptor in or near the cell membrane [10]. In addition, it has been shown by confocal microscopy that classical ER-alpha translocates into or near plasma membrane upon estradiol treatment of MCF-7 cells [8].

A still-debated question concerns the mechanism(s) responsible for the extra-nuclear localization of steroid receptors involved in rapid actions. Different possibilities have been proposed. They include interaction with caveolin-1 [11], palmitoylation of ER-alpha [12] or a truncated form of this receptor [13].

Recent evidence shows that GPR30 acts as an estradiol membrane receptor at plasma or intracellular membrane [14, 15].

Association of steroid receptors with different signaling effectors active at cell membrane, such as Src, PI3-K, FAK, or extra-nuclear scaffold proteins has been reported by different groups. It is possible that in the presence of steroids, these associations, which are crucial for signal transducing pathway activation by hormones, regulate in a dynamic way receptor localization at cell membrane. In this context, nuclear export of steroid receptors may also contribute to receptor localization outside the nucleus. Notably, the relative amount of steroid receptors responsible for non-genomic effects is very small, and according to different criteria has been evaluated at about 10% of the total receptor expressed in cells, or less. In contrast, receptor-dependent transcriptional activity depends on a larger amount of receptor [4], possibly required for receptor dimerization. These findings have interesting implications, since it is expected that cells expressing a low level of receptor may respond to steroids with signal transducing pathway activation in the absence of receptor transcriptional activity [4]. Because of the increased sensitivity of receptor assays, many cell types, previously thought to be non-hormone responsive, have been recently found to express a low amount of functional receptors. If our hypothesis is correct, non-genomic action in the absence of genomic activity might be a hallmark of different types of non-classical steroid hormone-responsive cells.

15.3. ROLE OF RAPID STEROID SIGNALING IN 'REPRODUCTIVE' CELLS

15.3.1. Estradiol Receptor

Estradiol very rapidly and transiently activates the Src/Ras/MAPK pathway in MCF-7 breast cancer cells [5]. This activation has a proliferative role as shown by experiments with small molecule inhibitors or dominant negative versions of signaling effectors [16]. In MCF-7 cells, estradiol triggers association of ER-alpha with Src. This occurs through interaction between the single tyrosine-phosphorylated ER-alpha at position 537 and the SH2 domain of Src [7]. In addition, a small fraction of ER and AR in mammary and prostate cancer cells is associated under basal conditions [17]. This probably explains why estradiol or androgen stimulation causes simultaneous interaction of both receptors with Src. Although association of either ER or AR with Src activates this kinase, the presence of both receptors induces stronger Src activation [7] and [17]. This is probably because each of the receptors interrupts only one of the two intra-molecular inhibitory interactions that keep Src in an inactive conformation under basal conditions [7]. Estradiol activation of the Src-dependent pathway occurs alongside PI3-K [2]. Hormone stimulation of MCF-7 cells induces the assembly of a multi-molecular complex made up of ER-alpha, Src p85 alpha, the regulatory subunit of PI3-K, and the atypical protein kinase C, PKC-zeta. This complex triggers simultaneous activation of Src- and PI3-K-dependent pathways. Estradiol-activated PI3-K targets Akt and PKC-zeta. Once activated, Akt increases cyclin D1 transcription, whereas PKC-zeta links Ras recruitment to the ER/Src/PI3-K complex, Erk-2 nuclear translocation and nuclear exclusion of p27. This interplay fosters cell cycle progression and DNA synthesis in MCF-7 cells [2, 18].

An ER-interacting 'adapter' protein, termed MNAR (modulator of non-genomic activity of estrogen receptor), containing both LXXLL (nuclear receptor binding) and PXXP (SH3-domain binding) motifs, is involved in ER-alpha/Src interaction [19]. MNAR phosphorylation in Tyr 920 also controls estrogen activation of the PI3-K/Akt pathway in MCF-7 cells [20].

Recently, it has been reported that arginine methylation of ERalpha triggered by estradiol is required for association of this receptor with Src and the regulatory subunit of PI3-K [21]. Another adapter protein, p130Cas, is associated with ER-alpha and Src in breast cancer T47D cells [22]. As p130Cas interacts with focal adhesion kinase (FAK), it is expected that estrogen-induced morphological changes can be mediated by the p130Cas/ER-alpha/Src complex in breast cancer cells.

The role of Src in estradiol action has been corroborated by findings showing that Src-null mice exhibit defects in mammary gland development and ER-alpha-mediated signaling [23]. Moreover, specific interference in the ER-alpha/Src interaction by a tyrosine phosphorylated peptide derived from an ER-alpha sequence inhibits the growth of mammary tumor cells in vitro and in nude mice. Noteworthy, inhibition of DNA synthesis by this compound occurs in the absence of ER transcriptional activity interference [24].

Several years ago, Szego and Davis reported a rapid, acute elevation of uterine cAMP by estradiol treatment of rats [25]. Subsequent findings have shown that estradiol treatment of breast cancer cells rapidly generates cAMP. This action results from ER-alpha-dependent G protein activation, and signaling is then transmitted to various effectors, including PKA, PKC, MAPK and PI3-K (reviewed by [26]. The importance of these signals in estradiol action in vitro and in vivo is well documented, and candidates in mediating these events are also represented by traditional G protein-coupled receptors (GPRs). As previously discussed, one of these receptors has been recently identified as GPR30, an orphan GPR [27, 15]. It binds estradiol and transiently regulates MAPK, since it is involved in both the rapid activation of MAPK and its subsequent inactivation. These findings indicate that estradiol controls MAPK even in the absence of classical ER [28].

There is now mounting interest in defining how signal transducing pathways contribute to steroid receptor localization and conversely how steroid receptor localization controls signaling activation. In addition to the link between estradiol-activated PI3-K/Akt pathway and nuclear export of ER-alpha discussed in one of the following sections in this chapter, it has been observed that expression of a shortened form of the metastatic tumor antigen 1 (MTA1s) sequesters ER-alpha in the cytoplasm and leads to malignant phenotypes by enhancing ER non-genomic functions in MCF-7 cells [29]. These and other findings imply that cytoplasmic localization of ER controls signal transducing-dependent functions, such as DNA synthesis and anchorage-dependent growth of target cells.

15.3.2. Androgen Receptor

Androgen stimulation of non-genomic signaling has been extensively studied in the LNCaP human prostate cancer cell line employed as a model of prostate cancer. It was initially observed that stimulation of these cells with androgen rapidly activates the Erk-2 pathway [30]. Subsequent studies showed that androgen or estradiol treatment of LNCaP cells promotes Erk-2 activation and proliferative effects by inducing rapid assembly of a cytosolic signaling complex containing classical AR, estrogen receptor beta (ER-beta) and Src [7]. This ternary complex follows direct interaction of a proline-rich motif of AR with the SH3-Src domain, and a phosphorylated tyrosine of ER-beta, possibly Tyr 443, with the SH2-Src domain [7]. Such an assembly fully activates Src and its dependent pathway in prostate and breast cancer cells [7], 2001). These results also underline the role of AR in breast cancer. In fact, this receptor is frequently expressed in breast cancer and its expression defines a subset of ER/PgR negative breast cancers [31].

Recently, it has been observed that interference in the AR/Src interaction by a peptide derived from the proline-rich AR motif inhibits the growth of prostate tumor cells in vitro and in nude mice [32]. Notably, this peptide does not affect the transcriptional activity of AR or signaling effects that are independent of AR [32].

The scaffolding protein MNAR is associated with the Src/AR signaling complex and its constitutively up-regulation correlates with transition of LNCaP cells to androgen independence [33]. More recently, it has been reported that androgen stimulation of LNCaP cells activates Akt-1 and increases AR/Akt-1 interaction in lipid rafts. This interaction is inhibited by the anti-androgen, bicalutamide, but is insensitive to PI3-K inhibition [34]. These findings suggest that in cells that have lost the expression of PTEN, like LNCaP cells, androgen activation of Akt may depend on kinases different from PI3-K, such as the *raptor*-mTOR complex [35].

Androgen activation of other signaling pathways leading to cAMP release and PKA activation has been described in prostate cells (reviewed by [36]. Treatment of LNCaP cells with testosterone or testosterone–BSA rapidly stimulates a rise in calcium, actin cytoskeleton reorganization and PSA secretion. Flow cytometric analysis of testosterone–BSA–FITC (fluoroscein isothiocyanate) binding studies has shown specific, saturable binding to LNCaP cells. Although the identity of the binding protein is unknown, it might not be the classical AR, since plasma membrane labeling is undetectable by immuno-cytochemistry for AR [36].

15.3.3. Progesterone Receptor

Progesterone activation of the Src/MAPK pathway was initially described in T47D breast cancer cells [6]. In these cells, which express PgR under basal conditions, progestin binds PgR-B and triggers association of ER-alpha with Src and consequent activation of the Src/Ras/MAPK pathway [6]. Interestingly, in Cos cells transfected with PgR-B and ER-alpha cDNAs, progestin stimulation induces much more robust Src activation than in cells transfected with PgR-B cDNA alone. These findings implicate that in these cells the PgR-B/ER-alpha cross talk either induces or amplifies the progestin signal. Microinjection of T47D cells with dominant negative forms of signaling effectors and chemical inhibitor experiments showed that progesterone activation of Src/Ras/MAPK pathway is required for DNA synthesis [16]. Subsequent studies further clarified the molecular mechanism underlying progesterone activation of the Src-dependent pathway by cross talk between PgR-B and ER-alpha [37]. This activation depends on the association of unliganded ER-alpha, with PgR-B via two domains of PgR [37]. Such a cross talk also occurs in endometrial stromal cells, and is responsible for progesterone-induced proliferation mediated by non-genomic pathway activation [38]. Activation of MAPK cascade by progesterone leads to phosphorylation of histone H3 with the consequent induction of progesterone target genes in T47D cells. This hormonal effect is prevented by pure antiestrogens, again implicating the action of PgR-B/ER cross talk [39]. It should be noted that at least two cross talks between sex steroid receptors are involved in non-genomic hormone action: AR/ER [7] and [17] and PgR-B/ER [6].

It has been shown that human PgR, through a proline-rich (PXXP) motif, directly interacts with the Src-SH3 domain in a ligand-dependent manner [40]. It is noteworthy the fact that the role of hormone-induced PgR-B association with Src has been mostly investigated in different PgR-B- as well as ER-negative cells ectopically expressing PgR-B but not ER. Depending on the cell line and experimental conditions, growth arrest [40] or stimulation of cell proliferation [41, 42] have been observed upon progestin stimulation. Altogether, these findings converge in supporting the regulatory role of Src in progestin-induced proliferation.

Progestin-activated MAPK leads to phosphorylation of Ser 294 of PgR and in this way mediates its nuclear export and degradation. Leptomycin B inhibits this event, which is involved in regulation of PgR transcriptional activity [43]. These and other findings involving ER-alpha [44] support the view that hormonal activation of signaling pathways plays a regulatory role in steroid receptor export and coupled functions.

15.4. ROLE OF RAPID STEROID SIGNALING IN 'NON-REPRODUCTIVE' CELLS

15.4.1. Estrogen Receptor

The role of non-genomic action of steroid receptors is of paramount importance in 'atypical' hormone targets such as non-reproductive cells. Among these, bone represents one of the best recognized models. Bone homeostasis results from a subtle balance between the activity of osteoblasts and osteoclasts, which are highly sensitive to sex-steroid hormones. Signaling of these hormones involves genomic and non-genomic pathways. Estrogens stimulate bone formation and inhibit bone resorption. ER-alpha in osteoclasts appears to inhibit bone resorption in women but not men [45]. A growing body of evidence indicates that rapid effects of estrogen in osteoblasts play a pivotal role in bone preservation. Such a view is supported by experiments with estren, a synthetic ligand for ER-alpha and AR, indicating that while this compound seems to lack the ability to induce gene transcription, it efficiently reduces bone loss [46]. Estren stimulates the activity of Elk-1, CCAAT enhancer binding protein-beta (C/EBPbeta), and cAMP-response element binding protein (CREB), or c-Jun/c-Fos by ER-alpha- (or AR-) dependent activation of Src/Shc/Erk pathway or down-regulation of JNK [47]. It is also worth noting that osteoblasts and osteoclasts derive from fibroblasts and macrophages respectively. These cells express detectable levels of ER and respond to a wide constellation of external signals, such as interleukins, TNF and CSF with activation of several signaling cascades, basically involving different isoforms of PI3-K. The observation that ER-alpha directly activates PI3-K in different tissues paves the way to the hypothesis that estrogens could affect bone growth and survival by activating specific PI3-K isoforms.

It is widely acknowledged that in endothelial cells estradiol induces a prompt vasodilatory action mediated by non-genomic mechanisms. This hormone triggers rapid eNOS activation. The process requires Src and MAPK signaling and eNOS phosphorylation by PI3-K/Akt kinase, with Src and PI3-K associating with ER-alpha upon ligand activation [48, 49]. Endothelial cells express an ER-alpha variant designated ER46, a truncated form of ER-alpha lacking the N-terminal A and B regions, and capable of modulating eNOS activation [13]. ER46 is devoid of full transcriptional activity but retains the properties of ligand binding and coupling to Erk and PI3-K. Different reports identify direct eNOS phosphorylation on Ser 1171 by Akt as the key event involved in estrogen-induced vasorelaxation. The role of the Src/MAPK pathway is less defined in this context, although ER mutants, unable to interact with Src and activate MAPK pathway, also fail to mediate estradiol-induced eNOS activation [50, 23]. These observations provide further support to the pioneering work of Simoncini et al. demonstrating that estrogen-induced vasodilatation is mediated by NO synthase stimulation following PI3-K/Akt activation by estrogen receptor triggered through direct interaction of ER with the p85 regulatory subunit of PI3-K [3].

Activation of the Src/PI3-K/MAPK pathway by estradiol through a G-protein-mediated mechanism has been observed in differentiated rat pre-adipocytes [51]. This report shows that MAPK activation by estrogens, which also involves PI3-K and classical PKCs, leads to activation of the transcriptional factor AP-1 in adipocytes. This factor acts synergistically with CREB, which is induced by cAMP-dependent pathways. This suggests that a non-genomic ER-alpha dependent mechanism contributes to modulation of the adipocyte metabolism by these hormones.

Several observations point to a primary role of the rapid signaling of estradiol receptor in nervous system. Estrogens, in fact, exert a clear neuro-protective effect that is essentially mediated by PI3-K/Akt activation. Akt prevents neural cell apoptosis via phosphorylation and consequent de-activation of GSK3ß and BAD [52], two well-known death effectors for neurons. Akt activation also inhibits the MLK3-MKK4/7-JNK1/2 pathway, and in this way might be responsible for estrogen-induced neuro-protection against transient global cerebral ischemia in hormone-treated male rats [53]. ER-alpha association with the p85 subunit of PI3-K triggers Akt activation and the following events. Estradiol action in nervous system is strictly dependent on the cellular environment. Estradiol significantly and rapidly stimulates PKC activity in the preoptic area of female rat brain but not in hypothalamus or in cortex, and this stimulation enhances estrogen receptor-dependent gene transcription. In adult rat brain, estrogen activates the MAPK pathway and has a synergistic effect with IGF-I on Akt activation [54, 55].

The range of non-reproductive cells analyzed for rapid estrogen effects also includes gastro-intestinal and liver cells. In the colon carcinoma cell line CaCo-2, estradiol stimulates two Src family kinases, Src and Yes, leading to Ras/Erk pathway activation and consequent cell proliferation. This effect is observed in sub-confluent, undifferentiated CaCo-2 cells that, unlike confluent cells, express significant ER-alpha levels [56]. In HepG2 liver cells expressing ER-alpha unable to stimulate gene transactivation, estrogen treatment activates PKC-alpha and triggers IP3 production as well as DNA synthesis [57].

15.4.2. Androgen Receptor

In different cell types androgen acts through non-transcriptional mechanisms [45]. Androgens play an important role in bone homeostasis maintenance. In this context, the role of androgens appears rather controversial. In MC3T3-E1 osteoblasts, androgens increase phosphorylation and nuclear translocation of Akt, protecting the cells from apoptosis. This effect is inhibited by both LY294002 and hydroxyl-flutamide, indicating that this hormone action is mediated by AR-dependent PI3-K activation [58]. In contrast, upon overexpression of AR in the same cell line, DHT inhibits Erk signaling, decreases serine phosphorylation of Bcl-2 and increases its ubiquitination. These results suggest that androgen stimulates osteoblast apoptosis through an increased Bax/Bcl-2 ratio [59].

Androgen seems to play a major regulatory role in stromal cells. In this regard, the action of androgens in NIH3T3 fibroblasts deserves particular consideration. In these cells the synthetic androgen R1881 at a concentration as low as 1 pM induces or increases association of AR with Src and PI3-K respectively, thereby triggering DNA synthesis. In the same cells, 10 nM R1881 stimulates Rac-1 activity and cytoskeleton changes in the absence of the AR/Src/PI3-kinase complex assembly, with a negligible effect on DNA synthesis. The anti-androgen Casodex and specific inhibitors of Src and PI3-K prevent all the above-reported effects induced by androgen. They are mediated by a non-transcriptional mechanism. In fact, neither low nor high R1881 concentration induces receptor nuclear translocation or receptor-dependent transcriptional activity in fibroblasts, although these cells harbor the classical mouse AR. The very low amount of AR in NIH3T3 cells (about 7% of that present in LNCaP cells) activates signaling pathways, but apparently it is not sufficient to stimulate gene transcription [4]. The mechanism underlying this dose-dependent biphasic effect of androgen in mouse fibroblasts remains to be elucidated.

Non-genomic action of androgen has been also described in a variety of other cells. Acute and chronic testosterone effects on calcium channels in rat cardiomyocytes have been reported. A non-classical androgen receptor responsible for these rapid hormone effects has been hypothesized. Furthermore, testosterone reduces anxiety in male mice with very rapid dynamics. For this activity it has been postulated that the hormone acts through a cross talk between AR and GABA receptors, even though the observed increase in neuro-steroid metabolites of testosterone (androsterone and 3alpha-androstandione) level leaves room for the possibility that these neuro-steroids directly interact with GABA (A) receptors [60].

15.4.3. Progesterone Receptor

Several progestin actions in non-reproductive organs such as the cardiovascular system are mediated by rapid non-genomic mechanisms. Progesterone exerts a powerful vasodilator action in arteries and regulates cardiac re-polarization. In isolated guinea pig ventricular myocytes, progesterone rapidly shortens action potential duration. This effect is mainly attributable to enhancement of the slow delayed rectifier K+ current (I(Ks)) under basal conditions and inhibition of L-type Ca2+ currents (I(Ca,L)) under cAMP-stimulated conditions. The effect of progesterone is mediated by nitric oxide released via non-genomic activation of endothelial nitric oxide synthase. This signal transduction seems to take place in the *caveolae* as sucrose density gradient fractionation experiments show co-localization of the progesterone receptor, c-Src, PI3-K, Akt, and endothelial nitric oxide synthase with repolarizing channels (KCNQ1, KCNE1), and Ca(V)1.2 in the *caveolae* fraction [61]. Rapid action of progesterone has also been described in the kidney. Concentration of progesterone as low as 10 pM enhances Ca^{2+} uptake by the distal tubule membranes through a receptor-mediated, non-genomic mechanism. However, the molecular machinery involved in this action has not been elucidated [62].

15.5. BIDIRECTIONAL INTEGRATION BETWEEN EXTRA-NUCLEAR AND NUCLEAR STEROID ACTION

Steroid regulation of gene transcription occurs through different pathways. It requires direct interaction of receptors with either steroid-regulated promoters or transcriptional factors, which interact with specific DNA sequences [63]. In addition, steroid-activated kinase cascades phosphorylate transcriptional factors or steroid receptors or their co-activators resulting in increased receptor transcriptional activity [64]. In such a way, steroid-activated pathways can regulate nuclear functions of steroid receptors.

One of the initial examples of this type of regulation is represented by the observation that estradiol-induced elevation of intracellular cAMP stimulates cAMP response element (CRE)-mediated gene expression in MCF-7 cells [65]. These results showed for the first time that signals resulting from activation of G-protein and cAMP-signaling pathways contribute to gene regulation by estradiol.

In recent years, it has been reported that estradiol rapidly induces phosphorylation of co-activator SRC-3 and that mutations of the ER-alpha DNA binding domain do not block this rapid event. Subsequent studies have shown that MAPK signaling activated by estradiol-coupled ER-alpha phosphorylates SRC-3. Once phosphorylated, SRC-3 translocates into the nucleus and activates ER-mediated transcriptional events. These results provide evidence for an early, non-genomic action of ER on SRC-3 that regulates the downstream genomic effects of estradiol ([66] and refs therein).

Another example of integration between non-genomic and genomic action of a steroid hormone has been described by Vicent and coworkers [39]. Rapid Erk activation by progestin leads to PgR phosphorylation, activation of Msk1 and recruitment of a complex of three proteins (PgR/Erk-2/Msk1) to a nucleosome on the MMTV promoter. Activated Msk1 then phosphorylates histone H3 and leads to RNA polymerase II recruitment. Thus, progestin activation of signaling cascades in the cytoplasm is essential for chromatin remodeling and transcriptional activation of a subset of target genes [39].

Very recently, a new link between estradiol-activated signaling and nuclear events has been described in MCF-7 cells. Estradiol activation of PI3-K/Akt pathway leads to phosphorylation of the gatekeeper Ser 256 of FKHR, a member of the Forkhead transcription factor family. This phosphorylation triggers FKHR/ER-alpha association which, in turn, is responsible for the FKHR nuclear export. According to this model, ER-alpha exerts two integrated actions: the first in the extra-nuclear compartment, where it activates the PI3-K/Akt pathway; the other in the nucleus, where it forms a complex with FKHR. Both receptor actions converge on FKHR nuclear export and subsequent release of DNA synthesis inhibition [44].

In turn, transcriptional activity of steroid receptors controls the activity of signaling effectors. In addition to the well-known rapid and transient activation of MAPK, the progestin-bound PgR-B induces in T47D cells sustained activation of MAPK. This activation results from PgR-mediated transcriptional up-regulation of the secreted glycoprotein Wnt1, which binds to the seven-transmembrane receptor

Frizzled and stimulates the MMP-dependent cleavage of EGFR ligands. In this way, EGFR transactivation occurs and sustained activation of the downstream Src and MAPK effectors follows [67].

15.6. REVERSIBLE CROSS TALK BETWEEN GROWTH FACTORS AND STEROID HORMONES

Some time ago, it was shown that estradiol stimulates synthesis or release of IGF-1, EGF and other growth factors from breast cancer cells and mouse uterus [68].

It was then demonstrated that cellular phosphorylation pathways could activate steroid receptors in the absence of ligand. Dopamine, by acting at its own D1 membrane receptor, activates PgR both in cultured cells and in living animals [69]. To activate a steroid receptor, this ligand-independent pathway occurs via signaling cascades from membrane regulatory molecules such as growth factors, cAMP, dopamine, cytokines, and possibly other cellular regulators acting at the membrane. This pathway also represents a mechanism by which the cellular environment can modulate steroid receptor functions. It is particularly important in growth factor and neurotransmitter action ([66] and refs therein).

We now appreciate that steroid receptors and growth-factor-signaling pathways intersect and directly interact at multiple levels of signal transduction. This synergism has been documented in normal development of target tissues as well as cancer progression and endocrine therapy resistance. Current understanding of the molecular mechanisms of cross talk between sex steroid receptor and growth-factor-signaling is illustrated by different findings:

- Sex steroids activate CREB and in such a way up-regulate IGF-1R expression in prostate cancer cells through non-genomic pathways [70].
- Estradiol increase in pancreatic insulin content involves ER-alpha-dependent Erk 1/2 activation [71].
- IGF-1R is a key element in the translocation of ER-alpha to the cell membrane and in facilitating ER-alpha-mediated rapid estradiol action in MCF-7 cells [8].
- Estrogen responsiveness of uterine cells is reduced by blocking the IGF-R pathway. In turn, IGF-1 action is lost in ER-knockout cells [72].
- EGF and IGF-1 treatment does not trigger DNA synthesis in uterine epithelial cells of ER-alpha-KO mouse [73].
- EGF induces phosphorylation of ER-alpha on serine 118 and in such a way increases nuclear receptor transcriptional activity [74].
- EGF signaling in MCF-7 cells induces ER-alpha tyrosine phosphorylation responsible for the assembly of a ternary complex made up of ER-alpha/AR/Src. In this complex, Src is activated and in turn phosphorylates EGFR and up-regulates EGF signaling. A similar cross talk between EGF and extra-nuclear ER-beta/AR is detectable in LNCaP cells [17].

It is clear from these and other reports that different physiologic actions of peptide growth factors depend on steroid receptors.

REFERENCES

1. Gorski, J., Wendell, D., Gregg, D., and Chun, T. Y. (1997). Estrogens and the genetic control of tumor growth. *Prog Clin Biol Res* 396, 233–243.
2. Castoria, G., Migliaccio, A., Bilancio, A., Di Domenico, M., de Falco, A., Lombardi, M., Fiorentino, R., Varricchio, L., Barone, M. V., and Auricchio, F. (2001). PI3-kinase in concert with Src promotes the S-phase entry of estradiol-stimulated MCF-7 cells. *EMBO J* 20, 6050–6059.
3. Simoncini, T., Hafezi-Moghadam, A., Brazil, D. P., Ley, K., Chin, W. W., and Liao, J. K. (2000). Interaction of oestrogen receptor with the regulatory subunit of phosphatidylinositol-3-OH kinase. *Nature* 407, 538–541.
4. Castoria, G., Lombardi, M., Barone, M. V., Bilancio, A., Di Domenico, M., Bottero, D., Vitale, F., Migliaccio, A., and Auricchio, F. (2003). Androgen-stimulated DNA synthesis and cytoskeletal changes in fibroblasts by a nontranscriptional receptor action. *J Cell Biol* 161, 547–556.
5. Migliaccio, A., Di Domenico, M., Castoria, G., de Falco, A., Bontempo, P., Nola, E., and Auricchio, F. (1996). Tyrosine kinase/p21ras/MAP-kinase pathway activation by estradiol-receptor complex in MCF-7 cells. *EMBO J* 15, 1292–1300.
6. Migliaccio, A., Piccolo, D., Castoria, G., Di Domenico, M., Bilancio, A., Lombardi, M., Gong, W., Beato, M., and Auricchio, F. (1998). Activation of the Src/p21ras/Erk pathway by progesterone receptor via cross-talk with estrogen receptor. *EMBO J* 17, 2008–2018.
7. Migliaccio, A., Castoria, G., Di Domenico, M., de Falco, A., Bilancio, A., Lombardi, M., Barone, M. V., Ametrano, D., Zannini, M. S., Abbondanza, C., and Auricchio, F. (2000). Steroid-induced androgen receptor-oestradiol receptor-beta Src complex triggers prostate cancer cell proliferation. *EMBO J* 19, 5406–5417.
8. Song, R., Barnes, C., Zhang, Z., Bao, Y., Kumar, R., and Santen, R. (2004). The role of Shc and insulin-like growth factor 1 receptor in mediating the translocation of estrogen receptor alpha to the plasma membrane. *Proc Natl Acad Sci U S A* 101, 2076–2081.
9. Abrahám, I. M., Todman, M. G., Korach, K. S., and Herbison, A. E. (2004). Critical in vivo roles for classical estrogen receptors in rapid estrogen actions on intracellular signaling in mouse brain. *Endocrinology* 145, 3055–3061.
10. Pappas, T. C., Gametchu, B., and Watson, C. S. (1995). Membrane estrogen receptors identified by multiple antibody labeling and impeded-ligand binding. *FASEB J* 9(5), 404–410.
11. Razandi, M., Pedram, A., Park, S. T., and Levin, E. R. (2003). Proximal events in signaling by plasma membrane estrogen receptors. *J Biol Chem* 278, 2701–2712.
12. Acconcia, F., Ascenzi, P., Bocedi, A., Spisni, E., Tomasi, V., Trentalance, A., Visca, P., and Marino, M. (2005). Palmitoylation-dependent estrogen receptor alpha membrane localization: Regulation by 17beta-estradiol. *Mol Biol Cell* 16, 231–237.
13. Li, L., Haynes, M. P., and Bender, J. R. (2003). Plasma membrane localization and function of the estrogen receptor α variant (ER46) in human endothelial cells. *Proc Natl Acad Sci USA* 100, 4807–4812.
14. Filardo, E. J. and Thomas, P. (2005). GPR30: A seven-transmembrane-spanning estrogen receptor that triggers EGF release. *Trends Endocrinol Metab* 16, 362–367.
15. Revankar, C. M., Cimino, D. F., Sklar, L. A., Arterburn, J. B., and Prossnitz, E. R. (2005). A transmembrane intracellular estrogen receptor mediates rapid cell signaling. *Science* 307, 1625–1630.
16. Castoria, G., Barone, M. V., Di Domenico, M., Bilancio, A., Ametrano, D., Migliaccio, A., and Auricchio, F. (1999). Non-transcriptional action of estrogen and progestin triggers DNA synthesis. *EMBO J* 18, 2500–2510.
17. Migliaccio, A., Di Domenico, M., Castoria, G., Nanayakkara, M., Lombardi, M., de Falco, A., Bilancio, A., Varricchio, L., Ciociola, A., and Auricchio, F. (2005). Steroid receptor regulation of EGF signaling through Src in breast and prostate cancer cells: Steroid antagonist action. *Cancer Res* 65, 10585–10593.
18. Castoria, G., Migliaccio, A., Di Domenico, M., Lombardi, M., de Falco, A., Varricchio, L., Bilancio, A., Barone, M. V., and Auricchio, F. (2004). Role of atypical PKC in estradiol-triggered G1/S progression of MCF-7 cells. *Mol Cell Biol* 24, 7643–7653.

19. Wong, C. W., McNally, C., Nickbarg, E., Komm, B. S., and Cheskis, B. J. (2002). Estrogen receptor-interacting protein that modulates its nongenomic activity-crosstalk with Src/Erk phosphorylation cascade. *Proc Natl Acad Sci USA* 99, 14783–14788.
20. Greger, J. G., Fursov, N., Cooch, N., McLarney, S., Freedman, L. P., Edwards, D. P., and Cheskis, B. J. (2007). Phosphorylation of MNAR promotes estrogen activation of phosphatidylinositol 3-kinase. *Mol Cell Biol* 27, 1904–1913.
21. Le Romancer, M., Treilleux, I., Leconte, N., Robin-Lespinasse, Y., Sentis, S., Bouchekioua-Bouzaghou, K., Goddard, S., Gobert-Gosse, S., and Corbo, L. (2008). Regulation of estrogen rapid signaling through arginine methylation by PRMT1. *Mol Cell* 31, 212–221.
22. Cabodi, S., Moro, L., Baj, G., Smeriglio, M., Di Stefano, P., Gippone, S., Surico, N., Silengo, L., Turco, E., Tarone, G., and Defilippi, P. (2004). p130Cas interacts with estrogen receptor alpha and modulates non-genomic estrogen signaling in breast cancer cells. *J Cell Sci* 117, 1603–1611.
23. Kim, H., Laing, M., and Muller, W. (2005). c-Src-null mice exhibit defects in normal mammary gland development and ERalpha signaling. *Oncogene* 24, 5629–5636.
24. Varricchio, L., Migliaccio, A., Castoria, G., Yamaguchi, H., de Falco, A., Di Domenico, M., Giovannelli, P., Farrar, W., Appella, E., and Auricchio, F. (2007). Inhibition of estradiol receptor/Src association and cell growth by an estradiol receptor {alpha} tyrosine-phosphorylated peptide. *Mol Cancer Res* 11, 1213–1221.
25. Szego, C. M. and Davis, J. S. (1967). Adenosine 3′,5′-monophosphate in rat uterus: Acute elevation by estrogen. *Proc Natl Acad Sci U S A* 58, 1711–1718.
26. Levin, E. R. (2005). Integration of the extranuclear and nuclear actions of estrogen. *Mol Endocrinol* 19, 1951–1959.
27. Thomas, P., Pang, Y., Filardo, E. J., and Dong, J. (2005). Identity of an estrogen membrane receptor coupled to a G protein in human breast cancer cells. *Endocrinology* 146, 624–632.
28. Filardo, E. J., Quinn, J. A., Bland, K. I., and Frackelton, A. R., Jr. (2000). Estrogen-induced activation of Erk-1 and Erk-2 requires the G protein-coupled receptor homolog, GPR30, and occurs via transactivation of the epidermal growth factor receptor through release of HB-EGF. *Mol Endocrinol* 14, 1649–1660.
29. Kumar, R., Wang, R. A., Mazumdar, A., Talukder, A. H., Mandal, M., Yang, Z., Bagheri-Yarmand, R., Sahin, A., Hortobagyi, G., Adams, L., Barnes, C. J., and Vadlamudi, R. K. (2002). A naturally occurring MTA1 variant sequesters oestrogen receptor-alpha in the cytoplasm. *Nature* 418, 654–657.
30. Peterziel, H., Mink, S., Schonert, A., Becker, M., Klocker, H., and Cato, A. C. (1999). Rapid signalling by androgen receptor in prostate cancer cells. *Oncogene* 18, 6322–6329.
31. Doane, A. S., Danso, M., Lal, P., Donaton, M., Zhang, L., Hudis, C., and Gerald, W. L. (2006). An estrogen receptor-negative breast cancer subset characterized by a hormonally regulated transcriptional program and response to androgen. *Oncogene* 25, 3994–4008.
32. Migliaccio, A., Varricchio, L., de Falco, A., Castoria, G., Arra, C., Yamaguchi, H., Ciociola, A., Lombardi, M., Di Stasio, R., Barbieri, A., Baldi, A., Barone, M. V., Appella, E., and Auricchio, F. (2007). Inhibition of the SH3 domain-mediated binding of Src to the androgen receptor and its effect on tumor growth. *Oncogene* 26, 6619–6629.
33. Unni, E., Sun, S., Nan, B., McPhaul, M. J., Cheskis, B., Mancini, M. A., and Marcelli, M. (2004). Changes in androgen receptor nongenotropic signaling correlate with transition of LNCaP cells to androgen independence. *Cancer Res* 64, 7156–7168.
34. Cinar, B., Mukhopadhyay, N. K., Meng, G., and Freeman, M. R. (2007). Phosphoinositide 3-kinase-independent non-genomic signals transit from the androgen receptor to Akt1 in membrane raft microdomains. *J Biol Chem* 282, 29584–29593.
35. Sarbassov, D. D., Guertin, D. A., Ali, S. M., and Sabatini, D. M. (2005). Phosphorylation and regulation of Akt/PKB by the rictor-mTOR complex. *Science* 307, 1098–1101.
36. Rahman, F. and Christian, H. C. (2007). Non classical actions of testosterone: An updata. *Trends Endocrinol Metab* 18, 371–378.
37. Ballaré, C., Uhrig, M., Bechtold, T., Sancho, E., Di Domenico, M., Migliaccio, A., Auricchio, F., and Beato, M. (2003). Two domains of the progesterone receptor interact with the estrogen receptor

and are required for progesterone activation of the c-Src/Erk pathway in mammalian cells. *Mol Cell Biol* 23, 1994–2008.
38. Vallejo, G., Ballaré, C., Barañao, J. L., Beato, M., and Saragüeta, P. (2005). Progestin activation of nongenomic pathways via cross talk of progesterone receptor with estrogen receptor ß induces proliferation of endometrial stromal cells. *Mol Endo* 19, 3023–3037.
39. Vicent, G. P., Ballarè, C., Nacht, A. S., Clausell, A., Subtil-Rodriguez, I., Quiles, A., Jordan, A., and Beato, M. (2006). Induction of progesterone target genes requires activation of Erk and Msk kinases and phosphorylation of histone H3. *Mol Cell* 24, 367–381.
40. Boonyaratanakornkit, V., Scott, M. P., Ribon, V., Sherman, L., Anderson, S. M., Maller, J. L., Miller, W. T., and Edwards, D. P. (2001). Progesterone receptor contains a proline-rich motif that directly interacts with SH3 domains and activates c-Src family tyrosine kinases. *Molecular Cell* 8, 269–280.
41. Boonyaratanakornkit, V., McGowan, E., Sherman, L., Mancini, M. A., Cheskis, B. J., and Edwards, D. P. (2007). The role of extranuclear signaling actions of progesterone receptor in mediating progesterone regulation of gene expression and the cell cycle. *Mol Endocrinol* 21, 359–375.
42. Carnevale, R. P., Proietti, C. J., Salatino, M., Urtreger, A., Peluffo, G., Edwards, D. P., Boonyaratanakornkit, V., Charreau, E. H., Bal de Kier Joffé, E., Schillaci, R., and Elizalde, P. V. (2007). Progestin effects on breast cancer cell proliferation, proteases activation, and in vivo development of metastatic phenotype all depend on progesterone receptor capacity to activate cytoplasmic signaling pathways. *Mol Endocrinol* 21, 335–358.
43. Qiu, M., Olsen, A., Faivre, E., Horwitz, K. B., and Lange, C. A. (2003). Mitogen-activated protein kinase regulates nuclear association of human progesterone receptors. *Mol Endocrinol* 17, 628–642.
44. Lombardi, M., Castoria, G., Migliaccio, A., Barone, M. V., Di Stasio, R., Ciociola, A., Bottero, D., Yamagushi, H., Appella, E., and Auricchio, F. (2008). Hormone-dependent nuclear export of estradiol receptor and DNA synthesis in breast cancer cells. *J Cell Biol* 182, 327–340.
45. Vanderschueren, D., Gaytant, J., Boonen, S., and Venken, K. (2008). Androgens and bone. *Curr Opin Endocrinol Diabetes Obes* 15, 250–254.
46. Kousteni, S., Chen, J. R., Bellido, T., Han, L., Ali, A. A., O'Brien, C. A., Plotkin, L., Fu, Q., Mancino, A. T., Wen, Y., Vertino, A. M., Powers, C. C., Stewart, S. A., Ebert, R., Parfitt, A. M., Weinstein, R. S., Jilka, R. L., and Manolagas, S. C. (2002). Reversal of bone loss in mice by nongenotropic signaling of sex steroids. *Science* 298, 843–846.
47. Kousteni, S., Han, L., Chen, J. R., Almeida, M., Plotkin, L. I., Bellido, T., and Manolagas, S. C. (2003). Kinase-mediated regulation of common transcription factors accounts for the bone-protective effects of sex steroids. *J Clin Invest* 111, 1651–1664.
48. Chambliss, K. L., Simon, L., Yuhanna, I. S., Mineo, C., and Shaul, P. W. (2005). Dissecting basis of nongenomic activation of eNOS by estradiol: Role of ERα domains with known nuclear functions. *Mol Endocrinol* 19, 277–289.
49. Marino, M., Acconcia, F., and Ascenzi, P. (2005). Estrogen receptor signaling: Bases for drug actions. *Curr Drug Targets Immune Endocr Metabol Disord* 5, 305–314.
50. Chambliss, K. L., Yuhanna, I. S., Anderson, R. G., Mendelsohn, M. E., and Shaul, P. W. (2002). ERß has non genomic action in caveolae. *Mol Endocrinol* 16, 938–946.
51. Dos Santos, E. G., Dieudonne, M. N., Pecquery, R., le Moal, V., Giudicelli, Y., and Lacasa, D. (2002). Rapid nongenomic E2 effects on p42/44 MAPK, activator-protein-1 and cAMP response element binding protein in rat white adipocytes. *Endocrinology* 143, 930–940.
52. Goodenough, S., Schleusner, D., Pietrzik, C., Skutella, T., and Behl, C. (2005). Glycogen synthase kinase 3ß links neuroprotection by 17ß estrogen to key Alzheimer processes. *Neuroscience* 132, 581–589.
53. Wang, R., Zhang, Q. Z., Han, D., Lu, Q., and Zhang, G. Y. (2006). Inhibition of MLK3-MKK4/7-JNK1/2 pathway by Akt-1 in exogenous estrogen induced neuro-protection against transient global cerebral ischemia by a non-genomic mechanism in male rats. *J Neurochem* 99, 1543–1554.
54. Cardona-Gomez, G. P., Mendez, P., and Garcia-Segura, L. M. (2002). Synergistic interaction of estradiol and insulin-like growth factor-I in the activation of PI3K/Akt signaling in the adult rat hypothalamus. *Brain Res Mol Brain Res* 107, 80–88.

55. Maggi, A., Ciana, P., Belcredito, S., and Vegeto, E. (2004). Estrogens in the nervous system: Mechanisms and nonreproductive functions. *Annu Rev Physiol* 66, 291–313.
56. Di Domenico, M., Castoria, G., Bilancio, A., Migliaccio, A., and Auricchio, F. (1996). Estradiol activation of human colon carcinoma-derived Caco-2 cell growth. *Cancer Res* 56, 4516–4521.
57. Marino, M., Distefano, E., Trentalance, A., and Smith, C. L. (2001). Estradiol induced IP3 mediate the estrogen receptor activity expressed in human cells. *Mol Cell Endocrinol* 182, 19–26.
58. Kang, N. Y., Cho, C. L., Huang, K. L., Wang, J. C., Hu, Y. C., Lin, H. K., Chang, C., and Huang, K. E. (2004). Nongenomic androgen activation of phosphatidylinositol 3-kinase/Akt signaling pathway in MC3T3-E1 osteoblasts. *J Bone Miner Res* 17, 1181–1190.
59. Wiren, K. M., Toombs, A. R., Semirale, A. A., and Zhang, X. (2006). Osteoblast and osteocyte apoptosis associated with androgen action in bone: Reqirement of Bax/Bcl-2 ratio. *Bone* 34, 637–651.
60. Aikey, J. L., Nyby, J. G., Anmuth, D. M., and James, P. J. (2002). Testosterone rapidly reduces anxiety in male house mice (*Mus musculus*). *Horm Behav* 42, 448–460.
61. Nakamura, H., Kurokawa, J., Bai, C. X., Asada, K., Xu, J., Oren, R. V., Zhu, Z. I., Clancy, C. E., Isobe, M., and Furukawa, T. (2007). Progesterone regulates cardiac repolarization through a nongenomic pathway: An in vitro patch-clamp and computational modeling study. *Circulation* 116, 2913–2922.
62. Brunette, M. G. and Leclerc, M. (2002). Renal action of progesterone: Effect on calcium reabsorption. *Mol Cell Endocrinol* 194, 183–190.
63. Lonard, D. M. and O'Malley, B. W. (2008). Gene transcription: Two worlds merged. *Nature* 452, 946–947.
64. Björnström, L. and Sjöberg, M. (2005). Mechanisms of estrogen receptor signaling: Convergence of genomic and nongenomic actions on target genes. *Mol Endocrinol* 19, 833–842.
65. Aronica, S. M., Kraus, W. L., and Katzenellenbogen, B. S. (1994). Estrogen action via the cAMP signaling pathway: Stimulation of adenylate cyclase and cAMP-regulated gene transcription. *Proc Natl Acad Sci U S A* 91, 8517–8521.
66. O'Malley, B. W. (2005). A life-long search for the molecular pathways of steroid hormone action. *Mol Endocrinol* 19, 1402–1411.
67. Faivre, E. J. and Lange, C. A. (2007). Progesterone receptors upregulate Wnt-1 to induce epidermal growth factor receptor transactivation and c-Src-dependent sustained activation of Erk1/2 mitogen-activated protein kinase in breast cancer cells. *Mol Cell Biol* 27, 466–480.
68. Dickson, R. B., McManaway, M. E., and Lippman, M. E. (1986). Estrogen-induced factors of breast cancer cells partially replace estrogen to promote tumor growth. *Science* 232, 1540–1543.
69. Denner, L. A., Weigel, N. L., Maxwell, B. L., Schrader, W. T., and O'Malley, B. W. (1990). Regulation of progesterone receptor-mediated transcription by phosphorylation. *Science* 250, 1740–1743.
70. Genva, M., Pandini, G., Sijci, D., Castoria, G., Maggioliui, M., Vigneri, R., and Belfiore, A. (2009). Role of cyclic AMP response element binding protein in insulin-like growth factor. I receptor upregulation by sex steroids in prostate cancer cells. Cancer Res. 69(18), 7270–7277.
71. Alonso-Magdalena, P., Ropero, A. B., Carrera, M. P., Cederroth, C. R., Baquié, M., Gauthier, B. R., Nef, S., Stefani, E., and Nadal, A. (2008). Pancreatic insulin content regulation by the estrogen receptor ER alpha. *PLoS ONE* 3, e2069.
72. Klotz, D. M., Hewitt, S. C., Ciana, P., Raviscioni, M., Lindzey, J. K., Foley, J., Maggi, A., Di Augustine, R. P., and Korach, K. S. (2002). Requirement of estrogen receptor-alpha in insulin-like growth factor-1 (IGF-1)-induced uterine responses and in vivo evidence for IGF-1/estrogen receptor cross-talk. *J Biol Chem* 277, 8531–8537.
73. Hewitt, S. C., Harrell, J. C., and Korach, K. S. (2005). Lessons in estrogen biology from knockout and transgenic animals. *Annu Rev Physiol* 67, 285–308.
74. Chen, D., Washbrook, E., Sarwar, N., Bates, G. J., Pace, P. E., Thirunuvakkarasu, V., Taylor, J., Epstein, R. J., Fuller-Pace, F. V., Egly, J. M., Coombes, R. C., and Ali, S. (2002). Phosphorylation of human estrogen receptor alpha at serine 118 by two distinct signal transduction pathways revealed by phosphorylation-specific antisera. *Oncogene* 21, 4921–4931.

CHAPTER 16

LIGAND REGULATION AND NUCLEAR RECEPTOR ACTION

MARTIN HEWISON
Department of Orthopaedic Surgery, David Geffen School of Medicine UCLA, Los Angeles, CA 90095, USA

Abstract: Nuclear receptors signal primarily via ligand-dependent mechanisms that are initiated when the ligand in question interacts with a specific receptor binding domain. Subsequent transcriptional responses to ligand-receptor interaction are then dependent on an array of downstream mechanisms such as chromatin remodeling, receptor binding to target gene promoter DNA and the modulation of receptor co-activators and co-repressors. However, it is also important to recognize that the efficacy of nuclear receptor signaling will also be influenced by more upstream pathways that modulate the ligand itself. These include mechanisms that control the cellular acquisition of ligand and its subcellular translocation, as well as metabolic systems that increase or decrease the local concentration of nuclear receptor ligands. The following chapter describes examples of these pathways for regulation of nuclear receptor ligands and their developing importance as therapeutic targets for human disease.

16.1. INTRODUCTION

Signaling pathways for nuclear receptors are complex, involving the integration of ligand binding, chromatin remodeling, DNA binding and the modulation of co-activator and co-repressor proteins. Given these complexities, it is understandable that studies of nuclear receptor signaling have focused primarily on molecular events that follow the binding of specific ligands to their cognate receptors. However, the ligand-dependency of most receptor responses means that the regulation of ligands themselves remains a pivotal component of this pathway. As illustrated in Figure 16.1a, this is commonly perceived as a passive mechanism involving the membrane diffusion of lipid-soluble molecules such as steroid hormones/sterols, followed by random association with cytosolic or nuclear receptors according to intracellular concentration gradients. Clearly this is likely to be an over-simplification and current data indicate that distinct mechanisms are involved in facilitating the delivery of ligands to their receptors. As outlined in Figure 16.1b, these include:

C.M. Bunce, M.J. Campbell (eds.), *Nuclear Receptors*, Proteins and Cell Regulation 8, 381–417. DOI 10.1007/978-90-481-3303-1_16,

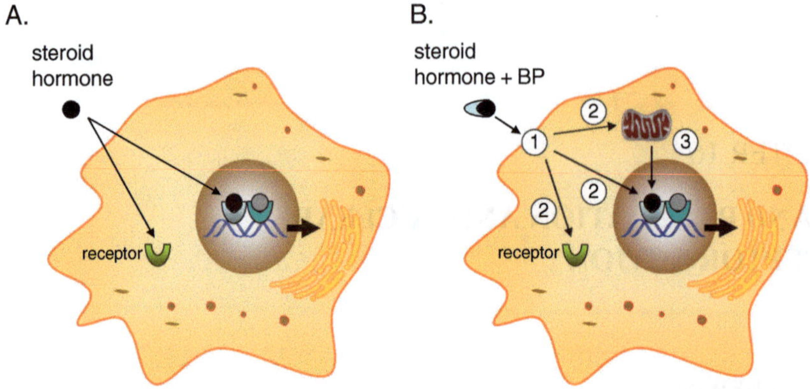

Figure 16.1. Ligand availability and nuclear receptor signalling. (**a**) Passive diffusion of active steroid hormones into target cells and subcellular translocation according to concentration gradients. (**b**) Facilitated mechanisms for: (*1*) cellular uptake of steroid hormones; (*2*) subcellular translocation; (*3*) intracrine activation of pro-hormone to active steroid hormone

(1) the cellular acquisition of lipophillic ligands from serum binding proteins
(2) intracellular trafficking of internalized ligands
(3) endogenous conversion to active or inactive metabolites

These adjuncts to nuclear signaling have been broadly termed 'Pre-receptor regulatory mechanisms' because of their 'upstream' role in modulating the 'downstream' actions of liganded nuclear receptors. The following chapter will review each of the three mechanisms using specific examples from the nuclear receptor superfamily to illustrate the importance of pre-receptor pathways in determining the initiation and efficacy of receptor action.

16.2. SERUM BINDING PROTEINS AND THE CELLULAR ACQUISITION OF NUCLEAR RECEPTOR LIGANDS

16.2.1. Introduction

Nuclear receptor ligands such as steroid hormones, along with retinoids and vitamin D share a structural relationship with cholesterol which means that they are highly lipophillic. Because of this they have a common requirement for serum carrier proteins that will ensure their appropriate delivery to target cells. Given the abundance of proteins in serum, much of the transport of nuclear receptor ligands is likely to be non-specific. Nevertheless, several ligand-specific serum carriers exist including corticosteroid-binding globulin (CBG) (glucocorticoids, mineralocorticoids), vitamin A (retinol)-binding protein (RBP), vitamin D-binding protein (DBP), sex hormone-binding globulin (SHBG) (estrogens, androgens), and thyroid hormone-binding globulin (TBG). To date these proteins have been studied primarily in the context of their impact on the analysis of serum levels of steroid hormones. However, recent reports have proposed a more active role for steroid binding globulins/proteins as a first step in the nuclear receptor signal transduction pathway [1, 2]. For example, CBG and SHBG act as relatively high affinity serum transporters for their respective groups of steroid hormones but they can also can bind to cell membranes in their liganded forms [3, 4], suggesting alternative actions for these proteins as signal transducers [5]. Membrane binding of liganded SHBG has been shown to stimulate intracellular cyclic AMP (cAMP) and an associated signaling cascade [6, 7]. In this way it appears that SHBG not only delivers estrogens and androgens to cells for classical receptor-mediated genomic signaling, but also provides its own non-genomic signal pathway with the potential to interact with more classical nuclear receptor responses.

Although membrane binding of proteins such as CBG and SHBG may in some instances act as an adjunct to nuclear signaling, the fundamental importance of serum binding proteins remains their role in delivering actual hormones to target cells. The mechanisms by which these molecules are then released and acquired by target cells are therefore crucial to nuclear receptor signaling and will be discussed in greater detail in the following sections with specific reference to vitamin D and retinol transport.

16.2.2. Cellular Acquisition of Receptor Ligands: Free Versus Bound

In serum, vitamin D metabolites are predominantly bound to the serum albumin and serum vitamin D binding protein (DBP). The affinity of albumin for 25OHD ($K_d = 1.7\ \mu M$) and $1,25(OH)_2D$ ($K_d = 19\ \mu M$) is substantially lower than that observed for DBP and 25OHD ($K_d = 1.4$ nM) and $1,25(OH)_2D$ ($K_d = 25$ nM) [8, 9]. Thus, the vast majority of vitamin D metabolites preferentially bind to DBP. However, because of the relative abundance of albumin in serum (650 μM) compared to DBP (5 μM) the potential remains for some vitamin D metabolites to be transported in the circulation by albumin. Additionally, in keeping with other steroid hormones, most of the DBP in serum is empty because of its molar abundance relative to the concentrations of vitamin D metabolites found in the periphery (25OHD = approximately 50 nM and $1,25(OH)_2D$ = approximately 0.1 nM). In view of these observations, it appears that most vitamin D metabolites in the circulation are bound to a carrier protein of some sort. However, despite this, the general assumption for vitamin D metabolites is that they are biological active when unbound, even though this fraction may be very small [9, 10]. Indeed the 'free-hormone hypothesis' has been proposed as a universal mechanism for the cellular uptake of nuclear receptor ligands [11, 12], largely because molecules such as vitamin D, sex hormones, corticosteroids, retinoids, and thyroid hormone are highly lipophillic and therefore have the potential to rapidly and passively diffuse across cell membranes. Nevertheless, in recent years the 'free-hormone hypothesis' has come under increasing scrutiny due, in part, to the disparity between likely amounts of ligand available for passive diffusion and the levels required to efficiently occupy target receptors. For example, it has been estimated that concentrations of free $1,25(OH)_2D$ in serum are as low as 10^{-13} M [9, 13], much less than the concentrations normally quoted for binding to the vitamin D receptor (VDR) (dissociation constant (K_d) = approximately 10^{-10} M).

Other objections to the free hormone hypothesis stem from recent studies which have shed new light on the manner in which steroid binding proteins such as DBP interact with target cells. Perhaps the most significant of these concern megalin, an endocytic protein which acts as a multi-ligand receptor at several sites, notably the apical surface of renal proximal tubule cells [14, 15]. In a seminal study, Willnow and colleagues showed that in the kidney megalin acts as a cell surface receptor for DBP, with the resulting complex being internalized by proximal tubule cells through endocytosis [16]. The presence of the vitamin D-activating enzyme 25-hydroxyvitamin D-1α-hydroxylase (1α-hydroxylase) in proximal tubules, coupled with DBP's relatively high affinity for 25OHD, means that the acquisition and internalization of DBP via its megalin receptor is a pivotal component of vitamin D metabolism. Consistent with this, megalin knockout mice were unable to recover DBP from the glomerulus of the kidney, and thus lose it and its vitamin D cargo in urine. As a consequence, megalin knockout mice are unable to adequately metabolize 25OHD to $1,25(OH)_2D$ resulting in a bone phenotype that resembles vitamin D deficient rickets [16].

The role of megalin as an endocytic receptor for steroid binding proteins is not restricted to vitamin D. Male and female megalin knockout mice show abnormalities in reproductive organs that are consistent with dysregulated sex hormone function [17]. However, this defect could not simply be rescued by increasing levels of androgens or estrogens in the knockout mice, suggesting that the sex hormone effect of megalin knockout was similar to that observed for vitamin D. Subsequent experiments confirmed that SHBG bound to ligands such as testosterone, dihydrotestosterone (DHT) or estradiol is taken up in an endocytic fashion by megalin-positive cells [17]. These observations, coupled with the expression of megalin in the urogenital tracts of male and female mice, support an additional role for this protein in sensitizing target tissues to the effects of androgens and estrogens through its effects on the acquisition of SHBG. Similar to its effects on vitamin D, megalin has also been linked to retinol homeostasis. Both the global and kidney-specific knockout of this gene have been shown to result in elevated excretion of retinol and its binding protein RBP [18, 19], leading to a concomitant depletion of liver stores of retinol metabolites [19].

16.2.3. Nuclear Receptor Responses in Serum Binding Protein Knockout Mice

In common with the megalin knockout mice, analysis of mice with ablation of genes for steroid hormone binding proteins/globulins has provided further insight on the role of these proteins in the cellular uptake of nuclear receptor ligands. For example, knockout mice for RBP are viable and fertile, but show retina dysfunction early in life [20]. However, this problem improved with age and the authors of this study concluded that the major consequence of RBP knockout was to impair mobilization of retinol stores under conditions of vitamin A restriction [20]. In a similar fashion, DBP knockout mice appear healthy and fertile despite having lower circulating levels of 25OHD and 1,25(OH)$_2$D [21]. The latter is presumably due to binding of vitamin D metabolites to albumin, which has a lower affinity for both 25OHD and 1,25(OH)$_2$D relative to DBP and will therefore be less effective in preventing urinary loss of vitamin D metabolites. Significantly, when DBP knockout mice were placed on a vitamin D deficient diet, they succumbed to bone mineralization abnormalities more rapidly than their wild type counterparts, underlining the importance of DBP in maintaining vitamin D bioavailability under conditions of dietary restriction. Conversely, loss of the DBP gene protected mice against potential toxic effects of vitamin D, decreased the half-life of vitamin D metabolites in the circulation and attenuated the timing of vitamin D-induced responses in peripheral tissues [21]. Collectively, these observations suggest that, like RBP, a key function of DBP is to maintain stable levels of 25OHD and/or 1,25(OH)$_2$D in serum whilst modulating their bioavailability to peripheral tissues.

More recent studies by Pike and colleagues confirmed the detrimental effect of DBP knockout on circulating levels of 1,25(OH)$_2$D but, paradoxically, showed that target tissue levels of the steroid hormone in DBP knockout mice were no different to those in tissue from heterozygous littermates [22]. Furthermore, parallel studies in vitro, showed that cells cultured in serum from DBP knockout mice were significantly more sensitive to 1,25(OH)$_2$D compared to cells cultured in serum from control mice [22]. These data suggest that for some VDR-expressing tissues, DBP is not an active facilitator of 1,25(OH)$_2$D uptake, but instead functions to limit its tissue bioavailability. Given the higher affinity DBP has for 25OHD compared to 1,25(OH)$_2$D it is likely that similar effects will also be observed for the precursor form of vitamin D.

Collectively, knockout mouse studies outlined above have helped to clarify the role of DBP as a mediator of the first stages of VDR signaling in vivo. On one level, DBP can act simply to sustain the bioavailability of vitamin D metabolites in the general circulation, particularly under conditions of vitamin D insufficiency. As outlined in Figure 16.2, this can be achieved, in part, by megalin-mediated cellular uptake of DBP and its bound metabolites. Such a mechanism is fundamental to the endocrinology of vitamin D and, specifically the renal synthesis of active 1,25(OH)$_2$D [23]. However, given the increasing numbers of extra-renal tissues documented as expressing CYP27b1, the question arises as to whether megalin-mediated uptake of DBP is also involved in delivering substrate to the enzyme in non-renal cells. Such a mechanism has been postulated for breast cancer and osteoblastic cells expressing CYP27b1 [24–27], but it may not be universal. Moreover, although the uptake of 25OHD by proximal tubule cells is clearly linked to their internalization of DBP, it is possible that related but different mechanisms apply to other

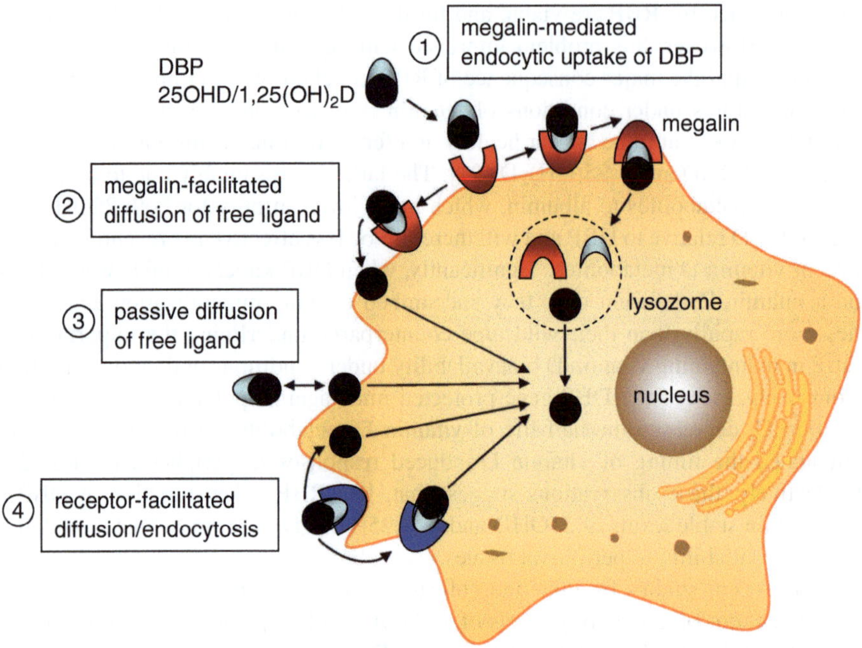

Figure 16.2. Membrane receptor-mediated and receptor-independent mechanisms for the cellular uptake of vitamin D. Vitamin D metabolites (25-hydroxyvitamin D [25OHD] and 1,25-dihydroxyvitamin D[$1,25(OH)_2D$]) are bound to vitamin D-binding protein (DBP) in serum and extracellular fluid. Intracellular uptake of vitamin D metabolites may occur via one of four different mechanisms outlined in the schemer

megalin-expressing tissues. Specifically, as shown in Figure 16.2, megalin-binding of DBP may simply act to increase localized concentrations of 'free' vitamin D metabolites at the cell membrane, thereby facilitating more effective passive diffusion of these molecules. It is also possible that for some cells internalization of DBP takes place via mechanisms that do not involve megalin. Uptake of DBP has been reported for B-lymphocytes, but this does not appear to involve the same clathrin-coated pits that are characteristic of the megalin-mediated pathway [28]. One possibility is that binding of DBP to this type of cell is mediated by the gamma Fc receptors that also associate with immunoglobulins [29]. Finally, in most cases, megalin-DBP interaction appears to facilitate transport of 25OHD and its subsequent intracrine activation via localized expression of CYP27b1. By contrast, studies using serum from wild type and knockout mice have shown that DBP acts to attenuate responses to its weaker ligand, $1,25(OH)_2D$, raising the possibility that endocrine responses to the active form of vitamin D are independent of its binding protein [30].

16.2.4. Mechanisms for the Transfer of Extracellular Ligands to Intracellular Binding Sites

A key advantage of membrane receptors for serum binding proteins is that it enables the targeting of nuclear receptor ligands to tissues where they are required. An additional potential benefit of this type of mechanism is that it may also help to deliver the ligand to corresponding proteins involved in their intracellular transport. In the case of vitamin D, it has been proposed that 25OHD or $1{,}25(OH)_2D$ traffic to either mitochondria or the nucleus via an intracellular vitamin D binding protein (IDBP) [31], and this is discussed in more detail later in the chapter. The precise manner by which this intracellular transport pathway interfaces with the extracellular vitamin D-DBP system has yet to be fully defined. However, a clearer picture of the membrane transfer of steroid hormones to intracellular targets is provided by studies of vitamin A and its corresponding serum binding protein, RBP.

Serum transport of vitamin A (all-*trans*-retinol, retinol) has many similarities to the vitamin D system. Retinol can associate non-specifically with serum albumin ($K_a = 45 \times 10^{-6}$ M), but its preferred carrier is RBP, a member of the lipocalin family of proteins ($K_a = 1.5 \times 10^{-6}$ M) [32]. In common with DBP, RBP is synthesized primarily in the liver [33], where it forms a complex with another protein transthyretin (TTR), which stabilizes binding of retinol to RBP and increases the half-life of the RBP-TTR complex [34]. The net effect of this is that for average circulating retinol concentrations of 2 μM, approximately 95.5% is carried as part of an RBP-TTR complex, 4.4% is bound to RBP in the absence of TTR and 0.1% is free retinol [35]. Thus, as with vitamin D, the question arises as to whether these binding proteins are central to the acquisition of retinol by target cells.

Membrane binding of RBP has been described for a variety of cell types (reviewed in [35]), suggesting that the intracellular uptake of retinol involves a specific plasma-membrane receptor for RBP [36]. As described earlier (Section 16.2.2), megalin can act as a binding site for RBP in the proximal tubules of the kidney, However, studies using bovine retina pigment epithelial cells have shown that the retinoic acid-induced protein STRA6 is a specific membrane receptor for RBP, which is expressed by a variety of tissue types including the brain, eye, testis, kidney, spleen and female reproductive tract [37, 38]. In contrast to the megalin-mediated endocytic uptake of RBP or DBP, the association between RBP and STRA6 involves membrane translocation and intracellular release of free retinol ligand only [35]. In a series of studies using chimeric proteins containing the RBP protein loops required for interaction with its membrane receptor, Findlay and colleagues showed that membrane binding of RBP results in conformational changes that lead to release of the retinol ligand [39]. As illustrated in Figure 16.3, this model system proposes that the RBP is only recognized by the membrane receptor in a liganded (holo) form. Binding of holo RBP to its receptor (STRA6) causes a conformational change in the receptor which releases the empty (apo) form of RBP, whilst simultaneously delivering free retinol for transfer to the cytoplasm.

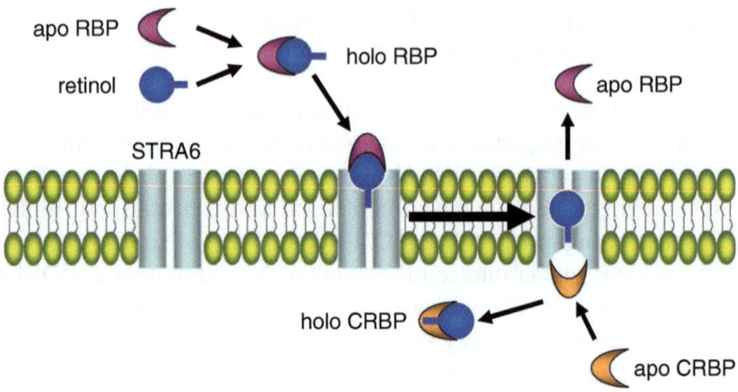

Figure 16.3. Membrane receptor-mediated cellular uptake of vitamin A. Vitamin A (retinol) is transported in serum and extracellular fluid by retinol binding protein (RBP). Liganded (holo) RBP binds to its membrane receptor STRA6 and in this form induces conformational changes in STRA6 that allow transfer of retinol to a hydrophobic channel. This in turn allows retinol to cross the cell membrane, with its hydrocarbon side-chain exposed. On the interior of the cell, unliganded (apo) type I cellular retinol-binding protein (CRBP I) interacts with the cytoplasmic surface of STRA6 and in this form induces conformational changes in STRA6 that allow acceptance of retinol by CRBP I

The conformational changes in the liganded- RBP receptor facilitate the membrane transfer of retinol and also provide a mechanism for subsequent transfer to intracellular proteins. The cellular retinol binding protein (CRBP) has structural similarities to lipophilic transport proteins and plays a pivotal role in both the cellular translocation and metabolism of retinoids [40] (see Sections 16.3.2 and 16.4.3). Prior to the identification of a specific membrane receptor for RBP, it was hypothesized that this protein was also involved in the acquisition of retinol by CRBP. Studies showed that transfer of retinol from RBP to CRBP was enhanced in the presence of cell membrane preparations, but this was substantially reduced when receptor-depleted membranes were used [36]. In these reports, it was also noted that: (1) transfer of retinol from RBP only occurs when CRBP is in its empty (apo) formation; (2) binding of retinol to CRBP (retinol hydrocarbon tail embedded in CRBP) is in the opposite orientation to retinol bound to RBP (retinol bound but hydrocarbon tail exposed) [36]. Thus, the RBP transfer model depicted in Figure 16.3 incorporates the intracellular acquisition of retinol by CRBP. Findlay and colleagues went on to show that only the apo form of CRBP was able to interact with holo RBP-membrane preparations. As RBP and CRBP do not directly associate with each other, it seems likely that transfer of retinol from one of these proteins to the other is mediated by the same receptor protein, suggesting that STRA6 is a receptor for both RBP and CRBP [39].

Thus, it has been proposed that binding of holo RBP to the extracellular surface of STRA6 introduces a conformational change in this receptor that enables the release of retinol for membrane translocation. The precise mechanism by which the retinol is then transferred to CRBP has yet to be defined. However, the channel that forms the retinol binding site within the membrane receptor is narrow but hydrophobic and, as such, may be ideal for the relatively water-insoluble retinol. Under these conditions retinol can move across the membrane with its hydrocarbon side-chain and hydroxyl group leading the way, and thus in the correct orientation for insertion in the CRBP, which is itself in the correct conformation (apo) for receiving retinol [39]. The downstream pathways that follow retinol-CRBP binding are discussed in more detail in subsequent sections of this chapter (Section 16.3.2), together with discussion of how related mechanisms may function for other nuclear receptors (Section 16.3.3).

16.3. SUBCELLULAR TRAFFICKING OF NUCLEAR RECEPTOR LIGANDS

16.3.1. Introduction

As illustrated in Figure 16.1a, the common view of nuclear receptor ligands is that as lipophillic molecules they enter cells by passive diffusion and then randomly migrate through the cytoplasm until they interact with their cognate intracellular receptor, localized either in the cytoplasm or nucleus. Until recently this perception restricted analysis of the subcellular movement of steroid hormones to a small number of studies focused primarily on the organelle-specific distribution of these molecules [41]. This is in stark contrast to the broad spectrum of publications associated with intracellular transport of sterols [42–46] and, in particular, the role of steroid acute regulatory protein (StAR) as a mitochondrial transporter for cholesterol during steroidogenesis [47–54]. Likewise, as detailed in several other chapters throughout this book, much is known about the mechanisms by which nuclear receptors undergo intracellular translocation [55–59]. However, the question arises at to whether there is a similar infrastructure for the subcellular movement of their ligands, and if so how might this interface with the membrane transport mechanisms outlined in the first part of this chapter?

16.3.2. Intracellular Transport of Vitamin A

In Section 16.2.4, the mechanism by which retinol is transferred from extracellular RBP to intracellular CRBP is described. Whilst this mechanism provides a detailed example of non-passive, receptor mediated, uptake of a sterol, it also illustrates the potential for involvement of intracellular binding proteins in the downstream trafficking of these molecules. As outlined in Figure 16.3, the transfer of holo RBP to STRA6 facilitates the membrane translocation of retinol, but this mechanism is then expanded by the binding of apo CRBP to the cytoplasmic surface of STRA6, which induces conformational changes in the membrane receptor that enable the transfer of retinol to CRBP. In this way, the membrane transport of retinol forms the first stage of a series of intracellular binding interactions that are central to both the metabolism of retinol and subsequent transfer of retinoic acid to its congnate nuclear retinoic acid receptor (RAR).

To date, four CRBPs (CRBPI-IV) belonging to the fatty acid-binding protein (FABP) family of proteins have been identified [60–62]. Although they share size (approximately 15 kDa) and sequence similarities, as well as common genetic and functional properties, the CRBPs exhibit distinct patterns of tissue distribution. Expression of CRBP I has been described for many tissues (except for the small intestine) where it functions as the principal cytoplasmic transporter for retinol [60]. CRBP I also appears to play a pivotal role in the metabolism of retinol. Knockout mice for CRBP I show none of the gross phenotypic abnormalities associated with retinoic acid deficiency but the livers of these animals exhibit a marked reduction in levels of retinyl ester, a retinol metabolite [63]. Moreover, when CRBP I knockout mice were placed on a vitamin A-deficient diet, the levels of retinyl esters decreased still further, leading to a rapid onset of symptoms of severe vitamin A deficiency in the animals [63, 64]. In contrast to CRBP I, the type II binding protein is almost exclusively expressed in the small intestine, where it plays a key role in the dietary uptake of vitamin A [60, 65, 66]. Ablation of the CRBP II gene in mice produces only moderate phenotypic changes when animals are vitamin A replete [67]. However, breeding of vitamin A-deficient knockout mice resulted in increased mortality of the resulting offspring, irrespective of the genotype of these pups [67]. Thus, in common with DBP [21] and RBP [20], CRBP I and CRBP II appear to be particularly important in maintaining the bioavailability of their ligands under conditions of dietary restriction. The physiological importance of CRBP I and CRBP II is underlined by the fact that they are highly conserved during evolution.

CRBP III is found mainly in the heart, skeletal muscle, adipose tissue and mammary gland [61]. The physiological significance of CRBP III has yet to be clarified, but knockout mouse studies suggest that it functions, in part, by facilitating the transfer of esterified retinol metabolites into breast milk [68]. CRBP IV is closely related to CRBP III [62]. Although its precise function remains uncertain, the relative abundance of CRBP IV in adult kidneys suggests that, along with megalin [18, 19], it plays a role in the renal homeostasis of retinol.

In addition to their association with retinol, CRBP I and CRBP II will also bind all-*trans*-retinal (retinal) which is generated from retinol in the presence of the enzyme retinol dehydrogenase. This is, in turn, leads to the generation of all-*trans*-retinoic acid (retinoic acid) following activity of the enzyme retinal dehydrogenase [69]. Retinoic acid is transported in the cytoplasm by one of two cellular retinoic acid binding proteins (CRABP I or CRABP II), and a related cellular retinal binding protein (CRALBP) is found in cells from the eye where it primarily binds 11-*cis*-retinal or 11-*cis*-retinol (reviewed in [70, 71]). CRABP I and CRABP II show 74% sequence similarity and, like CRBP I and II, they are highly conserved across species, underlining their potential importance during evolution [72–75]. In adults, CRABP I is virtually ubiquitous, while CRABP II is expressed only in the skin, uterus, ovary and choroid plexus (Reviewed in [70]). Knockout mice for either CRABP I or II are generally normal, but exhibit some minor limb abnormalities [76, 77], suggesting that the CRABPs facilitate the tissue retinoic acid gradients associated with normal limb bud development.

The diverse array of intracellular proteins with the capacity to bind retinoids facilitate three key steps in retinoid signaling: (1) modulation of the metabolism and storage of vitamin A metabolites; (2) subcellular translocation of retinoids; (3) delivery of retinoic acid to nuclear retinoic acid receptors (RARs) and, through further metabolism, to ligands for RXRs (see Figure 16.4). The extent to which these functions are operative at any given time depends on the cell type involved and the uptake of retinol. Under conditions of vitamin A sufficiency, retinol is stored in the liver and the model system illustrated in Figure 16.4 shows how CRBPs

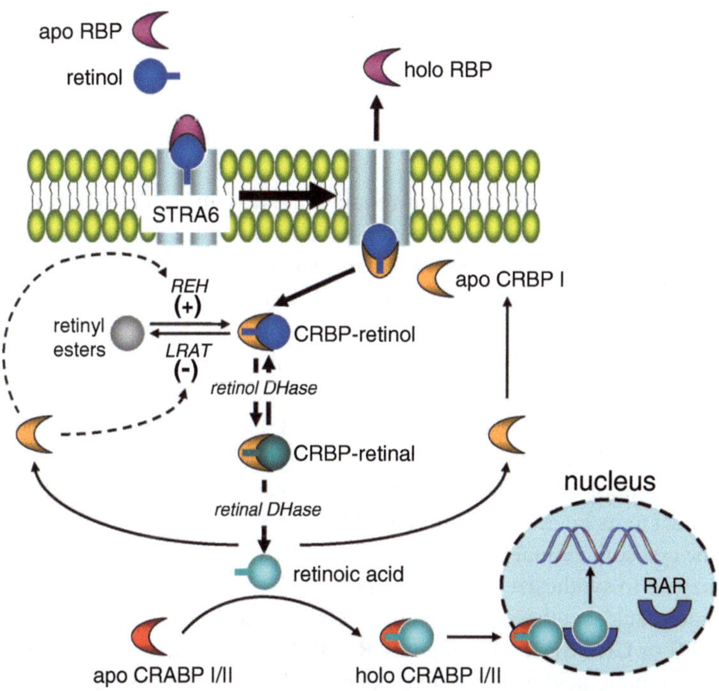

Figure 16.4. Intracellular binding proteins and the regulation of retinol metabolism and subcellular transport. Cellular uptake of retinol involving retinol-binding protein (RBP), STRA6 and type I cellular retinol-binding protein (CRBP I) is a key step in the regulation of retinol metabolism and cytoplasmic transport. Under conditions of vitamin A (retinol) sufficiency, Retinol bound to CRBP I is converted to storage retinyl esters catalyzed by lecithin:retinol acyl transferase (LRAT). Under conditions of retinol deficiency, excess of unliganded (apo) CRBP I suppresses LRAT and enhances retinyl ester hydrolase (REH) which catalyzes the synthesis of retinol from stored retinyl esters. CRBP-retinol is converted to retinal by the enzyme retinol dehydrogenase (retinol DHase). This, in turn, is converted to retinoic acid by retinal dehydrogenase (retinal DHase). Retinoic acid is bound in the cytoplasm by type I or type II cellular retinoic acid-binding proteins (CRABP I/CRABP II), which also interact with the nuclear retinoic acid receptor (RAR)

and CRABPs interact to facilitate retinol metabolism and signaling in a hepatic environment.

Retinol is stored in the form of retinyl esters following esterification of retinol in the presence of the enzyme lecithin: retinol acyltransferase (LRAT). This esterification step can also be carried out by the enzyme acyl-CoA retinyl acyl transferase (ARAT) but only LRAT can esterify CRBP-bound retinol (reviewed in [78]). Significantly the apparent K_m for LRAT and retinol is unaffected by presence of CRBP, suggesting that the binding protein participates in the delivery of substrate to LRAT [79]. In contrast to the facilitating effect of holo RBP, apo RBP acts as an inhibitor of LRAT activity [79]. Other studies have reported that apo CRBP also stimulates the enzyme retinyl ester hydrolase (REH), which opposes LRAT by catalyzing conversion of retinyl esters to retinol. In this way, CRBP I plays a key role in defining the dynamics of retinol/retinyl ester metabolites in liver cells, depending on the dietary availability of vitamin A. Under conditions of vitamin A sufficiency, holo RBP will promote storage of retinyl esters by promoting LRAT effects on retinol. By contrast, under conditions of vitamin A restriction, levels of apo CRBP will increase, leading to decreased levels of retinyl esters (either through inhibition of LRAT or induction of REH).

The conversion of retinol to retinoic acid requires the activity of dehydrogenase enzymes. Free retinol can be metabolized non-specifically by a variety of alcohol dehydrogenases. However, these enzymes are unable to access retinol when it is bound to CRBP I, so that another facet of CRBP I function is to facilitate the metabolism of retinol by a more specific enzyme, in this case retinol dehydrogenase (retinol DHase) [80]. As with LRAT, it appears that CRBP I can influence the conversion of retinol to retinal by retinol DHase, through direct interaction with the enzyme. As well as binding the substrate (retinol), CRBP I can also bind the product of retinol DHase, retinal. In doing so CRBP I has the potential to deliver retinal to the next enzyme in the retinol metabolic chain, retinal dehydrogenase (retinal DHase). However, in contrast to its interaction with retinol DHase, CRBP I does not appear to access readily the cytosolic retinal DHase but is more likely to utilize the microsomal forms of this enzyme to synthesize retinoic acid [80, 81].

All trans retinoic acid synthesized by retinal DHase can bind to two cytosolic proteins, CRABP I or CRABP II. Binding to CRABP I has been shown to facilitate the catabolism of retinoic acid to less active polar metabolites, suggesting that one of its functions is to rheostatically attenuate RAR signaling [82]. Both CRABP I and CRABP II are detectable in the nucleus and thus have the potential to interact with the RAR. The movement of retinoic acid from CRABP I to RAR appears to be a passive process which is dependent on retinoid concentration gradients. By contrast, CRABP II actively facilitates the delivery of retinoic acid to RAR through direct protein-protein interaction [83]. Thus, CRABP II appears to facilitate the formation of holo RAR. Indeed antisense inhibition of CRABP II has been shown to suppress RAR-mediated responses [84].

As illustrated in Figure 16.4, the complex family of intracellular binding proteins for retinol, retinal and retinoic acid act as pluripotent modulators of the metabolism, storage and subcellular translocation of vitamin A metabolites. The binding protein interactions outlined in Figure 16.4 play a central role in hepatic retinol homeostasis and in particular the metabolism and storage of retinol by hepatocytes and stellate cells [63]. However, the widespread distribution of CRBP I and CRABPs I and II means that these binding proteins can act as intracellular transporters of retinoids in many other tissues. It is likely that the intricate mechanisms for uptake and intracellular processing of retinol arose because of the evolutionary advantage they confer under conditions of vitamin A restriction. This raises the question as to whether similar mechanisms exist for other nuclear receptor ligands and whether these will also be linked in some way to improved homeostatic regulation of nuclear receptor action.

16.3.3. Intracellular Binding Proteins and the Cytoplasmic Trafficking of Vitamin D and Estrogen

New World primates (NWP) exhibit significantly higher circulating levels of steroid hormones than Old World primate (OWP) (including humans), but they are also relatively insensitive to these hormones [85]. End-organ resistance to glucocorticoids in NWPs is known to involve altered binding affinity of the glucocorticoid receptor (GR) for its preferred ligand cortisol [86]. However, more recent studies have implicated an additional mechanism involving differential interaction between GR and co-chaperones such as the immunophilins FKBP52 (humans and OWPs) and FKBP51 (NWP) [87]. NWP are also profoundly resistant to vitamin D and estrogens. As with glucocorticoids, insensitivity to $1,25(OH)_2D$ and estrogens such as 17β-estradiol (E_2) appears to be due to impaired signaling via cognate nuclear receptors, VDR and the estrogen receptor (ERα) respectively. In this instance, the blockade of VDR and ERα responses in NWPs has been shown to be associated with elevated levels of soluble proteins that act as competitors for binding of $1,25(OH)_2D$ or E_2 [88–90]. Purification of the NWP intracellular vitamin D-binding protein (IDBP) revealed similarity to the constitutively-expressed form of human heat-shock protein 70 (hsc70) [89], and also indicated that hsc70 was capable of binding both 25OHD and $1,25(OH)_2D$, as well as gonadal steroids such as estradiol [91]. However, a specific NWP intracellular estrogen-binding protein (IEBP) has also been cloned and has similarity to human heat-shock protein 27 (hsp27) [90].

Based on characterization of the IDBP and IEBP in NWP cells, it was assumed that hsc70 and hsp27 act as intracellular 'decoys' for the relatively high levels of $1,25(OH)_2D$ and E_2 present in NWPs, thereby attenuating over-exuberant VDR or ERα signaling. This is certainly true of hsp27 which suppresses E_2-ERα-mediated gene transcription when over-expressed in human cells [92]. By contrast, studies using various in vitro models showed that over-expression of hsc70-like IDBP cloned from NWP cells enhanced $1,25(OH)_2D$-VDR-mediated gene transcription [93]. The IDBP also enhanced metabolism of 25OHD to $1,25(OH)_2D$ via the enzyme 1α-hydroxylase, indicating that it is involved in the intracellular movement of both active and inactive vitamin D and can traffic these metabolites to both the nucleus and mitochondrion [93, 94]. Thus, it has been proposed that rather than acting as a simple VDR decoy, IDBP/hsc70 functions as a specific intracellular transporter for vitamin D [95, 96] (see Figure 16.5).

Hsc70 and hsp27 have ATP-binding sites which play a pivotal role in directing the conformational changes required for classical protein chaperone actions of heat-shock proteins [97, 98]. Recent studies have shown that binding of ATP to hsc70 alters the latters association with vitamin D metabolites and may thus facilitate the transfer of 25OHD or $1,25(OH)_2D$ to downstream effector proteins [99]. Potential targets include the VDR, CYP27b1 or other proteins such as the feedback control enzyme 24-hydroxylase (CYP24). However, given the protein chaperone functions of heat-shock proteins, it is also possible that hsc70 interacts with other proteins as part of the subcellular movement of vitamin D metabolites. One possible candidate for interaction with hsc70 is the co-chaperone Bcl-2–associated athanogene (BAG-1) which modulates the protein-folding activities of hsc70 [100]. BAG-1 can also act as

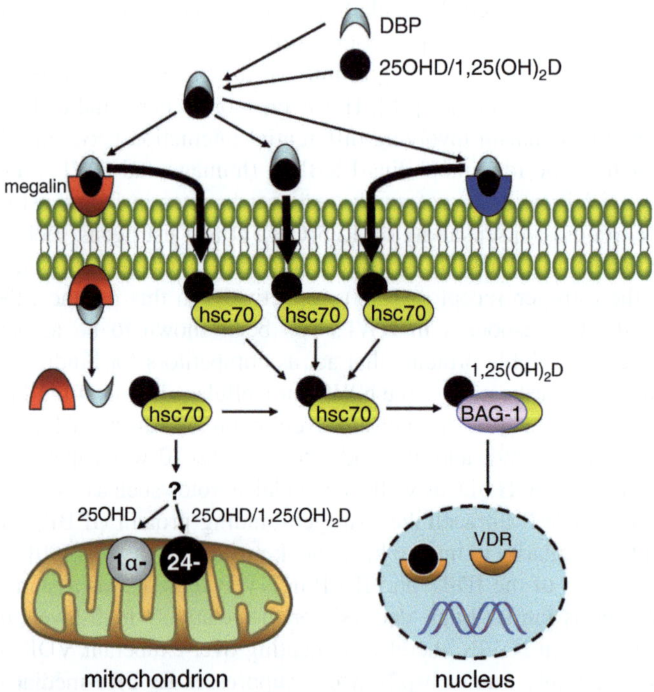

Figure 16.5. Intracellular binding proteins for vitamin D. Vitamin D metabolites such as 25-hydroxyvitamin D (25OHD) and 1,25-dihydroxyvitamin D ($1,25(OH)_2D$) are internalized by cells via megalin-dependent or –independent mechanisms. The constitutive form of heat-shock protein 70 (hsc70) acts as an intracellular binding protein for both metabolites, facilitating their metabolism by the mitochondrial enzymes 1α-hydroxylase (1α-) and 24-hydroxylase (24-) and the nuclear translocation of $1,25(OH)_2D$, the latter in conjunction with BAG-1

an intracellular binding site for vitamin D metabolites but, in contrast to hsc70, it has a higher affinity for $1,25(OH)_2D$ than 25OHD [101]. Furthermore, over-expression of BAG-1 in VDR-responsive cells increased $1,25(OH)_2D$-induced gene transcription but had no effect on either CYP27b1 or CYP24-metabolism in the same cells, suggesting that it acts as a nucleus-specific chaperone for $1,25(OH)_2D$ [101].

A key question arising from studies detailed above, is why are NWPs vitamin D resistant when they express high levels of an hsc70-like IDBP that acts as an efficient intracellular chaperone for vitamin D metabolites? The answer to this is that NWPs have high levels of another protein which efficiently suppresses VDR-mediated signaling at the level of gene promoter vitamin D response elements (VDRE). Adams and colleagues showed that, compared to OWPs, NWPs over-express a 38 kDa protein which actively competes with the VDR for binding to the half-sites of VDRE [102]. This VDRE-binding protein (VDRE-BP) was shown to be homologous with members of the heterogeneous nuclear ribonucleoprotein (hnRNP) family of RNA processing proteins, and acted to suppress $1,25(OH)_2D$-induced transcription when transfected into human cells [103]. Subsequent studies suggest that human VDRE-BP is hnRNPC1/C2 [104], and that aberrant expression of this protein can cause vitamin D resistant rickets in humans [105]. Thus, with respect to vitamin D, it appears that cells from NWPs express high levels of two distinct proteins with opposing effects on VDR-signaling. On the one hand the increased capacity for transport and metabolism of vitamin D via IDBP may have arisen in response to high levels of UV-driven synthesis of vitamin D in NWPs continually exposed to sunlight in their rain forest habitat. However, this scenario would also have detrimental hypercalcemic side-effects if allowed to operate unchecked, underlining the biological advantage conferred by coincidental expression of a protein that acts to attenuate VDR responses.

NWPs also exhibit high circulating concentrations of estrogens, in part because their diet includes substantial intake of phytoestogens. Thus, as with the vitamin D model, an IEBP is likely to confer a biological advantage by providing increased capacity for cytoplasmic handling of estrogens. However, in contrast to DBP, the IEBP acts to suppress ERα-mediated responses [90]. This appears to be due to the fact that hsp27 also binds to the estrogen equivalent of the VDRE-BP, the estrogen response element (ERE)-binding protein (ERE-BP) [106]. Cloning of the NWP ERE-BP revealed sequence similarity to a human hnRNPC-like protein and over-expression of this protein was shown to inhibit ERα-induced transcription by competing with ERα for ERE half-site binding [107, 108]. Similar ERE-BP activity was also observed in human cells transfected with hsp27 and yeast two hybrid and GST-pulldown analyses showed that this was due to dysregulated cytoplasmic-nuclear shuttling of the ERE-BP as a consequence of enhanced binding to the over-expressed IEBP [106]. Thus, in addition to its ability to bind ligand E_2, it appears that hsp27 can also influence estrogen signaling by binding to ERE-BP and modulating occupancy of the ERE. As yet, it is unclear whether similar protein-protein interaction also occurs with the IDBP and VDRE-BP but, in view of the enhancement of VDR signaling by IDBP, this seems unlikely.

16.4. PRE-RECEPTOR METABOLISM OF NUCLEAR RECEPTOR LIGANDS

16.4.1. Introduction

The simplistic model for nuclear receptor signaling outlined in Figure 16.1a not only makes the assumption that ligands access their cognate receptors via passive diffusion and subcellular translocation, it also assumes that this pathway involves only the specific ligand for the receptor in question. However, as documented for the retinol system in Section 16.3.1, endogenous enzyme activity can also play a significant role in nuclear receptor signaling by modulating the amount of intracellular ligand available for receptor binding. Thus, localized or intracrine metabolism provides the third mechanism for regulation of nuclear receptor ligand availability to be detailed in this chapter. Ligand metabolism itself represents a distinct component of endocrinology that has already been well documented in previous publications. As such, the remainder of this chapter will focus on the specific ways in which localized metabolism of nuclear receptor ligands can affect target receptor function. Broadly speaking these can be divided into four categories:

(1) Pre-receptor activation of adrenal steroids
(2) Pre-receptor activation of dietary/environmental pro-hormones
(3) Pre-receptor catabolism of steroid hormones
(4) Pre-receptor metabolism of thyroid hormone

16.4.2. Pre-receptor Activation of Adrenal Steroids

The network of enzymes involved in converting cholesterol to corticosteroids and sex hormones has been well documented in previous reviews [109]. As outlined in Figure 16.6, many of these reactions occur in the adrenal glands, resulting in the generation of adrenal steroids such as aldosterone and cortisol. However, some steroids generated by the adrenal glands function as precursors for other steroid hormones such as the sex hormones E_2, testosterone and dihydrotestosterone (DHT). In this case, the key reactions occur primarily at peripheral sites such as the ovaries (estradiol) or testes (testosterone). Despite the extra-adrenal nature of this metabolism, the systemic consequences of the products means that the enzymes in question (aromatase and 17β-hydroxysteroid dehydrogenases [17β-HSDs]) are still integral to the endocrine actions of estrogen and androgen receptors. However, in more recent years it has become clear that extra-adrenal steroid hormone metabolism is also central to the pre-receptor regulation of both corticosteroids and sex hormones. The most striking examples of this are the role of aromatase as a localized generator of E_2, and 11β-hydroxysteroid dehydrogenase type 1 (11β-HSD1) as a localized generator of cortisol.

Aromatase is a cytochrome P450 located in the endoplasmic reticulum which catalyzes the conversion of C19 androgens such as androstenedione and testosterone

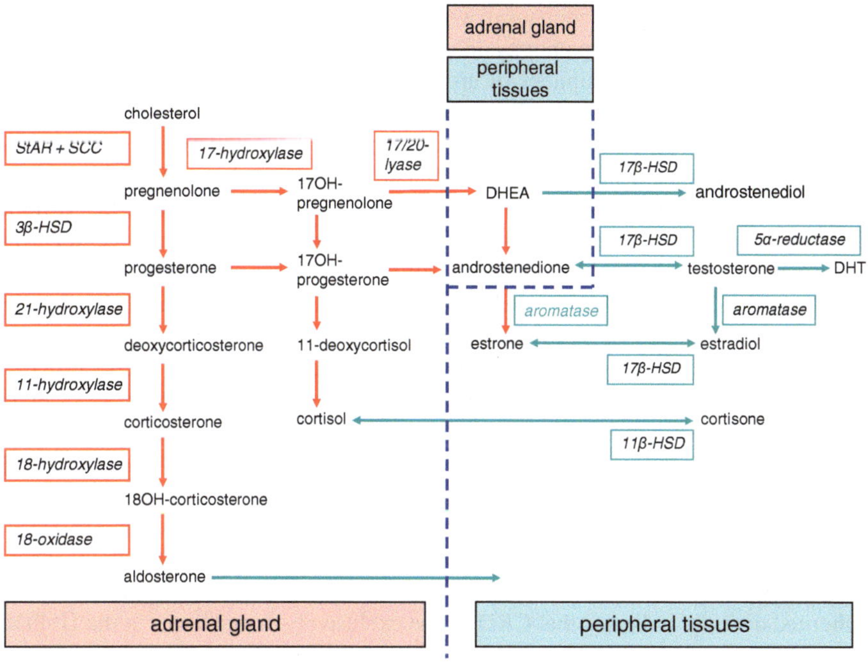

Figure 16.6. Steroidogenesis of sex hormones and corticosteroids in the adrenal gland and peripheral tissues

to their corresponding C18 estrogens (estrone and estradiol respectively) [110]. The enzyme plays a pivotal role in maintaining levels of sex hormones at both a systemic and local level, and this varies considerably with age and sex. For example, in men, testosterone acts as a systemic hormone, whereas in pre-menopausal women testosterone is synthesized within localized tissues. By contrast, in pre-menopausal women, estradiol acts in an endocrine fashion to support reproductive functions such as ovulation and implantation following fertilization, whilst in men and postmenopausal women, it is now clear that estradiol exerts its effects in an intracrine fashion [111]. Collectively, these observations, together with analysis of aromatase gene (CYP19) knockout (ArKO) mice [112], have underlined the importance of this enzyme as a pre-receptor regulator of estrogen and androgen receptor responses in both men and women. Furthermore, the expression of aromatase in a variety of ER-positive tissues outside the gonads, including breast and adipose tissue, bone, vasculature and brain indicates that the pre-receptor actions of aromatase have broad-ranging physiological consequences. Recent studies have linked the intracrine actions of aromatase with obesity [113], osteoporosis [114] and male sexual behaviour [115]. However, to date the most significant clinical facet of intracrine aromatase activity has been its effects on breast tumor development in post-menopausal women [116]. The high levels of estradiol detected in the neoplastic breast tissue from post-menopausal women, has implicated aberrant pre-receptor aromatase activity in the pathophysiology of these tumors. This, in turn, has led to the development of inhibitors of aromatase such as anastrozole, letrozole, and exemestane, which are now first-line agents for the treatment of breast cancer [117].

The other major group of steroid hormones that are subject to extensive peripheral metabolism are the glucocorticoids. As detailed later in this chapter (Section 16.4.4), cortisol produced by the adrenal glands is subject to sensitive intracrine inactivation in the kidney as part of a mechanism to protect the mineralocorticoid receptor from elicit occupancy. Although this activity fulfills a localized pre-receptor function, it also contributes to circulating levels of pro-hormone cortisone, which can then be re-converted into cortisol by the enzyme 11β-hydroxysteroid dehydrogenase type 1 (11β-HSD1) which, in contrast to 11β-HSD2, acts predominantly as a reductase (see Figure 16.7). Expression of 11β-HSD1 occurs primarily in glucocorticoid receptor (GR)-rich tissues such as liver and adipose tissue, so that it functions as a promoter of GR responses through the localized regeneration of cortisol levels [118, 119]. Defective activity of 11β-HSD1, known as cortisone reductase deficiency (CRD), was originally thought to be due to combined mutations in the gene for 11β-HSD1 (HSD11B1) and the enzyme hexose-6-phosphate dehydrogenase (H6PDH) that supplies NADPH co-factor to 11β-HSD1 [120]. However, a recent re-appraisal of this inherited disorder indicates that CRD is due exclusively to mutations in the H6PDH gene [121].

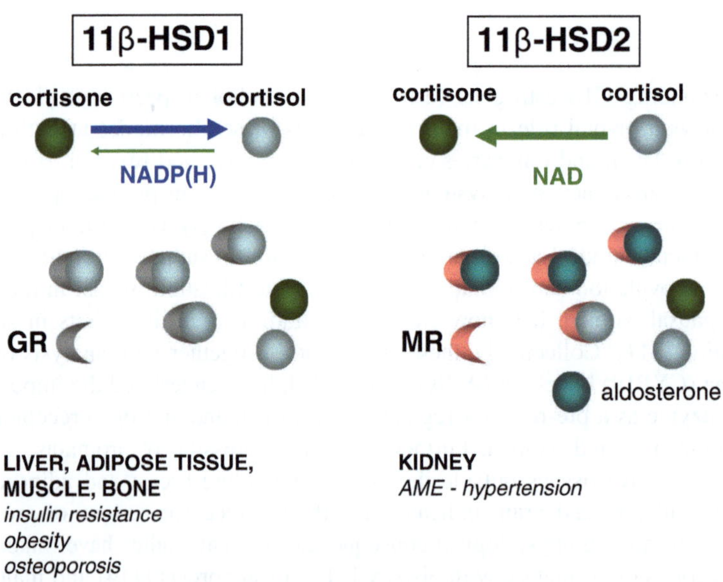

Figure 16.7. 11β-hydroxysteroid dehydrogenases, pre-receptor metabolism of corticosteroids and regulation of nuclear receptor function. Interconversion of cortisone and cortisol in glucocorticoid receptor (GR) and mineralocorticoid receptor (MR)-rich tissues by the enzymes 11β-hydroxysteroid dehydrogenase type 1(11β-HSD1) and 11β-HSD2

The abundant expression of 11β-HSD in tissues with strong GR-mediated responses such as adipose tissue [122–125], bone [126, 127] and the brain [128–131], indicates that the enzyme influences a wide-range of physiological responses at a pre-receptor level. This, in turn, has highlighted potential applications for inhibitors of 11β-HSD1 in the treatment of a variety of clinical disorders associated with aberrant glucocorticoid signaling. At present these studies are focused on major healthcare issue such as insulin resistance, obesity and type 2 diabetes [132–139]. The success of this pharmacological strategy will very much depend on the extent to which inhibitors of 11β-HSD1 can be applied to other glucocorticoid-related disorders such as steroid-induced osteoporosis. In addition, as with aromatase, further work is required to improve the tissue-specificity of 11β-HSD1 inhibitors.

Although intimately associated with glucocorticoid metabolism and GR signaling, the enzyme 11β-HSD1 has also been reported to catalyze metabolism of other steroidal substrates that may influence nuclear receptor signaling [140]. This observation underlines the potential for some enzyme to influence a broad spectrum of nuclear receptor signaling. Another example of this is the enzyme aldo-keto reductase 1C3 (AKR1C3) which exhibits two distinct types of metabolism. As a 17β-hydroxysteroid dehydrogenase (17β-HSD) it catalyzes conversion of Δ^4-androstene-3,17-dione to testosterone, increasing the accumulation of androgen receptor (AR) ligand in tissues such as the prostate [141]. The induction of AR signaling in the prostate is known to stimulate prostate cell proliferation and thus inhibition of AKRC13 represents an important target for the treatment of prostate cancer [142]. However, AKR1C3 may also play an indirect role in estrogen generation via coincident expression of aromatase which catalyzes conversion of testosterone to estradiol. In this way, inhibition of AKRC13 may provide a novel mechanism for modulating ER-mediated signaling in tissues such as breast and endometrium [143, 144].

In addition to its role as a 17β-HSD, AKR1C3 can also act as a prostaglandin (PG) F synthase by catalyzing the conversion of PGH_2 to $PGF_{2\alpha}$ or PGD_2 to $11\beta PGF_2$ [141, 145]. These metabolic pathways may promote nuclear receptor responses that are entirely independent of AKR1C3's role in androgen metabolism. For example, the reduction of PGD_2 to $11\beta PGF_2$ limits the amount of PGD_2 that is available for conversion to PGJ_2 metabolites, which can act as ligands for the nuclear peroxisomal proliferator-activated receptor (PPAR) γ. Signaling via PPARγ is known to promote anti-proliferative and pro-differentiative responses, and thus expression of AKR1C3 may act as an oncogenic mechanism by attenuating PGD_2 availability (and consequently PGJ_2) [146]. Thus, specific inhibitors of AKR1C3's prostaglandin metabolism have the potential to exert anti-cancer activity that is distinct from the enzyme's 17β-HSD activity [141, 146].

16.4.3. Pre-receptor Activation of Dietary/Environmental Ligands

As detailed in Section 16.3.1, the availability of retinoic acid for binding to RARs, is dependent on a complex array of extracellular and intracellular binding proteins as well as specific metabolic enzymes. The latter function firstly by providing a system for converting inactive dietary vitamin A (retinol) to active hormonal retinoic acid, catalyzed by sequential retinol and retinal dehydrogenase activities (see Figure 16.4). In addition, the enzymes LRAT and LEH act to modulate the amount of retinol that is available for metabolism by these dehydrogenases by catalyzing the interconversion of retinol and stored retinyl esters. Collectively these enzymes maximize the efficacy of RAR signaling by enabling localized, intracrine adaptation to the dietary availability of retinol.

This type of metabolic modulation of steroid hormone receptor action is also central to the vitamin D endocrine system. As with vitamin A, the amount of vitamin D present in serum varies considerably from one individual to another, depending on dietary intake of vitamin D_2 or exposure to UV light as a requirement for synthesis of vitamin D_3 in the skin [147]. Vitamin D also undergoes sequential metabolic steps to form the VDR ligand, $1,25(OH)_2D$, but in this instance the two enzymes involved are cytochrome P450 hydroxylases. The first of these, 25-hydroxylase, converts parental vitamin D to 25OHD, the major circulating form of vitamin D [148]. The precise identity of the cytochrome P450 associated with this activity is still unclear as several enzymes including CYP27A1, CYP2R1, CYP2C11 and CYP3A4 can catalyze this reaction. Of these, the most well characterized candidates are mitochondrial CYP27A1 [149] and microsomal CYP2R1 [150–152], with the latter likely to be the predominant 25-hydroxylase as mutations in its gene have been linked to vitamin D-deficient rickets [151].

Irrespective of whether cells produce their own 25OHD or utilize the abundant serum levels of this metabolite, a further hydroxylation step is required to produce active, hormonal $1,25(OH)_2D$. This is catalyzed by the enzyme 25-hydroxvitamin D-1α-hydroxylase (1α-hydroxylase) which, unlike, 25-hydroxylase, involves a single protein encoded by the CYP27b1 gene [153]. Expression of 1α-hydroxylase is abundant in the proximal tubules of the kidney where it acts to support endocrine production of $1,25(OH)_2D$ following stimulation of the enzyme by parathyroid hormone (PTH) [154, 155]. However, the fact that 1α-hydroxylase is also expressed by an array of extra-renal tissues [156], suggests that localized intracrine/paracrine/autocrine synthesis of $1,25(OH)_2D$ is an important facet of VDR signaling [157]. In this respect, the activity of extra-renal 1α-hydroxylase is similar to that previously described for the corresponding estrogenic cytochrome P450, aromatase, in that both enzymes act to increase tissue concentrations of nuclear receptor ligands independent of their effects on the circulating levels of these compounds (see Section 16.4.1).

16.4.4. Catabolism of Nuclear Receptor Ligands

Perhaps the most well-recognized mechanism for attenuation of nuclear receptor ligands involves the corticosteroid system and another of the 11β-HSD isozymes. In this instance catabolic enzyme activity not only functions to suppress the availability of ligand for receptor binding but in doing so preferentially enables binding of another steroid hormone to the same receptor. The receptor in question is the mineralocorticoid receptor (MR) which plays a pivotal role in regulating renal sodium-potassium homeostasis and is thus a key factor in the maintenance of extracellular water volume and normal blood pressure [158]. The natural ligand for MR is aldosterone but several related corticosteroids including the GR ligand cortisol can also bind to the MR with equal affinity [159, 160]. An over-abundance of these alternative MR ligands has the potential to mimic aldosterone excess, with concomitant hypercalcemic side effects. As a consequence, the MR is 'protected' from elicit occupancy by a dedicated enzyme, 11β-hydroxysteroid hydrogenase type 2 (11β-HSD2), which is related to the 11β-HSD1 detailed in earlier parts of this chapter (Section 16.4.1). However, whilst 11β-HSD1 has the capacity for bi-directional metabolism of glucocorticoids, 11β-HSD2 is only able to catalyze oxidative inactivation of cortisol to cortisone (reviewed in [118]). Expression of 11β-HSD2 has been reported for several tissues such as the colon [161], salivary gland [162], skin [163] and placenta [164] but it is most abundant in renal cortical collecting ducts that are essential for water and sodium-potassium homeostasis [165–167].

The importance of 11β-HSD2 as a modulator of ligand availability has been underlined by studies which have documented the consequences of impaired activity of this enzyme. Here the resulting aberrant MR function is manifested by a form of hypertension known as 'Apparent Mineralocorticoid Excess' (AME) [168–171]. One form of dysregulated 11β-HSD2 activity has been linked to ingestion of liquorice. A key chemical component of liquorice, glycyrrhetinic acid, is a potent inhibitor of 11β-HSD activity and thus promotes the elicit occupancy of MR by cortisol with concomitant induction of hypertension [172, 173]. However, AME is most commonly associated with mutations in the gene for 11β-HSD2 (HSD11B2), which are inherited in an autosomal recessive manner (reviewed in [118]). Approximately thirty different HSD11B2 mutations have been reported and the severity of AME associated with these defects varies depending on whether the mutation results in diminished or completely absent activity of 11β-HSD2. Similar observations have also been made with HSD11B2 knockout mice which appear normal at birth but rapidly develop symptoms of AME [174].

Although these studies have emphasized the importance of 11β-HSD2 as a regulator of corticosteroid signaling, the physiological advantage of this mechanism may be restricted to MR-rich tissues such as the kidney. Analysis of 11β-HSD expression in several types of neoplasms has demonstrated a shift from expression of 11β-HSD1 in normal cells to 11β-HSD2 in neoplastic cells [175–179]. In this instance, the elevated 11β-HSD2 activity was associated with predominant expression of the GR, suggesting that the impact of glucocorticoid metabolism by this enzyme is not restricted to the MR. The ability of glucocorticoids to mediate antiproliferative responses in GR-positive cells suggests that aberrant expression of 11β-HSD2 may play a role in tumor development through the attenuation of tissue-specific levels of active cortisol [180]. This has been endorsed by studies using GR-rich cells transfected with either 11β-HSD1 (cortisone to cortisol conversion) or

11β-HSD2 (cortisol to cortisone conversion) in which the latter potently stimulated cell proliferation whilst the former inhibited proliferation [181]. Significantly, the effects of 11β-HSD1 and 11β-HSD2 on cell proliferation occurred in the absence of any change in GR expression, indicating that the synthesis or catabolism of ligand (cortisol) was the determinant of GR-responsiveness in this model.

In common with other steroidogenic systems, the vitamin D metabolic pathway is subject to regulatory mechanisms aimed at maintaining optimal levels of $1,25(OH)_2D$ in target cells. In the kidney, this is achieved in part through modulation of 1α-hydroxylase activity, where PTH-induction of the enzyme is autoregulated by $1,25(OH)_2D$ itself as a consequence of direct VDR-mediated suppression of transcription [182]. However, this method of feedback control is not evident for most other sites expressing the VDR. In these tissues, as well as the kidney, local concentrations of $1,25(OH)_2D$ are modulated through expression of an additional mitochondrial cytochrome P450, vitamin D-24-hydroxylase (24-hydroxylase) whose primary function is to catalyze conversion of 25OHD or $1,25(OH)_2D$ to less active 24-hydroxylated metabolites. Thus, as illustrated in Figure 16.8, 24-hydroxylase plays a key role the regulation of VDR signaling by modulating the cellular concentrations of active $1,25(OH)_2D$. Expression of 24-hydroxylase is closely associated with VDR levels, and whilst most cells express a basal level of 24-hydroxylase this is profoundly enhanced following treatment with $1,25(OH)_2D$ [183]. Kinetic analyses for 24-hydroxylase activity are variable and very much depend on the nature of the enzyme preparations being studied. In studies using cell and detergent extracts Km values up to 1.5 μM have been described for $1,25(OH)_2D$ as a substrate whilst Km values for 25OHD were up to 30-times higher, suggesting that the active form of vitamin D is the preferred substrate for 24-hydroxylase (reviewed in [184]). By contrast, some studies using purified enzyme preparations have postulated that the opposite is true, with 24-hydroxylase exhibiting a Km of approximately 0.6 μM for 25OHD and 30 μM for $1,25(OH)_2D$ [185]. Based on analysis of mice in which the gene for 24-hydroxylase (CYP24A1) has been knocked out, the former of these two propositions appears to be the most relevant in an in vivo context. These CYP24A1-null animals show a remarkable accumulation of $1,25(OH)_2D$ in the circulation, providing further evidence that the preferred substrate for this enzyme is the active form of vitamin D [186]. Phenotypically, the CYP24 knockout mice were characterized by decreased bone mineralization, prompting suggestions that 24-hydroxylase activity contributes actively to skeletal homeostasis. In particular, it has been postulated that 24-hydroxylated vitamin D metabolites such as 24,25-dihydroxyvitamin D ($24,25(OH)_2D$) are not simply inactive catabolites, but act as physiological modulators of skeletal homeostasis [187].

There are no known diseases associated with mutations in the CYP24A1 gene. However, expression of this gene is frequently elevated in tumor cells [188], possibly as a result of amplification of the CYP24A1 gene [189]. In addition, studies of 24-hydroxylase mRNAs from various cell types have shown that expression of the enzyme is characterized by distinct gene splice variants. In addition to the full length 2.7 kb wild type 24-hydroxylase mRNA a truncated mRNA for the enzyme was

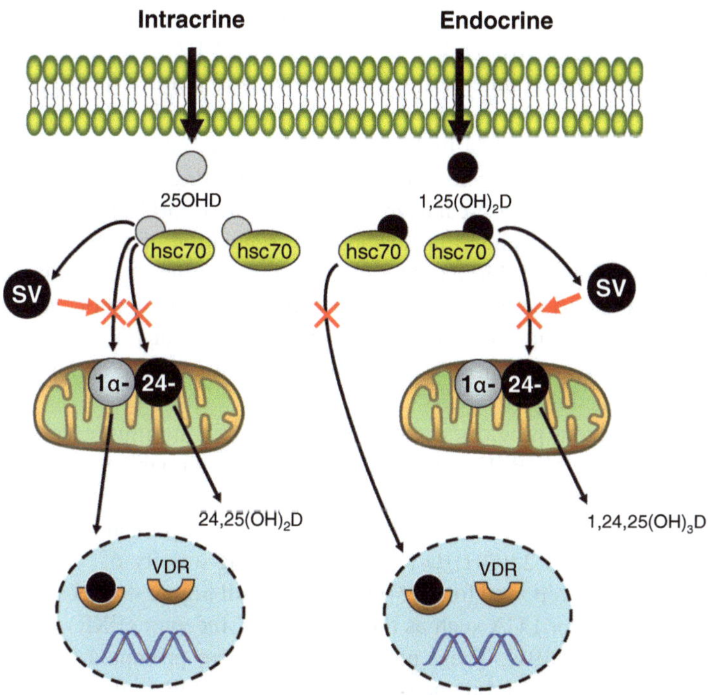

Figure 16.8. Pre-receptor metabolism and the regulation of endocrine and intracrine responses to vitamin D. Intracrine conversion of pro-hormone 25-hydroxyvitamin D (25OHD) to active 1, 25-dihydroxyvitamin D (1,25(OH)$_2$D) is catalyzed by the mitochondrial enzyme 25-hydroxyvitamin D-1α-hydroxylase (1α-hydroxylase [1α-]). 25OHD is also a substrate for another mitochondrial enzyme vitamin D-24-hydroxylase (24-hydroxylase [24-]) which can catalyze catabolism of 25OHD to 24,25(OH)$_2$D or 1,25(OH)$_2$D to 1,24,25(OH)$_3$D. 24-hydroxylase also exists as an amino terminus splice variant protein which lacks a mitochondrial-targeting domain. The resulting 24-hydroxylase splice variant (SV) is located in the cytoplasm and readily binds vitamin D metabolites despite showing no metabolic activity. Other abbreviations: Intracellular vitamin D binding heat-shock protein 70 (hsc70), vitamin D receptor (VDR)

also observed in several cell types, notably macrophages [190]. Subsequent studies showed that this was due to the 5' end of the wild-type CYP24A1, containing the mitochondrial targeting sequence, being deleted by alternative splicing of intron 2. This results in an in-frame, alternative start site of translation, encoding a protein that has normal sterol and heme binding domains but no amino-terminal mitochondrial targeting domain [190]. Based on these observations, it appears that the 24-hydroxylase splice variant (CYP24-SV) can act as a cytoplasmic 'decoy' for conventional mitochondrial 24-hydroxylase substrates such as 25OHD and 1,25(OH)$_2$D (see Figure 16.8). Significantly, when over-expressed in macrophage-like cells, CYP24-SV was a much more efficient attenuator of 1,25(OH)$_2$D synthesis than wild-type CYP24A1, for which 1,25(OH)$_2$D is the preferred substrate. By contrast, antisense inhibition of CYP24-SV enhanced 1,25$(OH)_2$D production, presumably by increasing the intracellular availability of 25OHD to CYP27b1. These data suggest that, like its wild-type counterpart, CYP24-SV acts to control local levels of 1,25$(OH)_2$D. Both are induced in cells following treatment with 1,25$(OH)_2$D. However, in contrast to wild type 24-hydroxylase, the splice variant appears to achieve its effect simply by binding 25OHD, thereby decreasing the availability of this metabolite for 1α-hydroxylase-catalyzed conversion to 1,25$(OH)_2$D. In this way CYP24-SV may provide a more efficient mode for regulation of VDR signaling by limiting the formation of 1,25$(OH)_2$D, rather than promoting its catabolism. Such a mechanism is likely to be particularly important in cells which show active synthesis of 1,25$(OH)_2$D, and it is therefore interesting to note that CYP24-SV is expressed in tissues such as the skin, placenta and kidney which have significant levels of 1α-hydroxylase activity.

The 24-hydroxylase system provides a potent mechanism for limiting access of vitamin D metabolites to the VDR. However, it is important to recognize that other ligands have the potential to signal via the VDR. Lithocholic acid (LCA), a toxic secondary bile acid is also able to bind to VDR and induce transcriptional responses similar to those observed for 1,25$(OH)_2$D [191]. The binding affinity of VDR for LCA is much lower than that reported for 1,25$(OH)_2$D, but still higher than observed for other putative receptors for LCA such as the farsenoid X receptor (FXR) or the pregnane X receptor (FXR) [191]. These effects of LCA have prompted suggestions that VDR also functions as a bile acid sensor [192], and this is discussed in greater detail in chapter 8. However, it is notable that a key target for VDR-LCA responses is the enzyme cytochrome P450 3A (CYP3A) which acts to detoxify LCA [193]. LCA is poorly reabsorbed into the enterohepatic circulation and can therefore accumulate at relatively high concentrations in the colon, where it has been shown to promote colorectal cancer. Induction of CYP3A is essential for protection against this LCA-induced gastrointestinal damage but by altering localized concentrations of an alternative VDR ligand, the enzyme can also as a novel regulator of LCA-mediated VDR signaling.

16.4.5. Tissue-Specific Metabolism and the Regulation of Thyroid Receptor Function

Thyroid hormone actions is dependent on target cell binding of active 3,5,3'-triiodothyironine (T3) to the nuclear thyroid hormone receptor (TR). However, the precursor to T3 thyroxine (T4) has a relatively long half-life (one week in humans) and thus has the potential to undergo tissue-specific metabolism that is independent of its classical endocrine regulation via the hypothalamus. Interconversion of T4 and T3 is catalyzed by three related selenodeiodinases – deiodinase 1 (D1), D2 and D3 [194, 195]. Although these enzymes play a role in determining the circulating levels of different thyroid hormones, it is clear that they are now also essential for localized tissue-specific metabolism [196]. As outlined in Figure 16.9a,b, the individual

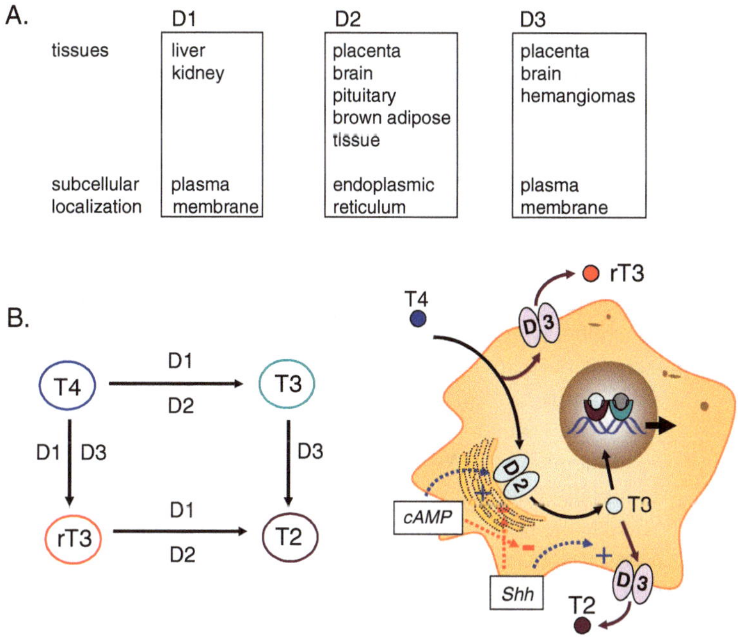

Figure 16.9. Extra-receptor regulation of thyroid hormone action. (**a**) Tissue and subcellular localization of thyroid deiodinases *D1*, *D2*, and *D3*. (**b**) Substrate and product specificity of thyroid deiodinases *D1*, *D2*, and *D3*. C schematic representation of the interaction between *D2* and *D3* in the regulation of intracrine conversion of pro-hormone thyroxine (*T4*) to active triiodothyironine (*T3*) and subsequent thyroid receptor (TR) signaling. Activation of *T4–T3* by D2 in the endoplasmic reticulum increases TR responses. Feedback control of this is maintained by the membrane catabolic enzyme D3 which converts *T3–T2* and *T4* to reverse *T3* (rT3). This feedback regulation is further modulated by membrane-mediated signal transduction pathways such as Sonic hedgehog (Shh) which attenuates TR signaling by suppressing *D2* activity and promoting *D3*, and by cAMP which promotes TR activity by enhancing expression of *D2* and by inhibiting Shh effects on *D3*

thyroid deiodinases display distinct substrate preferences, and tissue-specific localization to form a complex network of thyroid hormone metabolism that includes not only T3 and T4 but other thyroid hormones such as reverse T3 (rT3) and the less active catabolite T2.

Although circulating levels of T3 are much lower than those of T4, the latter is more tightly bound to thyroid binding globulins. Thus, the amount of free T3 and T4 for any given target cell is likely to be approximately the same [197]. Furthermore, both T3 and T4 enter target cells via the same mechanisms involving monocarboxylate transporter 8 and organic anion transporter polypeptide C1 so that the intracellular pool of thyroid hormone will include both T3 and T4 [198–200]. The net effect of this is similar to that previously documented for vitamin D (see Section 16.4.3), where both endocrine and intracrine synthesis of $1,25(OH)_2D$ may contribute to VDR signaling, depending on tissue-specific expression of the enzyme

1α-hydroxylase. In the case of thyroid hormone, the predominant deiodinase associated with extra-thyroidal conversion of T4–T3 appears to be D2 [201, 202], with D1 being important in hyperthyroidism [203], and D3 acting to catabolize T3 to less active metabolites (reviewed in [196]). The potential interaction between D2 and D3 as modulators of TR-mediated gene transcription is illustrated in Figure 16.9c. In this representative model, D2-expressing cells, internalized T4 can be converted in an intracrine fashion to the TR ligand T3. This pre-receptor activation of T4 can then be counteracted by pre-receptor catabolism by membrane-bound D3 which is able to convert substrate T4 or product T3 to less active metabolites. Several additional factors are known to influence this intracrine system. For example, D2 is targeted for degradation within its endoplasmic reticulum (ER) location and hence has a short half-life (approximately 40 minutes). Degradation of D2 is induced following its ubquitination [204, 205], a process which is enhanced by Sonic hedgehog (Shh) via the Shh-inducible gene WSB-1 which acts as a ubquitin ligase adapter for D2 [206]. By contrast, induction of membrane cAMP opposes Shh by enhancing D2 expression [206, 207] and coincidentally inhibiting Shh/WSB-1 function [208]. Thus the pre-receptor regulation of tissue-specific concentrations of T3 is fine-tuned by a series of pathways that are also essential in maintaining a balance between cell proliferation and differentiation [209]. This may be particularly important in regulating the established function of thyroid hormones during fetal development.

As yet, there have been no reported mutations in the genes for D1, D2 or D3. However, knockout mice models have been generated for these enzymes. As might be expected all three knockout models exhibited elevated serum levels of T4, and the D1 and D2. However, only the D3 knockout mice showed any change in T3 levels and both the D1 and D2 knockout mice remained physiologically euthyroid (reviewed in [196, 210]). This suggests that either compensatory mechanisms exist that utilize alternative deiodinases in the absence of the ablated enzyme, or that there may be altered secretion of T3.

16.5. SUMMARY

Compared with the wealth of information that has accumulated concerning other-nuclear receptor-regulatory events such as chromatin remodeling, the mechanisms that define availability of active ligand for binding to a specific nuclear receptor are amongst the least well characterized in endocrinology. Despite this, mechanisms involved in the regulation of ligand availability and metabolism clearly represent an important facet of nuclear receptor function. This is exemplified by studies that have defined an intracrine role for aromatase in breast cancer and 11β-HSD1 in type 2 diabetes. In both cases prevention of locally amplified receptor responses has substantial clinical benefits for prevalent clinical disorders. In future it is likely that similar strategies will be applied to the metabolism of many more nuclear receptor ligands. Other mechanisms involved in the regulation of nuclear receptor ligands such as cellular uptake and subcellular trafficking are less well characterized. However, in view of the increasing awareness of intracrine versus endocrine pathways, it is inevitable that future studies will increasingly look to these novel facets of nuclear receptor action.

REFERENCES

1. Watson, C. S. and Gametchu, B. (2003). Proteins of multiple classes may participate in nongenomic steroid actions. *Exp Biol Med (Maywood)* 228, 1272–1281.
2. de Wolf, F. A. and Brett, G. M. (2000). Ligand-binding proteins: Their potential for application in systems for controlled delivery and uptake of ligands. *Pharmacol Rev* 52, 207–236.
3. Orchinik, M., Hastings, N., Witt, D., and McEwen, B. S. (1997). High-affinity binding of corticosterone to mammalian neuronal membranes: Possible role of corticosteroid binding globulin. *J Steroid Biochem Mol Biol* 60, 229–236.
4. Khan, M. S., Hryb, D. J., Hashim, G. A., Romas, N. A., and Rosner, W. (1990). Delineation and synthesis of the membrane receptor-binding domain of sex hormone-binding globulin. *J Biol Chem* 265, 18362–18365.
5. Hammond, G. L. (1995). Potential functions of plasma steroid-binding proteins. *Trends Endocrinol Metab* 6, 298–304.
6. Rosner, W., Hryb, D. J., Khan, M. S., Nakhla, A. M., and Romas, N. A. (1999). Androgen and estrogen signaling at the cell membrane via G-proteins and cyclic adenosine monophosphate. *Steroids* 64, 100–106.
7. Rosner, W., Hryb, D. J., Khan, M. S., Nakhla, A. M., and Romas, N. A. (1999). Sex hormone-binding globulin mediates steroid hormone signal transduction at the plasma membrane. *J Steroid Biochem Mol Biol* 69, 481–485.
8. Bikle, D. D., Gee, E., Halloran, B., Kowalski, M. A., Ryzen, E., and Haddad, J. G. (1986). Assessment of the free fraction of 25-hydroxyvitamin D in serum and its regulation by albumin and the vitamin D-binding protein. *J Clin Endocrinol Metab* 63, 954–959.
9. Bikle, D. D., Gee, E., Halloran, B., and Haddad, J. G. (1984). Free 1,25-dihydroxyvitamin D levels in serum from normal subjects, pregnant subjects, and subjects with liver disease. *J Clin Invest* 74, 1966–1971.
10. Bouillon, R., Van Assche, F. A., Van Baelen, H., Heyns, W., and De Moor, P. (1981). Influence of the vitamin D-binding protein on the serum concentration of 1,25-dihydroxyvitamin D3. Significance of the free 1,25-dihydroxyvitamin D3 concentration. *J Clin Invest* 67, 589–596.
11. Mendel, C. M. (1989). The free hormone hypothesis: A physiologically based mathematical model. *Endocr Rev* 10, 232–274.
12. Hammond, G. L. (2002). Access of reproductive steroids to target tissues. *Obstet Gynecol Clin North Am* 29, 411–423.
13. Vieth, R. (1994). Simple method for determining specific binding capacity of vitamin D-binding protein and its use to calculate the concentration of "free" 1,25-dihydroxyvitamin D. *Clin Chem* 40, 435–441.
14. Moestrup, S. K. and Verroust, P. J. (2001). Megalin- and cubilin-mediated endocytosis of protein-bound vitamins, lipids, and hormones in polarized epithelia. *Annu Rev Nutr* 21, 407–428.
15. Christensen, E. I. and Birn, H. (2001). Megalin and cubilin: Synergistic endocytic receptors in renal proximal tubule. *Am J Physiol Renal Physiol* 280, F562–F573.
16. Nykjaer, A., Dragun, D., Walther, D., Vorum, H., Jacobsen, C., Herz, J., Melsen, F., Christensen, E. I., and Willnow, T. E. (1999). An endocytic pathway essential for renal uptake and activation of the steroid 25-(OH) vitamin D3. *Cell* 96, 507–515.
17. Hammes, A., Andreassen, T. K., Spoelgen, R., Raila, J., Hubner, N., Schulz, H., Metzger, J., Schweigert, F. J., Luppa, P. B., Nykjaer, A., and Willnow, T. E. (2005). Role of endocytosis in cellular uptake of sex steroids. *Cell* 122, 751–762.

18. Christensen, E. I., Moskaug, J. O., Vorum, H., Jacobsen, C., Gundersen, T. E., Nykjaer, A., Blomhoff, R., Willnow, T. E., and Moestrup, S. K. (1999). Evidence for an essential role of megalin in transepithelial transport of retinol. *J Am Soc Nephrol* 10, 685–695.
19. Raila, J., Willnow, T. E., and Schweigert, F. J. (2005). Megalin-mediated reuptake of retinol in the kidneys of mice is essential for vitamin A homeostasis. *J Nutr* 135, 2512–2516.
20. Quadro, L., Blaner, W. S., Salchow, D. J., Vogel, S., Piantedosi, R., Gouras, P., Freeman, S., Cosma, M. P., Colantuoni, V., and Gottesman, M. E. (1999). Impaired retinal function and vitamin A availability in mice lacking retinol-binding protein. *EMBO J* 18, 4633–4644.
21. Safadi, F. F., Thornton, P., Magiera, H., Hollis, B. W., Gentile, M., Haddad, J. G., Liebhaber, S. A., and Cooke, N. E. (1999). Osteopathy and resistance to vitamin D toxicity in mice null for vitamin D binding protein. *J Clin Invest* 103, 239–251.
22. Zella, L. A., Shevde, N. K., Hollis, B. W., Cooke, N. E., and Pike, J. W. (2008). Vitamin D-binding protein influences total circulating levels of 1,25-dihydroxyvitamin D3 but does not directly modulate the bioactive levels of the hormone in vivo. *Endocrinology* 149, 3656–3667.
23. White, P. and Cooke, N. (2000). The multifunctional properties and characteristics of vitamin D-binding protein. *Trends Endocrinol Metab* 11, 320–327.
24. Rowling, M. J., Kemmis, C. M., Taffany, D. A., and Welsh, J. (2006). Megalin-mediated endocytosis of vitamin D binding protein correlates with 25-hydroxycholecalciferol actions in human mammary cells. *J Nutr* 136, 2754–2759.
25. Chlon, T. M., Taffany, D. A., Welsh, J., and Rowling, M. J. (2008). Retinoids modulate expression of the endocytic partners megalin, cubilin, and disabled-2 and uptake of vitamin D-binding protein in human mammary cells. *J Nutr* 138, 1323–1328.
26. van Driel, M., Koedam, M., Buurman, C. J., Hewison, M., Chiba, H., Uitterlinden, A. G., Pols, H. A., and van Leeuwen, J. P. (2006). Evidence for auto/paracrine actions of vitamin D in bone: 1alpha-hydroxylase expression and activity in human bone cells. *FASEB J* 20, 2417–2419.
27. Atkins, G. J., Anderson, P. H., Findlay, D. M., Welldon, K. J., Vincent, C., Zannettino, A. C., O'Loughlin, P. D., and Morris, H. A. (2007). Metabolism of vitamin D3 in human osteoblasts: Evidence for autocrine and paracrine activities of 1 alpha,25-dihydroxyvitamin D3. *Bone* 40, 1517–1528.
28. Esteban, C., Geuskens, M., Ena, J. M., Mishal, Z., Macho, A., Torres, J. M., and Uriel, J. (1992). Receptor-mediated uptake and processing of vitamin D-binding protein in human B-lymphoid cells. *J Biol Chem* 267, 10177–10183.
29. Petrini, M., Galbraith, R. M., Emerson, D. L., Nel, A. E., and Arnaud, P. (1985). Structural studies of T lymphocyte Fc receptors. Association of Gc protein with IgG binding to Fc gamma. *J Biol Chem* 260, 1804–1810.
30. Zella, L. A., Shevde, N. K., Hollis, B. W., Cooke, N. E., and Pike, J. W. (2008). Vitamin D binding protein influences total circulating levels of 1,25-dihydroxyvitamin D3 but does not directly modulate the bioactive levels of the hormone in vivo. *Endocrinology* 149(7), 3656–3667.
31. Chun, R., Adams, J., and Hewison, M. (2008). Back to the future: A new look at 'old' vitamin D. *J Endocrinol* 198(2), 261–269.
32. Noy, N. and Xu, Z. J. (1990). Interactions of retinol with binding proteins: Implications for the mechanism of uptake by cells. *Biochemistry* 29, 3878–3883.
33. Bellovino, D., Morimoto, T., Tosetti, F., and Gaetani, S. (1996). Retinol binding protein and transthyretin are secreted as a complex formed in the endoplasmic reticulum in HepG2 human hepatocarcinoma cells. *Exp Cell Res* 222, 77–83.
34. Goodman, D. S. (1994). Plasma retinol binding protein. New York: Academic Press.
35. Redondo, C., Burke, B. J., and Findlay, J. B. (2006). The retinol-binding protein system: A potential paradigm for steroid-binding globulins?. *Horm Metab Res* 38, 269–278.
36. Sundaram, M., Sivaprasadarao, A., DeSousa, M. M., and Findlay, J. B. (1998). The transfer of retinol from serum retinol-binding protein to cellular retinol-binding protein is mediated by a membrane receptor. *J Biol Chem* 273, 3336–3342.

37. Kawaguchi, R., Yu, J., Honda, J., Hu, J., Whitelegge, J., Ping, P., Wiita, P., Bok, D., and Sun, H. (2007). A membrane receptor for retinol binding protein mediates cellular uptake of vitamin A. *Science* 315, 820–825.
38. Wolf, G. (2007). Identification of a membrane receptor for retinol-binding protein functioning in the cellular uptake of retinol. *Nutr Rev* 65, 385–388.
39. Redondo, C., Vouropoulou, M., Evans, J., and Findlay, J. B. (2008). Identification of the retinol-binding protein (RBP) interaction site and functional state of RBPs for the membrane receptor. *FASEB J* 22, 1043–1054.
40. Napoli, J. L. (1996). Biochemical pathways of retinoid transport, metabolism, and signal transduction. *Clin Immunol Immunopathol* 80, S52–S62.
41. Pertiwi, A. K., Kwan, T. K., and Gower, D. B. (2002). Pregnenolone metabolites in rat testis: Endogenous concentrations, and intracellular distribution in whole testes during incubation in vitro. *J Steroid Biochem Mol Biol* 81, 363–367.
42. Schroeder, F., Atshaves, B. P., McIntosh, A. L., Gallegos, A. M., Storey, S. M., Parr, R. D., Jefferson, J. R., Ball, J. M., and Kier, A. B. (2007). Sterol carrier protein-2: New roles in regulating lipid rafts and signaling. *Biochim Biophys Acta* 1771, 700–718.
43. Schneiter, R. (2007). Intracellular sterol transport in eukaryotes, a connection to mitochondrial function?. *Biochimie* 89, 255–259.
44. Maxfield, F. R. and Menon, A. K. (2006). Intracellular sterol transport and distribution. *Curr Opin Cell Biol* 18, 379–385.
45. Maxfield, F. R. and Mondal, M. (2006). Sterol and lipid trafficking in mammalian cells. *Biochem Soc Trans* 34, 335–339.
46. Seedorf, U., Ellinghaus, P., and Roch Nofer, J. (2000). Sterol carrier protein-2. *Biochim Biophys Acta* 1486, 45–54.
47. Miller, W. L. (2007). Steroidogenic acute regulatory protein (StAR), a novel mitochondrial cholesterol transporter. *Biochim Biophys Acta* 1771, 663–676.
48. Miller, W. L. (2007). Mechanism of StAR's regulation of mitochondrial cholesterol import. *Mol Cell Endocrinol* 265–266, 46–50.
49. Miller, W. L. (2007). StAR search–what we know about how the steroidogenic acute regulatory protein mediates mitochondrial cholesterol import. *Mol Endocrinol* 21, 589–601.
50. Stocco, D. M. (2001). Tracking the role of a star in the sky of the new millennium. *Mol Endocrinol* 15, 1245–1254.
51. Stocco, D. M. (2001). StAR protein and the regulation of steroid hormone biosynthesis. *Annu Rev Physiol* 63, 193–213.
52. Christenson, L. K. and Strauss, J. F., 3rd (2000). Steroidogenic acute regulatory protein (StAR) and the intramitochondrial translocation of cholesterol. *Biochim Biophys Acta* 1529, 175–187.
53. Stocco, D. M. (2000). The role of the StAR protein in steroidogenesis: Challenges for the future. *J Endocrinol* 164, 247–253.
54. Strauss, J. F., 3rd, Kallen, C. B., Christenson, L. K., Watari, H., Devoto, L., Arakane, F., Kiriakidou, M., and Sugawara, T. (1999). The steroidogenic acute regulatory protein (StAR): A window into the complexities of intracellular cholesterol trafficking. *Recent Prog Horm Res* 54, 369–394, discussion 394–365.
55. Kumar, S., Saradhi, M., Chaturvedi, N. K., and Tyagi, R. K. (2006). Intracellular localization and nucleocytoplasmic trafficking of steroid receptors: An overview. *Mol Cell Endocrinol* 246, 147–156.
56. Shank, L. C. and Paschal, B. M. (2005). Nuclear transport of steroid hormone receptors. *Crit Rev Eukaryot Gene Expr* 15, 49–73.
57. Pemberton, L. F. and Paschal, B. M. (2005). Mechanisms of receptor-mediated nuclear import and nuclear export. *Traffic* 6, 187–198.
58. Nagaich, A. K., Rayasam, G. V., Martinez, E. D., Becker, M., Qiu, Y., Johnson, T. A., Elbi, C., Fletcher, T. M., John, S., and Hager, G. L. (2004). Subnuclear trafficking and gene targeting by steroid receptors. *Ann N Y Acad Sci* 1024, 213–220.

59. Kawata, M., Matsuda, K., Nishi, M., Ogawa, H., and Ochiai, I. (2001). Intracellular dynamics of steroid hormone receptor. *Neurosci Res* 40, 197–203.
60. Li, E. and Norris, A. W. (1996). Structure/function of cytoplasmic vitamin A-binding proteins. *Annu Rev Nutr* 16, 205–234.
61. Folli, C., Calderone, V., Ottonello, S., Bolchi, A., Zanotti, G., Stoppini, M., and Berni, R. (2001). Identification, retinoid binding, and x-ray analysis of a human retinol-binding protein. *Proc Natl Acad Sci USA* 98, 3710–3715.
62. Folli, C., Calderone, V., Ramazzina, I., Zanotti, G., and Berni, R. (2002). Ligand binding and structural analysis of a human putative cellular retinol-binding protein. *J Biol Chem* 277, 41970–41977.
63. Ghyselinck, N. B., Bavik, C., Sapin, V., Mark, M., Bonnier, D., Hindelang, C., Dierich, A., Nilsson, C. B., Hakansson, H., Sauvant, P., Azais-Braesco, V., Frasson, M., Picaud, S., and Chambon, P. (1999). Cellular retinol-binding protein I is essential for vitamin A homeostasis. *EMBO J* 18, 4903–4914.
64. Napoli, J. L. (2000). A gene knockout corroborates the integral function of cellular retinol-binding protein in retinoid metabolism. *Nutr Rev* 58, 230–236.
65. Harrison, E. H. (2005). Mechanisms of digestion and absorption of dietary vitamin A. *Annu Rev Nutr* 25, 87–103.
66. Harrison, E. H. and Hussain, M. M. (2001). Mechanisms involved in the intestinal digestion and absorption of dietary vitamin A. *J Nutr* 131, 1405–1408.
67. Xueping, E., Zhang, L., Lu, J., Tso, P., Blaner, W. S., Levin, M. S., and Li, E. (2002). Increased neonatal mortality in mice lacking cellular retinol-binding protein II. *J Biol Chem* 277, 36617–36623.
68. Piantedosi, R., Ghyselinck, N., Blaner, W. S., and Vogel, S. (2005). Cellular retinol-binding protein type III is needed for retinoid incorporation into milk. *J Biol Chem* 280, 24286–24292.
69. Napoli, J. L. (1999). Retinoic acid: Its biosynthesis and metabolism. *Prog Nucleic Acid Res Mol Biol* 63, 139–188.
70. Noy, N. (2000). Retinoid-binding proteins: Mediators of retinoid action. *Biochem J* 348(Pt 3), 481–495.
71. Donovan, M., Olofsson, B., Gustafson, A. L., Dencker, L., and Eriksson, U. (1995). The cellular retinoic acid binding proteins. *J Steroid Biochem Mol Biol* 53, 459–465.
72. Fogh, K., Voorhees, J. J., and Astrom, A. (1993). Expression, purification, and binding properties of human cellular retinoic acid-binding protein type I and type II. *Arch Biochem Biophys* 300, 751–755.
73. Eller, M. S., Oleksiak, M. F., McQuaid, T. J., McAfee, S. G., and Gilchrest, B. A. (1992). The molecular cloning and expression of two CRABP cDNAs from human skin. *Exp Cell Res* 198, 328–336.
74. Astrom, A., Tavakkol, A., Pettersson, U., Cromie, M., Elder, J. T., and Voorhees, J. J. (1991). Molecular cloning of two human cellular retinoic acid-binding proteins (CRABP). Retinoic acid-induced expression of CRABP-II but not CRABP-I in adult human skin in vivo and in skin fibroblasts in vitro. *J Biol Chem* 266, 17662–17666.
75. Giguere, V., Lyn, S., Yip, P., Siu, C. H., and Amin, S. (1990). Molecular cloning of cDNA encoding a second cellular retinoic acid-binding protein. *Proc Natl Acad Sci USA* 87, 6233–6237.
76. Lampron, C., Rochette-Egly, C., Gorry, P., Dolle, P., Mark, M., Lufkin, T., LeMeur, M., and Chambon, P. (1995). Mice deficient in cellular retinoic acid binding protein II (CRABPII) or in both CRABPI and CRABPII are essentially normal. *Development* 121, 539–548.
77. Fawcett, D., Pasceri, P., Fraser, R., Colbert, M., Rossant, J., and Giguere, V. (1995). Postaxial polydactyly in forelimbs of CRABP-II mutant mice. *Development* 121, 671–679.
78. Ross, A. C. and Zolfaghari, R. (2004). Regulation of hepatic retinol metabolism: Perspectives from studies on vitamin A status. *J Nutr* 134, 269S–275S.
79. Herr, F. M. and Ong, D. E. (1992). Differential interaction of lecithin-retinol acyltransferase with cellular retinol binding proteins. *Biochemistry* 31, 6748–6755.

80. Napoli, J. L. (1993). Biosynthesis and metabolism of retinoic acid: Roles of CRBP and CRABP in retinoic acid: Roles of CRBP and CRABP in retinoic acid homeostasis. *J Nutr* 123, 362–366.
81. Boerman, M. H. and Napoli, J. L. (1996). Cellular retinol-binding protein-supported retinoic acid synthesis. Relative roles of microsomes and cytosol. *J Biol Chem* 271, 5610–5616.
82. Boylan, J. F. and Gudas, L. J. (1991). Overexpression of the cellular retinoic acid binding protein-I (CRABP-I) results in a reduction in differentiation-specific gene expression in F9 teratocarcinoma cells. *J Cell Biol* 112, 965–979.
83. Dong, D., Ruuska, S. E., Levinthal, D. J., and Noy, N. (1999). Distinct roles for cellular retinoic acid-binding proteins I and II in regulating signaling by retinoic acid. *J Biol Chem* 274, 23695–23698.
84. Vo, H. P. and Crowe, D. L. (1998). Transcriptional regulation of retinoic acid responsive genes by cellular retinoic acid binding protein-II modulates RA mediated tumor cell proliferation and invasion. *Anticancer Res* 18, 217–224.
85. Chrousos, G. P., Loriaux, D. L., Tomita, M., Brandon, D. D., Renquist, D., Albertson, B., and Lipsett, M. B. (1986). The new world primates as animal models of glucocorticoid resistance. *Adv Exp Med Biol* 196, 129–144.
86. Brandon, D. D., Markwick, A. J., Chrousos, G. P., and Loriaux, D. L. (1989). Glucocorticoid resistance in humans and nonhuman primates. *Cancer Res* 49, 2203s–2213s.
87. Fuller, P. J., Smith, B. J., and Rogerson, F. M. (2004). Cortisol resistance in the new world revisited. *Trends Endocrinol Metab* 15, 296–299.
88. Gacad, M. A. and Adams, J. S. (1991). Endogenous blockade of 1,25-dihydroxyvitamin D-receptor binding in New World primate cells. *J Clin Invest* 87, 996–1001.
89. Gacad, M. A., Chen, H., Arbelle, J. E., LeBon, T., and Adams, J. S. (1997). Functional characterization and purification of an intracellular vitamin D-binding protein in vitamin D-resistant new world primate cells. Amino acid sequence homology with proteins in the hsp-70 family. *J Biol Chem* 272, 8433–8440.
90. Chen, H., Hu, B., Huang, G. H., Trainor, A. G., Abbott, D. H., and Adams, J. S. (2003). Purification and characterization of a novel intracellular 17 beta-estradiol binding protein in estrogen-resistant New World primate cells. *J Clin Endocrinol Metab* 88, 501–504.
91. Gacad, M. A. and Adams, J. S. (1998). Proteins in the heat shock-70 family specifically bind 25-hydroxyvitamin D3 and 17beta-estradiol. *J Clin Endocrinol Metab* 83, 1264–1267.
92. Chen, H., Hewison, M., Hu, B., Sharma, M., Sun, Z., and Adams, J. S. (2004). An Hsp27-related, dominant-negative-acting intracellular estradiol-binding protein. *J Biol Chem* 279, 29944–29951.
93. Wu, S., Ren, S., Chen, H., Chun, R. F., Gacad, M. A., and Adams, J. S. (2000). Intracellular vitamin D binding proteins: Novel facilitators of vitamin D-directed transactivation. *Mol Endocrinol* 14, 1387–1397.
94. Wu, S., Chun, R., Gacad, M. A., Ren, S., Chen, H., and Adams, J. S. (2002). Regulation of 1, 25-dihydroxyvitamin d synthesis by intracellular vitamin d binding protein-1. *Endocrinology* 143, 4135.
95. Adams, J. S., Chen, H., Chun, R. F., Nguyen, L., Wu, S., Ren, S. Y., Barsony, J., and Gacad, M. A. (2003). Novel regulators of vitamin D action and metabolism: Lessons learned at the Los Angeles zoo. *J Cell Biochem* 88, 308–314.
96. Adams, J. S., Chen, H., Chun, R., Gacad, M. A., Encinas, C., Ren, S., Nguyen, L., Wu, S., Hewison, M., and Barsony, J. (2004). Response element binding proteins and intracellular vitamin D binding proteins: Novel regulators of vitamin D trafficking, action and metabolism. *J Steroid Biochem Mol Biol* 89–90, 461–465.
97. Young, J. C., Barral, J. M., and Ulrich Hartl, F. (2003). More than folding: Localized functions of cytosolic chaperones. *Trends Biochem Sci* 28, 541–547.
98. Jakob, U., Gaestel, M., Engel, K., and Buchner, J. (1993). Small heat shock proteins are molecular chaperones. *J Biol Chem* 268, 1517–1520.
99. Chun, R., Gacad, M. A., Hewison, M., and Adams, J. S. (2005). Adenosine 5′-triphosphate-dependent vitamin D sterol binding to heat shock protein-70 chaperones. *Endocrinology* 146, 5540–5544.

100. Takayama, S., Bimston, D. N., Matsuzawa, S., Freeman, B. C., Aime-Sempe, C., Xie, Z., Morimoto, R. I., and Reed, J. C. (1997). BAG-1 modulates the chaperone activity of Hsp70/Hsc70. *EMBO J* 16, 4887–4896.
101. Chun, R. F., Gacad, M., Nguyen, L., Hewison, M., and Adams, J. S. (2007). Co-chaperone potentiation of vitamin D receptor-mediated transactivation: A role for Bcl2-associated athanogene-1 as an intracellular-binding protein for 1,25-dihydroxyvitamin D3. *J Mol Endocrinol* 39, 81–89.
102. Arbelle, J. E., Chen, H., Gaead, M. A., Allegretto, E. A., Pike, J. W., and Adams, J. S. (1996). Inhibition of vitamin D receptor-retinoid X receptor-vitamin D response element complex formation by nuclear extracts of vitamin D-resistant New World primate cells. *Endocrinology* 137, 786–789.
103. Chen, H., Hu, B., Allegretto, E. A., and Adams, J. S. (2000). The vitamin D response element-binding protein. A novel dominant-negative regulator of vitamin D-directed transactivation. *J Biol Chem* 275, 35557–35564.
104. Chen, H., Hewison, M., and Adams, J. S. (2006). Functional characterization of heterogeneous nuclear ribonuclear protein C1/C2 in vitamin D resistance: A novel response element-binding protein. *J Biol Chem* 281, 39114–39120.
105. Chen, H., Hewison, M., Hu, B., and Adams, J. S. (2003). Heterogeneous nuclear ribonucleoprotein (hnRNP) binding to hormone response elements: A cause of vitamin D resistance. *Proc Natl Acad Sci USA* 100, 6109–6114.
106. Chen, H., Hewison, M., and Adams, J. S. (2008). Control of estradiol-directed gene transactivation by an intracellular estrogen-binding protein and an estrogen response element-binding protein. *Mol Endocrinol* 22, 559–569.
107. Chen, H., Arbelle, J. E., Gacad, M. A., Allegretto, E. A., and Adams, J. S. (1997). Vitamin D and gonadal steroid-resistant New World primate cells express an intracellular protein which competes with the estrogen receptor for binding to the estrogen response element. *J Clin Invest* 99, 669–675.
108. Chen, H., Hu, B., Gacad, M. A., and Adams, J. S. (1998). Cloning and expression of a novel dominant-negative-acting estrogen response element-binding protein in the heterogeneous nuclear ribonucleoprotein family. *J Biol Chem* 273, 31352–31357.
109. Payne, A. H. and Hales, D. B. (2004). Overview of steroidogenic enzymes in the pathway from cholesterol to active steroid hormones. *Endocr Rev* 25, 947–970.
110. Simpson, E. R., Mahendroo, M. S., Means, G. D., Kilgore, M. W., Hinshelwood, M. M., Graham-Lorence, S., Amarneh, B., Ito, Y., Fisher, C. R., Michael, M. D. et al. (1994). Aromatase cytochrome P450, the enzyme responsible for estrogen biosynthesis. *Endocr Rev* 15, 342–355.
111. Labrie, F., Luu-The, V., Labrie, C., Belanger, A., Simard, J., Lin, S. X., and Pelletier, G. (2003). Endocrine and intracrine sources of androgens in women: Inhibition of breast cancer and other roles of androgens and their precursor dehydroepiandrosterone. *Endocr Rev* 24, 152–182.
112. Simpson, E. R. (2004). Models of aromatase insufficiency. *Semin Reprod Med* 22, 25–30.
113. Simpson, E. R., Misso, M., Hewitt, K. N., Hill, R. A., Boon, W. C., Jones, M. E., Kovacic, A., Zhou, J., and Clyne, C. D. (2005). Estrogen–the good, the bad, and the unexpected. *Endocr Rev* 26, 322–330.
114. Simpson, E. R., Clyne, C., Rubin, G., Boon, W. C., Robertson, K., Britt, K., Speed, C., and Jones, M. (2002). Aromatase–a brief overview. *Annu Rev Physiol* 64, 93–127.
115. Cornil, C. A., Ball, G. F., and Balthazart, J. (2006). Functional significance of the rapid regulation of brain estrogen action: Where do the estrogens come from? *Brain Res* 1126, 2–26.
116. Pasqualini, J. R., Chetrite, G., Blacker, C., Feinstein, M. C., Delalonde, L., Talbi, M., and Maloche, C. (1996). Concentrations of estrone, estradiol, and estrone sulfate and evaluation of sulfatase and aromatase activities in pre- and postmenopausal breast cancer patients. *J Clin Endocrinol Metab* 81, 1460–1464.
117. Brueggemeier, R. W., Hackett, J. C., and Diaz-Cruz, E. S. (2005). Aromatase inhibitors in the treatment of breast cancer. *Endocr Rev* 26, 331–345.
118. Draper, N. and Stewart, P. M. (2005). 11beta-hydroxysteroid dehydrogenase and the pre-receptor regulation of corticosteroid hormone action. *J Endocrinol* 186, 251–271.

119. Tomlinson, J. W., Walker, E. A., Bujalska, I. J., Draper, N., Lavery, G. G., Cooper, M. S., Hewison, M., and Stewart, P. M. (2004). 11beta-hydroxysteroid dehydrogenase type 1: A tissue-specific regulator of glucocorticoid response. *Endocr Rev* 25, 831–866.
120. Draper, N., Walker, E. A., Bujalska, I. J., Tomlinson, J. W., Chalder, S. M., Arlt, W., Lavery, G. G., Bedendo, O., Ray, D. W., Laing, I., Malunowicz, E., White, P. C., Hewison, M., Mason, P. J., Connell, J. M., Shackleton, C. H., and Stewart, P. M. (2003). Mutations in the genes encoding 11beta-hydroxysteroid dehydrogenase type 1 and hexose-6-phosphate dehydrogenase interact to cause cortisone reductase deficiency. *Nat Genet* 34, 434–439.
121. Lavery, G. G., Walker, E. A., Tiganescu, A., Ride, J. P., Shackleton, C. H., Tomlinson, J. W., Connell, J. M., Ray, D. W., Biason-Lauber, A., Malunowicz, E. M., Arlt, W., and Stewart, P. M. (2008). Steroid biomarkers and genetic studies reveal inactivating mutations in hexose-6-phosphate dehydrogenase in patients with cortisone reductase deficiency. *J Clin Endocrinol Metab* 93, 3827–3832.
122. Tomlinson, J. W., Finney, J., Gay, C., Hughes, B. A., Hughes, S. V., and Stewart, P. M. (2008). Impaired glucose tolerance and insulin resistance are associated with increased adipose 11{beta}-hydroxysteroid dehydrogenase type 1 expression and elevated hepatic 5{alpha}-reductase activity. *Diabetes* 57(10), 2652–2660.
123. Valsamakis, G., Anwar, A., Tomlinson, J. W., Shackleton, C. H., McTernan, P. G., Chetty, R., Wood, P. J., Banerjee, A. K., Holder, G., Barnett, A. H., Stewart, P. M., and Kumar, S. (2004). 11beta-hydroxysteroid dehydrogenase type 1 activity in lean and obese males with type 2 diabetes mellitus. *J Clin Endocrinol Metab* 89, 4755–4761.
124. Walker, B. R. (2004). 11beta-hydroxysteroid dehydrogenase type 1 in obesity. *Obes Res* 12, 1–3.
125. Morton, N. M. and Seckl, J. R. (2008). 11beta-hydroxysteroid dehydrogenase type 1 and obesity. *Front Horm Res* 36, 146–164.
126. Cooper, M. S., Blumsohn, A., Goddard, P. E., Bartlett, W. A., Shackleton, C. H., Eastell, R., Hewison, M., and Stewart, P. M. (2003). 11beta-hydroxysteroid dehydrogenase type 1 activity predicts the effects of glucocorticoids on bone. *J Clin Endocrinol Metab* 88, 3874–3877.
127. Cooper, M. S., Rabbitt, E. H., Goddard, P. E., Bartlett, W. A., Hewison, M., and Stewart, P. M. (2002). Osteoblastic 11beta-hydroxysteroid dehydrogenase type 1 activity increases with age and glucocorticoid exposure. *J Bone Miner Res* 17, 979–986.
128. Chapman, K. E. and Seckl, J. R. (2008). 11beta-HSD1, inflammation, metabolic disease and age-related cognitive (dys)function. *Neurochem Res* 33, 624–636.
129. Holmes, M. C. and Seckl, J. R. (2006). The role of 11beta-hydroxysteroid dehydrogenases in the brain. *Mol Cell Endocrinol* 248, 9–14.
130. Seckl, J. R. (1997). 11beta-Hydroxysteroid dehydrogenase in the brain: A novel regulator of glucocorticoid action? *Front Neuroendocrinol* 18, 49–99.
131. Seckl, J. R., Yau, J., and Holmes, M. (2002). 11Beta-hydroxysteroid dehydrogenases: A novel control of glucocorticoid action in the brain. *Endocr Res* 28, 701–707.
132. Bujalska, I. J., Gathercole, L. L., Tomlinson, J. W., Darimont, C., Ermolieff, J., Fanjul, A. N., Rejto, P. A., and Stewart, P. M. (2008). A novel selective 11beta-hydroxysteroid dehydrogenase type 1 inhibitor prevents human adipogenesis. *J Endocrinol* 197, 297–307.
133. Hale, C. and Wang, M. (2008). Development of 11beta-HSD1 inhibitors for the treatment of type 2 diabetes. *Mini Rev Med Chem* 8, 702–710.
134. Saiah, E. (2008). The role of 11beta-hydroxysteroid dehydrogenase in metabolic disease and therapeutic potential of 11beta-hsd1 inhibitors. *Curr Med Chem* 15, 642–649.
135. Schnackenberg, C. G. (2008). 11beta-hydroxysteroid dehydrogenase type 1 inhibitors for metabolic syndrome. *Curr Opin Investig Drugs* 9, 295–300.
136. Tomlinson, J. W. and Stewart, P. M. (2007). Modulation of glucocorticoid action and the treatment of type-2 diabetes. *Best Pract Res Clin Endocrinol Metab* 21, 607–619.
137. Tomlinson, J. W. and Stewart, P. M. (2005). Mechanisms of disease: Selective inhibition of 11beta-hydroxysteroid dehydrogenase type 1 as a novel treatment for the metabolic syndrome. *Nat Clin Pract Endocrinol Metab* 1, 92–99.

138. Tomlinson, J. W. (2005). 11Beta-hydroxysteroid dehydrogenase type 1 in human disease: A novel therapeutic target. *Minerva Endocrinol* 30, 37–46.
139. Seckl, J. R. (2004). 11beta-hydroxysteroid dehydrogenases: Changing glucocorticoid action. *Curr Opin Pharmacol* 4, 597–602.
140. Nashev, L. G., Chandsawangbhuwana, C., Balazs, Z., Atanasov, A. G., Dick, B., Frey, F. J., Baker, M. E., and Odermatt, A. (2007). Hexose-6-phosphate dehydrogenase modulates 11beta-hydroxysteroid dehydrogenase type 1-dependent metabolism of 7-keto- and 7beta-hydroxy-neurosteroids. *PLoS ONE* 2, e561.
141. Penning, T. M. and Byrns, M. C. (2009). Steroid hormone transforming aldo-keto reductases and cancer. *Ann N Y Acad Sci* 1155, 33–42.
142. Penning, T. M., Steckelbroeck, S., Bauman, D. R., Miller, M. W., Jin, Y., Peehl, D. M., Fung, K. M., and Lin, H. K. (2006). Aldo-keto reductase (AKR) 1C3: Role in prostate disease and the development of specific inhibitors. *Mol Cell Endocrinol* 248, 182–191.
143. Smuc, T. and Rizner, T. L. (2009). Expression of 17beta-hydroxysteroid dehydrogenases and other estrogen-metabolizing enzymes in different cancer cell lines. *Chem Biol Interact* 178, 228–233.
144. Amin, S. A., Huang, C. C., Reierstad, S., Lin, Z., Arbieva, Z., Wiley, E., Saborian, H., Haynes, B., Cotterill, H., Dowsett, M., and Bulun, S. E. (2006). Paracrine-stimulated gene expression profile favors estradiol production in breast tumors. *Mol Cell Endocrinol* 253, 44–55.
145. Matsuura, K., Shiraishi, H., Hara, A., Sato, K., Deyashiki, Y., Ninomiya, M., and Sakai, S. (1998). Identification of a principal mRNA species for human 3alpha-hydroxysteroid dehydrogenase isoform (AKR1C3) that exhibits high prostaglandin D2 11-ketoreductase activity. *J Biochem* 124, 940–946.
146. Desmond, J. C., Mountford, J. C., Drayson, M. T., Walker, E. A., Hewison, M., Ride, J. P., Luong, Q. T., Hayden, R. E., Vanin, E. F., and Bunce, C. M. (2003). The aldo-keto reductase AKR1C3 is a novel suppressor of cell differentiation that provides a plausible target for the non-cyclooxygenase-dependent antineoplastic actions of nonsteroidal anti-inflammatory drugs. *Cancer Res* 63, 505–512.
147. Holick, M. F. (2007). Vitamin D deficiency. *N Engl J Med* 357, 266–281.
148. Sakaki, T., Kagawa, N., Yamamoto, K., and Inouye, K. (2005). Metabolism of vitamin D3 by cytochromes P450. *Front Biosci* 10, 119–134.
149. Guo, Y. D., Strugnell, S., Back, D. W., and Jones, G. (1993). Transfected human liver cytochrome P-450 hydroxylates vitamin D analogs at different side-chain positions. *Proc Natl Acad Sci USA* 90, 8668–8672.
150. Shinkyo, R., Sakaki, T., Kamakura, M., Ohta, M., and Inouye, K. (2004). Metabolism of vitamin D by human microsomal CYP2R1. *Biochem Biophys Res Commun* 324, 451–457.
151. Cheng, J. B., Levine, M. A., Bell, N. H., Mangelsdorf, D. J., and Russell, D. W. (2004). Genetic evidence that the human CYP2R1 enzyme is a key vitamin D 25-hydroxylase. *Proc Natl Acad Sci USA* 101, 7711–7715.
152. Cheng, J. B., Motola, D. L., Mangelsdorf, D. J., and Russell, D. W. (2003). De-orphanization of cytochrome P450 2R1: A microsomal vitamin D 25-hydroxilase. *J Biol Chem* 278, 38084–38093.
153. Fu, G. K., Lin, D., Zhang, M. Y., Bikle, D. D., Shackleton, C. H., Miller, W. L., and Portale, A. A. (1997). Cloning of human 25-hydroxyvitamin D-1 alpha-hydroxylase and mutations causing vitamin D-dependent rickets type 1. *Mol Endocrinol* 11, 1961–1970.
154. Bland, R., Zehnder, D., and Hewison, M. (2000). Expression of 25-hydroxyvitamin D3-1alpha-hydroxylase along the nephron: New insights into renal vitamin D metabolism. *Curr Opin Nephrol Hypertens* 9, 17–22.
155. Zehnder, D. and Hewison, M. (1999). The renal function of 25-hydroxyvitamin D3-1alpha-hydroxylase. *Mol Cell Endocrinol* 151, 213–220.
156. Zehnder, D., Bland, R., Williams, M. C., McNinch, R. W., Howie, A. J., Stewart, P. M., and Hewison, M. (2001). Extrarenal expression of 25-hydroxyvitamin d(3)-1 alpha-hydroxylase. *J Clin Endocrinol Metab* 86, 888–894.
157. Hewison, M., Zehnder, D., Chakraverty, R., and Adams, J. S. (2004). Vitamin D and barrier function: A novel role for extra-renal 1 alpha-hydroxylase. *Mol Cell Endocrinol* 215, 31–38.

158. Connell, J. M. and Davies, E. (2005). The new biology of aldosterone. *J Endocrinol* 186, 1–20.
159. Arriza, J. L., Simerly, R. B., Swanson, L. W., and Evans, R. M. (1988). The neuronal mineralocorticoid receptor as a mediator of glucocorticoid response. *Neuron* 1, 887–900.
160. Krozowski, Z. S. and Funder, J. W. (1983). Renal mineralocorticoid receptors and hippocampal corticosterone-binding species have identical intrinsic steroid specificity. *Proc Natl Acad Sci USA* 80, 6056–6060.
161. Whorwood, C. B., Ricketts, M. L., and Stewart, P. M. (1994). Epithelial cell localization of type 2 11 beta-hydroxysteroid dehydrogenase in rat and human colon. *Endocrinology* 135, 2533–2541.
162. Roland, B. L. and Funder, J. W. (1996). Localization of 11beta-hydroxysteroid dehydrogenase type 2 in rat tissues: In situ studies. *Endocrinology* 137, 1123–1128.
163. Kenouch, S., Lombes, M., Delahaye, F., Eugene, E., Bonvalet, J. P., and Farman, N. (1994). Human skin as target for aldosterone: Coexpression of mineralocorticoid receptors and 11 beta-hydroxysteroid dehydrogenase. *J Clin Endocrinol Metab* 79, 1334–1341.
164. Shams, M., Kilby, M. D., Somerset, D. A., Howie, A. J., Gupta, A., Wood, P. J., Afnan, M., and Stewart, P. M. (1998). 11Beta-hydroxysteroid dehydrogenase type 2 in human pregnancy and reduced expression in intrauterine growth restriction. *Hum Reprod* 13, 799–804.
165. Albiston, A. L., Obeyesekere, V. R., Smith, R. E., and Krozowski, Z. S. (1994). Cloning and tissue distribution of the human 11 beta-hydroxysteroid dehydrogenase type 2 enzyme. *Mol Cell Endocrinol* 105, R11–R17.
166. Agarwal, A. K., Mune, T., Monder, C., and White, P. C. (1994). NAD(+)-dependent isoform of 11 beta-hydroxysteroid dehydrogenase. Cloning and characterization of cDNA from sheep kidney. *J Biol Chem* 269, 25959–25962.
167. Stewart, P. M., Murry, B. A., and Mason, J. I. (1994). Type 2 11 beta-hydroxysteroid dehydrogenase in human fetal tissues. *J Clin Endocrinol Metab* 78, 1529–1532.
168. Hammer, F. and Stewart, P. M. (2006). Cortisol metabolism in hypertension. *Best Pract Res Clin Endocrinol Metab* 20, 337–353.
169. Palermo, M., Quinkler, M., and Stewart, P. M. (2004). Apparent mineralocorticoid excess syndrome: An overview. *Arq Bras Endocrinol Metabol* 48, 687–696.
170. Quinkler, M. and Stewart, P. M. (2003). Hypertension and the cortisol-cortisone shuttle. *J Clin Endocrinol Metab* 88, 2384–2392.
171. Edwards, C. R. and Stewart, P. M. (1991). The cortisol-cortisone shuttle and the apparent specificity of glucocorticoid and mineralocorticoid receptors. *J Steroid Biochem Mol Biol* 39, 859–865.
172. Stewart, P. M., Wallace, A. M., Valentino, R., Burt, D., Shackleton, C. H., and Edwards, C. R. (1987). Mineralocorticoid activity of liquorice: 11-beta-hydroxysteroid dehydrogenase deficiency comes of age. *Lancet* 2, 821–824.
173. Michael, A. E. (2003). Life after liquorice: The link between cortisol and conception. *Reprod Biomed Online* 7, 683–690.
174. Kotelevtsev, Y., Brown, R. W., Fleming, S., Kenyon, C., Edwards, C. R., Seckl, J. R., and Mullins, J. J. (1999). Hypertension in mice lacking 11beta-hydroxysteroid dehydrogenase type 2. *J Clin Invest* 103, 683–689.
175. Korbonits, M., Bujalska, I., Shimojo, M., Nobes, J., Jordan, S., Grossman, A. B., and Stewart, P. M. (2001). Expression of 11 beta-hydroxysteroid dehydrogenase isoenzymes in the human pituitary: Induction of the type 2 enzyme in corticotropinomas and other pituitary tumors. *J Clin Endocrinol Metab* 86, 2728–2733.
176. Rabbitt, E. H., Ayuk, J., Boelaert, K., Sheppard, M. C., Hewison, M., Stewart, P. M., and Gittoes, N. J. (2003). Abnormal expression of 11 beta-hydroxysteroid dehydrogenase type 2 in human pituitary adenomas: A prereceptor determinant of pituitary cell proliferation. *Oncogene* 22, 1663–1667.
177. Cooper, M. S., Walker, E. A., Bland, R., Fraser, W. D., Hewison, M., and Stewart, P. M. (2000). Expression and functional consequences of 11beta-hydroxysteroid dehydrogenase activity in human bone. *Bone* 27, 375–381.

178. Eyre, L. J., Rabbitt, E. H., Bland, R., Hughes, S. V., Cooper, M. S., Sheppard, M. C., Stewart, P. M., and Hewison, M. (2001). Expression of 11 beta-hydroxysteroid dehydrogenase in rat osteoblastic cells: Pre-receptor regulation of glucocorticoid responses in bone. *J Cell Biochem* 81, 453–462.
179. Bland, R., Worker, C. A., Noble, B. S., Eyre, L. J., Bujalska, I. J., Sheppard, M. C., Stewart, P. M., and Hewison, M. (1999). Characterization of 11beta-hydroxysteroid dehydrogenase activity and corticosteroid receptor expression in human osteosarcoma cell lines. *J Endocrinol* 161, 455–464.
180. Rabbitt, E. H., Gittoes, N. J., Stewart, P. M., and Hewison, M. (2003). 11beta-hydroxysteroid dehydrogenases, cell proliferation and malignancy. *J Steroid Biochem Mol Biol* 85, 415–421.
181. Rabbitt, E. H., Lavery, G. G., Walker, E. A., Cooper, M. S., Stewart, P. M., and Hewison, M. (2002). Prereceptor regulation of glucocorticoid action by 11beta-hydroxysteroid dehydrogenase: A novel determinant of cell proliferation. *FASEB J* 16, 36–44.
182. Kong, X. F., Zhu, X. H., Pei, Y. L., Jackson, D. M., and Holick, M. F. (1999). Molecular cloning, characterization, and promoter analysis of the human 25-hydroxyvitamin D3-1alpha-hydroxylase gene. *Proc Natl Acad Sci USA* 96, 6988–6993.
183. Ohyama, Y. and Yamasaki, T. (2004). Eight cytochrome P450s catalyze vitamin D metabolism. *Front Biosci* 9, 3007–3018.
184. Omdahl, J. M. B. (2005). The 25-hydroxyvitamin D-24-hydroxylase. New York: Elsevier.
185. Taniguchi, T., Eto, T. A., Shiotsuki, H., Sueta, H., Higashi, S., Iwamura, T., Okuda, K. I., and Setoguchi, T. (2001). Newly established assay method for 25-hydroxyvitamin D3 24-hydroxylase revealed much lower Km for 25-hydroxyvitamin D3 than for 1alpha,25-dihydroxyvitamin D3. *J Bone Miner Res* 16, 57–62.
186. St-Arnaud, R., Arabian, A., Travers, R., Barletta, F., Raval-Pandya, M., Chapin, K., Depovere, J., Mathieu, C., Christakos, S., Demay, M. B., and Glorieux, F. H. (2000). Deficient mineralization of intramembranous bone in vitamin D-24-hydroxylase-ablated mice is due to elevated 1,25-dihydroxyvitamin D and not to the absence of 24,25-dihydroxyvitamin D. *Endocrinology* 141, 2658–2666.
187. Schwartz, Z., Sylvia, V. L., Del Toro, F., Hardin, R. R., Dean, D. D., and Boyan, B. D. (2000). 24R,25-(OH)(2)D(3) mediates its membrane receptor-dependent effects on protein kinase C and alkaline phosphatase via phospholipase A(2) and cyclooxygenase-1 but not cyclooxygenase-2 in growth plate chondrocytes. *J Cell Physiol* 182, 390–401.
188. Cross, H. S., Kallay, E., Farhan, H., Weiland, T., and Manhardt, T. (2003). Regulation of extrarenal vitamin D metabolism as a tool for colon and prostate cancer prevention. *Recent Results Cancer Res* 164, 413–425.
189. Albertson, D. G., Ylstra, B., Segraves, R., Collins, C., Dairkee, S. H., Kowbel, D., Kuo, W. L., Gray, J. W., and Pinkel, D. (2000). Quantitative mapping of amplicon structure by array CGH identifies CYP24 as a candidate oncogene. *Nat Genet* 25, 144–146.
190. Ren, S., Nguyen, L., Wu, S., Encinas, C., Adams, J. S., and Hewison, M. (2005). Alternative splicing of vitamin D-24-hydroxylase: A novel mechanism for the regulation of extrarenal 1,25-dihydroxyvitamin D synthesis. *J Biol Chem* 280, 20604–20611.
191. Makishima, M., Lu, T. T., Xie, W., Whitfield, G. K., Domoto, H., Evans, R. M., Haussler, M. R., and Mangelsdorf, D. J. (2002). Vitamin D receptor as an intestinal bile acid sensor. *Science* 296, 1313–1316.
192. Nehring, J. A., Zierold, C., and DeLuca, H. F. (2007). Lithocholic acid can carry out in vivo functions of vitamin D. *Proc Natl Acad Sci USA* 104, 10006–10009.
193. Xie, W., Radominska-Pandya, A., Shi, Y., Simon, C. M., Nelson, M. C., Ong, E. S., Waxman, D. J., and Evans, R. M. (2001). An essential role for nuclear receptors SXR/PXR in detoxification of cholestatic bile acids. *Proc Natl Acad Sci USA* 98, 3375–3380.
194. St Germain, D. L. and Galton, V. A. (1997). The deiodinase family of selenoproteins. *Thyroid* 7, 655–668.
195. Kohrle, J. (2000). The deiodinase family: Selenoenzymes regulating thyroid hormone availability and action. *Cell Mol Life Sci* 57, 1853-1863.

196. Gereben, B., Zeold, A., Dentice, M., Salvatore, D., and Bianco, A. C. (2008). Activation and inactivation of thyroid hormone by deiodinases: Local action with general consequences. *Cell Mol Life Sci* 65, 570–590.
197. Bianco, A. C., Salvatore, D., Gereben, B., Berry, M. J., and Larsen, P. R. (2002). Biochemistry, cellular and molecular biology, and physiological roles of the iodothyronine selenodeiodinases. *Endocr Rev* 23, 38–89.
198. Jansen, J., Friesema, E. C., Milici, C., and Visser, T. J. (2005). Thyroid hormone transporters in health and disease. *Thyroid* 15, 757–768.
199. Friesema, E. C., Jansen, J., Milici, C., and Visser, T. J. (2005). Thyroid hormone transporters. *Vitam Horm* 70, 137–167.
200. Friesema, E. C., Jansen, J., and Visser, T. J. (2005). Thyroid hormone transporters. *Biochem Soc Trans* 33, 228–232.
201. Maia, A. L., Kim, B. W., Huang, S. A., Harney, J. W., and Larsen, P. R. (2005). Type 2 iodothyronine deiodinase is the major source of plasma T3 in euthyroid humans. *J Clin Invest* 115, 2524–2533.
202. Streckfuss, F., Hamann, I., Schomburg, L., Michaelis, M., Sapin, R., Klein, M. O., Kohrle, J., and Schweizer, U. (2005). Hepatic deiodinase activity is dispensable for the maintenance of normal circulating thyroid hormone levels in mice. *Biochem Biophys Res Commun* 337, 739–745.
203. Abuid, J. and Larsen, P. R. (1974). Triiodothyronine and thyroxine in hyperthyroidism. Comparison of the acute changes during therapy with antithyroid agents. *J Clin Invest* 54, 201–208.
204. Botero, D., Gereben, B., Goncalves, C., De Jesus, L. A., Harney, J. W., and Bianco, A. C. (2002). Ubc6p and ubc7p are required for normal and substrate-induced endoplasmic reticulum-associated degradation of the human selenoprotein type 2 iodothyronine monodeiodinase. *Mol Endocrinol* 16, 1999-2007.
205. Kim, B. W., Zavacki3, A. M., Curcio-Morelli, C., Dentice, M., Harney, J. W., Larsen, P. R., and Bianco, A. C. (2003). Endoplasmic reticulum-associated degradation of the human type 2 iodothyronine deiodinase (D2) is mediated via an association between mammalian UBC7 and the carboxyl region of D2. *Mol Endocrinol* 17, 2603–2612.
206. Dentice, M., Bandyopadhyay, A., Gereben, B., Callebaut, I., Christoffolete, M. A., Kim, B. W., Nissim, S., Mornon, J. P., Zavacki, A. M., Zeold, A., Capelo, L. P., Curcio-Morelli, C., Ribeiro, R., Harney, J. W., Tabin, C. J., and Bianco, A. C. (2005). The Hedgehog-inducible ubiquitin ligase subunit WSB-1 modulates thyroid hormone activation and PTHrP secretion in the developing growth plate. *Nat Cell Biol* 7, 698–705.
207. Silva, J. E. and Larsen, P. R. (1983). Adrenergic activation of triiodothyronine production in brown adipose tissue. *Nature* 305, 712–713.
208. Dentice, M., Luongo, C., Huang, S., Ambrosio, R., Elefante, A., Mirebeau-Prunier, D., Zavacki, A. M., Fenzi, G., Grachtchouk, M., Hutchin, M., Dlugosz, A. A., Bianco, A. C., Missero, C., Larsen, P. R., and Salvatore, D. (2007). Sonic hedgehog-induced type 3 deiodinase blocks thyroid hormone action enhancing proliferation of normal and malignant keratinocytes. *Proc Natl Acad Sci USA* 104, 14466–14471.
209. Sagar, G. D., Gereben, B., Callebaut, I., Mornon, J. P., Zeold, A., da Silva, W. S., Luongo, C., Dentice, M., Tente, S. M., Freitas, B. C., Harney, J. W., Zavacki, A. M., and Bianco, A. C. (2007). Ubiquitination-induced conformational change within the deiodinase dimer is a switch regulating enzyme activity. *Mol Cell Biol* 27, 4774–4783.
210. Bianco, A. C. and Kim, B. W. (2006). Deiodinases: Implications of the local control of thyroid hormone action. *J Clin Invest* 116, 2571–2579.

CHAPTER 17

NEW INSIGHTS TO NUCLEAR RECEPTOR GENE REGULATION FROM ANALYSIS OF THEIR RESPONSE ELEMENTS IN TARGET GENES

CARSTEN CARLBERG

Life Sciences Research Unit, University of Luxembourg, L-1511 Luxembourg City, Luxembourg; Department of Biosciences, University of Kuopio, FIN-70211 Kuopio, Finland

Abstract: Nuclear receptor (NR) target genes have key roles in cellular metabolism, cellular growth and differentiation and in controlling inflammation. Many NR target genes are also involved in dysregulated pathways that can lead to common human diseases, such as type 2 diabetes, atherosclerosis, Alzheimer's disease and cancer. On the genomic level these pathways converge on regulatory modules, some of which contain co-localizing NR binding sites, so-called response elements (REs). Recent advances in genomic techniques combined with computational analysis of binding modules are discussed in this chapter, primarily at the example of the NRs vitamin D receptor (VDR or NR1I1) and peroxisome proliferator-activated receptors (PPARα or NR1C1, PPARβ/δ or NR1C2 and PPARγ or NR1C3).

17.1. INTRODUCTION

Each individual human gene is under the control of a large set of transcription factors that can bind up- and downstream of its transcription start site (TSS) [1]. In order to directly activate a gene by a NR ligand at least one NR molecule has to bind in sufficient vicinity to the gene's TSS [2]. However, 'vicinity' could in some cases be a distance of up to 100 kB, irrespective if up- or downstream of the TSS. Moreover, there are a number of evidences that most primary NR target genes use multiple REs for their full functionality [3]. These REs typically arrange together with binding sites of other transcription factors into collections of neighboring sites, so-called modules or enhancers. Modules of transcription factors that act on focused genomic regions have been shown to be far more effective than individual factors on isolated locations. In an ideal case such transcription factor modules can be identified by parallel and comparative analysis of their binding sites. Here bioinformatics approaches can be of great help, in case they can predict the actions of the transcription factors precisely enough [4].

C.M. Bunce, M.J. Campbell (eds.), *Nuclear Receptors*, Proteins and Cell Regulation 8, 419–437. DOI 10.1007/978-90-481-3303-1_17,

The complete sequence of the human genome and also that of other mammalian species, such as chimp, dog, mouse and rat, is now available, so that we are able to screen for all putative REs. However, the constant packaging of genomic DNA into chromatin provides a repressive environment, which in most cases denies the access to putative REs [5]. Fortunately, new experimental techniques for genome-wide analyses of chromatin modifications and transcription factor binding, such as chromatin immunoprecipitation (ChIP)-chip and parallel sequencing (ChIP-Seq), are now available [6]. This will revolutionize our understanding of the genome-wide effects of NRs and of the diversity of their target genes as outlined in this chapter.

17.2. THE NR SUPERFAMILY

NRs form a superfamily with 48 human members, most of which have the special property to be ligand-activated [7, 8]. This property has attracted interest in the NR family as possible therapeutic targets. NRs belong to the best-characterized representatives of approximately 3,000 different mammalian proteins that are involved in transcriptional regulation in human tissues [9]. NRs modulate genes that affect processes as diverse as reproduction, development, inflammation and general metabolism. They can be classified based on ligand sensitivity [7], evolution of the NR genes [10] and their physiological role as interpreted from tissue-specific expression patterns [11].

The ligand sensitivity approach suggests three NR classes [7]. Class I contains the endocrine receptors with high-affinity hormonal lipids, such as the receptors for the steroid hormones estradiol (ER α and β or NR3A1 and 2), progesterone (PR or NR3C3), testosterone (AR or NR3C4), cortisol (GR or NR3C1) and aldosterol (MR or NR3C2), for thyroid hormones (TR α and β or NR1A1 and 2) and for the biologically active forms of the fat-soluble vitamins A and D, all-*trans* retinoic acid (RAR α, β and γ or NR1B1-3) and 1α,25-dihydroxyvitamin D_3 (1α,25$(OH)_2D_3$, VDR). These 12 NRs can be defined functionally as being able to bind their specific ligand with a K_d of 1 nM or less [7]. In class II are adopted orphan receptors that bind to dietary lipids and xenobiotics in the micro- to millimolar concentration range [12], such as PPARs α, β/δ and γ, liver X receptors (LXRs) α and β(NR1H3 and NR1H2), farnesoid X receptor (FXR or NR1H4), constitutive androstane receptor (CAR or NR1I3) and pregnane X receptor (PXR or NR1I2). Subsequently, these receptors have activation thresholds (in terms of K_d) in the same molar range, which functionally separates them from the endocrine receptors. Finally, in class III are orphan receptors, such as estrogen-related receptors (ERRs) α, β and γ (NR3B1-3), small heterodimerizing partner (SHP or NR0B2), dosage-sensitive sex reversal congenital adrenal hypoplasia critical region on the X chromosome (DAX-1 or NR0B1) and hepatocyte nuclear factors (HNFs) α, β and γ (NR2A1-3) for which a physiological ligand has not yet been identified. Some of these orphan NRs may shift eventually to the adopted orphan NR class, but others have such a tiny ligand-binding pocket within their ligand-binding domain (LBD) that they are very likely to be able to accommodate a ligand [13].

When the sequences of NRs are compared on DNA and protein level [10], the grouping significantly differs from the ligand-centered view. For example, VDR is in the same group with the PPARs, while the highly ligand-sensitive ERs and the orphan ERRs are together in another group. Again a different classification can be obtained on the basis of mRNA expression of all NR genes in 39 different tissues in two different mouse strains. Here the NRs are divided into clades with distinct physiological roles [11], where, for example, VDR is grouped to bile acid and xenobiotic metabolism based on its high expression in gastroentric tissues and PPARs are linked to lipid metabolism and energy homeostasis.

17.3. NRs AS MOLECULAR SWITCHES

NRs have a modular structure, onto which certain functions can be ascribed. The amino-terminus is of variable length and sequence in the different family members. It contains a transactivation domain, termed AF-1, which is recognized by co-activator proteins and/or other transcription factors, often in a ligand-independent fashion. The central DNA-binding domain (DBD) has two zinc-finger motifs that are common to the entire family. Interestingly, two orphan NRs, SHP and DAX-1, lack a DBD and function mainly as dominant negative repressors of other NRs [14].

The carboxy-terminal LBD, whose overall architecture is well conserved between the various family members, nonetheless diverges sufficiently to guarantee selective ligand recognition as well as accommodate the broad spectrum of NR ligand structures. The ligand-binding domain consists of 250–300 amino acids in 11–13 α-helices [15]. Ligand binding causes a conformational change within the LBD, whereby, at least in the case of endocrine NRs, helix 12, the most carboxy-terminal α-helix (also called AF-2 domain), closes the ligand-binding pocket via a 'mouse-trap like' intra-molecular folding event [16]. The LBD is also involved in a variety of interactions with nuclear proteins, such as other members of the NR superfamily and co-regulator proteins.

An essential prerequisite for the direct modulation of transcription by NR ligands is the location of at least one activated NR protein close to the TSS of the respective primary NR target gene. This is commonly achieved through the specific binding of NRs to a DNA binding site, a so-called RE, and DNA-looping towards the TSS [17]. Then the NRs recruit positive and negative co-regulatory proteins, referred to as co-activators [18] and co-repressors [19], respectively. In a simplified view of NR signaling, in the absence of ligand the NR interacts with co-repressor proteins, such as NCoR1, SMRT and Alien, which in turn associate with histone deacetylases leading to a locally increased chromatin packaging [20, 21]. The binding of ligand induces the dissociation of the co-repressor and the association of a co-activator of the p160-family, such as SRC-1, TIF2 or RAC3 [22]. Some co-activators have histone acetytransferase activity or are complexed with proteins harboring such activity and this results in the net effect of local chromatin relaxation [23]. In a subsequent step, ligand-activated NRs change rapidly from interacting with the co-activators of the p160-family to those of the mediator complex, such as Med1 [24]. The mediator complex, which consist of approximately 15–20 proteins, builds a bridge to the basal transcriptional machinery [25]. In this way ligand-activated NRs execute two tasks, the modification of chromatin and the regulation of transcription.

Cell- and time-specific patterns of relative protein expression levels of some co-regulators can distinctly modulate NR transcriptional activity. This aspect may have some diagnostic and therapeutic value in different types of cancer [26]. However, the switch between gene repression and activation is more complex than a simple alternative recruitment of two different regulatory complexes [27]. Most co-regulators are co-expressed in the same cell type at relatively similar levels, which raises the possibility of their concomitant recruitment to a specific promoter. This has been resolved by the mutually exclusive binding of co-activators and co-repressors to ligand-bound and -unbound NRs, respectively. Therefore, repression and activation is more likely achieved by a series of sequential multiple enzymatic reactions that are promoter and cell-type specific.

Transcriptional regulation is a highly dynamic event of rapid association and dissociation of proteins and their modifications, including proteolytic degradation and de novo synthesis. A pattern of recruitment and release of cohorts of co-regulatory complexes was demonstrated on a single ERα binding region of the *trefoil factor-1* promoter in breast cancer cells [28]. This study revealed detailed and coordinated patterns of co-regulator recruitment and preferential selectivity for factors that have similar enzymatic activities. Interestingly, similar cyclic behavior was also observed for VDR [29, 30] and PPAR [31]. Transcription cycles of 30–60 min length as described in these and in previous studies suggest that this phenomenon reflects the basal architecture of eukaryotic transcription across different cell types and species [28, 29, 32–35]. However, the extent of transcriptional cycling seems to depend on the REs that control the respective NR target gene.

17.4. CHROMATIN

The major protein constituents of chromatin are the four different histones that form a nucleosome, around which genomic DNA is wound. Chromatin carries numerous histone and DNA modifications, and many of these are associated with transcription. Covalent modifications of the lysines at the amino-terminal tails of histones neutralize their positive charge and thus their attraction for the negatively charged DNA backbone is diminished [36]. As a consequence, the association between the histone and the DNA becomes less stable. In addition, chemical moieties alter the nucleosome surface to promote the association of chromatin binding proteins. Both functions influence the degree of chromatin packaging and regulate the access of transcription factors to their potential binding sites [37]. Chromatin modifications are often termed epigenetic marks, although it is not clear how stable these alterations are during cell divisions and development. It seems that most, if not all, histone modifications are reversible, so it remains to be determined how epigenetic persistence of chromatin states is achieved, and which modifications are heritable.

A popular view within the field of chromatin research and in related areas has been that DNA methylation and histone post-translational modifications correlate with either positive or negative transcriptional states. These histone post-translational modifications are established across regulatory regions during gene activation through recruitment of the relevant enzymes by DNA-bound activators and RNA polymerase II. In turn, negative-acting marks are laid down across genes during repression by DNA-bound repressor recruitment or across heterochromatic regions of the genome.

Most co-regulators are not exclusive to NRs but also used in a similar manner by numerous other transcription factors. Based on their mode of action, co-regulators can be classified into two main groups [38]. The first group contains factors that covalently modify histones, such as acetylation/deacetylation and methylation/demethylation, a process that follows the precise and combinatorial histone code [36]. The second group of co-regulators includes ATP-dependent chromatin remodeling factors that modulate promoter accessibility to transcription factors and to the basal transcriptional machinery [39]. The complex network of co-regulators defines a co-regulator code characterized by distinct patterns of co-regulator recruitment and by their regulated enzymatic activities. The histone code can therefore be considered as a consequence as well as a determinant of this co-regulator code, as histones are crucial targets for the enzymatic activities of co-regulators but also have a key role in specifying co-regulator recruitment on the basis of the reading of the histone code.

17.5. NR REs

The DBDs of NRs contain two zinc finger motifs that enable them to bind to REs being composed of two hexameric DNA motifs of the consensus sequence RGKTSA (R = A or G, K = G or T, S = C or G), each of which are contacted by a short amino acid motif, the so-called P-box, within the NR DBD [40]. Some orphan NRs can also bind DNA as monomers, but they recognize additional specific nucleotides 5' flanking to the RGKTSA-motif [14]. Steroid NRs, such as AR and GR, and some orphan NRs form homodimers and recognize two hexameric arranged as inverted repeats (IRs) with one or three nucleotide spacing. Moreover, in contrast to all other NR superfamily members steroid NRs prefer to bind to RGAACA-like motifs [41].

Many NRs, which mainly include the sensors for micro- and macronutrients, preferentially form heterodimers with the NR retinoid X receptor (RXR) α, β or γ (NR2B1-3). Most heterodimers recognize two hexameric motifs in a direct repeat (DR) orientation, but recognition of everted repeat (ER) binding has been described as well for VDR, PXR, RARs and TRs [42]. Each heterodimer has individual preferences to the nucleotide composition and the tolerated variations from the consensus sequence, spacing and orientation. For example, PPAR-RXR heterodimers seem to bind more or less exclusively to DR1-type REs, while VDR-RXR heterodimers bind to DR3-type RE [43, 44] but to DR4-, ER6-, ER7-, ER8- and ER9-type REs [45, 46].

A RE classification according to the affinity for dimeric NR complexes suggests that the degree of deviation from the RGKTSA consensus sequence [47] is proportional to the loss of in vitro functionality [48]. Interestingly, the DR4-type RE of the rat *pit-1* gene [49], which contains two perfect core binding motifs, was found to be a higher affinity VDR-RXR heterodimer target than any known natural DR3-type RE [48]. However, one has to consider that a DR4-type REs is also recognized by the heterodimeric complexes of TRs, CAR, PXR and other orphan NRs with RXR [50, 51], whereas the same complexes bind to DR3-type REs far less tightly than VDR-RXR heterodimers. The competitive situation on DR4-type REs may therefore be the reason why in vivo VDR-RXR heterodimers still prefer DR3-type REs.

17.5.1. ChIP Analysis: The Concept of Multiple REs

For a detailed analysis of the regulatory regions of primary NR target genes and for the confirmation of the binding of a NR to a given RE in living cells, the method of ChIP became the golden standard. For example, for the VDR target genes *CYP24* [34], *CYP27B1* [52], *cyclin C* [53] and *p21* (*CDKN1A*) [30, 54] some 7–10 kB of their promoter regions were investigated by using in each case a set of 20–25 overlapping genomic regions. This approach identified four functional REs for both the *CYP24* and *cyclin C* genes, three in the *p21* promoter and two in the *CYP27B1* gene. Although the in vitro DNA binding affinity of VDR-RXR heterodimers to the REs described for these genes differs (compare [34, 53–55]), at the chromatin level all RE-containing promoter regions show comparable association strength with VDR and RXR. Each of the multiple $1\alpha,25(OH)_2D_3$-responsive promoter regions is able to contact independently the basal transcriptional machinery. This suggest that the simultaneous communication of the individual promoter regions with the RNA polymerase II complex occurs through a discrete 3-dimensional organization of the promoter within the nucleus of a cell and that this is achieved via a large protein conglomeration, such as the mediator complex. This arrangement would therefore allow the close contact of distant regions. Such a model could resemble the traditional 'DNA looping model' being discussed to explain the activity of upstream enhancer elements [56].

An alternative approach to the identification of primary NR target genes was performed with the six members of the *insulin-like growth factor binding protein (IGFBP)* gene family. Here first an *in silico* screen for VDR-RXR binding was performed, which was then followed by the analysis of candidate $1\alpha,25(OH)_2D_3$-responsive sequences in ChIP assays [55]. Induction of gene expression was confirmed independently using quantitative PCR. By using this approach, the genes *IGFBP1*, *3* and *5* were demonstrated to be primary VDR target genes. The *in silico* screening of the 174 kB of genomic sequence surrounding all six *IGFBP* genes identified 15 candidate REs, ten of which were shown to be functional in ChIP assays, i.e. also this approach confirmed the concept of multiple REs per NR target gene. Importantly, the *in silico* screening was not restricted to regulatory regions that comprise only 2 kB of sequence up- and downstream of the TSS, as in a recent whole genome screen for regulatory elements [57], but involved up to 10 kB of flanking sequences as well as intronic and intergenic sequences. In a comparable study the *5-lipoxygenase* (*5-LO*) gene has been analyzed and confirmed to be a primary 1α,25(OH)$_2$D$_3$ target gene. From the 22 putative REs identified in the whole *5-LO* gene sequence (–10 kB to +74 kB) by *in silico* screening, at least two have been validated to be functional in vitro and in the living cells. One of these REs is located far downstream of the TSS (+42 kB) and is one of the strongest known REs of the human genome [58]. No functional RE had been reported for *5-LO* before, since previous studies had been restricted to the proximal promoter region [59, 60]. Therefore, this study revealed candidate REs that are located more than 30 kB distant from their target gene's TSS. Based on the present understanding of enhancers, DNA looping and chromatin units being flanked by insulators or matrix attachment regions these distances are not limiting [61].

Interestingly, the number of REs within a promoter does not correlate with the inducibility of a NR target gene, since the average short-term transcriptional response of most primary NR target genes is only 2-fold or less [62]. However, most of them are simultaneously under the control of other transcription factors, such p53 in case of the *p21* gene [54], and therefore possess significant basal levels of transcription.

17.6. ChIP-CHIP AND ChIP-Seq ANALYSIS

The combination of ChIP assays with hybridization of the resulting chromatin fragments on microarrays, so-called ChIP-chip assays, provide an additional step for a larger scale analysis of NR target genes. For VDR target genes ChIP-chip technology has been applied for the analysis of the *VDR* gene itself [63], the intestinal calcium ion channel gene *TRPV6* [64], the Wnt signaling co-regulator *LRP5* [65] and the TNF-like factor *RANKL* that promotes the formation of calcium resorbing osteoclasts [66]. For all those genes a number of VDR-associated chromatin regions were identified, some of which were far upstream of the gene's TSS. These studies confirmed the concept that many, if not all, NR target genes have multiple NR-associated regions. However, not all of these REs may be functional, i.e. they may not contact the gene's TSS via DNA looping. Therefore, it is necessary to apply an additional method, the so-called chromosome-conformation-capture (3C) assay. So far, 3C assays confirmed the functionality of the REs in the *CYP27B1* [52] and the *p21* [27] gene.

Genome-wide ChIP-chip studies have been performed primarily with ERα [67]. These analyses suggest that the promoter-proximal regions, although important for some genes, do not constitute the majority of ERα target sites. Instead, it is apparent that a full definition of ERα binding to regulatory regions distinct from promoters is required to fully understand the estrogen response. Similar observations were made for genome-wide binding of the transcription factors c-Myc, p53 and Sp-1 [68]. In contrast, components of the basal transcriptional machinery, such as TFIID and RNA polymerase II, seem to be biased to promoter-proximal regions. This suggests that communication is often mediated at great distances between the transcription factors that initiate gene expression events and the transcription machinery that execute it. Based on Caroll et al. [67] less than one-third of early estrogen up-regulated genes have ERα binding sites within 50 kB of their TSS. Nevertheless, the complete set of ERα binding sites across the genome establishes a new resource for understanding estrogen action, for example in breast cancer. It correctly predicts the genes co-expressed with ERα in primary breast tumors and thus identifies important and previously unexplored regions of the genome that are the critical regulators of the estrogen dependence of breast cancer.

A technical improvement of the specificity and sensitivity of ChIP-chip assays was obtained by using the DNA selection and ligation (DSL) strategy [69]. This ChIP-DSL technology was also applied for ERα target genes in breast cancer cells and identified 578 highest confidence ERα-enriched promoters, which represents 3.3% of all reliably scored promoters. The majority of these ERα-bound promoters were also marked by RNA polymerase II and epigenetic markers associated with gene activation, but in microarray analyses only 54 of these were identified to be direct estrogen target genes. These 54 genes represent only 10% of all identified direct estrogen target genes. This suggests that the majority of the genes identified in microarray are only secondary estrogen targets and that only a fraction of the direct ERα target genes were highly responsive to estrogen.

'Big biology' projects, such as ENCODE [70], and other whole genome scans for histone modification and transcription factor association provide a large set of useful data for understanding the genome-wide action of transcription factors. In the ENCODE project from the NR superfamily only RARα and HNF4α are represented. RARα was studied in connection with neutrophil differentiation together with other transcription factors, co-regulators and histone modification in retinoic acid stimulated cells (Affymetrix ChIP-chip track [68]), while HNF4α was investigated together with other key liver transcription factors and histone 3 acetylation that marks active regions in liver cells (Uppsala ChIP track [71]).

The next step in genome-wide association studies is parallel sequencing of genomic fragments obtained after ChIP assays, also referred to as ChIP-Seq, with antibodies against NRs and their partner proteins. The first genome-wide ChIP-Seq study for a member of the NR superfamily was published for PPARγ and RXR binding sites and RNA polymerase II occupancy in adipocytes [72]. This study identified some 5,200 shared PPARγ-RXR binding sites in mature adipocytes, a number which was confirmed by a comparable ChIP-chip study [73]. Interestingly, the number of functional PPARγ-RXR heterodimers seem to have significantly increased during maturation of the cells starting with a very low number in undifferentiated cells. Some of these PPARγ-RXR binding regions (according to Lefterova et al. [73] 60% of all) also showed association the transcription factor C/EBP indicating a coordinated action of both transcription factors in adipocyte differentiation. This first ChIP-Seq study confirmed the finding from previous ChIP-chip studies that the majority of the NR binding sites are in intergenic regions. More ChIP-Seq data will be published soon.

17.7. REs IN THE CHROMATIN CONTEXT

It is assumed that matrix attachment regions subdivide genomic DNA into units of an average length of 100 kB containing the coding region of at least one gene [74]. DNA looping should be able to bring any DNA site within the same chromatin unit close to the basal transcriptional machinery that is assembled on the TSS. This model suggests that also very distant sequences can serve as REs and that even sequences downstream of the TSS could serve as functional VDR binding sites. Moreover, it fits well with the recent observation that the majority of functional NR binding sites are in intergenic regions. In fact the term 'intergenic region' may be defined more carefully and the respective genomic regions should be assigned to the genes, with which they share the same chromatin loop.

Due to its optimized 5'-flanking dinucleotide and core binding motif sequences the DR4-type RE of the rat *pit-1* gene is the most efficient known VDR binding site in vitro [48, 51]. However, the chromatin in the region of the *pit-1* gene promoter containing this RE seems to be closed in the adult rat, so that the responsiveness of the gene to $1\alpha,25(OH)_2D_3$ is lower than expected [23]. This indicates that a high in vitro binding affinity of VDR-RXR heterodimers for a RE is not sufficient for responsiveness to $1\alpha,25(OH)_2D_3$. When the promoter region that contains the RE is covered by condensed chromatin, VDR-RXR heterodimers are unable to bind there. This makes sufficiently decondensed chromatin an essential prerequisite for a functional RE. Chromatin decondensation is achieved by the activity of histone acetyltransferases, which are recruited to their local chromatin target by co-activator proteins. In turn, these co-activators are transiently attracted to a promoter region by ligand-activated NRs and other active transcription factors. Therefore, the more transcription factor binding sites a given genomic region has and the more of these transcription factors are expressed in the respective cell, the higher is the chance that this area of the promoter gets locally decondensed.

The above described co-localization of PPARγ and C/EBP binding sites in genes with impact in adipocyte differentiation may be an example for this concept. Another, more detailed example is the VDR binding site of the rat *osteocalcin* gene, which is flanked on both sides with a binding site for the transcription factor Runx2/Cbfa1 [75]. By contacting co-activator proteins and histone acetyltransferase Runx2/Cbfa1 seems to mediate the opening of chromatin locally, which allows efficient binding of VDR-RXR heterodimers to this decondensed region to occur. This paradigm suggests that REs are better targets for NR complexes, if other transcription factors are bound to the same chromatin region. In this respect, promoter context and cell-specific expression of other transcription factors may be of greater importance to a RE functionality and specificity than its in vitro binding profile.

Most models of transcriptional regulation still place the gene promoter and its REs central for recruiting transcription factors and co-regulators. However, it may also be possible that nuclear actin complexes pull promoters to pre-assembled transcription factories [76–78]. In this model the promoter-specific REs could represent an allosteric regulator of receptor activity and co-regulator recruitment. Similarly, the mode of interaction with DNA, either directly or through other transcription factors, could be an important regulator of specificity [79].

17.8. NEGATIVE REs

Expression profiling using microarray technology indicates that comparable numbers of genes are down-regulated by NR ligands as are up-regulated by the nuclear hormones [80]. In general, the mechanisms of the down-regulation of genes by NR ligands are much less understood, but they also seem to require the binding of an agonist to the NR. It is obvious that only genes, which show significant basal activity, can be down-regulated, i.e. these genes exhibit basal activity due to other transcription factors binding to their promoter. There are several different models that attempt to explain, how nuclear hormones and their ligands can mediate down-regulation of genes, but the common theme is that NR counteracts the activity of specific transcription factors [81]. For example, for the physiologically important down-regulation of the *CYP27B1* gene by $1\alpha,25(OH)_2D_3$ a negative RE located at position –0.5 kB has been proposed, where VDR-RXR heterodimers do not bind directly, but via the transcription factor VDR interacting repressor (VDIR, also called TCF3) [82]. In addition, two positive REs are located –2.6 and –3.2 kB upstream from the TSS and modulate the cell-specific activity of the negative RE [52]. Association of VDR-RXR heterodimers to TCF3 binding sites may also occur through ligand-dependent chromatin looping from more distal regions that directly bind the VDR [52]. In situations where these activating transcription factors are other NRs or transcription factors that bind to composite REs, the NR could simply compete for DNA binding sites [83, 84]. In a similar way, NRs can also compete for binding to partner proteins, such as RXR, or for common co-activators, such as SRC-1 or CBP [85]. In all these situations the down-regulating effects of the NR should be of general impact, i.e. the mechanism could apply to other genes in the same way.

The concept that multiple REs together with other transcription factor binding sites regulate primary NR target genes suggests that a promoter may contain both negative and positive REs. The activities of the different REs are determined by the promoter context and may not be simultaneously active. One might imagine that prior to stimulation with the NR ligand only the negative REs bind the NR and recruit co-repressors. This would actively condense the chromatin on a particular promoter region. The addition of ligand induces the release of co-repressor proteins and reduces chromatin density. The NR may then be transiently released from the negative RE and bind to a positive RE, which may be uncovered through NR ligand-dependent local nucleosome acetylation. The VDR then interacts with the mediator protein complex on this positive RE leading to transient transcriptional activation. After a certain period of time, newly synthesized, unliganded NR again binds to the negative RE, which initiates chromatin closing and inactivation of the positive RE [52]. In this or even more complex scenarios, the balance between negative and positive REs could explain the time course of the activation of primary NR target genes.

17.9. METHODS FOR *IN SILICO* SCREENING OF NR BINDING SITES

Characterization of REs from NR target gene promoters has resulted in a large collection of REs that deviate significantly from the consensus sequence. For example, an extensive binding data collection for PPARs was recently published [86], where more critical deviations and well-tolerated deviations from the consensus were identified. Statistically, a NR core binding motif, such as RGKTSA, should be found, on average, in every 256 bp of genomic DNA. Furthermore, dimeric assemblies of such hexamers should show up as DRs every 65,536 bp and as ERs every 32,768 bp in a random sequence. Therefore, an *in silico* screen of the human genome would identify for every NR on average of 50,000–100,000 putative REs. For example, Wang et al. combined microarray analysis and *in silico* genome-wide screens for DR3- and ER6-type VDR REs [87]. This approach identified several novel REs and VDR target genes, but most of the REs await a confirmation by ChIP and 3C assays.

Since NR proteins have an abundance of at most a few thousand molecules per cell and ChIP-Seq studies confirmed this order of functional NR binding sites per cell, a biologically realistic number of NR target genes per cell should be closer to this number. If one also considers the fact that many NR target genes appear to have more than one functional RE for any given NR, it could be expected that the real number of NR target genes in any cell type is much less than the number of NR molecules. These calculations make it obvious that not every putative NR binding site is used in nature in any cell at any given time.

The specificity of NRs for their binding sites allows constructing a model to describe the RE properties that can be used to predict potential binding sites in genomic sequences. For this the NR binding preference, often expressed as position weight matrix (PWM), has to be described on the basis of experimental data, such as series of gel shift assays with a large number of natural binding sites [88–91]. PWM descriptions lead to a prediction of REs every 1,000–10,000 bp of genomic sequence. This probably contains many false positive predictions, which is mainly due to scoring methodology and the limitations that are imposed by the available experimental data. For example, the quantitative characteristics of a transcription factor, i.e. its relative binding strength to a number of different binding sites, is neglected in a position frequency matrix, where simply the total number of observations of each nucleotide is recorded for each position. In a position frequency matrix the quantitative characteristics of a transcription factor, i.e. its relative binding strength to a number of different binding sites, is neglected, since simply the total number of observations of each nucleotide is recorded for each position. Moreover, in the past there was a positional bias of transcription factor binding sites upstream in close vicinity to the TSS. This would be apparent from the collection of identified REs [92], but is in contrast with a multi-genome comparison of NR binding site distribution [57] and other reports on wide-range associations of distal regulatory sites [93].

Internet-based software tools, such as TRANSFAC [94], screen DNA sequences with databases of matrix models. One approach used PWMs to describe the binding preferences of PPARs using all published PPAR binding sites [95]. The accuracy of such methods can be improved by taking the evolutionary conservation of the binding site and that of the flanking genomic region into account. Moreover, cooperative interactions between transcription factors, i.e. regulatory modules, can be taken into account by screening for binding site clusters. The combination of phylogenetic footprinting and PWM searches applied to orthologous human and mouse gene sequences reduces the rate of false predictions by an order of magnitude, but leads to some reduction in sensitivity [96]. Recent studies suggest that a surprisingly large fraction of regulatory sites may not be conserved but yet are functional, which suggests that sequence conservation revealed by alignments may not capture some relevant regulatory regions [97].

In effect, these approaches and tools are still insufficient and there has to be a focus on the creation of bioinformatics resources that include more directly the biochemical restrains to regulate gene transcription. One important aspect is that most putative REs are covered by nucleosomes, so that they are not accessible to the respective transcription factor. This repressive environment is found in particular for those sequences that are either contained within interspersed sequences, are positionally isolated from transcription factor modules or lie outside of insulator sequences marking the border of chromatin loops [98]. This perspective strongly discourages the idea that isolated, simple NR REs may be functional in vivo. In turn, this idea implies that the more transcription factor binding sites a given promoter region contains and the more of these transcription factors are expressed within a given cell, the higher is the chance that the chromatin on this area of the promoter becomes locally opened.

17.10. THE CLASSIFIER METHOD

Approaches for NR RE predictions have been based on a collection of disparate binding data and in general lack quantitative comparison of different experimental results. The recently published classifier method [86] used the in vitro binding preferences of the three PPAR subtypes on a panel of 39 systematic single nucleotide variations of the consensus DR1-type RE (**AGGTCAAAGGTCA**) [99] as an experimental data set. The single nucleotide variants were sorted into three classes, where in class I the PPAR subtypes are able to bind the sequence with a strength of $75 \pm 15\%$ of that of the consensus RE, in class II with $45 \pm 15\%$ and in class III with $15 \pm 15\%$. More than hundred additional DR1-type REs were sorted on the basis of counting increasing number of variations from the consensus and taking into account the single nucleotide variant binding strength. This led to only 11 RE categories, where combinations of class I, II and III variation still resulted in more than 1% relative binding. The main advantage, when comparing the classifier to PWM methods, is a clear separation between weak REs and those of medium and strong strength [86]. For the discovery of potential binding sites, this is extra information that could be especially of interest in processes considered context dependent, for example for REs that reside in genomic context of transcription factor modules.

All human genes that are known to be primary PPAR targets together with their mouse ortholog were clustered by predicted binding strength and evolutionary conservation of their REs [86]. This resulted in four clusters, where cluster I contains genes that carry multiple conserved REs, while genes in cluster II have only one or two strong or medium conserved RE in human, which are found in comparable strength and location in the mouse. Cluster III contains genes that have strong or medium REs in one species that are conserved only as weak REs in the other species. Finally, cluster IV contains more than 25% of all tested genes, which have the common property that they carry one or more REs, but none of them is conserved. These examples suggest that during evolution of gene regulation by PPARs apparently a number of different approaches were successful. However, minimal requirement for effective gene regulation by PPARs seems to be the presence of at least one strong REs or of several medium strength REs within the 20 kB surrounding the annotated TSS.

On this basis the gene-dense human chromosome 19 (63.8 MB, 1,445 known genes) and its syntenic mouse regions (956 genes have known orthologs) were selected for an *in silico* screening based on the above explained criteria; i.e. both species were investigated for medium and strong REs [86]. Interestingly, 20% of genes of chromosome 19 contain a co-localizing strong PPAR RE and additional 4% have more than two medium REs or a proximal medium RE. Experimentally, a complete evaluation of the selectivity of any such screen is complicated by the restricted expression profiles of the predicted genes, which prevents simple read-outs from individual target tissues. When requiring the detection in human and mouse, 12.1% of genes from chromosome 19 were predicted as PPAR targets.

Such a binding site screen will gain more power, when it can be integrated with other genomic screens, both experimental and bioinformatic. Experimental data, for example from microarray or ChIP-Seq origin, describing gene regulation in a disease state versus normal cells can be visualized by the same tool, such as the UCSC genome browser (http://genome.ucsc.edu), in order to detect overlap in functional binding sites. Given the high interest of the scientific community to better characterize binding profiles of different transcription factors and the improved experimental techniques to detect genome-wide binding events, such additional tracks combined with a RE binding track could be available in near future. Importantly, these datasets will motivate studies that aim to integrate the knowledge with systems biology methodology in order to model NR function in healthy versus disease state in various human tissues.

17.11. CONCLUSION

The sequencing of the complete human genome and the genome of other species, i.e. the availability of all regulatory sequences, enable a more mature understanding of the diversity of NR target genes. The identification of genes showing a primary response to NRs and their ligands, the so-called NR regulome, can be used as a prediction of their therapeutic potential as well as their possible side effects. Methods incorporating both experimental- and informatics-derived evidence to arrive at a more reliable prediction of NR targets and binding modules can bring all available data together with the aim to predict outcome in specific context. Taking the chromosome 19 *in silico* screening trial for PPAR REs as an example and extrapolating the results to the whole human genome, it suggests that approximately 10% of all human genes (an estimate of 2,000–2,500 genes) have the potential to be directly regulated by a specific NR by their RE content within 10 kB distance to their TSS. Translated to regulatory modules that co-localize with REs, an even larger number of genomic regions could be targeted by a given NR. It can be envisioned that in future the emphasis will shift from target genes to target regulatory modules to alter a physiological response and from individual genes to whole genome response. Therefore, a much larger challenge lies ahead when we will be confronted with the higher order of regulated networks of genes, where the sum effect of ligand treatment may reveal itself. In an effort to study this, we have started applying systems biology to the field of NR biology, through an EU-funded Marie Curie Research Training Network, NUCSYS (www.uku.fi/nucsys).

ACKNOWLEDGMENTS

The University of Luxembourg, the Academy of Finland, the Juselius Foundation and the EU (Marie Curie RTN NucSys) supported the work.

REFERENCES

1. Sperling, S. (2007). Transcriptional regulation at a glance. *BMC Bioinformatics* 8(Suppl 6), S2.
2. Carlberg, C. and Dunlop, T. W. (2006). The impact of chromatin organization of vitamin D target genes. *Anticancer Res* 26, 2637–2645.
3. Carlberg, C., Dunlop, T. W., Saramäki, A., Sinkkonen, L., Matilainen, M., and Väisänen, S. (2007). Controlling the chromatin organization of vitamin D target genes by multiple vitamin D receptor binding sites. *J Steroid Biochem Mol Biol* 103, 338–343.
4. Jones, S. J. (2006). Prediction of genomic functional elements. *Annu Rev Genomics Hum Genet* 7, 315–338.
5. Demeret, C., Vassetzky, Y., and Mechali, M. (2001). Chromatin remodelling and DNA replication: from nucleosomes to loop domains. *Oncogene* 20, 3086–3093.
6. Valouev, A., Johnson, D. S., Sundquist, A., Medina, C., Anton, E., Batzoglou, S., Myers, R. M., and Sidow, A. (2008). Genome-wide analysis of transcription factor binding sites based on ChIP-Seq data. *Nat Methods* 5, 829–834.
7. Chawla, A., Repa, J. J., Evans, R. M., and Mangelsdorf, D. J. (2001). Nuclear receptors and lipid physiology: opening the X-files. *Science* 294, 1866–1870.
8. Nuclear-Receptor-Committee (1999). A unified nomenclature system for the nuclear receptor superfamily. *Cell* 97, 161–163.
9. Maglich, J. M., Sluder, A., Guan, X., Shi, Y., McKee, D. D., Carrick, K., Kamdar, K., Willson, T. M., and Moore, J. T. (2001). Comparison of complete nuclear receptor sets from the human, Caenorhabditis elegans and Drosophila genomes. *Genome Biol* 2, R29.

10. Bertrand, S., Brunet, F. G., Escriva, H., Parmentier, G., Laudet, V., and Robinson-Rechavi, M. (2004). Evolutionary genomics of nuclear receptors: from twenty-five ancestral genes to derived endocrine systems. *Mol Biol Evol* 21, 1923–1937.
11. Bookout, A. L., Jeong, Y., Downes, M., Yu, R. T., Evans, R. M., and Mangelsdorf, D. J. (2006). Anatomical profiling of nuclear receptor expression reveals a hierarchical transcriptional network. *Cell* 126, 789–799.
12. Mohan, R. and Heyman, R. A. (2003). Orphan nuclear receptor modulators. *Curr Top Med Chem* 3, 1637–1647.
13. Nettles, K. W. and Greene, G. L. (2005). Ligand control of coregulator recruitment to nuclear receptors. *Annu Rev Physiol* 67, 309–333.
14. Giguere, V. (1999). Orphan nuclear receptors: from gene to function. *Endocr Rev* 20, 689–725.
15. Rochel, N., Wurtz, J. M., Mitschler, A., Klaholz, B., and Moras, D. (2000). Crystal structure of the nuclear receptor for vitamin D bound to its natural ligand. *Mol Cell* 5, 173–179.
16. Moras, D. and Gronemeyer, H. (1998). The nuclear receptor ligand-binding domain: structure and function. *Curr Opin Cell Biol* 10, 384–391.
17. Kersten, S., Desvergne, B., and Wahli, W. (2000). Roles of PPARs in health and disease. *Nature* 405, 421–424.
18. Aranda, A. and Pascual, A. (2001). Nuclear hormone receptors and gene expression. *Physiol Rev* 81, 1269–1304.
19. Burke, L. J. and Baniahmad, A. (2000). Co-repressors 2000. *FASEB J* 14, 1876–1888.
20. Polly, P., Herdick, M., Moehren, U., Baniahmad, A., Heinzel, T., and Carlberg, C. (2000). VDR-Alien: a novel, DNA-selective vitamin D_3 receptor-corepressor partnership. *FASEB J* 14, 1455–1463.
21. Privalsky, M. L. (2004). The role of corepressors in transcriptional regulation by nuclear hormone receptors. *Annu Rev Physiol* 66, 315–360.
22. Leo, C. and Chen, J. D. (2000). The SRC family of nuclear receptor coactivators. *Gene* 245, 1–11.
23. Castillo, A. I., Jimenez-Lara, A. M., Tolon, R. M., and Aranda, A. (1999). Synergistic activation of the prolactin promoter by vitamin D receptor and GHF-1: role of coactivators, CREB-binding protein and steroid hormone receptor coactivator-1 (SRC-1). *Mol Endocrinol* 13, 1141–1154.
24. Rachez, C., Suldan, Z., Ward, J., Chang, C. -P., Burakov, D., Erdjument-Bromage, H., Tempst, P., and Freedman, L. P. (1998). A novel protein complex that interacts with the vitamin D_3 receptor in a ligand-dependent manner and enhances transactivation in a cell-free system. *Genes Dev* 12, 1787–1800.
25. Rachez, C., Lemon, B. D., Suldan, Z., Bromleigh, V., Gamble, M., Näär, A. M., Erdjument-Bromage, H., Tempst, P., and Freedman, L. P. (1999). Ligand-dependent transcription activation by nuclear receptors requires the DRIP complex. *Nature* 398, 824–828.
26. Khanim, F. L., Gommersall, L. M., Wood, V. H., Smith, K. L., Montalvo, L., O'Neill, L. P., Xu, Y., Peehl, D. M., Stewart, P. M., Turner, B. M. et al. (2004). Altered SMRT levels disrupt vitamin D_3 receptor signalling in prostate cancer cells. *Oncogene* 23, 6712–6725.
27. Malinen, M., Saramäki, A., Ropponen, A., Degenhardt, T., Väisänen, S., and Carlberg, C. (2008). Distinct HDACs regulate the transcriptional response of human cyclin-dependent kinase inhibitor genes to Trichostatin A and 1a,25-dihydroxyvitamin D_3. *Nucleic Acids Res* 36, 121–132.
28. Metivier, R., Penot, G., Hubner, M. R., Reid, G., Brand, H., Kos, M., and Gannon, F. (2003). Estrogen receptor a directs ordered, cyclical, and combinatorial recruitment of cofactors on a natural target promoter. *Cell* 115, 751–763.
29. Kim, S., Shevde, N. K., and Pike, J. W. (2005). 1,25-Dihydroxyvitamin D_3 stimulates cyclic vitamin D receptor/retinoid X receptor DNA-binding, co-activator recruitment, and histone acetylation in intact osteoblasts. *J Bone Miner Res* 20, 305–317.
30. Saramäki, A., Diermeier, S., Kellner, R., Laitinen, H., Väisänen, S., and Carlberg, C. (2009). Cyclical chromatin looping and transcription factor association on the regulatory regions of the p21 (CDKN1A) gene in response to 1a,25-dihydroxyvitamin D_3. *J Biol Chem* 284, 8073–8082.
31. Degenhardt, T., Rybakova, K. N., Tomaszweska, A., Mone, M. J., Westerhoff, H. V., Bruggeman, F. J., and Carlberg, C. (2009). Population-level transcription cycles derive from stochastic timing of single-cell transcription. *Cell* 138(3), 489–501.

32. Kang, Z., Pirskanen, A., Jänne, O. A., and Palvimo, J. J. (2002). Involvement of proteasome in the dynamic assembly of the androgen receptor transcription complex. *J Biol Chem* 277, 48366–48371.
33. Karpova, T. S., Kim, M. J., Spriet, C., Nalley, K., Stasevich, T. J., Kherrouche, Z., Heliot, L., and McNally, J. G. (2008). Concurrent fast and slow cycling of a transcriptional activator at an endogenous promoter. *Science* 319, 466–469.
34. Väisänen, S., Dunlop, T. W., Sinkkonen, L., Frank, C., and Carlberg, C. (2005). Spatio-temporal activation of chromatin on the human CYP24 gene promoter in the presence of 1a,25-dihydroxyvitamin D_3. *J Mol Biol* 350, 65–77.
35. Sharma, D. and Fondell, J. D. (2002). Ordered recruitment of histone acetyltransferases and the TRAP/Mediator complex to thyroid hormone-responsive promoters in vivo. *Proc Natl Acad Sci U S A* 99, 7934–7939.
36. Jenuwein, T. and Allis, C. D. (2001). Translating the histone code. *Science* 293, 1074–1080.
37. Berger, S. L. (2007). The complex language of chromatin regulation during transcription. *Nature* 447, 407–412.
38. Perissi, V. and Rosenfeld, M. G. (2005). Controlling nuclear receptors: the circular logic of cofactor cycles. *Nat Rev Mol Cell Biol* 6, 542–554.
39. Acevedo, M. L. and Kraus, W. L. (2004). Transcriptional activation by nuclear receptors. *Essays Biochem* 40, 73–88.
40. Luisi, B. F., Xu, W. X., Otwinowski, Z., Freedman, L. P., Yamamoto, K. R., and Sigler, P. B. (1991). Crystallographic analysis of the interaction of the glucocorticoid receptor with DNA. *Nature* 352, 497–505.
41. Khorasanizadeh, S. and Rastinejad, F. (2001). Nuclear-receptor interactions on DNA-response element. *Trends Biochem Sci* 26, 384–390.
42. Carlberg, C. (1999). Lipid soluble vitamins in gene regulation. *Biofactors* 10, 91–97.
43. Carlberg, C., Bendik, I., Wyss, A., Meier, E., Sturzenbecker, L. J., Grippo, J. F., and Hunziker, W. (1993). Two nuclear signalling pathways for vitamin D. *Nature* 361, 657–660.
44. MacDonald, P. N., Dowd, D. R., Nakajima, S., Galligan, M. A., Reeder, M. C., Haussler, C. A., Ozato, K., and Haussler, M. R. (1993). Retinoid X receptors stimulate and 9-cis retinoic acid inhibits 1,25-dihydroxyvitamin D_3-activated expression of the rat osteocalcin gene. *Mol Cell Biol* 13, 5907–5917.
45. Schräder, M., Müller, K. M., Nayeri, S., Kahlen, J. P., and Carlberg, C. (1994). VDR-T_3R receptor heterodimer polarity directs ligand sensitivity of transactivation. *Nature* 370, 382–386.
46. Schräder, M., Nayeri, S., Kahlen, J. P., Müller, K. M., and Carlberg, C. (1995). Natural vitamin D_3 response elements formed by inverted palindromes: polarity-directed ligand sensitivity of vitamin D_3 receptor-retinoid X receptor heterodimer-mediated transactivation. *Mol Cell Biol* 15, 1154–1161.
47. Carlberg, C. (1995). Mechanisms of nuclear signalling by vitamin D_3. Interplay with retinoid and thyroid hormone signalling. *Eur J Biochem* 231, 517–527.
48. Toell, A., Polly, P., and Carlberg, C. (2000). All natural DR3-type vitamin D response elements show a similar functionality in vitro. *Biochem J* 352, 301–309.
49. Rhodes, S. J., Chen, R., DiMattia, G. E., Scully, K. M., Kalla, K. A., Lin, S. -C., Yu, V. C., and Rosenfeld, M. G. (1993). A tissue-specific enhancer confers Pit-1-dependent morphogen inducibility and autoregulation on the pit-1 gene. *Genes Dev* 7, 913–932.
50. Mangelsdorf, D. J. and Evans, R. M. (1995). The RXR heterodimers and orphan receptors. *Cell* 83, 841–850.
51. Quack, M. and Carlberg, C. (2000). Ligand-triggered stabilization of vitamin D receptor/retinoid X receptor heterodimer conformations on DR4-type response elements. *J Mol Biol* 296, 743–756.
52. Turunen, M. M., Dunlop, T. W., Carlberg, C., and Väisänen, S. (2007). Selective use of multiple vitamin D response elements underlies the 1a,25-dihydroxyvitamin D_3-mediated negative regulation of the human CYP27B1 gene. *Nucleic Acids Res* 35, 2734–2747.
53. Sinkkonen, L., Malinen, M., Saavalainen, K., Väisänen, S., and Carlberg, C. (2005). Regulation of the human cyclin C gene via multiple vitamin D_3-responsive regions in its promoter. *Nucleic Acids Res* 33, 2440–2451.

54. Saramäki, A., Banwell, C. M., Campbell, M. J., and Carlberg, C. (2006). Regulation of the human p21$^{(waf1/cip1)}$ gene promoter via multiple binding sites for p53 and the vitamin D_3 receptor. *Nucleic Acids Res* 34, 543–554.
55. Matilainen, M., Malinen, M., Saavalainen, K., and Carlberg, C. (2005). Regulation of multiple insulin-like growth factor binding protein genes by 1α,25-dihydroxyvitamin D_3. *Nucleic Acids Res* 33, 5521–5532.
56. Bulger, M. and Groudine, M. (1999). Looping versus linking: toward a model for long-distance gene activation. *Genes Dev* 13, 2465–2477.
57. Xie, X., Lu, J., Kulbokas, E. J., Golub, T. R., Mootha, V., Lindblad-Toh, K., Lander, E. S., and Kellis, M. (2005). Systematic discovery of regulatory motifs in human promoters and 3' UTRs by comparison of several mammals. *Nature* 434, 338–345.
58. Seuter, S., Väisänen, S., Radmark, O., Carlberg, C., and Steinhilber, D. (2007). Functional characterization of vitamin D responding regions in the human 5-lipoxygenase gene. *Biochim Biophys Acta* 1771, 864–872.
59. Klan, N., Seuter, S., Schnur, N., Jung, M., and Steinhilber, D. (2003). Trichostatin A and structurally related histone deacetylase inhibitors induce 5-lipoxygenase promoter activity. *Biol Chem* 384, 777–785.
60. Sorg, B. L., Klan, N., Seuter, S., Dishart, D., Radmark, O., Habenicht, A., Carlberg, C., Werz, O., and Steinhilber, D. (2006). Analysis of the 5-lipoxygenase promoter and characterization of a vitamin D receptor binding site. *Biochim Biophys Acta* 1761, 686–697.
61. Ogata, K., Sato, K., and Tahirov, T. H. (2003). Eukaryotic transcriptional regulatory complexes: cooperativity from near and afar. *Curr Opin Struct Biol* 13, 40–48.
62. Krishnan, A. V., Shinghal, R., Raghavachari, N., Brooks, J. D., Peehl, D. M., and Feldman, D. (2004). Analysis of vitamin D-regulated gene expression in LNCaP human prostate cancer cells using cDNA microarrays. *Prostate* 59, 243–251.
63. Zella, L. A., Kim, S., Shevde, N. K., and Pike, J. W. (2006). Enhancers located within two introns of the vitamin D receptor gene mediate transcriptional autoregulation by 1,25-dihydroxyvitamin D_3. *Mol Endocrinol* 20, 1231–1247.
64. Meyer, M. B., Watanuki, M., Kim, S., Shevde, N. K., and Pike, J. W. (2006). The human transient receptor potential vanilloid type 6 distal promoter contains multiple vitamin D receptor binding sites that mediate activation by 1,25-dihydroxyvitamin D_3 in intestinal cells. *Mol Endocrinol* 20, 1447–1461.
65. Fretz, J. A., Zella, L. A., Kim, S., Shevde, N. K., and Pike, J. W. (2007). 1,25-dihydroxyvitamin D3 induces expression of the Wnt signaling co-regulator LRP5 via regulatory elements located significantly downstream of the gene's transcriptional start site. *J Steroid Biochem Mol Biol* 103, 440–445.
66. Kim, S., Yamazaki, M., Zella, L. A., Shevde, N. K., and Pike, J. W. (2006). Activation of receptor activator of NF-kappaB ligand gene expression by 1,25-dihydroxyvitamin D_3 is mediated through multiple long-range enhancers. *Mol Cell Biol* 26, 6469–6486.
67. Carroll, J. S., Meyer, C. A., Song, J., Li, W., Geistlinger, T. R., Eeckhoute, J., Brodsky, A. S., Keeton, E. K., Fertuck, K. C., Hall, G. F. et al. (2006). Genome-wide analysis of estrogen receptor binding sites. *Nat Genet* 38, 1289–1297.
68. Cawley, S., Bekiranov, S., Ng, H. H., Kapranov, P., Sekinger, E. A., Kampa, D., Piccolboni, A., Sementchenko, V., Cheng, J., Williams, A. J. et al. (2004). Unbiased mapping of transcription factor binding sites along human chromosomes 21 and 22 points to widespread regulation of noncoding RNAs. *Cell* 116, 499–509.
69. Kwon, Y. S., Garcia-Bassets, I., Hutt, K. R., Cheng, C. S., Jin, M., Liu, D., Benner, C., Wang, D., Ye, Z., Bibikova, M. et al. (2007). Sensitive ChIP-DSL technology reveals an extensive estrogen receptor alpha-binding program on human gene promoters. *Proc Natl Acad Sci U S A* 104, 4852–4857.
70. ENCODE-Consortium (2004). The ENCODE (ENCyclopedia Of DNA Elements) Project. *Science* 306, 636–640.

71. Rada-Iglesias, A., Wallerman, O., Koch, C., Ameur, A., Enroth, S., Clelland, G., Wester, K., Wilcox, S., Dovey, O. M., Ellis, P. D. et al. (2005). Binding sites for metabolic disease related transcription factors inferred at base pair resolution by chromatin immunoprecipitation and genomic microarrays. *Hum Mol Genet* 14, 3435–3447.
72. Nielsen, R., Pedersen, T. A., Hagenbeek, D., Moulos, P., Siersbaek, R., Megens, E., Denissov, S., Borgesen, M., Francoijs, K. J., Mandrup, S. et al. (2008). Genome-wide profiling of PPARg:RXR and RNA polymerase II occupancy reveals temporal activation of distinct metabolic pathways and changes in RXR dimer composition during adipogenesis. *Genes Dev* 22, 2953–2967.
73. Lefterova, M. I., Zhang, Y., Steger, D. J., Schupp, M., Schug, J., Cristancho, A., Feng, D., Zhuo, D., Stoeckert, C. J., Jr., Liu, X. S. et al. (2008). PPARg and C/EBP factors orchestrate adipocyte biology via adjacent binding on a genome-wide scale. *Genes Dev* 22, 2941–2952.
74. Hancock, R. (2000). A new look at the nuclear matrix. *Chromosoma* 109, 219–225.
75. Sierra, J., Villagra, A., Paredes, R., Cruzat, F., Gutierrez, S., Javed, A., Arriagada, G., Olate, J., Imschenetzky, M., Van Wijnen, A. J. et al. (2003). Regulation of the bone-specific osteocalcin gene by p300 requires Runx2/Cbfa1 and the vitamin D_3 receptor but not p300 intrinsic histone acetyltransferase activity. *Mol Cell Biol* 23, 3339–3351.
76. Cook, P. R. (1999). The organization of replication and transcription. *Science* 284, 1790–1795.
77. Olave, I. A., Reck-Peterson, S. L., and Crabtree, G. R. (2002). Nuclear actin and actin-related proteins in chromatin remodeling. *Annu Rev Biochem* 71, 755–781.
78. Stein, G. S., Zaidi, S. K., Braastad, C. D., Montecino, M., van Wijnen, A. J., Choi, J. Y., Stein, J. L., Lian, J. B., and Javed, A. (2003). Functional architecture of the nucleus: organizing the regulatory machinery for gene expression, replication and repair. *Trends Cell Biol* 13, 584–592.
79. Narlikar, G. J., Fan, H. Y., and Kingston, R. E. (2002). Cooperation between complexes that regulate chromatin structure and transcription. *Cell* 108, 475–487.
80. White, J. H. (2004). Profiling 1,25-dihydroxyvitamin D_3-regulated gene expression by microarray analysis. *J Steroid Biochem Mol Biol* 89–90, 239–244.
81. Pascual, G. and Glass, C. K. (2006). Nuclear receptors versus inflammation: mechanisms of transrepression. *Trends Endocrinol Metab* 17, 321–327.
82. Murayama, A., Kim, M. S., Yanagisawa, J., Takeyama, K., and Kato, S. (2004). Transrepression by a liganded nuclear receptor via a bHLH activator through co-regulator switching. *EMBO J* 23, 1598–1608.
83. Towers, T. L. and Freedman, L. P. (1998). Granulocyte-macrophage colony-stimulating factor gene transcription is directly repressed by the vitamin D_3 receptor: implications for allosteric influences on nuclear receptor structure and function by a DNA element. *J Biol Chem* 273, 10338–10348.
84. Towers, T. L., Staeva, T. P., and Freedman, L. P. (1999). A two-hit mechanism for vitamin D_3-dediates transcriptional repression of the granulocyte-macrophage colony-stimulating factor gene: vitamin D receptor completes for DNA binding with NFAT1 and stabilizes c-Jun. *Mol Cell Biol* 19, 4191–4199.
85. Polly, P., Carlberg, C., Eisman, J. A., and Morrison, N. A. (1997). 1a,25-dihydroxyvitamin D_3 receptor as a mediator of transrepression of retinoid signaling. *J Cell Biochem* 67, 287–296.
86. Heinäniemi, M., Uski, J. O., Degenhardt, T., and Carlberg, C. (2007). Meta-analysis of primary target genes of peroxisome proliferator-activated receptors. *Genome Biol* 8, R147.
87. Wang, T. T., Tavera-Mendoza, L. E., Laperriere, D., Libby, E., MacLeod, N. B., Nagai, Y., Bourdeau, V., Konstorum, A., Lallemant, B., Zhang, R. et al. (2005). Large-scale in silico and microarray-based identification of direct 1,25-dihydroxyvitamin D_3 target genes. *Mol Endocrinol* 19, 2685–2695.
88. Mader, S., Leroy, P., Chen, J. -Y., and Chambon, P. (1993). Multiple parameters control the selectivity of nuclear receptors for their response elements. *J Biol Chem* 268, 591–600.
89. Schräder, M., Müller, K. M., Becker-André, M., and Carlberg, C. (1994). Response element selectivity for heterodimerization of vitamin D receptors with retinoic acid and retinoid X receptors. *J Mol Endocrinol* 12, 327–339.
90. Schräder, M., Müller, K. M., and Carlberg, C. (1994). Specificity and flexibility of vitamin D signaling. Modulation of the activation of natural vitamin D response elements by thyroid hormone. *J Biol Chem* 269, 5501–5504.

91. Schräder, M., Becker-Andre, M., and Carlberg, C. (1994). Thyroid hormone receptor functions as monomeric ligand-induced transcription factor on octameric half-sites. Consequences also for dimerization. *J Biol Chem* 269, 6444–6449.
92. Carlberg, C. and Polly, P. (1998). Gene regulation by vitamin D_3. *Crit Rev Eukaryot Gene Expr* 8, 19–42.
93. Barski, A., Cuddapah, S., Cui, K., Roh, T. Y., Schones, D. E., Wang, Z., Wei, G., Chepelev, I., and Zhao, K. (2007). High-resolution profiling of histone methylations in the human genome. *Cell* 129, 823–837.
94. Matys, V., Fricke, E., Geffers, R., Gossling, E., Haubrock, M., Hehl, R., Hornischer, K., Karas, D., Kel, A. E., Kel-Margoulis, O. V. et al. (2003). TRANSFAC: transcriptional regulation, from patterns to profiles. *Nucleic Acids Res* 31, 374–378.
95. Lemay, D. G. and Hwang, D. H. (2006). Genome-wide identification of peroxisome proliferator response elements using integrated computational genomics. *J Lipid Res* 47, 1583–1587.
96. Wasserman, W. W. and Sandelin, A. (2004). Applied bioinformatics for the identification of regulatory elements. *Nat Rev Genet* 5, 276–287.
97. Odom, D. T., Dowell, R. D., Jacobsen, E. S., Gordon, W., Danford, T. W., MacIsaac, K. D., Rolfe, P. A., Conboy, C. M., Gifford, D. K., and Fraenkel, E. (2007). Tissue-specific transcriptional regulation has diverged significantly between human and mouse. *Nat Genet* 39, 730–732.
98. Burns, J. L., Jackson, D. A., and Hassan, A. B. (2001). A view through the clouds of imprinting. *FASEB J* 15, 1694–1703.
99. Degenhardt, T., Matilainen, M., Herzig, K. H., Dunlop, T. W., and Carlberg, C. (2006). The insulin-like growth factor-binding protein 1 gene is a primary target of peroxisome proliferator-activated receptors. *J Biol Chem* 281, 39607–39619.

CHAPTER 18

SYSTEMS BIOLOGY: TOWARDS REALISTIC AND USEFUL MODELS OF MOLECULAR NETWORKS

F.J. BRUGGEMAN[1,2,3], A. KOLODKIN[3], K. RYBAKOVA[3], M. MONÉ[3], AND H.V. WESTERHOFF[3,4]

[1] *Regulatory Networks Group, Netherlands Institute for Systems Biology, University of Amsterdam, Amsterdam, The Netherlands*

[2] *Life Sciences, Institute for Mathematics and Computer Science (CWI), Amsterdam, 1098 SJ, The Netherlands*

[3] *Molecular Cell Physiology, Vrije Universiteit, Amsterdam, 1081 HV, The Netherlands*

[4] *Manchester Centre for Integrative Systems Biology, Manchester Interdisciplinary Biocentre, University of Manchester, Manchester, UK*

Abstract: Molecular biology is shifting focus from single molecules to networks of molecules. This development has changed our way of doing research and is challenging our thinking about cells. Cells turn out be complicated molecular systems displaying multivariate dynamics that can rarely be understood in terms of single molecules. One way to appreciate this complexity is to make mathematical models of signaling, gene, and metabolic network to assess the systemic consequences of specific molecular perturbations. This chapter gives a brief overview of some of the approach in mathematical modeling of molecular networks. We choose to keep the mathematical detail minimal and highlight a number of concepts and approaches that are emerging in the analysis of molecular networks.

18.1. THE SHIFT FROM MOLECULAR TO SYSTEMS BIOLOGY

In the last decade, biology has broadened its molecular focus more and more with a network perspective [1–3]. At hindsight, this was a logical progression. It has become increasingly apparent that cellular properties derive from the joint activity of macromolecules as part of dynamic networks; rarely, single molecules are responsible for cellular phenomena or diseases [4–6]. In the last decade, sophisticated screening methods have lead to the identification of many macromolecules, their functions and interaction partners. The emerging complexity of the resulting networks has inspired many scientists to re-evaluate their views on the nature of living cells and methods to study them. Reductionistic methods, aimed at studying single macromolecules at high detail, were quickly augmented with genome-wide exploratory studies relying on new technologies; such as yeast two hybrid screens

C.M. Bunce, M.J. Campbell (eds.), *Nuclear Receptors*, Proteins and Cell Regulation 8, 439–453. DOI 10.1007/978-90-481-3303-1_18,

to hunt for new protein interactions, mRNA microarrays, metabolomics, proteomics, and ChIP sequencing. Those studies typically leave the biologist with lots of data to mine for correlations and new hypotheses. Rarely such studies culminate in insight into how all those correlations emerge out of the dynamic molecular interactions nor about any rational strategies to specifically alter cellular behaviour by targeting particular molecular processes. Studies of the organizational principles of molecular networks, in terms of graph theory and statistical physics, indicated that biological networks share global organizational properties with networks as diverse as the internet and social networks [7, 5, 8].

All these studies leave us with one major challenge: how does all this organized molecular complexity underlie coherent and robust cellular and, ultimately, multicellular behaviour? To answer those questions, a new interdisciplinary approach, systems biology, developed. In its current form, systems biology is a field with experimental biologists, biochemists, engineers, mathematicians and physicists faced with the challenge to understand life in terms of the dynamic of molecular networks [4, 1, 9].

18.2. HIERARCHICAL NETWORKS

With current technologies, the dynamics of living cells can be measured simultaneously at the level of metabolism, signalling, and gene expression using a number of different methodologies. The dynamics of each of these levels arises from intra- and interlevel processes acting in concert. Functional concepts such robustness, redundancy, cross-talk, synergy and modularity immediately spring to mind and have been proved to be fruitful in understanding network organization [10, 11, 6, 12]. Classical approaches were mostly targeted to improving our understanding of intralevel phenomena. Nowadays we are faced often with problems related to interlevel phenomena and data integration. From a mechanistic perspective, a multitude of mechanisms have been identified by which levels communicate. The most prominent mechanisms are changes in protein level, covalent modification state, conformational state and cellular location [13]. For instance, transcription factors may become phosphorylated in the cytoplasm, move into the nucleus to regulate gene expression. Resultant changes in the protein levels, say in metabolism, could then alter the effector concentration of the kinase of the aforementioned transcription factor to further modulate signalling and gene expression. Many of such reverberating regulatory loops run through the entire molecular network. Networks composed out of interacting levels have been termed hierarchical networks. Much control and response theory has been developed to understand the design and functioning of hierarchical networks and to aid experimental data interpretation [14, 13, 15–19]. An example of a hierarchical network is shown in Figure 18.1. An example of such a system is the regulation of mammalian pyruvate dehydrogenase by a metabolite sensitive kinase. This kinase is encoded by PDK4, which is controlled by the nuclear receptor PPAR. The activity of the nuclear receptor depends on metabolite levels.

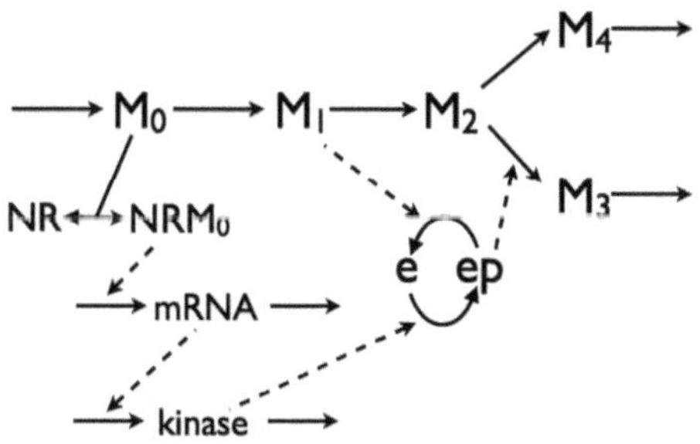

Figure 18.1. Example of a hierarchical network composed out of metabolism, gene expression, and covalent modification of a metabolic enzyme. In metabolism, metabolites (denoted by M_i) are converted by enzymes. The activity of the enzyme catalyzing the reaction of M_2 into M_3 is regulated by covalent modification. It has an altered activity upon phosphorylation. The level of the kinase is influenced by the level of metabolites through a nuclear receptor (NR). The activity of the phosphatase of the enzyme in addition depends on the level of a metabolite. For simplicity, the gene regulation of all other enzymes is not indicated

Hierarchical networks often display a natural modularity; as intra- and interlevel interactions differ in nature [20, 16, 17]. Intralevel molecular interactions can be of two types, either regulatory or mass flow interactions, whereas interlevel interactions are typically regulatory. The difference between the two is in their kinetic description. An effector involved in a mass flow interaction is converted in the processes whereas if merely act as a modifier, rather than a reactant, in a regulatory interaction. This has important consequences for the control properties of hierarchical networks (recently reviewed in Bruggeman et al. [20]).

18.3. FROM MOLECULAR INTERACTIONS AND REACTIONS TO NETWORKS

Different types of molecular networks are operative inside living cells. They can be distinguished on the basis of their biological function, physicochemical level of description and experimental methods for study. The function of metabolic networks is typically related to conversion fluxes. With current modeling approaches whole metabolic networks can be assessed for their metabolic capacities [21]. Signal transduction is better studied in terms of changes in the level of the (covalent modification) state and cellular location of proteins. Many systems biological studies address signal transduction using both experiment and modelling [22–29]. Metabolic networks are described in terms of reaction stoichiometry and enzyme kinetics whereas transcriptional regulation and signal transduction relies on the protein complex formation and modification. The latter systems are typically described in terms of mass action kinetics. Whereas stochastic phenomena do not play important roles in metabolism, they may prove very important in the regulation of gene expression [30]. Thus, understanding the functioning of hierarchical networks implies the usage

Table 18.1. Comparison of eukaryotic properties of metabolic, signalling and genetic networks

Aspect\network	Metabolic	Signalling	Genetic
Function	Metabolic flux	Information flow	Enzyme synthesis
Functional property	Metabolic flux	Protein states	Macromolecule levels
Stochastic phenomena	No	Yes/No	Yes
Diffusion dependent	No	Yes	Yes
Protein–protein interactions and complex formation	No	Yes	Yes
Enzyme (E) or mass action (MA) kinetics	E	MA	MA
Single molecule phenomenon	No	No	Yes[a]
Feedback regulation	Yes	Yes	Yes
Complex dynamics[b]	Yes[c]	Yes	Yes
Characteristic time scale	Seconds	Minutes	Hours
Characteristic intermediate concentration (molecules per cell)	10^3–10^7	50×10^4	1–5×10^3

[a]Response elements and transcription start site occur as one molecule per cell
[b]Steady states are frequently the functional states
[c]Examples: bistability and oscillations

of different techniques, concept and modelling approaches leaving us with the challenge to integrate it all into one consistent picture (see Table 18.1). This is a continuing challenge for systems biology, which probably has no single rigorous solution.

18.4. A KINETIC MODEL FOR TRANSCRIPTION FACTOR BINDING TO DNA

To illustrate the usage of the kinetic models to understand particular aspects of molecular networks, we will consider first transcription factor binding to the DNA. In the paragraphs that follow some other kinetic model aspects will be discussed. Good reviews exist about these approaches [31–34].

Rates of molecular processes cause the concentration of molecules to change over time. The description of the resultant network dynamics gets quickly complicated as the process rates depend in turn on those concentrations. We shall denote the rate of change in the concentration of a molecular intermediate, say molecule X, by dX/dt. The concentration increases – the rate of change will be positive – if the synthesis rate occurs at a higher activity than the consumption rates. This is reflected in the rate of change equation, the so-called mass balance for X, which gives the relationship between d*X*/d*t* and the rate of its synthesizing and degrading processes. For the reversible complex formation between *X* and *Y*, $X+Y \leftrightarrow XY$, we obtain,

$$\frac{dXY}{dt} = \underbrace{k^+ \cdot X \cdot Y}_{\text{association rate}} - \underbrace{k^- \cdot XY}_{\text{disociation rate}}$$

The two k's are rate constants; k^+ has as unit nM^{-1} min^{-1} and k^- min^{-1} if the concentration and time unit are nM and min, respectively. For protein complex formation, an estimate for k^+ is the diffusion-limited rate constant (see Table 18.2) [35]. It assumes that the protein complex formation time is predominantly determined by the time it takes for *X* and *Y* to encounter each other by diffusion. The time for the complex to be formed upon collision is assumed negligible. The diffusion-limited rate constant depends on the diffusion coefficients and the radii of the reactants, in the simplest case as,

$$k^+ = 4\pi(D_X + D_Y)(r_X + r_Y)$$

In this equation, *D* denotes a diffusion coefficient and *r* a molecular radius. The diffusion coefficients depend on the size of the protein and the viscosity of medium; in the simplest case for a spherical protein, $D = kT/(6\pi\eta r)$ (with *k* as the Boltzmann constant and η as the viscosity). The diffusion coefficient is an indication of the distance *l* a protein travels on average; $l = \sqrt{6Dt}$. This means that it takes a protein 0.13 s to travel the radius of an eukaryotic cell in distance. The time for two single proteins in a eukaryotic cell to find each other equals 26 min![1] For N of number molecules of each this time becomes $26/N^2$ min. Increasing the concentration of both molecules by a factor of 10 leads to 15 s for their encounter time. This means that a transcription factor, at a (free) concentration of 100 molecules per cell, finds a

Table 18.2. Useful physicochemical constants for signalling and gene expression models

Parameter	Value
Cell volume[a]	1.55 (cytoplasm) and 0.45 (nucleus) pl
Protein diffusion coefficient[b]	10 μm^2 s^{-1}
Diffusion limited rate constant	1 nM^{-1} s^{-1}
Average protein radius	5 nm
Cell radius	100 μm

[a]In this volume, the concentration of one molecule is 0.8 pM
[b]Eukaryotic cytoplasm (prokaryotic case, 1 μm^2 s^{-1})

[1] from $1/(k^+ \cdot X \cdot Y)$ with k^+ as 10^{-3} pM^{-1} min^{-1} and X and Y equal to 0.8 pM

specific target gene (occurring at 2 copies per cell) in 7.5 s. If the resulting protein-DNA complex would live for 10 s (a realistic value given FRAP experiments), the dissociation constant k^- equals 0.1 s^{-1}. A comparison with the lifetime of an mRNA, which ranges between tens of minutes to hours in eukaryotes, indicates that protein complexes on DNA are in a continuous dynamic equilibrium on the time scale of transcription.

Solving the mass balance given above yields,

$$\frac{XY(t)}{XY_{EQ}} = 1 - e^{-(k^- + k^+)\cdot Y \cdot t}$$

Here we kept Y, e.g. the transcription factor concentration, fixed, which amounts to the justified assumption that the transcription factor concentration is higher than the concentration of its DNA target. This equation shows that the complex concentration equals half the equilibrium concentration, XY_{EQ}, within $-\ln(0.5)Y^{-1}(k^+ + k^-)^{-1}$seconds, which is 0.09 s for the saturation of a gene by 50% by its transcription factor occurring at 100 molecules per cell. The equilibrium concentration of the complex equals,

$$\frac{XY_{EQ}}{X_T} = \frac{Y}{K_D + Y}$$

The dissociation constant K_D is defined as k^-/k^+ (units, pM). If Y denotes the free transcription factor and X the promotor concentration, we find as an estimate for the K_D ~6 nM, which is close to experimental observations. In the case that more than one molecule of transcription factor would bind to the DNA site, with either positive or negative cooperativity, we obtain for three molecules,

$$\frac{XY_{3,EQ}}{X_T} = \frac{\dfrac{Y^3}{K_{D,1}K_{D,2}K_{D,3}}}{1 + 3\dfrac{Y}{K_{D,1}} + 3\dfrac{Y^2}{K_{D,1}K_{D,2}} + \dfrac{Y^3}{K_{D,1}K_{D,2}K_{D,3}}}$$

In this case, the dependency of the complex depends on Y as a sigmoidal relationship rather than hyperbolic. In the absence of cooperativity, i.e. the binding affinity of Y for X does not depend on whether the binding state of X, the previous equation reduces to,

$$\frac{XY_{3,EQ}}{X_T} = \left(\frac{Y}{K_D{}^{+Y}}\right)^3$$

Irregardless of the complexity of the protein complex mechanism and the number of the protein constituents such equations can always be derived for thermodynamic equilibrium conditions. Below, we will show some of their consequences for the dynamics of genetic circuits. The incorporation of additional co-factors that determine transcription initiation frequencies is in principle possible but would quickly lead to complex mechanistic schemes. The dynamics of such system can be readily modelled but not much analytical insight can be obtained. This is recurrent problem in the modelling of the complex molecular network dynamics. Approximative theories, such as metabolic control analysis [36, 37], linear noise approximation [38, and biochemical systems theory [39, will be required to provide fundamental insight into the design and functioning of molecular networks. We anticipate that many of such theories will be developed in the next decades.

Typically, a transcription factor has many different genes as targets and the total concentration of transcription factor and target sites is comparable, i.e. of the order of a few thousands per cell. Whether all of these sites will be occupied will depend on the average dissociation (affinity) constants for transcription factors and their DNA sites. If we assume those target sites to all have a nearly identical affinity and that half of the sites are occupied, we can determine the waiting time for each gene to become bound by a transcription factor and the time that the transcription factor remains to the DNA. Every $\tau = (k^+ \cdot TF \cdot D + k^- \cdot TFD)^{-1}$ seconds a reaction occurs. The probability for an association reaction equals $p_A = k^+ \cdot TF \cdot D \cdot (k^+ \cdot TF \cdot D + k^- \cdot TFD)^{-1}$. The probability that on a specific DNA site an association reaction occurs equals $p_s = \mathrm{D}^{-1}$. Therefore, the waiting time for a particular unbound target site to become bound equals $\tau\,(p_S p_A)^{-1} = D \cdot \left(1 + K_D \frac{TFD}{TF \cdot D}\right) \cdot \tau$. Assuming, a thousand molecules of TF and DNA site each, this time equals $\sim 16\ \tau$, which amounts to 2.4 s. Such a transcription factor will stay on this site for $1/k^-$ second, i.e. 10 s. This analysis is an example of how simple kinetic models give insight into the dynamic nature of transcription initiation on the time scale of mRNA production, which takes about 10 min per mRNA.

So far, we have only considered kinetic models for the formation of molecular complexes. Inside cells, enzymes catalyze most of the molecular reactions. For instance, enzymes catalyze protein phosphorylation, histone modification, and metabolic conversions. Enzyme kinetics specifies how the rate of an enzyme depends on the concentrations of its reactants, effectors, and kinetic properties [40, 41]. The rate of an enzyme generally depends linearly on its concentration. For instance, the phosphorylation of a protein by a kinase (or in a similar fashion, histone acetylation by a acetylase),

$$v = \frac{V_{\mathrm{MAX}} \cdot \dfrac{\mathrm{ATP} \cdot E}{K_{\mathrm{ATP}} K_E} \cdot \left(1 - \dfrac{\mathrm{ATP} \cdot E}{\mathrm{ADP} \cdot E \cdot K_{\mathrm{EQ}}}\right)}{\left(1 + \dfrac{\mathrm{ATP}}{K_{\mathrm{ATP}}} + \dfrac{\mathrm{ADP}}{K_{\mathrm{ADP}}}\right)\left(1 + \dfrac{E}{K_E} + \dfrac{\mathrm{EP}}{K_{\mathrm{EP}}}\right)}$$

Here we assumed an enzyme with two binding sites in its catalytic site, one where ATP and ADP bind competitively and another such site for E and EP. The reaction is reversible in principle but typically the equilibrium constant K_{EQ} is so high that it is effectively irreversible. The other K constants are all affinity constants of the enzyme for its substrates. Effectors could in addition influence the affinity constants and the maximal rate of the enzyme, V_{MAX}. Typically, Km values lie in the mM range in central carbon and energy metabolism, such as glycolysis. This means that enzymes can be saturated with their substrate(–s) if their concentrations exceed ~1–10 mM.

Metabolic control analysis (MCA) [36, 37, 42] has been very insightful for understanding the control structures in metabolism (some of this is reviewed in, Bruggeman et al. [20]). An insightful book is 'Understanding the control of metabolism' by D. Fell [43]. MCA shows that flux controlling enzymes in metabolic pathways are typically insensitive to their reactants. This indicates that the fractional change in the flux J upon a 1% change in the activity v of an enzyme, i.e. dln$J$/dln$v$, is highest for those enzymes that are insensitive to their reactants. The latter insensitivity is quantified in terms of a low value of the elasticity coefficient ε_X^v in MCA, which gives the fractional change in the rate v of an enzyme upon a 1% change in the concentration of one of its reactants, e.g. X, while keeping all others fixed. Examples of controlling enzymes are saturated enzymes and feedback controlled enzymes. Whether an enzyme is flux controlling depend on the physiological state, at some conditions it may be at others not. For a simple linear metabolic pathway at steady state, i.e. S→X→P with S and P fixed, the flux control coefficient of the first enzyme is given by,

$$C_1^J = \frac{1}{1 - \frac{\varepsilon_X^1}{\varepsilon_X^2}}$$

As the effect of X on the rate of enzyme will be inhibitory (product inhibition), $\varepsilon_X^{v_1} = \partial \ln v_1 / \partial \ln X$, will be negative. MCA indicated that $C_1^J + C_2^J = 1$ (for the proof see for instance, Bruggeman et al. [20]). This equation indicates that the least sensitive enzyme will have the highest flux control. Another pronounced effect is that feedback in metabolic pathway causes the control on flux to shift to enzyme after the feedback metabolite [44].

18.5. GENE ACTIVITY

The activity of mammalian genes is under the control of large number of factors [45, 46]. The understanding of the dynamics of genes activity is a great challenge as they typically occur only at two copies per cell. This has as a consequence that gene activity becomes a process with a strongly stochastic component [47–52]. All these studies indicate that genes display stochastic dynamics between states that differ in activity, eventually leading to strong variability of mRNA and protein levels across isogenic cell populations. Also the waiting times for mRNA production events become strongly stochastic [53, 50, 54]. Depending on the dispersion across

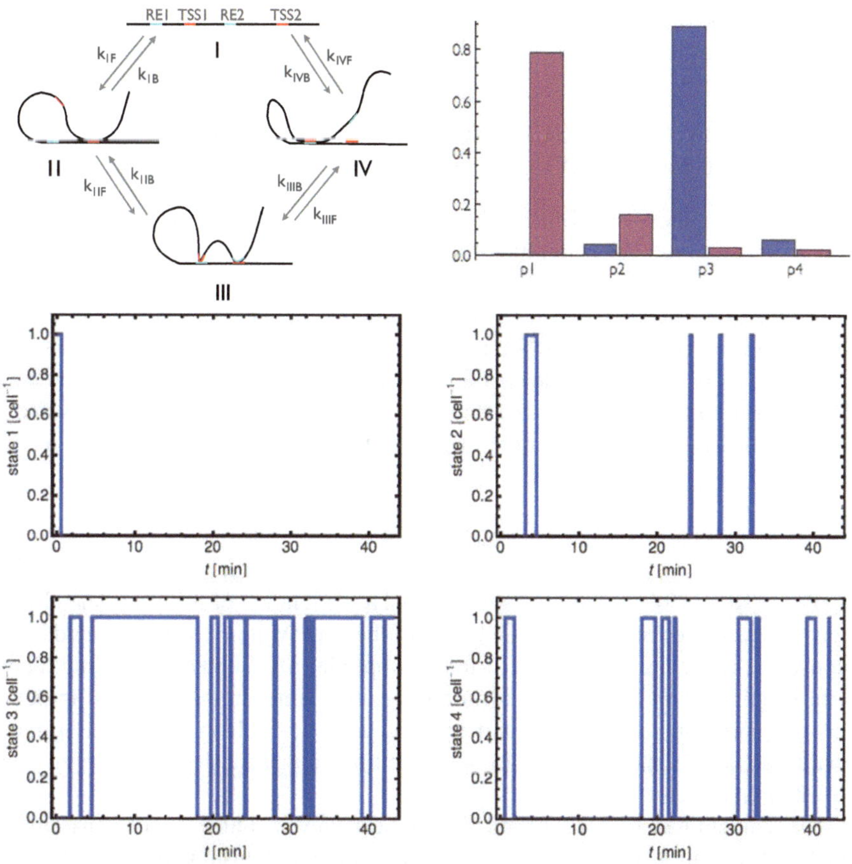

Figure 18.2. Different states of a gene promotor harboring two transcription start sites (TSSs) and two response elements (REs). The interaction between TSS_1-RE_2 and TSS_2-RE_1 has been assumed negligible (very short-lived). Rate constants quantify the transitions between the various states. Time scales are somewhat arbitrary in this case as presently kinetic parameters for such processes are poorly known. In the bar graph, the *red bars* denote gene state probabilities when the gene is not being activated by a transcription factor. In the presence of a high concentration of active transcription factor, the gene switches to another probability distribution over its states. The lower figures indicate the stochastic switching between states when the gene is turned on by a transcription factor (*blue bars*). The stochasticity of the process becomes immediately apparent. Cells in a population will display the same stationary probabilities but will shift between states at different moments in time. The life time of each state is given by the inverse of the sum of the rate constants for the reactions leaving the state

a population, studies of the population average loose important information [55, 30]. This phenomenon is illustrated for a gene with a complex promotor in Figure 18.2. The system stochastically switches between states. Cells in an isogenic population will shift at different times. The life time of each of the promotor states is given by the inverse of the sum of the escape rates from these states, e.g. for state 1, $\tau = (k_1^+ + k_4^-)^{-1}$.

18.6. NETWORK REGULATION OF GENE ACTIVITY AND mRNA LEVEL

At steady state, the level of mRNA is the result of the balance between the synthesis and degradation rate. The lifetime of an mRNA is given by the inverse of its degradation rate constant, k_{deg} (assuming the simplest degradation mechanism). The rate of degradation k_{deg} mRNA will equal the synthesis rate at steady state. The waiting time between consecutive mRNA productions at steady state, i.e. $(k_{\mathrm{deg}}\ \mathrm{mRNA})^{-1}$ (with mRNA in units molecules per cell), will equal the initiation time of the gene. The elongation time only becomes apparent when the gene is suddenly activated. If there is not any mRNA at the time of gene activation, the time for the first mRNA equals,

$$\tau = \tau_{\mathrm{initiation}} + \tau_{\mathrm{elongation}} + \tau_{\mathrm{processing}} + \tau_{\mathrm{export}}$$

(Processing involves capping, splicing, poly-adenylation of mRNA). The elongation time can be estimated from the length of the gene N_{GENE} (in kbp) divided by the elongation rate constant, $k_{\mathrm{elongation}}$ (in e.g. kbp/min, ~1.2 kpb/min). Taking a gene with length 12 kbp amounts to 10 min elongation time. In the case that initiation involves chromatin changes, RE and TSS preparation, and loop formation, initiation could also take 10 min or more [45, 46]. The time to make a single mRNA amounts to ~20 min (at least).

The regulation of the level of mRNA can be achieved at the level of transcription, mRNA degradation, or both. A recently discovered mechanism with a high potential for mRNA regulation are microRNA's [56–58]. In Figure 18.3 we show an example of how kinetic model can help to compare the influences of regulatory networks designs on dynamics. The simulation results indicate that microRNAs can speeds up both the onset and the decay of the dynamics.

Another aspect of gene regulation is the regulation of the localization of transcription factors [59–62]. Depending on their activity state, transcription factors may accumulate preferentially either in the nucleus or the cytoplasm. This shift in cellular compartment is related to their regulated affinity for the nucleocytoplasmic transport machinery [63]. In Figure 18.4, we illustrate an interesting consequence of this mechanism for nucleocytoplasmic transport. This figure shows that the parameters have an altered importance for gene regulation depending on the ligand level. In addition, some parameters are not important at all; gene expression is robust to change in their magnitude. The transport parameters prove to be among the least important parameters for the steady state level of transcription. They are among the most important ones for the regulation of the onset dynamics of transcription (not shown). Parameter sensitivity analysis is a useful method to identify important regulatory targets in kinetic models; for instance, for the identification of drug targets [24].

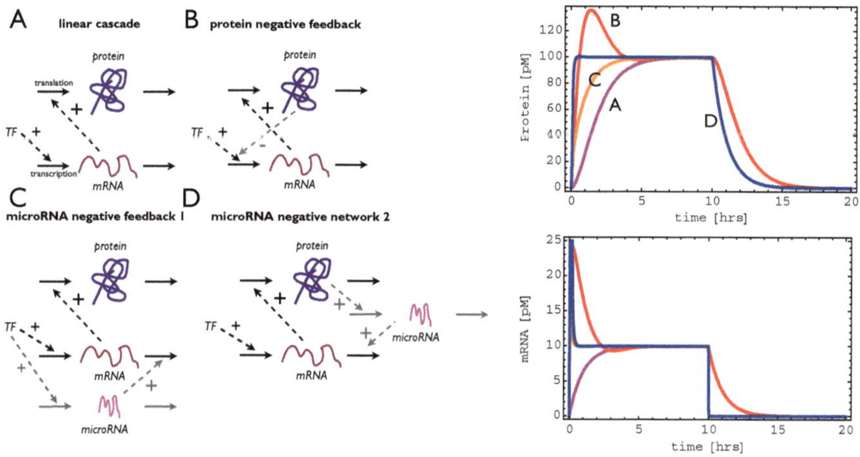

Figure 18.3. Four alternative gene regulatory designs. The first design (A) depends only on the level of a transcription factor. Network B has in addition a negative feedback effect of the gene product. This negative autoregulation motif speeds up the response of the network as shown the simulation curves (see also [64]). The addition of a miRNA in network C speeds up the degradation of the mRNA. Network A & B and C & D have the same degradation dynamics. Network D has the quickest onset and decay of the dynamics. This shows the additional regulatory capacity of mRNA regulation by miRNAs. To allow for model comparison all the models have the same steady state and basal kinetic parameters

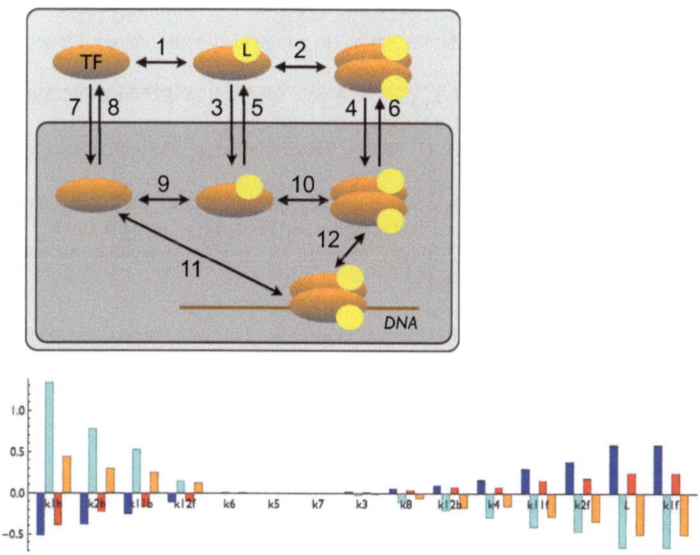

Figure 18.4. Parameter sensitivity analysis for a nucleocytoplasmic shuttling model of a transcription factor. The parameter sensitivity of the DNA-bound ligand-bound transcription factor at steady state is shown in the bar graph at two concentrations of ligand. The parameter sensitivity is quantified as the fractional change, (f(p+Δp)–f(p))/f(p). The parameters have been changed by 50% up (*blue and red*) and down (*orange and magenta*). *Blue and cyan bars* correspond to the low ligand concentration

18.7. CONCLUSION

Cell biology is becoming more and more reliant on interdisciplinary approaches. Experiments are being augmented with bioinformatics, kinetic modelling, and theories from engineering and physics. The object of focus has become more and more the network rather than different aspects of any its molecules. Many of the current challenges of cell biology lie at the interface between molecular properties and network dynamics. What are the systemic consequences of particular molecular properties or small network motifs? What is the best drug target and drug administration scheme? How to predict transcription factor binding sites or protein binding partners? Each of these questions requires the integration of a number of different approaches. One of these approaches will always be experimentation. The construction of kinetic models of molecular networks is another method that is frequently used in systems biology [22, 1, 3]. In this chapter we have given a short overview of some of the applications of kinetic models. Many of its applications have not been addressed such as spatial modelling, experimental design, and experiment-driven model discrimination. Kinetic models will continue to play a central role in systems biology and its aim to finally end up with a molecular understanding of cell behaviour and rational strategies for precise alterations of cellular behaviour.

REFERENCES

1. Bruggeman, F. J. and Westerhoff, H. V. (2007). The nature of systems biology. *Trends Microbiol* 15, 45–50.
2. Ideker, T., Galitski, T., and Hood, L. (2001). A new approach to decoding life: Systems biology. *Annu Rev Genomics Hum Genet* 2, 343–372.
3. Westerhoff, H. V. and Palsson, B. O. (2004). The evolution of molecular biology into systems biology. *Nat Biotechnol* 22, 1249–1252.
4. Alberghina, L. and Westerhoff, H. V. (Eds.) (2005). Systems biology: Definitions and perspectives (topics in current genetics). Springer-Verlag Berlin and Heidelberg GmbH & Co. KG.
5. Barabasi, A. L. and Oltvai, Z. N. (2004). Network biology: Understanding the cell's functional organization. *Nat Rev Genet* 5, 101–113.
6. Kitano, H. (2004). Biological robustness. *Nat Rev Genet* 5, 826–837.
7. Albert, R. and Barabasi, A. L. (2002). Statistical mechanics of complex networks. *Rev Modern Phys* 74, 47–97.
8. Newman, M. E. J. (2003). The structure and function of complex networks. *SIAM review* 45, 167–256.
9. Ideker, T., Winslow, L. R., and Lauffenburger, A. D. (2006). Bioengineering and systems biology. *Ann Biomed Eng* 34, 257–264.
10. Csete, M. E. and Doyle, J. C. (2002). Reverse engineering of biological complexity. *Science* 295, 1664–1669.
11. Del Vecchio, D., Ninfa, A. J., and Sontag, E. D. (2008). Modular cell biology: Retroactivity and insulation. *Mol Syst Biol* 4, 161.
12. Stelling, J., Sauer, U., Szallasi, Z., Doyle, F. J., 3rd, and Doyle, J. (2004). Robustness of cellular functions. *Cell* 118, 675–685.
13. Bruggeman, F. J., de Haan, J., Hardin, H., Bouwman, J., Rossell, S., van Eunen, K., Bakker, B. M., and Westerhoff, H. V. (2006). Time-dependent hierarchical regulation analysis: Deciphering cellular adaptation. *Syst Biol (Stevenage)* 153, 318–322.

14. Bruggeman, F. J., Westerhoff, H. V., Hoek, J. B., and Kholodenko, B. N. (2002). Modular response analysis of cellular regulatory networks. *J Theor Biol* 218, 507–520.
15. Daran-Lapujade, P., Rossell, S., van Gulik, W. M., Luttik, M. A., de Groot, M. J., Slijper, M., Heck, A. J., Daran, J. M., de Winde, J. H., Westerhoff, H. V., Pronk, J. T., and Bakker, B. M. (2007). The fluxes through glycolytic enzymes in Saccharomyces cerevisiae are predominantly regulated at posttranscriptional levels. *Proc Natl Acad Sci U S A* 104, 15753–15758.
16. Hofmeyr, J. H. and Westerhoff, H. V. (2001). Building the cellular puzzle: Control in multi-level reaction networks. *J Theor Biol* 208, 261–285.
17. Kahn, D. and Westerhoff, H. V. (1991). Control theory of regulatory cascades. *J Theor Biol* 153, 255–285.
18. Kholodenko, B. N., Kiyatkin, A., Bruggeman, F. J., Sontag, E., Westerhoff, H. V., and Hoek, J. B. (2002). Untangling the wires: A strategy to trace functional interactions in signaling and gene networks. *Proc Natl Acad Sci USA* 99, 12841–12846.
19. Rossell, S., van der Weijden, C. C., Lindenbergh, A., van Tuijl, A., Francke, C., Bakker, B. M., and Westerhoff, H. V. (2006). Unraveling the complexity of flux regulation: A new method demonstrated for nutrient starvation in Saccharomyces cerevisiae. *Proc Natl Acad Sci USA* 103, 2166–2171.
20. Bruggeman, F. J., Snoep, J. L., and Westerhoff, H. V. (2008). Control, responses and modularity of cellular regulatory networks: A control analysis perspective. *IET Syst Biol* 2, 397.
21. Price, N. D., Reed, J. L., and Palsson, B. O. (2004). Genome-scale models of microbial cells: Evaluating the consequences of constraints. *Nat Rev Microbiol* 2, 886–897.
22. Aldridge, B. B., Burke, J. M., Lauffenburger, D. A., and Sorger, P. K. (2006). Physicochemical modelling of cell signalling pathways. *Nat Cell Biol* 8, 1195–1203.
23. Bluthgen, N., Bruggeman, F. J., Legewie, S., Herzel, H., Westerhoff, H. V., and Kholodenko, B. N. (2006). Effects of sequestration on signal transduction cascades. *FEBS J* 273, 895–906.
24. Hornberg, J. J., Binder, B., Bruggeman, F. J., Schoeberl, B., Heinrich, R., and Westerhoff, H. V. (2005a). Control of MAPK signalling: From complexity to what really matters. *Oncogene* 24, 5533–5542.
25. Hornberg, J. J., Bruggeman, F. J., Binder, B., Geest, C. R., de Vaate, A. J., Lankelma, J., Heinrich, R., and Westerhoff, H. V. (2005b). Principles behind the multifarious control of signal transduction. *ERK phosphorylation and kinase/phosphatase control. FEBS J* 272, 244–258.
26. Kholodenko, B. N., Demin, O. V., Moehren, G., and Hoek, J. B. (1999). Quantification of short term signaling by the epidermal growth factor receptor. *J Biol Chem* 274, 30169–30181.
27. Schoeberl, B., Gaudet, S., Albeck, J. G., Janes, K., Sorger, P. K., and Lauffenburger, D. A. (2002). The apoptotic decision process in TNF alpha-stimulated HT-29 cells: A combined computational and experimental approach. *Molecular Biology of the Cell* 13, 11A–11A.
28. Suenaga, A., Kiyatkin, A. B., Hatakeyama, M., Futatsugi, N., Okimoto, N., Hirano, Y., Narumi, T., Kawai, A., Susukita, R., Koishi, T., Furusawa, H., Yasuoka, K., Takada, N., Ohno, Y., Taiji, M., Ebisuzaki, T., Hoek, J. B., Konagaya, A., and Kholodenko, B. N. (2004). Tyr-317 phosphorylation increases Shc structural rigidity and reduces coupling of domain motions remote from the phosphorylation site as revealed by molecular dynamics simulations. *J Biol Chem* 279, 4657–4662.
29. Swameye, I., Muller, T. G., Timmer, J., Sandra, O., and Klingmuller, U. (2003). Identification of nucleocytoplasmic cycling as a remote sensor in cellular signaling by databased modeling. *Proc Natl Acad Sci USA* 100, 1028–1033.
30. Raj, A. and van Oudenaarden, A. (2008). Nature, nurture, or chance: Stochastic gene expression and its consequences. *Cell* 135, 216–226.
31. Heinrich, R., Rapoport, S. M., and Rapoport, T. A. (1977). Metabolic regulation and mathematical models. *Prog Biophys Mol Biol* 32, 1–82.
32. Kholodenko, B. N. (2006). Cell-signalling dynamics in time and space. *Nat Rev Mol Cell Biol* 7, 165–176.
33. Szallazi, Z., Stelling, J., and Periwall, V. (2006). System modeling in cellular biology: From concepts to nuts and bolts. The MIT Press, Cambridge, MA.

34. Tyson, J. J., Chen, K. C., and Novak, B. (2003). Sniffers, buzzers, toggles and blinkers: dynamics of regulatory and signaling pathways in the cell. *Curr Opin Cell Biol* 15, 221–231.
35. von Hippel, P. H. and Berg, O. G. (1989). Facilitated target location in biological systems. *J Biol Chem* 264, 675–678.
36. Heinrich, R. and Rapoport, T. A. (1974). A linear steady-state treatment of enzymatic chains. General properties, control and effector strength. *Eur J Biochem* 42, 89–95.
37. Kacser, H. and Burns, J. A. (1973). The control of flux. *Symp Soc Exp Biol* 27, 65–104.
38. Paulsson, J. (2005). Models of stochastic gene expression. *Physics of Life Reviews* 2, 157–175.
39. Savageau, M. A. (1976). Biochemical systems analysis: A study of function and design in molecular biology. Addison-Wesley, New York.
40. Cornish-Bowden, A. (1995). Fundamentals of enzyme kinetics. Portland Press, London.
41. Segel, I. H. (1993). Enzyme kinetics: Behavior and analysis of rapid equilibrium and steady-state enzyme systems. John Wiley & Son, Inc., New York.
42. Westerhoff, H. V. and Van Dam, K. (1987). Thermodynamics and control of biological free-energy transduction. Elsevier Science Publishers B.V. (Biomedical Division), Amsterdam.
43. Fell, D. A. (1997). Understanding the control of metabolism. Portland Press, London and Miami.
44. Hofmeyr, J. S. and Cornish-Bowden, A. (2000). Regulating the cellular economy of supply and demand. *FEBS Lett* 476, 47–51.
45. Metivier, R., Reid, G., and Gannon, F. (2006). Transcription in four dimensions: Nuclear receptor-directed initiation of gene expression. *EMBO Rep* 7, 161–167.
46. Metivier, R., Penot, G., Hubner, M. R., Reid, G., Brand, H., Kos, M., and Gannon, F. (2003). Estrogen receptor-alpha directs ordered, cyclical, and combinatorial recruitment of cofactors on a natural target promoter. *Cell* 115, 751–763.
47. Chubb, J. R., Trcek, T., Shenoy, S. M., and Singer, R. H. (2006). Transcriptional pulsing of a developmental gene. *Curr Biol* 16, 1018–1025.
48. Elowitz, M. B., Levine, A. J., Siggia, E. D., and Swain, P. S. (2002). Stochastic gene expression in a single cell. *Science* 297, 1183–1186.
49. Levsky, J. M., Shenoy, S. M., Pezo, R. C., and Singer, R. H. (2002). Single-cell gene expression profiling. *Science* 297, 836–840.
50. Raj, A., Peskin, C. S., Tranchina, D., Vargas, D. Y., and Tyagi, S. (2006). Stochastic mRNA synthesis in mammalian cells. *PLoS Biol* 4, e309.
51. Rosenfeld, N., Young, J. W., Alon, U., Swain, P. S., and Elowitz, M. B. (2005). Gene regulation at the single-cell level. *Science* 307, 1962–1965.
52. Swain, P. S., Elowitz, M. B., and Siggia, E. D. (2002). Intrinsic and extrinsic contributions to stochasticity in gene expression. *Proc Natl Acad Sci USA* 99, 12795–12800.
53. Golding, I., Paulsson, J., Zawilski, S. M., and Cox, E. C. (2005). Real-time kinetics of gene activity in individual bacteria. *Cell* 123, 1025–1036.
54. Rando, O. J. and Paulsson, J. (2006). Noisy silencing of chromatin. *Dev Cell* 11, 134–136.
55. Kaufmann, B. B. and van Oudenaarden, A. (2007). Stochastic gene expression: From single molecules to the proteome. *Curr Opin Genet Dev* 17, 107–112.
56. Guarnieri, D. J. and DiLeone, R. J. (2008). MicroRNAs: A new class of gene regulators. *Ann Med* 40, 197–208.
57. Makeyev, E. V. and Maniatis, T. (2008). Multilevel regulation of gene expression by microRNAs. *Science* 319, 1789–90.
58. Selbach, M., Schwanhausser, B., Thierfelder, N., Fang, Z., Khanin, R., and Rajewsky, N. (2008). Widespread changes in protein synthesis induced by microRNAs. *Nature* 455, 58–63.
59. Kumar, S., Saradhi, M., Chaturvedi, N. K., and Tyagi, R. K. (2006). Intracellular localization and nucleocytoplasmic trafficking of steroid receptors: An overview. *Mol Cell Endocrinol* 246, 147–156.
60. Liu, J. and DeFranco, D. B. (2000). Protracted nuclear export of glucocorticoid receptor limits its turnover and does not require the exportin 1/CRM1-directed nuclear export pathway. *Mol Endocrinol* 14, 40–51.

61. Nicolas, F. J., De Bosscher, K., Schmierer, B., and Hill, C. S. (2004). Analysis of Smad nucleocytoplasmic shuttling in living cells. *J Cell Sci* 117, 4113–4125.
62. Schmierer, B. and Hill, C. S. (2005). Kinetic analysis of Smad nucleocytoplasmic shuttling reveals a mechanism for transforming growth factor beta-dependent nuclear accumulation of Smads. *Mol Cell Biol* 25, 9845–9858.
63. Stewart, M. (2007). Molecular mechanism of the nuclear protein import cycle. *Nat Rev Mol Cell Biol* 8, 195–208.
64. Alon, U. (2007). Network motifs: Theory and experimental approaches. *Nat Rev Genet* 8, 450–461.

INDEX

C.M. Bunce, M.J. Campbell (eds.), *Nuclear Receptors,* Proteins and Cell Regulation 8, 455–457. DOI 10.1007/978-90-481-3303-1,

MIX
Papier aus verantwortungsvollen Quellen
Paper from responsible sources
FSC® C105338

If you have any concerns about our products,
you can contact us on
ProductSafety@springernature.com

In case Publisher is established outside the EU,
the EU authorized representative is:
Springer Nature Customer Service Center GmbH
Europaplatz 3, 69115 Heidelberg, Germany

Printed by Libri Plureos GmbH
in Hamburg, Germany